This 11th Edition is dedicated to Dr. Clayton Crowe (1933–2012) and to our wonderful colleagues, students, friends, and families. We especially acknowledge our spouses Linda and Jim and Barbara's grandson Moses Pakootas for their patience and support.

CONTENTS

Preface   vii

CHAPTER**ONE Introduction**   **1**

1.1 Engineering Fluid Mechanics   2
1.2 How Materials are Idealized   3
1.3 Weight, Mass, and Newton's Law of Gravitation   8
1.4 Essential Math Topics   11
1.5 Density and Specific Weight   13
1.6 The Ideal Gas Law (IGL)   15
1.7 Units and Dimensions   18
1.8 Problem Solving   24
1.9 Summarizing Key Knowledge   27

CHAPTER**TWO Fluid Properties**   **31**

2.1 System, State, and Property   32
2.2 Looking Up Fluid Properties   33
2.3 Topics Related to Density   36
2.4 Pressure and Shear Stress   38
2.5 The Viscosity Equation   41
2.6 Surface Tension   47
2.7 Vapor Pressure   51
2.8 Characterizing Thermal Energy in Flowing Gases   52
2.9 Summarizing Key Knowledge   53

CHAPTER**THREE Fluid Statics**   **57**

3.1 Describing Pressure   58
3.2 The Hydrostatic Equations   63
3.3 Measuring Pressure   68
3.4 The Pressure Force on a Panel (Flat Surface)   72
3.5 Calculating the Pressure Force on a Curved Surface   78
3.6 Calculating Buoyant Forces   81
3.7 Predicting Stability of Immersed and Floating Bodies   83
3.8 Summarizing Key Knowledge   87

CHAPTER**FOUR The Bernoulli Equation and Pressure Variation**   **95**

4.1 Describing Streamlines, Streaklines, and Pathlines   95
4.2 Characterizing Velocity of a Flowing Fluid   98
4.3 Describing Flow   100
4.4 Acceleration   106
4.5 Applying Euler's Equation to Understand Pressure Variation   109
4.6 Applying the Bernoulli Equation along a Streamline   114
4.7 Measuring Velocity and Pressure   120
4.8 Characterizing the Rotational Motion of a Flowing Fluid   123
4.9 The Bernoulli Equation for Irrotational Flow   127
4.10 Describing the Pressure Field for Flow over a Circular Cylinder   128
4.11 Calculating the Pressure Field for a Rotating Flow   130
4.12 Summarizing Key Knowledge   132

CHAPTER**FIVE The Control Volume Approach and The Continuity Equation**   **141**

5.1 Characterizing the Rate of Flow   141
5.2 The Control Volume Approach   147
5.3 The Continuity Equation (Theory)   153
5.4 The Continuity Equation (Application)   154
5.5 Predicting Cavitation   161
5.6 Summarizing Key Knowledge   164

CHAPTER**SIX The Momentum Equation**   **171**

6.1 Understanding Newton's Second Law of Motion   171
6.2 The Linear Momentum Equation: Theory   175
6.3 The Linear Momentum Equation: Application   178
6.4 The Linear Momentum Equation for a Stationary Control Volume   180
6.5 Examples of the Linear Momentum Equation (Moving Objects)   189
6.6 The Angular Momentum Equation   194
6.7 Summarizing Key Knowledge   197

CHAPTER**SEVEN The Energy Equation**   **204**

7.1 Technical Vocabulary: Work, Energy, and Power   205
7.2 Conservation of Energy   207
7.3 The Energy Equation   209
7.4 The Power Equation   216
7.5 Mechanical Efficiency   218
7.6 Contrasting the Bernoulli Equation and the Energy Equation   221
7.7 Transitions   221
7.8 The Hydraulic and Energy Grade Lines   224
7.9 Summarizing Key Knowledge   227

CHAPTER **EIGHT Dimensional Analysis and Similitude**   **234**

8.1   The Need for Dimensional Analysis   234
8.2   Buckingham π Theorem   236
8.3   Dimensional Analysis   236
8.4   Common π-Groups   240
8.5   Similitude   243
8.6   Model Studies for Flows without Free-Surface Effects   247
8.7   Model-Prototype Performance   250
8.8   Approximate Similitude at High Reynolds Numbers   251
8.9   Free-Surface Model Studies   254
8.10 Summarizing Key Knowledge   257

CHAPTER **NINE Viscous Flow Over a Flat Surface**   **261**

9.1   The Navier-Stokes Equation for Uniform Flow   262
9.2   Couette Flow   263
9.3   Poiseuille Flow in a Channel   264
9.4   The Boundary Layer (Description)   266
9.5   Velocity Profiles in the Boundary Layer   267
9.6   The Boundary Layer (Calculations)   269
9.7   Summarizing Key Knowledge   273

CHAPTER **TEN Flow in Conduits**   **278**

10.1 Classifying Flow   279
10.2 Specifying Pipe Sizes   281
10.3 Pipe Head Loss   282
10.4 Stress Distributions in Pipe Flow   284
10.5 Laminar Flow in a Round Tube   286
10.6 Turbulent Flow and the Moody Diagram   289
10.7 A Strategy for Solving Problems   294
10.8 Combined Head Loss   298
10.9 Nonround Conduits   302
10.10 Pumps and Systems of Pipes   304
10.11 Summarizing Key Knowledge   309

CHAPTER **ELEVEN Drag and Lift**   **318**

11.1 Relating Lift and Drag to Stress Distributions   318
11.2 Calculating the Drag Force   320
11.3 Drag of Axisymmetric and 3-D Bodies   323
11.4 Terminal Velocity   328
11.5 Vortex Shedding   330
11.6 Reducing Drag by Streamlining   331
11.7 Drag in Compressible Flow   331
11.8 The Theory of Lift   332
11.9 Lift and Drag on Airfoils   336
11.10 Lift and Drag on Road Vehicles   342
11.11 Summarizing Key Knowledge   345

CHAPTER **TWELVE Compressible Flow**   **351**

12.1 Wave Propagation in Compressible Fluids   351

12.2 Mach Number Relationships   356
12.3 Normal Shock Waves   361
12.4 Isentropic Compressible Flow through a Duct with Varying Area   366
12.5 Summarizing Key Knowledge   377

CHAPTER **THIRTEEN Flow Measurements**   **380**

13.1 Measuring Velocity and Pressure   380
13.2 Measuring Flow Rate (Discharge)   387
13.3 Summarizing Key Knowledge   402

CHAPTER **FOURTEEN Turbomachinery**   **406**

14.1 Propellers   407
14.2 Axial-Flow Pumps   411
14.3 Radial-Flow Machines   415
14.4 Specific Speed   418
14.5 Suction Limitations of Pumps   420
14.6 Viscous Effects   422
14.7 Centrifugal Compressors   423
14.8 Turbines   426
14.9 Summarizing Key Knowledge   434

CHAPTER **FIFTEEN Flow in Open Channels**   **439**

15.1 Description of Open-Channel Flow   440
15.2 The Energy Equation for Steady Open-Channel Flow   442
15.3 Steady Uniform Flow   443
15.4 Steady Nonuniform Flow   451
15.5 Rapidly Varied Flow   451
15.6 Hydraulic Jump   461
15.7 Gradually Varied Flow   466
15.8 Summarizing Key Knowledge   473

CHAPTER **SIXTEEN Modeling of Fluid Dynamics Problems**   **477**

16.1 Models in Fluid Mechanics   478
16.2 Foundations for Learning Partial Differential Equations (PDEs)   482
16.3 The Continuity Equation   491
16.4 The Navier-Stokes Equation   497
16.5 Computational Fluid Dynamics (CFD)   501
16.6 Examples of CFD   506
16.7 A Path for Moving Forward   508
16.8 Summarizing Key Knowledge   509

Appendix   515

Answers   525

Index   529

ELEVENTH EDITION

ENGINEERING FLUID MECHANICS

International Student Version

DONALD F. ELGER
University of Idaho, Moscow

BARBARA A. LeBRET
University of Idaho, Moscow

CLAYTON T. CROWE
Washington State University, Pullman

JOHN A. ROBERSON
Washington State University, Pullman

WILEY

PREFACE

Audience

This book is written for engineering students of all majors who are taking a first or second course in fluid mechanics. Students should have background knowledge in physics (mechanics), chemistry, statics, and calculus.

Why We Wrote This Book

Our mission is to equip people to do engineering skillfully. Thus, we wrote this book to explain the main ideas of fluid mechanics at a level appropriate for a first or second college course. In addition, we have included selected engineering skills (e.g., critical thinking, problem solving, and estimation) because we believe that practicing these skills will help all students learn fluid mechanics better.

Approach

Knowledge. Each chapter begins with statements of what is important to learn. These learning outcomes are formulated in terms of what *students will be able to do.* Then, the chapter sections present the knowledge. Finally, the knowledge is summarized at the end of each chapter.

Practice with Feedback. The research of Dr. Anders Ericsson suggests that learning is brought about through *deliberate practice.* Deliberate practice involves doing something and then getting feedback. To provide opportunities for deliberate practice, this text contains end-of-chapter problems. The answers to selected, even-numbered problems are provided in the back of the book. Professors can gain access to the solution manual by contacting their Wiley representative.

Features of this Book

Learning Outcomes. Each chapter begins with learning outcomes so that students can identify what knowledge they should gain by studying the chapter.

Rationale. Each section describes what content is presented and why this content is relevant.

Visual Approach. This text uses sketches and photographs to help students learn more effectively by connecting images to words and equations.

Foundational Concepts. This text presents major concepts in a clear and concise format. These concepts form building blocks for higher levels of learning.

Seminal Equations. This text emphasizes technical derivations so that students can learn to do the derivations on their own, increasing their level of knowledge. Features include the following:

- Derivations of main equations are presented in a step-by-step fashion.

- The holistic meaning of main equations is explained using words.
- Main equations are named and listed in Table F.2.
- Main equations are summarized in tables in the chapters.
- A process for applying each main equation is presented in the relevant chapter.

Chapter Summaries. Each chapter concludes with a summary so that students can review the key knowledge in the chapter.

Process Approach. A process is a method for getting results. A process approach involves figuring out how experts do things and adapting this same approach. This textbook presents multiple processes.

Wales-Woods Model. The Wales-Woods Model represents how experts solve problems. This model is presented in Chapter 1 and used in example problems throughout the text.

Grid Method. This text presents a systematic process, called the grid method, for carrying and canceling units. Unit practice is emphasized because it helps engineers spot and fix mistakes and because it helps engineers put meaning on concepts and equations.

Traditional and SI Units. Examples and homework problems are presented using both SI and traditional unit systems. This presentation helps students gain familiarity with units that are used in professional practice.

Example Problems. Each chapter has examples to show how the knowledge is used in context and to present essential details needed for application.

Solutions Manual. The text includes a detailed solutions manual for instructors. Many solutions are presented with the Wales-Woods Model.

Image Gallery. The figures from the text are available in PowerPoint format, for easy inclusion in lecture presentations. This resource is available only to instructors. To request access to this and all instructor resources, please contact your local Wiley sale representative.

Interdisciplinary Approach. Historically, this text was written for the civil engineer. We are retaining this approach while adding material so that the text is also appropriate for other engineering disciplines. For example, the text presents the Bernoulli equation using both head terms (civil engineering approach) and terms with units of pressure (the approach used by chemical and mechanical engineers). We include problems that are relevant to product development as practiced by mechanical and electrical engineers. Some problems feature other disciplines, such as exercise physiology. The reason for this interdisciplinary approach is that the world of today's engineer is becoming more and more interdisciplinary.

What is New in the 11th Edition

1. Critical Thinking (CT) is introduced in Chapter 1. **Rationale**: When students apply CT, they learn fluid mechanics better. Also, they become better engineers.

2. Learn outcomes are organized into categories. **Rationale**: The grouping of outcomes increases the clarity about what is important.

3. New material was added in Chapter 1 describing force, mass, weight, Newton's law of universal gravitation, density, and specific weight. **Rationale**: We have seen many instances of student work indicating that these basic concepts are sometimes not in place. Also, introducing these topics in Chapter 1 provides a way to introduce engineering calculations earlier in the book.

4. We introduced the **Voice of the Engineer** in Chapter 1 as a way to present wisdom. **Rationale**: The **Voice of the Engineer** provides a structure for presenting an attitude that is widely shared in the professional engineering community.

5. In Chapter 1, new material was added about the ideal gas law (IGL). **Rationale**: The IGL section now has the right level of technical detail for engineering problems.

6. In Chapter 1, the material on problem solving was rewritten. Also, the Wales-Woods Model is now summarized on one page. **Rationale**: Solving problems and building math models are fundamental skills for the engineer. The ideas in Chapter 1 represent the best ideas that we have seen in the literature.

7. Chapter 2 has a new section on finding fluid properties. This new section, §2.2, contains the summary table that previously was situated at the end of the chapter. **Rationale**: Finding fluid properties is an important learning outcome for Chapter 2. The new section puts an emphasis on this outcome and organizes the ideas in one place. Previously, the knowledge needed to find fluid properties was scattered throughout Chapter 2.

8. Chapter 2 has a new section on stress, how to relate stress to force, and on common forces. **Rationale**: Stress and force are seminal ideas in mechanics. This section defines the relevant terms and shows how they are related.

9. The Chapter 2 discussions on the shear stress equation were edited to increase clarity and concision. **Rationale**: The shear stress equation is one of the seminal fluid mechanics equations.

10. The end-of-chapter problems include new or revised problems. **Rationale**: Both learning and assessment of learning are made easier by having problems available.

11. Chapter 9 was rewritten to make the chapter more suitable for students taking a first course in fluid mechanics.

Author Team

The book was originally written by Professor John Roberson, with Professor Clayton Crowe adding the material on compressible flow. Professor Roberson retired from active authorship after the 6th edition, Professor Donald Elger joined on the 7th edition, and Professor Barbara LeBret joined on the 9th edition. Professor Crowe retired from active authorship after the 9th edition. Professor Crowe passed away on February 5, 2012.

Acknowledgments

We acknowledge our colleagues and mentors. Donald Elger acknowledges his Ph.D. mentor, Ronald Adams, who always asked why and how. He also acknowledges Ralph Budwig, fluid mechanics researcher and colleague, who has provided many hours of delightful inquiry about fluid mechanics. Barbara

Donald Elger, Barbara LeBret, and Clayton Crowe (Photo by Archer Photography: www.archerstudio.com)

LeBret acknowledges Wilfried Brutsuert at Cornell University and George Bloomsburg at the University of Idaho, who inspired her passion for fluid mechanics. Last but not least, we thank technical reviewer Daniel Flick for his excellent work.

Contact Us

We welcome feedback and ideas for interesting end-of-chapter problems. Please contact us at the following email addresses:

Donald F. Elger (donelger@google.com)
Barbara A. LeBret (barbwill@uidaho.edu)

TABLE F.1 Formulas for Unit Conversions*

Name, Symbol, Dimensions			Conversion Formula
Length	L	L	$1\,\mathbf{m}$ = 3.281 ft = 1.094 yd = 39.37 in = km/1000 = $10^6\,\mu$m $1\,\mathbf{ft}$ = 0.3048 m = 12 in = mile/5280 = km/3281 $1\,\mathbf{mm}$ = m/1000 = in/25.4 = 39.37 mil = 1000 μm = 10^7Å
Speed	V	L/T	$1\,\mathbf{m/s}$ = 3.600 km/hr = 3.281 ft/s = 2.237 mph = 1.944 knots $1\,\mathbf{ft/s}$ = 0.3048 m/s = 0.6818 mph = 1.097 km/hr = 0.5925 knots
Mass	m	M	$1\,\mathbf{kg}$ = 2.205 lbm = 1000 g = slug/14.59 = (metric ton or tonne or Mg)/1000 $1\,\mathbf{lbm}$ = lbf·s^2/(32.17 ft) = kg/2.205 = slug/32.17 = 453.6 g = 16 oz = 7000 grains = short ton/2000 = metric ton (tonne)/2205
Density	ρ	M/L^3	$1000\,\mathbf{kg/m^3}$ = 62.43 lbm/ft^3 = 1.940 slug/ft^3 = 8.345 lbm/gal (US)
Force	F	ML/T^2	$1\,\mathbf{lbf}$ = 4.448 N = 32.17 lbm·ft/s^2 $1\,\mathbf{N}$ = kg·m/s^2 = 0.2248 lbf = 10^5 dyne

Pressure, shear stress	p, τ	M/LT^2	$1\,\mathbf{Pa}$ = N/m^2 = kg/m·s^2 = 10^{-5} bar = 1.450×10^{-4} lbf/in^2 = inch H$_2$O/249.1
			    = 0.007501 torr = 10.00 dyne/cm^2
			$1\,\mathbf{atm}$ = 101.3 kPa = 2116 psf = 1.013 bar = 14.70 lbf/in^2 = 33.90 ft of water
			    = 29.92 in of mercury = 10.33 m of water = 760 mm of mercury = 760 torr
			$1\,\mathbf{psi}$ = atm/14.70 = 6.895 kPa = 27.68 in H$_2$O = 51.71 torr

Volume	Ψ	L^3	$1\,\mathbf{m^3}$ = 35.31 ft^3 = 1000 L = 264.2 U.S. gal $1\,\mathbf{ft^3}$ = 0.02832 m^3 = 28.32 L = 7.481 U.S. gal = acre-ft/43,560 $1\,\mathbf{U.S.\ gal}$ = 231 in^3 = barrel (petroleum)/42 = 4 U.S. quarts = 8 U.S. pints = 3.785 L = 0.003785 m^3
Volume flow rate (discharge)	Q	L^3/T	$1\,\mathbf{m^3/s}$ = 35.31 ft^3/s = 2119 cfm = 264.2 gal (US)/s = 15850 gal (US)/m $1\,\mathbf{cfs}$ = 1 ft^3/s = 28.32 L/s = 7.481 gal (US)/s = 448.8 gal (US)/m
Mass flow rate	$\dot{m}$	M/T	$1\,\mathbf{kg/s}$ = 2.205 lbm/s = 0.06852 slug/s
Energy and work	E, W	ML^2/T^2	$1\,\mathbf{J}$ = kg·m^2/s^2 = N·m = W·s = volt·coulomb = 0.7376 ft·lbf = 9.478×10^{-4} Btu = 0.2388 cal = 0.0002388 Cal = 10^7 erg = kWh/3.600 × 10^6
Power	$P, \dot{E}, \dot{W}$	ML^2/T^3	$1\,\mathbf{W}$ = J/s = N·m/s = kg·m^2/s^3 = 1.341×10^{-3} hp = 0.7376 ft·lbf/s = 1.0 volt-ampere = 0.2388 cal/s = 9.478×10^{-4} Btu/s $1\,\mathbf{hp}$ = 0.7457 kW = 550 ft·lbf/s = 33,000 ft·lbf/min = 2544 Btu/h
Angular speed	ω	T^{-1}	$1.0\,\mathbf{rad/s}$ = 9.549 rpm = 0.1591 rev/s
Viscosity	μ	M/LT	$1\,\mathbf{Pa \cdot s}$ = kg/m·s = N·s/m^2 = 10 poise = 0.02089 lbf·s/ft^2 = 0.6720 lbm/ft·s
Kinematic viscosity	ν	L^2/T	$1\,\mathbf{m^2/s}$ = 10.76 ft^2/s = 10^6 cSt
Temperature	T	Θ	$\mathbf{K}$ = °C + 273.15 = °R/1.8 $\mathbf{°C}$ = (°F − 32)/1.8 $\mathbf{°R}$ = °F + 459.67 = 1.8 K $\mathbf{°F}$ = 1.8°C + 32

*Visit www.onlineconversion.com for a useful online reference.

TABLE F.2 Commonly Used Equations

Ideal gas law equations

$p = \rho R T$

$p \Psi = m R T$

$p \Psi = n R_u T$

$M = m/n; R = R_u/M$ (§1.6)

Specific weight

$\gamma = \rho g$ (Eq. 1.21)

Kinematic viscosity

$\nu = \mu/\rho$ (Eq. 2.1)

Specific gravity

$S = \dfrac{\rho}{\rho_{H_2O \text{ at } 4°C}} = \dfrac{\gamma}{\gamma_{H_2O \text{ at } 4°C}}$ (Eq. 2.3)

Definition of viscosity

$\tau = \mu \dfrac{dV}{dy}$ (Eq. 2.15)

Pressure equations

$p_{gage} = p_{abs} - p_{atm}$ (Eq. 3.3a)

$p_{vacuum} = p_{atm} - p_{abs}$ (Eq. 3.3b)

Hydrostatic equation

$\dfrac{p_1}{\gamma} + z_1 = \dfrac{p_2}{\gamma} + z_2 = \text{constant}$ (Eq. 3.10a)

$p_z = p_1 + \gamma z_1 = p_2 + \gamma z_2 = \text{constant}$ (Eq. 3.10b)

$\Delta p = -\gamma \Delta z$ (Eq. 3.10c)

Manometer equations

$p_2 = p_1 + \displaystyle\sum_{down} \gamma_i h_i - \sum_{up} \gamma_i h_i$ (Eq. 3.21)

$h_1 - h_2 = \Delta h (\gamma_B/\gamma_A - 1)$ (Eq. 3.22)

Hydrostatic force equations (flat panels)

$F_P = \bar{p} A$ (Eq. 3.28)

$y_{cp} - \bar{y} = \dfrac{\bar{I}}{\bar{y} A}$ (Eq. 3.33)

Buoyant force (Archimedes equation)

$F_B = \gamma \Psi_D$ (Eq. 3.41a)

The Bernoulli equation

$\left(\dfrac{p_1}{\gamma} + \dfrac{V_1^2}{2g} + z_1 \right) = \left(\dfrac{p_2}{\gamma} + \dfrac{V_2^2}{2g} + z_2 \right)$ (Eq. 4.21b)

$\left(p_1 + \dfrac{\rho V_1^2}{2} + \rho g z_1 \right) = \left(p_2 + \dfrac{\rho V_2^2}{2} + \rho g z_2 \right)$ (Eq. 4.21a)

Volume flow rate equation

$Q = \bar{V} A = \dfrac{\dot{m}}{\rho} = \displaystyle\int_A V dA = \int_A \mathbf{V} \cdot \mathbf{dA}$ (Eq. 5.10)

Mass flow rate equation

$\dot{m} = \rho A \bar{V} = \rho Q = \displaystyle\int_A \rho V dA = \int_A \rho \mathbf{V} \cdot \mathbf{dA}$ (Eq. 5.11)

Continuity equation

$\dfrac{d}{dt} \displaystyle\int_{cv} \rho \, d\Psi + \int_{cs} \rho \mathbf{V} \cdot \mathbf{dA} = 0$ (Eq. 5.28)

$\dfrac{d}{dt} M_{cv} + \displaystyle\sum_{cs} \dot{m}_o - \sum_{cs} \dot{m}_i = 0$ (Eq. 5.29)

$\rho_2 A_2 V_2 = \rho_1 A_1 V_1$ (Eq. 5.33)

Momentum equation

$\displaystyle\sum \mathbf{F} = \dfrac{d}{dt} \int_{cv} \mathbf{v} \rho \, d\Psi + \int_{cs} \mathbf{v} \rho \mathbf{V} \cdot \mathbf{dA}$ (Eq. 6.7)

$\displaystyle\sum \mathbf{F} = \dfrac{d(m_{cv} \mathbf{v}_{cv})}{dt} + \sum_{cs} \dot{m}_o \mathbf{v}_o - \sum_{cs} \dot{m}_i \mathbf{v}_i$ (Eq. 6.10)

Energy equation

$\left(\dfrac{p_1}{\gamma} + \alpha_1 \dfrac{\bar{V}_1^2}{2g} + z_1 \right) + h_p = \left(\dfrac{p_2}{\gamma} + \alpha_2 \dfrac{\bar{V}_2^2}{2g} + z_2 \right) + h_t + h_L$

(Eq. 7.29)

The power equation

$P = FV = T\omega$ (Eq. 7.3)

$P = \dot{m} g h = \gamma Q h$ (Eq. 7.31)

Efficiency of a machine

$\eta = \dfrac{P_{output}}{P_{input}}$ (Eq. 7.32)

Reynolds number (pipe)

$Re_D = \dfrac{VD}{\nu} = \dfrac{\rho VD}{\mu} = \dfrac{4Q}{\pi D \nu} = \dfrac{4\dot{m}}{\pi D \mu}$ (Eq. 10.1)

Combined head loss equation

$h_L = \displaystyle\sum_{pipes} f \dfrac{L}{D} \dfrac{V^2}{2g} + \sum_{components} K \dfrac{V^2}{2g}$ (Eq. 10.45)

Friction factor f (Resistance coefficient)

$f = \dfrac{64}{Re_D}$ $Re_D \leq 2000$ (Eq. 10.34)

$f = \dfrac{0.25}{\left[\log_{10} \left(\dfrac{k_s}{3.7D} + \dfrac{5.74}{Re_D^{0.9}} \right) \right]^2}$ $(Re_D \geq 3000)$ (Eq. 10.39)

Drag force equation

$F_D = C_D A \left(\dfrac{\rho V_0^2}{2} \right)$ (Eq. 11.5)

Lift force equation

$F_L = C_L A \left(\dfrac{\rho V_0^2}{2} \right)$ (Eq. 11.17)

TABLE F.3 Useful Constants

Name of Constant	Value
Acceleration of gravity	$g = 9.81 \text{ m/s}^2 = 32.2 \text{ ft/s}^2$
Universal gas constant	$R_u = 8.314 \text{ kJ/kmol·K} = 1545 \text{ ft·lbf/lbmol·°R}$
Standard atmospheric pressure	$p_{atm} = 1.0 \text{ atm} = 101.3 \text{ kPa} = 14.70 \text{ psi} = 2116 \text{ psf} = 33.90 \text{ ft of water}$ $p_{atm} = 10.33 \text{ m of water} = 760 \text{ mm of Hg} = 29.92 \text{ in of Hg} = 760 \text{ torr} = 1.013 \text{ bar}$

TABLE F.4 Properties of Air [$T = 20°C$ (68°F), $p = 1$ atm]

Property	SI Units	Traditional Units
Specific gas constant	$R_{air} = 287.0 \text{ J/kg·K}$	$R_{air} = 1716 \text{ ft·lbf/slug·°R}$
Density	$\rho = 1.20 \text{ kg/m}^3$	$\rho = 0.0752 \text{ lbm/ft}^3 = 0.00234 \text{ slug/ft}^3$
Specific weight	$\gamma = 11.8 \text{ N/m}^3$	$\gamma = 0.0752 \text{ lbf/ft}^3$
Viscosity	$\mu = 1.81 \times 10^{-5} \text{ N·s/m}^2$	$\mu = 3.81 \times 10^{-7} \text{ lbf·s/ft}^2$
Kinematic viscosity	$\nu = 1.51 \times 10^{-5} \text{ m}^2/\text{s}$	$\nu = 1.63 \times 10^{-4} \text{ ft}^2/\text{s}$
Specific heat ratio	$k = c_p/c_v = 1.40$	$k = c_p/c_v = 1.40$
Specific heat	$c_p = 1004 \text{ J/kg·K}$	$c_p = 0.241 \text{ Btu/lbm·°R}$
Speed of sound	$c = 343 \text{ m/s}$	$c = 1130 \text{ ft/s}$

TABLE F.5 Properties of Water [$T = 15°C$ (59°F), $p = 1$ atm]

Property	SI Units	Traditional Units
Density	$\rho = 999 \text{ kg/m}^3$	$\rho = 62.4 \text{ lbm/ft}^3 = 1.94 \text{ slug/ft}^3$
Specific weight	$\gamma = 9800 \text{ N/m}^3$	$\gamma = 62.4 \text{ lbf/ft}^3$
Viscosity	$\mu = 1.14 \times 10^{-3} \text{ N·s/m}^2$	$\mu = 2.38 \times 10^{-5} \text{ lbf·s/ft}^2$
Kinematic viscosity	$\nu = 1.14 \times 10^{-6} \text{ m}^2/\text{s}$	$\nu = 1.23 \times 10^{-5} \text{ ft}^2/\text{s}$
Surface tension (water–air)	$\sigma = 0.073 \text{ N/m}$	$\sigma = 0.0050 \text{ lbf/ft}$
Bulk modulus of elasticity	$E_v = 2.14 \times 10^9 \text{ Pa}$	$E_v = 3.10 \times 10^5 \text{ psi}$

TABLE F.6 Properties of Water [$T = 4°C$ (39°F), $p = 1$ atm]

Property	SI Units	Traditional Units
Density	$\rho = 1000 \text{ kg/m}^3$	$\rho = 62.4 \text{ lbm/ft}^3 = 1.94 \text{ slug/ft}^3$
Specific weight	$\gamma = 9810 \text{ N/m}^3$	$\gamma = 62.4 \text{ lbf/ft}^3$

CHAPTER**ONE**

Introduction

CHAPTER ROAD MAP   Our purpose is to equip you for success. Success means that you can do engineering skillfully. This chapter presents (a) fluid mechanics topics and (b) engineering skills. The engineering skills are optional. We included these skills because we believe that applying these skills while you are learning fluid mechanics will strengthen your fluid mechanics knowledge while also making you a better engineer.

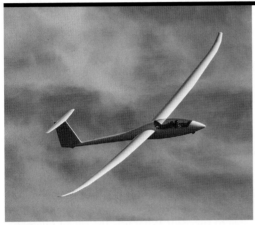

FIGURE 1.1
As engineers, we get to design fascinating systems like this glider. This is exciting! (© Ben Blankenburg/Corbis RF/Age Fotostock America, Inc.)

LEARNING OUTCOMES

ENGINEERING FLUID MECHANICS (§1.1*).
- Define engineering.
- Define fluid mechanics.

MATERIAL SCIENCE TOPICS (§1.2).
- Explain material behaviors using either a microscopic or a macroscopic approach or both.
- Know the main characteristics of liquids, gases, and fluids.
- Understand the concepts of body, material particle, body-as-a-particle, and the continuum assumption.

DENSITY AND SPECIFIC WEIGHT (§1.5).
- Know the main ideas about $W = mg$.
- Know the main ideas about density and specific weight.

THE IDEAL GAS LAW (IGL) (§1.6).
- Describe an ideal gas and a real gas.
- Convert temperature, pressure, and mole/mass units.
- Apply the IGL equations.

OPTIONAL ENGINEERING SKILLS (§1.1, §1.3, §1.4, §1.7, §1.8).
- Apply critical thinking to fluid mechanics problems.
- Make estimates when solving fluid mechanics problems.
- Apply ideas from calculus to fluid mechanics.
- Carry and cancel units when doing calculations.
- Check that an equation is DH (dimensionally homogeneous).
- Apply problem solving methods to fluid mechanics problems.

*The symbol § means "section"; e.g., the notation "§1.1" means Section 1.1.

1.1 Engineering Fluid Mechanics

In this section, we explain what engineering fluid mechanics means, and then we introduce *critical thinking* (CT), a method that is at the heart of doing engineering well.

About Engineering Fluid Mechanics

Why study engineering fluid mechanics? To answer this question, we'll start with some examples:

- When people started living in cities, they faced problems involving water. Those people who solved these problems were the engineers. For example, engineers designed aqueducts to bring water to the people. Engineers innovated technologies to remove waste water from the cities, thereby keeping the towns clean and free from effluent. Engineers developed technologies for treating water to remove waterborne diseases and to remove hazards such as arsenic.

- At one time, people had no flying machines. So, the Wright Brothers applied the engineering method to develop the world's first airplane. In the 1940s, engineers developed practical jet engines. More recently, the engineers at The Boeing Company developed the 787 Dreamliner.

- People have access to electrical power because engineers have developed technologies such as the water turbine, the wind turbine, the electric generator, the motor, and the electric grid system.

The preceding examples reveal that engineers solve problems and innovate in ways that lead to the development or improvement of technology. *How are engineers able to accomplish these difficult feats?* Why were the Wright Brothers able to succeed? What was the secret sauce that Edison had? The answer is that engineers have developed a method for success that is called the **engineering method**, which is actually a collection of submethods such as building math models, designing and conducting experiments, and designing and building physical systems.

Based on the ideas just presented, **engineering** is the body of knowledge that is concerned with solving problems by creating, designing, applying, and improving technology. **Engineering fluid mechanics** is engineering when a project involves substantial knowledge from the discipline of fluid mechanics.

Defining Mechanics

Mechanics is the branch of science that deals with motion and the forces that produce this motion. Mechanics is organized into two main categories: **solid mechanics** (materials in the solid state) and **fluid mechanics** (materials in the gas or liquid state). Note that many of the concepts of mechanics apply to both fluid mechanics and solid mechanics.

Critical Thinking (CT)

This section introduces critical thinking. Rationale. (1) The heart of the engineering method is critical thinking (CT); thus, skill with CT will give you the ability to do engineering well. (2) Applying CT while you are learning fluid mechanics will result in better learning.

Examples of CT are common. One example is seen when a police detective uses physical evidence and deductive reasoning to reach a conclusion about who committed a crime. A second example occurs when a medical doctor uses diagnostic test data and evidence from a physical examination to reach a conclusion about why a patient is ill. A third example exists when an engineering researcher gathers experimental data about groundwater flow, then reaches some conclusions and publishes these conclusions in a scientific journal. A fourth

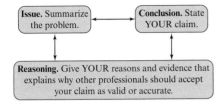

FIGURE 1.2

The Standard Structure of Critical Thinking (SSCT).

example arises when a practicing engineer uses experimental data and engineering calculations to conclude that Site ABC is a good choice for a wind turbine. These examples reveal some facts about critical thinking:

- CT is used by professionals in most fields (e.g., detectives, medical doctors, scientists, and engineers).
- Professionals apply CT to avoid major mistakes. No competent detective wants an innocent person convicted of a crime. No competent physician wants to make an incorrect diagnosis. No competent engineer wants a bridge to fail.
- CT involves methods that are agreed upon by a professional community. For example, the method of fingerprinting is accepted within the law enforcement community. Similarly, engineering has many agreed-upon methods (you can learn some of these methods in this book).

In summary, **critical thinking** is a collection of beliefs and methods that are accepted by a professional community for reaching a sound or strong conclusion. Some examples of the beliefs associated with CT are as follows:

- I want to find out the best idea or what is most correct (I have no interest in *being right*; I want to find out *what is right*).
- I want to make sure that my technical work is valid or correct (I don't want major mistakes or flaws; I bend over backwards to validate my findings).
- I am open to new beliefs and ideas, especially when these ideas are aligned with the knowledge and the beliefs of the professional community (I don't get stuck thinking that I am always right, my ideas are best, or that I know everything; by being open to new ideas, I open myself up to learning).

Regarding "how to do critical thinking," we teach and we apply the **Standard Structure of CT** (Fig. 1.2), which involves three methods:

1. **Issue**. Define the problem you are trying to solve so that is clear and unambiguous. Note that you will often need to rewrite or paraphrase the issue or question.

2. **Reasoning**. List the reasons that explain why professionals should accept your claim (i.e., your answer, your explanation, your conclusion, or your recommendation). To create your reasoning, take actions such as stating facts, citing references, defining terms, applying deductive logic, applying inductive logic, and building subconclusions.

3. **Conclusions**. State your claim. Make sure your claim addresses the issue. Recognize that a claim can be presented in multiple ways, such as an answer, a recommendation, or your stance on a controversial issue.

1.2 How Materials are Idealized

To understand the behavior of materials, engineers apply a few simple ideas. This section presents some of these ideas.

The Microscopic and Macroscopic Descriptions

Engineers strive to understand things. For example, an engineer might ask, why does Steel Alloy #1 fail given that Steel Alloy #2 does not fail in the same application? Or, an engineer might ask, why does water boil? Why does this boiling sometimes damage materials, as in cavitation*? To address questions about materials, engineers often apply the following ideas:

- **Microscopic Description**. Explain something about a material by describing what is happening at the atomic level (i.e., describing the atoms, molecules, electrons, etc.).

- **Macroscopic Description**. Explain something about a material without resorting to descriptions at the atomic level.

Forces between Molecules

One of the best ways to understand materials is to apply the idea that molecules attract one another if they are close together and repel if they are too close[†] (Fig. 1.3).

Defining the Liquid, Gas, and Fluid

In science, there are four states of matter: gas, liquid, solid, and plasma. A **gas** is a state of matter in which the molecules are on average far apart so that the forces between molecules (or atoms) is typically very small or zero. Consequently, a gas lacks a fixed shape, and it also lacks a fixed volume, because a gas will expand to fill its container.

A **liquid** is a state of matter in which the molecules are on average close together so that the forces between molecules (or atoms) are strong. In addition, the molecules are relatively

FIGURE 1.3

A description[‡] of the forces between molecules.

When two molecules are far apart (on average), there is no force between them. This is like the molecules in an *ideal gas*.

If two molecules are close enough, there is an attractive force (on average). This is like the molecules in a *gas that cannot be modeled with the ideal gas law*.

At a certain distance, there is a maximum attractive force between two molecules. This is like the molecules in a *liquid* or *solid*.

However, if two molecules are too close, there is a strong repulsive force between these molecules. This is why both liquids and solids are difficult to compress.

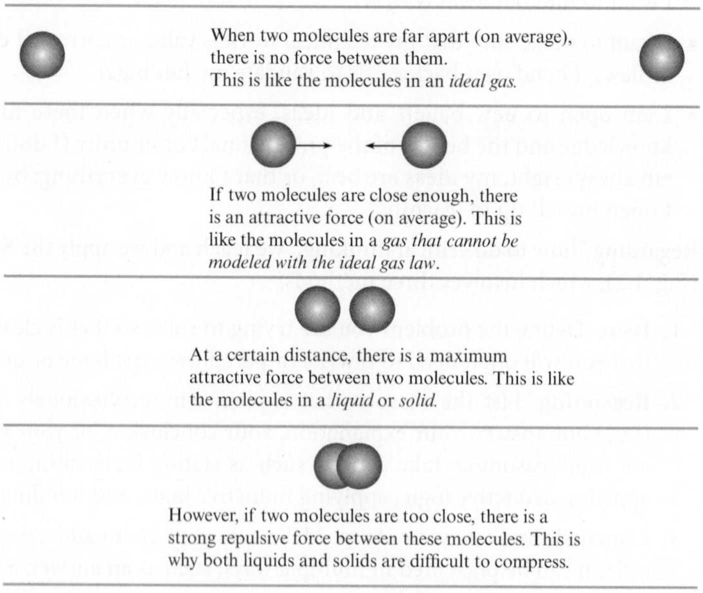

*Cavitation is explained in §5.5.

[†]Dr. Richard Feynman, who won the Nobel Prize in Physics, calls this the *single most important idea in science*. See the *Feynman Lectures on Physics*, Vol. 1, p. 2.

[‡]For additional details about forces between molecules, consult an expert source, such as a chemistry text or a professor who teaches material science.

TABLE 1.1 Comparison of Solids, Liquids, and Gases

Attribute	Solid	Liquid	Gas
Typical Visualization			
Description	Solids hold their shape; no need for a container	Liquids take the shape of the container and will stay in an open container	Gases expand to fill a closed container
Mobility of Molecules	Molecules have low mobility because they are bound in a structure by strong intermolecular forces	Molecules move around freely even though there are strong intermolecular forces between the molecules	Molecules move around freely with little interaction except during collisions; this is why gases expand to fill their container
Typical Density	Often high; e.g., the density of steel is 7700 kg/m^3	Medium; e.g., the density of water is 1000 kg/m^3	Small; e.g., the density of air at sea level is 1.2 kg/m^3
Molecular Spacing	Small—molecules are close together	Small—molecules are held close together by intermolecular forces	Large—on average, molecules are far apart
Effect of Shear Stress	Produces deformation	Produces flow	Produces flow
Effect of Normal Stress	Produces deformation that may associate with volume change; can cause failure	Produces deformation associated with volume change	Produces deformation associated with volume change
Viscosity	NA	High; decreases as temperature increases	Low; increases as temperature increases
Compressibility	Difficult to compress; bulk modulus of steel is 160×10^9 Pa	Difficult to compress; bulk modulus of liquid water is 2.2×10^9 Pa	Easy to compress; bulk modulus of a gas at room conditions is about 1.0×10^5 Pa

free to move around. In comparison, when a material is in the solid state, atoms tend to be fixed in place—for example, in a crystalline lattice. Thus, a liquid flows easily as compared to a solid. Due to the strong forces between molecules, a liquid has a fixed volume but not a fixed shape.

The term **fluid** refers to both a liquid and a gas and is generally defined as a state of matter in which the material flows freely under the action of a shear stress.*

Table 1.1 provides additional facts about solids, liquids, and gases. Notice that many features in this table can be explained by applying the ideas in Fig. 1.3. **Example**. The density of a liquid or a solid is much higher than the density of a gas because the strong attractive forces in a liquid or solid act to bring the molecules closer together. **Example**. A liquid is difficult to compress because the molecules will have strong repulsive forces if they are brought close together. In contrast, a gas is easy to compress because there are no forces (on average) between the molecules.

*Shear stress is explained in §2.4.

FIGURE 1.4

To find examples of material particles: (1) Select any body; for example, we selected a steel tank filled with water and air. (2) Select a small amount of matter and define this small chunk of matter as a material particle. This figure shows a material particle comprised of air, a material particle comprised of water, and a material particle comprised of steel.

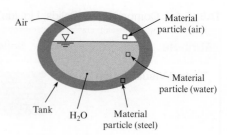

The Body, the Material Particle, the Body-as-a-Particle

Engineers have invented terms that can be used to describe *any material*. Learning this vocabulary will help you learn engineering.

In engineering, the term "body" or "material body" has a special meaning. Examples: A coffee cup can be a *body*. The air inside a basketball can be a *body*. A jet airplane can be a *body*. **Body** or **material body** is a label to identify objects or matter that exists in the real world, without specifying any specific object. It is like applying the term "sports" to identify many activities (e.g., soccer, tennis, golf, or swimming) without specifying a particular sport.

A **material particle** is a small region of matter within a *material body* (Fig. 1.4). Some useful facts about material particles are as follows:

- A material particle is often imagined to be *infinitesimal* in the calculus sense.
- A material particle can be selected or visualized so that it has any shape (e.g., spherical, cubical, cylindrical, or amorphous*).
- The term "fluid particle" refers to a material particle that is comprised of a liquid or a gas.

There is another way that engineers use the term "particle." For example, to model the motion of an airplane, an engineer can idealize the airplane as a *particle*. A physics book might state that Newton's second law of motion only applies to a *particle*. In this context, the term has a different meaning than *material particle*. This alternative concept is that the **particle (the body-as-a-particle)** is a way to idealize a material body as if all the mass of the body is concentrated at a point and the dimensions of the body are negligible.

Summary. There are two distinct concepts used in engineering: the *material particle* and the *body-as-a-particle*. However, it is common for the label "particle" to be used for both of these ideas. Engineers typically figure out which idea is meant by the context in which the term is being used.

The Continuum Assumption

Because a body of fluid is comprised of molecules, properties are due to average molecular behavior. That is, a fluid usually behaves as if it were comprised of continuous matter that is infinitely divisible into smaller and smaller parts. This idea is called the **continuum assumption**.

When the continuum assumption is valid, engineers can apply limit concepts from differential calculus. A limit concept typically involves letting a length, an area, or a volume approach zero. Because of the continuum assumption, fluid properties such as density and velocity can be considered continuous functions of position with a value at each point in space.

To gain insight into the validity of the continuum assumption, consider a hypothetical experiment to find density. Fig. 1.5a shows a container of gas in which a volume ΔV has been

---

*"Amorphous" means without a clearly defined shape or form.

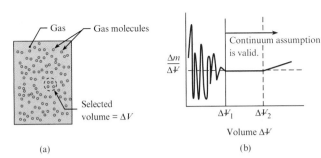

FIGURE 1.5

When a measuring volume ΔV is large enough for random molecular effects to average out, the continuum assumption is valid.

identified. The idea is to find the mass of the molecules Δm inside the volume and then to calculate density by

$$\rho = \frac{\Delta m}{\Delta V}$$

The calculated density is plotted in Fig. 1.5b. When the measuring volume ΔV is very small (approaching zero), the number of molecules in the volume will vary with time because of the random nature of molecular motion. Thus, the density will vary as shown by the wiggles in the blue line. As volume increases, the variations in calculated density will decrease until the calculated density is independent of the measuring volume. This condition corresponds to the vertical line at ΔV_1. If the volume is too large, as shown by ΔV_2, then the value of density may change due to spatial variations.

In most applications, the continuum assumption is valid, as shown by the next example.

EXAMPLE. Probability theory shows that including 10^6 molecules in a volume will allow the determination of density to within 1%. Thus, a cube that contains 10^6 molecules should be large enough to accurately estimate macroscopic properties such as density and velocity. Find the length of a cube that contains 10^6 molecules. Assume room conditions. Do calculations for (a) water and (b) air.

Solution. (a) The number of moles of water is $10^6/6.02 \times 10^{23} = 1.66 \times 10^{-18}$ mol. The mass of the water is $(1.66 \times 10^{-18}$ mol$)(0.0180$ kg/mol$) = 2.99 \times 10^{-20}$ kg. The volume of the cube is $(2.99 \times 10^{-20}$ kg$)/(999$ kg/m$^3) = 2.99 \times 10^{-23}$ m^3. Thus, the length of the side of a cube is 3.1×10^{-8} m. (b) Repeating this calculation with air gives a length of 3.5×10^{-7} m.

Review. For the continuum assumption to apply, the object being analyzed would need to be larger than the lengths calculated in the solution. If we select 100 times larger as our criteria, then the *continuum assumption applies* to objects with

- Length $(L) > 3.1 \times 10^{-6}$ m (for liquid water at room conditions)
- Length $(L) > 3.5 \times 10^{-5}$ m (for air at room conditions)

Given the two length scales just calculated, it is apparent that the *continuum assumption applies to most problems of engineering importance.* However, there are a few situations where the problem length scales are too small.

EXAMPLE. When air is in motion at a very low density, such as when a spacecraft enters the earth's atmosphere, the spacing between molecules is significant in comparison to the size of the spacecraft.

EXAMPLE. When a fluid flows through the tiny passages in nanotechnology devices, the spacing between molecules is significant compared to the size of these passageways.

1.3 Weight, Mass, and Newton's Law of Gravitation

This section reviews weight and mass and also introduces ideas (called the "voice of the engineer") that will help you learn fluid mechanics better.

Voice of the Engineer. *Build working knowledge in every subject that you learn.* Working knowledge is defined as knowledge that you have firmly locked into your brain (no need to look up anything) that is useful for engineering tasks. **Rationale.** Working knowledge is essential for estimation and validation, and these two skills are essential for doing engineering well. Examples of working knowledge are as follows:

- 1.0 pound of force (i.e., 1.0 lbf) is about 4.5 newtons.
- 1.0 horsepower is about 750 watts.
- The weight of water at typical room conditions is about 10,000 newtons for each cubic meter.

Voice of the Engineer. Learn the meaning of main concepts such as mass and force. **Rationale.** Understanding concepts and the relationships between these concepts is needed if you want to apply your knowledge.

Defining Mass

The *mass* of 1.0 liters of liquid water at room conditions is 1.0 kilograms. A body with a *mass* of 2.0 slugs has a *mass* of 29 kilograms. In Newton's second law, the sum-of-forces-vector is exactly balanced by the product of the *mass* and the acceleration. **Mass** is a property of a *body* that provides a measure of the amount of matter in the body. For example, Body *A*, which has a mass of 20 grams, has more matter than Body *B*, which has a mass of 5 grams.

Recommended working knowledge. Know four mass units (kilograms, grams, slugs, and pounds mass) and be able to convert between these units without the need of a calculator.* Regarding conversion formulas, see Table F.1, which is located on the inside cover of this text.

Defining Force

When water falls in a waterfall, we can say that the earth is pulling on the water with a force that is called the *gravity force*. When wind blows on a stop sign, we can say that the air is exerting a *drag force* on the sign. When water behind a dam pushes on the dam, we can say that the water is exerting a *hydrostatic force* on the face of the dam.

Some facts about force are as follows:

- Every force can be thought of as a push or a pull of one body on another.
- Force is a vector. In this text, we use a bold face roman font (e.g., **F**) to represent a vector. To represent the magnitude of a vector, we use an italic font (e.g., *F*).
- *Recommended working knowledge.* Know two force units: pounds-force (lbf) and newtons (N). Be able to convert units (i.e., make estimates) without the need of a calculator.
- Forces classify into two categories:
  - A **surface force** is any force that requires the two bodies to be touching. Most forces are surface forces. Some books use the term *contact force*.

*Your accuracy should be typical of an engineering estimate—for example, within 10% of the number you would get if you used a calculator.

- A **body force** is any force that does not require the two bodies to be touching. There are only a few types of body forces (e.g., the *gravity force*, the *electrostatic force*, and the *magnetic force*).

- Another way to describe forces is to talk about *action forces* (a force that acts to cause a body to accelerate) and *reaction forces* (a force that acts to prevent a body from accelerating; typically a force from a support). We do not use the concepts of action and reaction forces in this textbook.

In summary, a **force** is a push or pull between two bodies. A push or pull is an action that will cause a body to accelerate if the sum-of-forces vector in Newton's Second Law of Motion is nonzero.

Equation Literacy

Voice of the Engineer. Build equation literacy in all your engineering subjects. **Rationale.** Equation literacy is essential for building math models, and building math models is arguably the most important skill of the engineering method.

You have **equation literacy** for equation XYZ if you can do the following tasks: (1) You can explain how the equation was derived or where the equation came from. (2) You can explain the main ideas—that is, the physical interpretation—of the equation. (3) You can list the common equational forms, define each variable, and state the units and dimensions. (4) You can describe the assumptions and limitations of the equation and make correct choices about when to apply this equation or when to avoid applying this equation. (5) You have a systematic method for applying the equation correctly.

Newton's Law of Universal Gravitation (NLUG)

Newton's Law of Universal Gravitation (NLUG) reveals that any two bodies will attract each other with a force **F**, which is called the gravitational force (Fig. 1.6). Because this idea applies to any two bodies located anywhere in the universe, the equation is *universal* (hence the name).

The magnitude of the gravitational force F is given by

$$F = G \frac{m_1 m_2}{R^2} \tag{1.1}$$

where the term $G = 6.674 \times 10^{-11} \, \text{N·m}^2/\text{kg}^2$ is called the *gravitational constant*, m_1 is the mass of Body #1, m_2 is the mass of Body #2, and R is the distance between the center of mass of each body.

The law of gravity, like nearly all scientific laws, was developed by *inductive reasoning*. In particular, Newton examined data on planetary motion and found that the data were fit by Eq. (1.1). Newton concluded that the equation must be true in general.

To apply Eq. (1.1) on earth, start with Fig. 1.6 and let Body #1 represent the earth and Body #2 represent a body that is on or near the surface of the earth. Now, G and m_1 are constant and R is very nearly constant. Thus, define a new constant g that is given by $g \equiv Gm_E/R_E^2$, in

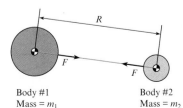

Body #1
Mass = m_1

Body #2
Mass = m_2

FIGURE 1.6

Any two bodies will attract one another. The corresponding force is called the **gravitational force**. Note that the magnitude of the gravitational force on Body #1 equals the magnitude of the gravitational force on Body #2.

which the subscript E denotes "earth." Also, rename the gravitational force F to be the weight of the body W. Then, Eq. (1.1) simplifies to

$$W = mg \tag{1.2}$$

where W is the weight of a body on a planet (typically Earth), m is the mass of the body, and g is a constant.

Useful Facts and Information

- The constant g is called **gravitational acceleration**. On earth, this parameter varies slightly with altitude; however, engineers commonly use the standard value, which is $g = 9.80665$ m/s^2 = 32.1740 ft/s^2.
- Gravitational acceleration (g) has a useful physical interpretation; g is the vertical component of acceleration that results when the vertical component of the sum-of-forces-vector in Newton's Second Law of Motion is exactly equal to the weight vector.
- In general, a falling body will not accelerate at a rate g because of the presence of additional forces, such as the lift force, the drag force, or the buoyant force.
- It is common for people to state that $W = mg$ is derived from $\Sigma \mathbf{F} = m\mathbf{a}$. However, it is more correct to say that $W = mg$ is derived from NLUG.
- **Weight** is the gravitational force acting on a body from a planet (typically Earth).
- Weight and mass are different concepts. For example, the mass of a body is the same at any location, whereas the weight can change with location. For example, if a body weighs 60 newtons on Earth, the same body will weigh about 10 newtons on the moon. Also, recognize that it is common (but incorrect) to report a weight using mass units. For example, to say that a body weighs 10 grams or that a body weighs 60 kg is incorrect.

Relating Force and Mass Units

We wrote this section because we have seen many mistakes involving force and mass units. Three useful ideas about units are (1) units were invented by people, (2) units are related to each other by equations, and (3) the definition of a given unit can be looked up.

The definition of a newton is "one newton of force is the quantity of force that will give one kilogram of mass an acceleration of one meter per second squared."

To relate force and mass units, engineers start with Newton's Second Law of Motion ($\Sigma \mathbf{F} = m\mathbf{a}$). Next, apply the definition of the newton to conclude that it must be true that

$$(1.0 \text{ N}) \equiv (1.0 \text{ kg})(1.0 \text{ m/s}^2) \tag{1.3}$$

Since Eq. (1.3) is true, it must also be true (by algebra) that

$$1.0 = \left(\frac{\text{kg} \cdot \text{m}}{\text{N} \cdot \text{s}^2} \right) \tag{1.4}$$

Thus, the weight of a 2.0 kilogram body must be 19.6 N because of the analysis shown in Eq. (1.5):

$$W = mg = \frac{2.0 \text{ kg}}{} \frac{9.81 \text{ m}}{\text{s}^2} \frac{\text{N} \cdot \text{s}^2}{\text{kg} \cdot \text{m}} = \boxed{19.6 \text{ N}} \tag{1.5}$$

Do you see the logic? Eq. (1.5) must be true because it is based on correct facts that are applied in a correct way. The main issue that we want to address is that many people become confused with English units. However, with English units you can apply the same logic. In particular, start with the definition of the pound-force (lbf). One pound of force is the amount of force that will accelerate one pound of mass (lbm) at a rate of 32.2 feet per second squared. Thus, it is true that

$$(1.0 \text{ lbf}) \equiv (1.0 \text{ lbm})(32.2 \text{ ft/s}^2) \tag{1.6}$$

Since Eq. (1.6) is true, it must also be true (by algebra) that

$$1.0 = \left(\frac{\text{lbf} \cdot \text{s}^2}{32.2 \text{ lbm} \cdot \text{ft}} \right) \tag{1.7}$$

Thus, the weight of a 2.0 lbm body must be 2.0 lbf because of the analysis shown in Eq. (1.8):

$$W = mg = \frac{2.0 \text{ lbm}}{} \left| \frac{32.2 \text{ ft}}{\text{s}^2} \right| \frac{\text{lbf} \cdot \text{s}^2}{32.2 \text{ lbm} \cdot \text{ft}} = \boxed{2.0 \text{ lbf}} \tag{1.8}$$

Eq. (1.8) shows that the magnitude of the weight (2.0) is the same as the magnitude of the mass (2.0). This occurs because of the way that English units are defined. It is correct to say that a body that has a mass of 2.0 lbm will have a weight of 2.0 lbf on earth. However, avoid generalizing this. For example, a body with a mass of 2.0 lbm will have a weight about 0.33 lbf on the moon. Also, avoid saying that 2.0 lbm equals 2.0 lbf, because mass and weight are different concepts.

The General Equation

A **general equation** is an equation that applies to many or to all problems. A **special-case equation** is an equation that is derived from a general equation but is more limited in scope because there are assumptions that must to be met in order to apply the special-case equation.

   Voice of the Engineer. *Learn the general equations and then derive each special-case equation on an as needed basis*. **Rationale**. (1) Given that there are only a few general equations, this approach will make your learning simpler. (2) You are less likely to make mistakes because general equations, by definition, apply more often than special-case equations. Examples of general and special case equations follow:

- NLUG is a general equation, and $W = mg$ is a special-case equation that is derived from NLUG.
- Newton's Second Law of Motion, $\Sigma \mathbf{F} = m\mathbf{a}$, is a general equation; note that this is a vector equation. Some special-case equations that can be derived from this equation are $\Sigma F_x = ma_x$ (a scalar equation) and $\Sigma F_z = 0$ (also a scalar equation).
- The general equation that defines mechanical work W is the line integral of the force vector dotted with the displacement vector $W = \oint_{x_1}^{x_2} \mathbf{F} \cdot \mathbf{dx}$. One associated special-case equation is $W = Fd$, where W is work, F is force, and d is displacement.

1.4 Essential Math Topics

Estimates

Voice of the Engineer. *Become skilled with pencil/paper estimates*. A **pencil/paper estimate** is defined as an estimate that you can do using only your brain, a pencil, and a sheet of paper (i.e., no books, calculators, or computers needed). **Rationale**: (1) All engineering calculations are estimates anyway; learning pencil/paper estimation skills will give you great insight into the nature of engineering estimates. (2) In the process of learning how to do pencil/paper estimates, you will acquire a great deal of practical knowledge. (3) You will save yourself huge amounts of time because you will do calculations much more quickly. (4) You will have strong methods for validating your technical work. (5) It is fun to figure out clever ways to estimate things.

Four Tips for Representing Numbers

To represent your numerical results in simple and effective ways, we have four recommendations:

1. Represent your result so that the digit term is between 0.1 and 1000; this makes your result easier to understand and remember. For example, 645798 can be represented as 646E3 or as 64.6E4 or as 6.46E5.

2. Use scientific or engineering notation to represent large and small numbers.

3. Use metric prefixes to represent numbers; for example, 142,711 pascals can be represented as 143 kPa.

4. Use a maximum of three significant figures to represent your final answers (unless you can justify more significant figures).

Scientific notation is a method of writing a number as a product of two numbers: a digit term and an exponential term. For example, the number 7600 is written as the product of 7.6 (the digit term) and 10^3 (the exponential term) to give 7.6×10^3. **Fact**. There are three common forms of scientific notation, which are as follows: $7.6 \times 10^3 = 7.6E3$ (upper case "E") $= 7.6e3$ (lower case "e"). Avoid mixing up the "e" that is used in scientific notation with Euler's number, which is $e = 2.718$.

Engineering notation is a version of scientific notation in which the powers of 10 are written as multiples of three. **Example**. $0.000475 = 4.75E\text{-}4$ (scientific notation) $= 0.475E\text{-}3$ (engineering notation) $= 475E\text{-}6$ (engineering notation). **Example**. $692000 = 6.92E5$ (scientific notation) $= 0.692e6$ (engineering notation).

Unit Prefixes (Metric System). In the SI system, it is common to use prefixes on units to multiply or divide by powers of 10. **Example**. 0.001 newtons $= 1.0$ mN. **Example**. 0.000475 m $= 0.475$ mm $= 475$ μm. **Example**. 1000 pascals $= 1.0$ kPa.

Significant Figures. When a number is reported with three significant figures (e.g., 1.97), this means that two of the digits are known with precision (i.e., the 1 and the 9), and one of the digits (i.e., the 7) is an approximation. The rationale for significant figures is that values in engineering (e.g., the density of water is about 998 kg/m^3) ultimately come from measurements, and measurements can only provide certain levels of precision. In this text, we report answers with three significant figures. Of course, during intermediate calculations, you should carry more than three significant digits to prevent rounding errors.

Thinking with the Derivative

We have seen many mistakes because the main idea of the derivative was not in place. Thus, we wrote this subsection to explain this idea in detail.

To describe a common mistake, we'll give an example of this mistake. Suppose you were asked to answer the following true/false question. **(T/F)**. If a car has traveled in a straight line for $\Delta x = 10.0$ meters during a time interval of $\Delta t = 2.5$ seconds, then its speed at the end of the time interval is $(10.0 \text{ m})/(2.5 \text{ s}) = 4.0$ m/s.

It seems like one could answer this question as true, because $V = (\Delta x)/(\Delta t) = (10.0 \text{ m})/(2.5 \text{ s}) = 4.0$ m/s. However, this answer is only valid if the speed of the car was constant with time. A better answer is to say false, because there is not enough information to reach the conclusion that the car is traveling at 4 m/s at the end of the time interval. The problem we are illustrating is the difference between *average speed* and *instantaneous speed*. The best way to think about speed is to apply the definition of the derivative. In words, speed is the ratio of distance traveled to the amount of time in the limit as the amount of time goes to zero. In equation form (more compact), speed V is defined by

$$V = \lim_{\Delta t \to 0} \frac{\Delta x}{\Delta t} \tag{1.9}$$

If speed is constant, then Eq. (1.9) will automatically simplify to give the equation for average speed. If speed is varying with time, then Eq. (1.9) will give a correct value of speed. Of course,

Eq. (1.9) is based on the definition of the derivative. Regarding this definition, calculus books give the definition in three ways:

$$\frac{dy}{dx} = \lim_{h \to 0} \frac{y(x + h) - y(x)}{h} \tag{1.10}$$

$$= \lim_{\Delta x \to 0} \frac{y(x + \Delta x) - y(x)}{\Delta x} \tag{1.11}$$

$$= \lim_{\Delta x \to 0} \frac{\Delta y}{\Delta x} \tag{1.12}$$

We apply the last definition, Eq. (1.12), in multiple places in this text. This definition shows that the derivative means the ratio of Δy to Δx in the limit as Δx goes to zero. Note that the delta symbol (i.e., the triangle) preceding the variable y denotes an amount or quantity of the variable y.

Thinking with the Integral

The integral was invented to solve problems in which rates change with time. To build up the definition of the integral, we note that it is tempting to state that the distance a car travels (Δx) is given by $\Delta x = V \Delta t$, where V is the speed and Δt is the time that the car has been traveling. The problem with this formula is that it does not apply in general, because speed can be changing. To modify the formula so that it is more general, one can do the following:

$$\Delta x = \sum_{i=1}^{N} V_i \, \Delta t_i \tag{1.13}$$

where the motion has been divided into time intervals. Here, Δt_i is a small time interval, V_i is the speed during this time interval, and N is the number of time intervals. To make this formula more accurate, we can let $N \to \infty$, and we arrive at a general formula for distance traveled:

$$\Delta x = \lim_{N \to \infty} \sum_{i=1}^{N} V_i \, \Delta t_i \tag{1.14}$$

Now, the summation on the right-hand side of Eq. (1.14) can be modified by applying the definition of the integral to give

$$\Delta x = \int_{0}^{t_f} V \, dt \tag{1.15}$$

In calculus texts, you will find the following definition of the integral:

$$\int_{a}^{b} f(x) \, dx = \lim_{N \to \infty} \sum_{i=1}^{N} f(x_i) \, \Delta x_i \tag{1.16}$$

Thus, the integral is an infinite sum of small terms that is applied when a dependent variable f is changing in response to changes in the independent variable x.

1.5 Density and Specific Weight

Solving most problems in fluids requires calculation of mass or weight. These calculations involve the properties of density and specific weight, which are presented in this section.

Defining Density

For a simple problem, density (ρ) can be found by taking the ratio of mass (Δm) to volume ($\Delta \forall$) as in

$$\rho = \frac{\Delta m}{\Delta \forall} \tag{1.17}$$

For example, if you were to take 1.0 liter of water at room conditions and measured the mass, the amount of mass would be (Δm) ≈ 1000 grams, and so the density would be

$$\rho = \Delta m / \Delta \forall = (1000 \text{ grams})/(1.0 \text{ liter}) = 1.0 \text{ kg/L}$$

> **EXAMPLE.** What is the mass of 2.5 liters of water? **Reasoning.** (1) The mass is given by $\Delta m = \rho(\Delta \forall)$. (2) The density of water at room conditions is about 1.0 kg/L. (3) Thus, the mass is $\Delta m = (1.0 \text{ kg/L})(2.5 \text{ L}) = 2.5 \text{ kg}$.

Eq. (1.17) defines average density, not the density at a point. To build a more general definition of density, apply the concept of the derivative (see §1.4). In general, density is defined using the derivative as shown in Eq. (1.18):

$$\rho \equiv \lim_{\Delta \forall \to 0} \frac{\Delta m}{\Delta \forall} \tag{1.18}$$

where $\Delta \forall$ denotes the volume of a tiny region of material surrounding a point (e.g., an x,y,z location) and Δm is the corresponding amount of mass that is contained within this region. Thus, density can be defined as the ratio of mass to volume at a point.

Some useful facts about density are as follows:

- You can look up density values in the front of the book (Tables F.4 to F.6) and in the appendices (Tables A.2 to A.5).

- In general, the value of the density will vary with the pressure and temperature of the material. For a liquid, the variation with pressure is usually negligible.

- The density of a gas is often calculated by applying the *density form* of the ideal gas law: $p = \rho RT$.

- To calculate the amount of mass in a given volume, it is tempting to apply $\Delta m = \rho \Delta \forall$. However, this equation is a special-case equation, not a general equation. The general equation that accounts for the fact that density can vary with position is

$$m = \int_\forall \rho \, d\forall \tag{1.19}$$

- *Recommended working knowledge.* Know the density of liquid water at typical room conditions in common units: $\rho = 1000 \text{ kg/m}^3 = 1.0 \text{ gram/mL} = 1.0 \text{ kg/L} = 62.4 \text{ lbm/ft}^3 = 1.94 \text{ slug/ft}^3$. Know the density of air at atmospheric pressure and 20°C: $\rho = 1.2 \text{ kg/m}^3 = 1.2 \text{ g/L}$.

Defining Specific Weight

Specific weight is the ratio of weight to volume at a point:

$$\gamma \equiv \lim_{\Delta \forall \to 0} \frac{\Delta W}{\Delta \forall} \tag{1.20}$$

where $\Delta \mathbb{V}$ denotes the volume of a tiny region of material surrounding a point (e.g., an x,y,z location) and ΔW is the corresponding weight of the mass that is contained within this region. Specific weight and density are related by this equation:

$$\gamma = \rho g \tag{1.21}$$

Thus, if you know one property, you can easily calculate the other. **Example**. The specific weight corresponding to a density of 800 kg/m³ is $\gamma = (800 \text{ kg/m}^3)(9.81 \text{ m/s}^2) = 7.85 \text{ kN/m}^3$.

The reasoning to show that Eq. (1.21) is true involves the following steps. (1) On Earth, NLUG simplifies to $W = mg$. (2) Divide $W = mg$ by volume to give $(\Delta W / \Delta \mathbb{V}) = (\Delta m / \Delta \mathbb{V})g$. (3) Take the limit as volume goes to zero. (4) Apply the definitions of γ and ρ to give $\gamma = \rho g$.

Some useful facts about specific weight are as follows:

- You can look values of γ in the front of the book (Tables F.4 to F.6) and in the back of the book (Tables A.3 to A.5).
- Since ρ and γ are related via Eq. (1.21), γ varies with temperature and pressure in a similar fashion to density.
- Specific weight is commonly used for liquids but is not commonly used for gases.
- *Recommended working knowledge.* Know the specific weight of liquid water at typical room conditions: $\gamma = 9800 \text{ N/m}^3 = 9.80 \text{ N/L} = 62.4 \text{ lbf/ft}^3$.

1.6 The Ideal Gas Law (IGL)

The IGL is commonly applied in fluid mechanics. For example, you will likely apply the IGL when you are designing products such as air bags, shock absorbers, combustion systems, and aircraft.

The IGL, the Ideal Gas, and the Real Gas

The IGL was developed by the logical method called induction. *Induction* involves making many experimental observations and then concluding that something is always true because every experiment indicates this truth. For example, if a person concludes that the sun will rise tomorrow because it has risen every day in the past, this is an example of inductive reasoning.

The IGL was developed by combining three empirical equations that had been discovered previously. The first of these equations, called Boyle's law, states that when temperature T is held constant, the pressure p and volume $\mathbb{V}$ of a fixed quantity of gas are related by

$$p\mathbb{V} = \text{constant} \quad \text{(Boyle's law)} \tag{1.22}$$

The second equation, Charles's law, states that when pressure is held constant, the temperature and volume $\mathbb{V}$ of a fixed quantity of gas are related by

$$\frac{\mathbb{V}}{T} = \text{constant} \quad \text{(Charles's law)} \tag{1.23}$$

The third equation was derived by a hypothesis formulated by Avogadro: *Equal volumes of gases at the same temperature and pressure contain equal number of molecules.* When Boyle's law, Charles's law, and Avogadro's law are combined, the result is the ideal gas equation in this form:

$$p\mathbb{V} = nR_u T \tag{1.24}$$

where n is the amount of gas measured in units of moles.

Eq. (1.24) is called the *$p\mathbb{V}T$ form* or the *mole form* of the IGL. **Tip**. There is no need to remember Charles's law or Boyle's law because they are both special cases of the IGL.

The ideal gas and the real gas can be defined as follows:

- An **ideal gas** refers to a gas that can be modeled using the ideal gas equation, Eq. (1.24), with an acceptable degree of accuracy; for example, calculations have less than a 5% deviation from the true values. Another way to define an ideal gas is to state that an ideal gas is any gas in which the molecules do not interact, except during collisions.

- A **real gas** refers to a gas that cannot be modeled using the ideal gas equation, Eq. (1.24), with an acceptable degree of accuracy because the molecules are close enough together (on average) that there are forces between the molecules. Although real gas behavior can be modeled, the equations are more complex than the IGL. Thus, the IGL is the preferred model if it provides an acceptable level of accuracy.

For every problem that we (the authors) have solved, the IGL has provided a valid model for gas behavior; that is, we have never needed to apply the equations used to model real gas behavior. However, there are a few instances in which you should be careful about applying the IGL:

- When a gas is very cold or under very high pressure, the molecules can move close enough together to invalidate the IGL.

- When both the liquid phase and the gas phase are present (e.g., propane in a tank used for a barbecue), you might want to be careful about applying the IGL to the gas phase.

- When a gas is very hot, such as the exhaust stream of a rocket, the gas can ionize or disassociate. Both of these effects can invalidate the ideal gas law.

Also, the IGL works well for modeling a mixture of gases. The classic example is air, which is a mixture of nitrogen, oxygen, and other gases.

Units in the IGL

Because we have seen many mistakes with units, we wrote this subsection to give you the essential facts so that you can avoid most of these mistakes and also save time.

Temperature in the IGL must be expressed using *absolute temperature*. Absolute temperature is measured relative to a temperature of absolute zero, which is the temperature at which (theoretically) all molecular motion ceases. The SI unit of absolute temperature is Kelvin (K with no degree symbol, as in 300 K). A temperature given in Celsius (°C) can be converted to Kelvin using this equation: $T(K) = T(°C) + 273.15$. For example, a temperature of 15°C will convert to 15°C + 273 = 288 K. The English unit of absolute temperature is Rankine; for example, a temperature of 70°F is the same as a temperature of 530°R. A temperature given in Fahrenheit (°F) may be converted to Rankine using this equation: $T(°R) = T(°F) + 459.67$ For example, a temperature of 65°F will convert to 65°F + 460 = 525°R.

Pressure in the IGL must be expressed using *absolute pressure*. Absolute pressure is measured relative to a perfect vacuum, such as outer space. Now, it is common in engineering to give a value of pressure that is measured relative to local atmospheric pressure; this is called *gage pressure*. To convert a gage pressure to absolute pressure, add the value of local atmospheric pressure. For example, if the gage pressure is 20 kPa and the local atmospheric pressure is 100 kPa, then the absolute pressure will be 100 kPa + 20 kPa = 120 kPa. If the local atmospheric pressure is unavailable, then use the standard value of atmospheric pressure, which is 101.325 kPa (14.696 psi or 2116.2 psf). More details about pressure are presented in §3.1.

The IGL also uses the **mole**, defined as the amount of material that has the same number of "entities" (atoms, molecules, ions, etc.) as there are atoms in 12 grams of carbon 12 (C^{12}). Think of the mole as a way to count *how many*. By analogy, the dozen is also a unit for counting how many; for example, three dozen donuts is a way of specifying 36 donuts. The number of

TABLE 1.2 Selected Values of Molar Mass

Substance	Molar Mass (grams/mole)
Hydrogen	1.0079
Helium	4.0026
Carbon	12.0107
Nitrogen N_2	14.0067
Oxygen O_2	15.9994
Dry Air	28.97

atoms in 12.0 grams of carbon 12 is equal to 1 mole of atoms. This number, called **Avogadro's number**, is 6.022×10^{23} entities. There are three different mole units in use:

- A gram mole (mol) has 6.022×10^{23} entities (atoms, molecules, etc.).
- A kilogram mole (kg-mol) has (6.022E2)(1000 grams/kg) = 6.022E26 entities.
- A pound-mass mole (lbm-mol) has (454.3 g/lbm)(6.022×10^{23}) = 2.732E26 entities.

Another unit issue arises because the amount of matter can be characterized by using either moles or mass. Moles and mass units are related by the **molar mass**, which is defined by

$$M = \frac{\text{amount of mass}}{\text{number of moles}} = \frac{m}{n} \tag{1.25}$$

Values of molar mass can be looked up on the Internet. Some typical values are also listed in Table 1.2.

EXAMPLE. What is the mass (in kg) of 2.7 moles of air?

Solution. $m = nM$ = (28.97E-3 kg/mol)(2.7 mol) = 78.2E-3 kg.

The Universal and Specific Gas Constant (R_u and R)

In the IGL, there are two gas constants: the universal gas constant and the specific gas constant. When you write the IGL like this, $p\forall = nR_uT$, the term R_u is called the universal gas constant. The word "universal" means that this gas constant is the same for every gas. The value of R_u in SI units is R_u = 8.314 462 J/mol·K. The value of R_u in traditional units is R_u = 1545.349 ft·lbf/lbm-mol·°R.

Often, engineers prefer to work with mass units instead of mole units. In this case, the IGL can be modified like this: (1) Start with Eq. (1.24) and substitute $n = m/M$ to give $p\forall = m(R_u/M)T$. (2) Define the specific gas constant (R) using this equation: $R \equiv R_u/M$. **Conclusion**. An alternative way to write the IGL is $p\forall = mRT$, where R is the *specific gas constant*.

Summary. Any time you are using the IGL, figure out whether you need to use R or R_u. As needed, you can relate R and R_u using this equation:

$$R = \frac{R_u}{M} \tag{1.26}$$

Also, you can find values of the specific gas constant (R) in Table A.2.

EXAMPLE. If 3.0 moles of a gas has a mass of 66 grams, what is the specific gas constant for this gas (SI units)?

TABLE 1.3 The Ideal Gas Law (IGL) and Related Equations

Description	Equation	Variables
Density form of the IGL	$p = \rho RT$	p = pressure (Pa) (use absolute pressure, not gage or vacuum pressure) ρ = density (kg/m^3) R = specific gas constant (J/(kg·K)) (look up R in Table A.2) T = temperature (K) (use absolute temperature)
Mass form of the IGL	$p\mathcal{V} = mRT$	$\mathcal{V}$ = volume (m^3) m = mass (kg)
Mole form of the IGL, or the $p\mathcal{V}T$ form	$p\mathcal{V} = nR_uT$	n = number of moles R_u = universal gas constant (R_u = 8.314 J/(mol·K) = 1545 (ft·lbf)/(lbmol·°R))
Apply this equation to relate R and R_u	$R = \dfrac{R_u}{M}$	M = molar mass (kg/mol)
Apply this equation to relate mass and moles	$M = m/n$	

Reasoning. (1) Since molar mass is the ratio of mass/moles, $M = (0.066 \text{ kg})/(3 \text{ mol}) = 0.022$ kg/mol. (2) Now that M is known, $R = R_u/M = (8.314 \text{ J/mol·K})/(0.022 \text{ kg/mol}) = 378$ J/kg·K.

Conclusion. $R = 378$ J/kg·K.

The IGL (Working Equations)

The purpose of this subsection is to (a) present three equations that are commonly used to represent the IGL and (b) explain the meaning of a working equation. Before we do this, we want to share an idea that we have found to be useful.

Voice of the Engineer. *Become skillful with the working equations in each engineering subject you study.* A **working equation** is defined as an equation that is often used in application. The benefit of using working equations is simplicity; in particular, each engineering subject has about 15 working equations. If you know these equations well, then you know a great deal about the subject. It is true that most engineering textbooks have hundreds of equations in them. This is because the authors are using these equations to explain things, but you do not need to remember most of these equations.

The working equations associated with the IGL are summarized in Table 1.3. Notice that there are three common IGL equations called the density form, the mass form, and the mole form. These equations are equivalent because you can start with one of these equations and derive the other two. Notice that the last column in Table 1.3 provides SI units and tips for application.

1.7 Units and Dimensions

Because math involves abstraction, units are uncommon. In contrast, engineering is about doing practical things, so units are essential because units make engineering calculations more concrete, understandable, and relevant. In addition, using units and dimensions will save you abundant amounts of time because errors can be identified and fixed. Units are so helpful that we carry and cancel units 100% of the time, and we encourage this practice for all engineers that we teach.

Definition of a Unit

Nearly everyone is familiar with units; for example:

- Mass units include the gram, the kilogram, and the pound mass.
- Length units include the meter, the centimeter, the inch, and the foot.
- Time units include the second, the hour, the day, the week, and the year.

In general, a **unit** is a quantity that is chosen as a standard so that one can describe an amount or quantity. That is, units allow quantification (i.e., describing "how much"); for example:

- If newtons are the standard (i.e., the unit) for force, then 5 N quantifies how much push or pull is applied.
- If pounds-mass are the standard (i.e., the unit) for mass, then 50 lbm describes a specific amount of matter.
- If seconds are the standard (i.e., the unit) for time, then 500 s describes a specific amount of time.

The combination of a number plus an associated unit (e.g., 5 N, 50 lbm, or 500 s) is called a *measurement* or a *value*.

The Grid Method

Of the various methods for carrying and canceling units, the *grid method* (Fig. 1.7) is the best method that we have seen. To learn how to apply the grid method, see the method and examples presented in Table 1.4

The essence of the grid method is to multiply the right side of the equation by 1.0 (i.e., the *multiplicative identity*) over and over until the units cancel in a way that gives you the desired unit. For example, in Fig 1.7, the right side of the equation was multiplied by 1.0 three times:

$$1.0 = \frac{1.0 \text{ m/s}}{2.237 \text{ mph}} \text{ (first time)}$$

$$1.0 = \frac{1.0 \text{ N}}{0.2248 \text{ lbf}} \text{ (second time)}$$

$$1.0 = \frac{1.0 \text{ W} \cdot \text{s}}{\text{N} \cdot \text{m}} \text{ (third time)}$$

As shown in the above three examples, a **conversion ratio** is an equation involving numbers and units that can be arranged so that the number 1.0 appears on one side of the equation. **Example.** 100 cm = 1.0 m is a conversion ratio because this equation can be written as 1.0 = (100 cm)/(1.0 m).

$$P = FV = \frac{4.0 \text{ lbf}}{} \frac{20 \text{ mph}}{} \left| \frac{1.0 \text{ m/s}}{2.237 \text{ mph}} \right| \frac{1.0 \text{ N}}{0.2248 \text{ lbf}} \left| \frac{\text{W} \cdot \text{s}}{\text{N} \cdot \text{m}} \right.$$

$$= \boxed{159 \text{ W}}$$

FIGURE 1.7

The *grid method*. This example shows a calculation of the power P required to ride a bicycle at a speed of $V = 20$ mph when the force to move against wind drag is $F = 4.0$ lbf.

TABLE 1.4 Applying the Grid Method (Two Examples)

Step	Example 1	Example 2
Problem Statement =>	**Situation:** Convert a pressure of 2.00 psi to pascals.	**Situation:** Find the force in newtons that is needed to accelerate a mass of 10 g at a rate of 15 ft/s².
Step 1. Write the equation down.	not applicable	$F = ma$
Step 2. Insert numbers and units.	$p = 2.00$ psi	$F = ma = (0.01 \text{ kg})(15 \text{ ft/s}^2)$
Step 3. Look up conversion ratios (see Table F.1).	$1.0 = \dfrac{1 \text{ Pa}}{1.45 \times 10^{-4} \text{ psi}}$	$1.0 = \dfrac{1.0 \text{ m}}{3.281 \text{ ft}} \quad 1.0 = \dfrac{\text{N} \cdot \text{s}^2}{\text{kg} \cdot \text{m}}$
Step 4. Multiply terms and cancel units.	$p = [2.00 \text{ psi}] \left[\dfrac{1 \text{ Pa}}{1.45 \times 10^{-4} \text{ psi}} \right]$	$F = [0.01 \text{ kg}] \left[\dfrac{15 \text{ ft}}{\text{s}^2} \right] \left[\dfrac{1.0 \text{ m}}{3.281 \text{ ft}} \right] \left[\dfrac{\text{N} \cdot \text{s}^2}{\text{kg} \cdot \text{m}} \right]$
Step 5. Do calculations.	$p = 13.8$ kPa	$F = 0.0457$ N

We recommend four methods for finding conversion ratios.

- **Method #1.** Derive the conversion ratio as shown in the following example:
  1. Power is defined as

  $$\text{power} = \frac{\text{work}}{\text{time}}$$

  2. Substituting SI units shows that

  $$1.0 \text{ W} = \frac{1.0 \text{ N} \cdot \text{m}}{1.0 \text{ s}}$$

  3. Algebra shows that

  $$1.0 = \frac{\text{W} \cdot \text{s}}{\text{N} \cdot \text{m}}$$

- **Method #2.** Derive the conversion ratio using data from Table F.1 (front of book). **Example.** To relate the speed units of m/s and mph, find the row labeled "Speed" in Table F.1 and extract the data that 1.0 m/s = 2.237 mph. Then, do algebra to show that 1.0 = (1.0 m/s)/ (2.237 mph). **Example.** To relate the pressure units of kPa and torr, find the row labeled "Pressure/Shear Stress" and extract the data that 6.895 kPa = 51.71 torr. Then, do algebra to show that 1.0 = (6.895 kPa)/(51.71 torr) = (1.0 kPa)/(7.50 torr).

- **Method #3.** Apply a fact; for example, because there are 30.48 centimeters in 1.0 foot, the conversion ratio from meters to feet is 1.0 = (0.3048 m)/(1.0 ft).

- **Method #4.** Use web resources. We recommend Google and www.onlineconversion.com. **Example.** A common way to measure the volume of water in hydrology is to use the unit of acre-feet. However, this unit is not in this textbook. Thus, go to Google and type in "acre-feet to cubic meters," and Google will output "1 acre-feet = 1 233.48184 cubic meters." Then, do algebra to show that 1.0 = (1233 m³)/(1.0 acre-feet).

Consistent Units

Voice of the Engineer. *Before you solve a problem, convert all your units to consistent units (SI preferred), do your analysis, and then report your answer in the units that are the most useful for* *your context*. We call this idea the **Consistent Unit Rule**. The rationale is that this will save you a lot of time, keep your documentation shorter and neater, and eliminate mistakes.

Consistent units are defined as any set of units for which the conversion factors only contain the number 1.0. This means, for example, that

- (1.0 unit of force) = (1.0 unit of mass)(1.0 unit of acceleration),
- (1.0 unit of power) = (1.0 unit of work)/(1.0 unit of time), and
- (1.0 unit of speed) = (1.0 unit of distance)/(1.0 units of time).

> **EXAMPLE.** If length is measured in millimeters and force in newtons, then what is the consistent unit of pressure? **Reasoning**. (1) The definition of consistent units means that (1.0 unit of pressure) = (1.0 unit of force)/(1.0 unit of area). (2) The unit of area in this case is millimeters squared. (3) Combining steps 1 and 2 gives (1.0 unit of pressure) = (1.0 N)/ (1.0 mm^2) = N/mm^2. (4) Because the unit of N/mm^2 is uncommon, it is best to covert this to more familiar units like this: (1.0 N/mm^2) = (1.0 N)/[(10^{-3})^2(1.0 m)^2] = 10^6 N/m^2 = 1.0 MPa. **Conclusion**. The consistent unit of pressure for the given units is MPa (mega pascal).

> **EXAMPLE.** Is the *given set* of units consistent (*given set*: force is in units of pounds-force (lbf), mass is in lbm, and acceleration is in ft/s^2)? **Reasoning**. (1) By definition, (1.0 lbf) = (1.0 lbm)(32.2 ft/s^2). (2) By the definition of consistent units, the only number that can appear is the number 1.0. (3) The number 32.2 is not the number 1.0. (4) Thus, the given set of units cannot be consistent. **Conclusion**. The given set of units is not consistent.

In principle, there are an infinite number of sets of consistent units. Fortunately, people before us have figured out an optimum set—that is, the SI unit system. The best method for using consistent units is to *convert all your units to SI units* (Fig. 1.8).

We recommend that do all your technical work in SI units. However, we also recommend that you become skilled with English units. This is like being able to speak two languages, as in I speak "SI units" and I speak "English units," but making one of the languages (i.e., SI units) your language of choice. Regarding English units, there are actually two systems of units in use. In this text, we combine these two systems and call them "traditional units" or "English units."

Consistent units for both the SI system and the English system are listed in Table 1.5. The way to use this table is to convert all variables in your problem so that they are expressed using only the units listed in Table 1.5. **Example**. Convert the following values so that they have consistent units: ρ = 50 lbm/ft^3, V = 200 ft/min, D = 12 in. **Reasoning**. The method is to convert the given units so that they match the units specified in Table 1.5. The conversions are straightforward, so we do not show these. **Conclusion**. Use ρ = 1.55 slug/ft^3, V = 3.33 ft/s, and D = 1.0 ft.

The Dimension: A Way to Organize Units

Because there are thousands of units, this section will show you a way to organize units into categories called *dimensions*. Dimensions will be used throughout this book and will be

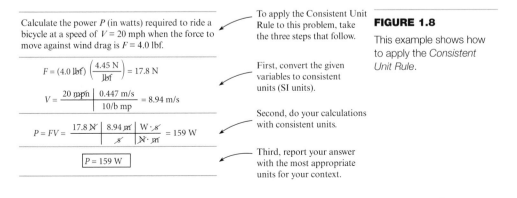

Calculate the power P (in watts) required to ride a bicycle at a speed of V = 20 mph when the force to move against wind drag is F = 4.0 lbf.

$$F = (4.0\ \text{lbf}) \left(\frac{4.45\ \text{N}}{\text{lbf}} \right) = 17.8\ \text{N}$$

$$V = \frac{20\ \text{mph}}{10/b\ \text{mp}} \left| \frac{0.447\ \text{m/s}}{} \right. = 8.94\ \text{m/s}$$

$$P = FV = \frac{17.8\ \text{N}}{} \left| \frac{8.94\ \text{m}}{\text{s}} \right| \frac{\text{W} \cdot \text{s}}{\text{N} \cdot \text{m}} = 159\ \text{W}$$

$$\boxed{P = 159\ \text{W}}$$

To apply the Consistent Unit Rule to this problem, take the three steps that follow.

First, convert the given variables to consistent units (SI units).

Second, do your calculations with consistent units.

Third, report your answer with the most appropriate units for your context.

FIGURE 1.8

This example shows how to apply the *Consistent Unit Rule*.

TABLE 1.5 Consistent Units

Dimension	SI System	English (Traditional) Units
length	meter (m)	foot (ft)
mass	kilogram (kg)	slug (slug)
time	second (s)	second (s)
force	newton (N)	pounds-force (lbf)
pressure	pascal (Pa)	pounds-force per square foot (psf)
density	kilogram per meter cubed (kg/m^3)	slug per foot cubed (slug/ft^3)
volume	cubic meters (m^3)	cubic feet (ft^3)
power	watt (W)	foot pounds-force per second (ft·lbf/s)

featured in Chapter 8, in which a powerful method of analysis (called dimensional analysis) is introduced.

Mass is an example of a *dimension*. To describe the amount of mass, engineers apply various units (e.g., slug, gram, kilogram, ounce, pound-mass). *Time* is an example of a dimension. To describe the amount of time, you can apply various units (seconds, minutes, hours, days, weeks, months, years, centuries, etc.). Other examples of dimensions are speed, volume, and energy. Each dimension has associated with it many possible units, but the dimension itself does not have a specified unit. As these examples show, a **dimension** is an entity that is measured using units. The relationship between dimensions and units is shown in Fig. 1.9. Notice that dimensions can be identified by asking this question: *What are we interested in measuring?* For example, engineers are generally interested in measuring force, power, energy, and time. Each of these entities is a dimension.

> **EXAMPLE.** Is temperature a dimension? **Reasoning.** (1) A dimension is an entity that is measured and quantified with units. (2) Temperature is something (i.e., an entity) that is measured and quantified with units such as Kelvin, Celsius, and Fahrenheit. (3) Thus, temperature aligns with the definition of dimension. **Conclusion.** Temperature is a dimension.

Dimensions can be related by using equations. For example, Newton's Second Law, $F = ma$, relates the dimensions of force, mass, and acceleration. Because dimensions can be related, engineers and scientists can express dimensions using a limited set of dimensions that are called **primary dimensions** (Table 1.6).

A **secondary dimension** is any dimension that can be expressed using primary dimensions. For example, the secondary dimension "force" is expressed in primary dimensions by using $\Sigma \mathbf{F} = m\mathbf{a}$. The primary dimensions of acceleration are L/T^2, so

$$[F] = [ma] = M\frac{L}{T^2} = \frac{ML}{T^2} \tag{1.27}$$

FIGURE 1.9

Dimensions describe *what is measured*. Units provide the method by which quantification is possible.

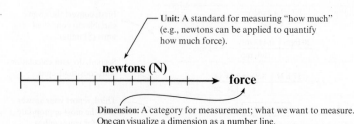

Unit: A standard for measuring "how much" (e.g., newtons can be applied to quantify how much force).

newtons (N)

force

Dimension: A category for measurement; what we want to measure. One can visualize a dimension as a number line.

TABLE 1.6 Primary Dimensions

Dimension	Symbol	Unit (SI)
Length	L	meter (m)
Mass	M	kilogram (kg)
Time	T	second (s)
Temperature	θ	kelvin (K)
Electric current	i	ampere (A)
Amount of light	C	candela (cd)
Amount of matter	N	mole (mol)

In Eq. (1.27), the square brackets mean "dimensions of." Thus $[F]$ means "the dimension of force." Similarly, $[ma]$ means "the dimensions of mass times acceleration." Notice that primary dimensions are not enclosed in brackets. For example, ML/T^2 is not enclosed in brackets.

To find the primary dimensions, we recommend two methods:

Method #1 (Primary Method). Figure out the primary dimensions by applying fundamental definitions on physical quantities.

Method #2 (Secondary Method). Look up the primary dimensions in Table F.1 (front of book) or in other engineering references. We recommend that you only use this method if you have not yet had enough practice using Method #1.

> **EXAMPLE.** If work is given the symbol W, what are $[W]$?
>
> **Reasoning.**
>
> 1. The symbol $[W]$ means "the primary dimensions of work." Thus, the question is asking *what are the primary dimensions of work?*
> 2. The definition of mechanical work reveals that (work) = (force)(distance).
> 3. Thus, $[W] = [F][d] = (ML/T^2)(L) = ML^2/T^2$.
>
> **Conclusion.** $[W] = ML^2/T^2$.

Dimensional Homogeneity (DH)

Voice of the Engineer. Routinely check each equation you encounter for dimensional homogeneity. **Reasoning.** (1) You can recognize and fix mistakes in equations. (2) This skill will help you make sense out of each equation you encounter and also make equations easier to remember.

An equation is **dimensionally homogenous** if each term in the equation has the same primary dimensions. The method for checking an equation for DH is to find the primary dimensions on each term and then check to see if each term has the same primary dimensions.* This method is illustrated in the next example.

> **EXAMPLE.** Show that the IGL (density form) is DH.
>
> **Reasoning.**
>
> 1. The density form of the IGL is $p = \rho RT$.
> 2. The secondary dimensions of pressure are $[p] = [\text{force}]/[\text{area}]$.
>   Thus, the primary dimensions are $[p] = M/LT^2$.

*Of course, one can also use secondary dimensions or units. However, we recommend using primary dimensions because this builds knowledge that will be useful when you learn dimensional analysis in Chapter 8.

FIGURE 1.10

This example shows how to analyze the Reynolds number to establish that the primary dimensions cancel out.

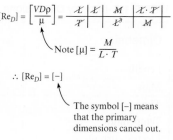

3. The SI units of the specific gas constant R are J/kg·K.

   Thus, the secondary dimensions are $[R] = $ [energy]/([mass][temperature]).

   Thus, the primary dimensions are $[R] = L^2/T^2\theta$.

4. The IGL can be analyzed as follows:

$$[p] = [\rho][R][T]$$

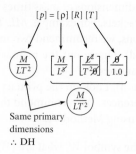

Conclusion. The density form of the IGL is dimensionally homogeneous, as shown by the analysis just presented.

The π-group (Dimensionless Group)

In fluid mechanics, it is common to arrange variables so that the primary dimensions cancel out. This group of variables is called a dimensionless group or a π-**group**. The reason for the use of pi (i.e., π) in the label is that the main theorem used in analysis is called the Buckingham Π theorem. This topic is presented in Chapter 8.

A common example of a π-group is the Reynolds number (Re_D). One equation for the Reynolds number is $\text{Re}_D = (\rho V D)/\mu$, where $\rho = $ fluid density, $V = $ velocity, D is pipe diameter, and $\mu = $ fluid viscosity. Analysis of the Re_D (Fig. 1.10) shows that the primary dimensions cancel out.

1.8 Problem Solving

Although people solve problems every day, not everyone is equally skilled at problem solving. To illustrate this idea, consider the game of golf. Nearly anyone can strike a golf ball with a golf club, but only a tiny percentage of the population can do this well. Golfers who have a high level of skill owe their abilities to many years of practicing. Problem solving is like this as well, but the good news is that the number of skills you need to master is small. These skills are explained in this section. We hope you practice these skills (they are fun!) and that over time you develop into a great problem solver.

Defining Problem Solving

A **problem** is a situation that you need to resolve, especially when you have no clear idea of how to effectively resolve the problem. Given that a problem is a situation that needs to be

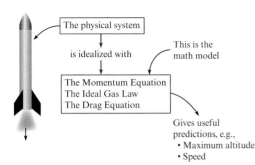

FIGURE 1.11

Example of a math model of a rocket.

resolved, **problem solving** is a label for the methods that empower you to solve problems. A person who is skilled at problem solving can create great solutions with minimal amounts of time, effort, and cost while also greatly enjoying the experience. In addition, the process of problem solving nearly always results in meaningful learning.

Applying Problem Solving to Building Math Models

There are general methods for solving problems. In this section, we will explain how to apply these methods in the context of engineering classes. Our logic can be explained by using an analogy: If you are going to spend a lot of time practicing the guitar, then you should apply methods that will help you develop as a great guitar player. In the same way, if you are going to spend a lot of time in engineering school doing calculation problems, then you should apply methods that will help you excel at problem solving in general and building math models in particular.

A math model (Fig. 1.11) comprises equations plus a method of solving these equations. The purpose of the math model is to help you predict variables that are useful for engineering a system.

On most engineering problems, a math model is useful. For example, suppose you are designing a pump and the associated piping system to deliver water from a lake to a building located 100 meters higher than the lake. A math model gives you the ability to predict useful parameters such as the optimum pipe diameter as well as the size and power requirements for the pump. If you did not have a math model, you would have to take a guess on sizing, then build something and take data. Then, you would repeat your steps until you had an acceptable design figured out. However, this guess, build, and repeat method is expensive and time-consuming.

In general, a **math model** can be defined as a collection of equations that you solve to give you values of parameters that are useful in the context of solving real-world problems. The main reason that a math model is useful is that it significantly reduces the cost and time you need to solve your problem.

The method that we use and that we teach is called the Wales-Woods Model (WWM), because it is based on the research of Professor Charles Wales and his colleagues (Anni Nardi and Robert Stager) and also based on the research of Professor Donald Woods (1-9).

If you apply the WWM, then you will learn more effectively and you will grow your problem-solving skills. There are several reasons that we say this: (1) These methods work for us and we can attest to their benefits, (2) we have observed many students become better problem solvers and have had many students report that these methods benefited them, and (3) the methods are backed up with research data* that show that the WWM is effective. In particular, Wales (3) analyzed five years of data and found that when students were taught the methods as

---

*The gains reported in the literature are far above the gains reported for nearly any other educational methods that we know of.

freshmen, the graduation rate increased by 32% and the average grade point average increased by 25%, as compared to the control group, the members of which were not taught these skills. Based on 20 years of data, Woods (9) reports that students taught problem-solving skills, as compared to control groups, showed significant gains in confidence, problem-solving ability, attitude toward lifetime learning, self-assessment, and recruiter response.

The WWM is explained in Table 1.7. Skills that are the most useful are marked with one or more check marks (✓). The best way to learn the WWM is to practice one or two skills at a time until you become good at them. Then, add a few more skills.

TABLE 1.7 The WWM for Problem Solving

	Example. This column lists a sample problem and then shows how the Wales-Woods Model of problem solving might be applied to this sample problem.	**Explanation.** This column describes the actions you can take to apply the problem-solving model. Check marks (✓) indicate how useful each action is in the context of an engineering course. *More check marks means that an item is more useful.*
Problem Definition	**Problem Statement** Find the total weight of a 17 ft³ tank of nitrogen if the nitrogen is pressurized to 500 psia, the tank itself weighs 50 lbf, and the temperature is 20°C. Work in SI units.	Figure out what you are being asked (while reading the problem): • (✓✓) Interpret the given problem statement. • (✓) Look up unfamiliar terms. • (✓✓) Figure out how the given system works. • (✓✓) Visualize the system as it might exist in the real world. • (✓✓✓) Identify ideas or equations that might apply.
	Define the Situation A tank holds compressed N_2 $\forall = 17\ \text{ft}^3 = 0.481\ \text{m}^3$ $P = 500\ \text{psia} = 3.45\text{E6 Pa}$ $T = 293\ \text{k}$ N_2 $W_{\text{tank}} = 50\ \text{lbf} = 222\ \text{N}$ Assume: IGL applies N_2 (A2): $R_{N_2} = 297\ \text{J/kg}\cdot\text{k}$	Document your interpretation of the problem: • Summarize the situation in one to two sentences. • (✓✓✓) Sketch a *system diagram*. • (✓) List values of known variables with units. • (✓) Convert units to *consistent units*. • List main assumptions. • List properties and other relevant data.
	State the Goal $W_T(\text{N}) \leftarrow$ Weight total (nitrogen + tank)	Describe your goal in a way that is unambiguous (the goal should be so clear that there will be no question about whether the goal is attained).
Figuring Out a Solution	**Generate Ideas** **1.** Weight total $$\overset{\boxed{?}}{W_T} = \overset{✓}{W_{\text{tank}}} + \overset{?}{W_{N_2}} \quad \text{(a)}$$ **2.** Newton's Law of Universal Gravity (applied to Earth) $$\overset{\boxed{?}}{W_{N_2}} = \overset{?}{(m_{N_2})}\overset{✓}{g} \quad \text{(b)}$$ **3.** The IGL (mass form) $$\overset{✓✓}{p}\overset{\boxed{?}}{\forall} = \overset{✓✓}{m_{N_2}R}T \quad \text{(c)}$$	Apply the GENI method (from Wales et al. (1)): **1.** (✓✓✓✓) Identify an equation that contains your goal. Mark your goal with a boxed question mark. Mark known variables with a check mark and unknown variables with a question mark (e.g., see line a). **2.** (✓✓✓✓) Make any unknown variable(s) your new goal. Repeat the marking process using checks and question marks (e.g., see lines b and c). **3.** (✓✓✓✓) Repeat steps 1 and 2 until the number of equations is equal to the number of unknowns. At this point, the problem is solvable (we say that the *problem is cracked*, which means it is now figured out). In this example, the *problem is cracked* because there are three equations (a, b, and c) and three unknown variables (weight of nitrogen, mass of nitrogen, and total weight of the tank).

Make a Plan: **1.** Calculate the mass of nitrogen using Eq. (c). **2.** Calculate the weight of nitrogen using Eq. (b). **3.** Calculate the total weight using Eq. (a).	As needed, list the set of steps that you can follow to reach your goal state (*most of the time, you can skip writing anything down*). In our examples, we often write out the plan steps so you can see what "making a plan" looks like.

Take Action (Execute the Plan)

1. IGL

$$m_{N_2} = \frac{p V}{RT} = \frac{3.45 \text{ E6 N}}{\text{m}^2} \left| \frac{0.481 \text{ m}^3}{} \right| \frac{\text{kg} \cdot \text{k}}{297 \text{ N} \cdot \text{m}} \left| 293 \text{ k} \right.$$

$$= 19.1 \text{ kg}$$

2. NLUG

$$W_{N_2} = (m_{N_2})(g) = \frac{19.1 \text{ kg}}{} \left| \frac{9.81 \text{ m}}{\text{s}^2} \right| \frac{\text{N} \cdot \text{s}^2}{\text{kg} \cdot \text{m}}$$

$$= 181 \text{ N}$$

3. Total Weight

$$W_T = W_{\text{tank}} + W_{N_2} = (222 + 187) \text{ N}$$
$$= \boxed{409 \text{ N}}$$

Build your solution:
- Do the calculations.
- (✓✓✓✓) Apply the grid method.
- Report your answer(s) with three significant figures.
- Box or mark your answer(s).

(Left margin label: Execution)

Review the Solution and the Process

1. When mass is the goal, the mass from of the IGL is the best equation to select.
2. To check the IGL assumption, I calculated the compressibility factor and found that the IGL was accurate to within about 98%.
4. For this problem, the weight of the gas is significant as compared to the weight of the tank.

Review your solution and your methods:
- (✓✓) Validate your solution.
- Figure out what recommendations you might make.
- (✓✓✓) Review your problem-solving methods:
  - What actions worked well for you?
  - What actions might you take in the future?
- Identify knowledge that was especially useful to you.
- Identify significant aspects of the solution.

(Left margin label: Reflective Thinking)

1.9 Summarizing Key Knowledge

Engineering Fluid Mechanics

- *Engineering* is the body of knowledge that equips individuals to solve problems by innovating, designing, applying, and improving technology.
- The *engineering method* is a label for the methods used to do engineering. The engineering method involves submethods such as critical thinking, building math models, the application of scientific experiments, and the application of existing technologies.
- *Mechanics* is the branch of science that deals with motion and with the forces that produce this motion. Mechanics is organized into two main categories: *solid mechanics* (materials in the solid state) and *fluid mechanics* (materials in the gas or liquid state).
- *Engineering fluid mechanics* is engineering when the project involves substantial knowledge from the discipline of fluid mechanics.

Fluids: Liquids and Gases

- Both liquids and gases are classified as fluids. A fluid is defined as a material that deforms continuously under the action of a shear stress.
- A significant difference between gases and liquids is that the molecules in a liquid experience strong intermolecular forces, whereas the molecules in a gas move about freely with little or no interactions except during collisions.
- Liquids and gases differ in many important respects. Example #1: Gases expand to fill their containers, whereas liquids will occupy a fixed volume. Example #2: Gases have much smaller values of density than liquids.

Ideas for Idealizing Materials

- A *microscopic viewpoint* involves understanding a material by understanding what the molecules are doing. A *macroscopic viewpoint* involves understanding a material without the need to consider what the molecules are doing.

- Much of material behavior can be explained by understanding the *forces between molecules*. Molecules far apart do not attract one another, but molecules close together have strong attractive forces between them. However, when molecules are too close, there is a very strong repulsive force.

- "Body" is a label to identify objects or matter that exists in the real world, without specifying any specific object.

- The term "particle" is used in two ways:
  - A *material particle* is a small chunk of a body.
  - A *body-as-a-particle* involves idealizing a body as if all the mass is concentrated at a single point and the dimensions of the body are not relevant. For example, to analyze an airplane, we can idealize the airplane as a particle.

- In the continuum assumption, matter is idealized as consisting of continuous material that can be broken into smaller and smaller parts. The continuum assumption applies to most problems that involve flowing fluids.

Weight and Mass

- *Mass* is a material property that characterizes the amount of matter of a body. *Weight* is a property that characterizes the gravitational force on a body from a nearby planet (e.g., Earth).

- Weight and mass are related to each other by Newton's Law of Universal Gravitation (NLUG). This law tells us that any two bodies anywhere in the universe will attract each other. The force of attraction depends on the mass of each body and inversely on the distance squared between the centers of mass of each body. In equation form,

NLUG is $F = (G\, m_1\, m_2)/R^2$. On Earth, NLUG simplifies to $W = mg$.

Density and Specific Weight

- Density is a material property that characterizes the ratio of mass/volume at a point; for example, the density of liquid water at room conditions is about $\rho = 1.0$ kg/L = 1000 kg/m^3.

- Specific weight is a material property that characterizes the ratio of weight/volume at a point; for example, the specific weight of liquid water at room conditions is about $\gamma = 9.8$ N/L = 9800 N/m^3.

- In general, ρ and γ vary with temperature and pressure. For liquids, ρ and γ are usually assumed to be constant with pressure but variable with temperature.

The Ideal Gas Law (IGL)

- Most gases can be idealized as an ideal gas.

- To apply the IGL, use correct temperature and pressure units.
  - Temperature must be in *absolute temperature* (Kelvin or Rankine), not Celsius or Fahrenheit.
  - Pressure must be in *absolute pressure*, not gage or vacuum pressure.

- Moles and mass are related by $m = nM$; similarly, the specific gas constant and the universal gas constant are related by $R = R_u/M$. The molar mass M has dimensions of (mass)/(mole).

- There are multiple ways to write equations that represent the IGL. Three of the most useful equations are $p = \rho RT$, $p\mathcal{V} = mRT$, and $p\mathcal{V} = nR_uT$.

REFERENCES

1. Wales, C.E., and Stager, R.A. *Thinking with Equations*. Center for Guided Design, West Virginia University, Morgantown, WV, 1990.

2. Wales, C.E. "Guided Design: Why & How You Should Use It," *Engineering Education*, vol. 62, no. 8 (1972).

3. Wales, C.E. "Does How You Teach Make a Difference?" *Engineering Education*, vol. 69, no. 5, 81–85 (1979).

4. Wales, C.E., Nardi, A.H., and Stager, R.A. *Thinking Skills: Making a Choice*. Center for Guided Design, West Virginia University, Morgantown, WV, 1987.

5. Wales, C.E., Nardi, A.H., and Stager, R.A. *Professional-Decision-Making*, Center for Guided Design, West Virginia University, Morgantown, WV, 1986.

6. Wales, C.E., and Stager, R.A., (1972b). "The Design of an Educational System." *Engineering Education*, vol. 62, no. 5 (1972).

7. Wales, C.E., and Stager, R.A. *Thinking with Equations*. Center for Guided Design, West Virginia University, Morgantown, WV, 1990.

8. Woods, D.R. "How I Might Teach Problem-Solving?" in J.E. Stice (ed.), *Developing Critical Thinking and Problem-Solving Abilities*. New Directions for Teaching and Learning, no. 30, San Francisco: Jossey-Bass, 1987.

9. Woods, D.R. "An Evidence-Based Strategy for Problem Solving." *Engineering Education*, vol. 89. no. 4, 443–459 (2000).

PROBLEMS

Engineering Fluid Mechanics (§1.1)

1.1 Apply critical thinking to an engineering-relevant issue that is important to you. Create a written document and use the CT (§1.1) process to justify your reasoning and your conclusions.

1.2 Read the definition of engineering in §1.1. How does this compare with your ideas of what engineering is? What is similar? What is different?

1.3 Given the definition of engineering in §1.1, what do you think that you should be learning? How do you know if you have learned it?

1.4 Should the definition of engineering in §1.1 include the idea that engineers also need to be very good with humanities and social sciences? What do you believe? Why?

1.5 Many engineering students believe that fixing a washing machine is an example of engineering because it involves solving a problem. Write a brief essay in which you address the following questions: Is fixing a washing machine an example of engineering? Why or why not? How do your ideas align or misalign with the definition of engineering given in §1.1?

How Materials Are Idealized (§1.2)

1.6 (T/F) A fluid is defined as a material that continuously deforms under the action of a shear stress.

Weight, Mass, and NLUG (§1.3)

1.7 Fill in the blanks. Show your work, using conversion factors found in Table F.1.

    a. 750 g is _____ slugs

    b. 36 lbm is _____ kg

    c. 150 slugs is _____ kg

    d. 6 lbm is _____ g

    e. 10 slugs is _____ lbm

1.8 Answer the following questions related to mass and weight. Show your work, and cancel and carry units.

    a. What is the weight on Earth (in N) of a 80-kg body?

    b. What is the mass (in lbm) of 35 lbf of water on Earth?

    c. What is the mass (in slugs) of 35 lbf of water on Earth?

    d. How many N are needed to accelerate 3 kg at 1 m/s²?

    e. How many lbf are needed to accelerate 3 lbm at 1 ft/s²?

    f. How many lbf are needed to accelerate 3 slugs at 1 ft/s²?

Essential Math Topics (§1.4)

1.9 An engineer measured the speed of a flowing fluid as a function of the distance y from a wall; the data are shown in the table. Show how to calculate the maximum value of dV/dy for this data set. Express your answer in SI units.

y (mm)	V (m/s)
0.0	0.00
1.0	1.00
2.0	1.99
3.0	2.96
4.0	3.91

Problem 1.9

Density and Specific Weight (§1.5)

1.10 If a gas has $\gamma = 13$ N/m³, what is its density? State your answers in SI units and in traditional units.

Ideal Gas Law (IGL) (§1.6)

1.11 Calculate the density and specific weight of carbon dioxide at a pressure of 100 kN/m² absolute and 90°C.

1.12 Determine the density of methane gas at a pressure of 300 kN/m² absolute and 30°C.

1.13 A spherical tank is being designed to hold 200 moles of methane gas at an absolute pressure of 5 bar and a temperature of 80°F. What diameter spherical tank should be used? The molecular weight of methane is 16 g/mole.

1.14 Natural gas is stored in a spherical tank at a temperature of 12°C. At a given initial time, the pressure in the tank is 75 kPa gage, and the atmospheric pressure is 100 kPa. Some time later, after considerably more gas is pumped into the tank, the pressure in the tank is 300 kPa gage, and the temperature is still 12°C. What will be the ratio of the mass of natural gas in the tank when $p = 300$ kPa gage to that when the pressure was 75 kPa gage?

1.15 Find the total weight of a 12 ft³ tank of oxygen if the oxygen is pressurized to 184 psia, the tank itself weighs 100 lbf, and the temperature is 95°F.

1.16 A 10 m³ oxygen tank is at 17°C and 800 kPa absolute. The valve is opened, and some oxygen is released until the pressure in the tank drops to 600 kPa. Calculate the mass of oxygen that has been released from the tank if the temperature in the tank does not change during the process.

1.17 What is the (a) specific weight and (b) density of air at an absolute pressure of 650 kPa and a temperature of 21°C?

1.18 Meteorologists often refer to air masses in forecasting the weather. Estimate the mass of 2 mi³ of air in slugs and kilograms. Make your own reasonable assumptions with respect to the conditions of the atmosphere.

1.19 Start with the ideal gas law and prove that

    a. Boyle's law is true.

    b. Charles's law is true.

1.20 A bicycle rider has several reasons to be interested in the effects of temperature on air density. The aerodynamic drag force decreases linearly with density. Also, a change in temperature will affect the tire pressure.

   a. To visualize the effects of temperature on air density, write a computer program that calculates the air density at atmospheric pressure for temperatures from $-10°C$ to $50°C$.

   b. Also assume that a bicycle tire was inflated to an absolute pressure of 450 kPa at $20°C$. Assume the volume of the tire does not change with temperature. Write a program to show how the tire pressure changes with temperature in the same temperature range, $-10°C$ to $50°C$.

Prepare a table or graph of your results for both problems. What engineering insights do you gain from these calculations?

1.21 A team is designing a helium-filled balloon that will fly to an altitude of 80,000 ft. As the balloon ascends, the upward force (buoyant force) will need to exceed the total weight. Thus, weight is critical. Estimate the weight (in newtons) of the helium inside the balloon. The balloon is inflated at a site where the atmospheric pressure is 0.89 bar and the temperature is $22°C$. When inflated prior to launch, the balloon is spherical (radius 1.3 m) and the inflation pressure equals the local atmospheric pressure.

Units and Dimensions (§1.7)

1.22 If the local atmospheric pressure is 120 kPa, use the grid method to find the pressure in units of

   a. psi

   b. psf

   c. bar

   d. atmospheres

   e. feet of water

   f. inches of mercury

Problem Solving (§1.8)

1.23 Apply the WWM and the grid method to find the acceleration for a force of 10 N acting on an object of mass 15 ounces. The relevant equation is Newton's Second Law of Motion, $F = ma$. Work in SI units, and provide the answer in meters per second squared (m/s^2).

CHAPTER**TWO**

Fluid Properties

CHAPTER ROAD MAP This chapter introduces ideas for idealizing real-world problems, introduces fluid properties, and presents the viscosity equation.

FIGURE 2.1
This photo shows engineers observing a **flume**, which is an artificial channel for conveying water. This flume is used to study sediment transport in rivers. (Photo courtesy of Professor Ralph Budwig of the Center for Ecohydraulics Research, University of Idaho.)

LEARNING OUTCOMES

SYSTEM, STATE, AND PROPERTY (§2.1).
- Define system, boundary, surroundings, state, steady state, process, and property.

FINDING FLUID PROPERTIES (§2.2).
- Look up appropriate values of fluid properties and document your work.
- Define each of the eight common fluid properties.

DENSITY TOPICS (§2.3).
- Know the main ideas about specific gravity.
- Explain the constant density assumption and make decisions about whether or not this assumption is valid.
- Determine changes in the density of water corresponding to a pressure change or a temperature change.

STRESS (§2.4).
- Define stress, pressure, and shear stress.
- Explain how to relate stress and force.
- Describe each of the seven common fluid forces.

THE VISCOSITY EQUATION (§2.5).
- Define the velocity gradient and find values of the velocity gradient.
- Describe the no-slip condition.
- Explain the main ideas of the viscosity equation.
- Solve problems that involve the viscosity equation.
- Describe a Newtonian and non-Newtonian fluid.

SURFACE TENSION (§2.6).
- Know the main ideas about surface tension.
- Solve problems that involve surface tension.

VAPOR PRESSURE (§2.7).
- Explain the main ideas of the vapor pressure curve.
- Find the pressure at which water will boil.

2.1 System, State, and Property

The vocabulary introduced in this section is useful for solving problems. In particular, these ideas allow engineers to describe problems in ways that are precise and concrete.

A **system** is the specific entity that is being studied or analyzed by the engineer. A system can be a collection of matter, or it can be a region in space. Anything that is not part of the system is considered to be part of the **surroundings**. The **boundary** is the imaginary surface that separates the system from its surroundings. For each problem you solve, it is your job as the engineer to select and identify the system that you are analyzing.

EXAMPLE. For the flume shown in Fig. 2.1, the water that is situated inside the flume could be defined as the system. For this system, the surroundings would be the flume walls, the air above the flume, and so on. Notice that *engineers are specific about what the system is, what the surroundings are, and what boundary is.*

EXAMPLE. Suppose an engineer is analyzing the air flow from a tank being used by a SCUBA diver. As shown in Fig. 2.2, the engineer might select a system comprised of the tank and the regulator. For this system, everything that is external to the tank and regulator is the surroundings. Notice that *the system is defined with a sketch* because this is sound professional practice.

If you make a wise choice when you select a system, you increase your probability of getting an accurate solution, and you minimize the amount of work you need to do. Although the choice of system must fit the problem at hand, there are often multiple possibilities for which system to select. This topic will be revisited throughout this textbook as various kinds of systems are introduced and applied.

Systems are described by specifying numbers that characterize the system. The numbers are called properties. A **property** is a measurable characteristic of a system that depends only on the present conditions within the system.

EXAMPLE. In Fig. 2.2, some examples of properties (i.e., measurable characteristics) are as follows:

- The pressure of the air inside the tank
- The density of air inside the tank
- The weight of the system (tank plus air plus regulator)

Some parameters in engineering are measurable, yet they are not properties. For example, work is not a property because the quantity of work depends on how a system interacts with its surroundings. Similarly, neither force nor torque are properties because these parameters depend on the interaction between a system and its surroundings. Heat transfer is not a property. Mass flow rate is not a property.

The **state** of a system means the condition of the system as defined by specifying its properties. When a system changes from one state to another state, this is called a **process**. When the properties of a system are constant with time, the system is said to be at **steady state**.

FIGURE 2.2

Example of a system, its surroundings, and the boundary.

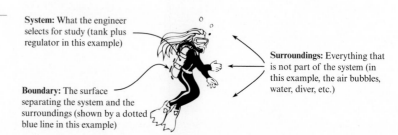

System: What the engineer selects for study (tank plus regulator in this example)

Surroundings: Everything that is not part of the system (in this example, the air bubbles, water, diver, etc.)

Boundary: The surface separating the system and the surroundings (shown by a dotted blue line in this example)

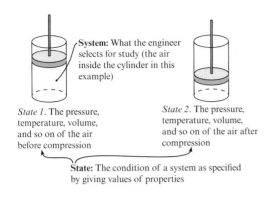

System: What the engineer selects for study (the air inside the cylinder in this example)

State 1. The pressure, temperature, volume, and so on of the air before compression

State 2. The pressure, temperature, volume, and so on of the air after compression

State: The condition of a system as specified by giving values of properties

FIGURE 2.3

Air in a cylinder being compressed by a piston. State 1 is a label for the conditions of the system prior to compression. State 2 is a label for the conditions of the system after compression.

EXAMPLE. Fig. 2.3 shows air being compressed by a piston in a cylinder. The air inside the cylinder is defined as the system. At state 1, the conditions of the system are defined by specifying properties such as pressure, temperature, and density. Similarly, state 2 is defined by specifying these same properties.

EXAMPLE. When air is compressed (Fig. 2.3), this is a process because the air (i.e., the system) has changed from one set of conditions (state 1) to another set of conditions (state 2). Engineers label processes that commonly occur. For example, an *isothermal process* is one in which the temperature of the system is held constant, and an *adiabatic process* is one in which there is no heat transfer between the system and the surroundings.

Properties are often classified into categories. Two examples of categories are as follows:

- **Kinematic properties**. These properties characterize the motion of your system. Examples include position, velocity, and acceleration.
- **Material properties**. These properties characterize the nature of the materials in your system. Examples include viscosity, density, and specific weight.

2.2 Looking Up Fluid Properties

One of the most common tasks that engineers perform is to look up material properties. This section presents ideas that help you perform this task well.

Overview of Properties

Although there are many fluid properties, there are only a few that you need often. These properties are summarized in Table 2.1. Notice that the properties are organized into three groups.

Group #1: Weight and Mass Properties. Three properties (ρ, γ, and SG) are used to characterize weight or mass. In general, you can find one of these properties and then calculate either of the other two using the following equations: $\gamma = \rho g$ and $SG = \rho/\rho_{H_2O, (4°C)} = \gamma/\gamma_{H_2O, (4°C)}$

Group #2: Properties for Characterizing Viscosity. To characterize friction-like effects in flowing fluids, engineers use viscosity, μ. Viscosity has two common synonyms: *dynamic viscosity* and *absolute viscosity*. In addition to viscosity, engineers use another term, kinematic viscosity, which is given the symbol ν. **Kinematic viscosity** is defined by

$$\nu = \frac{\mu}{\rho} \tag{2.1}$$

An easy way to distinguish between μ and ν is to check units or dimensions because $[\mu] = M/(L\cdot T)$ and $[\nu] = L^2/T$. Regarding viscosity, we recommend that you build a physical feel for this property by finding examples that make sense to you. In this spirit,

TABLE 2.1 Summary of Fluid Properties

Property	Units (SI)	Temperature Effects	Pressure Effects (common trends)	Notes
Density (ρ): Ratio of mass to volume at a point	$\dfrac{kg}{m^3}$	$\rho\downarrow$ as $T\uparrow$ if the gas is free to expand	$\rho\uparrow$ as $p\uparrow$ if a gas is compressed	• *Air.* Find ρ in Table F.4 or Table A.3. • *Other Gases.* Find ρ in Table A.2. • *Caution!* Tables for gases are for $p = 1$ atm. For other pressures, find ρ using the ideal gas law.
		$\rho\downarrow$ as $T\uparrow$ for liquids	A liquid is usually idealized with ρ independent of pressure	• *Water.* Find ρ in Table F.5 or Table A.5. • *Note.* For water, $\rho\uparrow$ as $T\uparrow$ for temperatures from 0 to about 4°C. Maximum density of water is at $T \approx 4$°C. • *Other Liquids.* Find ρ in Table A.4.
Specific Weight (γ): Ratio of weight to volume at a point	$\dfrac{N}{m^3}$	$\gamma\downarrow$ as $T\uparrow$ if fluid is free to expand	Gas: $\gamma\uparrow$ as $p\uparrow$ if a gas is compressed Liquid: a liquid is usually idealized with γ independent of pressure	• Use same tables as for density. • ρ and γ can be related using $\gamma = \rho g$. • *Caution!* Tables for gases are for $p = 1$ atm. For other pressures, find γ using the ideal gas law and $\gamma = \rho g$. • Typically, γ is not used for gases.
Specific Gravity (*S* or *SG*): Ratio of (density of a liquid) to (density of water at 4°C)	none	$SG\downarrow$ as $T\uparrow$	A liquid is usually idealized with SG independent of pressure	• Find SG data in Table A.4. • SG is used for liquids, not commonly used for gases. • Density of water (at 4°C) is listed in Table F.6. • $SG = \gamma/\gamma_{H_2O,\,4°C} = \rho/\rho_{H_2O,\,4°C}$.
Viscosity (μ): A property that characterizes resistance to shear stress and fluid friction	$\dfrac{N\cdot s}{m^2}$	$\mu\uparrow$ as $T\uparrow$ for a gas	A gas is usually idealized with μ independent of pressure	• *Air:* Find μ in Table F.4, Table A.3, Fig. A.2. • *Other gases:* Find properties in Table A.2, Fig. A.2. • *Hint:* Viscosity is also known as dynamic viscosity and absolute viscosity. • *Caution!* Avoid confusing viscosity and kinematic viscosity; these are different properties.
		$\mu\downarrow$ as $T\uparrow$ for a liquid	A liquid is usually idealized with μ independent of pressure	• *Water:* Find μ in Table F.5, Table A.5, Fig. A.2. • *Other Liquids.* Find μ in Table A.4, Fig. A.2.
Kinematic Viscosity (ν): A property that characterizes the mass and viscous properties of a fluid	$\dfrac{m^2}{s}$	$\nu\uparrow$ as $T\uparrow$ for a gas	$\nu\uparrow$ as $p\uparrow$ for a gas	• *Air:* Find μ in Table F.4, Table A.3, Fig. A.3. • *Other gases:* Find properties in Table A.2, Fig. A.3. • *Caution!* Avoid confusing viscosity and kinematic viscosity; these are different properties. • *Caution!* Gas tables are for $p = 1$ atm. For other pressures, look up $\mu = \mu(T)$, then find ρ using the ideal gas law, and calculate ν using $\nu = \mu/\rho$.
		$\nu\downarrow$ as $T\uparrow$ for a liquid	A liquid is usually idealized with ν independent of pressure	• *Water:* Find ν in Table F.5, Table A.5, Fig. A.3. • *Other liquids:* Find ν in Table A.4, Fig. A.3.
Surface Tension (σ): A property that characterizes the tendency of a liquid surface to behave as a stretched membrane	$\dfrac{N}{m},\dfrac{J}{m^2}$	$\sigma\downarrow$ as $T\uparrow$ for a liquid	A liquid is usually idealized with σ independent of pressure	• *Water:* Find σ in Fig. 2.18. • *Other liquids:* Find σ in Table A.4. • Surface tension is a property of liquids (not gases). • Surface tension is greatly reduced by contaminates or impurities.
Vapor Pressure p_v: The pressure at which a liquid will boil	Pa	$p_v\uparrow$ as $T\uparrow$ for a liquid	Not applicable	• *Water:* Find p_v in Table A-5.
Bulk Modulus of Elasticity E_v: A property that characterizes the compressibility of a fluid	Pa	Not presented here	Not presented here	• *Ideal gas (isothermal process):* $E_v = p = $ pressure. • *Ideal gas (adiabatic process):* $E_v = kp$; $k = c_p/c_v$. • *Water:* $E_v \approx 2.2 \times 10^9$ Pa.

(Left-margin group labels: Mass and Weight Properties — Density, Specific Weight, Specific Gravity; Properties Related to Viscosity — Viscosity, Kinematic Viscosity; Miscellaneous Properties — Surface Tension, Vapor Pressure, Bulk Modulus of Elasticity)

the following are two examples that we like: **Example**. Honey has a much higher value of viscosity than does liquid water. Thus, it is harder to push a spoon through a bowl of honey than it is to push a spoon through a bowl of water. **Example**. If you try to pour motor oil out of its container on a cold day, the oil will pour very slowly because the value of viscosity is high. If you heat the motor oil up, then the value of viscosity decreases and the motor oil is easier to pour.

Group #3: Miscellaneous Properties. The last three properties (σ, p_v, and E_v) are used for specialized problems. These properties are described later in this chapter.

Property Variation with Temperature and Pressure

In general, the value of a fluid property varies with both temperature and pressure. These variations are summarized in the third and fourth columns of Table 2.1. The notation $\rho \downarrow$ as $T \uparrow$ is a shorthand for saying that density goes down as temperature rises. The blue shading is used to distinguish between gases and liquids. For example, in the row for viscosity, the text in the blue shaded region indicates that the viscosity of a liquid decreases with a temperature rise. Similarly, the text that is not shaded indicates that the viscosity of a gas increases with a temperature rise.

Notice that the values of many properties (e.g., density of a liquid, viscosity of a gas) can be idealized as being independent of pressure. However, every property varies with temperature.

Finding Fluid Properties

We built Table 2.1 to summarize the details needed for looking up fluid properties. For example, the last column of Table 2.1 lists locations in the text where values of properties are tabulated. In the examples that follow, the key details used to solve the problems came from Table 2.1.

> **EXAMPLE.** What is the density of kerosene (SI units) at room conditions? **Reasoning**. (1) At room conditions, kerosene is a liquid. (2) Liquid properties can be found in Table A.4. **Conclusion**. $\rho = 814$ kg/m^3 (20°C and 1.0 atm).

> **EXAMPLE.** In traditional units, what is the dynamic viscosity of gasoline at 150°F? **Reasoning**. (1) Gasoline is a liquid. (2) Because the goal is to find "dynamic viscosity," note that this property is also called "viscosity" and "absolute viscosity." (3) Viscosity of liquids as a function of temperature can be found in Fig. A2.* (4) Read Fig. A.2[†] to find that μ is approximately 4E-6 lbf·ft/s^2. (5) *Note*: Given that the vertical scale on Fig. A.2 is hard to read, the value of μ was reported to one significant figure. **Conclusion**. $\mu = $ 4E-6 lbf·ft/s^2.

> **EXAMPLE.** What is the specific weight of air at 20°C and 3.0 atmospheres of pressure (gage)? **Reasoning**. (1) Specific weight is related to density via $\gamma = \rho g$. (2) Density can be calculated with the IGL: $\rho = p/RT = $ (4.053E5 Pa)/(287 J/kg·K)(293.2K) $ = $ 4.817 kg/m^3. (3) Thus, $\gamma = \rho g = $ (4.817 kg/m^3)(9.807 m/s^2) $ = $ 47.2 N/m^3. **Conclusion**. $\gamma = $ 47.2 N/m^3.

Quality in Documentation

Voice of the Engineer. *Document your technical work so well that you or a colleague could retrieve the work three years in the future and easily figure out what was done.* **Rationale**. (1) When you build effective documentation, this provides you with a structure that promotes good thinking. (2) In professional practice, you can use your documentation to recall the technical details

---

*Fig. A-2 has a semilog scale. As an engineer, you need to be skilled at reading data from a log scale and skilled at plotting on log scales. If you have not yet gained these skills, we recommend that you ask your teacher for assistance, or consult the Internet.

[†]We recommend using a ruler and drawing straight pencil lines whenever you are using a log scale. This allows you to read data more accurately.

FIGURE 2.4

An example of how to document fluid properties.

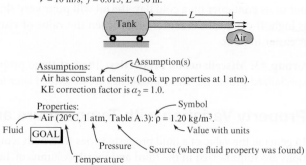

months or years after a project is completed. (3) Thorough documentation helps you protect your intellectual property and also helps protect your reputation (and your pocketbook) if you are involved in a legal conflict.

Most people (including us) dislike documentation, but well-crafted documentation saves abundant amounts of time and effort, so most professionals document their work well. We teach and practice a rule called the *5% rule,* which is this: *Document your technical work in real time (no rewriting allowed*) and do this so effectively that the maximum amount of extra time you need is 5% of your total time.*

Regarding quality in the documentation of fluid properties, we recommend six practices (Fig. 2.4):

1. List the name of the fluid.

2. List the temperature and pressure at which the property was reported by the source. **Rationale**. In general, fluid properties vary with both temperature and pressure, so these values need to be listed. Also, the state (gas, liquid, or solid) depends on temperature and pressure.

3. Cite the source of the fluid property. **Rationale**. Property data are often inaccurate; thus, citing your source is a way to provide evidence that your technical work is trustworthy.

4. List relevant assumptions.

5. List the value of and the units of the fluid property.

6. Be concise; write down the minimum amount of information required to get the job done.

2.3 Topics Related to Density

This section presents three topics (specific weight, the constant density assumption, and the bulk modulus) that are related to fluid density. The first two topics are very important; the last topic is of secondary importance.

Specific Gravity

Specific gravity is useful for characterizing the density or specific weight of a material. **Specific gravity** (represented by *S* or *SG*) is defined as the ratio of the density of a material to the density of a reference material. The reference material used in this text is liquid water at 4°C. Thus,

$$SG = \frac{\rho}{\rho_{H_2O \text{ at } 4°C}} \qquad (2.2)$$

---

*This is the goal; even the best of us need to rewrite our work every now and then. What you want to avoid is getting into the habit of being sloppy and then rewriting your work.

Because $\gamma = \rho g$, Eq. (2.2) can be multiplied by g to give

$$SG = \frac{\rho}{\rho_{H_2O \text{ at } 4°C}} = \frac{\gamma}{\gamma_{H_2O \text{ at } 4°C}} \qquad (2.3)$$

Useful Facts

- If $SG < 1$, then the material will often float on water (e.g., oil, gasoline, wood, and Styrofoam float on water). If $SG > 1$, the material will generally sink (e.g., a piece of potato, concrete, or steel will generally sink in water).
- If you add oil (e.g., $SG = 0.9$) to water ($SG = 1.0$), the oil will float on top of the water. This is because oil and water are *immiscible,* which means that they are not capable of being mixed. If you add alcohol (e.g., $SG = 0.8$) to water, the alcohol and water will mix; fluids that are capable of mixing are *miscible.*
- The properties ρ, SG, and γ are related. If you know one of these properties, you can easily calculate the other two by applying Eq. (2.3).
- Values of ρ and γ for water at 4°C are listed in Table F.6 (front pages).
- Values of SG for liquids are listed in Table A.4 (appendix).
- SG is commonly used for solids and liquids but is rarely used for gases. This textbook does not use SG for gases.

Recommended working knowledge:

- SG (petroleum products; e.g., gasoline or oil) ≈ 0.7 to 0.9
- SG (seawater) ≈ 1.03; SG (mercury) ≈ 13.6
- SG (steel) ≈ 7.8; SG (aluminum) ≈ 2.6, SG (concrete) ≈ 2.2 to 2.4

The Constant Density Assumption

If you can justify the assumption that a fluid has a constant density, this will make your analysis much simpler and faster. This section presents information about this assumption.

The **constant density assumption** means that you can idealize the fluid involved in your problem as if the density was constant with both position and time. Another way to state the assumption is to say that the density can be assumed to be constant even though temperature, pressure, or both are changing. To say that the constant density assumption is a "sound" or "valid" assumption means that the numbers you calculate in your problem are only impacted in a small way (e.g., by less than 5%) by this assumption.

Useful facts:

- Most of the topics and problems in this textbook and other fluid mechanics textbooks assume that density is constant. One notable exception is compressible flow (Chapter 12).
- To characterize a density change with respect to a pressure change, engineers often use the *bulk modulus*; this topic is presented in the next subsection.
- The variation of the density of liquid water with respect to temperature is given in Table A.5.
- When flow is steady,[*] engineers commonly make the following assumptions:
  - **Liquids.** In general, liquids in steady flow are assumed to have constant density.
  - **Gas.** For a gas in steady flow, the density is assumed to be constant if the Mach number[†] is less than about 0.3.

---

[*]Steady flow is defined in §4.3.
[†]The Mach number gives a ratio of the fluid speed to the speed of sound; see Chapter 12.

- When fluid temperatures are changing, it is common to look up a density at an average temperature and then assume that the density is constant. **Example**. If water enters a heat exchanger at 10°C and exits at 90°C, assume that the density is constant and look up the value of density at 50°C in Table A.5.

The Bulk Modulus of Elasticity

In practice, liquids are nearly always treated as if they are incompressible,* which means that the volume of a liquid will not go down if the pressure acting on the liquid is increased; that is, the liquid cannot be compressed. However, as an engineer, you want to understand that liquids are compressible but that the incompressible assumption is nearly always justified for liquids.

The fluid property called bulk modulus gives engineers a way to quantify the degree to which a liquid is compressible. For a liquid, the bulk modulus can be described using Eq. (2.4):

$$E_v = \frac{-\Delta p}{\Delta \forall / \forall} = \frac{\text{change in pressure}}{\text{fractional change in volume}} \qquad (2.4)$$

Given that the bulk modulus of elasticity for liquid water at room conditions is 2.2 GN/m², you can apply Eq. (2.4) to quantify the volume change of liquid water.

EXAMPLE. A 1.0 L volume of liquid water is subjected to an isothermal compression from atmospheric pressure to a pressure of 1.0 MPa absolute. What is the change in the volume of the water? **Reasoning**. (1) From Eq. (2.4), $\Delta \forall = \forall(-\Delta p)/E_v$. (2) Substituting numbers into this equation gives $(1E\text{-}3\ m^3)(-(1.0E6 - 1.0E5)\ Pa)/(2.2E9\ Pa) = -4.5E\text{-}7\ m^3$. **Conclusion**. The volume decreases by about 0.00045 liters, which is about 0.045%.

Summary. It is common but incorrect to say that a given liquid (e.g., water) is incompressible. A better statement is that liquids can usually be assumed to be incompressible. For certain types of problems (e.g., water hammer and acoustics), the compressibility of liquids must be modeled to produce accurate predictions.

For an ideal gas, E_v for an isothermal compression or expansion is given by

$$E_v = p \text{ (isothermal process)} \qquad (2.5)$$

where p is pressure. To apply Eq. (2.4) to a gas, you would want to let $E_v = p$ and then integrate the resulting equation because E_v is not a constant. In addition, E_v depends on the nature of the process. For example, if the compression or expansion were adiabatic, then

$$E_v = kp \text{ (adiabatic process)} \qquad (2.6)$$

where k, the specific heat ratio, is defined in §2.8.

2.4 Pressure and Shear Stress

When you understand stress, many topics in mechanics become easier. The big picture is that there are only two kinds of stress: *normal stress* and *shear stress*. In fluid mechanics, the normal stress† is nearly always just the fluid pressure; thus, the two kinds of stress are pressure and shear stress.

---

*Except for a few special types of problems, such as water hammer problems and modeling the fluid in an ink-jet print head.
†There is also a viscous component to the normal stress. However, this term is seldom important, and this topic is best left to more advanced fluids books.

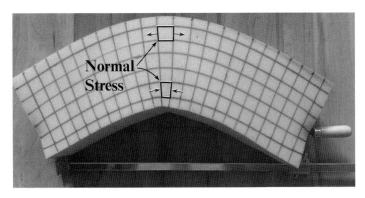

FIGURE 2.5

This figure shows our favorite way to visualize stress in a body. The method goes like this: (1) Select a *body* comprised of a *beam* made of foam. (2) Mark *material particles;* this example uses squares that are 25 mm on a side. (3) *Load* the beam; this example uses a clamp to exert a *bending moment*. (4) Observe how stress has deformed the material particles. (Photo by Donald Elger)

Definition of Stress

To define stress, we begin by noting that stress acts on material particles. For example, if you bend a beam, the material particles are deformed by normal stress (Fig. 2.5).

Thus, stress is caused by a load acting on a body. An example for a fluid body is shown in Fig. 2.6.

To build a definition of stress, we start by recognizing that the secondary dimensions of stress are force/area:

$$\text{stress} = \frac{\text{force}}{\text{area}} \tag{2.7}$$

Next, visualize the force on one face of a material particle. Resolve this force into a normal component of magnitude ΔF_n and a tangential component of magnitude ΔF_t (Fig. 2.7).

Then, the pressure is defined as the ratio of normal force to area:

$$p \equiv \lim_{\Delta A \to 0} \frac{\Delta F_n}{\Delta A} \tag{2.8}$$

And shear stress is defined as the ratio of shear force to area:

$$\tau \equiv \lim_{\Delta A \to 0} \frac{\Delta F_t}{\Delta A} \tag{2.9}$$

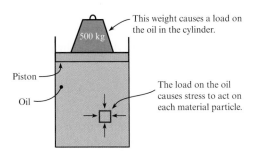

FIGURE 2.6

In this example, a load (i.e., a weight situated on piston) causes stress to act on the oil in a cylinder.

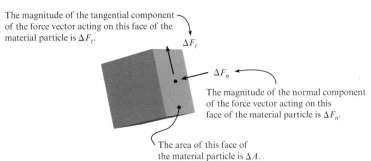

FIGURE 2.7

This sketch shows how the force on one face of a *material particle* can be resolved into a normal force and a tangential force.

FIGURE 2.8

This figure shows the pressure distribution associated with fluid flowing over a body that has a circular shape. This can represent, for example, how pressure varies around the outside of a round pier submerged in a river.

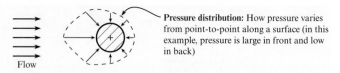

Pressure distribution: How pressure varies from point-to-point along a surface (in this example, pressure is large in front and low in back)

Summary. In mechanics, **stress** is an entity that expresses the forces that material particles exert on each other. Stress is the ratio of force to area at a point and is resolved into two components:

- **Pressure (normal stress).** The ratio of normal force to area.
- **Shear stress.** The ratio of shear force to area.

More advanced textbooks will present additional ideas about stress. For example, stress is often represented mathematically as a second-order tensor. However, these topics are beyond the scope of this text.

Relating Stress to Force

A common problem is how to relate the stress acting on an area to the associated force on the same area. The solution is to integrate the stress distribution as follows.

$$\text{force} = \int_{\text{Area}} \left(\frac{\text{force}}{\text{area}} \right) dA \qquad \overset{\text{definition of stress}}{} \tag{2.10}$$

To build the details of the integration, we'll start with a pressure distribution (Fig. 2.8). To represent force as a vector quantity, we select a small area and define a unit vector (Fig. 2.9). The force on the small area is $\mathbf{dF} = -p\mathbf{n}\, dA$. To obtain the force on the body, add up the small forces ($\mathbf{F} = \Sigma \mathbf{dF}$) while letting the size of the small area (i.e., dA) go towards zero. The summation of small terms is the definition of the integral (§1.4). Thus,

$$\mathbf{F}_p = \int_A -p\mathbf{n}\, dA \tag{2.11}$$

where $\mathbf{F}_p$, called the **pressure force**, represents the net force on the area A due to the pressure distribution. Eq. (2.11) has an important special case:

$$F_p = pA \tag{2.12}$$

The reasoning to prove that Eq. (2.12) is true goes like this: (1) Assume that the area A in Eq. (2.11) represents a flat surface. (2) Assume that the pressure in Eq. (2.11) is constant so that p comes out of the integral. (3) Thus, Eq. (2.11) can be simplified like this: $F_p = p\int_A dA = pA$. Note that the unit vector was omitted because the direction of a pressure force on a flat surface is normal to the surface and directed towards the surface.

FIGURE 2.9

The pressure force on a small section of area on a cylinder.

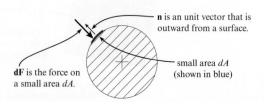

n is an unit vector that is outward from a surface.

small area dA (shown in blue)

dF is the force on a small area dA.

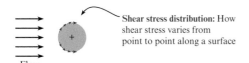

Shear stress distribution: How shear stress varies from point to point along a surface

Flow

Summary. The pressure force is *always given* by the integral of pressure over area, which is $\mathbf{F}_p = \int_A -p\mathbf{n}\, dA$. Only in the special case of uniform pressure acting on a flat surface can you calculate the pressure force by using $F_p = pA$. As always, we recommend that you remember the general equation (i.e., the integral) and then derive $F_p = pA$ whenever this equation is needed.

Now we can tackle the equation for the shear force, which is represented by the symbol F_s or sometimes by F_τ. To build an equation for F_s, we can apply the same logic that was used for the pressure force. Step 1 is to start with a stress distribution (Fig. 2.10). Step 2 is to define a small area and an associated unit vector (Fig. 2.11). Step 3 is to represent the force on the small area as $\mathbf{dF} = \tau\mathbf{t}\, dA$ and then to add up the small forces by using the integral. The final result is

$$\mathbf{F}_\tau = \int_A \tau\mathbf{t}\, dA \qquad (2.13)$$

where $\mathbf{F}_\tau$, the **shear force**, represents the net force on the area A due to the shear stress distribution. If shear stress is constant and the area of integration is a flat surface, then Eq. (2.13) reduces to

$$F_s = \tau A \qquad (2.14)$$

Summary. The shear force is *always given* by the integral of shear stress over area, which is $\mathbf{F}_s = \int_A \tau\mathbf{t}\, dA$. Only in the special case of a uniform shear stress acting on a flat surface can you calculate the shear force by using $F_s = \tau A$.

The Seven Common Fluid Forces

In fluid mechanics, correct analysis of forces is sometimes difficult. Thus, we'd like to share an idea that we have found to be helpful: *When a force acts between Body #1 (comprised of a fluid) and Body #2 (comprised of any material including another fluid), there are seven common forces that arise. Six of the seven forces are associated with the pressure distribution, the shear stress distribution, or both.*

The seven forces are summarized in Table 2.2. Notice the descriptions and tips presented in the third column. Notice in the fourth column that all of the forces are associated with the stress distribution except for the surface tension force.

2.5 The Viscosity Equation

The viscosity equation is used to represent viscous (i.e., frictional) effects in flowing fluids. This equation is important because viscous effects influence practical matters such as energy usage, pressure drop, and the fluid dynamic drag force.

t is an unit vector that is tangent to the surface.

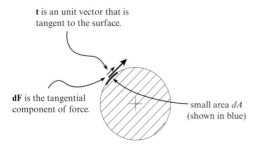

dF is the tangential component of force.

small area dA (shown in blue)

TABLE 2.2 The Seven Common Fluid Forces

#	Name	Description and Tips	Associated With
1	Pressure force	The force caused by a pressure distribution. Use gage pressure for most problems.	Pressure stress
2	Shear force (viscous force)	The force caused by a shear stress distribution. This force requires the fluid to be flowing.	Shear stress
3	Buoyant force	The force on a submerged or partially submerged body that is caused by the hydrostatic pressure distribution.	Pressure stress
4	Surface tension force	The force caused by surface tension. The common formula is $F = \sigma L$.	Forces between molecules
5	Drag force	When fluid flows over a body, the drag force is the component of the total force that is parallel to the fluid velocity.	Both the pressure stress and the shear stress
6	Lift force	When fluid flows over a body, the lift force is the component of the total force that is perpendicular to the fluid velocity.	Both the pressure stress and the shear stress (typically, the effect of shear stress is negligible as compared to the pressure stress)
7	Thrust force	The force associated with propulsion; that is, the force caused by a propeller, jet engine, rocket engine, etc.	Both the pressure stress and the shear stress (typically, the effect of shear stress is negligible as compared to the pressure stress)

The Viscosity Equation

The viscosity equation* is

$$\tau = \mu \frac{dV}{dy} \tag{2.15}$$

The viscosity equation relates shear stress τ to viscosity μ and velocity gradient dV/dy. The viscosity equation is called *Newton's Law of Viscosity* in many references.

The Velocity Gradient

The term (dV/dy) is called the **velocity gradient**.† The variable V represents the magnitude of the velocity vector. In mechanics, **velocity** is defined as the speed and direction of travel of a material particle. Thus, when a fluid is flowing, each material particle will have a different velocity (Fig. 2.12).

The variable y in dV/dy represents position as measured from a wall. Because dV/dy is an ordinary derivative, you can analyze this term by applying your knowledge of calculus. Three methods that we recommend are as follows:

Method #1. If you have a plot of $V(y)$, find dV/dy by sketching a tangent line and then finding the slope of the tangent line by using rise over run.

Method #2. If you have a table of experimental data (e.g., V versus y data), make an estimate based on the definition of the derivative from §1.4: $dV/dy \approx \Delta V/\Delta y$.

---

*There is a more general form of this equation that involves partial derivatives. However, Eq. (2.15) applies to many flows of engineering interest; thus, we leave the more general form to advanced courses.
†This is called the *velocity gradient* because the gradient operator from calculus reduces to the ordinary derivative dV/dy for most simple flows.

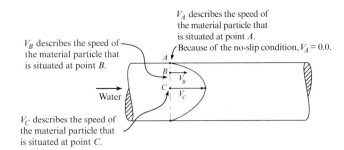

V_A describes the speed of the material particle that is situated at point A. Because of the no-slip condition, $V_A = 0.0$.

V_B describes the speed of the material particle that is situated at point B.

V_C describes the speed of the material particle that is situated at point C.

FIGURE 2.12

This sketch shows water flowing through a round pipe. A **velocity profile** is a sketch or an equation that shows how velocity varies with position.

Method #3. If you have an equation for $V(y)$, differentiate the equation using methods from calculus.

In the context of analyzing the velocity gradient, you will often need to apply the **no-slip condition**, which is this: *When fluid is in contact with a solid body, the velocity of the fluid at the point of contact is the same as the velocity of the solid body at the same point.* **Example.** When water flows in a pipe, the fluid velocity at the wall is equal to the velocity of the wall, which is zero. **Example.** When an airplane moves through the air, the fluid velocity at a point situated on the wing equals the wing velocity at this same point.

You will often see the *velocity gradient* called the *rate of strain* because one can start with the definition of strain and prove that the rate of strain of a fluid particle is given by the velocity gradient. However, this derivation is best left to advanced texts.

Newtonian versus Non-Newtonian Fluids

As an engineer, you need to make decisions about whether or not an equation applies to a situation that you are analyzing. One issue in making this decision is whether or not a fluid can be modeled as a Newtonian fluid. To define a Newtonian fluid, imagine using air or water, setting up an experiment that involves measuring shear stress as a function of velocity gradient, and then plotting your data. You will get a straight line (Fig. 2.13) because both air and water are Newtonian fluids.

If you were to select other fluids, run experiments, and plot the data, you would find that some of the datasets do not plot the same as a Newtonian fluid (Fig. 2.14).

Fig. 2.14 shows three categories of non-Newtonian fluids. For a *shear-thinning* fluid, the viscosity of the fluid decreases as the rate of shear strain (dV/dy) increases. Some common shear-thinning fluids are ketchup, paints, and printer's ink. For a *shear-thickening* fluid, the viscosity increases with shear rate. One example of a shear-thickening fluid is a mixture of starch and water. A *Bingham plastic* acts like a solid for small values of shear stress and then behaves as a fluid at higher shear stress. Some common fluids that are idealized as Bingham plastics are mayonnaise, toothpaste, and certain muds.

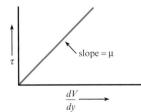

slope = μ

$\dfrac{dV}{dy}$

FIGURE 2.13

A fluid is defined as a **Newtonian fluid** when a plot of shear stress versus velocity gradient gives a straight line.* The slope will be equal to the value of the viscosity μ because the governing equation is $\tau = \mu(dV/dy)$.

*The curve also needs to pass through the origin to distinguish a Newtonian fluid from a Bingham plastic, which is a class of non-Newtonian fluids.

FIGURE 2.14

A **non-Newtonian fluid** (blue lines) is any fluid that does not follow the relationship between *shear stress* and *velocity gradient* that is followed by a Newtonian fluid.

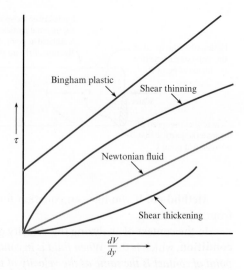

In general, non-Newtonian fluids have molecules that are more complex than Newtonian fluids. Thus, if you are working with a fluid that may be non-Newtonian, consider doing some research; many of the equations and math models presented in textbooks (including this one) only apply to Newtonian fluids. To learn more about non-Newtonian fluids, watch the film entitled *Rheological Behavior of Fluids* (1) or see references (2) and (3).

Reasoning with the Viscosity Equation

Notice that the viscosity equation for a Newtonian fluid is a *linear equation*. It is a linear equation because a plot of the equation (Fig. 2.13) is a straight line. In particular, the general equation for a straight line is $y = mx + b$, where m is the slope and b is the y intercept. Because the viscosity equation is $\tau = \mu(dV/dy)$, you can see that τ is the dependent variable, μ is the slope, dV/dy is the independent variable, and 0.0 is the y intercept.

By using the viscosity equation, you can assess the magnitude of the *velocity gradient* (i.e., dV/dy) and figure out things about the magnitude of the *shear stress* τ (e.g., see Fig. 2.15). The reasoning can be represented by using arrows, like this:

$$\tau\uparrow = \mu\left(\frac{dV}{dy}\uparrow\right) \tag{2.16}$$

In words, Eq. (2.16) says that if the slope (i.e., magnitude of dV/dy) increases, then the shear stress must increase. Similarly, the viscosity equation tells us that if slope decreases, then the shear stress must decrease. And, if the slope is constant (e.g., Couette flow, which is our next topic), then the shear stress must be constant.

FIGURE 2.15

This example shows the velocity profile associated with laminar flow in a round pipe. Notice how information about shear stress can be deduced from a velocity profile. Here, r is the radial position as measured from the centerline of the pipe.

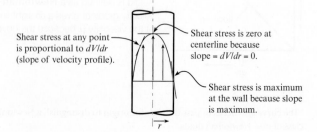

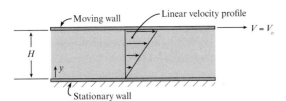

- Moving wall
- Linear velocity profile
- $V = V_o$
- H
- y
- Stationary wall

FIGURE 2.16

Couette flow is a flow that is driven by a moving wall. The velocity profile in the fluid is linear.

Couette Flow

Couette flow is used as a model for a variety of flows that involve lubrication. In Couette flow (e.g., see Fig. 2.16), a moving surface causes fluid to flow. Because of the no-slip condition, the velocity of the fluid at $y = H$ is equal to the velocity of the moving wall. Similarly, the velocity of the fluid at $y = 0$ is zero because the bottom plate is stationary. In the region between the plates, the velocity profile is linear.

When the viscosity equation is applied to Couette flow, the derivative can be replaced with a ratio because the velocity gradient is linear.

$$\tau = \mu\frac{dV}{dy} = \mu\frac{\Delta V}{\Delta y} \tag{2.17}$$

The terms on the right side of Eq. (2.17) can be analyzed as follows:

$$\tau = \mu\frac{\Delta V}{\Delta y} = \mu\frac{V_o - 0}{H - 0} = \mu\frac{V_o}{H}$$

Thus,

$$\tau\big|_{\text{Couette Flow}} = \text{constant} = \mu\frac{V_o}{H} \tag{2.18}$$

Eq. (2.18) reveals that the shear stress at all points in a Couette flow is constant with a magnitude of $\mu V_o/H$.

EXAMPLE 2.1

Applying the Viscosity Equation to Calculate Shear Stress in a Poiseuille Flow

Problem Statement

A famous solution in fluid mechanics, called Poiseuille flow, involves laminar flow in a round pipe (see Chapter 10 for details). Consider Poiseuille flow with a velocity profile in the pipe given by

$$V(r) = V_o(1 - (r/r_o)^2)$$

where r is radial position as measured from the centerline, V_o is the velocity at the center of the pipe, and r_o is the pipe radius. Find the shear stress at the center of the pipe, at the wall, and where $r = 1$ cm. The fluid is water (15°C), the pipe diameter is 4 cm, and $V_o = 1$ m/s.

Define the Situation

Water flows in a round pipe (Poiseuille flow).

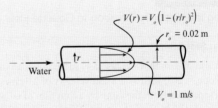

$V(r) = V_o\left(1 - (r/r_o)^2\right)$
$r_o = 0.02$ m
Water
r
$V_o = 1$ m/s

Water (15°C, 1 atm, Table A.5): $\mu = 1.14 \times 10^{-3}$ N·s/m².

State the Goal

Calculate the shear stress at three points:

$\tau(r = 0.00\text{ m}) \text{ (N/m}^2) \Leftarrow$ pipe centerline

$\tau(r = 0.01\text{ m}) \text{ (N/m}^2) \Leftarrow$ middle of the pipe

$\tau(r = 0.02\text{ m}) \text{ (N/m}^2) \Leftarrow$ the wall

Generate Ideas and Make a Plan

Because the goal is τ, select the *viscosity equation*. Let the position variable be r instead of y.

$$\tau = -\mu\frac{dV}{dr} \qquad \textbf{(a)}$$

Regarding the minus sign in Eq. (a), the y in the viscosity equation is measured from the wall. The coordinate r is in the opposite direction. The sign change occurs when the variable is changed from y to r.

To find the velocity gradient in Eq. (a), differentiate the given velocity profile.

$$\frac{dV(r)}{dr} = \frac{d}{dr}(V_o(1 - (r/r_o)^2)) = \frac{-2V_o r}{r_o^2} \qquad \textbf{(b)}$$

Now, the goal can be found. **Plan**. Apply Eq. (b) to find the velocity gradient. Then, substitute into Eq. (a).

Take Action (Execute the Plan)

1. Viscosity equation ($r = 0$ m):

$$\left.\frac{dV(r)}{dr}\right|_{r=0\,\text{m}} = \frac{-2V_o(0\,\text{m})}{r_o^2} = \frac{-2(1\,\text{m/s})(0\,\text{m})}{(0.02\,\text{m})^2} = 0.0\,\text{s}^{-1}$$

$$\tau(r = 0\,\text{m}) = \left.-\mu\frac{dV(r)}{dr}\right|_{r=0\,\text{m}}$$

$$= (1.14 \times 10^{-3}\,\text{N}\cdot\text{s/m}^2)(0.0\,\text{s}^{-1})$$

$$= \boxed{0.0\,\text{N/m}^2}$$

2. Viscosity equation ($r = 0.01$ m):

$$\left.\frac{dV(r)}{dr}\right|_{r=0.01\,\text{m}} = \frac{-2V_o(0.01\,\text{m})}{r_o^2}$$

$$\frac{-2(1\,\text{m/s})(0.01\,\text{m})}{(0.02\,\text{m})^2} = -50\,\text{s}^{-1}$$

Next, calculate shear stress:

$$\tau(r = 0.01\,\text{m}) = \left.-\mu\frac{dV(r)}{dr}\right|_{r=0.01\,\text{m}}$$

$$= (1.14 \times 10^{-3}\,\text{N}\cdot\text{s/m}^2)(50\,\text{s}^{-1})$$

$$= \boxed{0.0570\,\text{N/m}^2}$$

3. Viscosity equation ($r = 0.02$ m):

$$\left.\frac{dV(r)}{dr}\right|_{r=0.02\,\text{m}} = \frac{-2V_o(0.02\,\text{m})}{r_o^2}$$

$$= \frac{-2(1\,\text{m/s})(0.02\,\text{m})}{(0.02\,\text{m})^2} = -100\,\text{s}^{-1}$$

Next, calculate shear stress:

$$\tau(r = 0.02\,\text{m}) = \left.-\mu\frac{dV(r)}{dr}\right|_{r=0.02\,\text{m}}$$

$$= (1.14 \times 10^{-3}\,\text{N}\cdot\text{s/m}^2)(100\,\text{s}^{-1})$$

$$= \boxed{0.114\,\text{N/m}^2}$$

Review the Solution and the Process

1. *Tip*. On most problems, including this example, carrying and canceling units is useful, if not critical.

2. *Notice*. Shear stress varies with location. For this example, τ is zero on the centerline of the flow and nonzero everywhere else. The maximum value of shear stress occurs at the wall of the pipe.

3. *Notice*. For flow in a round pipe, the viscosity equation has a minus sign and uses the position coordinate r.

$$\tau = -\mu\frac{dV}{dr}$$

EXAMPLE 2.2

Applying the Viscosity Equation to Couette Flow

Problem Statement

A board 1 m by 1 m that weighs 25 N slides down an inclined ramp (slope = 20°) with a constant velocity of 2.0 cm/s. The board is separated from the ramp by a thin film of oil with a viscosity of 0.05 N·s/m². Assuming that the oil can be modeled as a Couette flow, calculate the space between the board and the ramp.

Define the Situation

A board slides down an oil film on a inclined plane.

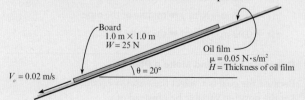

Assumptions. (1) Couette flow. (2) Board has constant speed.

State the Goal

H(mm) ← thickness of the film of oil

Generate Ideas and Make a Plan

Because the goal is H, apply the *viscosity equation* (Eq. 2.18):

$$H = \mu\frac{V_o}{\tau} \qquad \textbf{(a)}$$

To find the shear stress τ in Eq. (a), draw a *Free Body Diagram* (FBD) of the board. In the FBD, W is the weight, N is the normal force, and F_{shear} is shear force. Because shear stress is constant with x, the shear force can be expressed as $F_{\text{shear}} = \tau A$.

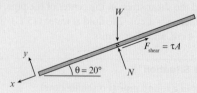

Because the board moves at constant speed, the forces are in balance. Thus, apply *force equilibrium*.

$$\Sigma F_x = 0 = W \sin\theta - \tau A \qquad \textbf{(b)}$$

Rewrite Eq. (b) as

$$\tau = (W\sin\theta)/A \qquad \textbf{(c)}$$

Eq. (c) can be solved for τ. The plan is as follows:

1. Calculate τ using force equilibrium (Eq. c).
2. Calculate H using the shear stress equation (Eq. a).

Take Action (Execute the Plan)

1. Force equilibrium:

$$\tau = (W\sin\theta)/A = (25\text{ N})(\sin 20°)/(1.0\text{ m}^2) = 8.55\text{ N/m}^2$$

2. Shear stress equation:

$$H = \mu\frac{V_o}{\tau} = (0.05\text{ N}\cdot\text{s/m}^2)\frac{(0.02\text{ m/s})}{(8.55\text{ N/m}^2)} = \boxed{0.117\text{ mm}}$$

Review the Solution and the Process

1. H is about 12% of a millimeter; this is quite small.
2. *Tip.* Solving this problem involved drawing an FBD. The FBD is useful for most problems involving Couette flow.

2.6 Surface Tension*

Engineers need to be able to predict and characterize surface tension effects because they affect many industrial problems. Some examples of surface tension effects include the following:

- *Wicking.* Water will wick into a paper towel. Ink will wick into paper. Polypropylene, an excellent fiber for cold-weather aerobic activity, wicks perspiration away from the body.
- *Capillary rise.* A liquid will rise in a small-diameter tube. Water will rise in soil.
- *Capillary instability.* A liquid jet will break up into drops.
- *Drop and bubble formation.* Water on a leaf beads up. A leaky faucet drips. Soap bubbles form.
- *Excess pressure.* The pressure inside a water drop is higher than ambient pressure. The pressure inside a vapor bubble during boiling is higher than ambient pressure.
- *Walking on water.* The water strider, an insect, can walk on water. Similarly, a metal paper clip or a metal needle can be positioned to float (through the action of surface tension) on the surface of water.
- *Detergents.* Soaps and detergents improve the cleaning of clothes because they lower the surface tension of water so that the water can more easily wick into the pores of the fabric.

Many experiments have shown that the surface of liquid behaves like a stretched membrane. The material property that characterizes this behavior is **surface tension**, σ (sigma). Surface tension can be expressed in terms of force:

$$\text{surface tension}(\sigma) = \frac{\text{force along an interface}}{\text{length of the interface}} \qquad \textbf{(2.19)}$$

Surface tension can also be expressed in terms of energy:

$$\text{surface tension}(\sigma) = \frac{\text{energy required to increase the surface area of a liquid}}{\text{unit area}} \qquad \textbf{(2.20)}$$

From Eq. (2.19), the unit of surface tension is the newton per meter (N/m). Surface tension typically has a magnitude ranging from 1 to 100 mN/m. The unit of surface tension can also be joule per meter squared (J/m^2) because

$$\frac{N}{m} = \frac{N\cdot m}{m\cdot m} = \frac{J}{m^2}$$

*The authors acknowledge and thank Dr. Eric Aston for his feedback and inputs on this section. Dr. Aston is a chemical engineering professor at the University of Idaho.

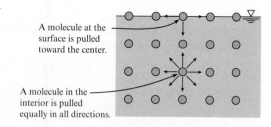

FIGURE 2.17

Forces between molecules in a liquid.

The physical mechanism of surface tension is based on **cohesive force**, which is the attractive force between like molecules. Because liquid molecules attract one another, molecules in the interior of a liquid (see Fig. 2.17) are attracted equally in all directions. In contrast, molecules at the surface are pulled toward the center because they have no liquid molecules above them. This pull on surface molecules draws the surface inward and causes the liquid to seek to minimize surface area. This is why a drop of water draws into a spherical shape.

Surface tension of water decreases with temperature (see Fig. 2.18) because thermal expansion moves the molecules farther apart, and this reduces the average attractive force between molecules (i.e., cohesive force goes down). Surface tension is strongly affected by the presence of contaminants or impurities. For example, adding soap to water decreases the surface tension. The reason is that impurities concentrate on the surface, and these molecules decrease the average attractive force between the water molecules. As shown in Fig. 2.18, the surface tension of water at 20°C is $\sigma = 0.0728 \approx 0.073$ N/m. This value is used in many of the calculations in this text.

In Fig. 2.18, surface tension is reported for an interface of air and water. It is common practice to report surface tension data based on the materials that were used during the measurement of the surface tension data.

To learn more about surface tension, we recommend the online film entitled *Surface Tension in Fluid Mechanics* (5) and Shaw's book (6).

Example Problems

Most problems involving surface tension are solved by drawing an FBD and applying force equilibrium. The force due to surface tension, from Eq. (2.19), is

$$\text{force due to surface tension} = F_\sigma = \sigma L \qquad \text{(2.21)}$$

where L is the length of a line that lies along the surface of the liquid. The use of force equilibrium to solve problems is illustrated in Examples 2.3 and 2.4.

Adhesion and Capillary Action

When a drop of water is placed on glass, the water will wet the glass (see Fig. 2.19) because water is strongly attracted to glass. This attractive force pulls the water outward as shown. The force between dissimilar surfaces is called **adhesion** (see Fig. 2.19b). Water will "wet out" on a surface when *adhesion is greater than cohesion*.

FIGURE 2.18

Surface tension of water for a water/air interface. Property values are from White (7).

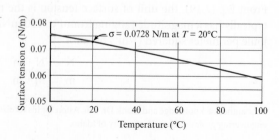

EXAMPLE 2.3

Applying Force Equilibrium to Calculate the Pressure Rise inside a Water Droplet

Problem Statement

The pressure inside a water drop is higher than the pressure of the surroundings. Derive a formula for this pressure rise. Then, calculate the pressure rise for a 2 mm diameter water drop. Use $\sigma = 73$ mN/m.

Define the Situation

Pressure inside a water drop is larger than ambient pressure. $d = 0.002$ m, $\sigma = 73$ mN/m.

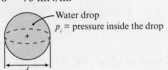

State the Goal

1. Derive an equation for p_i.
2. Calculate p_i in pascals.

Generate Ideas and Make a Plan

Because pressure is involved in a force balance, draw an FBD of the drop.

Force due to pressure = Force due to surface tension

$$F_p = F_\sigma \qquad \text{(a)}$$

From Eq. (2.19), the surface tension force is σ times the length of the interface:

$$F_\sigma = \sigma L = \sigma \pi d \qquad \text{(b)}$$

The pressure force is pressure times area:

$$F_p = p_i \frac{\pi d^2}{4} \qquad \text{(c)}$$

Combine Eqs. (a) to (c):

$$p_i \frac{\pi d^2}{4} = \sigma \pi d \qquad \text{(d)}$$

Solve for pressure:

$$\boxed{p_i = \frac{4\sigma}{d}} \qquad \text{(e)}$$

The first goal (equation for pressure) has been attained. The next goal (value of pressure) can be found by substituting numbers into Eq. (e).

Take Action (Execute the Plan)

$$p_i = \frac{4\sigma}{d} = \frac{4(0.073 \text{ N/m})}{(0.002 \text{ m})} = \boxed{146 \text{ Pa gage}}$$

Review the Results and the Process

1. *Notice.* The answer is expressed as gage pressure. Gage pressure in this context is the pressure rise above ambient.

2. *Physics.* The pressure rise inside a liquid drop is a consequence of the membrane effect of surface tension. One way to visualize this is make an analogy with a balloon filled with air. The pressure inside the balloon pushes outward against the membrane force of the rubber skin. In the same way, the pressure inside a liquid drop pushes outward against the membrane like the force of surface tension.

On some surfaces, such as Teflon and wax paper, a drop of water will bead up (Fig 2.20) because *adhesion between the water and teflon is less than cohesion of the water*. A surface on which water beads up is called *hydrophobic* (water hating). Surfaces such as glass on which water drops spread out are called *hydrophilic* (water loving).

Capillary action describes the tendency of a liquid to rise in narrow tubes or to be drawn into small openings. Capillary action is responsible for water being drawn into the narrow openings in soil or into the narrow openings between the fibers of a dry paper towel.

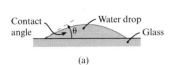

(a)

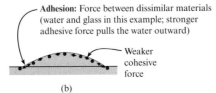

(b)

FIGURE 2.19

Water wets glass because adhesion is greater than cohesion. Wetting is associated with a contact angle less than 90°.

FIGURE 2.20

Water beads up a hydrophobic material such as Teflon because adhesion is less than cohesion. A nonwetting surface is associated with a contact angle greater than 90°.

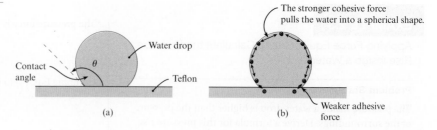

FIGURE 2.21

Water will rise up a glass tube (capillary rise), whereas mercury will move downward (capillary repulsion).

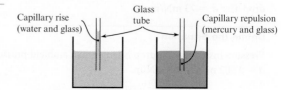

When a capillary tube is placed into a container of water, the water rises up the tube (Fig. 2.21) because the adhesive force between the water and the glass pulls the water up the tube. This is called capillary rise. Notice how the contact angle for the water is the same in Figs. 2.19 and 2.22. Alternatively, when a fluid is nonwetting (such as mercury on glass), then the liquid will display capillary repulsion.

To derive an equation for capillary rise (see Fig. 2.22), define a system comprised of the water inside the capillary tube. Then, draw an FBD. As shown, the pull of surface tension lifts the column of water. Applying force equilibrium gives

$$\text{weight} = \text{surface tension force}$$

$$\gamma \left(\frac{\pi d^2}{4} \right) \Delta h = \sigma \pi d \cos \theta \tag{2.22}$$

Assume the contact angle is nearly zero so that $\cos \theta \approx 1.0$. Note that this is a good assumption for a water/glass interface. Eq. (2.22) simplifies to

$$\Delta h = \frac{4\sigma}{\gamma d} \tag{2.23}$$

EXAMPLE. Calculate the capillary rise for water (20°C) in a glass tube of diameter $d = 1.6$ mm.

Solution. From Table A.5, $\gamma = 9790$ N/m³. From Fig. 2.18, $\sigma = 0.0728$ N/m. Now, calculate capillary rise using Eq. (2.23):

$$\Delta h = \frac{4(0.0728 \text{ N/m})}{(9790 \text{ N/m}^3)(1.6 \times 10^{-3} \text{ m})} = \boxed{18.6 \text{ mm}}$$

Example 2.4 shows a case involving a nonwetting surface.

FIGURE 2.22

Sketches used for deriving an equation for capillary rise.

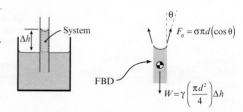

EXAMPLE 2.4

Applying Force Equilibrium to Determine the Size of a Sewing Needle that Can Be Supported by Surface Tension

Problem Statement

The Internet shows examples of sewing needles that appear to be "floating" on top of water. This effect is due to surface tension supporting the needle. Determine the largest diameter of sewing needle that can be supported by water. Assume that the needle material is stainless steel with $SG_{ss} = 7.7$.

Define the Situation

A sewing needle is supported by the surface tension of a water surface.

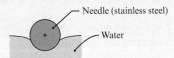

Needle (stainless steel)

Water

Assumptions
- Assume the sewing needle is a cylinder.
- Neglect end effects.

Properties
- Water (20°C, 1 atm, Fig. 2.18): $\sigma = 0.0728$ N/m
- Water (4°C, 1 atm, Table F.6): $\gamma_{H_2O} = 9810$ N/m^3
- SS: $\gamma_{ss} = (7.7)(9810$ N/m$^3) = 75.5$ kN/m^3

State the Goal

d(mm) ← diameter of the largest needle that can be supported by the water

Generate Ideas and Make a Plan

Because the weight of the needle is supported by the surface tension force, draw an FBD. Select a system comprised of the needle plus the surface layer of the water. The FBD is

Apply force equilibrium:

Force due to surface tension = Weight of needle

$$F_\sigma = W \qquad \text{(a)}$$

From Eq. (2.21)

$$F_\sigma = \sigma 2L \cos \theta \qquad \text{(b)}$$

where L is the length of the needle. The weight of the needle is

$$W = \left(\frac{\text{weight}}{\text{volume}} \right) [\text{volume}] = \gamma_{ss} \left[\left(\frac{\pi d^2}{4} \right) L \right] \qquad \text{(c)}$$

Combine Eqs. (a), (b), and (c). Also, assume the angle θ is zero because this gives the maximum possible diameter:

$$\sigma 2L = \gamma_{ss} \left(\frac{\pi d^2}{4} \right) L \qquad \text{(d)}$$

Plan. Solve Eq. (d) for d and then plug numbers in.

Take Action (Execute the Plan)

$$d = \sqrt{\frac{8\sigma}{\pi \gamma_{ss}}} = \sqrt{\frac{8(0.0728 \text{ N/m})}{\pi (75.5 \times 10^3 \text{ N/m}^3)}} = \boxed{1.57 \text{ mm}}$$

Review the Solution and the Process

Notice. When applying specific gravity, look up water properties at the reference temperature of 4°C.

2.7 Vapor Pressure

A liquid, even at a low temperature, can boil as it flows through a system. This boiling can reduce performance and damage equipment. Thus, engineers need to be able to predict when boiling will occur. This prediction is based on the vapor pressure.

Vapor pressure, p_v (kPa), is the pressure at which the liquid phase and the vapor phase of a material will be in thermal equilibrium. Vapor pressure is also called *saturation pressure,* and the corresponding temperature is called *saturation temperature.*

Vapor pressure can be visualized on a *phase diagram*. A phase diagram for water is shown in Fig. 2.23. As shown, water will exist in the liquid phase for any combination of temperature and pressure that lies above the blue line. Similarly, the water will exist in the vapor phase for points below the blue line. Along the blue line, the liquid and vapor phases are in thermal equilibrium. When boiling occurs, the pressure and temperature of the water will be given by one of the points on the blue line. In addition to Fig. 2.23, data for vapor pressure of water are tabulated in Table A.5.

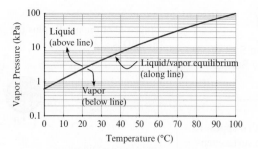

FIGURE 2.23

A phase diagram for water.

EXAMPLE. Water at 20°C flows through a venturi nozzle and boils. Explain why. Also, give the value of pressure in the nozzle.

Solution. The water is boiling because the pressure has dropped to the vapor pressure. Table 2.1 indicates that p_v can be looked up in Table A.5. Thus, the vapor pressure at 20°C (Table A.5) is $p_v = 2.34$ kPa absolute. This value can be validated by using Fig. 2.23.

Review. Vapor pressure is commonly expressed using *absolute pressure*. Absolute pressure is the value of pressure as measured relative to a pressure of absolute zero.

2.8 Characterizing Thermal Energy in Flowing Gases

Engineers characterize thermal energy changes using properties introduced in this section. **Thermal energy** is the energy associated with molecules in motion. This means that thermal energy is associated with temperature change (sensible energy change) and phase change (latent energy change). For most fluid problems, thermal properties are not important. However, thermal properties are used for compressible flow of gases (Chapter 12).

Specific Heat, c

Specific heat characterizes the amount of thermal energy that must be transferred to a unit mass of substance to raise its temperature by one degree. The dimensions of specific heat are energy per unit mass per degree temperature change, and the corresponding units are J/kg·K.

The magnitude of c depends on the process. For example, if a gas is heated at *constant volume*, less energy is required than if the gas is heated at *constant pressure*. This is because a gas that is heated at constant pressure must do work as it expands against its surroundings.

The *constant volume specific heat, c_v*, applies to a process carried out at constant volume. The *constant pressure specific heat, c_p*, applies to a process carried out at constant pressure. The ratio c_p/c_v is called the **specific heat ratio** and is given the symbol k. Values for c_p and k for various gases are given in Table A.2.

Internal Energy

Internal energy includes all the energy in matter except for kinetic energy and potential energy. Thus, internal energy includes multiple forms of energy, such as chemical energy, electrical energy, and thermal energy. Specific internal energy, u, has dimensions of energy per unit mass. The units are J/kg.

Enthalpy

When a material is heated at constant pressure, the energy balance is

$$(\text{Energy added}) = \begin{pmatrix} \text{Energy to increase} \\ \text{thermal energy} \end{pmatrix} + \begin{pmatrix} \text{Energy to do work} \\ \text{as the material expands} \end{pmatrix}$$

The work term is needed because the material is exerting a force over a distance as it pushes its surroundings away during the process of thermal expansion.

Enthalpy is a property that characterizes the amount of energy associated with a heating or cooling process. Enthalpy per unit mass is defined mathematically by

$$(\text{enthalpy}) = (\text{internal energy}) + (\text{pressure/density})$$

$$h = u + p/\rho$$

Ideal Gas Behavior

For an ideal gas, the properties h, u, c_p, and c_v depend only on temperature, not on pressure.

2.9 Summarizing Key Knowledge

Describing Your System

To describe what you are analyzing, apply three ideas:
- The *system* is the matter that you select for analysis.
- The *surroundings* are everything else that is not part of the system.
- The *boundary* is the surface that separates the system from its surroundings.

To describe the *conditions* of your system, apply four ideas:

- The *state* of a system is the condition of the system as specified by values of the properties of the system.
- A *property* is a measurable characteristic of a system that depends only on the present state.
- *Steady state* means that all properties of the system are constant with time.
- A *process* is a change of a system from one state to another state.

Finding Fluid Properties

- To characterize the weight or mass of a fluid, use ρ, γ, or SG. If you know one of these properties, then you can calculate the other two using these equations: $\gamma = \rho g$ and $SG = \rho/\rho_{H_2O,\,(4°C)} = \gamma/\gamma_{H_2O,\,(4°C)}$.
- To characterize viscous effects (i.e., frictional effects), you can use *viscosity* μ, which is also called *dynamic viscosity* or *absolute viscosity*. You also will often use a different property, called *kinematic viscosity*, which is defined by $\nu = \mu/\rho$.
- When looking up properties, make sure that you account for the variation in the value of the property as a function of temperature and pressure.
- Quality in documentation involves listing the name of the property, the source of the property data, the units, the temperature and pressure, and any assumptions that you make.

Density Topics

- Modeling a fluid as *constant density* means that you assume the density is constant with position and time. *Variable density* means the density can change with position, time, or both.
- A gas in steady flow can be idealized as a constant density if the Mach number is less than 0.3. Liquids for most flow situations can be idealized as constant density. Two notable exceptions are water hammer problems and acoustics problems.
- All fluids, including liquids, will compress (i.e., decrease in volume) if the pressure is increased. The amount of volume change can be calculated by using the bulk modulus of elasticity.
- Specific gravity (S or SG) gives the ratio of the density of a material to the density of liquid water at 4°C.

Stress

In mechanics, **stress** is an entity that expresses the internal forces that material particles exert on each other. Stress is the ratio of force to area at a point and is resolved into two components:

- Pressure (normal stress) is the ratio of normal force to area
- Shear stress is the ratio of shear force to area

To relate force to stress, integrate the stress over area.

- The general equation for the *pressure force* is $\mathbf{F}_p = \int_A -p\mathbf{n}\, dA$. For the special case of a uniform pressure acting on a flat surface, this equation simplifies to $F_p = pA$.

- The general equation for the *shear force* is $\mathbf{F}_s = \int_A \tau\, \mathbf{t}\, dA$. For the special case of a uniform shear stress acting on a flat surface, this equation simplifies to $F_s = \tau A$.

- When a force acts between Body #1 (a fluid body) and Body #2 (any other body), the force can usually be identified as one of seven forces: (1) the pressure force, (2) the shear force, (3) the buoyant force, (4) the drag force, (5) the lift force, (6) the surface tension force, or (7) the thrust force. Except for the surface tension force, each of these forces is associated with a pressure distribution, a shear stress distribution, or both.

The Viscosity Equation

- The viscosity equation is useful for calculating the shear stress in a flowing fluid. The equation is $\tau = \mu\,(dV/dy)$. If μ is constant, then τ is linearly related to dV/dy.

- For many flows, the velocity gradient is the first derivative of velocity with respect to distance (dV/dy).

- The no-slip condition means that the velocity of fluid in contact with a solid surface will equal the velocity of the surface.

- A *Newtonian fluid* is one in which a plot of τ versus dV/dy is a straight line. A *non-Newtonian* fluid is one in which a plot of τ versus dV/dy is not a straight line. In general, non-Newtonian fluids have more complex molecular structures than Newtonian fluids. Examples of non-Newtonian fluids include paint, toothpaste, and molten plastics. Equations developed for Newtonian fluids (i.e., many textbook equations) often do not apply to non-Newtonian fluids.

- Couette flow involves a flow through a narrow channel with the top plate moving at a speed of V_o. In Couette flow, the shear stress is constant at every point and is given by $\tau_o = (\mu V_o)/H$.

Surface Tension and Vapor Pressure

- A liquid flowing in a system will boil when the pressure drops to the vapor pressure. This boiling often is detrimental to a design.

- Surface tension problems are usually solved by drawing an FBD and summing forces.

- The formula for capillary rise of water in a round glass tube is $\Delta h = (4\sigma)/(\gamma d)$.

REFERENCES

1. Fluid Mechanics Films, downloaded 7/31/11 from http://web.mit.edu/hml/ncfmf.html

2. Harris, J. *Rheology and non-Newtonian Flow.* New York: Longman, 1977.

3. Schowalter, W. R. *Mechanics of Non-Newtonian Fluids.* New York: Pergamon Press, 1978.

4. White, F. M. *Fluid Mechanics*, 7th ed. New York: McGraw-Hill, 2011, p. 828.

5. Fluid Mechanics Films, downloaded 7/31/11 from http://web.mit.edu/hml/ncfmf.html

6. Shaw, D. J. *Introduction to Colloid and Surface Chemistry*, 4e, Maryland Heights, MO: Butterworth-Heinemann, 1992.

PROBLEMS

System, State and Property (§2.1)

2.1 A system is separated from its surrounding by a
   a. boundary
   b. divisor
   c. border
   d. fractionation line

Looking Up Fluid Properties (§2.2)

2.2 An open vat in a food processing plant contains 600 L of water at 20°C and atmospheric pressure. If the water is heated to 60°C, what will be the percentage change in its volume? If the vat has a diameter of 2 m, how much will the water level rise due to this temperature increase?

2.3 When looking up values for density, absolute viscosity, and kinematic viscosity, which statement is most true for <u>both</u> liquids and gases?
   a. all three of these properties vary with temperature and pressure
   b. all three of these properties vary with pressure
   c. all three of these properties vary with temperature

2.4 Kinematic viscosity (select all that apply)
   a. is viscosity/density
   b. is another name for absolute viscosity
   c. has dimensions of L^2/T
   d. is dimensionless because forces cancel out
   e. only applies to compressible fluids

2.5 What is the change in the viscosity and density of water between 10°C and 50°C? What is the change in the viscosity and density of air between 10°C and 50°C? Assume standard atmospheric pressure ($p = 101$ kN/m^2).

2.6 Determine the change in the kinematic viscosity of air that is heated from 10°C to 80°C. Assume standard atmospheric pressure.

2.7 Find the dynamic and kinematic viscosities of kerosene, SAE 10W-30 motor oil, and water at a temperature of 20°C.

Topics Related to Density (§2.3)

2.8 Specific gravity (select all that apply)

   a. is dimensionless

   b. can have units of N/m^3

   c. decreases with temperature increase

   d. increases with temperature increase

2.9 Dimensions of the bulk modulus of elasticity are

   a. the same as the dimensions of pressure/density

   b. the same as the dimensions of pressure/volume

   c. the same as the dimensions of pressure

   d. dimensionless

2.10 A pressure of 4×10^6 N/m^2 is applied to mass of water that initially filled a 3200 cm^3 volume. Estimate its volume after the pressure is applied.

2.11 Calculate the pressure increase that must be applied to liquid water to reduce its volume by 4.5%.

Pressure and Shear Stress (§2.4)

2.12 Shear stress has dimensions of

   a. force/area

   b. force/length

   c. dimensionless

The Viscosity Equation (§2.5)

2.13 The no-slip condition (select all that apply)

   a. only applies to ideal flow

   b. only applies to rough surfaces

   c. means velocity, V, is the velocity of the wall

   d. means velocity, V, is zero at the wall

2.14 At a point in a flowing fluid, the shear stress is 1.8×10^{-4} psi, and the velocity gradient is 1 s^{-1}.

   a. What is the viscosity in traditional units?

   b. Convert this viscosity to SI units.

   c. Is this fluid more or less viscous than water?

2.15 SAE 10W-30 oil with viscosity 1×10^{-4} lbf·s/ft^2 is used as a lubricant between two parts of a machine that slide past one

another with a velocity difference of 7 ft/s. What spacing, in inches, is required if you don't want a shear stress of more than 2 lbf/ft^2? Assume Couette flow.

2.16 Suppose that glycerin ($T = 20°$C) is flowing and that the pressure gradient dp/dx is -1.5 kPa/m. What are the velocity and shear stress at a distance of 8 mm from the wall if the space B between the walls is 5.0 cm? What are the shear stress and velocity at the wall? The velocity distribution for viscous flow between stationary plates is

$$u = -\frac{1}{2\mu}\frac{dp}{dx}(By - y^2)$$

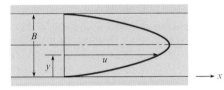

Problems 2.16

2.17 Two plates are separated by a 3/16 in. space. The lower plate is stationary; the upper plate moves at a velocity of 27 ft/s. Oil (SAE 10W-30, 150°F), which fills the space between the plates, has the same velocity as the plates at the surface of contact. The variation in velocity of the oil is linear. What is the shear stress in the oil?

2.18 The sliding plate viscometer shown below is used to measure the viscosity of a fluid. The top plate is moving to the right with a constant velocity of $V = 18$ m/s in response to a force of $F = 2$ N. The bottom plate is stationary. What is the viscosity of the fluid? Assume a linear velocity distribution.

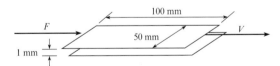

Problem 2.18

Surface Tension (§2.6)

2.19 Surface tension (select all that apply)

   a. only occurs at an interface, or surface

   b. has dimensions of force/area

   c. has dimensions of energy/area

   d. has dimensions of force/length

   e. depends on adhesion and cohesion

   f. varies as a function of temperature

2.20 A water bug is suspended on the surface of a pond by surface tension (water does not wet the legs). The bug has six legs, and each leg is in contact with the water over a length of 7 mm. What is the maximum mass (in grams) of the bug if it is to avoid sinking?

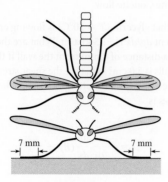

Problem 2.20

2.21 A water column in a glass tube is used to measure the pressure in a pipe. The tube is 1/4 in. in diameter. How much of the water column is due to surface-tension effects? What would be the surface-tension effects if the tube were 1/8 in. or 1/16 in. in diameter?

2.22 What is the gage pressure within a $d = 2.0$ mm spherical droplet of water?

Vapor Pressure (§2.7)

2.23 Water is at 20°C and the pressure is lowered until boiling is observed. What is the pressure?

2.24 A student in the laboratory plans to exert a vacuum in the air space above the surface of water in a closed tank. She plans for the absolute pressure in the air to be 12,300 Pa. The temperature in the water is 50°C. Will water boil under these circumstances?

CHAPTER**THREE**

Fluid Statics

CHAPTER ROAD MAP This chapter introduces concepts related to pressure and describes how to calculate forces associated with distributions of pressure. The emphasis is on fluids in hydrostatic equilibrium.

FIGURE 3.1

The first man-made structure to exceed the masonry mass of the Great Pyramid of Giza was Hoover Dam. The design of dams involves calculations of hydrostatic forces. (U.S. Bureau of Reclamation)

LEARNING OUTCOMES

PRESSURE (§3.1).
- Define pressure and convert pressure units.
- Describe atmospheric pressure and select an appropriate value.
- Define and apply gage, absolute, vacuum, and differential pressure.
- Know the main ideas about hydraulic machines and solve relevant problems.

THE HYDROSTATIC EQUATIONS (§3.2).
- Define hydrostatic equilibrium.
- Know the main ideas about the hydrostatic differential equation.
- Know the main ideas about the hydrostatic algebraic equation and solve relevant problems.

PRESSURE MEASUREMENT (§3.3).
- Explain how common scientific instruments work and do relevant calculations (this LO applies to the mercury barometer, piezometer, manometer, and Bourdon tube gage).

THE PRESSURE FORCE (§3.4).
- Define the center of pressure.
- Sketch a pressure distribution.
- Explain or apply the gage pressure rule.
- Calculate the force due to a uniform pressure distribution.
- Know the main ideas about the panel equations and be able to apply these equations.

CURVED SURFACES (§3.5).
- Solve problems that involve curved surfaces that are acted on by uniform or hydrostatic pressure distributions.

BUOYANCY (§3.6).
- Know the main ideas about buoyancy and be able to apply these ideas to solve problems.

3.1 Describing Pressure

Because engineers use pressure in the solution of nearly all fluid mechanics problems, this section introduces fundamental ideas about pressure.

Pressure

Pressure is the ratio of the normal force due to a fluid to the area that this force acts on, in the limit as this area shrinks to zero.

$$p = \left. \frac{\text{magnitude of normal force}}{\text{unit area}} \right|_{\substack{\text{at a point} \\ \text{due to a fluid}}} = \lim_{\Delta A \to 0} \frac{\Delta F_{normal}}{\Delta A} \qquad (3.1)$$

Pressure is defined at a point because pressure typically varies with each (x, y, z) location in a flowing fluid.

Pressure is a scalar that produces a resultant force by its action on an area. The resultant force is normal to the area and acts in a direction toward the surface (compressive).

Pressure is caused by the molecules of the fluid interacting with the surface. For example, when a soccer ball is inflated, the internal pressure on the skin of the ball is caused by air molecules striking the wall.

Units of pressure can be organized into three categories:

- *Force per area.* The SI unit is the newtons per square meter or pascals (Pa). The traditional units include psi, which is pounds-force per square inch, and psf, which is pounds-force per square foot.

- *Liquid column height.* Sometimes pressure units give an equivalent height of a column of liquid. For example, pressure in a balloon will push a water column upward about 8 inches (Fig. 3.2). Engineers state that the pressure in the balloon is 8 inches of water: $p = 8$ in-H_2O. When pressure is given in units of "height of a fluid column," the pressure value can be directly converted to other units using Table F.1. For example, the pressure in the balloon is

$$p = (8 \text{ in-}H_2O)(249.1 \text{ Pa/in-}H_2O) = 1.99 \text{ kPa}$$

- *Atmospheres.* Sometimes pressure units are stated in terms of atmospheres where 1.0 atm is the air pressure at sea level at standard conditions. Another common unit is the bar, which is very nearly equal to 1.0 atm (1.0 bar = 10^5 kPa).

Atmospheric Pressure

This subsection explains how to select an accurate value of atmospheric pressure (p_{atm}) because a value of p_{atm} is often needed in calculations.

The atmosphere of the earth is an extremely thin layer of air that extends from the surface of the earth to the edge of space. The atmosphere is held in place by gravitational force. According to NASA, "if the earth were the size of a basketball, a tightly held pillowcase would represent the thickness of the atmosphere."[*]

If you look at data, it is evident that p_{atm} is strongly influenced by elevation:[†]

- At London (EL = 35 m): $p_{atm} = 101$ kPa
- At Denver, Colorado, USA (EL = 1650 m), $p_{atm} = 83.4$ kPa
- Near the summit of Mount Everest, Nepal (EL = 8000 m): $p_{atm} = 35.6$ kPa
- At a typical cruise altitude of a jetliner (EL = 12,190 m): $p_{atm} = 18.8$ kPa

FIGURE 3.2

Pressure in a balloon causing a column of water to rise 8 inches.

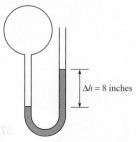

$\Delta h = 8$ inches

---

[*]http://www.grc.nasa.gov/WWW/k-12/airplane/atmosmet.html.
[†]The value of atmospheric pressure is an absolute pressure. Thus, engineers commonly say that $p_{atm} = 101$ kPa instead of saying that $p_{atm} = 101$ kPa abs.

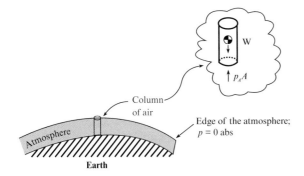

Column of air

Edge of the atmosphere; $p = 0$ abs

Atmosphere

Earth

FIGURE 3.3

Fact. Atmospheric pressure (p_{atm}) decreases as elevation increases. **Reasoning.** (1) Select a column of air that extends from the earth's surface to the upper edge of the atmosphere. (2) Idealize this column as stationary. (3) Because the column is stationary, the forces must sum to zero. (4) Thus, statics shows that atmospheric pressure equals the weight of the column divided by the section area. (5) At a higher elevation, the fluid column is shorter and thus has less weight. **Conclusion.** Elevation strongly influences the value of atmospheric pressure.

The reason that p_{atm} changes with elevation is explained in Fig. 3.3.

In addition to elevation, other variables influence p_{atm}. As elevation increases, the average temperature of the atmosphere decreases. For example, in the Alps, the average temperature on the summit of a mountain is lower than the average temperature in a town situated in a valley. Local weather influences p_{atm}. Good weather is associated with higher values of atmospheric pressure and bad weather with lower values. As the atmosphere is heated during the day and cooled during the night, the atmospheric pressure varies in response to temperature changes. Fortunately, it is simple to select an appropriate value of p_{atm}. Three methods that we recommend are as follows:

Method #1. If you lack information about elevation, select the standard value of atmospheric pressure at sea level,* which is

$$p_{atm}\text{(sea level)} = 1.000 \text{ atm} = 101.3 \text{ kPa} = 14.70 \text{ psi} = 2116 \text{ psf} = 33.90 \text{ ft-H}_2\text{O}$$
$$= 760.0 \text{ mm-Hg} = 29.92 \text{ in-Hg} = 1.013 \text{ bar}$$

Method #2. If you have information about elevation, you can calculate a typical value of atmospheric pressure using the *standard atmosphere*. The **U.S. standard atmosphere** is a math model that gives values of parameters such as temperature, density, and pressure corresponding to average conditions. The model, developed by NASA,[†] is valid from the earth's surface to an elevation of 1000 km. Regarding calculations, the equations of the math model are complex, so we recommend using the Digital Dutch online calculator.[‡]

Method #3. The most accurate way to find atmospheric pressure is to measure the value using a barometer. This method might be needed, for example, if you are processing experimental data and you want to know the exact value of atmospheric pressure at the time your data were recorded. As an alternative to using a barometer, you can look up a locally measured value on the Internet. Be careful when using the Internet as a resource, however, because many sites adjust the local atmospheric pressure to a value that the given location would have if it was situated at sea level.

EXAMPLE. What value of atmospheric pressure should be used for a project that will be built in Mexico City? **Reasoning.** (1) The elevation of Mexico City is 2250 m. (2) Using the U.S. standard atmosphere, as calculated with the Digital Dutch calculator,[§] shows that $p_{atm} = 77.1$ kPa at an elevation of 2250 m. **Conclusion.** Use $p_{atm} = 77$ kPa.

Absolute Pressure, Gage Pressure, Vacuum Pressure, and Differential Pressure

Absolute pressure is referenced to regions such as outer space, where the pressure is essentially zero because the region is devoid of gas. The pressure in a perfect vacuum is called absolute zero, and pressure measured relative to this zero pressure is termed **absolute pressure**.

*We recommend that you add these values to your *working knowledge*. As always, memorize the approximate values not the exact values. We recommend memorizing to two to three significant digits.
[†]The most recent version was published in 1976.
[‡]http://www.digitaldutch.com/atmoscalc.
[§]ibid.

FIGURE 3.4

Example of pressure relations.

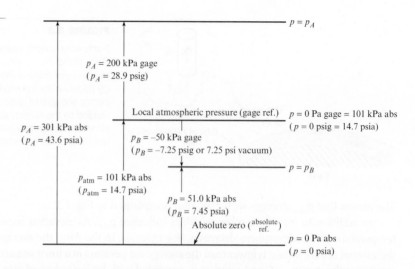

When pressure is measured relative to prevailing local atmospheric pressure, the pressure value is called **gage pressure**.* For example, when a tire pressure gage gives a value of 300 kPa (44 psi), this means that the absolute pressure in the tire is 300 kPa greater than local atmospheric pressure. To convert gage pressure to absolute pressure, add the local atmospheric pressure. For example, a gage pressure of 50 kPa recorded in a location where the atmospheric pressure is 100 kPa is expressed as either

$$p = 50 \text{ kPa gage} \quad \text{or} \quad p = 150 \text{ kPa abs} \tag{3.2}$$

In SI units, gage and absolute pressures are identified after the unit as shown in Eq. (3.2). In traditional units, gage pressure is identified by adding the letter g to the unit symbol. For example, a gage pressure of 10 pounds per square foot is designated as 10 psfg. Similarly, the letter a is used to denote absolute pressure. For example, an absolute pressure of 20 pounds force per square inch is designated as 20 psia.

When pressure is less than atmospheric, the pressure can be described using vacuum pressure. **Vacuum pressure** is defined as the difference between atmospheric pressure and actual pressure. Vacuum pressure is a positive number and equals the absolute value of gage pressure (which will be negative). For example, if $p_{atm} = 101$ kPa and a gage connected to a tank indicates a vacuum pressure of 31.0 kPa, this can also be stated as 70.0 kPa abs, or −31.0 kPa gage.

Fig. 3.4 provides a visual description of the three pressure scales. Notice that $p_B = 7.45$ psia is equivalent to −7.25 psig and +7.25 psi vacuum. Notice that $p_A =$ of 301 kPa abs is equivalent to 200 kPa gage. Gage, absolute, and vacuum pressure can be related using equations labeled as the "pressure equations."

$$p_{gage} = p_{abs} - p_{atm} \tag{3.3a}$$

$$p_{vacuum} = p_{atm} - p_{abs} \tag{3.3b}$$

$$p_{vacuum} = -p_{gage} \tag{3.3c}$$

> **EXAMPLE.** Convert 5 psi vacuum to absolute pressure in SI units.
>
> **Solution.** First, convert vacuum pressure to absolute pressure.
>
> $$p_{abs} = p_{atm} - p_{vacuum} = 14.7 \text{ psi} - 5 \text{ psi} = 9.7 \text{ psia}$$

*There are two correct spellings used in the literature: gage pressure and gauge pressure.

Second, convert units by applying a conversion ratio from Table F.1.

$$p = (9.7 \text{ psi})\left(\frac{101.3 \text{ kPa}}{14.7 \text{ psi}}\right) = 66,900 \text{ Pa absolute}$$

Recommendation. It is good practice, when writing pressure units, to specify whether the pressure is absolute, gage, or vacuum.

EXAMPLE. Suppose the pressure in a car tire is specified as 3 bar. Find the absolute pressure in units of kPa.

Solution. Recognize that tire pressure is commonly specified in gage pressure. Thus, convert the gage pressure to absolute pressure.

$$p_{abs} = p_{atm} + p_{gage} = (101.3 \text{ kPa}) + (3 \text{ bar})\frac{(101.3 \text{ kPa})}{(1.013 \text{ bar})} = 401 \text{ kPa absolute}$$

Another way to describe pressure is to use **differential pressure**, which is defined as the difference in pressure between two points and is given the symbol Δp (Fig. 3.5).

Some useful facts about differential pressure follow.

- The points (A and B) are typically selected so that differential pressure is positive; that is, $\Delta p > 0$.
- Differential pressure refers to the difference in pressure between two points, not to a "differential pressure" in the sense of a differential in calculus.
- The unit symbol psid stands for pounds-force per square inch differential. Similarly, psfd refers to a differential pressure.

Hydraulic Machines

A **hydraulic machine** uses a fluid to transmit forces or energy to assist in the performance of a human task. An example of a hydraulic machine is a hydraulic car jack in which a user can supply a small force to a handle and lift an automobile. Other examples of hydraulic machines include braking systems in cars, forklift trucks, power steering systems in cars, and airplane control systems.

The hydraulic machine provides a mechanical advantage (Fig. 3.6). **Mechanical advantage** is defined as the ratio of output force to input force:

$$(\text{mechanical advantage}) \equiv \frac{(\text{output force})}{(\text{input force})} \tag{3.4}$$

Mechanical advantage of a lever (Fig. 3.6) is found by summing moments about the fulcrum to give $F_1 L_1 = F_2 L_2$, where L denotes the length of the lever arm.

$$(\text{mechanical advantage; lever}) \equiv \frac{(\text{output force})}{(\text{input force})} = \frac{F_2}{F_1} = \frac{L_1}{L_2} \tag{3.5}$$

To find mechanical advantage of the hydraulic machine, apply force equilibrium to each piston (Fig. 3.6) to give $F_1 = p_1 A_1$ and $F_2 = p_2 A_2$, where p is pressure in the cylinder and A is face area of the piston. Next, recognize that $p_1 = p_2$ and then solve for the mechanical advantage

$$(\text{mechanical advantage; hydraulic machine}) \equiv \frac{(\text{output force})}{(\text{input force})} = \frac{F_2}{F_1} = \frac{A_2}{A_1} = \frac{D_2^2}{D_1^2} \tag{3.6}$$

The hydraulic machine is often used to illusrate Pascal's principle. This principle states that when there is an increase in pressure at any point in a confined fluid, there is an equal increase at every other point in the container. This principle is evident when a balloon is inflated because the balloon expands evenly in all directions. The principle is also evident in the hydraulic machine (Fig. 3.7).

FIGURE 3.5

An example of differential pressure for flow in a pipe. Points A and B are located on the centerline. The differential pressure (Δp) is the magnitude of the pressure at point A minus the magnitude of the pressure at point B.

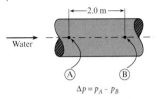

FIGURE 3.6

Both the lever and hydraulic machine provide a mechanical advantage.

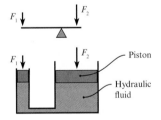

FIGURE 3.7

This figure shows how a hydraulic machine can be used to illustrate Pascal's principle.

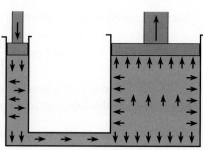

Pascal's principle. An applied force creates a pressure change that is transmitted to every point in the fluid and to the walls of the container.

EXAMPLE 3.1

Applying Force Equilibrium to a Hydraulic Jack

Problem Statement

A hydraulic jack has the dimensions shown. If one exerts a force F of 100 N on the handle of the jack, what load, F_2, can the jack support? Neglect lifter weight.

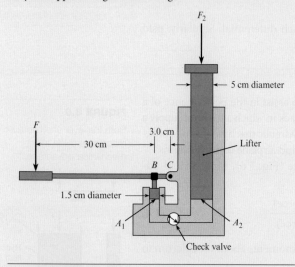

Define the Situation

A force of $F = 100$ N is applied to the handle of a jack.

Assumption: The weight of the lifter (see sketch) is negligible.

State the Goal

F_2(N) ⬅ load that the jack can lift

Generate Ideas and Make a Plan

Because the goal is F_2, apply force equilibrium to the lifter. Then, analyze the small piston and the handle. The plan is as follows:

1. Calculate force acting on the small piston by applying moment equilibrium.

2. Calculate pressure p_1 in the hydraulic fluid by applying force equilibrium.

3. Calculate the load F_2 by applying force equilibrium.

Take Action (Execute the Plan)

1. Moment equilibrium (handle):

$$\sum M_c = 0$$

$$(0.33 \text{ m}) \times (100 \text{ N}) - (0.03 \text{ m})F_1 = 0$$

$$F_1 = \frac{0.33 \text{ m} \times 100 \text{ N}}{0.03 \text{ m}} = 1100 \text{ N}$$

2. Force equilibrium (small piston):

$$\sum F_{\text{small piston}} = p_1 A_1 - F_1 = 0$$

$$p_1 A_1 = F_1 = 1100 \text{ N}$$

Thus,

$$p_1 = \frac{F_1}{A_1} = \frac{1100 \text{ N}}{\pi d^2/4} = 6.22 \times 10^6 \text{ N/m}^2$$

3. Force equilibrium (lifter):

$$\sum F_{\text{lifter}} = F_2 - p_1 A_2 = 0$$

$$F_2 = p_1 A_2 = \left(6.22 \times 10^6 \frac{\text{N}}{\text{m}^2}\right)\left(\frac{\pi}{4} \times (0.05 \text{ m})^2\right) = \boxed{12.2 \text{ kN}}$$

Note that $p_1 = p_2$ because they are at the same elevation (this fact will be established in the next section).

Review the Results and the Process

1. *Discussion.* The jack in this example, which combines a lever and a hydraulic machine, provides an output force of 12,200 N from an input force of 100 N. Thus, this jack provides a mechanical advantage of 122 to 1.

2. *Knowledge.* Hydraulic machines are analyzed by applying force and moment equilibrium. The force of pressure is typically given by $F = pA$.

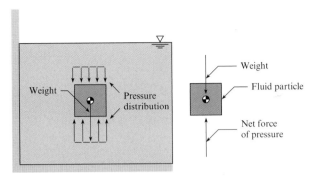

FIGURE 3.8

This example shows how to check to see if the *hydrostatic condition* applies. For this case, hydrostatic conditions do apply because the *weight* of the fluid particle is exactly balanced by the *pressure force*.

3.2 The Hydrostatic Equations

This section explains how to calculate the pressure for problems in which a fluid is in hydrostatic equilibrium. There are two main results:

- The hydrostatic *differential equation*, which is applied to problems in which density varies
- The hydrostatic *algebraic equation*, which is applied to problems in which density is constant

The Hydrostatic Condition

The equations in this section apply only if the fluid in your problem is in *hydrostatic equilibrium*. To tell if this condition applies, select a fluid particle, select a coordinate direction, and draw a free body diagram (FBD) that shows only the forces in the coordinate direction that you selected. If the acceleration of the fluid particle is zero in the coordinate direction you chose and if the only forces on the particle are the pressure force and the weight, then the hydrostatic condition applies on a plane that is parallel to your coordinate direction.

If a fluid is stationary (e.g., water in a lake as in Fig. 3.8), then the hydrostatic equation will always apply. The reason is that the acceleration of any fluid particle is zero and the only possible forces that can balance the weight of the fluid particle are the pressure force and the viscous force. However, the viscous force must be zero because of the definition of a fluid; that is, a fluid will deform continuously under the action of a viscous stress. Thus, the only force available to balance the weight of the fluid particle is the pressure force.

If a fluid is *flowing*, then the hydrostatic equation will sometimes apply (Fig. 3.9). For situations similar to those shown in the figure, you can apply the hydrostatic equation $\Delta p = -\rho g \Delta z$ to points situated in a plane.

The Hydrostatic Differential Equation (Variable Density)

This subsection shows how to derive $dp/dz = -\gamma$. This equation is important for understanding theory and for solving problems that involve varying density.

To begin the derivation, visualize any region of static fluid (e.g., water behind a dam), isolate a cylindrical body, and then sketch an FBD, as shown in Fig. 3.10. Notice that the cylindrical

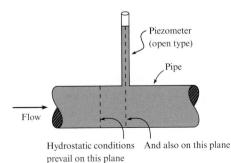

FIGURE 3.9

This sketch shows examples of when *hydrostatic conditions* apply to a flowing fluid. The reason why is that the pressure force balances the weight force for each fluid particle that is situated on one of the planes shown in the figure.

FIGURE 3.10

The system used to derive the hydrostatic differential equation.

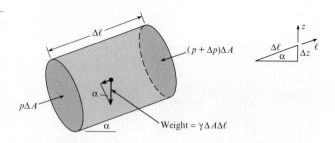

body is oriented so that its longitudinal axis is parallel to an arbitrary ℓ direction. The body is $\Delta\ell$ long, ΔA in cross-sectional area, and inclined at an angle α with the horizontal. Apply force equilibrium in the ℓ direction:

$$\sum F_\ell = 0$$

$$F_{\text{Pressure}} - F_{\text{Weight}} = 0$$

$$p\Delta A - (p + \Delta p)\Delta A - \gamma\Delta A\Delta\ell\sin\alpha = 0$$

Simplify and divide by the volume of the body $\Delta\ell\Delta A$ to give

$$\frac{\Delta p}{\Delta\ell} = -\gamma\sin\alpha$$

From Fig. 3.10, the sine of the angle is given by

$$\sin\alpha = \frac{\Delta z}{\Delta\ell}$$

Combining the previous two equations and letting Δz approach zero gives

$$\lim_{\Delta z\to 0}\frac{\Delta p}{\Delta z} = -\gamma$$

The final result is

$$\frac{dp}{dz} = -\gamma \qquad \text{(hydrostatic differential equation)} \qquad (3.7)$$

Eq. (3.7) means that changes in pressure correspond to changes in elevation. If one travels upward in the fluid (positive z direction), the pressure decreases; if one goes downward (negative z), the pressure increases; if one moves along a horizontal plane, the pressure remains constant. Of course, these pressure variations are exactly what a diver experiences when ascending or descending in a lake or pool.

The Hydrostatic Algebraic Equation (Constant Density)

Because modeling a fluid as if the density is constant is often well justified, it is useful to solve the hydrostatic differential equation for the special case of constant density. The resulting equation is called the hydrostatic algebraic equation, and we shorten this name to the *hydrostatic equation (HE)*. The hydrostatic equation is one of the most useful equations in fluid mechanics; thus, we recommend that you learn this equation well. To derive the equation, begin by integrating Eq. (3.7) for the case of constant density to give

$$p + \gamma z = p_z = \text{constant} \qquad (3.8)$$

where the term z is the elevation (vertical distance) above a fixed horizontal reference plane called a datum, and p_z is **piezometric pressure**. Dividing Eq. (3.8) by γ gives

$$\frac{p_z}{\gamma} = \left(\frac{p}{\gamma} + z\right) = h = \text{constant} \tag{3.9}$$

where h is the **piezometric head**. Because h is constant, Eq. (3.9) can be written as

$$\frac{p_1}{\gamma} + z_1 = \frac{p_2}{\gamma} + z_2 \tag{3.10a}$$

where the subscripts 1 and 2 identify any two points in a static fluid of constant density. Multiplying Eq. (3.10a) by γ gives

$$p_1 + \gamma z_1 = p_2 + \gamma z_2 \tag{3.10b}$$

In Eq. (3.10b), letting $\Delta p = p_2 - p_1$ and letting $\Delta z = z_2 - z_1$ gives

$$\Delta p = -\gamma \Delta z \tag{3.10c}$$

The hydrostatic equation is given by Eq. (3.10a), (3.10b), or (3.10c). These three equations are equivalent because any one of the equations can be used to derive the other two. The hydrostatic equation is valid for any constant density fluid in hydrostatic equilibrium.

Notice that the hydrostatic equation involves

$$\text{piezometric head} = h \equiv \left(\frac{p}{\gamma} + z\right) \tag{3.11}$$

$$\text{piezometric pressure} = p_z \equiv (p + \gamma z) \tag{3.12}$$

To calculate piezometric head or piezometric pressure, an engineer identifies a specific location in a body of fluid and then uses the value of pressure and elevation at that location. Piezometric pressure and head are related by

$$p_z = h\gamma \tag{3.13}$$

Piezometric head, h, a property that is widely used in fluid mechanics, characterizes hydrostatic equilibrium. When hydrostatic equilibrium prevails in a body of fluid of constant density, then h will be constant at all locations. For example, Fig. 3.11 shows a container with oil floating on water. Because piezometric head is constant in the water, $h_a = h_b = h_c$. Similarly, the piezometric head is constant in the oil: $h_d = h_e = h_f$. Notice that piezometric head is not constant when density changes. For example, $h_c \neq h_d$ because points c and d are in different fluids with different values of density.

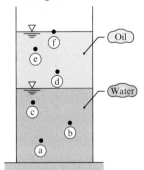

FIGURE 3.11

Oil floating on water.

Hydrostatic Equation (Working Equations)

To apply the hydrostatic equation, first check that the assumptions listed in Table 3.1 are valid. Then, select the most useful form of the hydrostatic equation. We recommend using the head form or the differential pressure form. We also recommend that you learn the meaning of the variables given in the third column because these names are used throughout fluid mechanics. For many problems, you will find the following two rules useful:

The **fluid interface rule** states that for a planar interface (e.g., Fig. 3.12) the pressure is constant across the interface (i.e., $p_1 = p_2$ at the interface). **Reasoning.** (1) The fluid interface is not moving, so $\Sigma\mathbf{F} = \mathbf{0}$. (2) Select an infinitesimally thin system so that the weight can be neglected. (3) Thus, the only forces on the interface are the pressure forces, and algebra shows that $p_1 = p_2$.

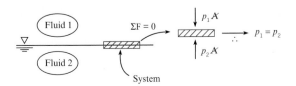

FIGURE 3.12

To prove the fluid interface rule (1) select an infinitesimally thin system on the interface and note that the weight of this system is negligible, (2) Apply $\Sigma F = 0$ to show that pressure is constant across the interface.

TABLE 3.1 The Hydrostatic Equation (Working Equations and Assumptions)

Name (Physical Interpretation)	Equation	Variables in the Equation
Head form (the piezometric head is constant at every point)	$\dfrac{p_1}{\gamma} + z_1 = \dfrac{p_2}{\gamma} + z_2$ Eq. (3.10a)	• p = pressure (N/m²) (use absolute or gage pressure; not vacuum pressure) • γ = specific weight (N/m³) • p/γ = pressure head (m) • z = elevation or elevation head (m) • $(p/\gamma + z)$ = piezometric head (m)
Differential pressure form (the differential pressure is linear with elevation change)	$\Delta p = \gamma \Delta z$ Eq. (3.10b)	• Δp = differential pressure (N/m²) • Δz = difference in elevation (m)
Piezometric pressure form (the piezometric pressure is constant at every point)	$p_1 + \gamma z_1 = p_2 + \gamma z_2$ Eq. (3.10c)	• $(p + \gamma z)$ = piezometric pressure (Pa)
Assumptions to check before you apply the hydrostatic equation		1. You can only apply the HE to a single fluid that has constant density. For problems that have multiple fluids (e.g., oil floating on water), the HE is applied successively to each fluid. 2. You can only apply the HE if the hydrostatic condition applies.

EXAMPLE 3.2

Applying the Hydrostatic Equation to Find Pressure in a Tank

Problem Statement

What is the water pressure at a depth of 35 ft in the tank shown?

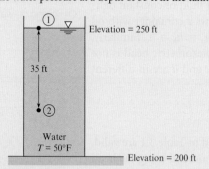

Elevation = 250 ft

35 ft

Water
$T = 50°F$

Elevation = 200 ft

Define the Situation

Water is contained in a tank that is 50 ft deep.

Properties: Water (50 °F, 1 atm, Table A.5): γ = 62.4 lbf/ft³

State the Goal

p_2 (psig) ← water pressure at point 2

Generate Ideas and Make a Plan

Apply the idea that piezometric head is constant. The plan steps are as follows:

1. Equate piezometric head at elevation 1 with piezometric head at elevation 2 (i.e., apply Eq. 3.10a).

2. Analyze each term in Eq. (3.10a).

3. Solve for the pressure at elevation 2.

Take Action (Execute the Plan)

1. Hydrostatic equation (Eq. 3.10a):

$$\frac{p_1}{\gamma} + z_1 = \frac{p_2}{\gamma} + z_2$$

2. Term-by-term analysis of Eq. (3.10a) yields:

   • $p_1 = p_{atm} = 0$ psig
   • $z_1 = 250$ ft
   • $z_2 = 215$ ft

3. Combine steps 1 and 2; solve for p_2:

$$\frac{p_1}{\gamma} + z_1 = \frac{p_2}{\gamma} + z_2$$

$$0 + 250 \text{ ft} = \frac{p_2}{62.4 \text{ lbf/ft}^3} + 215 \text{ ft}$$

$$p_2 = 2180 \text{ psfg} = \boxed{15.2 \text{ psig}}$$

Review the Solution and the Process

1. *Validation.* The calculated pressure change (15 psig) is slightly greater than 1 atm (14.7 psi). Because one atmosphere corresponds to a water column of 33.9 ft and this problem involves 35 feet of water column, the solution appears correct.

2. *Skill.* This example shows how to write down a governing equation and then analyze each term. This skill is called *term-by-term analysis.*

3. *Knowledge.* The gage pressure at the free surface of a liquid in contact with the atmosphere is zero ($p_1 = 0$ in this example).

4. *Skill.* Label a pressure as absolute or gage or vacuum. For this example, the pressure unit (psig) denotes a gage pressure.

5. *Knowledge.* The hydrostatic equation is valid when density is constant. This condition is met on this problem.

The **gas pressure change rule** states that the hydrostatic pressure change for a gas can usually be neglected. **Reasoning**. (1) The hydrostatic pressure change in a gas for a one-meter change of elevation is given by $\Delta p/\Delta z = \rho g$. (2) The given equation shows, for example, that the pressure change in air at room conditions is about 12 pascals/meter. (3) A pressure change of about 12 pascals/meter is typically negligible as compared to other relevant pressure changes. **Conclusion**. The hydrostatic pressure change in a gas can usually be neglected.

Example 3.3 shows how to find pressure by applying the idea of *constant piezometric head* to a problem involving several fluids. Notice the application of the fluid interface rule.

EXAMPLE 3.3

Applying the Hydrostatic Equation to Oil and Water in a Tank

Problem Statement

Oil with a specific gravity of 0.80 forms a layer 0.90 m deep in an open tank that is otherwise filled with water (10°C). The total depth of water and oil is 3 m. What is the gage pressure at the bottom of the tank?

Problem Definition

Oil and water are contained in a tank.

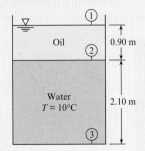

Properties:
- Water: (10°C, 1 atm, Table A.5): $\gamma_{\text{water}} = 9810 \text{ N/m}^3$
- Oil: $\gamma_{\text{oil}} = S\gamma_{\text{water, 4°C}} = 0.8(9810 \text{ N/m}^3) = 7850 \text{ N/m}^3$

State the Goal

p_3 (kPa gage) ◄ pressure at bottom of the tank

Generate Ideas and Make a Plan

Because the goal is p_3, apply the hydrostatic equation to the water. Then, analyze the oil. The plan steps are as follows:

1. Find p_2 by applying the hydrostatic equation (3.10a).
2. Equate pressures across the oil–water interface.
3. Find p_3 by applying the hydrostatic equation given in Eq. (3.10a).

Solution

1. Hydrostatic equation (oil):

$$\frac{p_1}{\gamma_{\text{oil}}} + z_1 = \frac{p_2}{\gamma_{\text{oil}}} + z_2$$

$$\frac{0 \text{ Pa}}{\gamma_{\text{oil}}} + 3 \text{ m} = \frac{p_2}{0.8 \times 9810 \text{ N/m}^3} + 2.1 \text{ m}$$

$$p_2 = 7.063 \text{ kPa}$$

2. Oil–water interface:

$$p_2\big|_{\text{oil}} = p_2\big|_{\text{water}} = 7.063 \text{ kPa}$$

3. Hydrostatic equation (water):

$$\frac{p_2}{\gamma_{\text{water}}} + z_2 = \frac{p_3}{\gamma_{\text{water}}} + z_3$$

$$\frac{7.063 \times 10^3 \text{ Pa}}{9810 \text{ N/m}^3} + 2.1 \text{ m} = \frac{p_3}{9810 \text{ N/m}^3} + 0 \text{ m}$$

$$\boxed{p_3 = 27.7 \text{ kPa gage}}$$

Review

Validation: Because oil is less dense than water, the answer should be slightly smaller than the pressure corresponding to a water column of 3 m. From Table F.1, a water column of 10 m ≈ 1 atm. Thus, a 3 m water column should produce a pressure of about 0.3 atm = 30 kPa. The calculated value appears correct.

FIGURE 3.13

A mercury barometer.

3.3 Measuring Pressure

When engineers design and conduct experiments, pressure nearly always needs to be measured. Thus, this section describes five scientific instruments for measuring pressure.

Barometer

An instrument that is used to measure atmospheric pressure is called a **barometer**. The most common types are the mercury barometer and the aneroid barometer. A mercury barometer is made by inverting a mercury-filled tube in a container of mercury, as shown in Fig. 3.13. The pressure at the top of the mercury barometer will be the vapor pressure of mercury, which is very small: $p_v = 2.4 \times 10^{-6}$ atm at 20°C. Thus, atmospheric pressure will push the mercury up the tube to a height h. The mercury barometer is analyzed by applying the hydrostatic equation:

$$p_{atm} = \gamma_{Hg}h + p_v \approx \gamma_{Hg}h \qquad (3.20)$$

Thus, by measuring h, local atmospheric pressure can be determined using Eq. (3.20).

An aneroid barometer works mechanically. An aneroid is an elastic bellows that has been tightly sealed after some air was removed. When atmospheric pressure changes, this causes the aneroid to change size, and this mechanical change can be used to deflect a needle to indicate local atmospheric pressure on a scale. An aneroid barometer has some advantages over a mercury barometer because it is smaller and allows data recording over time.

Bourdon-Tube Gage

A **Bourdon-tube** gage,* Fig. 3.14, measures pressure by sensing the deflection of a coiled tube. The tube has an elliptical cross section and is bent into a circular arc, as shown in Fig. 3.14b. When atmospheric pressure (zero gage pressure) prevails, the tube is undeflected, and for this condition the gage pointer is calibrated to read zero pressure. When pressure is applied to the gage, the curved tube tends to straighten (much like blowing into a party favor to straighten it out), thereby actuating the pointer to read a positive gage pressure. The Bourdon-tube gage is common because it is low cost, reliable, easy to install, and available in many different pressure ranges. Bourdon-tube gages have some disadvantages: dynamic pressures may not be measured accurately; accuracy of the gage can be lower than other instruments; and the gage can be damaged by excessive pressure pulsations.

Piezometer

A **piezometer** is a vertical tube, usually transparent, in which a liquid rises in response to a positive gage pressure. For example, Fig. 3.15 shows a piezometer attached to a pipe. Pressure

FIGURE 3.14

Bourdon-tube gage. (a) View of a typical gage (photo by Donald Elger). (b) Internal mechanism (schematic).

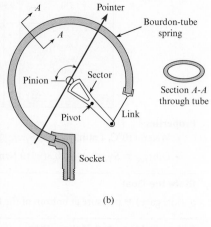

(a)

(b)

*Gage in this context means a scientific instrument. This word can also be correctly spelled as gauge.

in the pipe pushes the water column to a height h, and the gage pressure at the center of the pipe is $p = \gamma h$, which follows directly from the hydrostatic equation (3.10c). The piezometer has several advantages: simplicity, direct measurement (no need for calibration), and accuracy. However, a piezometer cannot easily be used for measuring pressure in a gas, and a piezometer is limited to low pressures because the column height becomes too large at high pressures.

FIGURE 3.15

Piezometer attached to a pipe.

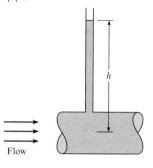

Manometer

A **manometer** (often shaped like the letter "U") is a device for measuring pressure by raising or lowering a column of liquid. For example, Fig. 3.16 shows a U-tube manometer that is being used to measure pressure in a flowing fluid. In the case shown, positive gage pressure in the pipe pushes the manometer liquid up a height Δh. To use a manometer, engineers relate the height of the liquid in the manometer to pressure, as illustrated in Example 3.4.

Once one is familiar with the basic principle of manometry, it is straightforward to write a single equation rather than separate equations as was done in Example 3.4. The single equation for evaluation of the pressure in the pipe of Fig. 3.16 is

$$0 + \gamma_m \Delta h - \gamma \ell = p_4$$

One can read the equation in this way: Zero pressure at the open end, plus the change in pressure from point 1 to 2 minus the change in pressure from point 3 to 4, equals the pressure in

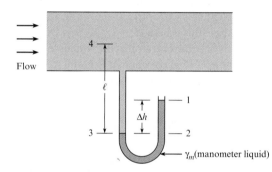

FIGURE 3.16

U-tube manometer.

EXAMPLE 3.4

Pressure Measurement (U-Tube Manometer)

Problem Statement

Water at 10°C is the fluid in the pipe of Fig. 3.16, and mercury is the manometer fluid. If the deflection Δh is 60 cm and ℓ is 180 cm, what is the gage pressure at the center of the pipe?

Define the Situation

Pressure in a pipe is being measured using a U-tube manometer.

Properties:

- Water (10°C), Table A.5: $\gamma = 9810$ N/m³
- Mercury, Table A.4: $\gamma = 133{,}000$ N/m³

State the Goal

Calculate gage pressure (kPa) in the center of the pipe.

Generate Ideas and Make a Plan

Start at point 1 and work to point 4 using ideas from Eq. (3.10c). When fluid depth increases, add a pressure change. When fluid depth decreases, subtract a pressure change.

Take Action (Execute the Plan)

1. Calculate the pressure at point 2 using the hydrostatic equation (3.10c):

$$p_2 = p_1 + \text{pressure increase between 1 and 2} = 0 + \gamma_m \Delta h_{12}$$
$$= \gamma_m (0.6 \text{ m}) = (133{,}000 \text{ N/m}^3)(0.6 \text{ m})$$
$$= 79.8 \text{ kPa}$$

2. Find the pressure at point 3:

   - The hydrostatic equation with $z_3 = z_2$ gives

   $$p_3|_\text{water} = p_2|_\text{water} = 79.8 \text{ kPa}$$

   - When a fluid-fluid interface is flat, pressure is constant across the interface. Thus, at the oil–water interface

   $$p_3|_\text{mercury} = p_3|_\text{water} = 79.8 \text{ kPa}$$

3. Find the pressure at point 4 using the hydrostatic equation given in Eq. (3.10c):

$$p_4 = p_3 - \text{pressure decrease between 3 and 4} = p_3 - \gamma_w \ell$$
$$= 79{,}800 \text{ Pa} - (9810 \text{ N/m}^3)(1.8 \text{ m})$$
$$= 62.1 \text{ kPa gage}$$

FIGURE 3.17

Apparatus for determining change in piezometric head corresponding to flow in a pipe.

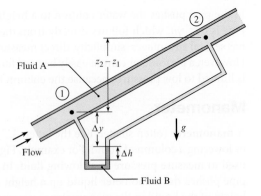

the pipe. The main concept is that pressure increases as depth increases and decreases as depth decreases.

The general equation for the pressure difference measured by the manometer is

$$p_2 = p_1 + \sum_{\text{down}} \gamma_i h_i - \sum_{\text{up}} \gamma_i h_i \qquad (3.21)$$

where γ_i and h_i are the specific weight and deflection in each leg of the manometer. It does not matter where one starts, that is, where one defines the initial point 1 and final point 2. When liquids and gases are both involved in a manometer problem, it is well within engineering accuracy to neglect the pressure changes due to the columns of gas. This is because $\gamma_{\text{liquid}} \gg \gamma_{\text{gas}}$. Example 3.5 shows how to apply Eq. (3.21) to perform an analysis of a manometer that uses multiple fluids.

Because the manometer configuration shown in Fig. 3.17 is common, it is useful to derive an equation specific to this application. To begin, apply the manometer equation (3.21) between points 1 and 2:

$$p_1 + \sum_{\text{down}} \gamma_i h_i - \sum_{\text{up}} \gamma_i h_i = p_2$$

$$p_1 + \gamma_A(\Delta y + \Delta h) - \gamma_B \Delta h - \gamma_A(\Delta y + z_2 - z_1) = p_2$$

EXAMPLE 3.5

Manometer Analysis

Problem Statement

What is the pressure of the air in the tank if $\ell_1 = 40$ cm, $\ell_2 = 100$ cm, and $\ell_3 = 80$ cm?

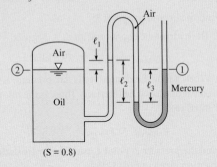

Define the Situation

A tank is pressurized with air.

Assumptions: Neglect the pressure change in the air column.

Properties:
- Oil: $\gamma_{\text{oil}} = S\gamma_{\text{water}} = 0.8 \times 9810 \text{ N/m}^3 = 7850 \text{ N/m}^3$
- Mercury, Table A.4: $\gamma = 133{,}000 \text{ N/m}^3$

State the Goal

Find the pressure (kPa gage) in the air.

Generate Ideas and Make a Plan

Apply the manometer equation (3.21) from location 1 to location 2.

Take Action (Execute the Plan)

Manometer equation:

$$p_1 + \sum_{\text{down}} \gamma_i h_i - \sum_{\text{up}} \gamma_i h_i = p_2$$

$$p_1 + \gamma_{\text{mercury}} \ell_3 - \gamma_{\text{air}} \ell_2 + \gamma_{\text{oil}} \ell_1 = p_2$$

$$0 + (133{,}000 \text{ N/m}^3)(0.8 \text{ m}) - 0 + (7850 \text{ N/m}^3)(0.4 \text{ m}) = p_2$$

$$\boxed{p_2 = p_{\text{air}} = 110 \text{ kPa gage}}$$

EXAMPLE 3.6

Change in Piezometric Head for Pipe Flow

Problem Statement

A differential mercury manometer is connected to two pressure taps in an inclined pipe as shown in Fig. 3.17. Water at 50°F is flowing through the pipe. The deflection of mercury in the manometer is 1 inch. Find the change in piezometric pressure and piezometric head between points 1 and 2.

Define the Situation

Water is flowing in a pipe.

Properties:

- Water (50 °F): Table A.5, $\gamma_{water} = 62.4$ lbf/ft^3.
- Mercury: Table A.4, $\gamma_{Hg} = 847$ lbf/ft^3.

State the Goal

Find the following:

- Change in piezometric head (ft) between points 1 and 2
- Change in piezometric pressure (psfg) between 1 and 2

Generate Ideas and Make a Plan

1. Find difference in the piezometric head using Eq. (3.22).
2. Relate piezometric head to piezometric pressure using Eq. (3.13).

Take Action (Execute the Plan)

1. Difference in piezometric head:

$$h_1 - h_2 = \Delta h\left(\frac{\gamma_{Hg}}{\gamma_{water}} - 1\right) = \left(\frac{1}{12}\,\text{ft}\right)\left(\frac{847\ \text{lbf/ft}^3}{62.4\ \text{lbf/ft}^3} - 1\right)$$

$$= \boxed{1.05\ \text{ft}}$$

2. Piezometric pressure:

$$p_z = h\gamma_{water}$$
$$= (1.05\ \text{ft})(62.4\ \text{lbf/ft}^3) = \boxed{65.5\ \text{psf}}$$

Simplifying gives

$$(p_1 + \gamma_A z_1) - (p_2 + \gamma_A z_2) = \Delta h(\gamma_B - \gamma_A)$$

Dividing through by γ_A gives

$$\left(\frac{p_1}{\gamma_A} + z_1\right) - \left(\frac{p_2}{\gamma_A} + z_2\right) = \Delta h\left(\frac{\gamma_B}{\gamma_A} - 1\right)$$

Recognize that the terms on the left side of the equation are piezometric head and rewrite to give the final result:

$$h_1 - h_2 = \Delta h\left(\frac{\gamma_B}{\gamma_A} - 1\right) \tag{3.22}$$

Equation (3.22) is valid when a manometer is used to measure differential pressure. Example 3.6 shows how this equation is used.

Summary of the Manometer Equations

These manometer equations are summarized in Table 3.2. Because the equations were derived from the hydrostatic equation, they have the same assumptions: constant fluid density and hydrostatic conditions. The process for applying the manometer equations is as follows:

Step 1. For measurement of pressure at a point, select Eq. (3.21). For measurement of pressure or head change between two points in a pipe, select Eq. (3.22).

Step 2. Select points 1 and 2 where you know information or where you want to find information.

Step 3. Write the general form of the manometer equation.

Step 4. Perform a term-by-term analysis.

TABLE 3.2 Summary of the Manometer Equations

Description	Equation	Terms
Gage pressure analysis. Use this equation for a manometer that is being applied to measure gage pressure (e.g., see Fig. 3.16).	$$p_2 = p_1 + \sum_{down} \gamma_i h_i - \sum_{up} \gamma_i h_i \quad (3.21)$$	p_1 = pressure at point 1 (N/m^2) p_2 = pressure at point 2 (N/m^2) γ_i = specific weight of fluid i (N/m^3) h_i = deflection of fluid in leg i (m)
Differential pressure analysis. Use this equation for a manometer that is being applied to measure differential pressure in a pipe with a flowing fluid (e.g., see Fig. 3.17).	$$h_1 - h_2 = \Delta h \left(\frac{\gamma_B}{\gamma_A} - 1 \right) \quad (3.22)$$	$h_1 = p_1/\gamma_A + z_1$ = piezometric head at point 1 (m) $h_2 = p_2/\gamma_A + z_2$ = piezometric head at point 2 (m) Δh = deflection of the manometer fluid (m) γ_A = specific weight of the flowing fluid (N/m^3) γ_B = specific weight of the manometer fluid (N/m^3)

The Pressure Transducer

A **pressure transducer** (PT) is a device that converts pressure to an electrical signal. For example, Fig. 3.18 shows a strain-gage pressure transducer. Pressure transducers have many advantages, such as the following:

- In general, PTs have high levels of accuracy as compared to other devices, such as Bourdon-tube gages and manometers.
- A PT can be used to measure gage pressure, absolute pressure, vacuum pressure, or differential pressure.
- Most PTs can measure pressure as a function of time and can be applied to electronic data logging.
- A PT is available for almost any pressure range you want to measure.

Pressure transducers also have some disadvantages, such as the following:

- Higher costs.
- Longer setup times because they are more complicated.
- In general, PTs need to be calibrated and used carefully.

3.4 The Pressure Force on a Panel (Flat Surface)

Many problems require a calculation of the pressure force on a panel. Thus, this section explains how to do this calculation for two cases:

- A *uniform* pressure distribution
- A *hydrostatic* pressure distribution

A **panel** is any surface that is flat or that can be idealized as if it were flat (e.g., face of a dam, a surface on an airplane wing, or the cross section inside a pressure vessel).

FIGURE 3.18

A strain gage pressure transducer operates as follows: (1) Pressure deforms a diaphragm. (2) The diaphragm deflection is sensed with a strain gage. (3) The voltage from the strain gage is amplified and then converted to a pressure value via software. (4) The pressure value is displayed.

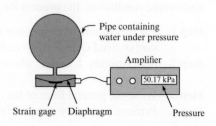

Pipe containing water under pressure

Amplifier

50.17 kPa

Strain gage   Diaphragm   Pressure

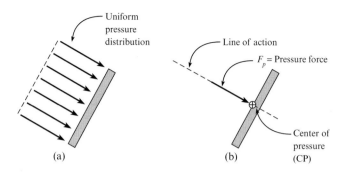

(a)   (b)   Center of pressure (CP)

FIGURE 3.19

This example shows (a) a uniform pressure distribution, and (b) the associated pressure force.

The Uniform Pressure Distribution

Fig. 3.19 shows a uniform pressure distribution and the associated pressure force $\mathbf{F}_p$. The value of F_p is calculated using

$$F_p = pA \tag{3.23}$$

where p is the *gage pressure* and A is the surface area of the panel. The pressure force acts at a location called the *center of pressure* (CP). For a uniform pressure, the CP is located at the centroid of the panel. The direction of the pressure force is normal to the panel. The reasoning for why Eq. (3.23) is true is as follows: (1) The pressure force on *any surface* is given by $\mathbf{F}_p = \int_A -p\,\mathbf{n}\,dA$. (2) Because the pressure is constant for a uniform pressure distribution, $\mathbf{F}_p = p\int_A -\mathbf{n}\,dA = pA(-\mathbf{n})$. **Conclusion**: The magnitude of $\mathbf{F}_p$ is $F_p = pA$. The direction of $\mathbf{F}_p$ is the $(-\mathbf{n})$ direction. Thus, Eq. (3.23) is true.

Some useful facts about pressure distributions follow.

- A uniform pressure distribution is commonly used to idealize the pressure distribution due to a gas and the pressure distribution due to a liquid when a panel is horizontal.

- Gage pressure (not absolute pressure) is used in Eq. (3.23) because of the *gage pressure rule*. This rule is explained in Fig. 3.20.

- To analyze a pressure vessel, apply the *pressure vessel force balance method*. This method is explained in Fig. 3.21.

The Hydrostatic Pressure Distribution

A **hydrostatic pressure distribution** (Fig. 3.22) describes the distribution of pressure when pressure varies only with elevation z according to $dp/dz = -\gamma$. When hydrostatic conditions prevail, any panel that is not horizontal is subjected to a hydrostatic pressure distribution.

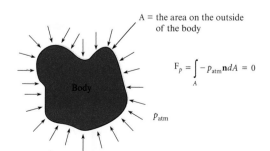

A = the area on the outside of the body

$$F_p = \int_A -p_{atm}\mathbf{n}\,dA = 0$$

FIGURE 3.20

Gage pressure rule: When a uniform atmospheric pressure acts on a body, integrating this pressure over area shows that the net pressure force is zero. Thus, use gage pressure when analyzing the pressure force.

FIGURE 3.21

The pressure vessel force balance is a method for analyzing the force (F_c) needed to clamp a pressure vessel together. To derive an equation, take the following steps: (1) Imagine cutting the tank where it is clamped. (2) Sketch an FBD of the cut portion of the tank. (3) Balance the pressure force with the clamping force to show that $F_c = p_i A_c$.

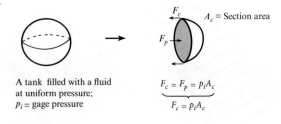

A tank filled with a fluid at uniform pressure; p_i = gage pressure

$F_c = F_p = p_i A_c$

$\underbrace{}$

$F_c = p_i A_c$

FIGURE 3.22

An example showing (a) a hydrostatic pressure distribution on a rectangular panel and (b) the corresponding pressure force.

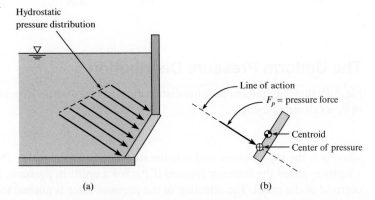

Hydrostatic pressure distribution

Line of action

F_p = pressure force

Centroid

Center of pressure

(a)   (b)

A pressure force acts at a point called the **center of pressure**, which is calculated so that the torque due to the pressure force is exactly the same as the torque due to the pressure distribution. In other words, if you want to replace the pressure distribution with a statically equivalent force that acts at a point, the correct point is the center of pressure. In this text, the symbol for the CP is a circle with a plus symbol inside: $\oplus$.

The **centroid** of an area can be thought of as the balance point of an area (see Fig. 3.23). In general, the equations for finding the centroid are integrals such as $x_c = (\int x\,dA)/A$. For common shapes, the equations have been solved, and engineers look up the value. In this text, centroid formulas are presented in the appendix, Fig. A.1.

Sketching a Pressure Distribution

As an engineer, you should be able to sketch a pressure distribution. Some guidelines are as follows: (1) draw each arrow so that its length represents the magnitude of the pressure, (2) sketch gage pressure, not absolute pressure, (3) draw each arrow so that the arrow is normal to the surface, and (4) draw each arrow to represent compression.

Theory: Force Caused by a Hydrostatic Pressure Distribution

Next, we will show how to find the force on one face of a panel that is acted on by a hydrostatic pressure distribution. To begin, sketch a panel of arbitrary shape submerged in a liquid

FIGURE 3.23

An example of the *centroid* for a triangular panel. The idea here is to (1) imagine making a model of the panel, then (2) the centroid is the point at which the model would balance on the tip of a pencil. This example assumes that the model has a uniform density and that the gravity field is uniform.

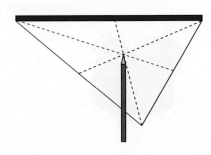

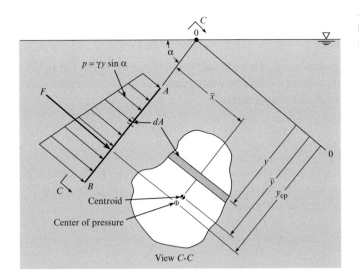

FIGURE 3.24

Distribution of hydrostatic pressure on a plane surface.

(Fig. 3.24). Line AB is the edge view of a panel. The plane of the panel intersects the horizontal liquid surface at axis 0-0 with an angle α. The distance from the axis 0-0 to the horizontal axis through the centroid of the area is given by $\bar{y}$. The distance from 0-0 to the differential area dA is y.

The force due to pressure is given by

$$F_p = \int_A p\,dA \tag{3.24}$$

In Eq. (3.24), the pressure can be found with the hydrostatic equation:

$$p = -\gamma\Delta z = \gamma y \sin \alpha \tag{3.25}$$

Combine Eqs. (3.24) and (3.25) to give

$$F_p = \int_A p\,dA = \int_A \gamma y \sin \alpha\, dA = \gamma \sin \alpha \int_A y\,dA \tag{3.26}$$

Because the integral on the right side of Eq. (3.26) is the first moment of the area, replace the integral by its equivalent, $\bar{y}A$. Therefore

$$F_p = \gamma \bar{y} A \sin \alpha = (\gamma \bar{y} \sin \alpha)A \tag{3.27}$$

Apply the hydrostatic equation to show that the variables within the parentheses on the right side of Eq. (3.27) are the pressure at the centroid of the area. Thus,

$$F_p = \bar{p}A \tag{3.28}$$

Equation (3.28) shows that the hydrostatic force on a panel of arbitrary shape (e.g., rectangular, round, elliptical) is given by the product of the panel area and the pressure at the elevation of the centroid.

Theory: The Center of Pressure for a Hydrostatic Pressure Distribution

This subsection shows how to derive an equation for the vertical location of the CP. For the panel shown in Fig. 3.24 to be in moment equilibrium, the torque due to the resultant force F_p must balance the torque due to each differential force:

$$y_{cp} F_p = \int y \, dF$$

Note that y_{cp} is the "*slant*" distance from the center of pressure to the surface of the liquid. The label "slant" denotes that the distance is measured in the plane that runs through the panel. The differential force dF is given by $dF = p \, dA$; therefore,

$$y_{cp} F = \int_A yp \, dA$$

Also, $p = \gamma y \sin \alpha$, so

$$y_{cp} F = \int_A \gamma y^2 \sin \alpha \, dA \qquad (3.29)$$

Because γ and $\sin \alpha$ are constants,

$$y_{cp} F = \gamma \sin \alpha \int_A y^2 \, dA \qquad (3.30)$$

The integral on the right-hand side of Eq. (3.30) is the second moment of the area (often called the area moment of inertia). This shall be identified as I_0. However, for engineering applications it is convenient to express the second moment with respect to the horizontal centroidal axis of the area. Hence by the parallel-axis theorem,

$$I_0 = \bar{I} + \bar{y}^2 A \qquad (3.31)$$

Substitute Eq. (3.31) into Eq. (3.30) to give

$$y_{cp} F = \gamma \sin \alpha (\bar{I} + \bar{y}^2 A)$$

However, from Eq. (3.25), $F = \gamma \bar{y} \sin \alpha A$. Therefore,

$$y_{cp}(\gamma \bar{y} \sin \alpha A) = \gamma \sin \alpha (\bar{I} + \bar{y}^2 A) \qquad (3.32)$$

$$y_{cp} = \bar{y} + \frac{\bar{I}}{\bar{y} A}$$

$$y_{cp} - \bar{y} = \frac{\bar{I}}{\bar{y} A} \qquad (3.33)$$

In Eq. (3.33), the area moment of inertia $\bar{I}$ is taken about a horizontal axis that passes through the centroid of area. Formulas for $\bar{I}$ are presented in Fig. A.1. The slant distance $\bar{y}$ measures the length from the surface of the liquid to the centroid of the panel along an axis that is aligned with the "slant of the panel," as shown in Fig. 3.24.

Equation (3.33) shows that the CP will be situated below the centroid. The distance between the CP and the centroid depends on the depth of submersion, which is characterized by $\bar{y}$, and on the panel geometry, which is characterized by $\bar{I}/A$.

TABLE 3.3 Summary of the Panel Equations

Purpose of the Equation	Equation	Variables
Predict the magnitude of the hydrostatic force	$F_p = \bar{p}A$ (3.28)	F_p = pressure force (N) $\bar{p}$ = gage pressure evaluated at the depth of the centroid (Pa) A = surface area of the plate (m²)
Calculate the location of the center of pressure (CP)	$y_{cp} - \bar{y} = \dfrac{\bar{I}}{\bar{y}A}$ (3.33)	$(y_{cp} - \bar{y})$ = slant distance from the centroid to the CP (m) $\bar{I}$ = area moment of inertia of the panel about its centroidal axis (m⁴; for formulas, see Fig. A.1 in the appendix) $\bar{y}$ = slant distance from the centroid to the liquid surface (m)
This figure defines variables		
Check these assumptions:		

1. The problem involves only one fluid. This fluid has a constant density.
2. The pressure distribution is hydrostatic.
3. The pressure at the free surface is zero gage.
4. The panel is symmetric about an axis parallel to the slant distance.

Due to assumptions in the derivations, Eqs. (3.28) and (3.33) have several limitations. First, they only apply to a single fluid of constant density. Second, the pressure at the liquid surface needs to be $p = 0$ gage to correctly locate the CP. Third, Eq. (3.33) gives only the vertical location of the CP, not the lateral location.

Panel Force Working Equations (Summary)

In Table 3.3, we have summarized information that is useful for applying the panel equations. Notice that this table gives the equations, the variables, and the main assumptions. These equations are applied in Examples 3.7 and 3.8.

EXAMPLE 3.7

Hydrostatic Force Due to Concrete

Problem Statement

Determine the force acting on one side of a concrete form 2.44 m high and 1.22 m wide (8 ft by 4 ft) that is used for pouring a basement wall. The specific weight of concrete is 23.6 kN/m³ (150 lbf/ft³).

Define the Situation

Concrete in a liquid state acts on a vertical surface.

The vertical wall is 2.44 m high and 1.22 m wide.

Assumptions: Freshly poured concrete can be represented as a liquid.

Properties: Concrete: $\gamma = 23.6$ kN/m³

State the Goal

Find the resultant force (kN) acting on the wall.

Plan

Apply the panel equation (3.28).

Solution

1. Panel equation:
$$F = \bar{p}A$$

2. Term-by-term analysis:
   - $\bar{p}$ = pressure at depth of the centroid
   $$\bar{p} = (\gamma_{concrete})(z_{centroid}) = (23.6 \text{ kN/m}^3)(2.44/2 \text{ m})$$
   $$= 28.79 \text{ kPa}$$
   - A = area of panel
   $$A = (2.44 \text{ m})(1.22 \text{ m}) = 2.977 \text{ m}^2$$

3. Resultant force:
$$F = \bar{p}A = (28.79 \text{ kPa})(2.977 \text{ m}^2) = \boxed{85.7 \text{ kN}}$$

EXAMPLE 3.8

Force to Open an Elliptical Gate

Problem Statement

An elliptical gate covers the end of a pipe 4 m in diameter. If the gate is hinged at the top, what normal force F is required to open the gate when water is 8 m deep above the top of the pipe and the pipe is open to the atmosphere on the other side? Neglect the weight of the gate.

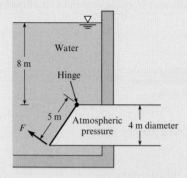

Define the Situation

Water pressure is acting on an elliptical gate.

Properties: Water (10°C): Table A.5, $\gamma = 9810$ N/m³

Assumptions:

1. Neglect the weight of the gate.
2. Neglect friction between the bottom on the gate and the pipe wall.

State the Goal

F(N) ← force needed to open gate

Generate Ideas and Make a Plan

1. Calculate resultant hydrostatic force using $F = \bar{p}A$.
2. Find the location of the center of pressure using Eq. (3.33).
3. Draw an FBD of the gate.
4. Apply moment equilibrium about the hinge.

Take Action (Execute the Plan)

1. Hydrostatic (resultant) force:
   - $\bar{p}$ = pressure at depth of the centroid

   $$\bar{p} = (\gamma_{water})(z_{centroid}) = (9810 \text{ N/m}^3)(10 \text{ m}) = 98.1 \text{ kPa}$$

   - A = area of elliptical panel (using Fig. A.1 to find formula)

   $$A = \pi ab$$
   $$= \pi(2.5 \text{ m})(2 \text{ m}) = 15.71 \text{ m}^2$$

   - Calculate resultant force:

   $$F_p = \bar{p}A = (98.1 \text{ kPa})(15.71 \text{ m}^2) = \boxed{1.54 \text{ MN}}$$

2. Center of pressure:
   - $\bar{y} = 12.5$ m, where $\bar{y}$ is the slant distance from the water surface to the centroid
   - Area moment of inertia $\bar{I}$ of an elliptical panel using a formula from Fig. A.1:

   $$\bar{I} = \frac{\pi a^3 b}{4} = \frac{\pi(2.5 \text{ m})^3(2 \text{ m})}{4} = 24.54 \text{ m}^4$$

   - Finding center of pressure:

   $$y_{cp} - \bar{y} = \frac{\bar{I}}{\bar{y}A} = \frac{25.54 \text{ m}^4}{(12.5 \text{ m})(15.71 \text{ m}^2)} = 0.125 \text{ m}$$

3. FBD of the gate:

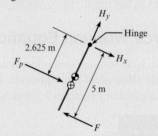

4. Moment equilibrium:

$$\sum M_{hinge} = 0$$
$$1.541 \times 10^6 \text{ N} \times 2.625 \text{ m} - F \times 5 \text{ m} = 0$$
$$F = \boxed{809 \text{ kN}}$$

3.5 Calculating the Pressure Force on a Curved Surface

As engineers, we calculate pressure forces on curved surfaces when we are designing components such as tanks, pipes, and curved gates. Thus, this topic is described in this section.

Consider the curved surface AB in Fig. 3.25a. The goal is to represent the pressure distribution with a resultant force that passes through the center of pressure. One approach is to integrate the pressure force along the curved surface and find the equivalent force. However, it is easier to sum forces for the free body shown in the upper part of Fig. 3.25b. The lower sketch in Fig. 3.25b shows how the force acting on the curved surface relates to the force

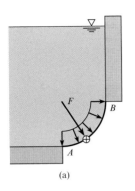

(a)

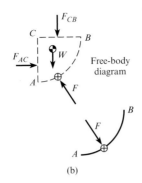

Free-body diagram

(b)

FIGURE 3.25

(a) Pressure distribution and equivalent force.
(b) Free body diagram and action-reaction force pair.

F acting on the free body. Using the FBD and summing forces in the horizontal direction shows that

$$F_x = F_{AC} \tag{3.34}$$

The line of action for the force F_{AC} is through the center of pressure for side AC.

The vertical component of the equivalent force is

$$F_y = W + F_{CB} \tag{3.35}$$

where W is the weight of the fluid in the free body and F_{CB} is the force on the side CB.

The force F_{CB} acts through the centroid of surface CB, and the weight acts through the center of gravity of the free body. The line of action for the vertical force may be found by summing the moments about any convenient axis.

Example 3.9 illustrates how curved surface problems can be solved by applying equilibrium concepts together with the panel force equations.

The central idea of this section is that *forces on curved surfaces may be found by applying equilibrium concepts to systems comprised of the fluid in contact with the curved surface*. Notice how equilibrium concepts are used in each of the situations discussed ahead.

Consider a sphere holding a gas pressurized to a gage pressure p_i, as shown in Fig. 3.26. The indicated forces act on the fluid in volume ABC. Applying equilibrium in the vertical direction gives

$$F = p_i A_{AC} + W$$

Because the specific weight for a gas is quite small, engineers usually neglect the weight of the gas:

$$F = p_i A_{AC} \tag{3.36}$$

Another example is finding the force on a curved surface submerged in a reservoir of liquid, as shown in Fig. 3.27a. If atmospheric pressure prevails above the free surface and on the outside of surface AB, then force caused by atmospheric pressure cancels out, and equilibrium gives

$$F = \gamma V_{ABCD} = W \downarrow \tag{3.37}$$

Hence, the force on surface AB equals the weight of liquid above the surface, and the arrow indicates that the force acts downward.

Now consider the situation in which the pressure distribution on a thin, curved surface comes from the liquid underneath, as shown in Fig. 3.27b. If the region above the surface, volume $abcd$, were filled with the same liquid, then the pressure acting at each point on the upper surface of ab would equal the pressure acting at each point on the lower surface. In other words, there would be no net force on the surface. Thus, the equivalent force on surface ab is given by

$$F = \gamma V_{abcd} = W \downarrow \tag{3.38}$$

where W is the weight of liquid needed to fill a volume that extends from the curved surface to the free surface of the liquid.

FIGURE 3.26

Pressurized spherical tank showing forces that act on the fluid inside the marked region.

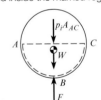

FIGURE 3.27

Curved surface with (a) liquid above and (b) liquid below. In (a), arrows represent forces acting on the liquid. In (b), arrows represent the pressure distribution on surface *ab*.

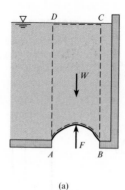

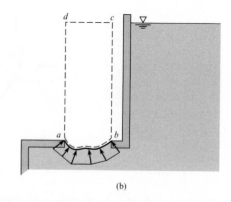

(a)   (b)

EXAMPLE 3.9

Hydrostatic Force on a Curved Surface

Problem Statement

Surface *AB* is a circular arc with a radius of 2 m and a width of 1 m into the paper. The distance *EB* is 4 m. The fluid above surface *AB* is water, and atmospheric pressure prevails on the free surface of the water and on the bottom side of surface *AB*. Find the magnitude and line of action of the hydrostatic force acting on surface *AB*.

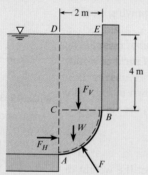

Define the Situation

Situation: A body of water is contained by a curved surface.

Properties: Water (10°C): Table A.5, $\gamma = 9810$ N/m³

State the Goal

Find:

1. Hydrostatic force (in newtons) on the curved surface *AB*
2. Line of action of the hydrostatic force

Generate Ideas and Make a Plan

Apply equilibrium concepts to the body of fluid *ABC*:

1. Find the horizontal component of *F* by applying Eq. (3.34).
2. Find the vertical component of *F* by applying Eq. (3.35).
3. Find the line of action of *F* by finding the lines of action of components and then using a graphical solution.

Take Action (Execute the Plan)

1. Force in the horizontal direction:

$$F_x = F_H = \bar{p}A = (5\text{ m})(9810\text{ N/m}^3)(2 \times 1\text{ m}^2)$$
$$= 98.1\text{ kN}$$

2. Force in the vertical direction:
   - Vertical force on side *CB*:

$$F_V = \bar{p}_0 A = 9.81\text{ kN/m}^3 \times 4\text{ m} \times 2\text{ m} \times 1\text{ m} = 78.5\text{ kN}$$

   - Weight of the water in volume *ABC*:

$$W = \gamma\Psi_{ABC} = (\gamma)(\tfrac{1}{4}\pi r^2)(w)$$
$$= (9.81\text{ kN/m}^3) \times (0.25 \times \pi \times 4\text{ m}^2)(1\text{ m}) = 30.8\text{ kN}$$

   - Summing forces:

$$F_y = W + F_V = 109.3\text{ kN}$$

3. Line of action (horizontal force):

$$y_{cp} = \bar{y} + \frac{\bar{I}}{\bar{y}A} = (5\text{ m}) + \left(\frac{1 \times 2^3/12}{5 \times 2 \times 1}\text{ m}\right)$$
$$y_{cp} = 5.067\text{ m}$$

4. The line of action (x_{cp}) for the vertical force is found by summing moments about point *C*:

$$x_{cp}F_y = F_V \times 1\text{ m} + W \times \bar{x}_w$$

The horizontal distance from point *C* to the centroid of the area *ABC* is found using Fig. A.1: $\bar{x}_w = 4r/3\pi = 0.849$ m. Thus,

$$x_{cp} = \frac{78.5\text{ kN} \times 1\text{ m} + 30.8\text{ kN} \times 0.849\text{ m}}{109.3\text{ kN}} = 0.957\text{ m}$$

5. The resultant force that acts on the curved surface is shown in the following figure:

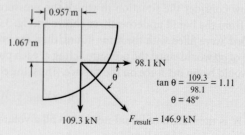

$$\tan\theta = \frac{109.3}{98.1} = 1.11$$
$$\theta = 48°$$

3.6 Calculating Buoyant Forces

Engineers calculate buoyant forces for applications such as the design of ships, sediment transport in rivers, and fish migration. Buoyant forces are sometimes significant in problems involving gases (e.g., a weather balloon). This section describes how to calculate the buoyant force on an object.

A **buoyant force** is defined as an upward force (with respect to gravity) on a body that is totally or partially submerged in a fluid, either a liquid or gas. Buoyant forces are caused by the hydrostatic pressure distribution.

The Buoyant Force Equation

To derive an equation, consider a body $ABCD$ submerged in a liquid of specific weight γ (Fig. 3.28). The sketch on the left shows the pressure distribution acting on the body. As shown by Eq. (3.38), pressures acting on the lower portion of the body create an upward force equal to the weight of liquid needed to fill the volume above surface ADC. The upward force is

$$F_{up} = \gamma(V_b + V_a)$$

where V_b is the volume of the body (i.e., volume $ABCD$) and V_a is the volume of liquid above the body (i.e., volume $ABCFE$). As shown by Eq. (3.37), pressures acting on the top surface of the body create a downward force equal to the weight of the liquid above the body:

$$F_{down} = \gamma V_a$$

Subtracting the downward force from the upward force gives the net or buoyant force F_B acting on the body:

$$F_B = F_{up} - F_{down} = \gamma V_b \qquad \text{(3.39)}$$

Hence, the net force or buoyant force (F_B) equals the weight of liquid that would be needed to occupy the volume of the body.

Consider a body that is floating as shown in Fig. 3.29. The marked portion of the object has a volume V_D. Pressure acts on curved surface ADC, causing an upward force equal to the weight of liquid that would be needed to fill volume V_D. The buoyant force is given by

$$F_B = F_{up} = \gamma V_D \qquad \text{(3.40)}$$

Hence, the buoyant force equals the weight of liquid that would be needed to occupy the volume V_D. This volume is called the displaced volume. Comparison of Eqs. (3.39) and (3.40) shows that one can write a single equation for the buoyant force:

$$F_B = \gamma V_D \qquad \text{(3.41a)}$$

In Eq. (3.41a), V_D is the volume that is displaced by the body. If the body is totally submerged, the displaced volume is the volume of the body. If a body is partially submerged, the displaced volume is the portion of the volume that is submerged.

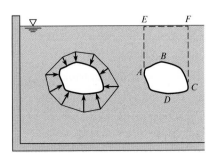

FIGURE 3.28

Two views of a body immersed in a liquid.

FIGURE 3.29

A body partially submerged in a liquid.

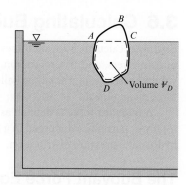

FIGURE 3.30

Hydrometer

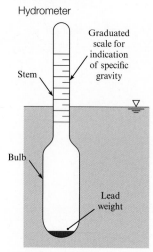

Graduated scale for indication of specific gravity

Stem

Bulb

Lead weight

Eq. (3.41b) is only valid for a single fluid of uniform density. The general principle of buoyancy is called **Archimedes' principle**:

$$\text{(buoyant force)} = F_B = \text{(weight of the displaced fluid)} \qquad \textbf{(3.41b)}$$

The buoyant force acts at a point called the center of buoyancy, which is located at the center of gravity of the displaced fluid.

The Hydrometer

A **hydrometer** (Fig. 3.30) is an instrument for measuring the specific gravity of liquids. It is typically made of a glass bulb that is weighted on one end so the hydrometer floats in an upright position. A stem of constant diameter is marked with a scale, and the specific weight of the liquid is determined by the depth at which the hydrometer floats. The operating principle of the hydrometer is buoyancy. In a heavy liquid (i.e., high γ), the hydrometer will float more shallowly because a lesser volume of the liquid must be displaced to balance the weight of the hydrometer. In a light liquid, the hydrometer will float deeper.

EXAMPLE 3.10

Buoyant Force on a Metal Part

Problem Statement

A metal part (object 2) is hanging by a thin cord from a floating wood block (object 1). The wood block has a specific gravity $S_1 = 0.3$ and dimensions of $50 \times 50 \times 10$ mm. The metal part has a volume of 6600 mm³. Find the mass m_2 of the metal part and the tension T in the cord.

Define the Situation

A metal part is suspended from a floating block of wood.

Properties:

- Water (15°C): Table A.5, $\gamma = 9800$ N/m³
- Wood: $S_1 = 0.3$

State the Goal

- Find the mass (in grams) of the metal part.
- Calculate the tension (in newtons) in the cord.

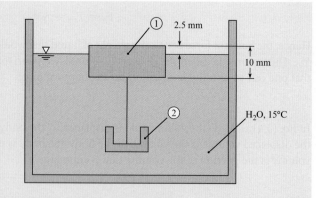

Generate Ideas and Make a Plan

1. Draw FBDs of the block and the part.
2. Apply equilibrium to the block to find the tension.
3. Apply equilibrium to the part to find the weight of the part.
4. Calculate the mass of the metal part using $W = mg$.

Take Action (Execute the Plan)

1. FBDs:

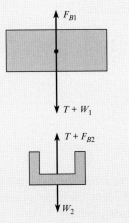

2. Force equilibrium (vertical direction) applied to block:

$$T = F_{B1} - W_1$$

- Buoyant force $F_{B1} = \gamma V_{D1}$, where V_{D1} is the submerged volume:

$$F_{B1} = \gamma V_{D1}$$
$$= (9800 \text{ N/m}^3)(50 \times 50 \times 7.5 \text{ mm}^3)(10^{-9} \text{ m}^3/\text{mm}^3)$$
$$= 0.184 \text{ N}$$

- Weight of the block:

$$W_1 = \gamma S_1 V_1$$
$$= (9800 \text{ N/m}^3)(0.3)(50 \times 50 \times 10 \text{ mm}^3)(10^{-9} \text{ m}^3/\text{mm}^3)$$
$$= 0.0735 \text{ N}$$

- Tension in the cord:

$$T = (0.184 - 0.0735) = \boxed{0.110 \text{ N}}$$

3. Force equilibrium (vertical direction) applied to metal part:

- Buoyant force:

$$F_{B2} = \gamma V_2 = (9800 \text{ N/m}^3)(6600 \text{ mm}^3)(10^{-9}) = 0.0647 \text{ N}$$

- Equilibrium equation:

$$W_2 = T + F_{B2} = (0.110 \text{ N}) + (0.0647 \text{ N})$$

4. Mass of metal part:

$$m_2 = W_2/g = \boxed{17.8 \text{ g}}$$

Review the Solution and the Process

Discussion. Notice that tension in the cord (0.11 N) is less than the weight of the metal part (0.18 N). This result is consistent with the common observation that an object will weigh less in water than in air.

Tip. When solving problems that involve buoyancy, draw an FBD.

3.7 Predicting Stability of Immersed and Floating Bodies

Engineers need to calculate whether an object will tip over or remain in an upright position when placed in a liquid (e.g., for the design of ships and buoys). Thus, stability is presented in this section.

Immersed Bodies

When a body is completely immersed in a liquid, its stability depends on the relative positions of the center of gravity of the body and the centroid of the displaced volume of fluid, which is called the **center of buoyancy**. If the center of buoyancy is above the center of gravity (see Fig. 3.31a), any tipping of the body produces a righting couple, and consequently the body is stable. Alternatively, if the center of gravity is above the center of buoyancy, any tipping produces

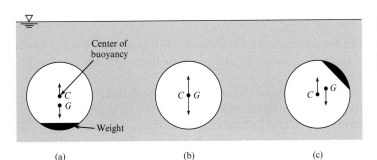

(a) (b) (c)

FIGURE 3.31

Conditions of stability for immersed bodies.
(a) Stable. (b) Neutral.
(c) Unstable.

FIGURE 3.32

Ship stability relations.

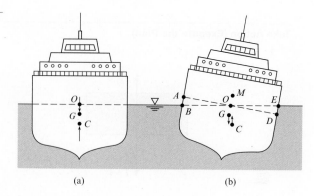

(a)   (b)

an overturning moment, thus causing the body to rotate through 180° (see Fig. 3.31c). If the center of buoyancy and center of gravity are coincident, the body is neutrally stable—that is, it lacks a tendency for righting itself or for overturning (see Fig. 3.31b).

Floating Bodies

The question of stability is more involved for floating bodies than for immersed bodies because the center of buoyancy may take different positions with respect to the center of gravity, depending on the shape of the body and the position in which it is floating. For example, consider the cross section of a ship shown in Fig. 3.32a. Here, the center of gravity G is above the center of buoyancy C. Therefore, at first glance it would appear that the ship is unstable and could flip over. However, notice the position of C and G after the ship has taken a small angle of heel. As shown in Fig. 3.32b, the center of gravity is in the same position, but the center of buoyancy has moved outward from the center of gravity, thus producing a righting moment. A ship having such characteristics is stable.

The reason for the change in the center of buoyancy for the ship is that part of the original buoyant volume, as shown by the wedge shape AOB, is transferred to a new buoyant volume EOD. Because the buoyant center is at the centroid of the displaced volume, it follows that for this case the buoyant center must move laterally to the right. The point of intersection of the lines of action of the buoyant force before and after heel is called the *metacenter* (M), and the distance GM is called the *metacentric height*. If GM is positive—that is, if M is above G—the ship is stable; however, if GM is negative, the ship is unstable. Quantitative relations involving these basic principles of stability are presented in the next paragraph.

Consider the ship shown in Fig. 3.33, which has taken a small angle of heel α. First, evaluate the lateral displacement of the center of buoyancy, CC'; then, it will be easy by simple trigonometry to solve for the metacentric height GM or to evaluate the righting moment. Recall that the center of buoyancy is at the centroid of the displaced volume. Therefore, resort to the fundamentals of centroids to evaluate the displacement CC'. From the definition of the centroid of a volume,

$$\bar{x}\,V = \Sigma x_i \Delta V_i \tag{3.42}$$

where $\bar{x} = CC'$, which is the distance from the plane about which moments are taken to the centroid of V; V is the total volume displaced; ΔV_i is the volume increment; and x_i is the moment arm of the increment of volume.

Take moments about the plane of symmetry of the ship. Recall from mechanics that volumes to the left produce negative moments and volumes to the right produce positive moments. For the right side of Eq. (3.42), write terms for the moment of the submerged volume about the plane of symmetry. A convenient way to do this is to consider the moment of the volume before heel, subtract the moment of the volume represented by the wedge AOB,

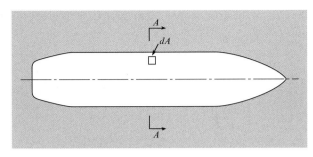

(a)

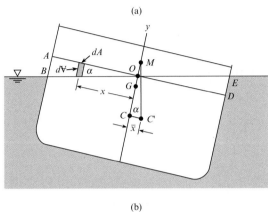

(b)

FIGURE 3.33

(a) A plan view of a ship. (b) Section *A-A* of the ship.

and add the moment represented by the wedge *EOD*. In a general way, this is given by the following equation:

$$\bar{x}\,V = \text{moment of } V \text{ before heel } - \text{ moment of } V_{AOB} + \text{ moment of } V_{EOD} \quad \textbf{(3.43)}$$

Because the original buoyant volume is symmetrical with *y-y*, the moment for the first term on the right is zero. Also, the sign of the moment of V_{AOB} is negative; therefore, when this negative moment is subtracted from the right-hand side of Eq. (3.43), the result is

$$\bar{x}\,V = \sum x_i \Delta V_{iAOB} + \sum x_i \Delta V_{iEOD} \quad \textbf{(3.44)}$$

Now, express Eq. (3.44) in integral form:

$$\bar{x}\,V = \int_{AOB} x\,dV + \int_{EOD} x\,dV \quad \textbf{(3.45)}$$

However, it may be seen from Fig. 3.33b that dV can be given as the product of the length of the differential volume, $x \tan \alpha$, and the differential area, dA. Consequently, Eq. (3.45) can be written as

$$\bar{x}\,V = \int_{AOB} x^2 \tan \alpha\, dA + \int_{EOD} x^2 \tan \alpha\, dA$$

Here, $\tan \alpha$ is a constant with respect to the integration. Also, because the two terms on the right-hand side are identical except for the area over which integration is to be performed, combine them as follows:

$$\bar{x}\,V = \tan \alpha \int_{A_{\text{waterline}}} x^2\, dA \quad \textbf{(3.46)}$$

The second moment, or moment of inertia of the area defined by the waterline, is given the symbol I_{00}, and the following is obtained:

$$\bar{x}\,\mathrm{V} = I_{00}\tan\alpha$$

Next, replace $\bar{x}$ by CC' and solve for CC':

$$CC' = \frac{I_{00}\tan\alpha}{\mathrm{V}}$$

From Fig. 3.33b,

$$CC' = CM\tan\alpha$$

Thus, eliminating CC' and $\tan\alpha$ yields

$$CM = \frac{I_{00}}{\mathrm{V}}$$

However,

$$GM = CM - CG$$

Therefore, the *metacentric height* is

$$GM = \frac{I_{00}}{\mathrm{V}} - CG \qquad (3.47)$$

Equation (3.47) is used to determine the stability of floating bodies. As already noted, if GM is positive, the body is stable; if GM is negative, the body is unstable.

Note that for small angles of heel α, the righting moment or overturning moment is given as follows:

$$RM = \gamma\,\mathrm{V}GM\alpha \qquad (3.48)$$

However, for large angles of heel, direct methods of calculation based on these same principles would have to be employed to evaluate the righting or overturning moment.

EXAMPLE 3.11

Stability of a Floating Block

Problem Statement

A block of wood 30 cm square in cross section and 60 cm long weighs 318 N. Will the block float with sides vertical as shown?

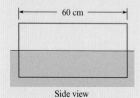

Side view

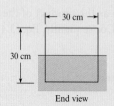

End view

Define the Situation

A block of wood is floating in water.

State the Goal

Determine the stable configuration of the block of wood.

Generate Ideas and Make a Plan

1. Apply force equilibrium to find the depth of submergence.
2. Determine if the block is stable about the long axis by applying Eq. (3.47).
3. If the block is not stable, repeat steps 1 and 2.

Take Action (Execute the Plan)

1. Equilibrium (vertical direction):

$$\sum F_y = 0$$

$$-\text{weight} + \text{buoyant force} = 0$$

$$-318\ \text{N} + 9810\ \text{N/m}^3 \times 0.30\ \text{m} \times 0.60\ \text{m} \times d = 0$$

$$d = 0.18\ \text{m} = 18\ \text{cm}$$

2. Stability (longitudinal axis):

$$GM = \frac{I_{00}}{\mathrm{V}} - CG = \frac{\frac{1}{12} \times 60 \times 30^3}{18 \times 60 \times 30} - (15 - 9)$$

$$= 4.167 - 6 = -1.833\ \text{cm}$$

Because the metacentric height is negative, the block is not stable about the longitudinal axis. Thus, a slight disturbance will make it tip to the orientation shown below. Note: calculations to find the dimensions (2.26 and 5.73 cm) are not shown in this example.

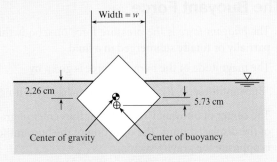

3. Equilibrium (vertical direction):

$$-\text{weight} + \text{buoyant force} = 0$$

$$-(318 \text{ N}) + (9810 \text{ N/m}^3)(V_D) = 0$$

$$V_D = 0.0324 \text{ m}^3$$

4. Find the dimension w:

   (Displaced volume)

   $$= (\text{Block volume}) - (\text{Volume above the waterline})$$

   $$V_D = 0.0324 \text{ m}^3 = (0.3^2)(0.6) \text{ m}^3 - \frac{w^2}{4}(0.6 \text{ m})$$

   $$w = 0.379 \text{ m}$$

5. Moment of inertia at the waterline:

   $$I_{00} = \frac{bh^3}{12} = \frac{(0.6 \text{ m})(0.379 \text{ m})^3}{12} = 0.00273 \text{ m}^4$$

6. Metacentric height:

   $$GM = \frac{I_{00}}{V} - CG = \frac{0.00273 \text{ m}^4}{0.0324 \text{ m}^3} - 0.0573 \text{ m} = 0.027 \text{ m}$$

Because the metacentric height is positive, the block will be stable in this position.

3.8 Summarizing Key Knowledge

Pressure

- *Pressure p* is the ratio of (magnitude of normal force due to a fluid) to (area) at a point.
  - Pressure always acts to compress the material that is in contact with the fluid exerting the pressure.
  - Pressure is a scalar not a vector.
- Engineers express pressure with gage pressure, absolute pressure, vacuum pressure, and differential pressure.
  - Absolute pressure is measured relative to absolute zero.
  - Gage pressure gives the magnitude of pressure relative to atmospheric pressure.

    $$p_{\text{abs}} = p_{\text{atm}} + p_{\text{gage}}$$

  - Vacuum pressure gives the magnitude of the pressure below atmospheric pressure.

    $$p_{\text{vacuum}} = p_{\text{atm}} - p_{\text{abs}}$$

  - Differential pressure (Δp) gives the difference in pressure between two points (e.g., A and B).

Hydrostatic Equilibrium

- A *hydrostatic condition* means that the weight of each fluid particle is balanced by the net pressure force.
- The weight of a fluid causes pressure to increase with increasing depth, giving the *hydrostatic differential equation*. The equations that are used in hydrostatics are derived from this equation. The hydrostatic differential equation is

  $$\frac{dp}{dz} = -\gamma = -\rho g$$

- If density is constant, the hydrostatic differential equation can be integrated to give the hydrostatic equation. The meaning (i.e., physics) of the hydrostatic equation is that piezometric head (or piezometric pressure) is constant everywhere in a static body of fluid.

  $$\frac{p}{\gamma} + z = \text{constant}$$

Pressure Distributions and Forces Due to Pressure

- A fluid in contact with a surface produces a *pressure distribution*, which is a mathematical or visual description of how the pressure varies along the surface.
- A pressure distribution is often represented as a statically equivalent force $\mathbf{F}_p$ acting at the *center of pressure* (CP).
- A *uniform pressure distribution* means that the pressure is the same at every point on a surface. Pressure distributions due to gases are typically idealized as uniform pressure distributions.
- A *hydrostatic pressure distribution* means that the pressure varies according to $dp/dz = -\gamma$.

Force on a Flat Surface

- For a panel subjected to a hydrostatic pressure distribution, the hydrostatic force is

$$F_p = \bar{p}A$$

- This hydrostatic force
  - Acts *at* the centroid of area for a uniform pressure distribution.
  - Acts *below* the centroid of area for a hydrostatic pressure distribution. The slant distance between the center of pressure and the centroid of area is given by

$$y_{cp} - \bar{y} = \frac{I}{\bar{y}A}$$

Hydrostatic Forces on a Curved Surface

- When a surface is curved, one can find the pressure force by applying force equilibrium to a free body comprised of the fluid in contact with the surface.

The Buoyant Force

- The *buoyant force* is the pressure force on a body that is partially or totally submerged in a fluid.
- The magnitude of the buoyant force is given by

  Buoyant force $= F_B =$ weight of the displaced fluid

- The center of buoyancy is located at the center of gravity of the displaced fluid. The direction of the buoyant force is opposite the gravity vector.
- When the buoyant force is due to a single fluid with constant density, the magnitude of the buoyant force is

$$F_B = \gamma \forall_D$$

Hydrodynamic Stability

- Hydrodynamic stability means that if an object is displaced from equilibrium, then there is a moment that causes the object to return to equilibrium.
- The criteria for stability are as follows:
  - *Immersed object.* The body is stable if the center of gravity is below the center of buoyancy.
  - *Floating object.* The body is stable if the metacentric height is positive.

PROBLEMS

Describing Pressure (§3.1)

3.1 Apply the grid method (§1.7) to each situation.

a. If the pressure is 21 inches of water (vacuum), what is the gage pressure in kPa?

b. If the pressure is 175 kPa abs, what is the gage pressure in psi?

c. If a gage pressure is 0.8 bar, what is absolute pressure in psi?

d. If a person's blood pressure is 122 mm Hg gage, what is their blood pressure in kPa abs?

3.2 A 93 mm diameter sphere contains an ideal gas at 20°C. Apply the grid method (§1.7) to calculate the density in units of kg/m³.

a. The gas is helium. The gage pressure is 15 in-H_2O.

b. The gas is methane. The vacuum pressure is 5.5 psi.

3.3 For the questions below, assume standard atmospheric pressure.

a. For a vacuum pressure of 52 kPa, what is the absolute pressure? Gage pressure?

b. For a pressure of 20.5 psig, what is the pressure in psia?

c. For a pressure of 210 kPa gage, what is the absolute pressure in kPa?

d. Give the pressure 100 psfg in psfa.

3.4 The local atmospheric pressure is 98 kPa. A gage on an oxygen tank reads a pressure of 225 kPa gage. What is the pressure in the tank in kPa abs?

3.5 The gage tester shown in the figure is used to calibrate or to test pressure gages. When the weights and the piston together weigh 94 N, the gage being tested indicates 197 kPa. If the piston diameter is 25 mm, what percentage of error exists in the gage?

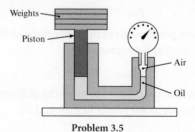

Problem 3.5

3.6 Using §3.1 and other resources, answer the following questions. Strive for depth, clarity, and accuracy while also combining sketches, words, and equations in ways that enhance the effectiveness of your communication.

a. What are five important facts that engineers need to know about pressure?

b. What are five common instances in which people use gage pressure?

c. What are the most common units for pressure?

d. Why is pressure defined using a derivative?

e. How is pressure similar to shear stress? How does pressure differ from shear stress?

The Hydrostatic Equation (§3.2)

3.7 Apply the grid method (§1.7) with the hydrostatic equation ($\Delta p = \gamma \Delta z$) to each of the following cases:

a. Predict the pressure change Δp in kPa for an elevation change Δz of 9.5 ft in a fluid with a density of 90 lbm/ft^3.

b. Predict the pressure change in psf for a fluid with $SG = 0.9$ and an elevation change of 16 m.

c. Predict pressure change in inches of water for a fluid with a density of 1.3 kg/m^3 and an elevation change of 1800 ft.

d. Predict the elevation change in millimeters for a fluid with $SG = 13.0$ that corresponds to a change in pressure of 1/4 atm.

3.8 Apply the grid method to each of the following situations:

a. What is the change in air pressure in pascals between the floor and the ceiling of a room with walls that are 12 ft tall?

b. A diver in the ocean ($SG = 1.03$) records a pressure of 1.2 atm on her depth gage. How deep is she?

c. A hiker starts a hike at an elevation where the air pressure is 960 mbar, and he ascends 1000 ft to a mountain summit. Assuming the density of air is constant, what is the pressure in mbar at the summit?

d. Lake Pend Oreille, in northern Idaho, is one of the deepest lakes in the world, with a depth of 380 m in some locations. This lake is used as a test facility for submarines. What is the maximum gage pressure that a submarine could experience in this lake?

e. A 60 m tall standpipe (a vertical pipe that is filled with water and open to the atmosphere) is used to supply water for firefighting. What is the maximum gage pressure in the standpipe?

3.9 A field test is used to measure the density of crude oil recovered during a fracking* operation. The crude oil recovered is mixed with brine. The oil and brine mixture are placed in an open tank and allowed to separate. After separation, a 1.0 m layer

*Hydraulic fracturing (or "fracking") is a method that is used to recover gas and oil. Fracking creates fractures in rocks by injecting high-pressure liquids containing particulate additives into smaller cracks and forcing the cracks to widen. The larger cracks allow more petroleum products to flow through the formation to the well. A density test as described here could be performed to make a preliminary determination of the approximate makeup of the oil. The brine must be disposed of after fracking.

of oil floats on top of 0.50 m of brine. The density of the brine is 1040 kg/m^3, and the pressure at the bottom of the tank is 13.8 kPa gage. Find the density of the oil.

3.10 If a 290 N force F_1 is applied to the piston with the 4 cm diameter, what is the magnitude of the force F_2 that can be resisted by the piston with the 10 cm diameter? Neglect the weights of the pistons.

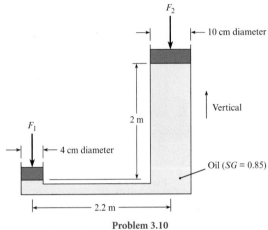

Problem 3.10

3.11 Water occupies the bottom 2.3 m of a cylindrical tank. On top of the water is 0.7 m of kerosene, which is open to the atmosphere. If the temperature is 20°C, what is the gage pressure at the bottom of the tank?

3.12 An engineer is designing a hydraulic lift with a capacity of 10 tons. The moving parts of this lift weigh 1000 lbf. The lift should raise the load to a height of 6 ft in 20 seconds. This will be accomplished with a hydraulic pump that delivers fluid to a cylinder. Hydraulic cylinders with a stroke of 72 inches are available with bore sizes from 2 to 8 inches. Hydraulic piston pumps with an operating pressure range from 200 to 3000 psig are available with pumping capacities of 5, 10, and 15 gallons per minute. Select a hydraulic pump size and a hydraulic cylinder size that can be used for this application.

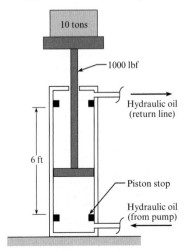

Problem 3.12

3.13 The steel pipe and steel chamber shown in the figure together weigh 680 lbf. What force will have to be exerted on the chamber by all the bolts to hold it in place? The dimension ℓ is equal to 3 ft. *Note:* There is no bottom on the chamber—only a flange bolted to the floor.

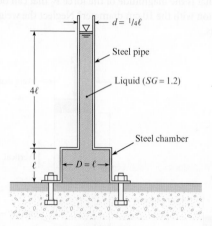

Problem 3.13

3.14 What force must be exerted through the bolts to hold the dome in place? The metal dome and pipe weigh 6 kN. The dome has no bottom. Here $\ell = 80$ cm and the specific weight of the water is $\gamma = 9810$ N/m³.

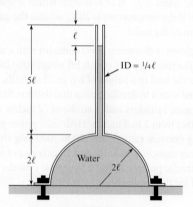

Problem 3.14

3.15 Find the vertical component of force in the metal at the base of the spherical dome shown when gage A reads 5 psig. Indicate whether the metal is in compression or tension. The specific gravity of the enclosed fluid is 1.5. The dimension L is 2 ft. Assume the dome weighs 1000 lbf.

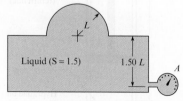

Problem 3.15

3.16 The piston shown weighs 5.5 lbf. In its initial position, the piston is restrained from moving towards the bottom of the cylinder by means of the metal stop. Assuming there is neither friction nor leakage between piston and cylinder, what volume of oil ($SG = 0.85$) would have to be added to the 1 in. tube to cause the piston to rise 1 in. from its initial position?

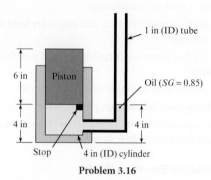

Problem 3.16

Measuring Pressure (§3.3)

3.17 Match the following pressure-measuring devices with the correct name. The device names are: barometer, Bourdon gage, piezometer, manometer, and pressure transducer.

a. Typically contains a diaphragm, a strain gage, and conversion to an electric signal.

b. A U-shaped tube in which changes in pressure cause changes in relative elevation of a liquid that is usually denser than the fluid in the system measured; can be used to measure vacuum.

c. A round face with a scale to measure needle deflection, in which the needle is deflected by changes in extension of a coiled hollow tube.

d. An instrument used to measure atmospheric pressure; can be of various designs.

e. A vertical tube in which a liquid rises in response to a positive gage pressure.

Applying the Manometer Equations (§3.3)

3.18 Using the Internet and other resources, answer the following questions:

a. What are three common types of manometers? For each type, make a sketch and give a brief description.

b. How would you build a manometer from materials that are commonly available? Sketch your design concept.

3.19 As shown, gas at pressure p_g raises a column of liquid to a height h. The gage pressure in the gas is given by $p_g = \gamma_{\text{liquid}}h$. Apply the grid method (§1.7) to each situation that follows.

a. The manometer uses a liquid with $SG = 1.5$. Calculate pressure in psia for $h = 0.3$ ft.

b. The manometer uses mercury. Calculate the column rise in mm for a gage pressure of 0.8 atm.

c. The liquid has a density of 18 lbm/ft³. Calculate pressure in psfg for $h = 5$ inches.

d. The liquid has a density of 850 kg/m³. Calculate the gage pressure in bar for $h = 4.0$ m.

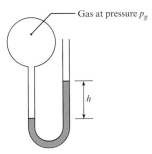

Problem 3.19

3.20 A manometer is used to measure the pressure difference between points A and B in a pipe as shown. Water flows in the pipe, and the specific gravity of the manometer fluid is 2.8. The distances and manometer deflection are indicated on the figure. Find (a) the pressure differences $p_A - p_B$, and (b) the difference in piezometric pressure, $p_{z,A} - p_{z,B}$. Express both answers in kPa.

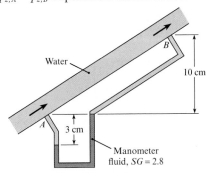

Problem 3.20

Pressure Forces on Panels (Flat Surfaces) (§3.4)

3.21 What is the force acting on the gate of an irrigation ditch if the ditch and gate are 4 ft wide, 3 ft deep, and the ditch is completely full of water? There is no water on the other side of the gate. The weather has been hot for weeks, so the water is 70°F.

3.22 An irrigation ditch is full, with slack water ($V = 0$ m/s; $T = 5°C$) restrained by a closed gate. The ditch and gate are both 2 m wide by 1.9 m. Find the force acting on the gate and the location of center of pressure on the gate as measured from the bottom of the ditch. There is no water on the downstream side of the gate.

3.23 As shown, water (15°C) is in contact with a square panel; $d = 1.2$ m and $h = 0.8$ m.

a. Calculate the depth of the centroid.

b. Calculate the resultant force on the panel.

c. Calculate the distance from the centroid to the CP.

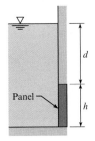

Problem 3.23

3.24 As shown, a round viewing window of diameter $D = 0.8$ m is situated in a large tank of seawater ($SG = 1.03$). The top of the window is 1.0 m below the water surface, and the window is angled at 60° with respect to the horizontal. Find the hydrostatic force acting on the window, and locate the corresponding CP.

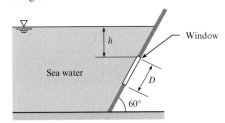

Problem 3.24

3.25 Find the force of the gate on the block as shown, where $d = 10$ m, $h = 2$ m, and $w = 2$ m.

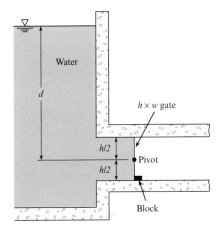

Problem 3.25

3.26 A rectangular gate is hinged at the water line, as shown. The gate has $h = 5$ ft of its length below the waterline, $L = 1$ ft above the waterline, and is 4 ft wide. The specific weight of water is 62.4 lbf/ft³. Find the force (lbf) applied at the bottom of the gate that is necessary to keep the gate closed.

3.27 A rectangular gate is hinged at the water line as shown. The gate has $h = 1$ m of its length below the waterline, $L = 0.3$ m above the waterline, and is 1 m wide. The specific weight of water is 9810 N/m³. Find the necessary force (in N) applied at the bottom of the gate to keep it closed.

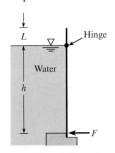

Problem 3.26, 3.27

3.28 This butterfly valve ($D = 8$ ft) is used to control the flow in an 8 ft diameter outlet pipe in a dam. In the position shown, the valve is closed. The valve is supported by a horizontal shaft through its center. The shaft is located $H = 40$ ft below the water surface. What torque would have to be applied to the shaft to hold the valve in the position shown?

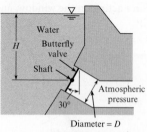

Problem 3.28

3.29 In constructing dams, the concrete is poured in lifts of approximately 1 m ($y_1 = 1$ m). The forms for the face of the dam are reused from one lift to the next. The figure shows one such form, which is bolted to the already cured concrete. For the new pour, what moment will occur at the base of the form per meter of length (normal to the page)? Assume that concrete acts as a liquid when it is first poured and has a specific weight of 23 kN/m³.

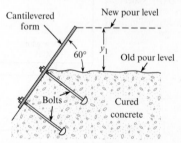

Problem 3.29

3.30 Assume that wet concrete ($\gamma = 150$ lbf/ft³) behaves as a liquid. Determine the force per unit foot of length exerted on the forms. If the forms are held in place as shown, with ties between vertical braces spaced every 2 ft, what force is exerted on the bottom tie?

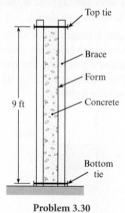

Problem 3.30

Pressure Force on a Curved Surface (§3.5)

3.31 If exactly 20 bolts of 2.5 cm diameter are needed to hold the air chamber together at *A-A* as a result of the high pressure

within, how many bolts will be needed at *B-B*? Here $D = 40$ cm and $d = 20$ cm.

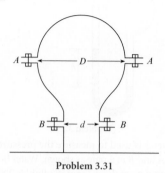

Problem 3.31

3.32 For the plane rectangular gate ($\ell \times w$ in size), figure (*a*), what is the magnitude of the reaction at *A* in terms of γ_w and the dimensions ℓ and *w*? For the cylindrical gate, figure (*b*), will the magnitude of the reaction at *A* be greater than, less than, or the same as that for the plane gate? Neglect the weight of the gates.

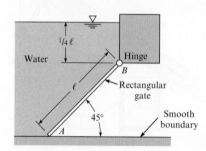

(a) Plane gate

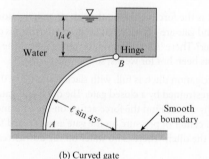

(b) Curved gate

Problem 3.32

3.33 Water is held back by this radial gate. Does the resultant of the pressure forces acting on the gate pass above the pin, through the pin, or below the pin?

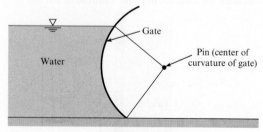

Problem 3.33

3.34 For the curved surface AB:

  a. Determine the magnitude, direction, and line of action of the vertical component of hydrostatic force acting on the surface. Here $\ell = 1$ m.

  b. Determine the magnitude, direction, and line of action of the horizontal component of hydrostatic force acting on the surface.

  c. Determine the resultant hydrostatic force acting on the surface.

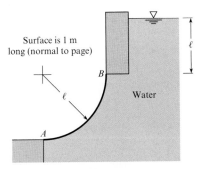

Problem 3.34

3.35 Determine the hydrostatic force acting on the radial gate if the gate is 40 ft long (normal to the page). Show the line of action of the hydrostatic force acting on the gate.

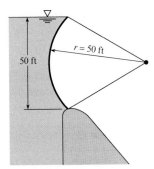

Problem 3.35

3.36 Consider the dome shown. This dome is 10 ft in diameter, but now the dome is not submerged. The water surface is at the level of the center of curvature of the dome. For these conditions, determine the magnitude and direction of the resultant hydrostatic force acting on the dome.

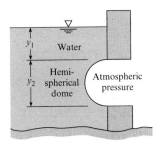

Problem 3.36

Calculating Buoyant Forces (§3.6)

3.37 Apply the grid method (§1.7 in Ch. 1) to each situation below.

  a. Determine the buoyant force in newtons on a basketball that is floating in a lake (10°C).

  b. Determine the buoyant force in newtons on a 1 mm diameter copper sphere that is immersed in kerosene.

  c. Determine the buoyant force in newtons on a 12 inch–diameter balloon. The balloon is filled with helium and situated in ambient air (20°C).

3.38 Using §3.6 and other resources, answer the following questions. Strive for depth, clarity, and accuracy while also combining sketches, words, and equations in ways that enhance the effectiveness of your communication.

  a. Why learn about buoyancy? That is, what are important technical problems that involve buoyant forces?

  b. For a buoyant force, where is the CP? Where is the line of action?

  c. What is a displaced volume? Why is it important?

  d. What is the relationship between a pressure distribution and the associated buoyant force?

3.39 A rock weighs 950 N in air and 720 N in water. Find its volume.

3.40 You are at an estate sale and trying to decide whether to bid on a gold pendant that is said to be 24-carat (pure) gold. The pendant looks like gold, but you would like to check. You are permitted to make some measurements, and you collect the following data: The pendant has a mass of 102 g in air and an apparent mass of 96 g when submerged in water. You know that the SG of 24-carat gold is 19.3, and the SG of 22-carat gold is 17.8; you decide to bid on anything that has $SG > 19.0$. Find the SG of the pendant, and decide whether you will bid.

3.41 As shown, a cube ($L = 75$ mm) suspended in carbon tetrachloride is exactly balanced by an object of mass $m_1 = 650$ g. Find the mass m_2 of the cube.

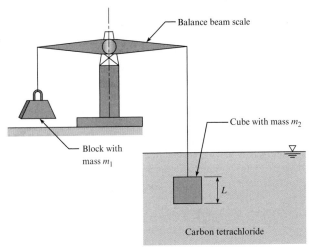

Problem 3.41

3.42 A 600 ft ship has a weight of 20,000 tons, and an area defined by the waterline of 18,000 ft². Will the ship take more or less draft when steaming from salt water to fresh water? How much will it settle or rise?

3.43 An 150 m long freighter weighs 250×10^6 N, and the area defined by its waterline is 2000 m². Will the ship ride higher or deeper in the water when traveling from fresh water to salt water as it leaves the harbor for the open ocean? How much (in meters) will it settle or rise?

3.44 A submerged spherical steel buoy that is 1.4 m in diameter and weighs 1200 N is to be anchored in salt water 30 m below the surface. Find the weight of scrap iron that should be sealed inside the buoy in order that the force on its anchor chain will not exceed 5 kN.

3.45 A block of material of unknown volume is submerged in water and found to weigh 350 N (in water). The same block weighs 705 N in air. Determine the specific weight and volume of the material.

3.46 A 1 ft diameter cylindrical tank is filled with water to a depth of 2 ft. A cylinder of wood 5 in. in diameter and 4 in. long is set afloat on the water. The weight of the wood cylinder is 3.2 lbf. Determine the change (if any) in the depth of the water in the tank.

3.47 A buoy is designed with a hemispherical bottom and conical top as shown in the figure. The diameter of the hemisphere is 1 m, and the half angle of the cone is 30°. The buoy has a mass of 460 kg. Find the location of the water line on the buoy floating in sea water ($\rho = 1010$ kg/m³).

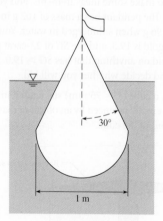

Problem 3.47

3.48 A 90° inverted cone contains water as shown. The volume of the water in the cone is given by $\Psi = (\pi/3)h^3$. The original depth of the water is 10 cm. A block with a volume of 200 cm³ and a specific gravity of 0.6 is floated in the water. What will be the change (in cm) in water surface height in the cone?

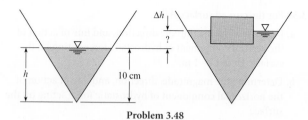

Problem 3.48

3.49 Determine the minimum volume of concrete ($\gamma = 23.6$ kN/m³) needed to keep the gate (1 m wide) in a closed position, with $\ell = 4$ m. Note the hinge at the bottom of the gate.

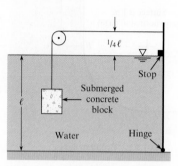

Problem 3.49

Measuring ρ, γ, and SG with Hydrometers (§3.6)

3.50 The hydrometer shown weighs 0.01 N. If the stem sinks 2.2 cm in oil ($z = 2.2$ cm), what is the specific gravity of the oil?

3.51 The hydrometer shown sinks 9 cm ($z = 9$ cm) in water (15°C). The bulb displaces 1.0 cm³, and the stem area is 0.1 cm². Find the weight of the hydrometer.

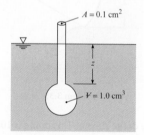

Problems 3.50, 3.51

Predicting Stability (§3.7)

3.52 A cylindrical block of wood 1 m in diameter and 1 m long has a specific weight of 6800 N/m³. Will it float in water with its axis vertical?

The Bernoulli Equation and Pressure Variation

CHAPTER ROAD MAP This chapter describes flowing fluids, introduces the Bernoulli equation, and describes pressure variations in flowing fluids.

FIGURE 4.1
This photo shows flow over a model truck in a wind tunnel. The purpose of the study was to compare the drag force on various designs of tonneau covers. The study was done by Stephen Lyda while he was an undergraduate engineering student. (Photo by Stephen Lyda.)

LEARNING OUTCOMES

DESCRIBING FLOW (§4.1 to 4.3, §4.12).
- Explain streamlines, streaklines, and pathlines.
- Explain the Eulerian and Lagrangian approaches.
- Know the terms defined in Table 4.4.

KINEMATIC PROPERTIES (§4.2, §4.4).
- Define velocity and the velocity field.
- Define acceleration.

EULER'S EQUATION (§4.5).
- Explain how Euler's equation is derived and the meaning of the terms that appear in the equation.

THE BERNOULLI EQUATION (§4.6).
- Know the main ideas about the Bernoulli equation.
- Solve problems that involve the Bernoulli equation.

VELOCITY MEASUREMENT (§4.7).
- Explain how the piezometer, the stagnation tube, and the Pitot-static tube work; do calculations.
- Define static pressure and kinetic pressure.

PRESSURE (§4.10, §4.12).
- Describe the pressure field for flow past a circular cylinder.
- Explain the three causes of pressure variation.

4.1 Describing Streamlines, Streaklines, and Pathlines

To visualize and describe flowing fluids, engineers use the streamline, streakline, and pathline. Hence, these topics are introduced in this section.

Pathlines and Streaklines

The **pathline** is the path of a fluid particle as it moves through a flow field. For example, when the wind blows a leaf, this provides an idea about what the flow is doing. If we imagine that the leaf is tiny and attached to a particle of air as this particle moves, then the motion of the leaf will reveal the motion of the particle. Another way to think of a pathline is to imagine attaching a light to a fluid particle. A time exposure photograph taken of the moving light would be the pathline. One way to reveal pathlines in a flow of water is to add tiny beads that are neutrally buoyant so that bead motion is the same as the motion of fluid particles. Observing these beads as they move through the flow reveals the pathline of each particle.

The **streakline** is the line generated by a tracer fluid, such as a dye, continuously injected into a flow field at a starting point. For example, if smoke is introduced into a flow of air, the resulting lines are streaklines. Streaklines are shown in Fig. 4.1. These streaklines were produced by vaporizing mineral oil on a vertical wire that was heated by passing an electrical current through the wire.

Streamlines

The **streamline** is defined as a line that is tangent everywhere to the local velocity vector.

> **EXAMPLE.** The flow pattern for water draining through an opening in a tank (Fig. 4.2a) can be visualized by examining streamlines. Notice that velocity vectors at points *a*, *b*, and *c* are tangent to the streamlines. Also, the streamlines adjacent to the wall follow the contour of the wall because the fluid velocity is parallel to the wall. The generation of a flow pattern is an effective way of illustrating the flow field.

Streamlines for flow around an airfoil (Fig. 4.2b) reveal that part of the flow goes over the airfoil and part goes under. The flow is separated by the **dividing streamline**. At the location where the dividing streamline intersects the body, the velocity will be zero with respect to the body. This is called the stagnation **point**.

Streamlines for flow over a Volvo ECC prototype (Fig. 4.3) allow engineers to assess aerodynamic features of the flow and possibly change the shape to achieve better performance, such as reduced drag.

Comparing Streamlines, Streaklines, and Pathlines

When flow is steady, the pathline, streakline, and streamline look the same so long as they all pass through the same point. Thus, the streakline, which can be revealed by experimental means, will show what the streamline looks like. Similarly, a particle in the flow will follow a line traced out of a streakline.

FIGURE 4.2

(a) Flow through an opening in a tank. (b) Flow over an airfoil section.

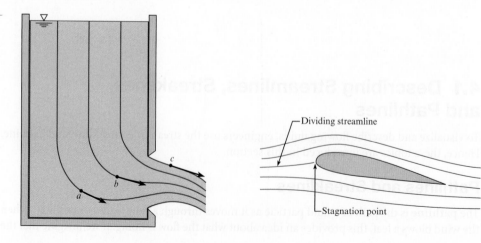

(a)   (b)

FIGURE 4.3

Predicted streamline pattern over the Volvo ECC prototype. (Courtesy of Analytical Methods, VSAERO software, Volvo Concept Center.)

When flow is unsteady, then the streamline, streakline, and pathline look different. A captivating film entitled *Flow Visualization* (1) shows how and why the streamline, streakline, and pathline differ in unsteady flow.

EXAMPLE. To show how pathlines, streaklines, and streamlines differ in unsteady flow, consider a two-dimensional flow that initially has horizontal streamlines (Fig. 4.4). At a given time, t_0, the flow instantly changes direction, and the flow moves upward to the right at 45° with no further change. A fluid particle is tracked from the starting point, and up to time t_0, the pathline is the horizontal line segment shown on Fig. 4.4a. After time t_0, the particle continues to follow the streamline and moves up the right as shown in Fig. 4.4b. Both line segments constitute the pathline. Notice in Fig. 4.4b that the pathline (black dotted line) differs from a streamline for $t < t_0$ and any streamline for $t > t_0$. Thus, the pathline and the streamline are not the same.

Next, consider the streakline by introducing black tracer fluid, as shown in Figures 4.4c and 4.4d. As shown, the streakline in Fig. 4.4d differs from the pathline and from any streamline.

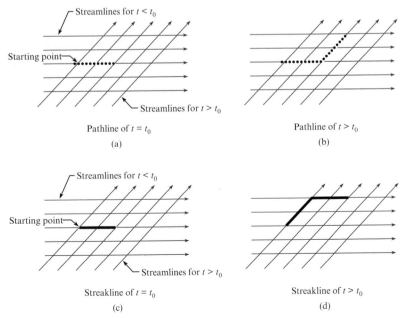

FIGURE 4.4

Streamlines, pathlines, and streaklines for an unsteady flow field.

4.2 Characterizing Velocity of a Flowing Fluid

This section introduces *velocity* and the *velocity field*. Then, these ideas are used to introduce two alternative methods for describing motion:

- *Lagrangian approach.* Describe motion of matter.
- *Eulerian approach.* Describe motion at locations in space.

Describing Velocity

Velocity, a property of a fluid particle, gives the speed and direction of travel of the particle at an instant in time. The mathematical definition of velocity is

$$\mathbf{V}_A = \frac{d\mathbf{r}_A}{dt} \qquad (4.1)$$

where $\mathbf{V}_A$ is the velocity of particle A, and $\mathbf{r}_A$ is the position of particle A at time t.

> **EXAMPLE.** When water drains from a tank (Fig. 4.5a), $\mathbf{V}_A$ gives the speed and direction of travel of the particle at point A. The velocity $\mathbf{V}_A$ is the time rate of change of the vector $\mathbf{r}_A$.

Velocity Field

A description of the velocity of each fluid particle in a flow is called a **velocity field**. In general, each fluid particle in a flow has a different velocity. For example, particles A and B in Fig. 4.5a have different velocities. Thus, the velocity field describes how the velocity varies with position (see Fig. 4.5b).

A velocity field can be described visually (Fig. 4.5b) or mathematically. For example, a steady, two-dimensional velocity field in a corner is given by

$$\mathbf{V} = (2x \, \text{s}^{-1})\mathbf{i} - (2y \, \text{s}^{-1})\mathbf{j} \qquad (4.2)$$

where x and y are position coordinates measured in meters and $\mathbf{i}$ and $\mathbf{j}$ are unit vectors in the x and y directions, respectively.

When a velocity field is given by an equation, a plot can help one visualize the flow. For example, select the location $(x, y) = (1, 1)$ and then substitute $x = 1.0$ meter and $y = 1.0$ meter into Eq. (4.2) to give the velocity as

$$\mathbf{V} = (2 \text{ m/s})\mathbf{i} - (2 \text{ m/s})\mathbf{j} \qquad (4.3)$$

Plot this point and repeat this process at other points to create Fig. 4.6a. Finally, use the definition of the streamline (the line that is tangent everywhere to the velocity vector) to create a streamline pattern (Fig. 4.6b).

Summary. The velocity field describes the velocity of each fluid particle in a spatial region. The velocity field can be shown visually, as in Figs. 4.5 and 4.6, or described mathematically, as in Eq. (4.2).

FIGURE 4.5

Water draining out of a tank.
(a) The velocity of particle A is the time derivative of the position.
(b) The velocity field represents the velocity of each fluid particle throughout the region of flow.

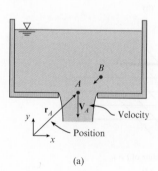

(a)

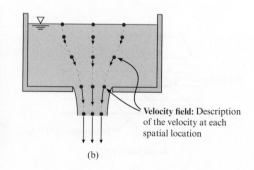

(b)

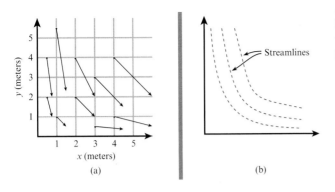

FIGURE 4.6

The velocity field specified by Eq. (4.2): (a) velocity vectors and (b) the streamline pattern.

The concept of a field can be generalized. A **field** is a mathematical or visual description of a variable as a function of position and time.

EXAMPLES. A pressure field describes the distribution of pressure at various points in space and time. A temperature field describes the distribution of temperature at various points in space and time.

A field can be scalar valued (e.g., temperature field, pressure field) or vector valued (e.g., velocity field, acceleration field).

The Eulerian and Lagrangian Approaches

In solid mechanics, it is straightforward to describe the motion of a particle or a rigid body. In contrast, the particles in a flowing fluid move in more complicated ways, and it is not practical to track the motion of each particle. Thus, researchers invented a second way to describe motion.

The first way to describe motion (called the **Lagrangian approach**) involves selecting a body and then describing the motion of this body. The second way (called the **Eulerian approach**) involves selecting a region in space and then describing the motion that is occurring at points in space. In addition, the Eulerian approach allows properties to be evaluated at spatial locations as a function of time because the Eulerian approach uses fields.

EXAMPLE. Consider falling particles (Fig. 4.7). The Lagrangian approach uses equations that describe an individual particle. The Eulerian approach uses an equation for the *velocity field*. Although the equations of the two approaches are different, they predict the same values of velocity. Note that the equation $v = \sqrt{2g|z|}$ in Fig. 4.7 was derived by letting the kinetic energy of the particle equal the change in gravitational potential energy.

When the ideas in Fig. 4.7 are generalized, the independent variables of the Lagrangian approach are initial position and time. The independent variables of the Eulerian approach are position in the field and time. Table 4.1 compares the Lagrangian and the Eulerian approaches.

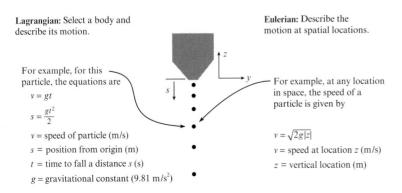

FIGURE 4.7

This figure shows small particles released from rest and falling under the action of gravity. Equations on the left side of the image show how motion is described using a Lagrangian approach. Equations on the right side show a Eulerian approach.

TABLE 4.1 Comparison of the Lagrangian and the Eulerian Approaches

Feature	Lagrangian Approach	Eulerian Approach
Basic idea	Observe or describe the motion of matter of fixed identity.	Observe or describe the motion of matter at spatial locations.
Solid mechanics (application)	Used in dynamics.	Used in elasticity. Can be used to model the flow of materials.
Fluid mechanics (application)	Fluid mechanics uses many Eulerian ideas (e.g., fluid particle, streakline, acceleration of a fluid particle). Equations in fluid mechanics are often derived from a Lagrangian viewpoint.	Nearly all mathematical equations in fluid mechanics are written using the Eulerian approach.
Independent variables	Initial position (x_0, y_0, z_0) and time (t).	Spatial location (x, y, z) and time (t).
Mathematical complexity	Simpler.	More complex; for example, partial derivatives and nonlinear terms appear.
Field concept	Not used in the Lagrangian approach.	The field is a Eulerian concept. When fields are used, the mathematics often includes the divergence, gradient, and curl.
Types of systems used	Closed systems, particles, rigid bodies, system of particles.	Control volumes.

Representing Velocity Using Components

When the velocity field is represented in Cartesian components, the mathematical form is

$$\mathbf{V} = u(x, y, z, t)\mathbf{i} + v(x, y, z, t)\mathbf{j} + w(x, y, z, t)\mathbf{k} \qquad (4.4)$$

where $u = u(x, y, z, t)$ is the x-component of the velocity vector and $\mathbf{i}$ is a unit vector in the x direction. The coordinates (x, y, z) give the spatial location in the field and t is time. Similarly, the components v and w give the y- and z-components of the velocity vector.

Another way to represent a velocity is to use *normal and tangential components*. In this approach (Fig. 4.8), unit vectors are attached to the particle and move with the particle. The tangential unit vector $\mathbf{u}_t$ is tangent to the path of the particle, and the normal unit vector $\mathbf{u}_n$ is normal to the path and directed inward toward the center of curvature. The position coordinate s measures distance traveled along the path. The velocity of a fluid particle is represented as $\mathbf{V} = V(s, t)\mathbf{u}_t$, where V is the speed of the particle and t is time.

4.3 Describing Flow

Engineers use many words to describe flowing fluids. Speaking and understanding this language is seminal to professional practice. Thus, this section introduces concepts for describing flowing fluids. Because there are many ideas, a summary table is presented (see Table 4.4).

FIGURE 4.8

Describing the motion of a fluid particle using normal and tangential components.

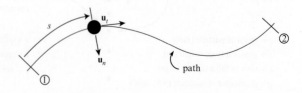

Uniform and Nonuniform Flow

To introduce uniform flow, consider a velocity field of the form

$$\mathbf{V} = \mathbf{V}(s, t)$$

where s is distance traveled by a fluid particle along a path and t is time (Fig. 4.9). This mathematical representation is called *normal and tangential components*. This approach is useful when the path of a particle is known.

In a **uniform flow**, the velocity is constant in magnitude and direction along a streamline at each instant in time. In uniform flow, the streamlines must be rectilinear, which means straight and parallel (see Fig. 4.10). Uniform flow can be described by an equation:

$$\left(\frac{\partial \mathbf{V}}{\partial s}\right)_t = \frac{\partial \mathbf{V}}{\partial s} = 0 \quad \text{(uniform flow)} \tag{4.5}$$

Regarding notation in this text, we omit the variables that are held constant when writing partial derivatives. For example, in Eq. (4.5), the leftmost terms show the formal way to write a partial derivative, and the middle term shows a simpler notation. The rationale for the simpler notation is that variables that are held constant can be inferred from the context.

In **nonuniform flow**, the velocity changes along a streamline in magnitude, direction, or both. It follows that any flow with streamline curvature is nonuniform. Any flow in which the speed of the flow is changing spatially is also nonuniform. Nonuniform flow can be described with an equation.

$$\frac{\partial \mathbf{V}}{\partial s} \neq 0 \quad \text{(nonuniform flow)}$$

EXAMPLES. Nonuniform flow occurs in the converging duct in Fig. 4.11a because the speed increases as the duct converges. Nonuniform flow occurs in the vortex in Fig. 4.11b because the streamlines are curved.

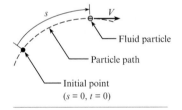

FIGURE 4.9

A fluid particle moving along a pathline.

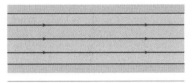

FIGURE 4.10

Uniform flow in a pipe.

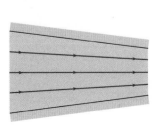

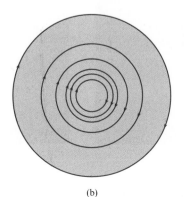

(a)   (b)

FIGURE 4.11

Flow patterns for nonuniform flow:
(a) converging flow, (b) vortex flow.

Steady and Unsteady Flow

In general, a velocity field $\mathbf{V}$ depends on position $\mathbf{r}$ and time t: $\mathbf{V} = \mathbf{V}(\mathbf{r}, t)$. However, in many situations, the velocity is constant with time, so $\mathbf{V} = \mathbf{V}(\mathbf{r})$. This is called steady flow. **Steady flow** means that velocity at each location in space is constant with time. This idea can be written mathematically as

$$\left.\frac{\partial \mathbf{V}}{\partial t}\right|_{\text{all points in velocity field}} = 0$$

In an **unsteady flow**, the velocity is changing, at least at some points, in the velocity field. This idea can be represented with an equation:

$$\frac{\partial \mathbf{V}}{\partial t} \neq 0$$

> **EXAMPLE.** If the flow in a pipe changed with time due to a valve opening or closing, then the flow would be unsteady; that is, the velocity at locations in the velocity field would be increasing or decreasing with time.

Laminar and Turbulent Flow

In a famous experiment, Osborne Reynolds showed that two different kinds of flow that can occur in a pipe.* The first type, called **laminar flow,** is a well-ordered state of flow in which adjacent fluid layers move smoothly with respect to each other. The flow occurs in layers or laminae. An example of laminar flow is the flow of thick syrup (Fig. 4.12a).

The second type of flow identified by Reynolds is called **turbulent flow,** which is an unsteady flow characterized by eddies of various sizes and intense cross-stream mixing. Turbulent flow can be observed in the wake of a ship. Also, turbulent flow can be observed for a smokestack (Fig. 4.12b). Notice that the mixing of the turbulent flow is apparent because the plume widens and disperses.

Laminar flow in a pipe (Fig. 4.13a) has a smooth parabolic velocity distribution. Turbulent flow (Fig. 4.13b) has a plug-shaped velocity distribution because eddies mix the flow, which tends to keep the distribution uniform. In both laminar and turbulent flow, the no-slip condition applies.

FIGURE 4.12

Examples of laminar and turbulent flow: (a) The flow of maple syrup is laminar. (Lauri Patterson/The Agency Collection/Getty Images.) (b) The flow out of a smokestack is turbulent. (Photo by Donald Elger.)

(a)   (b)

*Reynolds' experiment is described in Chapter 10.

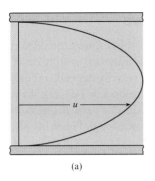 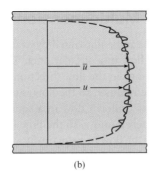

FIGURE 4.13

Laminar and turbulent flow in a straight pipe.
(a) Laminar flow.
(b) Turbulent flow.
Both sketches assume fully developed flow.

(a)   (b)

Time-Averaged Velocity

Turbulent flow is unsteady, so the standard approach is to represent the velocity as a time-averaged velocity $\bar{u}$ plus a fluctuating component u'. Thus, the velocity is expressed as $u = \bar{u} + u'$ (see Fig. 4.13b). The fluctuating component is defined as the difference between the local velocity and the time-averaged velocity. A turbulent flow is designated as "steady" if the time-averaged velocity is unchanging with time. For an interesting look at turbulent flows, see the film entitled *Turbulence* (3). Table 4.2 compares laminar and turbulent flows.

TABLE 4.2 Comparison of Laminar and Turbulent Flows

Feature	Laminar Flow	Turbulent Flow
Basic description	Smooth flow in layers (laminae).	The flow has many eddies of various sizes. The flow appears random, chaotic, and unsteady.
Velocity profile in a pipe	Parabolic; ratio of mean velocity to centerline velocity is 0.5 for fully developed flow.	Pluglike; ratio of mean velocity to centerline velocity is between 0.8 and 0.9.
Mixing of materials added to the flow	Low levels of mixing. Difficult to get a material to mix with a fluid in laminar flow.	High levels of mixing. Easy to get a material to mix; for example, visualize cream mixing with coffee.
Variation with time	Can be steady or unsteady.	Always unsteady.
Dimensionality of flow	Can be 1-D, 2-D, or 3-D.	Always 3-D.
Availability of mathematical solutions	In principle, any laminar flow can be solved with an analytical or computer solution. There are many existing analytical solutions. Solutions are very close to what would be measured with an experiment.	There is no complete theory of turbulent flow. There are a limited number of semiempirical solution approaches. Many turbulent flows cannot be accurately predicted with computer models or analytical solutions. Engineers often rely on experiments to characterize turbulent flow.
Practical importance	A small percentage of practical problems involve laminar flow.	The majority of practical problems involve turbulent flow. Typically, the flow of air and water in piping systems is turbulent. Most flows of water in open channels are turbulent.
Occurrence (Reynolds number)	Occurs at lower values of Reynolds numbers. (The Reynolds number is introduced in Chapter 8.)	Occurs at higher values of Reynolds numbers.

One-Dimensional and Multidimensional Flows

The dimensionality of a flow field can be illustrated by example. Fig. 4.14a shows the velocity distribution for an axisymmetric flow in a circular duct. The flow is uniform, or fully developed, so the velocity does not change in the flow direction (z). The velocity depends on only one spatial dimension, namely the radius r, so the flow is one-dimensional or 1-D. Fig. 4.14b shows the velocity distribution for uniform flow in a square duct. In this case, the velocity depends on two dimensions, namely x and y, so the flow is two-dimensional. Fig. 4.14c also shows the velocity distribution for the flow in a square duct, but the duct cross-sectional area is expanding in the flow direction, so the velocity will be dependent on z as well as x and y. This flow is three-dimensional, or 3-D.

Turbulence is another good example of three-dimensional flow because the velocity components at any one time depend on the three coordinate directions. For example, the velocity component u at a given time depends on x, y, and z; that is, $u(x, y, z)$. Turbulent flow is unsteady, so the velocity components also depend on time.

Another definition frequently used in fluid mechanics is quasi-one-dimensional flow. By this definition, it is assumed that there is only one component of velocity in the flow direction and that the velocity profiles are uniformly distributed; there is constant velocity across the duct cross section.

Viscous and Inviscid Flow

In a **viscous flow**, the forces associated with viscous shear stresses are large enough to effect the dynamic motion of the particles that comprise the flow. For example, when a fluid flows in a pipe, as shown in Fig. 4.13, this is a viscous flow. Indeed, both laminar and turbulent flows are types of viscous flows.

In an **inviscid flow**, the forces associated with viscous shear stresses are small enough that they do not affect the dynamic motion of the particles that comprise the flow. Thus, in an inviscid flow, the viscous stresses can be neglected in the equations for motion.

FIGURE 4.14

Flow dimensionality: (a) one-dimensional flow, (b) two-dimensional flow, and (c) three-dimensional flow.

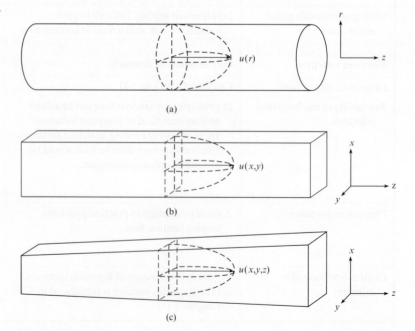

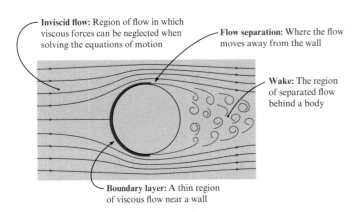

Inviscid flow: Region of flow in which viscous forces can be neglected when solving the equations of motion

Flow separation: Where the flow moves away from the wall

Wake: The region of separated flow behind a body

Boundary layer: A thin region of viscous flow near a wall

FIGURE 4.15

Flow pattern around a sphere when the Reynolds number is high. The sketch shows the regions of flow.

Boundary Layer, Wake, and Potential Flow Regions

To idealize many complex flows, engineers use ideas that can be illustrated by flow over a sphere (Fig. 4.15). As shown, the flow is divided into three regions: an inviscid flow region, a wake, and a boundary layer.

Flow Separation

Flow separation (Fig. 4.15) occurs when the fluid particles adjacent to a body deviate from the contours of the body. Fig. 4.16 shows flow separation behind a square rod. Notice that the flow separates from the shoulders of the rod and that the wake region is large. In both Figs. 4.15 and 4.16, the flow follows the contours of the body on the upstream sides of the objects. The region in which a flow follows the body contour is called **attached flow**.

When flow separates (Fig. 4.16), the drag force on the body is usually large. Thus, designers strive to reduce or eliminate flow separation when designing products such as automobiles and airplanes. In addition, flow separation can lead to structural failure because the wake is unsteady due to vortex shedding, and this creates oscillatory forces. These forces cause structural vibrations, which can lead to failure when the structure's natural frequency is closely matched to the vortex-shedding frequency. In a famous example, vortex shedding associated with flow separation caused the Tacoma Narrows Bridge near Seattle, Washington, to oscillate wildly and to fail catastrophically.

Fig. 4.17 shows flow separation for an airfoil (an airfoil is a body with the cross-sectional shape of a wing). Flow separation occurs when the airfoil is rotated to an angle of attack that is too high. Flow separation in this context causes an airplane to stall, which means that the lifting force drops dramatically and the wings can no longer keep the airplane in level flight. Stall is to be avoided.

Flow separation can occur inside pipes. For example, flow passing through an orifice in a pipe will separate (see Fig. 13.14). In this case, the zone of separated flow is usually called a recirculating zone. Separating flow within a pipe is usually undesirable because it causes energy losses, low-pressure zones that can lead to cavitation and vibrations.

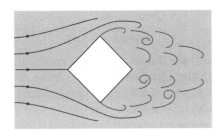

FIGURE 4.16

Flow pattern past a square rod illustrating separation at the edges.

FIGURE 4.17

Separated flow behind an airfoil section at a large angle of attack.

Summary. *Attached flow* means that flow is moving parallel to the walls of a body. *Flow separation*, which occurs in both internal and external flows, means the flow moves away from the wall. Flow separation is related to phenomenon of engineering interests such as drag, structural vibrations, and cavitation.

4.4 Acceleration

Predicting forces is important to the designer. Because forces are related to acceleration, this section describes what acceleration means in the context of a flowing fluid.

Definition of Acceleration

Acceleration is a property of a fluid particle that characterizes the change in speed of the particle and the change in the direction of travel at an instant in time. The mathematical definition of acceleration is

$$\mathbf{a} = \frac{d\mathbf{V}}{dt} \tag{4.6}$$

where $\mathbf{V}$ is the velocity of the particle and t is time.

Physical Interpretation of Acceleration

Acceleration occurs when *a fluid particle is changing its speed, changing its direction of travel, or both.*

EXAMPLE. As a particle moves along the straight streamline in Fig. 4.18, it is slowing down. Because the particle is changing speed, it is accelerating (actually, decelerating in this case). Whenever a particle is changing speed, there must be a component of the acceleration vector tangent to the path. This component of acceleration is called the *tangential component of acceleration*.

EXAMPLE. As a particle moves along a curved streamline (see Fig. 4.19), the particle must have a component of acceleration directed inward as shown. This component is called the *normal component of the acceleration vector*. In addition, if the particle is changing speed, then the tangential component will also be present.

FIGURE 4.18

This figure shows flow over a sphere. The blue sphere is a fluid particle that is moving along the stagnation streamline.

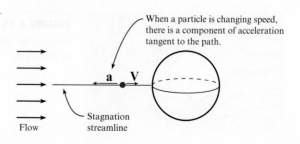

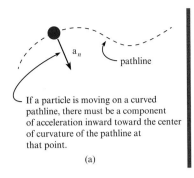

If a particle is moving on a curved pathline, there must be a component of acceleration inward toward the center of curvature of the pathline at that point.

(a)

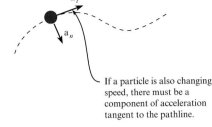

If a particle is also changing speed, there must be a component of acceleration tangent to the pathline.

FIGURE 4.19

This figure shows a particle moving on a curved streamline.

Summary. Acceleration is a property of a fluid particle. The tangential component of the acceleration vector is associated with a change in speed. The normal component is associated with a change in direction. The normal component will be nonzero anytime a particle is moving on a curved streamline because the particle is continually changing its direction of travel.

Describing Acceleration Mathematically

Because the velocity of a flowing fluid is described with a velocity field (i.e., an Eulerian approach), the mathematical representation of acceleration is different from what is presented in courses on subjects such as physics and dynamics. This subsection develops the ideas about fluid acceleration.

To begin, picture a fluid particle on a streamline, as shown in Fig. 4.20. Write the velocity using normal-tangential components:

$$\mathbf{V} = V(s, t)\mathbf{u}_t$$

In this equation, the speed of the particle V is a function of position s and time t. The direction of travel of the particle is given by the unit vector $\mathbf{u}_t$, which by definition is tangent to the streamline.

Use the definition of acceleration:

$$\mathbf{a} = \frac{d\mathbf{V}}{dt} = \left(\frac{dV}{dt}\right)\mathbf{u}_t + V\left(\frac{d\mathbf{u}_t}{dt}\right) \tag{4.7}$$

To evaluate the derivative of speed in Eq. (4.7), the chain rule for a function of two variables can be used:

$$\frac{dV(s, t)}{dt} = \left(\frac{\partial V}{\partial s}\right)\left(\frac{ds}{dt}\right) + \frac{\partial V}{\partial t} \tag{4.8}$$

In a time dt, the fluid particle moves a distance ds, so the derivative ds/dt corresponds to the speed V of the particle, and Eq. (4.8) becomes

$$\frac{dV}{dt} = V\left(\frac{\partial V}{\partial s}\right) + \frac{\partial V}{\partial t} \tag{4.9}$$

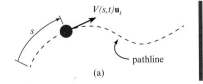

$V(s,t)\mathbf{u}_t$

pathline

(a)

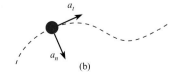

a_t

a_n

(b)

FIGURE 4.20

Particle moving on a pathline: (a) velocity, (b) acceleration.

In Eq. (4.7), the derivative of the unit vector $d\mathbf{u}_t/dt$ is nonzero because the direction of the unit vector changes with time as the particle moves along the pathline. The derivative is

$$\frac{d\mathbf{u}_t}{dt} = \frac{V}{r}\mathbf{u}_n \tag{4.10}$$

where $\mathbf{u}_t$ is the unit vector perpendicular to the pathline and pointing inward toward the center of curvature (2).

Substituting Eqs. (4.9) and (4.10) into Eq. (4.7) gives the acceleration of the fluid particle:

$$\mathbf{a} = \left(V\frac{\partial V}{\partial s} + \frac{\partial V}{\partial t}\right)\mathbf{u}_t + \left(\frac{V^2}{r}\right)\mathbf{u}_n \tag{4.11}$$

The interpretation of this equation is as follows: The acceleration on the left side is acceleration of the fluid particle. The terms on the right side represent a way to evaluate this acceleration by using the velocity, the velocity gradient, and the velocity change with time.

Eq. (4.11) shows that the magnitude of the normal component of acceleration is V^2/r. The direction of this acceleration component is normal to the streamline and inward toward the center of curvature of the streamline. This term is sometimes called the **centripetal acceleration**, where "centripetal" means *center seeking*.

Convective and Local Acceleration

In Eq. (4.11), the term $\partial V/\partial t$ means the time rate of change of speed while holding position (x, y, z) constant. Time-derivative terms in Eulerian formulation for acceleration are called **local acceleration** because position is held constant. All other terms are called **convective acceleration** because they typically involve variables associated with fluid motion.

> **EXAMPLE.** The concepts of Eq. (4.11) can be illustrated by use of the cartoon in Fig. 4.21. The carriage represents the fluid particle, and the track represents the pathline. A direct way to measure the acceleration is to ride on the carriage and read the acceleration off an accelerometer. This gives the acceleration on the left side of Eq. (4.11). The Eulerian approach is to record data so that terms on the right side of Eq. (4.11) can be calculated. One would measure the carriage velocity at two locations separated by a distance Δs and calculate the convective term using
>
> $$V\frac{\partial V}{\partial s} \approx V\frac{\Delta V}{\Delta s}$$
>
> Next, one would measure V and r and then calculate V^2/r. The local acceleration, for this example, would be zero. When one performs the calculations on the right side of Eq. (4.11), the calculated value will match the value recorded by the accelerometer.

FIGURE 4.21

Measuring convective acceleration by two different approaches. (Cartoon by Chad Crowe.)

Summary. The physics of acceleration are described by considering changing speed and changing direction of a fluid particle. Local and convective acceleration are labels for the mathematical terms that appear in the Eulerian formulation of acceleration.

When a velocity field is specified, this denotes an Eulerian approach, and one can calculate the acceleration by substituting numbers into the equation. Example 4.1 illustrates this method.

EXAMPLE 4.1

Calculating Acceleration when a Velocity Field is Specified

Problem Statement

A nozzle is designed such that the velocity in the nozzle varies as

$$u(x) = \frac{u_0}{1.0 - 0.5x/L}$$

where the velocity u_0 is the entrance velocity and L is the nozzle length. The entrance velocity is 10 m/s, and the length is 0.5 m. The velocity is uniform across each section. Find the acceleration at the station halfway through the nozzle ($x/L = 0.5$).

Define the Situation

A velocity distribution is specified in a nozzle.

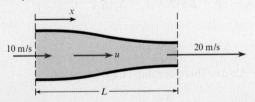

Assumptions: Flow field is quasi-one-dimensional (negligible velocity normal to nozzle centerline).

State the Goal

Calculate the acceleration at the nozzle midpoint.

Generate Ideas and Make a Plan

1. Select the pathline along the centerline of the nozzle.
2. Evaluate the terms in Eq. (4.11).

Take Action (Execute the Plan)

The distance along the pathline is x, so s in Eq. (4.11) becomes x and V becomes u. The pathline is straight, so $r \to \infty$.

1. Term-by-term analysis:
   - Convective acceleration:

$$\frac{\partial u}{\partial x} = -\frac{u_0}{(1 - 0.5x/L)^2} \times \left(-\frac{0.5}{L}\right)$$

$$= \frac{1}{L} \frac{0.5u_0}{(1 - 0.5x/L)^2}$$

$$u\frac{\partial u}{\partial x} = 0.5 \frac{u_0^2}{L} \frac{1}{(1 - 0.5x/L)^3}$$

   Evaluation at $x/L = 0.5$:

$$u\frac{\partial u}{\partial x} = 0.5 \times \frac{10^2}{0.5} \times \frac{1}{0.75^3}$$

$$= 237 \text{ m/s}^2$$

   - Local acceleration:

$$\frac{\partial u}{\partial t} = 0$$

   - Centripetal acceleration (also a convective acceleration):

$$\frac{u^2}{r} = 0$$

2. Combine the terms:

$$a_x = 237 \text{ m/s}^2 + 0$$

$$= \boxed{237 \text{ m/s}^2}$$

$$a_n \text{ (normal to pathline)} = \boxed{0}$$

Review the Solution and the Process

Knowledge. Because a_x is positive, the direction of the acceleration is positive; that is, the velocity increases in the x direction, as expected. Even though the flow is steady, the fluid particles still accelerate.

4.5 Applying Euler's Equation to Understand Pressure Variation

Euler's equation, the topic of this section, is used by engineers to understand pressure variation.

Derivation of Euler's Equation

Euler's equation is derived by applying $\Sigma\mathbf{F} = m\mathbf{a}$ to a fluid particle. The derivation is similar to the derivation of the hydrostatic differential equation (Chapter 3).

FIGURE 4.22

(a) Forces acting on a fluid particle and (b) a sketch showing the geometry.

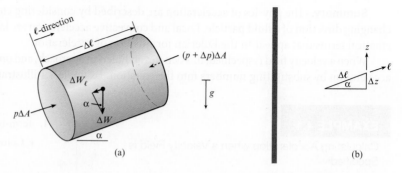

(a)

(b)

To begin, select a fluid particle (Fig. 4.22a) and orient the particle in an arbitrary direction ℓ and at an angle α with respect to the horizontal plane (Fig. 4.22b). Assume that viscous forces are zero. Assume the particle is in a flow and that the particle is accelerating. Now, apply Newton's second law in the ℓ-direction:

$$\sum F_\ell = ma_\ell$$

$$F_{\text{pressure}} + F_{\text{gravity}} = ma_\ell \tag{4.12}$$

The mass of the particle is

$$m = \rho \Delta A \Delta \ell$$

The net force due to pressure in the ℓ-direction is

$$F_{\text{pressure}} = p \Delta A - (p + \Delta p)\Delta A = -\Delta p \Delta A$$

The force due to gravity is

$$F_{\text{gravity}} = -\Delta W_\ell = -\Delta W \sin \alpha \tag{4.13}$$

From Fig. 4.22b, note that $\sin \alpha = \Delta z / \Delta \ell$. Therefore, Eq. (4.13) becomes

$$F_{\text{gravity}} = -\Delta W \frac{\Delta z}{\Delta \ell}$$

The weight of the particle is $\Delta W = \gamma \Delta \ell \Delta A$. Substituting the mass of the particle and the forces on the particle into Eq. (4.12) yields

$$-\Delta p \Delta A - \gamma \Delta \ell \Delta A \frac{\Delta z}{\Delta \ell} = \rho \Delta \ell \Delta A a_\ell$$

Dividing through by the volume of the particle $\Delta A \Delta \ell$ results in

$$-\frac{\Delta p}{\Delta \ell} - \gamma \frac{\Delta z}{\Delta \ell} = \rho a_\ell$$

Taking the limit as $\Delta \ell$ approaches zero (reduce the particle to an infinitesimal size) leads to

$$-\frac{\partial p}{\partial \ell} - \gamma \frac{\partial z}{\partial \ell} = \rho a_\ell \tag{4.14}$$

Assume a constant density flow, so γ is constant, and Eq. (4.14) reduces to

$$-\frac{\partial}{\partial \ell}(p + \gamma z) = \rho a_\ell \tag{4.15}$$

Equation (4.15) is a scalar form of Euler's equation. Because this equation is true in any scalar direction, one can write it in an equivalent vector form:

$$-\nabla p_z = \rho \mathbf{a} \tag{4.16}$$

where ∇p_z is the gradient of the piezometric pressure, and $\mathbf{a}$ is the acceleration of the fluid particle.

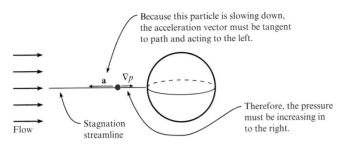

FIGURE 4.23

This figure shows flow over a sphere. The blue object is a fluid particle moving along the stagnation streamline.

Physical Interpretation of Euler's Equation

Euler's equation shows that the pressure gradient is colinear with the acceleration vector and opposite in direction.

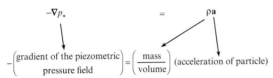

Thus, by using knowledge of acceleration, one can make inferences about the pressure variation. Three important cases are presented next. At this point, we recommend the film entitled *Pressure Fields and Fluid Acceleration* (4), which illustrates fundamental concepts using laboratory experiments.

Case 1: Pressure Variation Due to Changing Speed of a Particle

When a fluid particle is speeding up or slowing down as it moves along a streamline, pressure will vary in a direction tangent to the streamline. For example, Fig. 4.23 shows a fluid particle moving along a stagnation streamline. Because the particle is slowing down, the acceleration vector points to the left. Therefore, the pressure gradient must point to the right. Thus, the pressure is increasing along the streamline, and the direction of increasing pressure is to the right. *Summary.* When a particle is changing speed, pressure will vary in a direction that is tangent to the streamline.

Case 2: Pressure Variation Normal to Rectilinear Streamlines

When streamlines are straight and parallel (Fig. 4.24), then *piezometric pressure will be constant along a line that is normal to the streamlines.*

Case 3: Pressure Variation Normal to Curved Streamlines

When streamlines are curved (Fig. 4.25), then *piezometric pressure will increase along a line that is normal to the streamlines.* The direction of increasing pressure will be outward from the center of curvature of the streamlines. Fig. 4.25 shows why pressure will vary. A fluid particle

FIGURE 4.24

Flow with rectilinear streamlines. The numbered steps give the logic to show that the pressure variation normal to rectilinear streamlines is hydrostatic.

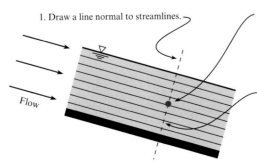

1. Draw a line normal to streamlines.

2. Recognize that the normal component of acceleration for this particle must be zero.

3. Because the acceleration is zero, the gradient of the piezometric pressure along this line must be zero.

4. Conclude that the piezometric pressure must be constant along this line. Therefore, *the pressure variation normal to rectilinear streamlines is hydrostatic.*

FIGURE 4.25

Flow with curved streamlines. Assume that the fluid particle has constant speed. Thus, the acceleration vector points inward towards the center of curvature.

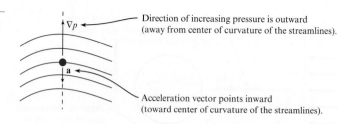

Direction of increasing pressure is outward (away from center of curvature of the streamlines).

Acceleration vector points inward (toward center of curvature of the streamlines).

on a curved streamline must have a component of acceleration inward. Therefore, the gradient of the pressure will point outward. Because the gradient points in the direction of increasing pressure, we conclude that pressure will increase along the line drawn normal to the streamlines. **Summary**. When streamlines are curved, pressure increases outward from the center of curvature* of the streamlines.

Calculations Involving Euler's Equation

In most cases, calculations involving Euler's equation are beyond the scope of this book. However, when a fluid is accelerating as a rigid body, then Euler's equation can be applied in a simple way. Examples 4.2 and 4.3 show how to do this.

EXAMPLE 4.2

Applying Euler's Equation to a Column of Fluid Being Accelerated Upward

Problem Statement

A column of water in a vertical tube is being accelerated by a piston in the vertical direction at 100 m/s². The depth of the water column is 10 cm. Find the gage pressure on the piston. The water density is 10^3 kg/m³.

Define the Situation

A column of water is being accelerated by a piston.

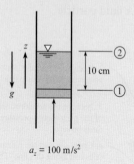

$a_z = 100$ m/s²

Assumptions:
- Acceleration is constant.
- Viscous effects are unimportant.
- Water is incompressible.

Properties: $\rho = 10^3$ kg/m³

State the Goal

Find: The gage pressure on the piston.

Generate Ideas and Make a Plan

1. Apply Euler's equation, Eq. (4.15), in the z direction.
2. Integrate between locations 1 and 2.
3. Set pressure equal to zero (gage pressure) at section 2.
4. Calculate the pressure on the piston.

Take Action (Execute the Plan)

1. Because the acceleration is constant, there is no dependence on time, so the partial derivative in Euler's equation can be replaced by an ordinary derivative. Euler's equation becomes

$$\frac{d}{dz}(p + \gamma z) = -\rho a_z$$

---

*Each streamline has a center of curvature at each point along the streamline. There is not a single center of curvature of a group of streamlines.

2. Integration between sections 1 and 2:

$$\int_1^2 d(p + \gamma z) = \int_1^2 (-\rho a_z)\, dz$$

$$(p_2 + \gamma z_2) - (p_1 + \gamma z_1) = -\rho a_z (z_2 - z_1)$$

3. Algebra:

$$p_1 = (\gamma + \rho a_z)\Delta z = \rho(g + a_z)\Delta z$$

4. Evaluation of pressure:

$$p_1 = 10^3\ \text{kg/m}^3 \times (9.81 + 100)\ \text{m/s}^2 \times 0.1\ \text{m}$$

$$p_1 = \boxed{10.9 \times 10^3\ \text{Pa} = 10.9\ \text{kPa, gage}}$$

EXAMPLE 4.3

Applying Euler's Equation to Gasoline in a Decelerating Tanker

Problem Statement

The tank on a trailer truck is filled completely with gasoline, which has a specific weight of 42 lbf/ft³ (6.60 kN/m³). The truck is decelerating at a rate of 10 ft/s² (3.05 m/s²).

1. If the tank on the trailer is 20 ft (6.1 m) long and if the pressure at the top rear end of the tank is atmospheric, what is the pressure at the top front?

2. If the tank is 6 ft (1.83 m) high, what is the maximum pressure in the tank?

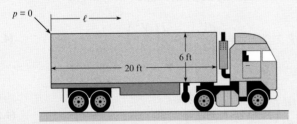

Define the Situation

A decelerating tank of gasoline has pressure equal to a zero gage at the top-rear end.

Assumptions:

1. Deceleration is constant.

2. Gasoline is incompressible.

Properties: $\gamma = 42$ lbf/ft³ (6.60 kN/m³)

State the Goal

Find:

1. The pressure (psfg and kPa, gage) at the top front of the tank.

2. The maximum pressure (psfg and kPa, gage) in the tank.

Make a Plan

1. Apply Euler's equation, Eq. (4.15), along the top of the tank. Elevation, z, is constant.

2. Evaluate pressure at the top front.

3. Maximum pressure will be at the front bottom. Apply Euler's equation from top to bottom at the front of the tank.

4. Using the result from step 2, evaluate the pressure at the front bottom.

Take Action (Execute the Plan)

1. Euler's equation along the top of the tank:

$$\frac{dp}{d\ell} = -\rho a_\ell$$

Integration from back (1) to front (2):

$$p_2 - p_1 = -\rho a_\ell \Delta \ell = -\frac{\gamma}{g} a_\ell \Delta \ell$$

2. Evaluation of p_2 with $p_1 = 0$:

$$p_2 = -\left(\frac{42\ \text{lbf/ft}^3}{32.2\ \text{ft/s}^2}\right) \times (-10\ \text{ft/s}^2) \times 20\ \text{ft}$$

$$= \boxed{261\ \text{psfg}}$$

In SI units:

$$p_2 = -\left(\frac{6.60\ \text{kN/m}^3}{9.81\ \text{m/s}^2}\right) \times (-3.05\ \text{m/s}^2) \times 6.1\ \text{m}$$

$$= \boxed{12.5\ (\text{kPa gage})}$$

3. Euler's equation in the vertical direction:

$$\frac{d}{dz}(p + \gamma z) = -\rho a_z$$

4. For the vertical direction, $a_z = 0$. Integration from the top of the tank (2) to the bottom (3):

$$p_2 + \gamma z_2 = p_3 + \gamma z_3$$

$$p_3 = p_2 + \gamma(z_2 - z_3)$$

$$p_3 = 261\ \text{lbf/ft}^2 + 42\ \text{lbf/ft}^3 \times 6\ \text{ft} = \boxed{513\ \text{psfg}}$$

In SI units:

$$p_3 = 12.5\ \text{kN/m}^2 + 6.6\ \text{kN/m}^3 \times 1.83\ \text{m}$$

$$p_3 = \boxed{24.6\ \text{kPa gage}}$$

4.6 Applying the Bernoulli Equation along a Streamline

The Bernoulli equation is used frequently in fluid mechanics; this section introduces this topic.

Derivation of the Bernoulli Equation

Select a particle on a streamline (Fig. 4.26). The position coordinate s gives the particle's position. The unit vector $\mathbf{u}_t$ is tangent to the streamline, and the unit vector $\mathbf{u}_n$ is normal to the streamline. Assume steady flow so that the velocity of the particle depends on position only. That is, $V = V(s)$.

Assume that viscous forces on the particle can be neglected. Then, apply Euler's equation (Eq. 4.15) to the particle in the $\mathbf{u}_t$ direction:

$$-\frac{\partial}{\partial s}(p + \gamma z) = \rho a_t \qquad (4.17)$$

Acceleration is given by Eq. (4.11). Because the flow is steady, $\partial V/\partial t = 0$, and Eq. (4.11) gives

$$a_t = V\frac{\partial V}{\partial s} + \frac{\partial V}{\partial t} = V\frac{\partial V}{\partial s} \qquad (4.18)$$

Because p, z, and V in Eqs. (4.17) and (4.18) depend only on position s, the partial derivatives become ordinary derivatives (i.e., functions only of a single variable). Thus, write the these derivatives as ordinary derivatives, and combine Eqs. (4.17) and (4.18) to give

$$-\frac{d}{ds}(p + \gamma z) = \rho V\frac{dV}{ds} = \rho\frac{d}{ds}\left(\frac{V^2}{2}\right) \qquad (4.19)$$

Move all the terms to one side:

$$\frac{d}{ds}\left(p + \gamma z + \rho\frac{V^2}{2}\right) = 0 \qquad (4.20)$$

When the derivative of an expression is zero, the expression is equal to a constant. Thus, rewrite Eq. (4.20) as

$$p + \gamma z + \rho\frac{V^2}{2} = C \qquad (4.21a)$$

where C is a constant. Eq. (4.21a) is the *pressure form of the Bernoulli equation*. This is called the pressure form because all terms have units of pressure. Dividing Eq. (4.21a) by the specific weight yields the *head form of the Bernoulli equation,* which is given as Eq. (4.21b). In the head form, all terms have units of length.

$$\frac{p}{\gamma} + z + \frac{V^2}{2g} = C \qquad (4.21b)$$

FIGURE 4.26

Sketch used for the derivation of the Bernoulli equation.

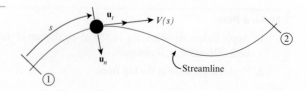

Physical Interpretation #1 (Energy Is Conserved)

One way to interpret the Bernoulli equation leads to the idea that *when the Bernoulli equation applies, the total head of the flowing fluid is a constant along a streamline*. To develop this interpretation, recall that the piezometric head, introduced in Chapter 3, is defined as

$$\text{piezometric head} = h \equiv \frac{p}{\gamma} + z \tag{4.22}$$

Introduce Eq. (4.22) into Eq. (4.21b):

$$h + \frac{V^2}{2g} = \text{Constant} \tag{4.23}$$

Now, the velocity head is defined by

$$\text{velocity head} \equiv \frac{V^2}{2g} \tag{4.24}$$

Combine Eqs. (4.22) to (4.24) to give

$$\left(\begin{array}{c}\text{Piezometric}\\ \text{head}\end{array}\right) + \left(\begin{array}{c}\text{Velocity}\\ \text{head}\end{array}\right) = \left(\begin{array}{c}\text{Constant along}\\ \text{streamline}\end{array}\right) \tag{4.25}$$

Eq. (4.25) is shown visually in Fig. 4.27. Notice that the piezometric head (blue lines) and the velocity head (gray lines) are changing, but the sum of the piezometric head plus the velocity head is constant everywhere. Thus, the total head is constant for all points along a streamline when the Bernoulli equation applies.

The previous discussion introduced head. **Head** *is a concept that is used to characterize the balance of work and energy in a flowing fluid*. As shown in Fig. 4.27, *head can be visualized as the height of a column of liquid*. Each type of head describes a work or energy term. Velocity head characterizes the kinetic energy in a flowing fluid, elevation head characterizes the gravitational potential energy of a fluid, and pressure head is related to work done by the pressure force. As shown in Fig. 4.27, the total head is constant. This means that when the Bernoulli equation applies, the fluid is not losing energy as it flows. The reason is that viscous effects are the cause of energy losses, and viscous effects are negligible when the Bernoulli equation applies.

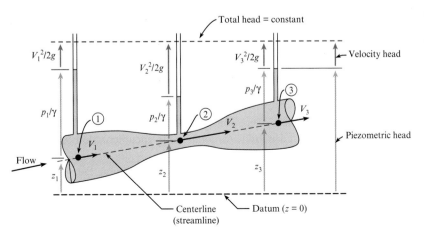

FIGURE 4.27

Water flowing through a Venturi nozzle. The piezometers show the piezometric head at locations 1, 2, and 3.

FIGURE 4.28

(a) The Vinturi wine aerator and (b) a sketch illustrating the operating principle. (Photo courtesy of Vinturi Inc.)

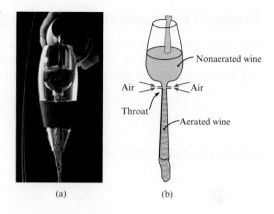

(a)   (b)

Physical Interpretation #2 (Velocity and Pressure Vary Inversely)

A second way to interpret the Bernoulli equation leads to the idea that *when velocity increases, then pressure will decrease.* To develop this interpretation, recall that piezometric pressure, introduced in Chapter 3, is defined as

$$\text{piezometric pressure} = p_z \equiv p + \gamma z \qquad (4.26)$$

Introduce Eq. (4.26) into Eq. (4.21a):

$$p_z + \frac{\rho V^2}{2} = \text{Constant} \qquad (4.27)$$

For Eq. (4.27) to be true, piezometric pressure and velocity must vary inversely so that the sum of p_z and $(V^2/2g)$ is a constant. Thus, the pressure form of the Bernoulli equation shows that *piezometric pressure varies inversely with velocity.* In regions of high velocity, piezometric pressure will be low; in regions of low velocity, piezometric pressure will be high.

> **EXAMPLE.** Fig. 4.28 shows a Vinturi™ red wine aerator, which is a product that is used to add air to wine. When wine flows through the Vinturi, the shape of the device causes an increase in the wine's velocity and a corresponding decrease in its pressure. At the throat, the pressure is below atmospheric pressure, so air flows inward through two inlet ports and mixes with the wine to create aerated wine, which tastes better to most people.

Working Equations and Process

Table 4.3 summarizes the Bernoulli equation. Example 4.4 shows how to apply the Bernoulli equation to a draining tank of water. A systematic method for applying the Bernoulli equation is as follows.

Step 1:   Selection. Select the head form or the pressure form. Check that the assumptions are satisfied.

Step 2:   Sketch. Select a streamline. Then, select points 1 and 2 where you know information or where you want to find information. Annotate your documentation to show the streamline and points.

Step 3:   General equation. Write the general form of the Bernoulli equation. Perform a term-by-term analysis to simplify the *general equation* to a *reduced equation* that applies to the problem at hand.

Step 4:   Validation. Check the reduced equation to ensure that it makes physical sense.

TABLE 4.3 Summary of the Bernoulli Equation

Description	Equation	Terms
Bernoulli equation (head form) Recommended form to use for liquids	$\left(\dfrac{p_1}{\gamma} + \dfrac{V_1^2}{2g} + z_1\right) = \left(\dfrac{p_2}{\gamma} + \dfrac{V_2^2}{2g} + z_2\right)$ Eq. (4.21b)	p = static pressure (Pa) (use gage pressure or abs pressure; avoid vacuum pressure) γ = specific weight (N/m^3) V = speed (m/s) g = gravitational constant = 9.81 m/s^3 z = elevation or elevation head (m)
Bernoulli equation (pressure form) Recommended form to use for gases	$\left(p_1 + \dfrac{\rho V_1^2}{2} + \rho g z_1\right) = \left(p_2 + \dfrac{\rho V_2^2}{2} + \rho g z_2\right)$ Eq. (4.21a)	$\dfrac{p}{\gamma}$ = pressure head (m) $\dfrac{V^2}{2g}$ = velocity head (m) $\dfrac{p}{\gamma} + z$ = piezometric head (m) $p + \gamma z$ = piezometric pressure (Pa) $\dfrac{\rho V^2}{2}$ = kinetic pressure (Pa)

EXAMPLE 4.4

Applying the Bernoulli Equation to Water Draining out of a Tank

Problem Statement

Water in an open tank drains through a port at the bottom of the tank. The elevation of the water in the tank is 10 m above the drain. Find the velocity of the liquid in the drain port.

Define the Situation

Water flows out of a tank.

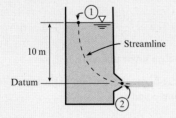

Assumptions:
- Steady flow.
- Viscous effects are negligible.

State the Goal

V_2 (m/s) ← velocity at the exit port

Generate Ideas and Make a Plan

Selection: Select the head form of the Bernoulli equation because the fluid is a liquid. Document assumptions (see above).

Sketching: Select point 1 where information is known and point 2 where information is desired. On the situation diagram (see above), sketch the streamline, label points 1 and 2, and label the datum.

General equation:

$$\left(\frac{p_1}{\gamma} + \frac{V_1^2}{2g} + z_1\right) = \left(\frac{p_2}{\gamma} + \frac{V_2^2}{2g} + z_2\right) \quad \text{(a)}$$

Term-by-term analysis:
- $p_1 = p_2 = 0$ kPa gage
- Let $V_1 = 0$ because $V_1 \ll V_2$
- Let $z_1 = 10$ m and $z_2 = 0$ m

Reduce Eq. (a) so that it applies to the problem at hand:

$$(0 + 0 + 10 \text{ m}) = \left(0 + \frac{V_2^2}{2g} + 0\right) \quad \text{(b)}$$

Simplify Eq. (b):

$$V_2 = \sqrt{2g(10 \text{ m})} \quad \text{(c)}$$

Because Eq. (c) has only one unknown, the plan is to use this equation to solve for V_2.

Take Action (Execute the Plan)

$$V_2 = \sqrt{2g(10 \text{ m})}$$
$$V_2 = \sqrt{2(9.81 \text{ m/s}^2)(10 \text{ m})}$$
$$\boxed{V_2 = 14.0 \text{ m/s}}$$

Review the Solution and the Process

1. *Knowledge.* Notice that the same answer would be calculated for an object dropped from the same elevation as the water in the tank. This is because both problems involve equating gravitational potential energy at 1 with kinetic energy at 2.

2. *Validate.* The assumption of the small velocity at the liquid surface is generally valid. It can be shown

(Chapter 5) that

$$\frac{V_1}{V_2} = \frac{D_2^2}{D_1^2}$$

For example, a diameter ratio of 10 to 1 ($D_2/D_1 = 0.1$) results in the velocity ratio of 100 to 1 ($V_1/V_2 = 1/100$).

When the Bernoulli equation is applied to a gas, it is common to neglect the elevation terms because these terms are negligibly small as compared to the pressure and velocity terms. An example of applying the Bernoulli equation to a flow of air is presented in Example 4.5.

EXAMPLE 4.5

Applying the Bernoulli Equation to Air Flowing around a Bicycle Helmet

Problem Statement

The problem is to estimate the pressure at locations A and B so that these values can be used to estimate the ventilation in a bicycle helmet currently being designed. Assume an air density of $\rho = 1.2$ kg/m^3 and an air speed of 12 m/s relative to the helmet. Point A is a stagnation point, and the velocity of air at point B is 18 m/s.

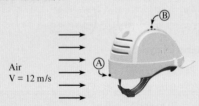

Define the Situation

Idealize flow around a bike helmet as flow around the upper half of a sphere. Assume steady flow. Assume that point B is outside the boundary layer. Relabel the points as shown in the situation diagram; this makes application of the Bernoulli equation easier.

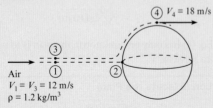

State the Goal

p_2(Pa gage) ← pressure at the forward stagnation point
p_4(Pa gage) ← pressure at the shoulder

Generate Ideas and Make a Plan

Selection: Select the pressure form of the Bernoulli equation because the flow is air. Then, write the Bernoulli equation along the stagnation streamline (i.e., from point 1 to point 2):

$$\left(p_1 + \frac{\rho V_1^2}{2} + \rho g z_1 \right) = \left(p_2 + \frac{\rho V_2^2}{2} + \rho g z_2 \right) \quad \text{(a)}$$

Term-by-term analysis:

- $p_1 = 0$ kPa gage because the external flow is at atmospheric pressure.
- $V_1 = 12$ m/s.
- let $z_1 = z_2 = 0$ because elevation terms are negligibly small for a gas flow such as a flow of air.
- let $V_2 = 0$ because this is a stagnation point.

Now, simplify Eq. (a):

$$0 + \frac{\rho V_1^2}{2} + 0 = p_2 + 0 + 0 \quad \text{(b)}$$

Eq. (b) has only a single unknown (p_2).

Next, apply the Bernoulli equation to the streamline that connects points 3 and 4:

$$\left(p_3 + \frac{\rho V_3^2}{2} + \rho g z_3 \right) = \left(p_4 + \frac{\rho V_4^2}{2} + \rho g z_4 \right) \quad \text{(c)}$$

Do a term-by-term analysis to give:

$$\left(0 + \frac{\rho V_3^2}{2} + 0 \right) = \left(p_4 + \frac{\rho V_4^2}{2} + 0 \right) \quad \text{(d)}$$

Eq. (d) has only one unknown (p_4). The plan is as follows:

1. Calculate p_2 using Eq. (b).
2. Calculate p_4 using Eq. (d).

Take Action (Execute the Plan)

1. Bernoulli equation (point 1 to point 2):

$$p_2 = \frac{\rho V_1^2}{2} = \frac{(1.2\ \text{kg/m}^3)(12\ \text{m/s})^2}{2}$$

$$\boxed{p_2 = 86.4\ \text{Pa gage}}$$

2. Bernoulli equation (point 3 to point 4):

$$p_4 = \frac{\rho(V_3^2 - V_4^2)}{2} = \frac{(1.2\ \text{kg/m}^3)(12^2 - 18^2)(\text{m/s})^2}{2}$$

$$\boxed{p_4 = -108\ \text{Pa gage}}$$

Review the Solution and the Process

1. *Discussion.* Notice that where the velocity is high (i.e., point 4), the pressure is low (negative gage pressure).

2. *Knowledge.* Remember to specify pressure units in gage pressure or absolute pressure.

3. *Knowledge.* Theory shows that the velocity at the shoulder of a sphere is 3/2 times the velocity in the free stream.

Example 4.6 involves a venturi. A **venturi** (also called a venturi nozzle) is a constricted section, as shown in this example. As fluid flows through a venturi, the pressure is reduced in the narrow area, called the throat. This drop in pressure is called the venturi effect.

The venturi can be used to entrain liquid drops into a flow of gas, as in a carburetor. The venturi can also be used to measure the flow rate. The venturi is commonly analyzed with the Bernoulli equation.

EXAMPLE 4.6

Applying the Bernoulli Equation to Flow through a Venturi Nozzle

Problem Statement

Piezometric tubes are tapped into a venturi section as shown in the figure. The liquid is incompressible. The upstream piezometric head is 1 m, and the piezometric head at the throat is 0.5 m. The velocity in the throat section is twice as large as in the approach section. Find the velocity in the throat section.

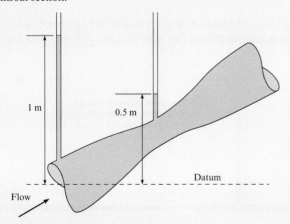

Define the Situation

A liquid flows through a venturi nozzle.

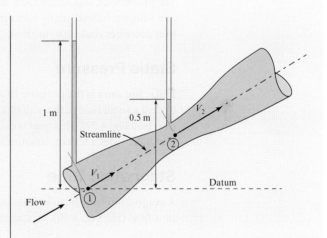

State the Goal

V_2(m/s) ← velocity at point 2

Generate Ideas and Make a Plan

Select the Bernoulli equation because the problem involves flow through a nozzle. Select the head form because a liquid is involved. Select a streamline and points 1 and 2.
Sketch these choices on the situation diagram.

Write the general form of the Bernoulli equation:

$$\frac{p_1}{\gamma} + z_1 + \frac{V_1^2}{2g} = \frac{p_2}{\gamma} + z_2 + \frac{V_2^2}{2g} \tag{a}$$

Introduce piezometric head because this is what the piezometer measures:

$$h_1 + \frac{V_1^2}{2g} = h_2 + \frac{V_2^2}{2g}$$

$$(1.0 \text{ m}) + \frac{V_1^2}{2g} = (0.5 \text{ m}) + \frac{V_2^2}{2g}$$

Let $V_1 = 0.5 \, V_2$

$$(1.0 \text{ m}) + \frac{(0.5 \, V_2)^2}{2g} = (0.5 \text{ m}) + \frac{V_2^2}{2g} \qquad \textbf{(b)}$$

Plan: Use Eq. (b) to solve for V_2.

Take Action (Execute the Plan)

Bernoulli equation (i.e., Eq. b):

$$(0.5 \text{ m}) = \frac{0.75 \, V_2^2}{2g}$$

Thus,

$$V_2 = \sqrt{\frac{2g(0.5 \text{ m})}{0.75}}$$

$$V_2 = \sqrt{\frac{2(9.81 \text{ m/s}^2)(0.5 \text{ m})}{0.75}}$$

$$V_2 = \boxed{3.62 \text{ m/s}}$$

Review the Solution and the Process

1. *Knowledge.* Notice how a piezometer is used to measure the piezometric head in the nozzle.

2. *Knowledge.* A piezometer could not be used to measure the piezometric head if the pressure anywhere in the line were subatmospheric. In this case, pressure gages or manometers could be used.

4.7 Measuring Velocity and Pressure

The piezometer, stagnation tube, and Pitot-static tube have long been used to measure pressure and velocity. Indeed, many concepts in measurement are based on these instruments. This section describes these instruments.

Static Pressure

Static pressure is the pressure in a flowing fluid. A common way to measure static pressure is to drill a small hole in the wall of a pipe and then connect a piezometer or pressure gage to this port (see Fig. 4.29). This port is called a **pressure tap**. The reason that a pressure tap is useful is that it provides a way to measure static pressure that does not disturb the flow.

Stagnation Tube

A **stagnation tube** (also known as a total head tube) is an open-ended tube directed upstream in a flow (see Fig. 4.30). A stagnation tube measures the sum of static pressure and kinetic pressure.

FIGURE 4.29

This figure defines a pressure port and shows how a piezometer is connected to a wall and used to measure static pressure.

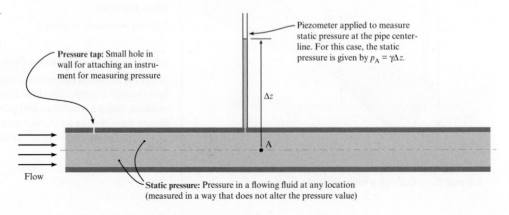

Pressure tap: Small hole in wall for attaching an instrument for measuring pressure

Piezometer applied to measure static pressure at the pipe centerline. For this case, the static pressure is given by $p_A = \gamma \Delta z$.

Δz

A

Flow

Static pressure: Pressure in a flowing fluid at any location (measured in a way that does not alter the pressure value)

FIGURE 4.30

Stagnation tube.

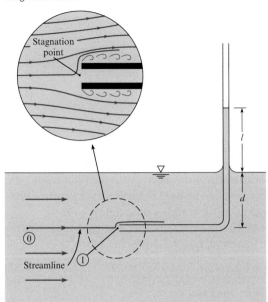

FIGURE 4.31

Pitot-static tube.

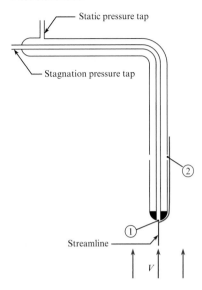

Kinetic pressure is defined at an arbitrary point A as:

$$\left(\begin{array}{c} \text{kinetic pressure} \\ \text{at point A} \end{array} \right) = \frac{\rho V_A^2}{2}$$

Next, we will derive an equation for velocity in an open channel flow. For the stagnation tube in Fig. 4.30, select points 0 and 1 on the streamline, and let $z_0 = z_1$. The Bernoulli equation reduces to

$$p_1 + \frac{\rho V_1^2}{2} = p_0 + \frac{\rho V_0^2}{2} \tag{4.28}$$

The velocity at point 1 is zero (stagnation point). Hence, Eq. (4.28) simplifies to

$$V_0^2 = \frac{2}{\rho}(p_1 - p_0) \tag{4.29}$$

Next, apply the hydrostatic equation: $p_0 = \gamma d$ and $p_1 = \gamma(l + d)$. Therefore, Eq. (4.29) can be written as

$$V_0^2 = \frac{2}{\rho}(\gamma(l + d) - \gamma d)$$

which reduces to

$$V_0 = \sqrt{2gl} \tag{4.30}$$

Pitot-Static Tube

The **Pitot-static tube,** named after the eighteenth-century French hydraulic engineer who invented it, is based on the same principle as the stagnation tube, but it is much more versatile. The Pitot-static tube, shown in Fig. 4.31, has a pressure tap at the upstream end of the tube for sensing the kinetic pressure. There are also ports located several tube diameters downstream of the front end of the tube for sensing the static pressure in the fluid, in which the velocity is essentially the same as the approach velocity. When the Bernoulli equation, Eq. (4.21a), is applied between points 1 and 2 along the streamline shown in Fig. 4.31, the result is

$$p_1 + \gamma z_1 + \frac{\rho V_1^2}{2} = p_2 + \gamma z_2 + \frac{\rho V_2^2}{2}$$

However, $V_1 = 0$, so solving that equation for V_2 gives an equation for velocity:

$$V_2 = \left[\frac{2}{\rho}(p_{z,1} - p_{z,2}) \right]^{1/2} \tag{4.31}$$

Here $V_2 = V$, where V is the velocity of the stream and $p_{z,1}$ and $p_{z,2}$ are the piezometric pressures at points 1 and 2, respectively.

By connecting a pressure gage or manometer between the pressure taps shown in Fig. 4.31 that lead to points 1 and 2, one can easily measure the flow velocity with the Pitot-static tube. A major advantage of the Pitot-static tube is that it can be used to measure velocity in a pressurized pipe; a stagnation tube is not convenient to use in such a situation.

If a differential pressure gage is connected across the taps, then the gage measures the difference in piezometric pressure directly. Therefore, Eq. (4.31) simplifies to

$$V = \sqrt{2\Delta p/\rho} \tag{4.32}$$

where Δp is the pressure difference measured by the gage.

More information on Pitot-static tubes and flow measurement is available in the *Flow Measurement Engineering Handbook* (5). Example 4.7 illustrates the application of the Pitot-static tube with a manometer. Then, Example 4.8 illustrates application with a pressure gage.

EXAMPLE 4.7

Applying a Pitot-Static Tube (Pressure Measured with a Manometer)

Problem Statement

A mercury manometer is connected to the Pitot-static tube in a pipe transporting kerosene as shown. If the deflection on the manometer is 7 in., what is the kerosene velocity in the pipe? Assume that the specific gravity of the kerosene is 0.81.

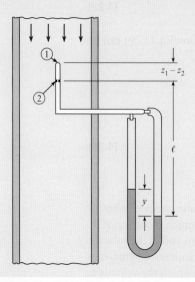

Define the Situation

A Pitot-static tube is mounted in a pipe and connected to a manometer.

Assumptions: Pitot-static tube equation is applicable.

Properties:

- $S_{kero} = 0.81$; from Table A.4
- $S_{Hg} = 13.55$

State the Goal

Find flow velocity (m/s).

Generate Ideas and Make a Plan

1. Find difference in piezometric pressure using the manometer equation.
2. Substitute in the Pitot-static tube equation.
3. Evaluate velocity.

Take Action (Execute the Plan)

1. Manometer equation between points 1 and 2 on Pitot-static tube:

$$p_1 + (z_1 - z_2)\gamma_{kero} + \ell\gamma_{kero} - y\gamma_{Hg} - (\ell - y)\gamma_{kero} = p_2$$

or

$$p_1 + \gamma_{kero}z_1 - (p_2 + \gamma_{kero}z_2) = y(\gamma_{Hg} - \gamma_{kero})$$

$$p_{z,1} - p_{z,2} = y(\gamma_{Hg} - \gamma_{kero})$$

2. Substitution into the Pitot-static tube equation:

$$V = \left[\frac{2}{\rho_{kero}} y(\gamma_{Hg} - \gamma_{kero})\right]^{1/2}$$

$$= \left[2gy\left(\frac{\gamma_{Hg}}{\gamma_{kero}} - 1\right)\right]^{1/2}$$

3. Velocity evaluation:

$$V = \left[2 \times 32.2 \text{ ft/s}^2 \times \frac{7}{12}\text{ ft}\left(\frac{13.55}{0.81} - 1\right)\right]^{1/2}$$

$$= \left[2 \times 32.2 \times \frac{7}{12}(16.7 - 1)\text{ ft}^2/\text{s}^2\right]^{1/2}$$

$$= \boxed{24.3 \text{ ft/s}}$$

Review the Solution and the Process

Discussion. The −1 in the quantity (16.7 − 1) reflects the effect of the column of kerosene in the right leg of the manometer, which tends to counterbalance the mercury in the left leg. Thus, with a gas–liquid manometer, the counterbalancing effect is negligible.

EXAMPLE 4.8

Applying a Pitot-Static Tube (Pressure Measured with a Pressure Gage)

Problem Statement

A differential pressure gage is connected across the taps of a Pitot-static tube. When this Pitot-static tube is used in a wind tunnel test, the gage indicates a Δp of 730 Pa. What is the air velocity in the tunnel? The pressure and temperature in the tunnel are 98 kPa absolute and 20°C, respectively.

Define the Situation

A differential pressure gage is attached to a Pitot-static tube for velocity measurement in a wind tunnel.

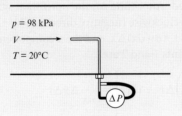

p = 98 kPa

V ⟶

T = 20°C

ΔP

Assumptions:

- Airflow is steady.
- The Pitot-tube equation is applicable.

Properties: From Table A.2: R_{air} = 287 J/kg K.

State the Goal

Find the air velocity (in m/s).

Generate Ideas and Make a Plan

1. Using the ideal gas law, calculate air density.
2. Using the Pitot-static tube equation, calculate the velocity.

Take Action (Execute the Plan)

1. Density calculation:

$$\rho = \frac{p}{RT} = \frac{98 \times 10^3 \text{ N/m}^2}{(287 \text{ J/kg K}) \times (20 + 273 \text{ K})} = 1.17 \text{ kg/m}^3$$

2. Pitot-static tube equation with differential pressure gage:

$$V = \sqrt{2\Delta p/\rho}$$

$$V = \sqrt{(2 \times 730 \text{ N/m}^2)/(1.17 \text{ kg/m}^3)} = \boxed{35.3 \text{ m/s}}$$

4.8 Characterizing the Rotational Motion of a Flowing Fluid

In addition to velocity and acceleration, engineers also describe the rotation of a fluid. This topic is introduced in this section. At this point, we recommend the online film *Vorticity* (6) because it shows the concepts in this section using laboratory experiments.

Concept of Rotation

Rotation of a fluid particle is defined as the average rotation of two initially mutually perpendicular faces of a fluid particle. The test is to look at the rotation of the line that bisects both faces (*a-a* and *b-b* in Fig. 4.32). The angle between this line and the horizontal axis is the rotation, θ.

The general relationship between θ and the angles defining the sides is shown in Fig. 4.33, where θ_A is the angle of one side with the *x*-axis, and the angle θ_B is the angle of the other side

FIGURE 4.32

Rotation of a fluid particle in flow between moving and stationary parallel plates.

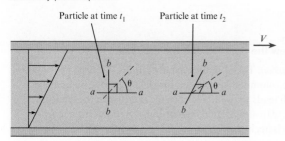

FIGURE 4.33

Orientation of a rotated fluid particle.

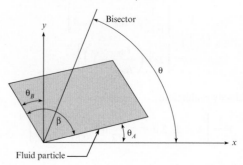

with the y-axis. The angle between the sides is $\beta = \dfrac{\pi}{2} + \theta_B - \theta_A$, so the orientation of the particle with respect to the x-axis is

$$\theta = \frac{1}{2}\beta + \theta_A = \frac{\pi}{4} + \frac{1}{2}(\theta_A + \theta_B)$$

The rotational rate of the particle is

$$\dot{\theta} = \frac{1}{2}(\dot{\theta}_A + \dot{\theta}_B) \tag{4.33}$$

If $\dot{\theta} = 0$, then the flow is **irrotational**, which means that the rotation rate (as defined by Eq. 4.33) is zero for all points in the velocity field.

Next, we derive an equation for $\dot{\theta}$ in terms of the velocity field. Consider the particle shown in Fig. 4.34. The sides of the particle are initially perpendicular, with lengths Δx and Δy. Then, the particle moves with time and deforms as shown, with point 0 going to $0'$, point 1 to $1'$, and point 2 to $2'$. The lengths of the sides are unchanged. After time Δt, the horizontal side has rotated counterclockwise by $\Delta\theta_A$ and the vertical side clockwise (negative direction) by $-\Delta\theta_B$.

The y velocity component of point 1 is $v + (\partial v/\partial x)\Delta x$, and the x component of point 2 is $u + (\partial u/\partial y)\Delta y$. The net displacements of points 1 and 2 are*

$$\Delta y_1 \sim \left[\left(v + \frac{\partial v}{\partial x}\Delta x\right)\Delta t - v\Delta t\right] = \frac{\partial v}{\partial x}\Delta x\Delta t$$

$$\Delta x_2 \sim \left[\left(u + \frac{\partial u}{\partial y}\Delta y\right)\Delta t - u\Delta t\right] = \frac{\partial u}{\partial y}\Delta y\Delta t \tag{4.34}$$

FIGURE 4.34

Translation and deformation of a fluid particle.

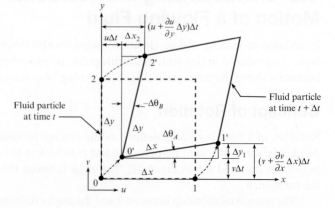

*The symbol ~ means that the quantities are approximately equal but become exactly equal as the quantities approach zero.

Referring to Fig. 4.34, the angles $\Delta\theta_A$ and $\Delta\theta_B$ are given by

$$\Delta\theta_A = \text{asin}\left(\frac{\Delta y_1}{\Delta x}\right) \sim \frac{\Delta y_1}{\Delta x} \sim \frac{\partial v}{\partial x}\Delta t$$

$$-\Delta\theta_B = \text{asin}\left(\frac{\Delta x_2}{\Delta y}\right) \sim \frac{\Delta x_2}{\Delta y} \sim \frac{\partial u}{\partial y}\Delta t$$

(4.35)

Dividing the angles by Δt and taking the limit as $\Delta t \to 0$,

$$\dot{\theta}_A = \lim_{\Delta t \to 0} \frac{\Delta\theta_A}{\Delta t} = \frac{\partial v}{\partial x}$$

$$\dot{\theta}_B = \lim_{\Delta t \to 0} \frac{\Delta\theta_B}{\Delta t} = -\frac{\partial u}{\partial y}$$

(4.36)

Substituting these results into Eq. (4.33) gives the rotational rate of the particle about the z-axis (normal to the page):

$$\dot{\theta} = \frac{1}{2}\left(\frac{\partial v}{\partial x} - \frac{\partial u}{\partial y}\right)$$

This component of rotational velocity is defined as Ω_z, so

$$\Omega_z = \frac{1}{2}\left(\frac{\partial v}{\partial x} - \frac{\partial u}{\partial y}\right)$$

(4.37a)

Likewise, the rotation rates about the other axes are

$$\Omega_x = \frac{1}{2}\left(\frac{\partial w}{\partial y} - \frac{\partial v}{\partial z}\right)$$

(4.37b)

$$\Omega_y = \frac{1}{2}\left(\frac{\partial u}{\partial z} - \frac{\partial w}{\partial x}\right)$$

(4.37c)

The rate-of-rotation vector is

$$\Omega = \Omega_x\mathbf{i} + \Omega_y\mathbf{j} + \Omega_z\mathbf{k}$$

(4.38)

An irrotational flow ($\Omega = 0$) requires that

$$\frac{\partial v}{\partial x} = \frac{\partial u}{\partial y}$$

(4.39a)

$$\frac{\partial w}{\partial y} = \frac{\partial v}{\partial z}$$

(4.39b)

$$\frac{\partial u}{\partial z} = \frac{\partial w}{\partial x}$$

(4.39c)

The most extensive application of these equations is in ideal flow theory. An ideal flow is the flow of an irrotational, incompressible fluid. Flow fields in which viscous effects are small can often be regarded as irrotational. In fact, if a flow of an incompressible, inviscid fluid is initially irrotational, then it will remain irrotational.

Vorticity

The most common way to describe rotation is to use **vorticity**, which is a vector equal to twice the rate-of-rotation vector. The magnitude of the vorticity indicates the rotationality of a flow and is very important in flows in which viscous effects dominate, such as boundary layer, separated, and wake flows. The vorticity equation is

$$\omega = 2\Omega$$

$$= \left(\frac{\partial w}{\partial y} - \frac{\partial v}{\partial z}\right)\mathbf{i} + \left(\frac{\partial u}{\partial z} - \frac{\partial w}{\partial x}\right)\mathbf{j} + \left(\frac{\partial v}{\partial x} - \frac{\partial u}{\partial y}\right)\mathbf{k} \qquad (4.40)$$

$$= \nabla \times \mathbf{V}$$

where $\nabla \times \mathbf{V}$ is the curl of the velocity field.

An irrotational flow signifies that the vorticity vector is zero everywhere. Example 4.9 illustrates how to evaluate the rotationality of a flow field, and Example 4.10 evaluates the rotation of a fluid particle.

EXAMPLE 4.9

Evaluating Rotation

Problem Statement

The vector $\mathbf{V} = 10x\mathbf{i} - 10y\mathbf{j}$ represents a two-dimensional velocity field. Is the flow irrotational?

Define the Situation

Velocity field is given.

State the Goal

Determine if the flow is irrotational.

Generate Ideas and Make a Plan

Because $w = 0$ and $\dfrac{\partial}{\partial z} = 0$, apply Eq. (4.39a) to evaluate rotationality.

Take Action (Execute the Plan)

Velocity components and derivatives:

$$u = 10x \qquad \frac{\partial u}{\partial y} = 0$$

$$v = -10y \qquad \frac{\partial v}{\partial x} = 0$$

Thus, flow is irrotational.

EXAMPLE 4.10

Rotation of a Fluid Particle

Problem Definition

A fluid exists between stationary and moving parallel flat plates, and the velocity is linear as shown. The distance between the plates is 1 cm, and the upper plate moves at 2 cm/s. Find the amount of rotation that the fluid particle located at 0.5 cm will undergo after it has traveled a distance of 1 cm.

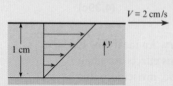

Define the Situation

This problem involves Couette flow.

Assumptions: Planar flow ($w = 0$ and $\dfrac{\partial}{\partial z} = 0$).

State the Goal

Find the rotation of a fluid particle (in radians) at the midpoint after traveling 1 cm.

Generate Ideas and Make a Plan

1. Use Eq. (4.37a) to evaluate rotational rate with $v = 0$.
2. Find time for a particle to travel 1 cm.
3. Calculate the amount of rotation.

Take Action (Execute the Plan)

1. Velocity distribution:

$$u = 0.02 \text{ m/s} \times \frac{y}{0.01 \text{ m}} = 2y \text{ (l/s)}$$

Rotational rate:

$$\Omega_z = \frac{1}{2}\left(\frac{\partial v}{\partial x} - \frac{\partial u}{\partial y}\right) = -1 \text{ rad/s}$$

2. Time to travel 1 cm:

$$u = 2 \ (\text{l/s}) \times 0.005 \ \text{m} = 0.01 \ \text{m/s}$$

$$\Delta t = \frac{\Delta x}{u} = \frac{0.01 \ \text{m}}{0.01 \ \text{m/s}} = 1 \ \text{s}$$

3. Amount of rotation:

$$\Delta \theta = \Omega_z \times \Delta t = -1 \times 1 = -1 \ \text{rad}$$

Review the Solution and the Process

Discussion. Note that the rotation is negative (in the clockwise direction).

4.9 The Bernoulli Equation for Irrotational Flow

When flow is irrotational, the Bernoulli equation can be applied between any two points in this flow. That is, the points do not need to be on the same streamline. This *irrotational form* of the Bernoulli equation is used extensively in applications such as classical hydrodynamics, the aerodynamics of lifting surfaces (wings), and atmospheric winds. Thus, this section describes how to derive the Bernoulli equation for an irrotational flow.

To begin the derivation, apply the Euler equation, Eq. (4.15), in the *n* direction (normal to the streamline):

$$-\frac{d}{dn}(p + \gamma z) = \rho a_n \tag{4.41}$$

where the partial derivative of *n* is replaced by the ordinary derivative because the flow is assumed to be steady (no time dependence). Two adjacent streamlines and the direction *n* is shown in Fig. 4.35. The local fluid speed is *V*, and the local radius of curvature of the streamline is *r*. The acceleration normal to the streamline is the centripetal acceleration, so

$$a_n = -\frac{V^2}{r} \tag{4.42}$$

where the negative sign occurs because the direction *n* is outward from the center of curvature and the centripetal acceleration is toward the center of curvature. Using the irrotationality condition, the acceleration can be written as

$$a_n = -\frac{V^2}{r} = -V\left(\frac{V}{r}\right) = V\frac{dV}{dr} = \frac{d}{dr}\left(\frac{V^2}{2}\right) \tag{4.43}$$

Also, the derivative with respect to *r* can be expressed as a derivative with respect to *n* by

$$\frac{d}{dr}\left(\frac{V^2}{2}\right) = \frac{d}{dn}\left(\frac{V^2}{2}\right)\frac{dn}{dr} = \frac{d}{dn}\left(\frac{V^2}{2}\right)$$

because the direction of *n* is the same as *r*, so $dn/dr = 1$. Eq. (4.43) can be rewritten as

$$a_n = \frac{d}{dn}\left(\frac{V^2}{2}\right) \tag{4.44}$$

Substituting the expression for acceleration into Euler's equation, Eq. (4.41), and assuming constant density results in

$$\frac{d}{dn}\left(p + \gamma z + \rho\frac{V^2}{2}\right) = 0 \tag{4.45}$$

FIGURE 4.35

Two adjacent streamlines showing direction *n* between lines.

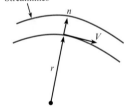

Streamlines

or

$$p + \gamma z + \rho \frac{V^2}{2} = C \qquad (4.46)$$

which is the Bernoulli equation, and C is constant in the n direction (across streamlines).

Summary. For an irrotational flow, the constant C in the Bernoulli equation is the same across streamlines as well as along streamlines, so it is the same everywhere in the flow field. Thus, *when applying the Bernoulli equation for irrotational flow, one can select points 1 and 2 at any locations, not just along a streamline.*

4.10 Describing the Pressure Field for Flow over a Circular Cylinder

Flow over a circular cylinder is a paradigm (i.e., model) for external flow over many objects. This flow is described in this section.

The Pressure Coefficient

To describe the pressure field, engineers often use a dimensionless group called the **pressure coefficient**:

$$C_p = \frac{p_z - p_{zo}}{\rho V_o^2/2} = \frac{h - h_o}{V_o^2/(2g)} \qquad (4.47)$$

Pressure Distribution for an Ideal Fluid

An **ideal fluid** is defined as a fluid that is nonviscous and that has constant density. If we assume an irrotational flow of an ideal fluid, then calculations reveal the results shown in Fig. 4.36a. Features to notice in this figure are as follows:

- The pressure distribution is symmetric on the front and back of the cylinder.
- The pressure coefficient is sometimes negative (plotted outward), which corresponds to negative gage pressure.
- The pressure coefficient is sometimes positive (plotted inward), which corresponds to positive gage pressure.

FIGURE 4.36

Irrotational flow past a cylinder: (a) streamline pattern, (b) pressure distribution.

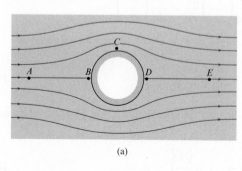

(a)

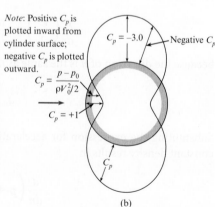

Note: Positive C_p is plotted inward from cylinder surface; negative C_p is plotted outward.

$$C_p = \frac{p - p_0}{\rho V_0^2/2}$$

$C_p = -3.0$

Negative C_p

$C_p = +1$

C_p

(b)

- The maximum pressure ($C_p = +1.0$) occurs on the front and back of the cylinder at the stagnation points (points B and D).
- The minimum pressure ($C_p = -3.0$) occurs at the midsection, where the velocity is highest (point C).

Next, we introduce the concepts of favorable and adverse pressure gradients. To begin, apply Euler's equation while neglecting gravitational effects:

$$\rho a_t = -\frac{\partial p}{\partial s}$$

Note that $a_t > 0$ if $\partial p/\partial s < 0$; that is, the fluid particle accelerates if the pressure decreases with distance along a pathline. This is a **favorable pressure gradient**. On the other hand, $a_t < 0$ if $\partial p/\partial s > 0$, so the fluid particle decelerates if the pressure increases along a pathline. This is an **adverse pressure gradient**. The definitions of pressure gradient are summarized as follows.

Favorable pressure gradient	$\partial p/\partial s < 0$	$a_t > 0$ (acceleration)
Adverse pressure gradient	$\partial p/\partial s > 0$	$a_t < 0$ (deceleration)

Visualize the motion of a fluid particle in Fig. 4.36a as it travels around the cylinder from A to B to C to D and finally to E. Notice that it first decelerates from the free-stream velocity to zero velocity at the forward stagnation point as it travels in an adverse pressure gradient. Then, as it passes from B to C, it is in a favorable pressure gradient, and it accelerates to its highest speed. From C to D, the pressure increases again toward the rearward stagnation point, and the particle decelerates but has enough momentum to reach D. Finally, the pressure decreases from D to E, and this favorable pressure gradient accelerates the particle back to the free-stream velocity.

Pressure Distribution for a Viscous Flow

Consider the flow of a real (viscous) fluid past a cylinder, as shown in Fig. 4.37. The flow pattern upstream of the midsection is very similar to the pattern for an ideal fluid. However, in a viscous fluid the velocity at the surface is zero (no-slip condition), whereas with the flow of an inviscid fluid the surface velocity need not be zero. Because of viscous effects, a boundary layer forms next to the surface. The velocity changes from zero at the surface to the free-stream velocity across the boundary layer. Over the forward section of the cylinder, where the pressure gradient is favorable, the boundary layer is quite thin.

Downstream of the midsection, the pressure gradient is adverse, and the fluid particles in the boundary layer, slowed by viscous effects, can only go so far and then are forced to detour away from the surface. The particle is pushed off the wall by pressure force associated with the adverse pressure gradient. The point where the flow leaves the wall is called the separation point.

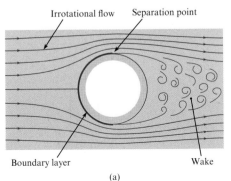

(a)

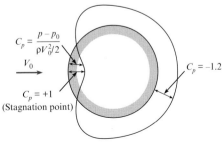

(b)

FIGURE 4.37

Flow of a real fluid past a circular cylinder: (a) flow pattern, (b) pressure distribution.

A recirculatory flow called a wake develops behind the cylinder. The flow in the wake region is called separated flow. The pressure distribution on the cylinder surface in the wake region is nearly constant, as shown in Fig. 4.37b. The reduced pressure in the wake leads to increased drag.

4.11 Calculating the Pressure Field for a Rotating Flow

This section describes how to relate pressure and velocity for a *fluid in a solid body rotation*. To understand solid body rotation, consider a cylindrical container of water (Fig. 4.38a) that is stationary. Imagine that the container is placed into rotational motion about an axis (Fig. 4.38b) and allowed to reach steady state with an angular speed of ω. At steady state, the fluid particles will be at rest with respect to each other. That is, the distance between any two fluid particles will be constant. This condition also describes the rotation of a rigid body; thus, this type of motion is defined as a **fluid in a solid body rotation**.

Situations in which a fluid rotates as a solid body are found in many engineering applications. One common application is the centrifugal separator. The centripetal accelerations resulting from rotating a fluid separate the heavier particles from the lighter particles as the heavier particles move toward the outside and the lighter particles are displaced toward the center. A milk separator operates in this fashion, as does a cyclone separator for removing particulates from an airstream.

Derivation of an Equation for a Fluid in Solid Body Rotation

To begin, apply Euler's equation in the direction normal to the streamlines and outward from the center of rotation. In this case, the fluid particles rotate as the spokes of a wheel, so the direction ℓ in Euler's equation, Eq. (4.15), is replaced by r, giving

$$-\frac{d}{dr}(p + \gamma z) = \rho a_r \qquad (4.48)$$

where the partial derivative has been replaced by an ordinary derivative because the flow is steady and a function only of the radius r. From Eq. (4.11), the acceleration in the radial direction (away from the center of curvature) is

$$a_r = -\frac{V^2}{r}$$

and Euler's equation becomes

$$-\frac{d}{dr}(p + \gamma z) = -\rho\frac{V^2}{r} \qquad (4.49)$$

For solid body rotation about a fixed axis,

$$V = \omega r$$

FIGURE 4.38

Sketch used to define a fluid in solid body rotation.

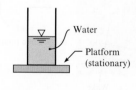

(a)

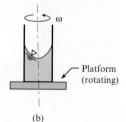

(b)

Substituting this velocity distribution into Euler's equation results in

$$\frac{d}{dr}(p + \gamma z) = \rho r \omega^2 \qquad\qquad (4.50)$$

Integrating Eq. (4.50) with respect to r gives

$$p + \gamma z = \frac{\rho r^2 \omega^2}{2} + \text{const} \qquad\qquad (4.51)$$

or

$$\frac{p}{\gamma} + z - \frac{\omega^2 r^2}{2g} = C \qquad\qquad (4.52a)$$

This equation can also be written as

$$p + \gamma z - \rho \frac{\omega^2 r^2}{2} = C \qquad\qquad (4.52b)$$

These equivalent equations describe the *pressure variation in rotating flow*. Example 4.11 shows how to apply the equations, and Example 4.12 illustrates the analysis of a rotating flow in a manometer.

EXAMPLE 4.11

Calculating the Surface Profile of a Rotating Liquid

Problem Statement

A cylindrical tank of liquid shown in the figure is rotating as a solid body at a rate of 4 rad/s. The tank diameter is 0.5 m. The line AA depicts the liquid surface before rotation, and the line $A'A'$ shows the surface profile after rotation has been established. Find the elevation difference between the liquid at the center and the wall during rotation.

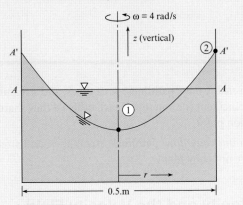

Define the Situation

A liquid is rotating in a cylindrical tank.

State the Goal

Calculate the elevation difference (in meters) between liquid at the center and at the wall.

Generate Ideas and Make a Plan

1. Apply Eq. (4.52a), between points 1 and 2.
2. Calculate the elevation difference.

Take Action (Execute the Plan)

1. Equation (4.52a):

$$\frac{p_1}{\gamma} + z_1 - \frac{\omega^2 r_1^2}{2g} = \frac{p_2}{\gamma} + z_2 - \frac{\omega^2 r_2^2}{2g}$$

The pressure at both points is atmospheric, so $p_1 = p_2$ and the pressure terms cancel out. At point 1, $r_1 = 0$, and at point 2, $r = r_2$. The equation reduces to

$$z_2 - \frac{\omega^2 r_2^2}{2g} = z_1$$

$$z_2 - z_1 = \frac{\omega^2 r_2^2}{2g}$$

2. Elevation difference:

$$z_2 - z_1 = \frac{(4 \text{ rad/s})^2 \times (0.25 \text{ m})^2}{2 \times 9.81 \text{ m/s}^2}$$

$$= \boxed{0.051 \text{ m or 5.1 cm}}$$

Review the Solution and the Process

Notice that the surface profile is parabolic.

EXAMPLE 4.12

Evaluating a Rotating Manometer Tube

Problem Statement

When the U-tube is not rotated, the water stands in the tube as shown. If the tube is rotated about the eccentric axis at a rate of 8 rad/s, what are the new levels of water in the tube?

Define the Situation

A manometer tube is rotated around an eccentric axis.

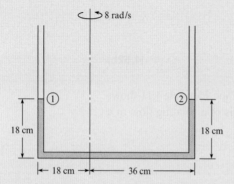

Assumptions: Liquid is incompressible.

State the Goal

Find the levels of water in each leg.

Generate Ideas and Make a Plan

The total length of the liquid in the manometer must be the same before and after rotation—namely, 90 cm. Assume, to start with, that liquid remains in the bottom leg. The pressure at the top of the liquid in each leg is atmospheric.

1. Apply the equation for pressure variation in rotating flows, Eq. (4.52a), to evaluate the difference in elevation in each leg.

2. Using the constraint of total liquid length, find the level in each leg.

Take Action (Execute the Plan)

1. Apply Eq. (4.52a) between the top of the leg on left (1) and on the right (2):

$$z_1 - \frac{r_1^2 \omega^2}{2g} = z_2 - \frac{r_2^2 \omega^2}{2g}$$

$$z_2 - z_1 = \frac{\omega^2}{2g}(r_2^2 - r_1^2)$$

$$= \frac{(8 \text{ rad/s})^2}{2 \times 9.81 \text{ m/s}^2}(0.36^2 \text{ m}^2 - 0.18^2 \text{ m}^2) = 0.317 \text{ m}$$

2. The sum of the heights in each leg is 36 cm.

$$z_2 + z_1 = 0.36 \text{ m}$$

Solution for the leg heights:

$$z_2 = 0.338 \text{ m}$$
$$z_1 = 0.022 \text{ m}$$

Review the Solution and the Process

Discussion. If the result was a negative height in one leg, it would mean that one end of the liquid column would be in the horizontal leg, and the problem would have to be reworked to reflect this configuration.

4.12 Summarizing Key Knowledge

Pathline, Streamlines, and Streaklines

- To visualize flow, engineers use the streamline, streakline, and the pathline:
  - The *streamline* is a curve that is tangent everywhere to the local velocity vector.
  - The *streamline* is a mathematical entity that cannot be observed in the physical world.
  - The configuration of streamlines in a flow field is called the *flow pattern*.
  - The *pathline* is the line (straight or curved) that a particle follows.
  - A *streakline* is the line produced by a dye or other marker fluid introduced at a point.

- In *steady flow*, pathlines, streaklines, and streamlines are coincident (i.e., on top of each other) if they share a common point.

- In *unsteady flow*, pathlines, streaklines, and streamlines are not coincident.

Velocity and the Velocity Field

- In a flowing fluid, *velocity* is defined as the speed and direction of travel of a fluid particle.

- A *velocity field* is a mathematical or graphical description that shows the velocity at each point (i.e., spatial location) within a flow.

Eulerian and Lagrangian Descriptions

There are two ways to describe motion (Lagrangian and Eulerian):

- In the *Lagrangian approach*, the engineer identifies a specified collection of matter and describes its motion. For example, when an engineer is describing the motion of a fluid particle, this is a Lagrangian-based description.
- In the *Eulerian approach*, the engineer identifies a region in space and describes the motion of matter that is passing by in terms of what is happening at various spatial locations. For example, the velocity field is an Eulerian-based concept.

- The Eulerian approach uses fields. A *field* is a mathematical or graphical description that shows how a variable is distributed spatially. A field can be a *scalar field* or a *vector field*.
- The Eulerian approach uses the divergence, gradient, and curl operators.
- The Eulerian approach uses more complicated mathematics (e.g., partial derivatives) than the Lagrangian approach.

Describing Flow

Engineers describe flowing fluids using the ideas summarized in Table 4.4.

TABLE 4.4 How Engineers Describe Flowing Fluids

Description	Key Knowledge
Engineers classify flows as *uniform* or *nonuniform*.	• Uniform and nonuniform flow describe how velocity varies spatially. • *Uniform flow* means that the velocity at each point on a given streamline is the same. Uniform flow requires rectilinear streamlines (straight and parallel). • *Nonuniform flow* means that velocity at various points on a given streamline differs.
Engineers classify flows as *steady* or *unsteady*.	• *Steady flow* means the velocity is constant with respect to time at every point in space. • *Unsteady flow* means the velocity is changing with time at some or all points in space. • Engineers often idealize unsteady flows as steady flow. For example, draining tank of water is commonly assumed to be a steady flow.
Engineers classify flows as *laminar* or *turbulent*.	• *Laminar flow* involves flow in smooth layers (laminae), with low levels of mixing between layers. • *Turbulent flow* involves flow that is dominated by eddies of various size. Flow is chaotic, unsteady, and 3-D. There are high levels of mixing. • Occasionally, engineers describe a flow as *transitional*. This means that the flow is changing from a laminar flow to a turbulent flow.
Engineers classify flows as *1-D, 2-D,* or *3-D*.	• *One-dimensional (1-D) flow* means the velocity depends on one spatial variable; for example, velocity depends on radius r only. • *Three-dimensional (3-D) flow* means the velocity depends on three spatial variables; for example, velocity depends on three position coordinates: $\mathbf{V} = \mathbf{V}(x, y, z)$.
Engineers classify flows as *viscous flow* or *inviscid flows*.	• In a *viscous flow,* the forces associated with viscous shear stresses are significant. Thus, viscous terms are included when solving the equations of motion. • In an *inviscid flow,* the forces associated with viscous shear stresses are insignificant. Thus, viscous terms are neglected when solving the equations of motion. The fluid behaves as if its viscosity were zero.
Engineers describe flows by describing an *inviscid flow region*, a *boundary layer*, and a *wake*.	• In the *inviscid flow region,* the streamlines are smooth, and the flow can be analyzed with Euler's equation. • The *boundary layer* is a thin region of fluid next to wall. Viscous effects are significant in the boundary layer. • The *wake* is the region of separated flow behind a body.
Engineers describe flows as separated or attached.	• *Flow separation* occurs when fluid particles move away from the wall. • *Attached flow* occurs when fluid particles are moving along a wall or boundary. • The region of separated flow inside a pipe or duct is often called a *recirculation zone*.

Acceleration

- *Acceleration* is a property of a fluid particle that characterizes
  - The change in speed of the particle or
  - The change in direction of travel of the particle.
- *Acceleration* is defined mathematically as the derivative of the velocity vector.
- *Acceleration* of a fluid particle can be described qualitatively. Guidelines:
  - If a particle is traveling on a curved streamline, there will be a component of acceleration that is normal to the streamline and directed inwards toward the center of curvature.
  - If the particle is changing speed, there will be a component of acceleration that is tangent to the streamline.
- In an *Eulerian representation of acceleration*
  - Terms that involve derivatives with respect to time are *local acceleration* terms and
  - All other terms are *convective acceleration* terms. Most of these terms involve derivatives with respect to position.

Euler's Equation

- *Euler's equation* is *Newton's second law of motion* applied to a fluid particle when the flow is inviscid and incompressible.
- Euler's equation can be written as a *vector equation*:

$$-\nabla p_z = \rho \mathbf{a}$$

- This vector form can be also be written as a *scalar equation* in an arbitrary ℓ direction:

$$-\frac{\partial}{\partial \ell}(p + \gamma z) = -\left(\frac{\partial p_z}{\partial \ell}\right) = \rho a_\ell$$

- *Physics of Euler's equation*: The gradient of piezometric pressure is colinear with acceleration and opposite in direction. This reveals how pressure varies:
  - When streamlines are curved, pressure will increase outward from the center of curvature.
  - When a streamline is rectilinear and a particle on the streamline is changing speed, then the pressure will change in a direction tangent to the streamline. The direction of increasing pressure is opposite the acceleration vector.
  - When streamlines are rectilinear, pressure variation normal to the streamlines is hydrostatic.

The Bernoulli Equation

- The *Bernoulli equation* is *conservation of energy* applied to a fluid particle. It is derived by integrating Euler's equation for steady, inviscid, and constant density flow.
- For the assumptions just stated, the Bernoulli equation is applied between any two points on the same streamline.
- The Bernoulli equations has two forms:
  - *Head form*: $p/\gamma + z + V^2/(2g) =$ constant
  - *Pressure form*: $p + \rho gz + (\rho V^2)/2 =$ constant
- There are two equivalent ways to describe the physics of the Bernoulli equation:
  - When speed increases, piezometric pressure decreases (along a streamline).
  - The total head (velocity head plus piezometric head) is constant along a streamline. This means that energy is conserved as a fluid particle moves along a streamline.

Measuring Velocity and Pressure

- When pressure is measured at a *pressure tap* on the wall of a pipe, this provides a measurement of static pressure. This same measurement can also be used to determine pressure head or piezometric head.
- *Static pressure* is defined as the pressure in a flowing fluid. Static pressure must be measured in a way that does not change the value of the measured pressure.
- *Kinetic pressure* is $(\rho V^2)/2$.
- A *stagnation tube* provides a measurement of (static pressure) + (kinetic pressure):

$$p + (\rho V^2)/2$$

- The *Pitot-static* tube provides a method to measure both static pressure and kinetic pressure at a point in a flowing fluid and thus provides a way to measure fluid velocity.

Fluid Rotation, Vorticity, and Irrotational Flow

- Rate of rotation Ω
  - Is a property of a fluid particle that describes how fast the particle is rotating,
  - Is defined by placing two perpendicular lines on a fluid particle and then averaging the rotational rate of these lines, and
  - Is a vector quantity with the direction of the vector given by the right-hand rule.
- A common way to describe rotation is to use the vorticity vector ω, which is twice the rotation vector: $\omega = 2\Omega$.

- In Cartesian coordinates, the vorticity is given by

$$\omega = \left(\frac{\partial w}{\partial y} - \frac{\partial v}{\partial z}\right)\mathbf{i} + \left(\frac{\partial u}{\partial z} - \frac{\partial w}{\partial x}\right)\mathbf{j} + \left(\frac{\partial v}{\partial x} - \frac{\partial u}{\partial y}\right)\mathbf{k}$$

- An irrotational flow is one in which vorticity is zero everywhere.
- When applying the Bernoulli equation for irrotational flow, one can select points 1 and 2 at any locations, not just along a streamline.

Describing the Pressure Field

- The pressure field is often described using a π-group called the pressure coefficient.
- The pressure gradient near a body is related to flow separation:
  - An adverse pressure gradient is associated with flow separation.
  - A positive pressure gradient is associated with attached flow.
- The pressure field for flow over a circular cylinder is a paradigm for understanding external flows. The pressure along the front of the cylinder is high, and the pressure in the wake is low.

- When flow is rotating as a solid body, the pressure field p can be described using

$$p + \gamma z - \rho\frac{\omega^2 r^2}{2} = C$$

where ω is the rotational speed and r is the distance from the axis of rotation to the point in the field.

Describing the Pressure Field (Summary)

Pressure variations in a flowing fluid are associated with three phenomenon:

- **Weight.** Due to the weight of a fluid, pressure increases with increasing depth (i.e., decreasing elevation). This topic is presented in Chapter 3 (Hydrostatics).
- **Acceleration.** When fluid particles are accelerating, there are usually pressure variations associated with the acceleration. In inviscid flow, the gradient of the pressure field is aligned in a direction opposite the acceleration vector.
- **Viscous effects.** When viscous effects are significant, there can be associated pressure changes. For example, there are pressure drops associated with flows in horizontal pipes and ducts. This topic is presented in Chapter 10 (Conduit Flow).

REFERENCES

1. *Flow Visualization*, Fluid Mechanics Films, downloaded 7/31/11 from http://web.mit.edu/hml/ncfmf.html.

2. Hibbeler, R.C. *Dynamics*. Englewood Cliffs, NJ: Prentice Hall, 1995.

3. *Turbulence*, Fluid Mechanics Films, downloaded 7/31/11 from http://web.mit.edu/hml/ncfmf.html.

4. *Pressure Fields and Fluid Acceleration*, Fluid Mechanics Films, downloaded 7/31/11 from http://web.mit.edu/hml/ncfmf.html.

5. Miller, R.W. (ed.) *Flow Measurement Engineering Handbook*, New York: McGraw-Hill, 1996.

6. *Vorticity, Part 1, Part 2*, Fluid Mechanics Films, downloaded 7/31/11 from http://web.mit.edu/hml/ncfmf.html.

PROBLEMS

Streamlines, Streaklines, and Pathlines (§4.1)

4.1 A windsock is a sock-shaped device attached to a swivel on top of a pole. Windsocks at airports are used by pilots to see instantaneous shifts in the direction of the wind. If one drew a line colinear with a windsock's orientation at any instant, the line would best approximate (a) a streamline, (b) a streakline, or (c) a pathline.

4.2 For streamlines, streaklines, and streamlines to all be colinear, the flow must be

  a. dividing

  b. steady

  c. stagnant

  d. a tracer

4.3 For a given hypothetical flow, the velocity from time $t = 0$ to $t = 5$ s was $u = 2$ m/s, $v = 0$. Then, from time $t = 5$ s to $t = 10$ s, the velocity was $u = +3$ m/s, $v = -4$ m/s. A dye streak was started at a point in the flow field at time $t = 0$, and the path of a particle in the fluid was also traced from that same point starting at the same time. Draw to scale the streakline, pathline of the particle, and streamlines at time $t = 10$ s.

Velocity and the Velocity Field (§4.2)

4.4 A velocity field is given mathematically as $\mathbf{V} = (3x + 2y)\mathbf{i}$. The velocity field is

  a. 1-D in x-direction

  b. 1-D in y-direction

  c. 2-D in x and y directions

The Eulerian and Lagrangian Approaches (§4.2)

4.5 Many 3-D groundwater flow problems involve the unsteady transport of fluid moving between adjacent control volumes, referred to as fluid elements. Groundwater flow equations are solved for each 3-D fluid element that is defined by (x, y, z, t) variables. Fluid flows from upstream to downstream locations because of a pressure distribution, which can be viewed as a field. Is the given approach best classified as (a) a Lagrangian approach, or (b) an Eulerian approach. What is your rationale?

Describing Flow (§4.3)

4.6 A velocity field is given by $\mathbf{V} = 10xt\mathbf{i} + 2y\mathbf{j}$. It is

a. 1-D and steady

b. 1-D and unsteady

c. 2-D and steady

d. 2-D and unsteady

4.7 Correctly match the items in column A with those in column B.

A	B
Steady flow	$\partial V_s/\partial s = 0$
Unsteady flow	$\partial V_s/\partial t = 0$
Uniform flow	$\partial V_s/\partial t \neq 0$
Nonuniform flow	$\partial V_s/\partial s \neq 0$

Acceleration (§4.4)

4.8 In a flowing fluid, acceleration means that a fluid particle is

a. changing speed

b. changing direction

c. changing both speed and direction

d. any of the above

4.9 The flow passing through a nozzle is steady. The speed of the fluid increases between the entrance and the exit of the nozzle. The acceleration halfway between the entrance and the nozzle is

a. local

b. convective

c. both

4.10 Local acceleration

a. is close to the origin

b. is always nonuniform

c. occurs in unsteady flow

4.11 The velocity along a pathline is given by V (m/s) $= s^2 t^{1/2}$ where s is in meters and t is in seconds. The radius of curvature is 0.3 m. Evaluate the acceleration tangent and normal to the path at $s = 1$ m and $t = 1$ seconds.

4.12 In this flow passage, the velocity is varying with time. The velocity varies with time at section A-A as

$$V = 6 \text{ m/s} - 2.25\frac{t}{t_0} \text{ m/s}$$

At time $t = 0.50$ s, it is known that at section A-A the velocity gradient in the s direction is $+2.4$ m/s per meter. Given that t_0 is 0.2 s and assuming quasi-one-dimensional flow, answer the following questions for time $t = 0.5$ s:

a. What is the local acceleration at A-A?

b. What is the convective acceleration at A-A?

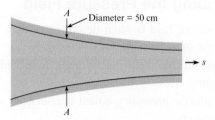

Diameter = 50 cm

Problem 4.12

4.13 The nozzle in the figure is shaped such that the velocity of the fluid varies linearly from the base of the nozzle to its tip. Assuming quasi-one-dimensional flow, what is the convective acceleration midway between the base and the tip if the velocity is 1 ft/s at the base and 6 ft/s at the tip? Nozzle length is 25 inches.

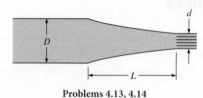

Problems 4.13, 4.14

4.14 In Prob. 4.13 the velocity varies linearly with time throughout the nozzle. The velocity at the base is $1t$ (ft/s) and at the tip is $5t$ (ft/s). What is the local acceleration midway along the nozzle when $t = 2$ s?

4.15 Liquid flows through this two-dimensional slot with a velocity of $V = 2(q_0/b)(t/t_0)$, where q_0 and t_0 are reference values. What will be the local acceleration at $x = 2B$ and $y = 0$ in terms of B, t, t_0, and q_0?

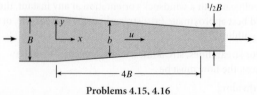

Problems 4.15, 4.16

4.16 What will be the convective acceleration for the conditions of Prob. 4.15?

Euler's Equation and Pressure Variation (§4.5)

4.17 If the piston and water ($\rho = 62.4$ lbm/ft³) are accelerated upward at a rate of $0.4g$, what will be the pressure at a depth of 2 ft in the water column?

Problem 4.17

4.18 Water ($\rho = 62.4$ lbm/ft³) stands at a depth of 10 ft in a vertical pipe that is closed at the bottom by a piston. Assuming that the vapor pressure is zero (abs), determine the maximum downward acceleration that can be given to the piston without causing the water immediately above it to vaporize.

4.19 A pipe slopes upward in the direction of liquid flow at an angle of $30°$ with the horizontal. What is the pressure gradient in the flow direction along the pipe in terms of the specific weight of the liquid, γ, if the liquid is decelerating (accelerating opposite to flow direction) at a rate of $0.2\ g$?

4.20 What pressure gradient is required to accelerate kerosene ($SG = 0.81$) vertically upward in a vertical pipe at a rate of $0.3\ g$?

4.21 The hypothetical liquid in the tube shown in the figure has zero viscosity and a specific weight of 10 kN/m³. If $p_B - p_A$ is equal to 11 kPa, one can conclude that the liquid in the tube is being accelerated (a) upward, (b) downward, or (c) neither: acceleration = 0.

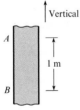

Problem 4.21

4.22 Water ($\rho = 62.4$ lbm/ft³) stands at a depth of 6 ft in a vertical pipe that is open at the top and closed at the bottom by a piston. What upward acceleration of the piston is necessary to create a pressure of 12 psig immediately above the piston?

Problem 4.22

4.23 What pressure gradient is required to accelerate water ($\rho = 998$ kg/m³) in a horizontal pipe at a rate of 9.2 m/s²?

Applying the Bernoulli Equation (§4.6)

4.24 An engineer is designing a fountain, as shown, and will install a nozzle that can produce a vertical jet. How high (h) will the water in the fountain rise if $V_n = 21$ m/s at $h = 0$?

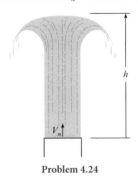

Problem 4.24

4.25 The tank shown is used to pressurize a water-fertilizer solution for delivery from a sprayer. The tank is pressurized at $p = 20$ kPa gage. Height h is 0.9 m. What is the velocity (m/s) of the fertilizer at the outlet?

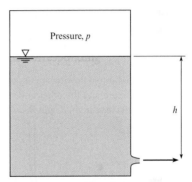

Problem 4.25

4.26 Water flows through a vertical contraction (venturi) section. Piezometers are attached to the upstream pipe and minimum area section as shown. The mean velocity in the pipe is $V = 8$ ft/s. The difference in elevation between the two water levels in the piezometers is $\Delta z = 6$ inches. The water temperature is $68°F$. What is the velocity (ft/s) at the minimum area?

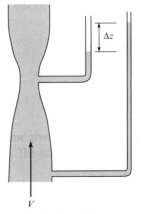

Problem 4.26

4.27 Kerosene at 20°C flows through a contraction section as shown. A pressure gage connected between the upstream pipe and throat section shows a pressure difference of 25 kPa. The gasoline velocity in the throat section is 12 m/s. What is the velocity (m/s) in the upstream pipe?

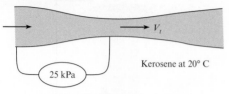

Kerosene at 20° C

25 kPa

Problem 4.27

Stagnation Tubes and Pitot-Static Tubes (§4.7)

4.28 A stagnation tube placed in a river (select all that apply)

 a. can be used to determine water velocity

 b. can be used to determine air pressure

 c. measures kinetic pressure + static pressure

4.29 A Pitot-static tube is mounted on an airplane to measure airspeed. At an altitude of 10,000 ft, where the temperature is 23°F and the pressure is 9 psia, a pressure difference corresponding to 6.5 in. of water is measured. What is the airspeed?

4.30 A Pitot tube is placed in an open channel as shown. What is the velocity V_A (m/s) when height h is 21 cm?

4.31 A glass tube is inserted into a flowing stream of water with one opening directed upstream and the other end vertical. If the water velocity, V_A is 3.2 m/s, how high, h, will the water rise?

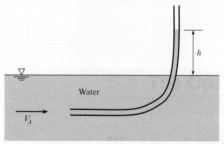

Water

V_A

h

Problems 4.30, 4.31

4.32 To measure air velocity in a food-drying plant ($T = 150°F$, $p = 14$ psia), an air-water manometer is connected to a Pitot-static tube. When the manometer deflects 3 in., what is the velocity?

4.33 A Bourdon-tube gage is taped into the center of a disk as shown. Then for a disk that is about 1 ft in diameter and for an approach velocity of air (V_0) of 40 ft/s, the gage would read a pressure intensity that is (a) less than $\rho V_0^2/2$, (b) equal to $\rho V_0^2/2$, or (c) greater than $\rho V_0^2/2$.

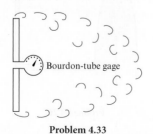

Bourdon-tube gage

Problem 4.33

4.34 A Pitot-static tube is used to measure the velocity at the center of a 8 in. pipe. If kerosene at 68°F is flowing and the deflection on a mercury-kerosene manometer connected to the Pitot tube is 3.9 in., what is the velocity?

4.35 A Pitot-static tube used to measure air velocity is connected to a pressure gage. If the air temperature is 10°C at standard atmospheric pressure at sea level, and if the gage reads a differential pressure of 3.5 kPa, what is the air velocity?

4.36 A Pitot-static tube used to measure air velocity is connected to a pressure gage. If the air temperature is 200°F at standard atmospheric pressure, and if the gage reads a differential pressure of 18 psf, what is the air velocity?

4.37 A Pitot-static tube is used to measure the gas velocity in a duct. A pressure transducer connected to the Pitot tube registers a pressure difference of 3.0 psi. The density of the gas in the duct is 0.25 lbm/ft³. What is the gas velocity in the duct?

4.38 The "spherical" Pitot probe shown is used to measure the flow velocity in hot water ($\rho = 965$ kg/m³). Pressure taps are located at the forward stagnation point and at 90° from the forward stagnation point. The speed of fluid next to the surface of the sphere varies as $1.5\,V_0 \sin \theta$, where V_0 is the free-stream velocity and θ is measured from the forward stagnation point. The pressure taps are at the same level; that is, they are in the same horizontal plane. The piezometric pressure difference between the two taps is 7 kPa. What is the free-stream velocity V_0?

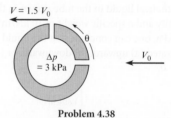

$V = 1.5\,V_0$

θ

$\Delta p = 3$ kPa

V_0

Problem 4.38

4.39 Explain how you might design a spherical Pitot-static probe to provide the direction and velocity of a flowing stream. The Pitot-static probe will be mounted on a string that can be oriented in any direction.

4.40 This navy surveillance sphere is being tested for the pressure field that will be induced in front of it as a function of velocity. Velocimeters in the test basin show that when $V_A = 16$ m/s, the velocity at B is 8 m/s and at C is 3 m/s. What is $p_B - p_C$? (Velocities are measured with respect to a stationary, i.e., lab, reference frame.)

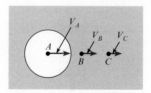

V_A

V_B

V_C

A

B

C

Problem 4.40

4.41 A rugged instrument used frequently for monitoring gas velocity in smokestacks consists of two open tubes oriented to the flow direction as shown and connected to a manometer. The pressure coefficient is 1.0 at *A* and −0.2 at *B*. Assume that water, at 20°C, is used in the manometer and that a 7 mm deflection is noted. The pressure and temperature of the stack gases are 101 kPa abs and 250°C. The gas constant of the stack gases is 229 J/kg·K. Determine the velocity of the stack gases.

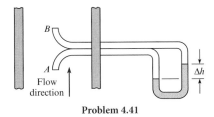

Problem 4.41

4.42 A Pitot-static tube is used to measure the airspeed of an airplane. The Pitot tube is connected to a pressure-sensing device calibrated to indicate the correct airspeed when the temperature is 17°C and the pressure is 101 kPa. The airplane flies at an altitude of 3000 m, where the pressure and temperature are 80 kPa and −8.5°C. The indicated airspeed is 56 m/s. What is the true airspeed?

4.43 A sphere moves horizontally through still water at a speed of 11 ft/s. A short distance directly ahead of the sphere (call it point *A*), the velocity, with respect to the earth, induced by the sphere is 1 ft/s in the same direction as the motion of the sphere. If p_0 is the pressure in the undisturbed water at the same depth as the center of the sphere, then the value of the ratio p_A/p_0 will be (a) less than unity, (b) equal to unity, or (c) greater than unity.

4.44 The apparatus shown in the figure is used to measure the velocity of air at the center of a duct having a 10 cm diameter. A tube mounted at the center of the duct has a 2 mm diameter and is attached to one leg of a slant-tube manometer. A pressure tap in the wall of the duct is connected to the other end of the slant-tube manometer. The well of the slant-tube manometer is sufficiently large that the elevation of the fluid in it does not change significantly when fluid moves up the leg of the manometer. The air in the duct is at a temperature of 20°C, and the pressure is 150 kPa. The manometer liquid has a specific gravity of 0.7, and the slope of the leg is 30°. When there is no flow in the duct, the liquid surface in the manometer lies at 2.3 cm on the slanted scale. When there is flow in the duct, the liquid moves up to 6.7 cm on the slanted scale. Find the velocity of the air in the duct. Assuming a uniform velocity profile in the duct, calculate the rate of flow of the air.

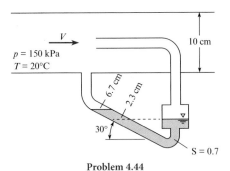

Problem 4.44

4.45 An aircraft flying at 10,000 feet uses a Pitot-static tube to measure speed. The instrumentation on the aircraft provides the differential pressure as well as the local static pressure and the local temperature. The local static pressure is 9.8 psig, and the air temperature is 25°F. The differential pressure is 0.5 psid. Find the speed of the aircraft in mph.

Characterizing Rotational Motion of a Fluid (§4.8)

4.46 The vector $\mathbf{V} = 4x\mathbf{i} - 10y\mathbf{j}$ represents a two-dimensional velocity field. Is the flow irrotational?

4.47 A two-dimensional flow field is defined by $u = x^4 - y^4$ and $v = -4xy$. Is the flow rotational or irrotational?

4.48 Fluid flows between two parallel stationary plates. The distance between the plates is 1 cm. The velocity profile between the two plates is a parabola with a maximum velocity at the centerline of 2 cm/s. The velocity is given by

$$u = 2(1 - 4y^2)$$

where *y* is measured from the centerline. The cross-flow component of velocity, *v*, is zero. There is a reference line located 1 cm downstream. Find an expression, as a function of *y*, for the amount of rotation (in radian) a fluid particle will undergo when it travels a distance of 1 cm downstream.

4.49 A combination of a forced and a free vortex is represented by the velocity distribution

$$v_\theta = \frac{1}{r}[1 - \exp(-r^2)]$$

For $r \to 0$ the velocity approaches a rigid body rotation, and as *r* becomes large, a free-vortex velocity distribution is approached. Find the amount of rotation (in radians) that a fluid particle will experience in completing one circuit around the center as a function of *r*. *Hint:* The rotation rate in a flow with concentric streamlines is given by

$$2\dot\theta = \frac{dv_\theta}{dr} + \frac{v_\theta}{r} = \frac{1}{r}\frac{d}{dr}(v_\theta r)$$

Evaluate the rotation for $r = 0.5, 1.0,$ and 1.5.

The Bernoulli Equation (Irrotational Flow) (§4.9)

4.50 Euler's equations for a planar (two-dimensional) flow in the *xy*-plane are

$$u\frac{\partial u}{\partial x} + v\frac{\partial u}{\partial y} = -g\frac{\partial h}{\partial x} \quad x = \text{direction}$$

$$u\frac{\partial v}{\partial x} + v\frac{\partial v}{\partial y} = -g\frac{\partial h}{\partial y} \quad y = \text{direction}$$

a. The slope of a streamline is given by

$$\frac{dy}{dx} = \frac{v}{u}$$

Using this relation in Euler's equation, show that

$$d\left(\frac{u^2 + v^2}{2g} + h\right) = 0$$

or

$$d\left(\frac{V^2}{2g} + h\right) = 0$$

which means that $V^2/2g + h$ is constant along a streamline.

b. For an irrotational flow,

$$\frac{\partial u}{\partial y} = \frac{\partial v}{\partial x}$$

Substituting this equation into Euler's equation, show that

$$\frac{\partial}{\partial x}\left(\frac{V^2}{2g} + h\right) = 0$$

$$\frac{\partial}{\partial y}\left(\frac{V^2}{2g} + h\right) = 0$$

which means that $V^2/2g + h$ is constant in all directions.

Pressure Field for a Circular Cylinder (§4.10)

4.51 Figure 4.36 shows irrotational flow past a circular cylinder. Assume that the approach velocity at A is constant (does not vary with time).

a. Is this a case of one-dimensional, two-dimensional, or three-dimensional flow?

b. Is the flow past the cylinder steady or unsteady?

c. Are there any regions of the flow where local acceleration is present? If so, show where they are and show vectors representing the local acceleration in the regions where it occurs.

d. Are there any regions of flow where convective acceleration is present? If so, show vectors representing the convective acceleration in the regions where it occurs.

Pressure Field for a Rotating Flow (§4.11)

4.52 Take a spoon and rapidly stir a cup of liquid. Report on the contour of the surface. Provide an explanation for the observed shape.

4.53 A U-tube is rotated at 50 rev/min about one leg. The fluid at the bottom of the U-tube has a specific gravity of 3.0. The distance between the two legs of the U-tube is 1 ft. A 6 in. height of another fluid is in the outer leg of the U-tube. Both legs are open to the atmosphere. Calculate the specific gravity of the other fluid.

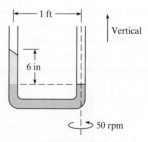

Problem 4.53

4.54 Water stands in these tubes as shown when no rotation occurs. Derive a formula for the angular speed at which water will just begin to spill out of the small tube when the entire system is rotated about axis A-A.

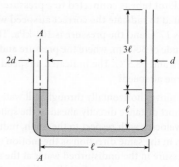

Problem 4.54

4.55 The tank shown is 4 ft in diameter and 12 ft long and is closed and filled with water ($\rho = 62.4$ lbm/ft³). It is rotated about its horizontal-centroidal axis, and the water in the tank rotates with the tank ($V = r\omega$). The maximum velocity is 25 ft/s. What is the maximum difference in pressure in the tank? Where is the point of minimum pressure?

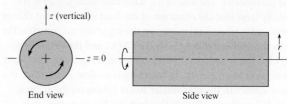

Problem 4.55

CHAPTER FIVE

The Control Volume Approach and The Continuity Equation

CHAPTER ROAD MAP This chapter describes how conservation of mass can be applied to a flowing fluid. The resulting equation is called the *continuity equation*. The continuity equation is applied to a spatial region called a control volume, which is also introduced.

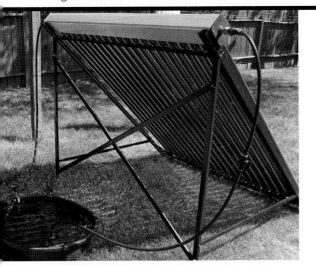

FIGURE 5.1
The photo shows an evacuated-tube solar collector that is being tested to measure its efficiency. This project was run by undergraduate engineering students. The team applied the control volume concept, the continuity equation, the flow rate equations, and knowledge from thermodynamics and heat transfer. (Photo by Donald Elger.)

LEARNING OUTCOMES

FLOW RATE (§5.1).
- Know the main ideas about mass and volume flow rate.
- Define mean velocity and know typical values.
- Solve problems that involve the flow rate equations.

THE CONTROL VOLUME APPROACH (§5.2).
- Describe the six types of systems.
- Distinguish between intensive and extensive properties.
- Explain how to use the dot product to characterize net outflow.
- Know the main ideas of the Reynolds Transport Theorem.

THE CONTINUITY EQUATION (§5.3, §5.4).
- Know the main ideas about the continuity equation.
- Solve problems that involve the continuity equation.

CAVITATION (§5.7).
- Know the main ideas about cavitation—for example, why cavitation happens, why cavitation matters, how to spot potential cavitation sites, and how to design to reduce the possibility of cavitation.

5.1 Characterizing the Rate of Flow

Engineers characterize the rate of flow using the (a) mass flow rate, $\dot{m}$, and (b) the volume flow rate, Q. These concepts and the associated equations are introduced in this section.

Volume Flow Rate (Discharge)

Volume flow rate Q is the *ratio of volume to time at an instant in time*. In equation form,

$$Q = \left(\frac{\text{volume of fluid passing through a cross-sectional area}}{\text{interval of time}} \right)_{\substack{\text{instant} \\ \text{in time}}} = \lim_{\Delta t \to 0} \frac{\Delta \forall}{\Delta t} \quad \textbf{(5.1)}$$

> **EXAMPLE.** To describe volume flow rate (Q) for a gas pump (Fig. 5.2a), select a cross-sectional area. Then, Q is the volume of gasoline that flowed across the specified section during a specified time interval (say, one second) divided by the time interval. The units could be gallons per minute or liters per second.

FIGURE 5.2

Sketches used to define volume flow rate:
(a) gasoline flowing out of a valve at a filling station,
(b) air flowing inward to a person during inhalation.

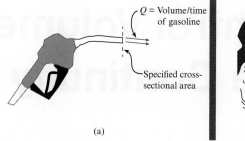

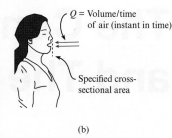

(a)   (b)

EXAMPLE. To describe volume flow rate (Q) for a person inhaling while doing yoga (Fig. 5.2b), select a cross-sectional area as shown. Then, Q is the volume of air that flowed across the specified section during a specified time interval (say, $\Delta t = 0.01$ s) divided by the time interval. Notice that the time interval should be short because the flow rate is continuously varying during breathing. The idea is to let $\Delta t \to 0$ so that the flow rate is characterized at an instant in time.

Volume flow rate is often called *discharge*. Because these two terms are synonyms, this text uses both terms interchangeably.

The SI units of discharge are cubic meters of volume per second (m³/s). In traditional units, the consistent unit is cubic feet of volume per second (ft³/s). Often, this unit is written as cfs, which stands for cubic feet per second.

Deriving Equations for Volume Flow Rate (Discharge)

This subsection shows how to derive useful equations for discharge Q in terms of fluid velocity and section area A.

To relate Q to velocity V, select a flow of fluid (Fig. 5.3) in which velocity is assumed to be constant across the pipe cross section. Suppose a marker is injected over the cross section at section A-A for a period of time Δt. The fluid that passes A-A in time Δt is represented by the marked volume. The length of the marked volume is $V\Delta t$, so the volume is $\Delta \forall = AV\Delta t$. Apply the definition of Q:

$$Q = \lim_{\Delta t \to 0} \frac{\Delta \forall}{\Delta t} = \lim_{\Delta t \to 0} \frac{AV\Delta t}{\Delta t} = VA \qquad (5.2)$$

In Eq. (5.2), notice how the units work out:

$$Q = VA$$

$$\text{flow rate (m}^3\text{/s)} = \text{velocity (m/s)} \times \text{area (m}^2\text{)}$$

FIGURE 5.3

Volume of fluid in flow, with uniform velocity distribution that passes section A-A in time Δt.

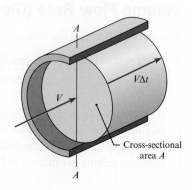

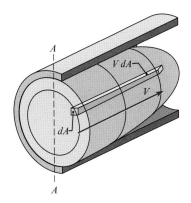

FIGURE 5.4

Volume of fluid that passes section *A-A* in time Δt.

Because Eq. (5.2) is based on a uniform velocity distribution, consider a flow in which the velocity varies across the section (see Fig. 5.4). The blue shaded region shows the volume of fluid that passes across a differential area of the section. Using the idea of Eq. (5.2), let $dQ = V\, dA$. To obtain the total flow rate, add up the volume flow rate through each differential element and then apply the definition of the integral:

$$Q = \sum_{\text{section}} V_i \, dA_i = \int_A V \, dA \qquad (5.3)$$

Eq. (5.3) means that *velocity integrated over section area gives discharge*. To develop another useful result, divide Eq. (5.3) by area A to give

$$\overline{V} = \frac{Q}{A} = \frac{1}{A} \int_A V \, dA \qquad (5.4)$$

Eq. (5.4) provides a definition of $\overline{V}$, which is called the **mean velocity**. As shown, the mean velocity is an area-weighted average velocity. For this reason, mean velocity is sometimes called *area-averaged velocity*. This label is useful for distinguishing an area-averaged velocity from a *time-averaged velocity*, which is used for characterizing turbulent flow (see §4.3). Some useful values of mean velocity are summarized in Table 5.1.

Eq. (5.4) can be generalized by using the concept of the dot product. The dot product is useful when the velocity vector is aligned at an angle with respect to the section area (Fig. 5.5). The only component of velocity that contributes to the flow through the differential area dA is the component normal to the area, V_n. The differential discharge through area dA is

$$dQ = V_n \, dA$$

TABLE 5.1 Values of Mean Velocity

Situation	Equation for Mean Velocity
Fully developed laminar flow in a round pipe. For more information, see §10.5.	$\overline{V}/V_{max} = 0.5$, where V_{max} is the value of the maximum velocity in the pipe. Note that V_{max} is the value of the velocity at the center of the pipe.
Fully developed laminar flow in a rectangular channel (channel has infinite width).	$\overline{V}/V_{max} = 2/3 = 0.667$.
Fully developed turbulent flow in a round pipe. For more information, see §10.6.	$\overline{V}/V_{max} \approx 0.79$ to 0.86, where the ratio depends on the Reynolds number.

FIGURE 5.5

Velocity vector oriented at angle θ with respect to normal.

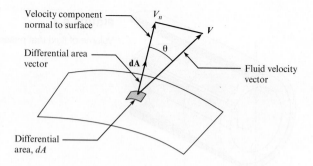

If the vector, **dA**, is defined with magnitude equal to the differential area, dA, and direction normal to the surface, then $V_n dA = |\mathbf{V}| \cos \theta \, dA = \mathbf{V} \cdot \mathbf{dA}$, where $\mathbf{V} \cdot \mathbf{dA}$ is the dot product of the two vectors. Thus, a more general equation for the discharge or volume flow rate through a surface A is

$$Q = \int_A \mathbf{V} \cdot \mathbf{dA} \tag{5.5}$$

If the velocity is constant over the area and the area is a planar surface, then the discharge is

$$Q = \mathbf{V} \cdot \mathbf{A}$$

If, in addition, the velocity and area vectors are aligned, then

$$Q = VA$$

which reverts to the original equation developed for discharge, Eq. (5.2).

Mass Flow Rate

Mass flow rate $\dot{m}$ is the *ratio of mass to time at an instant in time*. In equation form,

$$\dot{m} = \left(\frac{\text{mass of fluid passing through a cross sectional area}}{\text{interval of time}} \right)_{\substack{\text{instant} \\ \text{in time}}} = \lim_{\Delta t \to 0} \frac{\Delta m}{\Delta t} \tag{5.6}$$

The common units for mass flow rate are kg/s, lbm/s, and slugs/s.

Using the same approach as for volume flow rate, the mass of the fluid in the marked volume in Fig. 5.3 is $\Delta m = \rho \Delta \forall$, where ρ is the average density. Thus, one can derive several useful equations:

$$\dot{m} = \lim_{\Delta t \to 0} \frac{\Delta m}{\Delta t} = \rho \lim_{\Delta t \to 0} \frac{\Delta \forall}{\Delta t} = \rho Q$$
$$= \rho A V \tag{5.7}$$

The generalized form of the mass flow equation corresponding to Eq. (5.5) is

$$\dot{m} = \int_A \rho \mathbf{V} \cdot \mathbf{dA} \tag{5.8}$$

where both the velocity and fluid density can vary over the cross-sectional area. If the density is constant, then Eq. (5.7) is recovered. Also, if the velocity vector is aligned with the area vector, then Eq. (5.8) reduces to

$$\dot{m} = \int_A \rho V \, dA \tag{5.9}$$

TABLE 5.2 Summary of the Flow Rate Equations

Description	Equation	Terms
Volume flow rate equation	$Q = \overline{V}A = \dfrac{\dot{m}}{\rho} = \displaystyle\int_A V\,dA = \int_A \mathbf{V} \cdot \mathbf{dA}$ (5.10)	Q = volume flow rate = discharge (m³/s) $\overline{V}$ = mean velocity = area averaged velocity (m/s) A = cross-sectional area (m²) $\dot{m}$ = mass flow rate (kg/s) V = speed of a fluid particle (m/s) dA = differential area (m²) $\mathbf{V}$ = velocity of a fluid particle (m/s) $\mathbf{dA}$ = differential area vector (m²) (points outward from the control surface)
Mass flow rate equation	$\dot{m} = \rho A \overline{V} = \rho Q = \displaystyle\int_A \rho V\,dA = \int_A \rho \mathbf{V} \cdot \mathbf{dA}$ (5.11)	$\dot{m}$ = mass flow rate (kg/s) ρ = mass density (kg/m³)

Working Equations

Table 5.2 summarizes the flow rate equations. Notice that multiplying Eq. (5.10) by density gives Eq. (5.11).

Example Problems

For most problems, application of the flow rate equation involves substituting numbers into the appropriate equation; see Example 5.1 for this case.

EXAMPLE 5.1

Applying the Flow Rate Equations to a Flow of Air in a Pipe

Problem Statement

Air that has a mass density of 1.24 kg/m³ (0.00241 slugs/ft³) flows in a pipe with a diameter of 30 cm (0.984 ft) at a mass rate of flow of 3 kg/s (0.206 slugs/s). What are the mean velocity and discharge in this pipe for both systems of units?

Define the Situation

Air flows in a pipe.

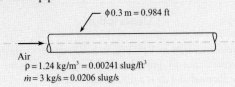

$\phi\,0.3\,\text{m} = 0.984\,\text{ft}$

Air
$\rho = 1.24\ \text{kg/m}^3 = 0.00241\ \text{slug/ft}^3$
$\dot{m} = 3\ \text{kg/s} = 0.0206\ \text{slug/s}$

State the Goal

$Q\,(\text{m}^3/\text{s and ft}^3/\text{s})$ ◄ volume flow rate (discharge)
$\overline{V}\,(\text{m/s and ft/s})$ ◄ mean velocity

Generate Ideas and Make a Plan

Because Q is the goal and $\dot{m}$ and ρ are known, apply the mass flow rate equation (Eq. 5.11):

$$\dot{m} = \rho Q \qquad \text{(a)}$$

To find the last goal ($\overline{V}$), apply the volume flow rate equation (Eq. 5.10):

$$Q = \overline{V}A \qquad \textbf{(b)}$$

The plan is as follows:

1. Calculate Q using Eq. (a).
2. Calculate $\overline{V}$ using Eq. (b).

Take Action (Execute the Plan)

1. Mass flow rate equation:

$$Q = \frac{\dot{m}}{\rho} = \frac{3\ \text{kg/s}}{1.24\ \text{kg/m}^3} = \boxed{2.42\ \text{m}^3/\text{s}}$$

$$Q = 2.42\ \text{m}^3/\text{s} \times \left(\frac{35.31\ \text{ft}^3}{1\ \text{m}^3}\right) = \boxed{85.5\ \text{cfs}}$$

2. Volume flow rate equation:

$$\overline{V} = \frac{Q}{A} = \frac{2.42\ \text{m}^3/\text{s}}{(\frac{1}{4}\pi) \times (0.30\ \text{m})^2} = \boxed{34.2\ \text{m/s}}$$

$$\overline{V} = 34.2\ \text{m/s} \times \left(\frac{1\ \text{ft}}{0.3048\ \text{m}}\right) = \boxed{112\ \text{ft/s}}$$

EXAMPLE 5.2

Calculating the Volume Flow Rate by Applying
the Dot Product

Problem Statement

Water flows in a channel that has a slope of 30°. If the velocity
is assumed to be constant, 12 m/s, and if a depth of 60 cm is
measured along a vertical line, what is the discharge per meter
of width of the channel?

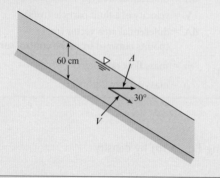

State the Goal

$Q(m^3/s)$ ⟵ discharge per meter of width of the channel

Generate Ideas and Make a Plan

Because V and A are not at right angles, apply
$Q = \mathbf{V} \cdot \mathbf{A} = VA \cos \theta$. Because all variables are known except
Q, the plan is to substitute in values.

Take Action (Execute the Plan)

$$Q = \mathbf{V} \cdot \mathbf{A} = V(\cos 30°)A$$
$$= (12 \text{ m/s})(\cos 30°)(0.6 \text{ m})$$
$$= \boxed{6.24 \text{ m}^3/\text{s per meter}}$$

Review the Solution and the Process

1. *Knowledge.* This example involves a channel flow. A flow
   is a *channel flow* when a liquid (usually water) flows
   with an open surface exposed to air under the action of
   gravity.

2. *Knowledge.* The discharge per unit width is usually
   designated as q.

Define the Situation

Water flows in an open channel.

When fluid passes across a control surface and the velocity vector is at an angle with
respect to the surface normal vector, use the dot product. This case is illustrated by Example 5.2.
Another important case is when velocity varies at different points on the control surface.
In this case, use an integral to determine flow rate, as specified by Eq. (5.10):

$$Q = \int_A V \, dA$$

In this integral, the differential area dA depends on the geometry of the problem. Two common
cases are shown in Table 5.3. Analyzing a variable velocity is illustrated by Example 5.3.

TABLE 5.3 Differential Areas for Determining Flow Rate

Label	Sketch	Description
Channel flow	*dy*, *dA = wdy*, Channel wall, *y*, *w*	When velocity varies as $V = V(y)$ in a rectangular channel, use a differential area dA, given by $dA = wdy$, where w is the width of the channel and dy is a differential height.
Pipe flow	*dA = 2πrdr*, *r*, Pipe wall	When velocity varies as $V = V(r)$ in a round pipe, use a differential area dA, given by $dA = 2\pi r dr$, where r is the radius of the differential area and dr is a differential radius.

EXAMPLE 5.3

Determining Flow Rate by Integration

Problem Statement

The water velocity in the channel shown in the accompanying figure has a velocity distribution across the vertical section equal to $u/u_{max} = (y/d)^{1/2}$. What is the discharge in the channel if the water is 2 m deep ($d = 2$ m), the channel is 5 m wide, and the maximum velocity is 3 m/s?

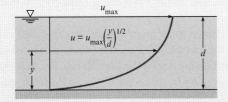

Define the Situation

Water flows in a channel.

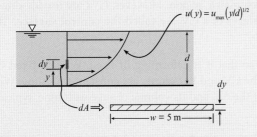

State the Goal

$Q(\text{m}^3/\text{s}) \Leftarrow$ discharge (volume flow rate)

Generate Ideas and Make a Plan

Because velocity is varying over the cross-sectional area, apply Eq. (5.10):

$$Q = \int_A V \, dA \qquad \text{(a)}$$

Because Eq. (a) has two unknowns (V and dA), find equations for these unknowns. The velocity is given:

$$V = u(y) = u_{max}(y/d)^{1/2} \qquad \text{(b)}$$

From Table 5.3, the differential area is

$$dA = w \, dy \qquad \text{(c)}$$

Notice that the differential area is sketched in the situation diagram. Substitute Eqs. (b) and (c) into Eq. (a):

$$Q = \int_0^d u_{max}(y/d)^{1/2} w \, dy \qquad \text{(d)}$$

The plan is to integrate Eq. (d) and then plug numbers in.

Take Action (Execute the Plan)

$$Q = \int_0^d u_{max}(y/d)^{1/2} w \, dy$$

$$= \frac{w u_{max}}{d^{1/2}} \int_0^d y^{1/2} \, dy$$

$$= \frac{w u_{max}}{d^{1/2}} \frac{2}{3} y^{3/2} \Big|_0^d = \frac{w u_{max}}{d^{1/2}} \frac{2}{3} d^{3/2}$$

$$= \frac{(5 \text{ m})(3 \text{ m/s})}{(2 \text{ m})^{1/2}} \times \frac{2}{3} \times (2 \text{ m})^{3/2} = \boxed{20 \text{ m}^3/\text{s}}$$

5.2 The Control Volume Approach

Engineers solve problems in fluid mechanics using the *control volume approach*. Equations for this approach are derived using the *Reynolds transport theorem*. These topics are presented in this section.

The Closed System and the Control Volume

As introduced in Section 2.1, a *system* is whatever the engineer selects for study. The *surroundings* are everything that is external to the system, and the *boundary* is the interface between the system and the surroundings. Systems can be classified into two categories: the closed system and the open system (also known as a control volume).

The **closed system** (also known as a *control mass*) is a fixed collection of matter that the engineer selects for analysis. By definition, mass cannot cross the boundary of a closed system. The boundary of a closed system can move and deform.

> **EXAMPLE.** Consider air inside a cylinder (see Fig. 5.6). If the goal is to calculate the pressure and temperature of the air during compression, then engineers select a closed system comprised of the air inside the cylinder. The system boundaries would deform as the piston moves so that the closed system always contains the same matter. This is an example of a closed system because the mass within the system is always the same.

FIGURE 5.6

Example of a closed system.

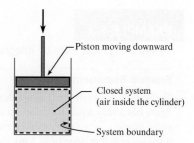

Because the closed system involves selection and analysis of a specific collection of matter, the closed system is a Lagrangian concept.

The **control volume** (CV or cv; also known as an open system) is a specified volumetric region in space that the engineer selects for analysis. The matter inside a control volume is usually changing with time because mass is flowing across the boundaries. Because the control volume involves selection and analysis of a region in space, the CV is an Eulerian concept.

EXAMPLE. Suppose water is flowing through a tank (Fig. 5.7) and the goal is to calculate the depth of water *h* as a function of time. A key to solving this problem is to select a system, and the best choice of a system is a CV surrounding the tank. Note that the CV is always three-dimensional because it is a volumetric region. However, CVs are usually drawn in two dimensions. The boundary surfaces of a CV are called the **control surface**. This is abbreviated as CS or cs.

A control volume can be defined so that it is deforming or fixed. When a **fixed CV** is defined, this means that the shape of the CV and its volume are constant with time. When a **deforming CV** is defined, the shape of the CV and its volume change with time, typically to mimic the volume of a region of fluid.

EXAMPLE. To model a rocket made from a balloon suspended on a string, one can define a deforming CV that surrounds the deflating balloon and follow the shape of the balloon during the process of deflation.

Summary. When engineers analyze a problem, they select the type of system that is most useful (see Fig. 5.8). There are two approaches. Using the *control volume approach*, the engineer selects a region in space and analyzes flow through this region. Using the *closed system approach*, the engineer selects a body of matter of fixed identity and analyzes this matter.

Table 5.4 compares the two approaches.

Intensive and Extensive Properties

Properties, which are measurable characteristics of a system, can be classified into two categories. An **extensive property** is any property that depends on the amount of matter present. An **intensive property** is any property that is independent of the amount of matter present.

FIGURE 5.7

Water entering a tank through the top and exiting through the bottom.

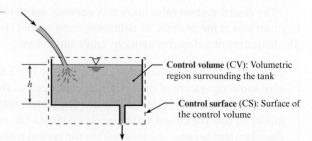

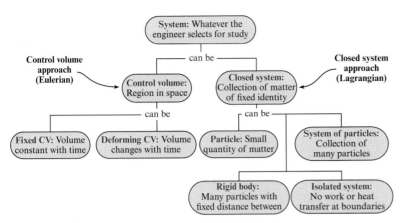

FIGURE 5.8

When engineers select a system, they choose either the *control volume approach* or the *closed system approach*. Then, they select the specific type of system from a choice of six possibilities.

Examples (extensive). Mass, momentum, energy, and weight are extensive properties because each of these properties depends on the amount of matter present. **Examples (intensive).** Pressure, temperature, and density are intensive properties because each of these properties are independent on the amount of matter present.

Many intensive properties are obtained by taking the ratio of two extensive properties. For example, density is the ratio of mass to volume. Similarly, specific energy e is the ratio of energy to mass.

To develop a general equation to relate intensive and extensive properties, define a generic extensive property, B. Also, define a corresponding intensive property b.

$$b = \left(\frac{B}{mass}\right)_{point\ in\ space}$$

The amount of extensive property B contained in a control volume at a given instant is

$$B_{cv} = \int_{cv} b\,dm = \int_{cv} b\rho\,d\forall \qquad (5.12)$$

where dm and $d\forall$ are the differential mass and differential volume, respectively, and the integral is carried out over the control volume.

TABLE 5.4 Comparison of the Control Volume and the Closed System Approaches

Feature	Closed System Approach	Control Volume Approach
Basic idea	Analyze a body or a fixed collection of matter.	Analyze a spatial region.
Lagrangian vs. Eulerian	Lagrangian approach.	Eulerian approach.
Mass crossing the boundaries	Mass cannot cross the boundaries.	Mass is allowed to cross the boundaries.
Mass (quantity)	The mass of the closed system must stay constant with time; always the same number of kilograms.	The mass of the materials inside the CV can stay constant or can change with time.
Mass (identity)	Always contains the same matter.	Can contain the same matter at all times, or the identity of the matter can vary with time.
Application	Solid mechanics, fluid mechanics, thermodynamics, and other thermal sciences.	Fluid mechanics, thermodynamics, and other thermal sciences.

FIGURE 5.9

Flow through a control volume in a duct.

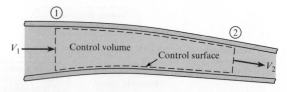

FIGURE 5.10

Control surfaces are represented by area vectors and velocities by velocity vectors.

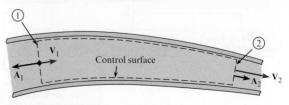

Property Transport across the Control Surface

Because a flowing fluid transports mass, momentum, and energy across a control surface, the next step is to describe this transport. Consider flow through a duct (Fig. 5.9) and assume that the velocity is uniformly distributed across the control surface. Then, the mass flow rate through each section is given by

$$\dot{m}_1 = \rho_1 A_1 V_1 \qquad \dot{m}_2 = \rho_2 A_2 V_2$$

The rate of outflow minus the rate of inflow is

$$\text{(outflow minus inflow)} = \text{(net mass outflow rate)} = \dot{m}_2 - \dot{m}_1 = \rho_2 A_2 V_2 - \rho_1 A_1 V_1$$

Next, we'll introduce velocity. The same control volume is shown in Fig. 5.10, with each control surface area represented by a vector **A** oriented outward from the control volume and with magnitude equal to the cross-sectional area. The velocity is represented by a vector **V**. Taking the dot product of the velocity and area vectors at both stations gives

$$\mathbf{V}_1 \cdot \mathbf{A}_1 = -V_1 A_1 \qquad \mathbf{V}_2 \cdot \mathbf{A}_2 = V_2 A_2$$

The negative value at station 1 occurs because the velocity and area vectors are in opposite directions. Similarly, the positive value at station 2 occurs because these vectors are in the same direction. Now, the net mass outflow rate can be written as

$$
\begin{aligned}
\text{Net mass outflow rate} &= \rho_2 V_2 A_2 - \rho_1 V_1 A_1 \\
&= \rho_2 \mathbf{V}_2 \cdot \mathbf{A}_2 + \rho_1 \mathbf{V}_1 \cdot \mathbf{A}_1 \qquad \textbf{(5.13)} \\
&= \sum_{cs} \rho \mathbf{V} \cdot \mathbf{A}
\end{aligned}
$$

Equation (5.13) states that if the dot product $\rho \mathbf{V} \cdot \mathbf{A}$ is summed for all flows into and out of the control volume, the result is the net mass flow rate out of the control volume, or the net mass efflux (*efflux* means outflow). If the summation is positive, then the net mass flow rate is out of the control volume. If it is negative, then the net mass flow rate is into the control volume. If the inflow and outflow rates are equal, then

$$\sum_{cs} \rho \mathbf{V} \cdot \mathbf{A} = 0$$

To obtain the net rate of flow of an extensive property B across a section, write

$$\underbrace{b}_{\left(\dfrac{B}{\text{mass}}\right)} \quad \underbrace{\dot{m}}_{\left(\dfrac{\text{mass}}{\text{time}}\right)} \quad = \quad \underbrace{\dot{B}}_{\left(\dfrac{B}{\text{time}}\right)}$$

Next, include all inlet and outlet ports:

$$\dot{B}_{\text{net}} = \sum_{\text{cs}} b \overbrace{\rho \mathbf{V} \cdot \mathbf{A}}^{\dot{m}}$$

(5.14)

Equation (5.14) is applicable for all flows in which the properties are uniformly distributed across the flow area. To account for property variation, replace the sum with an integral:

$$\dot{B}_{\text{net}} = \int_{\text{cs}} b \rho \mathbf{V} \cdot \mathbf{dA}$$

(5.15)

Eq. (5.15) will be used in the derivation of the Reynolds transport theorem.

The Reynolds Transport Theorem

The Reynolds transport theorem is an equation that relates a derivative term for a *closed system* to the corresponding terms for a *control volume*. The reason for the theorem is that the conservation laws of science were originally formulated for closed systems. Over time, researchers figured out how to modify the equations so that they apply to a control volume. The result is the Reynolds transport theorem.

To derive the Reynolds transport theorem, consider a flowing fluid; see Fig. 5.11. The darker shaded region is a *closed system*. As shown, the boundaries of the closed system change with time so that the system always contains the same matter. Also, define a CV as identified by the dashed line. At time t, the closed system consists of the material inside the control volume and the material going in, so the property B of the system at this time is

$$B_{\text{closed system}}(t) = B_{\text{cv}}(t) + \Delta B_{\text{in}}$$

(5.16)

At time $t + \Delta t$, the closed system has moved and now consists of the material in the control volume and the material passing out, so B of the system is

$$B_{\text{closed system}}(t + \Delta t) = B_{\text{cv}}(t + \Delta t) + \Delta B_{\text{out}}$$

(5.17)

The rate of change of the property B is

$$\frac{dB_{\text{closed system}}}{dt} = \lim_{\Delta t \to 0} \left[\frac{B_{\text{closed system}}(t + \Delta t) - B_{\text{closed system}}(t)}{\Delta t} \right]$$

(5.18)

Substituting in Eqs. (5.16) and (5.17) results in

$$\frac{dB_{\text{closed system}}}{dt} = \lim_{\Delta t \to 0} \left[\frac{B_{\text{cv}}(t + \Delta t) - B_{\text{cv}}(t) + \Delta B_{\text{out}} - \Delta B_{\text{in}}}{\Delta t} \right]$$

(5.19)

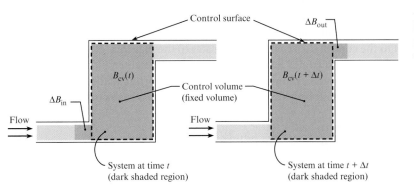

FIGURE 5.11

Progression of a closed system through a control volume.

Rearranging terms yields

$$\frac{dB_{\text{closed system}}}{dt} = \lim_{\Delta t \to 0} \left[\frac{B_{\text{cv}}(t + \Delta t) - B_{\text{cv}}(t)}{\Delta t} \right] + \lim_{\Delta t \to 0} \frac{\Delta B_{\text{out}}}{\Delta t} - \lim_{\Delta t \to 0} \frac{\Delta B_{\text{in}}}{\Delta t} \qquad \textbf{(5.20)}$$

The first term on the right side of Eq. (5.20) is the rate of change of the property B inside the control volume, or

$$\lim_{\Delta t \to 0} \left[\frac{B_{\text{cv}}(t + \Delta t) - B_{\text{cv}}(t)}{\Delta t} \right] = \frac{dB_{\text{cv}}}{dt} \qquad \textbf{(5.21)}$$

The remaining terms are

$$\lim_{\Delta t \to 0} \frac{\Delta B_{\text{out}}}{\Delta t} = \dot{B}_{\text{out}} \qquad \text{and} \qquad \lim_{\Delta t \to 0} \frac{\Delta B_{\text{in}}}{\Delta t} = \dot{B}_{\text{in}}$$

These two terms can be combined to give

$$\dot{B}_{\text{net}} = \dot{B}_{\text{out}} - \dot{B}_{\text{in}} \qquad \textbf{(5.22)}$$

or the net efflux, or net outflow rate, of the property B through the control surface. Equation (5.20) can now be written as

$$\frac{dB_{\text{closed system}}}{dt} = \frac{d}{dt} B_{\text{cv}} + \dot{B}_{\text{net}}$$

Substituting in Eq. (5.15) for $\dot{B}_{\text{net}}$ and Eq. (5.12) for B_{cv} results in the general form of the Reynolds transport theorem:

$$\underbrace{\frac{dB_{\text{closed system}}}{dt}}_{\text{Lagrangian}} = \underbrace{\frac{d}{dt} \int_{\text{cv}} b\rho \, d\Psi + \int_{\text{cs}} b\rho \mathbf{V} \cdot d\mathbf{A}}_{\text{Eulerian}} \qquad \textbf{(5.23)}$$

Eq. (5.23) may be expressed in words as

$$\left\{ \begin{array}{c} \text{Rate of change} \\ \text{of property } B \\ \text{in closed system} \end{array} \right\} = \left\{ \begin{array}{c} \text{Rate of change} \\ \text{of property } B \\ \text{in control volume} \end{array} \right\} + \left\{ \begin{array}{c} \text{Net outflow} \\ \text{of property } B \\ \text{through control surface} \end{array} \right\}$$

The left side of the equation is the Lagrangian form—that is, the rate of change of property B for the closed system. The right side is the Eulerian form—that is, the change of property B evaluated in the control volume and the flux measured at the control surface. This equation applies at the instant the system occupies the control volume and provides the connection between the Lagrangian and Eulerian descriptions of fluid flow. The velocity $\mathbf{V}$ is always measured with respect to the control surface because it relates to the mass flux across the surface.

A simplified form of the Reynolds transport theorem can be written if the mass crossing the control surface occurs through a number of inlet and outlet ports, and the velocity, density and intensive property b are uniformly distributed (constant) across each port. Then

$$\frac{dB_{\text{closed system}}}{dt} = \frac{d}{dt} \int_{\text{cv}} b\rho \, d\Psi + \sum_{\text{cs}} \rho b \mathbf{V} \cdot \mathbf{A} \qquad \textbf{(5.24)}$$

where the summation is carried out for each port crossing the control surface.

An alternative form can be written in terms of the mass flow rates:

$$\frac{dB_{\text{closed system}}}{dt} = \frac{d}{dt} \int_{cv} \rho b \, d\forall + \sum_{cs} \dot{m}_o b_o - \sum_{cs} \dot{m}_i b_i \qquad (5.25)$$

where the subscripts i and o refer to the inlet and outlet ports, respectively, located on the control surface. This form of the equation does not require that the velocity and density be uniformly distributed across each inlet and outlet port, but the property b must be.

5.3 The Continuity Equation (Theory)

The continuity equation is the law of *conservation of mass* applied to a control volume. Because this equation is commonly used by engineers, this section presents the relevant topics.

Derivation

The law of conservation of mass for a closed system can be written as

$$\frac{d(\text{mass of a closed system})}{dt} = \frac{dm_{\text{closed system}}}{dt} = 0 \qquad (5.26)$$

To transform Eq. (5.26) into an equation for a control volume, apply the Reynolds transport theorem, Eq. (5.23). In Eq. (5.23), the extensive property is $B =$ mass. The corresponding intensive property is

$$b = \frac{B}{\text{mass}} = \frac{\text{mass}}{\text{mass}} = 1.0$$

Substituting for B and b in Eq. (5.23) gives

$$\frac{dm_{\text{closed system}}}{dt} = \frac{d}{dt} \int_{cv} \rho \, d\forall + \int_{cs} \rho \mathbf{V} \cdot \mathbf{dA} \qquad (5.27)$$

Combining Eq. (5.26) to Eq. (5.27) gives the *general form of the continuity equation*.

$$\frac{d}{dt} \int_{cv} \rho \, d\forall + \int_{cs} \rho \mathbf{V} \cdot \mathbf{dA} = 0 \qquad (5.28)$$

If mass crosses the boundaries at a number of inlet and exit ports, then Eq. (5.28) reduces to give the *simplified form of the continuity equation*:

$$\frac{d}{dt} m_{cv} + \sum_{cs} \dot{m}_o - \sum_{cs} \dot{m}_i = 0 \qquad (5.29)$$

Physical Interpretation of the Continuity Equation

Fig. 5.12 shows the meaning of the terms in the continuity equation. The top row gives the general form (Eq. 5.28), and the second row gives the simplified form (Eq. 5.29). The arrows show which terms have the same conceptual meaning.

The **accumulation** term describes the changes in the quantity of mass inside the control volume (CV) with respect to time. Mass inside a CV can increase with time (accumulation is positive), decrease with time (accumulation is negative), or stay the same (accumulation is zero).

FIGURE 5.12

This figure shows the conceptual meaning of the continuity equation.

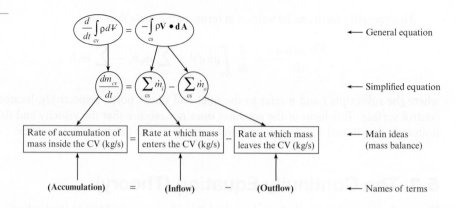

The **inflow and outflow** terms describe the rates at which mass is flowing across the surfaces of the control volume. Sometimes, inflow and outflow are combined to give **efflux**, which is defined as the net positive rate at which is mass is flowing out of a CV. That is, (efflux) = (outflow) − (inflow). When efflux is positive, there is a net flow of mass out of the CV, and accumulation is negative. When efflux is negative, accumulation is positive.

As shown in Fig. 5.12, the physics of the continuity equation can be summarized as:

$$\text{accumulation} = \text{inflow} - \text{outflow} \qquad (5.30)$$

where all terms in Eq. (5.30) are rates (see Fig. 5.12)

Eq. (5.30) is called a balance equation because the ideas relate to our everyday experiences with how things balance. For example, the accumulation of cash in a bank account equals the inflows (deposits) minus the outflows (withdrawals). Because the continuity equation is a balance equation, it is sometimes called the *mass balance equation.*

The continuity equation is applied at an instant in time and the units are kg/s. Sometimes, the continuity equation is integrated with respect to time and the units are kg. To recognize a problem that will involve integration, *look for a change in state during a time interval.*

5.4 The Continuity Equation (Application)

This section describes how to apply the continuity equation and presents example problems.

Working Equations

Three useful forms of the continuity equations are summarized in Table 5.5.

The process for applying the continuity equation is as follows:

Step 1: **Selection.** Select the continuity equation when flow rates, velocity, or mass accumulation are involved in the problem.

Step 2: **Sketching.** Select a CV by locating CSs that cut through where (a) you know information or (b) you want information. Sketch the CV and label it appropriately. Note that it is common to label the inlet port as section 1 and the outlet port as section 2.

Step 3: **Analysis.** Write the continuity equation and perform a term-by-term analysis to simplify the general equation to the reduced equation.

Step 4: **Validation.** Check units. Check the basic physics; that is, check that (inflow) − (outflow) = (accumulation).

Example Problems

The first example problem (Example 5.4) shows how continuity is applied to a problem that involves accumulation of mass.

TABLE 5.5 Summary of the Continuity Equation

Description	Equations	Terms
General form. Valid for any problem.	$$\frac{d}{dt}\int_{cv} \rho \, d\mathcal{V} + \int_{cs} \rho \mathbf{V} \cdot \mathbf{dA} = 0 \qquad \text{(Eq. 5.28)}$$	t = time (s) ρ = density (kg/m³) $d\mathcal{V}$ = differential volume (m³)
Simplified form. Useful when there are well defined inlet and exit ports.	$$\frac{d}{dt}m_{cv} + \sum_{cs} \dot{m}_o - \sum_{cs} \dot{m}_i = 0 \qquad \text{(Eq. 5.29)}$$	$\mathbf{V}$ = fluid velocity vector (m/s) (reference frame is the control surface) $\mathbf{dA}$ = differential area vector (m²) (positive direction of $\mathbf{dA}$ is outward from CS)
Pipe flow form. Valid for flow in a pipe. *For gases*: Density can vary but the density must be uniform across sections 1 and 2. *For liquids*: The equation reduces to $A_2 V_2 = A_1 V_1$ for a constant density assumption.	$$\rho_2 A_2 V_2 = \rho_1 A_1 V_1 \qquad \text{(Eq. 5.33)}$$	m_{cv} = mass inside the control volume (kg) $\dot{m} = \rho A V$ = mass/time crossing CS (kg/s) A = area of flow (m²) V = mean velocity (m/s)

EXAMPLE 5.4

Applying the Continuity Equation to a Tank with an Inflow and an Outflow

Problem Statement

A stream of water flows into an open tank. The speed of the incoming water is $V = 7$ m/s, and the section area is $A = 0.0025$ m². Water also flows out of the tank at a rate of $Q = 0.003$ m³/s. Water density is 1000 kg/m³. What is the rate at which water is being stored (or removed from) the tank?

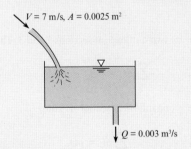

$V = 7$ m/s, $A = 0.0025$ m²

$Q = 0.003$ m³/s

Define the Situation

Water flows into a tank at the top and out at the bottom.

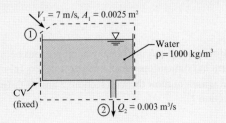

$V_1 = 7$ m/s, $A_1 = 0.0025$ m²

① Water $\rho = 1000$ kg/m³

CV (fixed)

② $Q_2 = 0.003$ m³/s

State the Goal

(dm_{cv}/dt) (kg/s) ⬅ rate of accumulation of water in tank

Generate Ideas and Make a Plan

Selection: Select the simplified form of the continuity equation (Eq. 5.29).

Sketching: Modify the situation diagram to show the CV and sections 1 and 2. Notice that the CV in the upper left corner is sketched so that it is at a right angle to the inlet flow.

Analysis: Write the continuity equation (simplified form):

$$\frac{d}{dt}m_{cv} + \sum_{cs} \dot{m}_o - \sum_{cs} \dot{m}_i = 0 \qquad \text{(a)}$$

Analyze the outflow and inflow terms:

$$\sum_{cs} \dot{m}_o = \rho Q_2 \qquad \text{(b)}$$

$$\sum_{cs} \dot{m}_i = \rho A_1 V_1 \qquad \text{(c)}$$

Combine Eqs. (a), (b), and (c):

$$\frac{d}{dt}m_{cv} = \rho A_1 V_1 - \rho Q_2 \qquad \text{(d)}$$

Validate: Each term has units of kilograms per second. Eq. (d) makes physical sense; (rate of accumulation of mass) = (rate of mass flow in) − (rate of mass flow out).

Because variables on the right side of Eq. (d) are known, the problem can be solved. The plan is as follows:

1. Calculate the flow rates on the right side of Eq. (d).

2. Apply Eq. (d) to calculate the rate of accumulation.

Take Action (Execute the Plan)

1. Mass flow rates (inlet and outlet):

$$\rho A_1 V_1 = (1000 \text{ kg/m}^3)(0.0025 \text{ m}^2)(7 \text{ m/s}) = 17.5 \text{ kg/s}$$
$$\rho Q_2 = (1000 \text{ kg/m}^3)(0.003 \text{ m}^3/\text{s}) = 3 \text{ kg/s}$$

2. Accumulation:

$$\frac{dm_{cv}}{dt} = 17.5 \text{ kg/s} - 3 \text{ kg/s}$$
$$= \boxed{14.5 \text{ kg/s}}$$

Review the Solution and the Process

1. *Discussion.* Because the accumulation is positive, the quantity of mass within the control volume is increasing with time.

2. *Discussion.* The rising level of water in the tank causes air to flow out of the CV. Because air has a density that is about 1/1000 of the density of water, this effect is negligible.

Example 5.5 shows how to solve a problem that involves accumulation by using a fixed CV.

EXAMPLE 5.5

Applying the Continuity Equation to Calculate the Rate of Water Rise in a Reservoir

Problem Statement

A river discharges into a reservoir at a rate of 400,000 ft³/s (cfs), and the outflow rate from the reservoir through the flow passages in a dam is 250,000 cfs. If the reservoir surface area is 40 mi², what is the rate of rise of water in the reservoir?

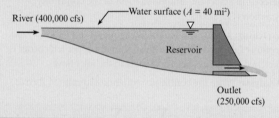

Define the Situation

A reservoir is filling with water.

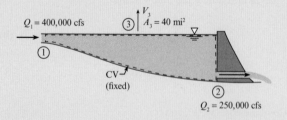

State the Goal

V_3(ft/h) ← speed at which the water surface is rising

Generate Ideas and Make a Plan

Selection: Select the continuity equation because the problem involves flow rates and accumulation of mass in a reservoir.

Sketching: Select a fixed control volume and sketch this CV on the situation diagram. The control surface at section 3 is just below the water surface and is stationary. Mass passes through control surface 3 as the water level in the reservoir rises (or falls). The mass within the control volume is constant because the volume of the CV is constant.

Analysis: Write the continuity equation (simplified form):

$$\frac{d}{dt}m_{cv} + \sum_{cs} \dot{m}_o - \sum_{cs} \dot{m}_i = 0 \qquad \text{(a)}$$

Next, analyze each term:

• Mass in the control volume is constant. Thus,

$$dm_{cv}/dt = 0. \qquad \text{(b)}$$

• There are two outflows, at sections 2 and 3. Thus,

$$\sum_{cs} \dot{m}_o = \rho Q_2 + \rho A_3 V_3 \qquad \text{(c)}$$

• There is one inflow, at section 1. Thus,

$$\sum_{cs} \dot{m}_i = \rho Q_1. \qquad \text{(d)}$$

Substitute Eqs. (b), (c), and (d) into Eq. (a). Then, divide each term by density:

$$Q_2 + A_3 V_3 = Q_1 \qquad \text{(e)}$$

Validation: Eq. (e) is dimensionally homogeneous because each term has dimensions of volume per time. Eq. (e) makes physical sense: (outflow through sections 2 and 3) = (inflow from section 1).

Because Eq. (e) contains the problem goal and all other variables are known, the problem is cracked. The plan is as follows:

1. Use Eq. (e) to derive an equation for V_3.
2. Solve for V_3.

Take Action (Execute the Plan)

1. Continuity equation:

$$V_3 = \frac{Q_1 - Q_2}{A_3}$$

2. Calculations:

$$V_{rise} = \frac{400,000 \text{ cfs} - 250,000 \text{ cfs}}{40 \text{ mi}^2 \times (5280 \text{ ft/mi})^2}$$

$$= 1.34 \times 10^{-4} \text{ ft/s} = \boxed{0.482 \text{ ft/hr}}$$

Example 5.6 shows (a) how to use a deforming CV and (b) how to integrate the continuity equation.

EXAMPLE 5.6

Applying the Continuity Equation to Predict the Time for a Tank to Drain

Problem Statement

A 10 cm jet of water issues from a 1.0 m diameter tank. Assume the Bernoulli equation applies, so the velocity in the jet is $\sqrt{2gh}$ m/s, where h is the elevation of the water surface above the outlet jet. How long will it take for the water surface in the tank to drop from $h_o = 2$ m to $h_f = 0.50$ m?

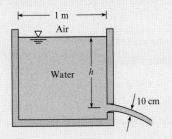

Define the Situation

Water is draining from a tank.

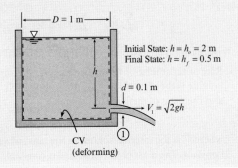

Initial State: $h = h_o = 2$ m
Final State: $h = h_f = 0.5$ m

$d = 0.1$ m

$V_1 = \sqrt{2gh}$

CV (deforming)

State the Goal

$t_f(s)$ ← time for the tank to drain from h_o to h_f

Generate Ideas and Make a Plan

Selection: Select the continuity equation by recognizing that the problem involves outflow and accumulation of mass in a tank.

Also note that the continuity equation will need to be integrated because this problem involves time and a defined initial state and final state.

Sketching: Select a deforming CV that is defined so that the top surface area is coincident with the surface level of the water. Sketch this CV in the situation diagram.

Analysis: Write the continuity equation:

$$\frac{d}{dt}m_{cv} + \sum_{cs} \dot{m}_o - \sum_{cs} \dot{m}_i = 0 \qquad \text{(a)}$$

Analyze each term in a step-by-step fashion:

- Mass in the control volume is given by*

$$m_{cv} = (\text{density})(\text{volume}) = \rho \left(\frac{\pi D^2}{4} \right) h \qquad \text{(b)}$$

- Differentiate Eq. (b) with respect to time. Note that the only variable that changes with time is water depth h, so the other variables can come out of the derivative.

$$\frac{dm_{cv}}{dt} = \frac{d}{dt} \left(\rho \left(\frac{\pi D^2}{4} \right) h \right) = \rho \left(\frac{\pi D^2}{4} \right) \frac{dh}{dt} \qquad \text{(c)}$$

- The inflow is zero and the outflow is

$$\sum_{cs} \dot{m}_o = \rho A_1 V_1 = \rho \left(\frac{\pi d^2}{4} \right) \sqrt{2gh} \qquad \text{(d)}$$

---

*The mass in the CV also includes the mass of the water below the outlet. However, when dm_{cv}/dt is evaluated, this term will go to zero.

Substitute Eqs. (b), (c), and (d) into Eq. (a):

$$\rho\left(\frac{\pi D^2}{4}\right)\frac{dh}{dt} = -\rho\left(\frac{\pi d^2}{4}\right)\sqrt{2gh} \qquad \text{(e)}$$

Validation: In Eq. (e), each term has units of kg/s. Also, this equation makes physical sense; (accumulation rate) = (the negative of the outflow rate).

Integration: To begin, simplify Eq. (e)

$$\left(\frac{D}{d}\right)^2\frac{dh}{dt} = -\sqrt{2gh} \qquad \text{(f)}$$

Next, apply the method of *separation of variables*. Put the variables involving h on the left side and the other variables on the right side. Integrate using definite integrals:

$$-\int_{h_o}^{h_f}\frac{dh}{\sqrt{2gh}} = \int_0^{t_f}\left(\frac{d}{D}\right)^2 dt \qquad \text{(g)}$$

Perform the integration to give

$$\frac{2(\sqrt{h_o} - \sqrt{h_f})}{\sqrt{2g}} = \left(\frac{d}{D}\right)^2 t_f \qquad \text{(h)}$$

Because Eq. (h) contains the problem goal (t_f) and all other variables in this equation are known, the plan is to use Eq. (h) to calculate (t_f).

Take Action (Execute the Plan)

$$t_f = \left(\frac{D}{d}\right)^2\left(\frac{2(\sqrt{h_o} - \sqrt{h_f})}{\sqrt{2g}}\right)$$

$$= \left(\frac{1\ \text{m}}{0.1\ \text{m}}\right)^2\left(\frac{2(\sqrt{2\ \text{m}} - \sqrt{0.5\ \text{m}})}{\sqrt{2(9.81\ \text{m/s}^2)}}\right)$$

$$\boxed{t_f = 31.9\ \text{s}}$$

Example 5.7 shows another instance in which the continuity equation is integrated with respect to time.

EXAMPLE 5.7

Depressurization of Gas in Tank

Problem Definition

Methane escapes through a small ($10^{-7}\ \text{m}^2$) hole in a 10 m³ tank. The methane escapes so slowly that the temperature in the tank remains constant at 23°C. The mass flow rate of methane through the hole is given by $\dot{m} = 0.66\ pA/\sqrt{RT}$, where p is the pressure in the tank, A is the area of the hole, R is the gas constant, and T is the temperature in the tank. Calculate the time required for the absolute pressure in the tank to decrease from 500 to 400 kPa.

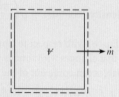

Define the Situation

Methane leaks through a $10^{-7}\ \text{m}^2$ hole in 10 m³ tank.

Assumptions:

1. Gas temperatures constant at 23°C during leakage.
2. Ideal gas law is applicable.

Properties: Table A.2: $R = 518\ \text{J/kgK}$.

State the Goal

Find: Time (in seconds) for pressure to decrease from 500 kPa to 400 kPa.

Generate Ideas and Make a Plan

Select a CV that encloses the whole tank:

1. Apply the continuity equation, Eq. (5.29).
2. Analyze term by term.
3. Solve the equation for elapsed time.
4. Calculate time.

Take Action (Execute the Plan)

1. Continuity equation:

$$\frac{d}{dt}m_{cv} + \sum_{cs}\dot{m}_o - \sum_{cs}\dot{m}_i = 0$$

2. Term-by-term analysis:
   - Rate of accumulation term. The mass in the control volume is the sum of the mass of the tank shell, M_{shell}, and the mass of methane in the tank,

$$m_{cv} = m_{shell} + \rho V$$

   where V is the internal volume of the tank, which is constant. The mass of the tank shell is constant, so

$$\frac{dm_{cv}}{dt} = V\frac{d\rho}{dt}$$

- There is no mass inflow:

$$\sum_{cs} \dot{m}_i = 0$$

- Mass out flow rate is

$$\sum_{cs} \dot{m}_o = 0.66\,\frac{pA}{\sqrt{RT}}$$

Substituting terms into the continuity equation gives

$$\forall\,\frac{d\rho}{dt} = -0.66\,\frac{pA}{\sqrt{RT}}$$

3. Equation for elapsed time:
   - Use ideal gas law for ρ:

$$\forall\,\frac{d}{dt}\left(\frac{p}{RT}\right) = -0.66\,\frac{pA}{\sqrt{RT}}$$

- Because R and T are constant,

$$\frac{dp}{dt} = -0.66\,\frac{pA\sqrt{RT}}{\forall}$$

- Next, separate variables:

$$\frac{dp}{p} = -0.66\,\frac{A\sqrt{RT}\,dt}{\forall}$$

- Integrating the equation and substituting limits for initial and final pressure gives

$$t = \frac{1.52\,\forall}{A\sqrt{RT}}\,\ln\frac{p_0}{p_f}$$

4. Elapsed time:

$$t = \frac{1.52\,(10\text{ m}^3)}{(10^{-7}\text{ m}^2)\left(518\,\dfrac{\text{J}}{\text{kg}\cdot\text{K}}\times 300\text{ K}\right)^{1/2}}\,\ln\frac{500}{400} = \boxed{8.6\times 10^4\text{ s}}$$

Review the Solution and the Process

1. *Discussion.* The time corresponds to approximately one day.

2. *Knowledge.* Because the ideal gas law is used, the pressure and temperature have to be in absolute values.

Continuity Equation for Flow in a Conduit

A conduit is a pipe or duct or channel that is completely filled with a flowing fluid. Because flow in conduits is common, it is useful to derive an equation that applies to this case. To begin the derivation, recognize that in a conduit (see Fig. 5.13), there is no place for mass to accumulate,* so Eq. (5.28) simplifies to

$$\int_{cs} \rho\mathbf{V}\cdot d\mathbf{A} = 0 \tag{5.31}$$

Mass is crossing the control surface at sections 1 and 2, so Eq. (5.31) simplifies to

$$\int_{\text{section 2}} \rho V dA - \int_{\text{section 1}} \rho V dA = 0 \tag{5.32}$$

If density is assumed to be constant across each section, Eq. (5.32) simplifies to

$$\rho_1 A_1 V_1 = \rho_2 A_2 V_2 \tag{5.33}$$

Eq. (5.33), which is called the *pipe flow form* of the continuity equation, is the final result. The meaning of this equation is (rate of inflow of mass at section 1) = (rate of outflow of mass at section 2).

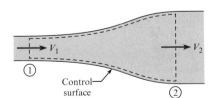

FIGURE 5.13

Flow through a conduit.

---

*The mass accumulation term in a conduit can be nonzero for some unsteady flow problems, but this is rare. This topic is left to advanced textbooks.

There are other useful ways of writing the continuity equation. For example, Eq. (5.33) can be written in several equivalent forms:

$$\rho_1 Q_1 = \rho_2 Q_2 \tag{5.34}$$

$$\dot{m}_1 = \dot{m}_2 \tag{5.35}$$

If density is assumed to be constant, then Eq. (5.34) reduces to

$$Q_2 = Q_1 \tag{5.36}$$

Eq. (5.34) is valid for both steady and unsteady incompressible flow in a pipe. If there are more than two ports and the accumulation term is zero, then Eq. (5.29) can be reduced to

$$\sum_{cs} \dot{m}_i = \sum_{cs} \dot{m}_o \tag{5.37}$$

If the flow is assumed to have constant density, Eq. (5.37) can be written in terms of discharge:

$$\sum_{cs} Q_i = \sum_{cs} Q_o \tag{5.38}$$

Summary. Depending on the assumptions of the problem, there are many ways to write the continuity equation. However, one can analyze any problem using the three equations summarized in Table 5.5. Thus, we recommend starting with one of these three equations because this is simpler than remembering many different equations.

Example 5.8 shows how to apply continuity to flow in a pipe.

EXAMPLE 5.8

Applying the Continuity Equation to Flow in a Variable Area Pipe

Problem Statement

A 120 cm pipe is in series with a 60 cm pipe. The speed of the water in the 120 cm pipe is 2 m/s. What is the water speed in the 60 cm pipe?

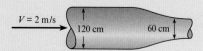

Define the Situation

Water flows through a contraction in a pipe.

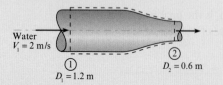

State the Goal

V_2(m/s) ← mean velocity at section 2

Generate Ideas and Make a Plan

Selection: Select the continuity equation because the problem variables are velocity and pipe diameter.

Sketching: Select a fixed CV. Sketch this CV on the situation diagram. Label the inlet as section 1 and outlet as section 2.

Analysis: Select the pipe flow form of continuity (i.e., Eq. 5.33) because the problem involves flow in a pipe:

$$\rho A_1 V_1 = \rho A_2 V_2 \tag{a}$$

Assume density is constant (this is standard practice for steady flow of a liquid). The continuity equation reduces to

$$A_1 V_1 = A_2 V_2 \tag{b}$$

Validate: To validate Eq. (b), notice that the primary dimensions of each term are L^3/T. Also, this equation makes physical sense because it can be interpreted as (inflow) = (outflow).

Plan: Eq. (b) contains the goal (V_2), and all other variables are known. Thus, the plan is to substitute numbers into this equation.

Take Action (Execute the Plan)

Continuity equation:

$$V_2 = V_1 \frac{A_1}{A_2} = V_1 \left(\frac{D_1}{D_2} \right)^2$$

$$V_2 = (2 \text{ m/s}) \left(\frac{1.2 \text{ m}}{0.6 \text{ m}} \right)^2 = \boxed{8 \text{ m/s}}$$

Example 5.9 shows how the continuity equation can be applied together with the Bernoulli equation.

EXAMPLE 5.9

Applying the Bernoulli and Continuity Equations to Flow through a Venturi

Problem Statement

Water with a density of 1000 kg/m^3 flows through a vertical venturimeter as shown. A pressure gage is connected across two taps in the pipe (station 1) and the throat (station 2). The area ratio $A_{\text{throat}}/A_{\text{pipe}}$ is 0.5. The velocity in the pipe is 10 m/s. Find the pressure difference recorded by the pressure gage. Assume the flow has a uniform velocity distribution and that viscous effects are not important.

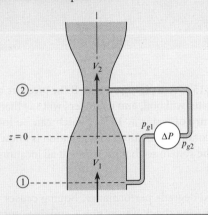

Define the Situation

Water flows in venturimeter. Area ratio = 0.5. V_1 = 10 m/s.

Assumptions:

1. Velocity distribution is uniform.
2. Viscous effects are unimportant.

Properties: $\rho = 1000$ kg/m^3.

State the Goal

Find: Pressure difference measured by gage.

Generate Ideas and Make a Plan

1. Because viscous effects are unimportant, apply the Bernoulli equation between stations 1 and 2.
2. Combine the continuity equation (5.33) with the results of step 1.
3. Find the pressure on the gage by applying the hydrostatic equation.

Take Action (Execute the Plan)

1. The Bernoulli equation:

$$p_1 + \gamma z_1 + \rho \frac{V_1^2}{2} = p_2 + \gamma z_2 + \rho \frac{V_2^2}{2}$$

Rewrite the equation in terms of piezometric pressure:

$$p_{z_1} - p_{z_2} = \frac{\rho}{2}(V_2^2 - V_1^2)$$

$$= \frac{\rho V_1^2}{2}\left(\frac{V_2^2}{V_1^2} - 1\right)$$

2. Continuity equation $V_2/V_1 = A_1/A_2$:

$$p_{z_1} - p_{z_2} = \frac{\rho V_1^2}{2}\left(\frac{A_1^2}{A_2^2} - 1\right)$$

$$= \frac{1000 \text{ kg/m}^3}{2} \times (10 \text{ m/s})^2 \times (2^2 - 1)$$

$$= 150 \text{ kPa}$$

3. Apply the hydrostatic equation between the gage attachment point where the pressure is p_{g_1} and station 1, where the gage line is tapped into the pipe:

$$p_{z_1} = p_{g_1}$$

Also, $p_{z_2} = p_{g_2}$, so

$$\Delta p_{\text{gage}} = p_{g_1} - p_{g_2} = p_{z_1} - p_{z_2} = \boxed{150 \text{ kPa}}$$

5.5 Predicting Cavitation

Designers can encounter a phenomenon called cavitation, in which a liquid starts to boil due to low pressure. This situation is beneficial for some applications, but it is usually a problem that should be avoided by thoughtful design. This section describes cavitation and discusses how to design systems to minimize the possibility of harmful cavitation.

Description of Cavitation

Cavitation takes place when fluid pressure at a given point in a system drops to the vapor pressure and boiling occurs.

FIGURE 5.14

Cavitation damage to a propeller. (Photo by Erik Axdahl.)

EXAMPLE. Consider water flowing at 15°C in a piping system. If the pressure of the water drops to the vapor pressure, then the water will boil, and engineers will say that the system is cavitating. Because the vapor pressure of water at 15°C (which can be looked up in Appendix A.5) is $p_v = 1.7$ kPa abs, the condition required for cavitation is known. To avoid cavitation, the designer can configure the system so that pressures at all locations are above 1.7 kPa absolute.

Cavitation can damage equipment and degrade performance. Boiling causes vapor bubbles to form, grow, and then collapse, producing shock waves, noise, and dynamic effects that lead to decreased equipment performance and, frequently, equipment failure. Cavitation damage to a propeller (see Fig. 5.14) occurs because the spinning propeller creates low pressures near the tips of the blades where the velocity is high. In 1983, cavitation caused major damage to spillway tunnels at the Glen Canyon Dam. Thus, engineers found a solution and implemented this solution for several dams in the U.S.; see Fig. 5.15.

Cavitation degrades materials because of the high pressures associated with the collapse of vapor bubbles. Experimental studies reveal that very high intermittent pressure, as high as 800 MPa (115,000 psi), develops in the vicinity of the bubbles when they collapse (1). Therefore, if bubbles collapse close to boundaries such as pipe walls, pump impellers, valve casings, and dam slipway floors, they can cause considerable damage. Usually this damage occurs in the form of fatigue failure brought about by the action of millions of bubbles impacting (in effect, imploding) against the material surface over a long period of time, thus producing a material pitting in the zone of cavitation.

In some applications, cavitation is beneficial. Cavitation is responsible for the effectiveness of ultrasonic cleaning. Supercavitating torpedoes have been developed in which a large bubble envelops the torpedo, significantly reducing the contact area with the water and leading to significantly faster speeds. Cavitation plays a medical role in shock wave lithotripsy for the destruction of kidney stones.

The world's largest and most technically advanced water tunnel for studying cavitation is located in Memphis, Tennessee—the William P. Morgan Large Cavitation Tunnel. This facility is used to test large-scale models of submarine systems and full-scale torpedoes as well as applications in the maritime shipping industry. More detailed discussions of cavitation can be found in Brennen (2) and Young (3).

Identifying Cavitation Sites

To predict cavitation, engineers look for locations with low pressures. For example, when water flows through a pipe restriction (Fig. 5.16), the velocity increases according to the continuity equation, and the pressure decreases in turn, as dictated by the Bernoulli equation. For low flow rates, there is a relatively small drop in pressure at the restriction, so the water remains well above the vapor pressure and boiling does not occur. However, as the flow rate increases, the pressure at the restriction becomes progressively lower until a flow rate is reached at which the pressure is equal to the vapor pressure, as shown in Fig. 5.16. At this point, the liquid boils to form bubbles and cavitation ensues. The onset of cavitation can also be affected by the presence of contaminant gases, turbulence, and viscous effects.

The formation of vapor bubbles at the restriction in Fig. 5.16 is shown in Fig. 5.17a. The vapor bubbles form and then collapse as they move into a region of higher pressure and are swept downstream with the flow. When the flow velocity is increased further, the minimum

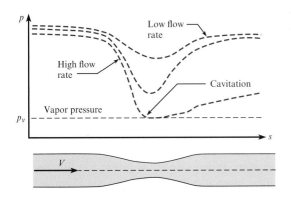

FIGURE 5.16

Flow through pipe restriction: variation of pressure for three different flow rates.

FIGURE 5.17

Formation of vapor bubbles in the process of cavitation: (a) cavitation, (b) cavitation—higher flow rate.

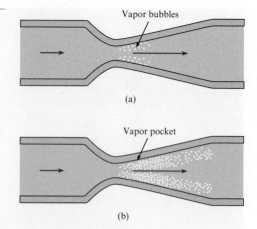

Vapor bubbles

(a)

Vapor pocket

(b)

pressure is still the local vapor pressure, but the zone of bubble formation is extended as shown in Fig. 5.17b. In this case, the entire vapor pocket may intermittently grow and collapse, producing serious vibration problems.

Summary. Cavitation, which is caused by the boiling of liquids at low pressures, is usually problematic in an engineered system. Cavitation is most likely to occur at locations with low pressures, such as the following:

- High elevation points
- Locations with high velocities (e.g., constrictions in pipes, tips of propeller blades)
- The suction (inlet) side of pumps

5.6 Summarizing Key Knowledge

Characterizing Flow Rate ($\dot{m}$ and Q)

- *Volume flow rate,* Q (m³/s), is defined by

$$Q = \left(\frac{\begin{array}{c}\text{volume of fluid passing through}\\\text{a cross-sectional area}\end{array}}{\text{interval of time}}\right)_{\substack{\text{Instant}\\\text{in time}}} = \lim_{\Delta t \to 0} \frac{\Delta V}{\Delta t}$$

- Volume flow rate is also called *discharge.*
- Q can be calculated with four equations:

$$Q = \overline{V}A = \frac{\dot{m}}{\rho} = \int_A V\, dA = \int_A \mathbf{V} \cdot \mathbf{dA}$$

- *Mass flow rate,* $\dot{m}$ (kg/s), is defined as

$$\dot{m} = \left(\frac{\begin{array}{c}\text{mass of fluid passing through}\\\text{a cross-sectional area}\end{array}}{\text{interval of time}}\right)_{\substack{\text{Instant}\\\text{in time}}} = \lim_{\Delta t \to 0} \frac{\Delta m}{\Delta t}$$

- $\dot{m}$ can be calculated with four equations:

$$\dot{m} = \rho A\overline{V} = \rho Q = \int_A \rho V\, dA = \int_A \rho \mathbf{V} \cdot \mathbf{dA}$$

- *Mean velocity,* $\overline{V}$ or V, is the value of velocity averaged over the section area at an instant in time. This concept is different than *time-averaged velocity,* which involves velocity averaged over time at a point in space.
- Typical values of mean velocity:
  - $\overline{V}/V_{\max} = 0.5$ for laminar flow in a round pipe
  - $\overline{V}/V_{\max} = 2/3 = 0.667$ for laminar flow in a rectangular conduit
  - $\overline{V}/V_{\max} \approx 0.79$ to 0.86 for turbulent flow in a round pipe
- Problems solvable with the flow rate equations can be organized into three categories:
  - *Algebraic equations.* Problems in this category are solved by straightforward application of the equations (see Example 5.1).
  - *Dot product.* When the area is not aligned with the velocity vector, apply the dot product ($\mathbf{V} \cdot \mathbf{A}$; see Example 5.2).
  - *Integration.* When velocity is given as a function of position, integrate velocity over area (see Example 5.3).

The Control Volume Approach and Reynolds Transport Theorem

- A *system* is what the engineer selects to analyze. Systems can be classified into two categories: the closed system and the control volume.

  - A *closed system* is a given quantity of matter of fixed identity. Fixed identity means the closed system is always comprised of the same matter. Thus, mass cannot cross the boundary of a closed system.

  - A *control volume* (cv or CV) is a geometric region defined in space and enclosed by a *control surface* (cs or CS).

  - The Reynolds transport theorem is a mathematical tool for converting a derivative written for a closed system to terms that apply to a control volume.

The Continuity Equation

- The *law of conservation of mass* for a control volume is called the *continuity equation*.

- The physics of the continuity equation are

$$\begin{pmatrix} \text{rate of} \\ \text{accumulation of mass} \end{pmatrix} = \begin{pmatrix} \text{rate of} \\ \text{inflow of mass} \end{pmatrix} - \begin{pmatrix} \text{rate of} \\ \text{outflow of mass} \end{pmatrix}$$

- The continuity equation can be applied at an instant in time and the units are kg/s. Also, the continuity equation

can be integrated and applied over a finite time interval (e.g., 5 minutes), in which case the units are kg.

- Three useful forms of the continuity equation (see Table 5.5) are as follows:

  - The general equation (always applies)

  - The simplified form (useful when there are well-defined inlet and outlet ports)

  - The pipe flow form (applies to flow in a conduit)

Cavitation

- Cavitation occurs in a flowing liquid when the pressure drops to the local vapor pressure of the liquid.

- Vapor pressure is discussed in Chapter 2. Data for water are presented in Table A.5.

- Cavitation is usually undesirable because it can reduce performance. Cavitation can also cause erosion or pitting of solid materials, noise, vibrations, and structural failures.

- Cavitation is most likely to occur in regions of high velocity, in inlet regions of centrifugal pumps, and at locations of high elevations.

- To reduce the probability of cavitation, designers can specify that components that are susceptible to cavitation (e.g., valves and centrifugal pumps) be situated at low elevations.

REFERENCES

1. Knapp, R. T., J. W. Daily, and F. G. Hammitt. *Cavitation*. New York: McGraw-Hill, 1970.

2. Brennen, C. E. *Cavitation and Bubble Dynamics*. New York: Oxford University Press, 1995.

3. Young, F. R. *Cavitation*. New York: McGraw-Hill, 1989.

PROBLEMS

Characterizing Flow Rates (§5.1)

5.1 Consider filling the gasoline tank of an automobile at a gas station. (a) Estimate the discharge in gpm. (b) Using the same nozzle, estimate the time to put 50 gallons in the tank. (c) Estimate the cross-sectional area of the nozzle and calculate the velocity at the nozzle exit.

5.2 The discharge of water in a 28-cm-diameter pipe is $0.07 \ \text{m}^3/\text{s}$. What is the mean velocity?

5.3 A pipe with a 15 in. diameter carries water having a velocity of 5 ft/s. What is the discharge in cubic feet per second and in gallons per minute (1 cfs equals 449 gpm)?

5.4 Natural gas (methane) flows at 20 m/s through a pipe with a 0.7 m diameter. The temperature of the methane is 15°C, and the pressure is 200 kPa gage. Determine the mass flow rate.

5.5 An aircraft engine test pipe is capable of providing a flow rate of 180 kg/s at altitude conditions corresponding to an absolute pressure of 50 kPa and a temperature of −18°C. The velocity of air through the duct attached to the engine is 255 m/s. Calculate the diameter of the duct.

5.6 Water flows in a pipe that has a 3 ft diameter and the following hypothetical velocity distribution: The velocity is maximum at the centerline and decreases linearly with r to a minimum at the pipe wall. If $V_{max} = 15$ ft/s and $V_{min} = 9$ ft/s, what is the discharge in cubic feet per second and in gallons per minute?

5.7 Air enters this square duct at section 1 with the velocity distribution as shown. Note that the velocity varies in the y direction only (for a given value of y, the velocity is the same for all values of z).

  a. What is the volume rate of flow?

  b. What is the mean velocity in the duct?

  c. What is the mass rate of flow if the mass density of the air is 1.1 kg/m³?

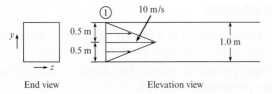

Problem 5.7

5.8 The velocity at section A-A is 15 ft/s, and the vertical depth y at the same section is 3 ft. If the width of the channel is 30 ft, what is the discharge in cubic feet per second?

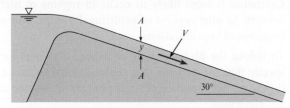

Problem 5.8

5.9 Water from a pipe is diverted into a weigh tank for exactly 1 min. The increased weight in the tank is 60 kN. What is the discharge in cubic meters per second? Assume $T = 20°C$.

5.10 Engineers are developing a new design of a jet engine for an unmanned aerial vehicle (i.e., a drone). During testing, to simulate flight, air is supplied to the jet engine through an attached round duct. In the duct, just upstream of the jet engine, the following specifications are required: $V = 200$ m/s, $\dot{m} = 415$ kg/s, $p = 60$ kPa abs, and $T = -17°C$. What diameter pipe is needed to meet these specs?

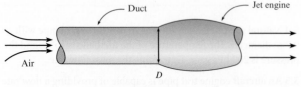

Problem 5.10

5.11 An empirical equation for the velocity distribution in a horizontal, rectangular, open channel is given by $u = u_{max} (y/d)^n$, where u is the velocity at a distance y meters above the floor of the channel. If the depth d of flow is 1.4 m, $u_{max} = 5$ m/s, and $n = 1/6$, what is the discharge in cubic meters per second per meter of width of channel? What is the mean velocity?

5.12 Plot the velocity distribution across the pipe, and determine the discharge of a fluid flowing through a pipe 1 m in diameter that has a velocity distribution given by $V = 12(1 - r^2/r_0^2)$ m/s. Here r_0 is the radius of the pipe, and r is the radial distance from the centerline. What is the mean velocity?

5.13 Water flows through a 2.0 in. diameter pipeline at 100 lbm/min. Calculate the mean velocity. Assume $T = 60°F$.

5.14 The cross section of a heat exchanger consists of three circular pipes inside a larger pipe. The internal diameter of the three smaller pipes is $D_S = 2.0$ cm, and their pipe wall thicknesses are each 3 mm. The inside diameter of the larger pipe is $D_L = 11$ cm. If the mean velocity of the fluid in the region between the smaller pipes and larger pipe is 18 m/s, what is the corresponding discharge in m³/s?

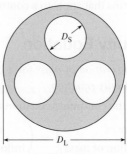

Problem 5.14

5.15 The mean velocity of water in a 6-in. pipe is 10 ft/s. Determine the flow rate in slugs per second, gallons per minute, and cubic feet per second if $T = 60°F$.

Lagrangian and Eulerian Approaches (§5.2)

5.16 State whether each of the following quantities is extensive or intensive:

  a. density

  b. mass

  c. volume

  d. energy

  e. specific energy

The Control Volume Approach (§5.2)

5.17 For cases a and b shown in the figure, respond to the following questions and statements concerning the application of the Reynolds transport theorem to the continuity equation.

  a. What is the value of b?

  b. Determine the value of dB_{sys}/dt.

  c. Determine the value of $\sum_{cs} b\rho \mathbf{V} \cdot \mathbf{A}$.

  d. Determine the value of $d/dt \int_{cv} b\rho \, d\forall$.

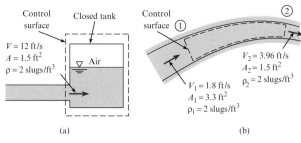

(a)   (b)

Problem 5.17

The Continuity Equation (Theory) (§5.3)

5.18 The continuity equation is sometimes called:

   a. the conservative equation

   b. the mass balance equation

Applying the Continuity Equation (§5.4)

5.19 Two pipes are connected together in series. The diameter of one pipe is three (3) times the diameter of the second pipe. With liquid flowing in the pipes, the velocity in the large pipe is 6 m/s. What is the velocity in the smaller pipe?

5.20 Two streams discharge into a pipe as shown. The flows are incompressible. The volume flow rate of stream A into the pipe is given by $Q_A = 0.01t$ m³/s and that of stream B by $Q_B = 0.006\ t^2$ m³/s, where t is in seconds. The exit area of the pipe is 0.01 m². Find the velocity and acceleration of the fluid at the exit at $t = 2$ s.

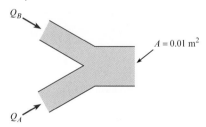

Problem 5.20

5.21 During the production of biodiesel, glycerin is a waste product and is collected by gravity separation because its SG (1.26) is much higher than that of biodiesel. Glycerin is withdrawn from the bottom of the separation tank through a pipe. At section 1, the pressure is $p_1 = 320$ kPa gage, the diameter is $D = 80$ cm, and the flow rate is 700 L/s. Find the pressure at section 2, which is 0.8 m higher and where the diameter has been reduced to $d = 35$ cm. Assume that the flow is inviscid.

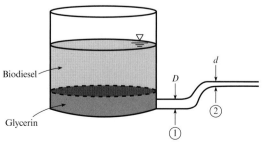

Problem 5.21

5.22 In a food-drying facility, a constant-diameter heated pipe is used to raise the temperature of air. At the pipe entrance, the velocity is 10 m/s, the pressure is 100 kPa absolute, and the temperature is 20°C. At the outlet, the pressure is 78 kPa absolute and the temperature is 80°C. What is the velocity at the outlet? Can the Bernoulli equation be used to relate the pressure and velocity changes? Explain.

5.23 Two parallel disks of diameter D are brought together, each with a normal speed of V. When their spacing is h, what is the radial component of convective acceleration at the section just inside the edge of the disk (section A) in terms of V, h, and D? Assume uniform velocity distribution across the section.

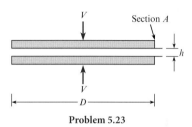

Problem 5.23

5.24 A 6-in.-diameter cylinder falls at a rate of 4 ft/s in an 8-in.-diameter tube containing an incompressible liquid. What is the mean velocity of the liquid (with respect to the tube) in the space between the cylinder and the tube wall?

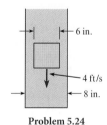

Problem 5.24

5.25 A sphere 8 inches in diameter falls at 4 ft/s downward axially through water in a 1-ft-diameter container. Find the upward speed of the water with respect to the container wall at the midsection of the sphere.

5.26 A mechanical pump is used to pressurize a bicycle tire. The inflow to the pump is 0.4 cfm. The density of the air entering the pump is 0.075 lbm/ft³. The inflated volume of a bicycle tire is 0.050 ft³. The density of air in the inflated tire is 0.5 lbm/ft³. How many seconds does it take to pressurize the tire if there initially was no air in the tire?

5.27 A 30 cm pipe divides into a 20 cm branch and a 10 cm branch. If the total discharge is 0.40 m³/s and if the same mean velocity occurs in each branch, what is the discharge in each branch?

5.28 Water flows in a 11 in. pipe that is connected in series with a 8 in. pipe. If the rate of flow is 950 gpm (gallons per minute), what is the mean velocity in each pipe?

5.29 For a steady flow of gas in the conduit shown, what is the mean velocity at section 2?

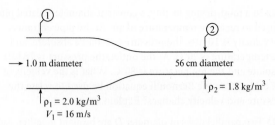

Problem 5.29

5.30 A lake with no outlet is fed by a river with a constant flow of 1700 ft^3/s. Water evaporates from the surface at a constant rate of 11 ft^3/s per square mile surface area. The area varies with depth h (feet) as A (square miles) = 4.5 + 5.5h. What is the equilibrium depth of the lake? Below what river discharge will the lake dry up?

5.31 Oxygen and methane are mixed at 204 kPa absolute pressure and 100°C. The velocity of the gases into the mixer is 6 m/s. The density of the gas leaving the mixer is 2.0 kg/m^3. Determine the exit velocity of the gas mixture.

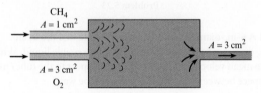

Problem 5.31

5.32 A pipe with a series of holes, as shown in the figure, is used in many engineering systems to distribute gas into a system. The volume flow rate through each hole depends on the pressure difference across the hole and is given by

$$Q = 0.67 \, A_o \left(\frac{2 \Delta p}{\rho} \right)^{1/2}$$

where A_o is the area of the hole, Δp is the pressure difference across the hole, and ρ is the density of the gas in the pipe. If the pipe is sufficiently large, the pressure will be uniform along the pipe. A distribution pipe for air at 20° C is 0.5 meters in diameter and 8 m long. The gage pressure in the pipe is 100 Pa. The pressure outside the pipe is atmospheric at 1 bar. The hole diameter is 2.5 cm, and there are 35 holes per meter length of pipe. The pressure is constant in the pipe. Find the velocity of the air entering the pipe.

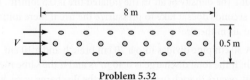

Problem 5.32

5.33 A compressor supplies gas to a 5 m^3 tank. The inlet mass flow rate is given by $\dot{m} = 0.5 \, \rho_0/\rho$ (kg/s), where ρ is the density in the tank and ρ_0 is the initial density. Find the time it would take to increase the density in the tank by a factor of 2 if the initial density is 1.8 kg/m^3. Assume the density is uniform throughout the tank.

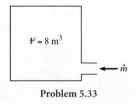

Problem 5.33

5.34 A slow leak develops in a tire (assume constant volume), in which it takes 3 hr for the pressure to decrease from 30 psig to 25 psig. The air volume in the tire is 0.5 ft^3, and the temperature remains constant at 60°F. The mass flow rate of air is given by $\dot{m} = 0.68 \, pA/\sqrt{RT}$. Calculate the area of the hole in the tire. Atmospheric pressure is 14 psia.

5.35 A cylindrical drum of water, lying on its side, is being emptied through a 2 in.–diameter short pipe at the bottom of the drum. The velocity of the water out of the pipe is $V = \sqrt{2gh}$, where g is the acceleration due to gravity and h is the height of the water surface above the outlet of the tank. The tank is 4 ft long and 2 ft in diameter. Initially the tank is half full. Find the time for the tank to empty.

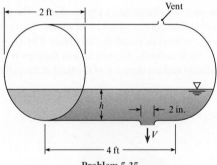

Problem 5.35

5.36 For the type of tank shown, the tank diameter is given as $D = d + C_1 h$, where d is the bottom diameter and C_1 is a constant. Derive a formula for the time of fall of liquid surface from $h = h_0$ to $h = h$ in terms of d_j, d, h_0, h, and C_1. Solve for t if $h_0 = 1$ m, $h = 20$ cm, $d = 20$ cm, $C_1 = 0.3$, and $d_j = 5$ cm. The velocity of water in the liquid jet exiting the tank is $V_e = \sqrt{2gh}$.

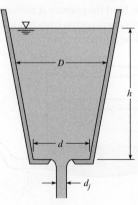

Problem 5.36

5.37 A tank containing oil is to be pressurized to decrease the draining time. The tank, shown in the figure, is 2 m in diameter and 6 m high. The oil is originally at a level of 5 m. The oil has a density of 880 kg/m^3. The outlet port has a diameter of 2 cm, and the velocity at the outlet is given by

$$V_e = \sqrt{2gh + \frac{2p}{\rho}}$$

where p is the gage pressure in the tank, ρ is the density of the oil, and h is the elevation of the surface above the hole. Assume during the emptying operation that the temperature of the air in the tank is constant. The pressure will vary as

$$p = (p_0 + p_{atm})\frac{(L - h_0)}{(L - h)} - p_{atm}$$

where L is the height of the tank, p_{atm} is the atmospheric pressure, and the subscript 0 refers to the initial conditions. The initial pressure in the tank is 300 kPa gage, and the atmospheric pressure is 100 kPa.

Applying the continuity equation to this problem, one finds

$$\frac{dh}{dt} = -\frac{A_e}{A_T}\sqrt{2gh + \frac{2p}{\rho}}$$

Integrate this equation to predict the depth of the oil with time for a period of one hour.

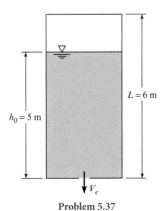

Problem 5.37

5.38 Water is draining from a pressurized tank as shown in the figure. The exit velocity is given by

$$V_e = \sqrt{\frac{2p}{\rho} + 2gh}$$

where p is the pressure in the tank, ρ is the water density, and h is the elevation of the water surface above the outlet. The depth of the water in the tank is 2.2 m. The tank has a cross-sectional area of 1.2 m^2, and the exit area of the pipe is 9 cm^2. The pressure in the tank is maintained at 10 kPa. Find the time required to empty the tank. Compare this value with the time required if the tank is not pressurized.

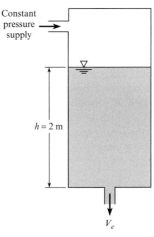

Problem 5.38

5.39 Gas is flowing from section 1 to 2 in the pipe expansion shown. The inlet density, diameter and velocity are ρ_1, D_1, and V_1 respectively. If D_2 is $2D_1$ and V_2 is half of V_1, what is the magnitude of ρ_2?

    a. $\rho_2 = 4\,\rho_1$

    b. $\rho_2 = 2\,\rho_1$

    c. $\rho_2 = \tfrac{1}{2}\,\rho_1$

    d. $\rho_2 = \rho_1$

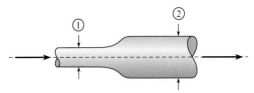

Problem 5.39

5.40 Air with a density of 0.06 lbm/ft^3 is flowing upward in the vertical duct, as shown. The velocity at the inlet (station 1) is 70 ft/s, and the area ratio between stations 1 and 2 is 0.3 ($A_2/A_1 = 0.3$). Two pressure taps, 10 ft apart, are connected to a manometer, as shown. The specific weight of the manometer liquid is 120 lbf/ft^3. Find the deflection, Δh, of the manometer.

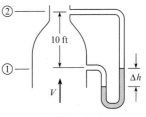

Problem 5.40

5.41 The piston shown is moving up during the exhaust stroke of a four-cycle engine. Mass escapes through the exhaust port at a rate given by

$$\dot{m} = 0.65\frac{p_c A_v}{\sqrt{RT_c}}$$

where p_c and T_c are the cylinder pressure and temperature, A_v is the valve opening area, and R is the gas constant of the exhaust gases. The bore of the cylinder is 10 cm, and the piston is moving upward at 30 m/s. The distance between the piston and the head is 10 cm. The valve opening area is 1 cm², the chamber pressure is 300 kPa abs, the chamber temperature is 600°C, and the gas constant is 350 J/kg K. Applying the continuity equation, determine the rate at which the gas density is changing in the cylinder. Assume the density and pressure are uniform in the cylinder and the gas is ideal.

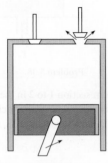

Problem 5.41

5.42 Water flows through a rigid contraction section of circular pipe in which the outlet diameter is one-half the inlet diameter. The velocity of the water at the inlet varies with time as $V_{in} = (10 \text{ m/s})[1 - \exp(-t/10)]$. How will the velocity vary with time at the outlet?

5.43 Air flows through a constant-area heated pipe. At the entrance, the velocity is 10 m/s, the pressure is 100 kPa absolute, and the temperature is 20°C. At the outlet, the pressure is 80 kPa absolute, and the temperature is 50°C. What is the velocity at the outlet? Can the Bernoulli equation be used to relate the pressure and velocity changes? Explain.

Predicting Cavitation (§5.5)

5.44 When the hydrofoil shown was tested, the minimum pressure on the surface of the foil was found to be 2.7 psi vacuum when the foil was submerged 3.5 ft and towed at a speed of 15 ft/s. At the same depth, at what speed will cavitation first occur? Assume irrotational flow and $T = 50°F$.

5.45 For the conditions of Prob. 5.44 , at what speed will cavitation begin if the depth is increased to 10 ft?

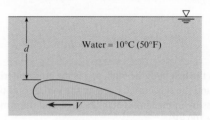

Problems 5.44, 5.45

5.46 A sphere is moving in water at a depth where the absolute pressure is 18 psia. The maximum velocity on a sphere occurs 90° from the forward stagnation point and is 1.5 times the free-stream velocity. The density of water is 62.4 lbm/ft³. Calculate the speed of the sphere at which cavitation will occur. $T = 50°F$.

5.47 The minimum pressure on a cylinder moving horizontally in water ($T = 10°C$) at 5 m/s at a depth of 1 m is 80 kPa absolute. At what velocity will cavitation begin? Atmospheric pressure is 100 kPa absolute.

CHAPTER SIX

The Momentum Equation

CHAPTER ROAD MAP This chapter presents (a) the linear momentum equation and (b) the angular momentum equation. Both equations are derived from Newton's second law of motion.

FIGURE 6.1
Engineers design systems by using a small set of fundamental equations, such as the momentum equation. (Photo courtesy of NASA.)

LEARNING OUTCOMES

NEWTON'S SECOND LAW (§6.1).
- Know the main ideas about Newton's second law of motion.
- Solve problems that involve Newton's second law by applying the visual solution method.

THE LINEAR MOMENTUM EQUATION (§6.2 to §6.4).
- List the steps to derive the momentum equation and explain the physics.
- Draw a force diagram and a momentum diagram.
- Explain or calculate the momentum flow.
- Apply the linear momentum equation to solve problems.

MOVING CONTROL VOLUMES (§6.5).
- Distinguish between an inertial and noninertial reference frame.
- Solve problems that involve moving control volumes.

6.1 Understanding Newton's Second Law of Motion

Because Newton's second law is the theoretical foundation of the momentum equation, this section reviews relevant concepts.

Body and Surface Forces

A **force** is an interaction between two bodies that can be idealized as a push or pull of one body on another body. A push/pull interaction is one that can cause acceleration.

Newton's third law tells us that forces must involve the interaction of *two bodies* and that *forces occur in pairs*. The two forces are equal in magnitude, opposite in direction, and colinear.

> **EXAMPLE.** To give examples of force, consider an airplane that is flying in a straight path at constant speed (Fig. 6.2). Select the airplane as the *system* for analysis. Idealize the airplane as a *particle*. Newton's first law (i.e., force equilibrium) tells us that the sum of forces must balance. There are four forces on the airplane:
>
> - The *lift force* is the net upward push of the air (body 1) on the airplane (body 2).
>
> - The *weight* is the pull of the earth (body 1) on the airplane (body 2) through the action of gravity.

171

FIGURE 6.2

When an airplane is flying in straight and level flight, the forces sum to zero.

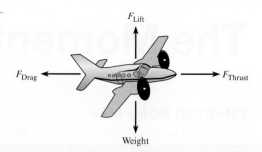

- The *drag force* is the net resistive force of the air (body 1) on the airplane (body 2).
- The *thrust force* is the net horizontal push of the air (body 1) on the surfaces of the propeller (body 2).

Notice that each of the four interactions just described can be classified as a force because (a) they involve a push or pull and (b) they involve the interaction of two bodies of matter.

Forces can be classified into two categories: body force and surface force. A **surface force** (also known as a contact force) is a force that requires physical contact or touching between the two interacting bodies. The lift force (Fig. 6.2) is a surface force because the air (body 1) must touch the wing (body 2) to create the lift force. Similarly, the thrust and drag forces are surface forces.

A **body force** is a force that can act without physical contact. For example, the weight force is a body force because the airplane (body 1) does not need to touch the earth (body 2) for the weight force to act.

A body force acts on every particle within a system. In contrast, a surface force acts only on the particles that are in physical contact with the other interacting body. For example, consider a system comprised of a glass of water sitting on a table. The weight force is pulling on every particle within the system, and we represent this force as a vector that passes through the center of gravity of the system. In contrast, the normal force on the bottom of the cup acts only on the particles of glass that are touching the table.

Summary. Forces can be classified into two categories: body forces and surface forces (see Fig. 6.3). Most forces are surface forces.

Newton's Second Law of Motion

In words, Newton's second law is this: *The sum of forces on a particle is proportional to the acceleration, and the constant of proportionality is the mass of the particle.* Notice that this law

FIGURE 6.3

Forces can be classified as body forces or surface forces.

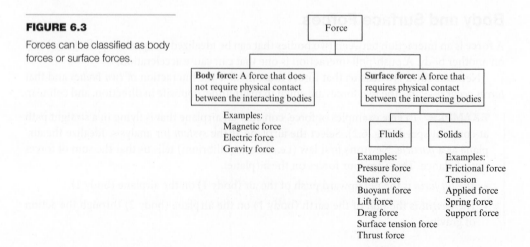

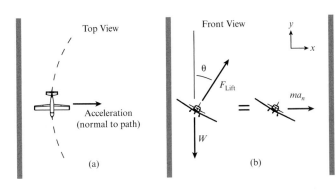
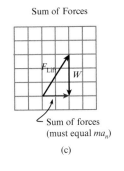

FIGURE 6.4

An airplane flying with a steady speed on curved path in a horizontal plane: (a) top view, (b) front view, (c) a sketch showing how the $\Sigma \mathbf{F}$ vector balances the $m\mathbf{a}$ vector.

applies only to a particle. The second law asserts that *acceleration and unbalanced forces are proportional*. This means, for example, that

- If a particle is accelerating, then the sum of forces on the particle is nonzero, and
- If the sum of forces on a particle is nonzero, then the particle will be accelerating.

Newton's second law can be written as an equation:

$$\left(\sum \mathbf{F} \right)_{ext} = m\mathbf{a} \tag{6.1}$$

where the subscript "ext" is a reminder to sum only external forces.

EXAMPLE. To illustrate the relationship between unbalanced forces and acceleration, consider an airplane that is turning left while flying at a constant speed in a horizontal plane (Fig. 6.4a). Select the airplane as a *system*. Idealize the airplane as a *particle*. Because the airplane is traveling in a circular path at constant speed, the acceleration vector must point inward. Fig. 6.4b shows the vectors that appear in Newton's second law. For Newton's second law of motion to be satisfied, the sum of the force vectors (Fig. 6.4c) must be equal to the $m\mathbf{a}$ vector.

The airplane example illustrates a method for visualizing and solving a vector equation called the *Visual Solution Method* (VSM). This method was adapted from Hibbeler (1) and is presented in the next subsection.

Solving a Vector Equation with the Visual Solution Method (VSM)

The VSM is an approach for solving a vector equation that reveals the physics while also showing visually how the equation can be solved. Thus, the VSM simplifies problem solving. The VSM has three steps:

Step 1. Identify the vector equation in its general form.

Step 2. Draw a diagram that shows the vectors that appear in the left side of the equation. Then, draw a second diagram that shows the vectors that appear on the right side of the equation. Add an equal sign between the diagrams.

Step 3. From the diagrams, apply the general equation and simplify the results to create the reduced equation(s). The reduced equation(s) can be written as a vector equation or as one or more scaler equations.

EXAMPLE. This example shows how to apply the VSM to the airplane problem (see Fig. 6.4).

Step 1. The general equation is Newton's second law $(\Sigma \mathbf{F})_{\text{ext}} = m\mathbf{a}$.

Step 2. The two diagrams separated by an equal sign are shown in Fig. 6.4b.

Step 3. By looking at the diagrams, one can write the reduced equation using scalar equations:

$$(x \text{ direction}) \qquad F_{\text{lift}} \sin \theta = ma_n$$

$$(y \text{ direction}) \qquad -W + F_{\text{lift}} \cos \theta = 0$$

Alternatively, you can look at the diagrams and then write the reduced equation using a vector equation:

$$F_{\text{Lift}}(\sin \theta \mathbf{i} + \cos \theta \mathbf{j}) - W\mathbf{j} = (ma_n)\mathbf{i}$$

EXAMPLE. This example shows how to apply the VSM to a generic vector equation.

Step 1. Suppose the general equation is $\Sigma \mathbf{x} = \mathbf{y}_2 - \mathbf{y}_1$.

Step 2. Suppose the vectors are known. Then, sketch the diagrams (Fig. 6.5).

Step 3. By looking at the diagrams, write the reduced equations. To get the signs correct, notice that the general equation shows that vector $\mathbf{y}_1$ is subtracted. The reduced equations are

$$(x \text{ direction}) \qquad x_2 + x_3 - x_4 \cos 30° = y_2 \cos 30° - y_1$$

$$(y \text{ direction}) \qquad x_1 + x_4 \sin 30° = -y_2 \sin 30°$$

Newton's Second Law (System of Particles)

Newton's second law (Eq. 6.1) applies to one particle. Because a flowing fluid involves many particles, the next step is to modify the second law so that it applies to a system of particles. To begin the derivation, note that the mass of a particle must be constant. Then, modify Eq. (6.1) to give

$$\left(\sum \mathbf{F}\right)_{\text{ext}} = \frac{d(m\mathbf{v})}{dt} \tag{6.2}$$

where $m\mathbf{v}$ is the momentum of one particle.

To extend Eq. (6.2) to multiple particles, apply Newton's second law to each particle, and then add the equations together. Internal forces, which are defined as forces between the particles of the system, cancel out; the result is

$$\left(\sum \mathbf{F}\right)_{\text{ext}} = \frac{d}{dt} \sum_{i=1}^{N} (m_i \mathbf{v}_i) \tag{6.3}$$

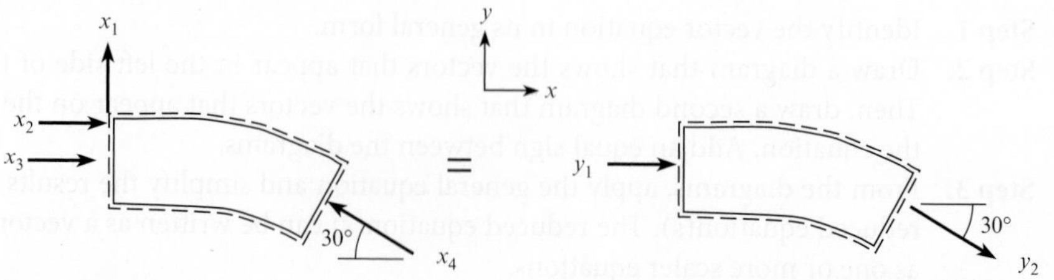

FIGURE 6.5

Vectors used to illustrate how to solve a vector equation.

where $m_i\mathbf{v}_i$ is the momentum of the ith particle and $(\Sigma\mathbf{F})_{\text{ext}}$ are forces that are external to the system. Next, let

$$(\text{Total momentum of the system}) \equiv \mathbf{M} = \sum_{i=1}^{N} (m_i\mathbf{v}_i) \tag{6.4}$$

Combine Eqs. (6.3) and (6.4):

$$\left(\sum\mathbf{F}\right)_{\text{ext}} = \left.\frac{d(\mathbf{M})}{dt}\right|_{\text{closed system}} \tag{6.5}$$

The subscript "closed system" reminds us that Eq. (6.5) is for a closed system.

6.2 The Linear Momentum Equation: Theory

This section shows how to derive the linear momentum equation and explains the physics.

Derivation

Start with Newton's second law for a system of particles (Eq. 6.5). Next, apply the Reynolds transport theorem (Eq. 5.23) to the right side of the equation. The extensive property is momentum, and the corresponding intensive property is the momentum per unit mass, which ends up being the velocity. Thus, the Reynolds transport theorem gives

$$\left.\frac{d\mathbf{M}}{dt}\right|_{\text{closed system}} = \frac{d}{dt}\int_{\text{cv}} \mathbf{v}\rho\,dV + \int_{\text{cs}} \mathbf{v}\rho\mathbf{V}\cdot d\mathbf{A} \tag{6.6}$$

Combining Eqs. (6.5) and (6.6) gives the *general form* of the *momentum equation*,

$$\left(\sum\mathbf{F}\right)_{\text{ext}} = \frac{d}{dt}\int_{\text{cv}} \mathbf{v}\rho\,dV + \int_{\text{cs}} \rho\mathbf{v}(\mathbf{V}\cdot d\mathbf{A}) \tag{6.7}$$

where $(\Sigma\mathbf{F})_{\text{ext}}$ is the sum of external forces acting on the matter in the control volume, $\mathbf{v}$ is fluid velocity relative to an inertial reference frame, and $\mathbf{V}$ is velocity relative to the control surface.

Eq. (6.7) can be simplified. To begin, assume that each particle inside the CV has the same velocity. Thus, the first term on the right side of Eq. (6.7) can be written as

$$\frac{d}{dt}\int_{\text{cv}} \mathbf{v}\rho\,dV = \frac{d}{dt}\left[\mathbf{v}\int_{\text{cv}}\rho\,dV\right] = \frac{d(m_{\text{cv}}\mathbf{v}_{\text{cv}})}{dt} \tag{6.8}$$

Next, assume that velocity is uniformly distributed as it crosses the control surface. Then, the last term in Eq. (6.7) can be written as

$$\int_{\text{cs}} \mathbf{v}\rho\mathbf{V}\cdot d\mathbf{A} = \mathbf{v}\int_{\text{cs}}\rho\mathbf{V}\cdot d\mathbf{A} = \sum_{\text{cs}}\dot{m}_o\mathbf{v}_o - \sum_{\text{cs}}\dot{m}_i\mathbf{v}_i \tag{6.9}$$

Combining Eqs. (6.7) to (6.9) gives the final result,

$$\left(\sum\mathbf{F}\right)_{\text{ext}} = \frac{d(m_{\text{cv}}\mathbf{v}_{\text{cv}})}{dt} + \sum_{\text{cs}}\dot{m}_o\mathbf{v}_o - \sum_{\text{cs}}\dot{m}_i\mathbf{v}_i \tag{6.10}$$

where m_{cv} is the mass of the matter that is inside the control volume. The subscripts o and i refer to the outlet and inlet ports, respectively. Eq. (6.10) is the *simplified form* of the momentum equation.

FIGURE 6.6

The conceptual meaning of the momentum equation.

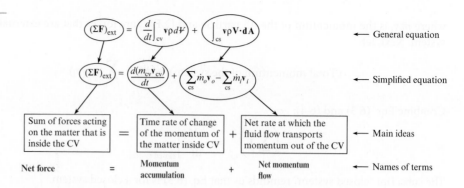

$$\left(\Sigma \mathbf{F}\right)_{\text{ext}} = \left(\frac{d}{dt}\int_{\text{cv}} \mathbf{v}\rho d\mathcal{V}\right) + \left(\int_{\text{cs}} \mathbf{v}\rho \mathbf{V}\cdot\mathbf{dA}\right)$$ ⟵ General equation

$$\left(\Sigma \mathbf{F}\right)_{\text{ext}} = \frac{d(m_{\text{cv}}\mathbf{v}_{\text{cv}})}{dt} + \sum_{\text{cs}} \dot{m}_o\mathbf{v}_o - \sum_{\text{cs}} \dot{m}_i\mathbf{v}_i$$ ⟵ Simplified equation

| Sum of forces acting on the matter that is inside the CV | = | Time rate of change of the momentum of the matter inside CV | + | Net rate at which the fluid flow transports momentum out of the CV | ⟵ Main ideas |

Net force $=$ Momentum accumulation $+$ Net momentum flow ⟵ Names of terms

Physical Interpretation of the Momentum Equation

The momentum equation asserts that the sum of forces is exactly balanced by the momentum terms; see Fig. 6.6.

Momentum Flow (Physical Interpretation)

To understand what momentum flow means, select a cylindrical fluid particle passing across a CS (see Fig. 6.7). Let the particle be long enough that it travels across the CS during a time interval Δt. Then, the particle's length is

$$L = (\text{length}) = \left(\frac{\text{length}}{\text{time}}\right)(\text{time}) = (\text{speed})(\text{time}) = v\Delta t$$

and the particle's volume is $\mathcal{V} = (v\Delta t)\Delta A$. The momentum of the particle is

$$\text{momentum of one particle} = (\text{mass})(\text{velocity}) = (\rho\Delta\mathcal{V})\mathbf{v} = (\rho v\Delta t\Delta A)\mathbf{v}$$

Next, add up the momentum of all particles that are crossing the control surface through a given face:

$$\text{momentum of all particles} = \sum_{\text{cs}} (\rho v\Delta t\Delta A)\mathbf{v} \qquad (6.11)$$

Now, let the time interval Δt and the area ΔA approach zero, and replace the sum with the integral. Eq. (6.11) becomes

$$\left(\frac{\text{momentum of all particles crossing the CS}}{\text{interval of time}}\right)_{\text{instant in time}} = \int_{\text{cs}} (\rho v)\mathbf{v}\, dA$$

Summary. Momentum flow describes the rate at which the flowing fluid transports momentum across the control surface.

Momentum Flow (Calculations)

When fluid crosses the control surface, it transports momentum across the CS. At section 1 (Fig. 6.8), momentum is transported into the CV. At section 2, momentum is transported out of the CV.

FIGURE 6.7

A fluid particle passing across the control surface during a time interval Δt.

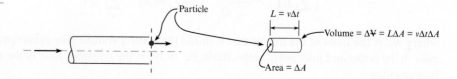

Particle

$L = v\Delta t$

Volume $= \Delta\mathcal{V} = L\Delta A = v\Delta t\Delta A$

Area $= \Delta A$

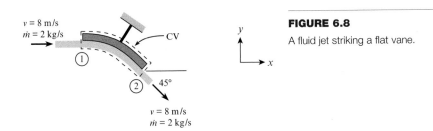

FIGURE 6.8

A fluid jet striking a flat vane.

When the velocity is uniformly distributed across the CS, Eq. (6.10) indicates

$$\left(\begin{array}{c} \text{magnitude of} \\ \text{momentum flow} \end{array}\right) = \dot{m}v = \rho A v^2 \tag{6.12}$$

Thus, at section 1, the momentum flow has a magnitude of

$$\dot{m}v = (2\text{ kg/s})(8\text{ m/s}) = 16\text{ kg} \cdot \text{m/s}^2 = 16\text{ N}$$

and the direction of the vector is to the right. Similarly, at section 2, the momentum flow has a magnitude of 16 newtons and a direction of 45° below horizontal. From Eq. (6.10), the net momentum flow term is

$$\dot{m}\mathbf{v}_2 - \dot{m}\mathbf{v}_1 = \{(16\text{ N})\cos(45°\mathbf{i} - \sin 45°\mathbf{j})\} - \{(16\text{ N})\mathbf{i}\}$$

Summary. For uniform velocity, momentum flow terms have a magnitude $\dot{m}v = \rho A v^2$ and a direction parallel to the velocity vector. The net momentum flow is calculated by subtracting the inlet momentum flow vector(s) from the outlet momentum flow vector(s).

Momentum Accumulation (Physical Interpretation)

To understand what accumulation means, consider a control volume around a nozzle (Fig. 6.9). Then, divide the control volume into many small volumes. Pick one of these small volumes, and note that the momentum inside this volume is $(\rho \Delta V)\mathbf{v}$.

To find the total momentum inside the CV, add up the momentum for all the small volumes that comprise the CV. Then, let $\Delta V \to 0$, and use the fact that an integral is the sum of many small terms:

$$\left(\begin{array}{c} \text{total momentum} \\ \text{inside the CV} \end{array}\right) = \sum (\rho \Delta V)\mathbf{v} = \sum \mathbf{v}\rho \Delta V = \int_{cv} \mathbf{v}\rho\, dV \tag{6.13}$$

Taking the time derivative of Eq. (6.13) gives the final result:

$$\left(\begin{array}{c} \text{momentum} \\ \text{accumulation} \end{array}\right) = \left(\begin{array}{c} \text{rate of change of the} \\ \text{total momentum} \\ \text{inside the CV} \end{array}\right) = \frac{d}{dt}\int_{cv} \mathbf{v}\rho\, dV \tag{6.14}$$

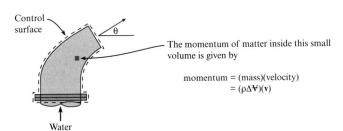

Control surface

The momentum of matter inside this small volume is given by

momentum = (mass)(velocity)
= $(\rho\Delta V)(\mathbf{v})$

Water

FIGURE 6.9

Water flowing through a nozzle.

TABLE 6.1 Summary of the Linear Momentum Equation

Description	Equation	Terms
General equation	$$\left(\sum \mathbf{F}\right)_{\text{ext}} = \frac{d}{dt}\int_{\text{cv}} \mathbf{v}\rho\,d\mathcal{V} + \int_{\text{cs}} \rho\mathbf{v}(\mathbf{V}\cdot d\mathbf{A})$$ Eq. (6.7)	$(\sum\mathbf{F})_{\text{ext}}$ = sum of external forces (N) t = time (s) $\mathbf{v}$ = velocity measured from the selected ref. frame (m/s) (must select a reference frame that is inertial)
Simplified equation Use this equation for most problems. Assumptions: (a) All particles inside the CV have the same velocity, and (b) when flow crosses the CS, the velocity is uniformly distributed.	$$\left(\sum \mathbf{F}\right)_{\text{ext}} = \frac{d(m_{\text{cv}}\mathbf{v}_{\text{cv}})}{dt} + \sum_{\text{cs}} \dot{m}_o\mathbf{v}_o - \sum_{\text{cs}} \dot{m}_i\mathbf{v}_i$$ Eq. (6.10)	$\mathbf{v}_{\text{cv}}$ = velocity of CV from selected ref. frame (m/s) $\mathbf{V}$ = velocity measured from the control surface (m/s) ρ = density of fluid (kg/m³) m_{cv} = mass of the matter inside the control volume (kg) $\dot{m}_o$ = mass flow rate out of the control volume (kg/s) $\dot{m}_i$ = mass flow rate into the control volume (kg/s)

Summary. Momentum accumulation describes the time rate of change of the momentum inside the CV. For most problems, the accumulation term is zero or negligible. To analyze the momentum accumulation term, one can ask two questions: *Is the momentum of the matter inside the CV changing with time? Is this change significant?* If the answers to both questions are yes, then the momentum accumulation term should be analyzed. Otherwise, the accumulation term can be set to zero.

6.3 The Linear Momentum Equation: Application

Working Equations

Table 6.1 summarizes the linear momentum equation.

Force and Momentum Diagram

The recommended method for applying the momentum equation, the VSM, is illustrated in the next example.

> **EXAMPLE.** This example explains how to apply the VSM for water flowing out of a nozzle (Fig. 6.10a). The water enters at section 1 and jets out at section 2.
>
> **Step 1.** Write the momentum equation (see Fig. 6.10b). Select a control volume that surrounds the nozzle.
>
> **Step 2a.** To represent the force terms, sketch a *force diagram* (Fig. 6.10c). A **force diagram** illustrates the forces that are acting on the matter that is inside the CV. A force diagram is similar to a *free body diagram* in terms of how it is drawn and how it looks. However, a free body diagram is a Lagrangian idea, whereas a force diagram is a Eulerian idea. This is why different names are used.
>
> To draw the force diagram, sketch the CV, then sketch the external forces acting on the CV. In Fig. 6.10c, the weight vector, W, represents the weight of the water plus the weight of the nozzle material. The pressure vector, symbolized with p_1A_1, represents the water in the pipe pushing the water through the nozzle. The force vector, symbolized with F_x and F_y, represents the force of the support that is holding the nozzle stationary.

FIGURE 6.10

The recommended way to apply the momentum equation is to sketch force and momentum diagrams and then to write the reduced form of the momentum equation in the x and y directions.

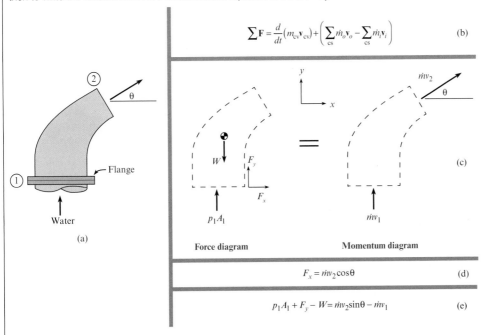

$$\sum \mathbf{F} = \frac{d}{dt}(m_{cv}\mathbf{v}_{cv}) + \left(\sum_{cs} \dot{m}_o \mathbf{v}_o - \sum_{cs} \dot{m}_i \mathbf{v}_i\right) \qquad (b)$$

(a)

Force diagram   Momentum diagram   (c)

$$F_x = \dot{m}v_2\cos\theta \qquad (d)$$

$$p_1 A_1 + F_y - W = \dot{m}v_2\sin\theta - \dot{m}v_1 \qquad (e)$$

Step 2b. To represent the momentum terms, sketch a **momentum diagram** (Fig. 6.10c). This diagram shows the momentum terms from the right side of the momentum equation. The momentum outflow is represented with $\dot{m}v_2$ and momentum inflow is represented with $\dot{m}v_1$. The momentum accumulation term is zero because the total momentum inside the CV is constant with time.

Step 3. Using the diagrams, write the reduced equations (see Figs. 6.10d and 6.10e).

Summary. The *force diagram* shows forces on the CV, and the *momentum diagram* shows the momentum terms. We recommend drawing these diagrams and using the VSM.

A Process for Applying the Momentum Equation

Step 1: Selection. Select the linear momentum equation when the problem involves forces and accelerating fluid particles and torque does not need to be considered.

Step 2: Sketching. Select a CV so that control surfaces cut through where (a) you know information or (b) you want information. Then, sketch a force diagram and a momentum diagram.

Step 3: Analysis. Write scalar or vector equations by using the VSM.

Step 4: Validation. Check that all forces are external forces. Check the signs on vectors. Check the physics. For example, if accumulation is zero, then the sum of forces should balance the momentum flow out minus the momentum flow in.

A Road Map for Problem Solving

Fig. 6.11 shows a classification scheme for problems. Like a road map, the purpose of this diagram is to help navigate the terrain. The next two sections present the details of each category of problems.

FIGURE 6.11

A classification scheme for problems that are solvable by application of the momentum equation.

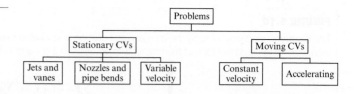

FIGURE 6.12

A problem involving a fluid jet.

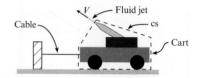

6.4 The Linear Momentum Equation for a Stationary Control Volume

When a CV is stationary with respect to the earth, the accumulation term is nearly always zero or negligible. Thus, the momentum equation simplifies to

$$(\text{sum of forces}) = (\text{rate of momentum out}) - (\text{rate of momentum in})$$

Fluid Jets

Problems in the category of **fluid jets** involve a free jet leaving a nozzle. However, analysis of the nozzle itself is not part of the problem. An example of a fluid jet problem is shown in Fig. 6.12. This problem involves a water cannon on a cart. The water leaves the nozzle with velocity V, and the goal is to find the tension in the cable.

Each category of problems has certain facts that make problem solving easier. These facts will be presented in the form of tips. **Tips** for fluid jet problems are as follows:

- **Tip 1.** When a free jet crosses the control surface, the jet does not exert a force. Thus, do not draw a force on the force diagram. The reason is that the pressure in the jet is ambient pressure, so there is no net force. This can be proven by applying Euler's equation.

- **Tip 2.** The momentum flow of the fluid jet is $\dot{m}v$.

Example 6.1 shows a problem in the fluid jet category.

EXAMPLE 6.1

Momentum Equation Applied to a Stationary Rocket

Problem Statement

The following sketch shows a 40 g rocket, of the type used for model rocketry, being fired on a test stand to evaluate thrust. The exhaust jet from the rocket motor has a diameter of $d = 1$ cm, a speed of $v = 450$ m/s, and a density of $\rho = 0.5$ kg/m^3. Assume the pressure in the exhaust jet equals ambient pressure. Find the force F_s acting on the support that holds the rocket stationary.

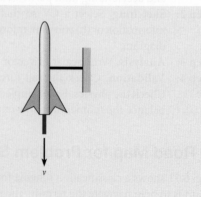

Define the Situation

A small rocket is fired on a test stand.

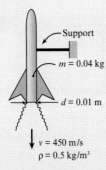

Support
$m = 0.04$ kg
$d = 0.01$ m

$v = 450$ m/s
$\rho = 0.5$ kg/m³

Assumptions: Pressure is 0.0 kPa gage at the nozzle exit plane.

State the Goal

F_s (N) ← force that acts on the support

Generate Ideas and Make a Plan

Selection: Select the momentum equation because fluid particles are accelerating due to pressures generated by combustion and because force is the goal.

Sketching: Select a CV surrounding the rocket because the control surface cuts

- through the support (where we want information) and
- across the rocket nozzle (where information is known).

Then, sketch a *force diagram* and a *momentum diagram*. Notice that the diagrams include an arrow to indicate the positive y direction. This is important because the momentum equation is a vector equation.

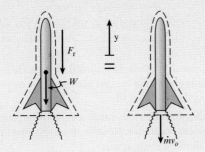

In the force diagram, the body force is the weight (W). The force (F_r) represents the downward push of the support on the rocket. There is no pressure force at the nozzle exit plane because pressure is atmospheric.

Analysis: Apply the momentum equation in the vertical direction by selecting terms off the diagrams:

$$F_r + W = \dot{m}v_o \qquad \text{(a)}$$

In Eq. (a), the only unknown is F_r. Thus, the plan is as follows:

1. Calculate momentum flow: $\dot{m}v_o = \rho A v_o^2$.
2. Calculate weight.
3. Solve for force F_r. Then, apply Newton's third law.

Take Action (Execute the Plan)

1. Momentum flow:

$$\rho A v^2 = (0.5 \text{ kg/m}^3)(\pi \times 0.01^2 \text{ m}^2/4)(450^2 \text{ m}^2/\text{s}^2)$$
$$= 7.952 \text{ N}$$

2. Weight:

$$W = mg = (0.04 \text{ kg})(9.81 \text{ m/s}^2) = 0.3924 \text{ N}$$

3. Force on the rocket (from Eq. (a)):

$$F_r = \rho A v_o^2 - W = (7.952 \text{ N}) - (0.3924 \text{ N}) = 7.56 \text{ N}$$

By Newton's third law, the force on the support is equal in magnitude to F_r and opposite in direction:

$$\boxed{F_s = 7.56 \text{ N (upward)}}$$

Review

1. *Knowledge.* Notice that forces acting on the rocket do not sum to zero. This is because the fluid is accelerating.

2. *Knowledge.* For a rocket, the term $\dot{m}v$ is sometimes called a "thrust force." For this example, $\dot{m}v = 7.95$ N (1.79 lbf); this value is typical of a small motor used for model rocketry.

3. *Knowledge.* Newton's third law tells us that forces always occur in pairs, equal in magnitude and opposite in direction. In the sketch ahead, F_r and F_s are equal in magnitude and opposite in direction.

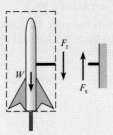

Example 6.2 gives another problem in the fluid jet category.

EXAMPLE 6.2

Momentum Equation Applied to a Fluid Jet

Problem Statement

As shown in the sketch, concrete flows into a cart sitting on a scale. The stream of concrete has a density of $\rho = 150$ lbm/ft^3, an area of $A = 1$ ft^2, and a speed of $v = 10$ ft/s. At the instant shown, the weight of the cart plus the concrete is 800 lbf. Determine the tension in the cable and the weight recorded by the scale. Assume steady flow.

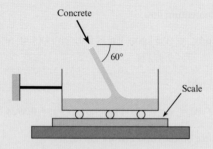

Define the Situation

Concrete is flowing into a cart that is being weighed.

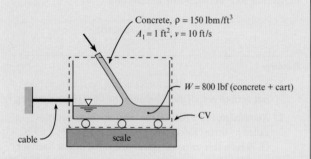

State the Goal

T(lbf) ◂ tension in cable
W_s(lbf) ◂ weight recorded by the scale

Generate Ideas and Make a Plan

Select the momentum equation. Then, select a CV and sketch this in the situation diagram. Next, sketch a force diagram and momentum diagram.

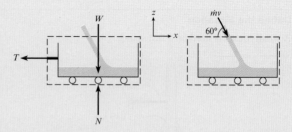

Notice in the force diagram that the liquid jet does not exert a force at the control surface. This is because the pressure in the jet equals atmospheric pressure.

To apply the momentum equation, use the force and momentum diagrams to visualize the vectors:

$$\sum \mathbf{F} = \dot{m}_o \mathbf{v}_o - \dot{m}_i \mathbf{v}_i$$

$$-T\mathbf{i} + (N - W)\mathbf{k} = -\dot{m}v((\cos 60°)\mathbf{i} - (\sin 60°)\mathbf{j})$$

Next, write scalar equations:

$$-T = -\dot{m}v\cos 60° \tag{a}$$
$$(N - W) = \dot{m}v\sin 60° \tag{b}$$

Now, the goals can be solved for. The plan is as follows:

1. Calculate T using Eq. (a).
2. Calculate N using Eq. (b). Then, let $W_s = -N$.

Take Action (Execute the Plan)

1. Momentum equation (horizontal direction):

$$T = \dot{m}v\cos 60° = \rho A v^2 \cos 60°$$

$$T = (150 \text{ lbm/ft}^3)\left(\frac{\text{slugs}}{32.2 \text{ lbm}}\right)(1 \text{ ft}^2)(10 \text{ ft/s})^2 \cos 60°$$

$$= \boxed{233 \text{ lbf}}$$

2. Momentum equation (vertical direction):

$$N - W = \dot{m}v\sin 60° = \rho A v^2 \sin 60°$$
$$N = W + \rho A v^2 \sin 60°$$
$$= 800 \text{ lbf} + 403 \text{ lbf} = \boxed{1200 \text{ lbf}}$$

Review

1. *Discussion.* The weight recorded by the scale is larger than the weight of the cart because of the momentum carried by the fluid jet.

2. *Discussion.* The momentum accumulation term in this problem is nonzero. However, it was assumed to be small and was neglected.

Vanes

A **vane** is a structural component, typically thin, that is used to turn a fluid jet (Fig. 6.13). A vane is used to idealize many components of engineering interest. Examples include a blade in a turbine, a sail on a ship, and a thrust reverser on an aircraft engine.

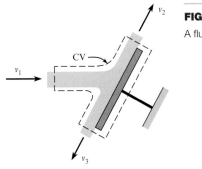

FIGURE 6.13

A fluid jet striking a flat vane.

To make solving vane problems easier, we offer the following **tips**:

- **Tip 1.** Assume that $v_1 = v_2 = v_3$. This assumption can be justified with the Bernoulli equation. In particular, assume inviscid flow and neglect elevation changes, and the Bernoulli equation can be used to prove that the velocity of the fluid jet is constant.

- **Tip 2.** Let each momentum flow equal $\dot{m}\mathbf{v}$. For example, in Fig. 6.13, the momentum inflow is $\dot{m}_1\mathbf{v}_1$. The momentum outflows are $\dot{m}_2\mathbf{v}_2$ and $\dot{m}_3\mathbf{v}_3$.

- **Tip 3.** If the vane is flat, as in Fig. 6.13, assume that the force to hold the vane stationary is normal to the vane because viscous stresses are small relative to pressure stresses. Thus, the load on the vane can assumed to be due to pressure, which acts normal to the vane.

- **Tip 4.** When the jet is a free jet, as in Fig. 6.13, recognize that the jet does not cause a net force at the control surface because the pressure in the jet is atmospheric. Only pressures different than atmospheric cause a net force.

EXAMPLE 6.3

Momentum Equation Applied to a Vane

Problem Statement

A water jet ($\rho = 1.94$ slug/ft^3) is deflected 60° by a stationary vane as shown in the figure. The incoming jet has a speed of 100 ft/s and a diameter of 1 in. Find the force exerted by the jet on the vane.

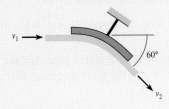

Define the Situation

A water jet is deflected by a vane.

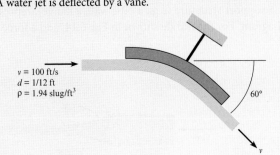

Assumptions:

- Jet velocity is constant: $v_1 = v_2 = v$.
- Jet diameter is constant: $d_1 = d_2 = d$.
- Neglect gravitational effects.

State the Goal

$\mathbf{F}_{jet}(N)$ ⟵ force of the fluid jet on the vane

Generate Ideas and Make a Plan

Select: Because force is a parameter and fluid particles accelerate as the jet turns, select the linear momentum equation.

Sketch: Select a CV that cuts through support so that the force of the support can be found. Then, sketch a force diagram and a momentum diagram.

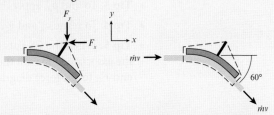

In the force and momentum diagrams, notice the following:

- Pressure forces are zero because pressures in the water jet at the control surface are zero gage.

- Each momentum flow is represented with $\dot{m}v$.

Analysis: To apply the momentum equation, use the force and momentum diagrams to write a vector equation:

$$\sum \mathbf{F} = \dot{m}_o \mathbf{v}_o - \dot{m}_i \mathbf{v}_i$$
$$(-F_x)\mathbf{i} + (-F_y)\mathbf{j} = \dot{m}v(\cos 60°\,\mathbf{i} - \sin 60°\,\mathbf{j}) - \dot{m}v\,\mathbf{i}$$

Now, write scalar equations:

$$-F_x = \dot{m}v(\cos 60° - 1) \quad\quad \textbf{(a)}$$
$$-F_y = -\dot{m}v(\sin 60°) \quad\quad \textbf{(b)}$$

Because there is enough information as follows: to solve Eqs. (a) and (b), the problem is cracked. The plan is

1. Calculate $\dot{m}v$.
2. Apply Eq. (a) to calculate F_x.
3. Apply Eq. (b) to calculate F_y.
4. Apply Newton's third law to find the force of the jet.

Take Action (Execute the Plan)

1. Momentum flow rate:

$$\dot{m}v = (\rho Av)v$$
$$= (1.94\ \text{slug/ft}^3)(\pi \times 0.0417^2\ \text{ft}^2)(100\ \text{ft/s})^2$$
$$= 105.8\ \text{lbf}$$

2. Linear momentum equation (x direction):

$$F_x = \dot{m}v(1 - \cos 60°)$$
$$= (105.8\ \text{lbf})(1 - \cos 60°)$$
$$F_x = 53.0\ \text{lbf}$$

3. Linear momentum equation (y direction):

$$F_y = \dot{m}v \sin 60°$$
$$= (105.8\ \text{lbf})\sin 60°$$
$$F_y = 91.8\ \text{lbf}$$

4. Newton's third law:

The force of the jet on the vane ($\mathbf{F}_{jet}$) is opposite in direction to the force required to hold the vane stationary ($\mathbf{F}$). Therefore,

$$\mathbf{F}_{jet} = (53.0\ \text{lbf})\mathbf{i} + (91.8\ \text{lbf})\mathbf{j}$$

Review

1. *Discussion.* Notice that the problem goal was specified as a vector. Thus, the answer was given as a vector.
2. *Skill.* Notice how the common assumptions for a vane were applied in the "define the situation" portion.

Nozzles

Nozzles are flow devices used to accelerate a fluid stream by reducing the cross-sectional area of the flow (Fig. 6.14). Problems in this category involve analysis of the nozzle itself, not analysis of the free jet.

To make solving nozzle problems easier, we offer the following **tips**:

- **Tip 1.** Let each momentum flow equal $\dot{m}v$. For the nozzle in Fig. 6.14, the momentum inflow is $\dot{m}v_A$ and the outflow is $\dot{m}v_B$.
- **Tip 2.** Include a pressure force where the nozzle connects to a pipe. For the nozzle in Fig. 6.14, include a pressure force of magnitude $p_A A_A$ on the force diagram. This pressure force, like all pressure forces, is compressive.
- **Tip 3.** To find p_A, apply the Bernoulli equation between A and B.
- **Tip 4.** To relate v_A and v_B, apply the continuity equation.
- **Tip 5.** When the CS cuts through a support structure (e.g., a pipe wall, a flange), represent the associated force on the force diagram. For the nozzle shown in Fig. 6.14, add a force F_{Ax} and F_{Ay} to the force diagram.

FIGURE 6.14

A fluid jet exiting a nozzle.

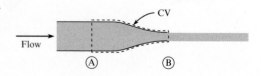

EXAMPLE 6.4
Momentum Equation Applied to a Nozzle

Problem Statement

The sketch shows air flowing through a nozzle. The inlet pressure is $p_1 = 105$ kPa abs, and the air exhausts into the atmosphere, where the pressure is 101.3 kPa abs. The nozzle has an inlet diameter of 60 mm and an exit diameter of 10 mm, and the nozzle is connected to the supply pipe by flanges. Find the force required to hold the nozzle stationary. Assume the air has a constant density of 1.22 kg/m³. Neglect the weight of the nozzle.

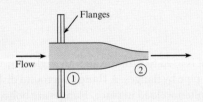

Define the Situation

Air flows through a nozzle.

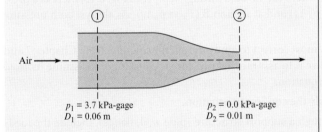

$p_1 = 3.7$ kPa-gage $p_2 = 0.0$ kPa-gage
$D_1 = 0.06$ m $D_2 = 0.01$ m

Properties: $\rho = 1.22$ kg/m³
Assumptions:
- The weight of the nozzle is negligible.
- Steady flow, constant density flow, inviscid flow.

State the Goals

$F(N)$ ← force required to hold nozzle stationary

Generate Ideas and Make a Plan

Select: Because force is a parameter and fluid particles are accelerating in the nozzle, select the momentum equation.

Sketch: Sketch a force diagram (FD) and momentum diagram (MD):

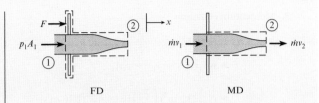

FD   MD

Write the momentum equation (x direction):

$$F + p_1 A_1 = \dot{m}(v_2 - v_1) \tag{a}$$

To solve for F, we need v_2 and v_1, which can be found using the Bernoulli equation. Thus, the plan is as follows:

1. Derive an equation for v_2 by applying the Bernoulli equation and the continuity equation.
2. Calculate v_2 and v_1.
3. Calculate F by applying Eq. (a).

Take Action (Execute the Plan)

1. *Bernoulli equation* (apply between 1 and 2):

$$p_1 + \gamma z_1 + \frac{1}{2}\rho v_1^2 = p_2 + \gamma z_2 + \frac{1}{2}\rho v_2^2$$

Term-by-term analysis:
- $z_1 = z_2 = 0$
- $p_1 = 3.7$ kPa; $p_2 = 0.0$

The Bernoulli equation reduces to

$$p_1 + \rho v_1^2/2 = \rho v_2^2/2$$

Continuity equation. Select a CV that cuts through sections 1 and 2. Neglect the mass accumulation terms. Continuity simplifies to

$$v_1 A_1 = v_2 A_2$$
$$v_1 d_1^2 = v_2 d_2^2$$

Substitute into the Bernoulli equation and solve for v_2:

$$v_2 = \sqrt{\frac{2p_1}{\rho(1 - (d_2/d_1)^4)}}$$

2. Calculate v_2 and v_1:

$$v_2 = \sqrt{\frac{2 \times 3.7 \times 1000 \text{ Pa}}{(1.22 \text{ kg/m}^3)(1 - (10/60)^4)}} = 77.9 \text{ m/s}$$

$$v_1 = v_2\left(\frac{d_2}{d_1}\right)^2$$

$$= 77.9 \text{ m/s} \times \left(\frac{1}{6}\right)^2 = 2.16 \text{ m/s}$$

3. Momentum equation:

$$F + p_1A_1 = \dot{m}(v_2 - v_1)$$

$$F = \rho A_1 v_1(v_2 - v_1) - p_1 A_1$$

$$= (1.22 \text{ kg/m}^3)\left(\frac{\pi}{4}\right)(0.06 \text{ m})^2(2.16 \text{ m/s})$$

$$\times (77.9 - 2.16)(\text{m/s})$$

$$-3.7 \times 1000 \text{ N/m}^2 \times \left(\frac{\pi}{4}\right)(0.06 \text{ m})^2$$

$$= 0.564 \text{ N} - 10.46 \text{ N} = -9.90 \text{ N}$$

Because F is negative, the direction is opposite to the direction assumed on the force diagram. Thus,

> Force to hold nozzle = 9.90 N ($\leftarrow$ direction)

Review

1. *Knowledge.* The direction initially assumed for the force on a force diagram is arbitrary. If the answer for the force is negative, then the force acts in a direction opposite the chosen direction.

2. *Knowledge.* Pressures were changed to gage pressure in the "define the situation" operation because it is the pressures' differences as compared to atmospheric pressure that cause net pressure forces.

Pipe Bends

A **pipe bend** is a structural component that is used to turn through an angle (Fig. 6.15). A pipe bend is often connected to straight runs of pipe by flanges. A **flange** is a round disk with a hole in the center that slides over a pipe and is often welded in place. Flanges are bolted together to connect sections of pipe.

The following tips are useful for solving problems that involve pipe bends.

- **Tip 1.** Let each momentum flow equal $\dot{m}\mathbf{v}$. For the bend in Fig. 6.15, the momentum inflow is $\dot{m}\mathbf{v}_A$ and the outflow is $\dot{m}\mathbf{v}_B$.

- **Tip 2.** Include pressure forces where the CS cuts through a pipe. In Fig. 6.15, there is a pressure force at section A ($F_A = p_A A_A$) and at section B ($F_B = p_B A_B$). As always, both pressure forces are compressive.

- **Tip 3.** To relate p_A and p_B, it is most correct to apply the *energy equation* from Chapter 7 and include head loss. An alternative is to assume that pressure is constant or to assume inviscid flow and apply the Bernoulli equation.

- **Tip 4.** To relate v_A and v_B, apply the continuity equation.

- **Tip 5.** When the CS cuts through a support structure (pipe wall, flange), include the loads caused by the support on the force diagram.

FIGURE 6.15

Pipe bend.

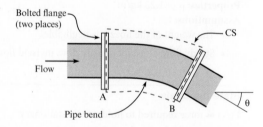

EXAMPLE 6.5

Momentum Equation Applied to a Pipe Bend

Problem Statement

A 1 m diameter pipe bend shown in the diagram is carrying crude oil ($S = 0.94$) with a steady flow rate of 2 m³/s. The bend

has an angle of 30° and lies in a horizontal plane. The volume of oil in the bend is 1.2 m³, and the empty weight of the bend is 4 kN. Assume the pressure along the centerline of the bend is constant with a value of 75 kPa gage. Find the force required to hold the bend in place.

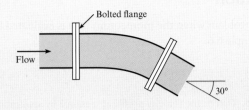

Bolted flange

Flow

30°

Define the Situation

Crude oil flows through a pipe bend:

- The bend lies in a horizontal plane.
- $V_{oil} = 1.2$ m^3 = volume of oil in bend.
- $W_{bend} = 4000$ N = empty weight of bend.
- $p = 75$ kPa-gage = pressure along the centerline.

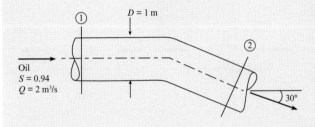

① $D = 1$ m

②

Oil
$S = 0.94$
$Q = 2$ m^3/s

30°

State the Goal

$\mathbf{F}$(N) ⬅ force to hold the bend stationary.

Generate Ideas and Make a Plan

Select: Because force is a parameter and fluid particles accelerate in the pipe bend, select the momentum equation.

Sketch: Select a CV that cuts through the support structure and through sections 1 and 2. Then, sketch the force and momentum diagrams.

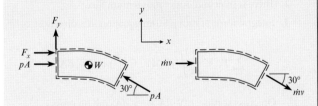

y

F_y

x

F_x
pA

W

$\dot{m}v$

30°

30°
pA

30°
$\dot{m}v$

Analysis: Using the diagrams as guides, write the momentum equation in each direction:

- x direction:

$$F_x + p_1 A_1 - p_2 A_2 \cos 30° = \dot{m}v_2 \cos 30° - \dot{m}v_1 \quad \text{(a)}$$

- y direction:

$$F_y + p_2 A_2 \sin 30° = -\dot{m}v_2 \sin 30° \quad \text{(b)}$$

- z direction:

$$F_z - W_{total} = 0 \quad \text{(c)}$$

Review these equations; notice that there is enough information to solve for the goals F_x, F_y, and F_z. Thus, create a plan:

1. Calculate the momentum flux $\dot{m}v$.
2. Calculate the pressure force pA.
3. Solve Eq. (a) for F_x.
4. Solve Eq. (b) for F_y.
5. Solve Eq. (c) for F_z.

Take Action (Execute the Plan)

1. Momentum flow:
   - Apply the volume flow rate equation:

$$v = Q/A = \frac{(2 \text{ m}^3/\text{s})}{(\pi \times 0.5^2 \text{ m}^2)} = 2.55 \text{ m/s}$$

   - Next, calculate the momentum flow:

$$\dot{m}v = \rho Q v = (0.94 \times 1000 \text{ kg/m}^3)(2 \text{ m}^3/\text{s})(2.55 \text{ m/s})$$
$$= 4.79 \text{ kN}$$

2. Pressure force:

$$pA = (75 \text{ kN/m}^2)(\pi \times 0.5^2 \text{ m}^2) = 58.9 \text{ kN}$$

3. Momentum equation (x direction):

$$F_x + p_1 A_1 - p_2 A_2 \cos 30° = \dot{m}v_2 \cos 30° - \dot{m}v_1$$
$$F_x = -pA(1 - \cos 30°) - \dot{m}v(1 - \cos 30°)$$
$$= -(pA + \dot{m}v)(1 - \cos 30°)$$
$$= -(58.9 + 4.79)(\text{kN})(1 - \cos 30°)$$
$$= -8.53 \text{ kN}$$

4. Momentum equation (y direction):

$$F_y + p_2 A_2 \sin 30° = -\dot{m}v_2 \sin 30°$$
$$F_y = -(pA + \dot{m}v) \sin 30°$$
$$= -(58.9 + 4.79)(\text{kN})(\sin 30°) = -31.8 \text{ kN}$$

5. Momentum equation (z direction) (the bend weight includes the oil plus the empty pipe):

$$-F_z - W_{total} = 0$$

$W = \gamma V + 4 \text{ kN}$

$$= (0.94 \times 9.81 \text{ kN/m}^3)(1.2 \text{ m}^3) + 4 \text{ kN} = 15.1 \text{ kN}$$

6. Force to hold the bend:

$$\boxed{\mathbf{F} = (-8.53 \text{ kN})\mathbf{i} + (-31.8 \text{ kN})\mathbf{j} + (15.1 \text{ kN})\mathbf{k}}$$

Variable Velocity Distribution

This subsection shows how to solve a problem when the momentum flow is evaluated by integration. This case is illustrated by Example 6.6.

EXAMPLE 6.6

Momentum Equation Applied with a Variable Velocity Distribution

Problem Statement

The drag force of a bullet-shaped device may be measured using a wind tunnel. The tunnel is round with a diameter of 1 m, the pressure at section 1 is 1.5 kPa gage, the pressure at section 2 is 1.0 kPa gage, and air density is 1.0 kg/m³. At the inlet, the velocity is uniform with a magnitude of 30 m/s. At the exit, the velocity varies linearly as shown in the sketch. Determine the drag force on the device and support vanes. Neglect viscous resistance at the wall, and assume pressure is uniform across sections 1 and 2.

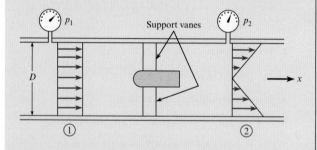

Define the Situation

Data are supplied for the wind tunnel test (see above).

Properties: Air: $\rho = 1.0$ kg/m³.

Assumptions: Steady flow.

State the Goal

Find: Drag force (in newtons) on the model

Make a Plan

1. Select a control volume that encloses the model.
2. Sketch the force diagram.
3. Sketch the momentum diagram.
4. The downstream velocity profile is not uniformly distributed. Apply the integral form of the momentum equation, Eq. (6.7).
5. Evaluate the sum of forces.

6. Determine velocity profile at section 2 by application of continuity equation.
7. Evaluate the momentum terms.
8. Calculate drag force on the model.

Take Action (Execute the Plan)

1. The control volume selected is shown. The control volume is stationary.

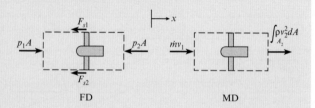

FD   MD

2. The forces consist of the pressure forces and the force on the model support struts cut by the control surface. The drag force on the model is equal and opposite to the force on the support struts: $F_D = F_{s1} + F_{s2}$.

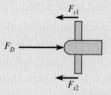

3. There is inlet and outlet momentum flux.
4. Integral form of momentum equation in x direction:

$$\sum F_x = \frac{d}{dt}\int_{cv}\rho v_x d\forall + \int_{cs}\rho v_x(\mathbf{V} \cdot d\mathbf{A})$$

On cross section 1, $\mathbf{V}\cdot d\mathbf{A} = -v_x dA$, and on cross section 2, $\mathbf{V}\cdot d\mathbf{A} = v_x dA$, so

$$\sum F_x = \frac{d}{dt}\int_{cv}\rho v_x d\forall - \int_1\rho v_x^2 dA + \int_2\rho v_x^2 dA$$

5. Evaluation of force terms:

$$\sum F_x = p_1 A - p_2 A - (F_{s1} + F_{s2})$$
$$= p_1 A - p_2 A - F_D$$

6. Velocity profile at section 2:

Velocity is linear in radius, so choose $v_2 = v_1 K(r/r_o)$, where r_o is the tunnel radius and K is a proportionality factor to be determined:

$$Q_1 = Q_2$$

$$A_1 v_1 = \int_{A_2} v_2(r)\, dA = \int_0^{r_o} v_1 K(r/r_o) 2\pi r\, dr$$

$$\pi r_o^2 v_1 = 2\pi v_1 K \frac{1}{3} r_o^2$$

$$K = \frac{3}{2}$$

7. Evaluation of momentum terms:
   - Accumulation term for steady flow is

$$\frac{d}{dt}\int_{cv} \rho v_x\, d\mathcal{V} = 0$$

- Momentum at cross section 1 with $v_x = v_1$ is

$$\int_1 \rho v_x^2\, dA = \rho v_1^2 A = \dot{m} v_1$$

- Momentum at cross section 2 is

$$\int_2 \rho v_x^2\, dA = \int_0^{r_o} \rho \left[\frac{3}{2} v_1 \left(\frac{r}{r_o}\right)\right]^2 2\pi r\, dr = \frac{9}{8}\dot{m} v_1$$

8. Drag force:

$$p_1 A - p_2 A - F_D = \dot{m} v_1 \left(\frac{9}{8} - 1\right)$$

$$F_D = (p_1 - p_2)A - \frac{1}{8}\rho A v_1^2$$

$$= (\pi \times 0.5^2\, \text{m}^2)(1.5 - 1.0)(10^3)\,\text{N/m}^2$$

$$-\frac{1}{8}(1\,\text{kg/m}^3)(\pi \times 0.5^2\,\text{m}^2)(30\,\text{m/s})^2$$

$$F_D = \boxed{304\,\text{N}}$$

6.5 Examples of the Linear Momentum Equation (Moving Objects)

This section describes how to apply the linear momentum equation to problems that involve moving objects, such as carts in motion and rockets. When an object is moving, let the CV move with the object. As shown ahead (repeated from Fig. 6.11), problems that involve moving CVs classify into two categories: objects moving with constant velocity and objects that are accelerating. Both categories involve selection of a reference frame, which is the next topic.

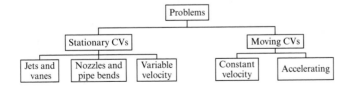

Reference Frame

When an object is moving, it is necessary to specify a reference frame. A **reference frame** is a three-dimensional framework from which an observer takes measurements. For example, Fig. 6.16 shows a rocket in flight. For this situation, one possible reference frame is fixed to the earth. Another possible reference frame is fixed to the rocket. Observers in these two frames of reference would report different values of the rocket velocity V_{Rocket} and the velocity of the fluid jet V_{jet}. The ground-based reference frame is inertial. An **inertial reference frame** is any reference frame that is stationary or moving with constant velocity with respect to the earth. Thus, an inertial reference frame is a nonaccelerating reference frame. Alternatively, a **noninertial reference frame** is any reference frame that is accelerating.

FIGURE 6.16

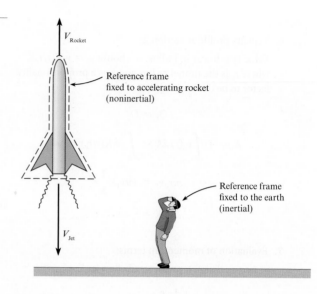

Regarding the linear momentum equation as presented in this text, this equation is only valid for an inertial frame. Thus, when objects are moving, the engineer should specify an inertial reference frame.

Analyzing a Moving Body (Constant Velocity)

When an object is moving with constant velocity, the reference frame can be placed on the moving object or fixed to the earth. However, most problems are simpler if the frame is fixed to the moving object. Example 6.7 shows how to solve a problem involving an object moving with constant velocity.

EXAMPLE 6.7

Momentum Equation Applied to a Moving CV

Problem Statement

A stationary nozzle produces a water jet with a speed of 50 m/s and a cross-sectional area of 5 cm^2. The jet strikes a moving block and is deflected 90° relative to the block. The block is sliding with a constant speed of 25 m/s on a surface with friction. The density of the water is 1000 kg/m^3. Find the frictional force F acting on the block.

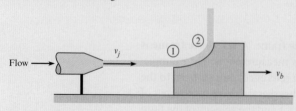

Define the Situation

A block slides at constant velocity due to a fluid jet.

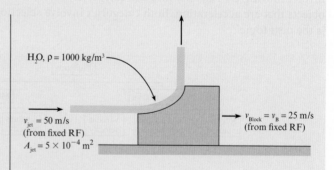

State the Goal

$F_f(\text{N})$ ← the frictional force on the block

Solution Method I (Moving RF)

When a body is moving at constant velocity, the easiest way to solve the problem is to put the RF on the moving body. This solution method is shown first.

Generate Ideas and Make a Plan

Select the linear momentum equation because force is the goal and fluid particles accelerate as they interact with the block.

Select a moving CV that surrounds the block because this CV involves known parameters (i.e., the two fluid jets) and the goal (frictional force).

Because the CV is moving at a constant velocity, select a reference frame (RF) that is fixed to the moving block. This RF makes analysis of the problem simpler.

Sketch the force and momentum diagrams and the RF.

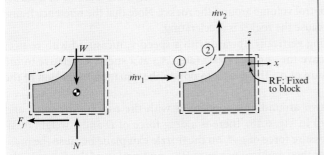

To apply the momentum equation, use the force and momentum diagrams to visualize the vectors. The momentum equation in the x direction is

$$-F_f = -\dot{m}v_1 \tag{a}$$

In Eq. (a), the mass flow rate describes the rate at which mass is crossing the control surface. Because the CS is moving away from the fluid jet, the mass flow rate term becomes

$$\dot{m} = \rho A V = \rho A_{jet}(v_{jet} - v_{block}) \tag{b}$$

In Eq. (a), the velocity v_1 is the velocity as measured from the selected reference frame. Thus,

$$v_1 = v_{jet} - v_{block} \tag{c}$$

Combining Eqs. (a), (b), and (c) gives

$$F_f = \dot{m}v_1 = \rho A_{jet}(v_{jet} - v_{block})^2 \tag{d}$$

Because all variables on the right side of Eq. (d) are known, we can solve the problem. The plan is simple: plug numbers into Eq. (d).

Take Action (Execute the Plan)

$$F_f = \rho A_{jet}(v_{jet} - v_{block})^2$$
$$F_f = (1000 \text{ kg/m}^2)(5 \times 10^{-4} \text{ m}^2)(50 - 25)^2(\text{m/s})^2$$
$$\boxed{F_f = 312 \text{ N}}$$

Solution Method II (Fixed RF)

Another way to solve this problem is to use a fixed reference frame. To implement this approach, sketch the force diagram, the momentum diagram, and the selected RF.

Notice that $\dot{m}v_2$ shows a vertical and horizontal component. This is because an observer in the selected RF would see these velocity components.

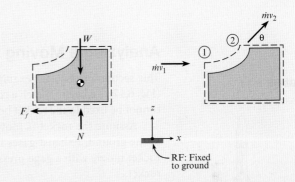

From the diagrams, one can write the momentum equation in the x direction:

$$-F_f = \dot{m}v_2\cos\theta - \dot{m}v_1$$
$$F_f = \dot{m}(v_1 - v_2\cos\theta) \tag{e}$$

In the momentum equation, the mass flow rate is measured relative to the control surface. Thus, $\dot{m}$ is independent of the RF, and one can use Eq. (b), which is repeated here:

$$\dot{m} = \rho A V = \rho A_{jet}(v_{jet} - v_{block}) \tag{f}$$

In Eq. (e), the velocity v_1 is the velocity as measured from the selected reference frame. Thus,

$$v_1 = v_{jet} \tag{g}$$

To analyze v_2, relate velocities by using a relative-velocity equation from a dynamics text:

$$v_{jet} = v_{block} + v_{jet/block} \tag{h}$$

where
- $v_2 = v_{jet}$ is the velocity of the jet at section 2 as measured from the fixed RF,
- v_{block} is the velocity of the moving block as measured from the fixed RF, and
- $v_{jet/block}$ is the velocity of the jet at section as measured from a RF fixed to the moving block.

Substitute numbers into Eq. (h) to give

$$\mathbf{v}_2 = (25 \text{ m/s})\mathbf{i} + (25 \text{ m/s})\mathbf{j} \tag{i}$$

Thus,

$$v_2\cos\theta = v_{2x} = 25 \text{ m/s} = v_{block} \tag{j}$$

Substitute Eqs. (f), (g), and (j) into Eq. (e):

$$F_f = \{\dot{m}\}(v_1 - v_2 \cos\theta)$$
$$= \{\rho A_{jet}(v_{jet} - v_{block})\}(v_{jet} - v_{block}) \qquad \textbf{(k)}$$
$$= \rho A_{jet}(v_{jet} - v_{block})^2$$

Eq. (k) is identical to Eq. (d). Thus, *solution method I* is equivalent to *solution method II*.

Review the Solution and the Process

1. *Knowledge.* When an object moves with constant velocity, select an RF fixed to the moving object because this is much easier than selecting an RF fixed to the earth.

2. *Knowledge.* Specifying the control volume and specifying the reference frame are independent decisions.

Analyzing a Moving Body (Accelerating)

FIGURE 6.17

Vertical launch of rocket.

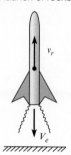

This section presents an example of an accelerating object—namely, the analysis of a rocket (Fig. 6.17). To begin, sketch a control volume around the rocket. Note that the reference frame cannot be fixed to the rocket because the rocket is accelerating.

Assume the rocket is moving vertically upward with a speed v_r measured with respect to the ground. Exhaust gases leave the engine nozzle (area A_e) at a speed V_e relative to the rocket nozzle with a gage pressure of p_e. The goal is to obtain the equation of motion of the rocket.

The control volume is drawn around and accelerates with the rocket. The force and momentum diagrams are shown in Fig. 6.18. There is a drag force of D and a weight of W acting downward. There is a pressure force of $p_e A_e$ on the nozzle exit plane because the pressure in a supersonic jet is greater than ambient pressure. The summation of the forces in the z direction is

$$\sum F_z = p_e A_e - W - D \qquad \textbf{(6.15)}$$

There is only one momentum flux out of the rocket nozzle, $\dot{m}v_o$. The speed v_o must be referenced to an inertial reference frame, which in this case is chosen as the ground. The speed of the exit gases with respect to the ground is

$$v_o = (V_e - v_r) \qquad \textbf{(6.16)}$$

because the rocket is moving upward with speed v_r with respect to the ground, and the exit gases are moving downward at speed V_e with respect to the rocket.

FIGURE 6.18

Force and momentum diagrams for rocket.

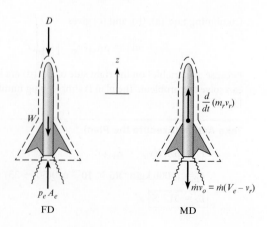

The momentum equation in the z direction is

$$\sum F_z = \frac{d}{dt}\int_{cv} v_z \rho \, d\forall + \sum_{cs} \dot{m}_o v_{oz} - \sum_{cs} \dot{m}_i v_{iz}$$

The velocity inside the control volume is the speed of the rocket, v_r, so the accumulation term becomes

$$\frac{d}{dt}\left(\int_{cv} v_z \rho \, d\forall\right) = \frac{d}{dt}\left[v_r \int_{cv}\rho\, d\forall\right] = \frac{d}{dt}(m_r v_r)$$

Substituting the sum of the forces and momentum terms into the momentum equation gives

$$p_e A_e - W - D = \frac{d}{dt}(m_r v_r) - \dot{m}(V_e - v_r) \tag{6.17}$$

Next, apply the product rule to the accumulation term. This gives

$$p_e A_e - W - D = m_r \frac{dv_r}{dt} + v_r\left(\frac{dm_r}{dt} + \dot{m}\right) - \dot{m}V_e \tag{6.18}$$

The continuity equation can now be used to eliminate the second term on the right. Applying the continuity equation to the control surface around the rocket leads to

$$\frac{d}{dt}\int_{cv}\rho\, d\forall + \sum \dot{m}_o - \sum \dot{m}_i = 0$$

$$\frac{dm_r}{dt} + \dot{m} = 0 \tag{6.19}$$

Substituting Eq. (6.19) into Eq. (6.18) yields

$$\dot{m}V_e + p_e A_e - W - D = m_r \frac{dv_r}{dt} \tag{6.20}$$

The sum of the momentum outflow and the pressure force at the nozzle exit is identified as the thrust of the rocket,

$$T = \dot{m}V_e + p_e A_e = \rho_e A_e V_e^2 + p_e A_e$$

so Eq. (6.20) simplifies to

$$m_r \frac{dv_r}{dt} = T - D - W \tag{6.21}$$

which is the equation used to predict and analyze rocket performance.

Integration of Eq. (6.21) leads to one of the fundamental equations for rocketry: the burnout velocity or the velocity achieved when all the fuel is burned. Neglecting the drag and weight, the equation of motion reduces to

$$T = m_r \frac{dv_r}{dt} \tag{6.22}$$

The instantaneous mass of the rocket is given by $m_r = m_i - \dot{m}t$, where m_i is the initial rocket mass and t is the time from ignition. Substituting the expression for mass into Eq. (6.22) and integrating with the initial condition $v_r(0) = 0$ results in

$$v_{bo} = \frac{T}{\dot{m}} \ln \frac{m_i}{m_f} \tag{6.23}$$

where v_{bo} is the burnout velocity and m_f is the final (or payload) mass. The ratio $T/\dot{m}$ is known as the specific impulse, I_{sp}, and has units of velocity.

6.6 The Angular Momentum Equation

This section presents the *angular momentum equation,* which is also called the *moment-of-momentum equation.* The angular momentum equation is very useful for situations that involve torques. Examples include analyses of rotating machinery such as pumps, turbines, fans, and blowers.

Derivation of the Equation

Newton's second law of motion can be used to derive an equation for the rotational motion of a system of particles:

$$\sum \mathbf{M} = \frac{d(\mathbf{H}_{sys})}{dt} \tag{6.24}$$

where $\mathbf{M}$ is a moment and $\mathbf{H}_{sys}$ is the total angular momentum of all mass forming the system.

To convert Eq. (6.24) to a Eulerian equation, apply the Reynolds transport theorem, Eq. (5.23). The extensive property B_{sys} becomes the angular momentum of the system: $B_{sys} = \mathbf{H}_{sys}$. The intensive property b becomes the angular momentum per unit mass. The angular momentum of an element is $\mathbf{r} \times m\mathbf{v}$, and so $b = \mathbf{r} \times \mathbf{v}$. Substituting for B_{sys} and b in Eq. (5.23) gives

$$\frac{d(\mathbf{H}_{sys})}{dt} = \frac{d}{dt} \int_{cv} (\mathbf{r} \times \mathbf{v}) \rho \, d\forall + \int_{cs} (\mathbf{r} \times \mathbf{v}) \rho \mathbf{V} \cdot d\mathbf{A} \tag{6.25}$$

Combining Eqs. (6.24) and (6.25) gives the integral form of the *moment-of-momentum equation:*

$$\sum \mathbf{M} = \frac{d}{dt} \int_{cv} (\mathbf{r} \times \mathbf{v}) \rho \, d\forall + \int_{cs} (\mathbf{r} \times \mathbf{v}) \rho \mathbf{V} \cdot d\mathbf{A} \tag{6.26}$$

where $\mathbf{r}$ is a position vector that extends from the moment center, $\mathbf{V}$ is flow velocity relative to the control surface, and $\mathbf{v}$ is flow velocity relative to the inertial reference frame selected.

If the mass crosses the control surface through a series of inlet and outlet ports with uniformly distributed properties across each port, then the moment-of-momentum equation becomes

$$\sum \mathbf{M} = \frac{d}{dt} \int_{cv} (\mathbf{r} \times \mathbf{v}) \rho \, d\forall + \sum_{cs} \mathbf{r}_o \times (\dot{m}_o \mathbf{v}_o) - \sum_{cs} \mathbf{r}_i \times (\dot{m}_i \mathbf{v}_i) \tag{6.27}$$

The moment-of-momentum equation has the following physical interpretation:

$$\begin{pmatrix} \text{sum of} \\ \text{moments} \end{pmatrix} = \begin{pmatrix} \text{angular momentum} \\ \text{accumulation} \end{pmatrix} + \begin{pmatrix} \text{angular momentum} \\ \text{outflow} \end{pmatrix} - \begin{pmatrix} \text{angular momentum} \\ \text{inflow} \end{pmatrix}$$

Application

The process for applying the angular momentum equation is similar to the process for applying the linear momentum equation. To illustrate this process, Example 6.8 shows how to apply the angular momentum equation to a pipe bend.

EXAMPLE 6.8

Applying the Angular Momentum Equation to Calculate the Moment on a Reducing Bend

Problem Statement

The reducing bend shown in the figure is supported on a horizontal axis through point A. Water (20°C) flows through the bend at 0.25 m³/s. The inlet pressure at cross section 1 is 150 kPa gage, and the outlet pressure at section 2 is 59.3 kPa gage. A weight of 1420 N acts 20 cm to the right of point A. Find the moment the support system must resist. The diameters of the inlet and outlet pipes are 30 cm and 10 cm, respectively.

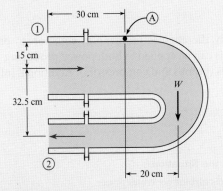

Define the Situation

Water flows through a pipe bend.

Assumptions: Steady flow.

Properties: Water (Table A.5, 20°C, $p = 1$ atm): $\rho = 998$ kg/m³

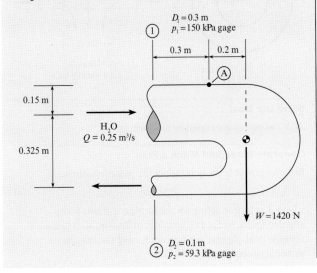

State the Goal

$M_A(N) \leftarrow$ moment acting on the support structure

Generate Ideas and Make a Plan

Select the moment-of-momentum equation (Eq. 6.27) because (a) torque is a parameter and (b) fluid particles are accelerating as they pass through the pipe bend.

Select a control volume surrounding the reducing bend. The reason is that this CV cuts through point A (where we want to know the moment) and also cuts through sections 1 and 2 where information is known.

Sketch the force and momentum diagrams. Add dimensions to the sketches so that it is easier to evaluate cross products.

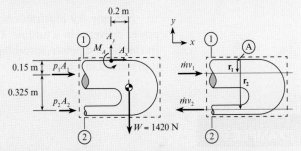

Select point "A" to sum moments about. Because the flow is steady, the accumulation of momentum term is zero. Also, there is one inflow of angular momentum and one outflow. Thus, the angular momentum equation (Eq. 6.27) simplifies to

$$\sum \mathbf{M}_A = \{\mathbf{r}_2 \times (\dot{m}\mathbf{v}_2)\} - \{\mathbf{r}_1 \times (\dot{m}\mathbf{v}_1)\} \quad \text{(a)}$$

Sum moments in the z direction:

$$\sum M_{A,z} = (p_1 A_1)(0.15 \text{ m}) + (p_2 A_2)(0.475 \text{ m}) + M_A - W(0.2 \text{ m}) \quad \text{(b)}$$

Next, analyze the momentum terms in Eq. (a):

$$\{\mathbf{r}_2 \times (\dot{m}\mathbf{v}_2)\} - \{\mathbf{r}_1 \times (\dot{m}\mathbf{v}_1)\}_z = \{-r_2 \dot{m}v_2\} - \{r_1 \dot{m}v_1\} \quad \text{(c)}$$

Substitute Eqs. (b) and (c) into Eq. (a):

$$(p_1 A_1)(0.15 \text{ m}) + (p_2 A_2)(0.475 \text{ m}) + M_A - W(0.2 \text{ m})$$
$$= \{-r_2 \dot{m}v_2\} - \{r_1 \dot{m}v_1\} \quad \text{(d)}$$

All the terms in Eq. (d) are known, so M_A can be calculated. Thus, the plan is as follows:

1. Calculate torques to due to pressure: $r_1 p_1 A_1$ and $r_2 p_2 A_2$.
2. Calculate momentum flow terms: $r_2 \dot{m}v_2 + r_1 \dot{m}v_1$.
3. Calculate M_A.

Take Action (Execute the Plan)

1. Torques due to pressure:

$$r_1 p_1 A_1 = (0.15 \text{ m})(150 \times 1000 \text{ N/m}^2)(\pi \times 0.3^2/4 \text{ m}^2)$$
$$= 1590 \text{ N} \cdot \text{m}$$
$$r_2 p_2 A_2 = (0.475 \text{ m})(59.3 \times 1000 \text{ N/m}^2)(\pi \times 0.15^2/4 \text{ m}^2)$$
$$= 498 \text{ N} \cdot \text{m}$$

2. Momentum flow terms:

$$\dot{m} = \rho Q = (998 \text{ kg/m}^3)(0.25 \text{ m}^3/\text{s})$$
$$= 250 \text{ kg/s}$$

$$v_1 = \frac{Q}{A_1} = \frac{0.25 \text{ m}^3/\text{s}}{\pi \times 0.15^2 \text{ m}^2} = 3.54 \text{ m/s}$$

$$v_2 = \frac{Q}{A_2} = \frac{0.25 \text{ m}^3/\text{s}}{\pi \times 0.075^2 \text{ m}^2} = 14.15 \text{ m/s}$$

$$\dot{m}(r_2 v_2 + r_1 v_1) = (250 \text{ kg/s})$$
$$\times (0.475 \times 14.15 + 0.15 \times 3.54)(\text{m}^2/\text{s})$$
$$= 1813 \text{ N} \cdot \text{m}$$

3. Moment exerted by support:

$$M_A = -0.15 p_1 A_1 - 0.475 p_2 A_2 + 0.2W - \dot{m}(r_2 v_2 + r_1 v_1)$$
$$= -(1590 \text{ N} \cdot \text{m}) - (498 \text{ N} \cdot \text{m})$$
$$+ (0.2 \text{ m} \times 1420 \text{ N}) - (1813 \text{ N} \cdot \text{m})$$
$$M_A = -3.62 \text{ kN} \cdot \text{m}$$

Thus, a moment of 3.62 kN·m acting in the clockwise direction is needed to hold the bend stationary.

> By Newton's third law, the moment acting on the support structure is $M_A = 3.62$ kN·m (counterclockwise).

Review the Solution and the Process

Tip. Use the "right-hand rule" to find the correct direction of moments.

Example 6.9 illustrates how to apply the angular momentum equation to predict the power delivered by a turbine. This analysis can be applied to both power-producing machines (turbines) and power-absorbing machines (pumps and compressors). Additional information is presented in Chapter 14.

EXAMPLE 6.9

Applying the Angular Momentum Equation to Predict the Power Delivered by a Francis Turbine

Problem Statement

A Francis turbine is shown in the diagram. Water is directed by guide vanes into the rotating wheel (runner) of the turbine. The guide vanes have a 70° angle from the radial direction. The water exits with only a radial component of velocity with respect to the environment. The outer diameter of the wheel is 1 m, and the inner diameter is 0.5 m. The distance across the runner is 4 cm. The discharge is 0.5 m³/s, and the rotational rate of the wheel is 1200 rpm. The water density is 1000 kg/m³. Find the power (kW) produced by the turbine.

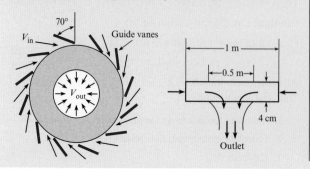

Define the Situation

A Francis turbine generates power.

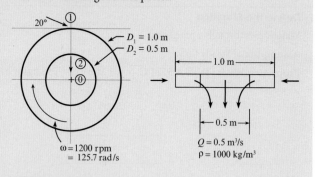

State the Goal

$P(\text{W}) \leftarrow$ power generated by the turbine

Generate Ideas and Make a Plan

Because power is the goal, select the *power equation*:

$$P = T\omega \qquad \text{(a)}$$

where T is torque acting on the turbine and ω is turbine angular speed. In Eq. (a), torque is unknown, so it becomes the new goal. Torque can be found using the angular momentum equation.

Sketch: To apply the angular momentum equation, select a control volume surrounding the turbine. Then, sketch a force and momentum diagram:

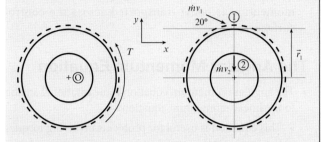

In the force diagram, the torque T is the external torque from the generator. Because this torque opposes angular acceleration, its direction is counterclockwise. The flow is idealized by using one inlet momentum flow at section 1 and one outlet momentum flow at section 2.

Select point "O" to sum moments about. Because the flow is steady, the accumulation of momentum is zero. Thus, the angular momentum equation (Eq. 6.26) simplifies to

$$\sum \mathbf{M}_A = \{\mathbf{r}_2 \times (\dot{m}\mathbf{v}_2)\} - \{\mathbf{r}_1 \times (\dot{m}\mathbf{v}_1)\} \qquad \textbf{(b)}$$

Apply Eq. (b) in the z direction. Also, recognize that the flow at section 2 has no angular momentum. That is, $\{\mathbf{r}_2 \times (\dot{m}\mathbf{v}_2)\} = 0$. Thus, Eq. (b) simplifies to

$$T = 0 - \{-r_1 \dot{m} v_1 \cos 20°\}$$

which can be written as

$$T = r_1 \dot{m} v_1 \cos 20° \qquad \textbf{(c)}$$

In Eq. (c), the velocity v_1 can be calculated using the flow rate equation. Because velocity is not perpendicular to area, use the dot product:

$$Q_1 = \mathbf{V}_1 \cdot \mathbf{A}_1$$
$$Q = v_1 A_1 \sin 20°$$

which can be rewritten as

$$v_1 = \frac{Q}{A_1 \sin 20°} \qquad \textbf{(d)}$$

Now, the number of equations equals the number of unknowns. Thus, the plan is as follows:

1. Calculate inlet velocity v_1 using Eq. (d).
2. Calculate mass flow rate using $\dot{m} = \rho Q$.
3. Calculate torque using Eq. (c).
4. Calculate power using Eq. (a).

Take Action (Execute the Plan)

1. Volume flow rate equation:

$$v_1 = \frac{Q}{A_1 \sin 20°} = \frac{(0.5 \text{ m}^3/\text{s})}{\pi(1.0 \text{ m})(0.04 \text{ m}) \sin 20°} = 11.63 \text{ m/s}$$

2. Mass flow rate equation:

$$\dot{m} = \rho Q = (1000 \text{ kg/m}^3)(0.5 \text{ m}^3/\text{s}) = 500 \text{ kg/s}$$

3. Angular momentum equation:

$$\begin{aligned} T &= r_1 \dot{m} v_1 \cos 20° \\ &= (0.5 \text{ m})(500 \text{ kg/s})(11.63 \text{ m/s}) \cos 20° \\ &= 2732 \text{ N} \cdot \text{m} \end{aligned}$$

4. Power equation:

$$P = T\omega = (2732 \text{ N} \cdot \text{m})(125.7 \text{ rad/s})$$
$$\boxed{P = 343 \text{ kW}}$$

6.7 Summarizing Key Knowledge

Newton's Second Law of Motion

- A *force* is a push or pull of one body on another. A push/pull is an interaction that can cause a body to accelerate. A force always requires the interaction of two bodies.
- Forces can be classified into two categories:
  - *Body forces.* Forces in this category do not require that the interacting bodies be touching. Common body forces include weight, the magnetic force, and the electrostatic force.
  - *Surface forces.* Forces in this category require that the two interacting bodies are touching. Most forces are surface forces.
- Newton's second law $\Sigma \mathbf{F} = m\mathbf{a}$ applies to a fluid particle; other forms of this law are derived from this equation.
- Newton's second law asserts that forces are related to accelerations:
  - Thus, if $\Sigma \mathbf{F} > \mathbf{0}$, the particle must accelerate.
  - Thus, if $\mathbf{a} > \mathbf{0}$, the sum of forces must be nonzero.

Solving Vector Equations

- A vector equation is one for which the terms are vectors.
- A vector equation can be written as one or more equivalent scalar equations.

- The Visual Solution Method (VSM) is an approach for solving a vector equation that makes problem solving easier. The process for the VSM is as follows:
  - **Step 1.** Identify the vector equation in its general form.
  - **Step 2.** Sketch a diagram that shows the vectors on the left side of the equation. Sketch an equal sign. Sketch a diagram that shows the vectors on the right side of the equation.
  - **Step 3.** From the diagrams, apply the general equation, write the final results, and simplify the results to create the reduced equation(s).

The Linear Momentum Equation

- The linear momentum equation is Newton's second law in a form that is useful for solving problems in fluid mechanics.
- To derive the momentum equation, proceed as follows:
  - Begin with Newton's second law for a single particle.
  - Derive Newton's second law for a system of particles.
  - Apply the Reynolds transport theorem to give the final result.
- Physical interpretation:

$$\begin{pmatrix} \text{sum of} \\ \text{forces} \end{pmatrix} = \begin{pmatrix} \text{momentum} \\ \text{accumulation} \end{pmatrix} + \begin{pmatrix} \text{momentum} \\ \text{outflow} \end{pmatrix} \\ - \begin{pmatrix} \text{momentum} \\ \text{inflow} \end{pmatrix}$$

- The *momentum accumulation* term gives the rate at which the momentum inside the control volume is changing with time.
- The *momentum flow* terms give the rate at which momentum is being transported across the control surfaces.

The Angular Momentum Equation

- The angular momentum equation is the rotational analog to the linear momentum equation:
  - This equation is useful for problems involving torques (i.e., moments).
  - This equation is commonly applied to rotating machinery such as pumps, fans, and turbines.
- The physics of the angular momentum equation are

$$\begin{pmatrix} \text{sum of} \\ \text{moments} \end{pmatrix} = \begin{pmatrix} \text{angular momentum} \\ \text{accumulation} \end{pmatrix} \\ + \begin{pmatrix} \text{angular momentum} \\ \text{outflow} \end{pmatrix} - \begin{pmatrix} \text{angular momentum} \\ \text{inflow} \end{pmatrix}$$

- To apply the angular momentum equation, use the same process as that used for the linear momentum equation.

REFERENCES

1. Hibbeler, R.C. *Dynamics*. Englewood Cliffs, NJ: Prentice Hall, 1995.

PROBLEMS

Newton's Second Law of Motion (§6.1)

6.1 Identify the surface and body forces acting on a glider in flight. Also, sketch a free body diagram and explain how Newton's laws of motion apply.

The Linear Momentum Equation: Theory (§6.2)

6.2 Which of the following are correct with respect to the derivation of the momentum equation? (Select all that apply.)

a. Reynold's transport theorem is applied to Fick's law.

b. The extensive property is momentum.

c. The intensive property is mass.

d. The net momentum flow is the "ins" minus the "outs."

e. The net force is the sum of forces acting on the matter inside the CV.

f. The velocity is assumed to be uniformly distributed across each inlet and outlet.

The Linear Momentum Equation: Application (§6.3)

6.3 When making a force diagram (FD) and its partner momentum diagram (MD) to set up the equations for a momentum equation problem (see Fig. 6.10), which of the following elements should be in the FD and which should be in the MD? (Classify each of the following as either "FD" or "MD.")

a. Forces required to hold walls, vanes, or pipes in place

b. Each mass stream with product $\dot{m}_o \mathbf{v}_o$ or product $\dot{m}_i \mathbf{v}_i$ crossing a control surface boundary

c. Weight of the fluid

d. Weight of a tank that contains the fluid

e. Pressure force associated with a fluid flowing across a control surface boundary

Applying the Momentum Equation to Fluid Jets (§6.4)

6.4 A "balloon rocket" is a balloon suspended from a taut wire by a hollow tube (drinking straw) and string. The nozzle is formed of a 1.1-cm-diameter tube, and an air jet exits the nozzle with a speed of 65 m/s and a density of 1.2 kg/m³. Find the force F needed to hold the balloon stationary. Neglect friction.

6.5 The balloon rocket is held in place by a force F. The pressure inside the balloon is 10 in-H_2O gage, the nozzle diameter is 0.5 cm, and the air density is 1.2 kg/m³. Find the exit velocity v and the force F. Neglect friction and assume the air flow is inviscid and irrotational.

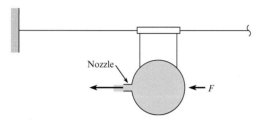

Problems 6.4, 6.5

6.6 A water jet of diameter 40 mm and speed $v = 15$ m/s is filling a tank. The tank has a mass of 27 kg and contains 30 liters of water at the instant shown. The water temperature is 15°C. Find the force acting on the exterior bottom of the tank and the force acting on the stop block. Neglect friction.

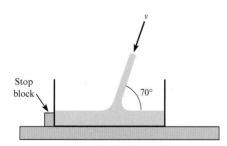

Problem 6.6

6.7 This tank provides a water jet (70°F) to cool a vertical metal surface during manufacturing. Calculate V when a horizontal force of 220 lbf is required to hold the metal surface in place. $Q = 5$ cfs.

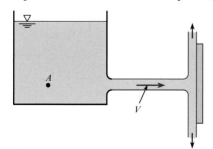

Problem 6.7

6.8 An engineer who is designing a water toy is making preliminary calculations. A user of the product will apply a force F_1 that moves a piston ($D = 90$ mm) at a speed of $V_{\text{piston}} = 250$ mm/s.

Water at 20°C jets out of a converging nozzle of diameter $d = 15$ mm. To hold the toy stationary, the user applies a force F_2 to the handle. Which force (F_1 versus F_2) is larger? Explain your answer using concepts of the momentum equation. Then calculate F_1 and F_2. Neglect friction between the piston and the walls.

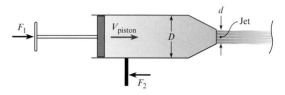

Problem 6.8

6.9 A firehose on a boat is producing a 2.5-in.-diameter water jet with a speed of $V = 70$ mph. The boat is held stationary by a cable attached to a pier, and the water temperature is 50°F. Calculate the tension in the cable.

6.10 A boat is held stationary by a cable attached to a pier. A firehose directs a spray of 5°C water at a speed of $V = 30$ m/s. If the allowable load on the cable is 4 kN, calculate the maximum allowable mass flow rate of the water jet. What is the corresponding diameter of the water jet?

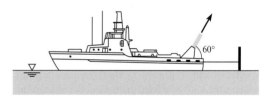

Problems 6.9, 6.10

6.11 A jet of water (60°F) is discharging at a constant rate of 2.0 cfs from the upper tank. If the jet diameter at section 1 is 3 in., what forces will be measured by scales A and B? Assume the empty tank weighs 150 lbf, the cross-sectional area of the lower tank is 4 ft², $h = 1$ ft, and $H = 9$ ft.

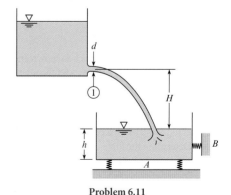

Problem 6.11

Applying the Momentum Equation to Vanes (§6.4)

6.12 Determine the forces in the x and y-directions needed to hold this fixed vane, which turns the oil jet ($SG = 0.9$) in a horizontal plane. Here $V_1 = 30$ m/s, $V_2 = 29$ m/s, and $Q = 0.8$ m³/s.

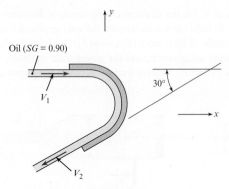

Problem 6.12

6.13 This planar water jet (60°F) is deflected by a fixed vane. What are the x- and y-components of force per unit width needed to hold the vane stationary? Neglect gravity.

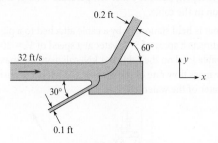

Problem 6.13

6.14 A water jet with a speed of 35 ft/s and a mass flow rate of 40 lbm/s is turned 30° by a fixed vane. Find the force of the water jet on the vane. Neglect gravity.

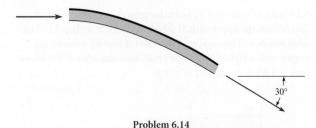

Problem 6.14

6.15 Water ($\rho = 1000$ kg/m³) strikes a block as shown and is deflected 30°. The flow rate of the water is 2.0 kg/s, and the inlet velocity is $V = 11$ m/s. The mass of the block is 1 kg. The coefficient of static friction between the block and the surface is 0.1. (The coefficient of static friction is defined as the ratio of friction force/normal force.) If the force parallel to the surface exceeds the frictional force, the block will move. Determine the force on the block and whether the block will move. Neglect the weight of the water.

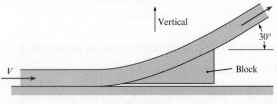

Problem 6.15

6.16 A two-dimensional liquid jet impinges on a vertical wall. Assuming that the incoming jet speed is the same as the exiting jet speed ($V_1 = V_2$), derive an expression for the force per unit width of jet exerted on the wall. What form do you think the upper liquid surface will take next to the wall? Sketch the shape you think it will take, and explain your reasons for drawing it that way.

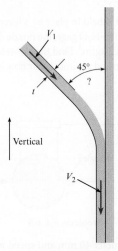

Problem 6.16

6.17 The "clam shell" thrust reverser sketched in the figure is often used to decelerate aircraft on landing. The sketch shows normal operation (a) and when deployed (b). The vanes are oriented 20° with respect to the vertical. The mass flow through the engine is 150 lbm/s, the inlet velocity is 300 ft/s, and the exit velocity is 1400 ft/s. Assume that when the thrust reverser is deployed, the exit velocity of the exhaust is unchanged. Assume the engine is stationary. Calculate the thrust under normal operation (lbf) and when the thrust reverser is deployed.

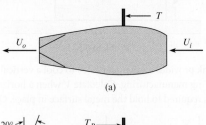

(a)

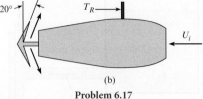

(b)

Problem 6.17

Applying the Momentum Equation to Nozzles (§6.4)

6.18 Firehoses are fitted with special nozzles. Use the Internet or contact your local fire department to find information on operational conditions and typical hose and nozzle sizes used.

6.19 High-speed water jets are used for specialty cutting applications. The pressure in the chamber is approximately 40,000 psig.

Using the Bernoulli equation, estimate the water speed exiting the nozzle exhausting to atmospheric pressure. Neglect compressibility effects and assume a water temperature of 60°F.

6.20 Water at 60°F flows through a nozzle that contracts from a diameter of 4 in. to 1 in. The pressure at section 1 is 2200 psfg, and atmospheric pressure prevails at the exit of the jet. Calculate the speed of the flow at the nozzle exit and the force required to hold the nozzle stationary. Neglect weight.

6.21 Water at 15°C flows through a nozzle that contracts from a diameter of 12 cm to 3 cm. The exit speed is $v_2 = 30$ m/s, and atmospheric pressure prevails at the exit of the jet. Calculate the pressure at section 1 and the force required to hold the nozzle stationary. Neglect weight.

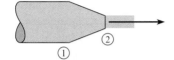

Problems 6.20, 6.21

6.22 Water (at 50°F) flows through this nozzle at a rate of 25 cfs and discharges into the atmosphere. $D_1 = 18$ in., and $D_2 = 9$ in. Determine the force required at the flange to hold the nozzle in place. Assume irrotational flow. Neglect gravitational forces.

6.23 Solve Prob. 6.22 using the following values: $Q = 0.30$ m³/s, $D_1 = 20$ cm, and $D_2 = 12$ cm. ($\rho = 1000$ kg/m³.)

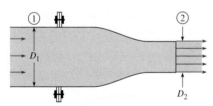

Problems 6.22, 6.23

6.24 A 15 cm nozzle is bolted with six bolts to the flange of a 30 cm pipe. If water ($\rho = 1000$ kg/m³) discharges from the nozzle into the atmosphere, calculate the tension load in each bolt when the pressure in the pipe is 200 kPa. Assume irrotational flow.

6.25 This spray head discharges water ($\rho = 62.4$ lbm/ft³) at a rate of 5 ft³/s. Assuming irrotational flow and an efflux speed of 60 ft/s in the free jet, determine what force acting through the bolts of the flange is needed to keep the spray head on the 6 in. pipe. Neglect gravitational forces.

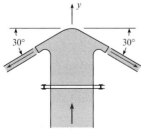

Problem 6.25

Applying the Momentum Equation to Pipe Bends (§6.4)

6.26 A hot gas stream enters a uniform-diameter return bend as shown. The entrance velocity is 75 ft/s, the gas density is 0.02 lbm/ft³, and the mass flow rate is 3 lbm/s. Water is sprayed into the duct to cool the gas down. The gas exits with a density of 0.05 lbm/ft³. The mass flow of water into the gas is negligible. The pressures at the entrance and exit are the same and equal to the atmospheric pressure. Find the force required to hold the bend.

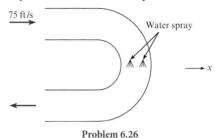

Problem 6.26

6.27 The pipe shown has a 180° horizontal bend in it as shown, and D is 30 cm. The discharge of water ($\rho = 1000$ kg/m³) in the pipe and bend is 0.45 m³/s, and the pressure in the pipe and bend is 100 kPa gage. If the bend volume is 0.10 m³, and the bend itself weighs 300 N, what force must be applied at the flanges to hold the bend in place?

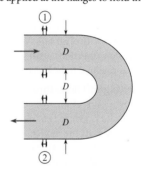

Problem 6.27

6.28 Water (at 50°F) flows in the 90° horizontal bend at a rate of 20 cfs and discharges into the atmosphere past the downstream flange. The pipe diameter is 1 ft. What force must be applied at the upstream flange to hold the bend in place? Assume that the volume of water downstream of the upstream flange is 3 ft³ and that the bend and pipe weigh 80 lbf. Assume the pressure at the upstream flange is 6 psig.

6.29 The gage pressure throughout the horizontal 90° pipe bend is 250 kPa. If the pipe diameter is 1.5 m and the water (at 10°C) flow rate is 15 m³/s, what x-component of force must be applied to the bend to hold it in place against the water action?

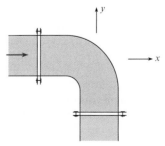

Problems 6.28 and 6.29

6.30 This 30° vertical bend in a pipe with a 2.5 ft diameter carries water ($\rho = 62.4$ lbm/ft^3) at a rate of 31.4 cfs. If the pressure p_1 is 10 psig at the lower end of the bend, where the elevation is 100 ft, and p_2 is 8.5 psig at the upper end, where the elevation is 103 ft, what will be the vertical component of force that must be exerted by the "anchor" on the bend to hold it in position? The bend itself weighs 300 lbf, and the length L is 6 ft.

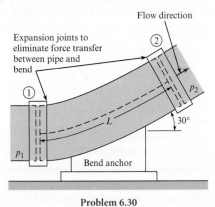

Problem 6.30

6.31 A horizontal reducing bend turns the flow of water ($\rho = 1000$ kg/m^3) through 60°. The inlet area is 0.001 m^2, and the outlet area is 0.0001 m^2. The water from the outlet discharges into the atmosphere with a velocity of 80 m/s. What horizontal force (parallel to the initial flow direction) acting through the metal of the bend at the inlet is required to hold the bend in place?

6.32 Water ($\rho = 1000$ kg/m^3) flows through a horizontal bend and T section as shown. At section a the flow enters with a velocity of 8 m/s, and the pressure is 6.5 kPa gage. At both sections b and c the flow exits the device with a velocity of 3 m/s, and the pressure at these sections is atmospheric ($p = 0$). The cross-sectional areas at a, b, and c are all the same: 0.20 m^2. Find the x- and y-components of force necessary to restrain the section.

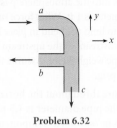

Problem 6.32

The Momentum Equation: Other Situations (§6.4)

6.33 An engineer is measuring the lift and drag on a wind turbine blade section mounted in a two-dimensional wind tunnel. The wind tunnel is 0.5 m high and 0.5 m deep (into the paper). The upstream wind velocity is uniform at 10 m/s, and the downstream velocity is 12 m/s and 8 m/s as shown. The vertical component of velocity is zero at both stations. The test section is 1 m long. The engineer measures the pressure distribution in the tunnel along the upper and lower walls and finds

$$p_u = 100 - 10x - 20x(1-x)(\text{Pa gage})$$
$$p_l = 100 - 10x + 20x(1-x)(\text{Pa gage})$$

where x is the distance in meters measured from the beginning of the test section. The gas density is uniform throughout and equal to 1.2 kg/m^3. The lift and drag are the vectors indicated on the figure. Find the lift and drag forces acting on the wind turbine blade section. Hint: The forces acting on the fluid are in the opposite direction to the lift and drag vectors drawn in the figure.

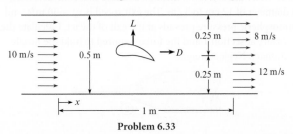

Problem 6.33

6.34 A modern turbofan engine in a commercial jet takes in air, part of which passes through the compressors, combustion chambers, and turbine, and the rest of which bypasses the compressor and is accelerated by the fans. The mass flow rate of bypass air to the mass flow rate through the compressor-combustor-turbine path is called the "bypass ratio." The total flow rate of air entering a turbofan is 300 kg/s with a velocity of 300 m/s. The engine has a bypass ratio of 2.5. The bypass air exits at 600 m/s, whereas the air through the compressor–combustor–turbine path exits at 1000 m/s. What is the thrust of the turbofan engine? Clearly show your control volume and application of momentum equation.

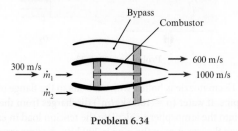

Problem 6.34

Applying the Momentum Equation to Moving CVs (§6.5)

6.35 A large tank of liquid is resting on a frictionless plane as shown. Explain in a qualitative way what will happen after the cap is removed from the short pipe.

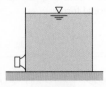

Problem 6.35

6.36 Consider a tank of water ($\rho = 1000$ kg/m^3) in a container that rests on a sled that is situated on a horizontal ice sheet. A high pressure is maintained by a compressor so that a jet of water leaving the tank horizontally from an orifice does so at a constant speed of 25 m/s relative to the tank. If there is 0.10 m³ of water in the tank at time t and the diameter of the jet is 15 mm, what will be the acceleration of the sled at time t if the empty tank and compressor have a weight of 350 N and the coefficient of friction between the sled and the ice is 0.05?

6.37 It is common practice in rocket trajectory analyses to neglect the body-force term and drag, so the velocity at burnout is given by

$$v_{bo} = \frac{T}{\dot{m}} \ln \frac{m_0}{m_f}$$

Assuming a thrust-to-mass-flow ratio of 3000 N · s/kg and a final mass of 50 kg, calculate the initial mass needed to establish the rocket in an earth orbit at a velocity of 7200 m/s.

The Angular Momentum Equation (§6.6)

6.38 Two small liquid-propellant rocket motors are mounted at the tips of a helicopter rotor to augment power under emergency conditions. The diameter of the helicopter rotor is 7 m, and it rotates at 1 rev/s. The air enters at the tip speed of the rotor, and

exhaust gases exit at 500 m/s with respect to the rocket motor. The intake area of each motor is 20 cm², and the air density is 1.2 kg/m³. Calculate the power provided by the rocket motors. Neglect the mass rate of flow of fuel in this calculation.

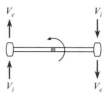

Problem 6.38

6.39 Design a rotating lawn sprinkler to deliver 0.25 in. of water per hour over a circle of 50 ft radius. Make the simplifying assumptions that the pressure to the sprinkler is 50 psig and that frictional effects involving the flow of water through the sprinkler flow passages are negligible (the Bernoulli equation is applicable). However, do not neglect the friction between the rotating element and the fixed base of the sprinkler.

CHAPTER**SEVEN**

The Energy Equation

CHAPTER ROAD MAP This chapter describes how conservation of energy can be applied to a flowing fluid. The resulting equation is called the *energy equation*.

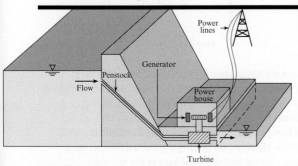

FIGURE 7.1
The energy equation can be applied to hydroelectric power generation. In addition, the energy equation can be applied to thousands of other applications. It is one of the most useful equations in fluid mechanics.

LEARNING OUTCOMES

WORK AND ENERGY (§7.1).
- Define energy, work, and power.
- Define a pump and a turbine.
- Classify energy into categories.
- Know common units.

CONSERVATION OF ENERGY FOR A CLOSED SYSTEM (§7.2).
- Know the main ideas about conservation of energy for a closed system.
- Apply the equation(s) to solve problems and answer questions.

THE ENERGY EQUATION (§7.3).
- Know the most important ideas about the energy equation.
- Calculate α.
- Define flow work and shaft work.
- Define head and know the various types of head.
- Apply the energy equation to solve problems.

THE POWER EQUATION (§7.4).
- Know the concepts associated with each of the power equations.
- Solve problems that involve the power equation.

EFFICIENCY (§7.4).
- Define mechanical efficiency.
- Solve problems that involve efficiency of components such as pumps and turbines.

THE SUDDEN EXPANSION (§7.7).
- Calculate the head loss for a sudden expansion.

THE EGL/HGL (§7.8).
- Explain the main ideas about the EGL and HGL.
- Sketch the EGL and HGL.
- Solve problems that involve the EGL and HGL.

7.1 Technical Vocabulary: Work, Energy, and Power

Conservation of energy is perhaps the *single most useful equation in all of engineering*. The key to applying this equation is to have solid knowledge of the foundational concepts of energy, work, and power. In addition to reviewing these topics, this section also defines pumps and turbines.

Energy

Energy is the property of a system that characterizes the amount of work that this system can do on its environment. In simple terms, if matter (i.e., the system) can be used to lift a weight, then that matter has energy.

Examples

- Water behind a dam has energy because the water can be directed through a pipe (i.e., a penstock), then used to rotate a wheel (i.e., a water turbine) that lifts a weight. Of course, this work can also rotate the shaft of an electrical generator, which is used to produce electrical power.
- Wind has energy because the wind can pass across a set of blades (e.g., a windmill), rotate the blades, and lift a weight that is attached to a rotating shaft. This shaft can also do work to rotate the shaft of an electrical generator.
- Gasoline has energy because it can be placed into a cylinder (e.g., a gas engine), burned, and expanded to move a piston in a cylinder. This moving cylinder can then be connected to a mechanism that is used to lift a weight.

The SI unit of energy, the *joule*, is the energy associated with a force of one newton acting through a distance of one meter. For example, if a person with a weight of 700 newtons travels up a 10-meter flight of stairs, then their gravitational potential energy has changed by $\Delta PE = (700 \text{ N})(10 \text{ m}) = 700 \text{ N·m} = 700 \text{ J}$. In traditional units, the unit of energy the *foot-pound force* (lbf) is defined as energy associated with a force of 1.0 lbf moving through a distance of 1.0 foot.

Another way to define a unit of energy is to describe the heating of water. A *small calorie* (cal) is the approximate amount of energy required to increase the temperature of 1.0 gram of water by 1°C. The unit conversion between small calories and joules is 1.0 cal = 4.187 J.* The *large calorie* (Cal) is the amount of energy to raise 1.0 kg of water by 1°C. Thus, 1.0 Cal = 4187 J. The large calorie is used in the United States to characterize the energy in food. Thus, a food item with 100 calories has an energy content of 0.4187 MJ. Energy in the traditional system is often measured using the British thermal unit (Btu). One Btu is the amount of energy required to raise the temperature of 1.0 lbm of water by 1.0°F.

Energy can be classified into categories:

- **Mechanical energy.** This is the energy associated with motion (i.e., *kinetic energy*) plus the energy associated with position in a field. Regarding position in a field, this refers to position in a gravitational field (i.e., gravitational potential energy) and to deflection of an elastic object such as a spring (i.e., spring potential energy).
- **Thermal energy.** This is energy associated with temperature changes and phase changes. For example, select a system comprised of 1 kg of ice (about 1 liter). The energy to melt the ice is 334 kJ. The energy to raise the temperature of the liquid water from 0°C to 100°C is 419 kJ.
- **Chemical energy.** This is the energy associated with chemical bonds between elements. For example, when methane (CH_4) is burned, there is a chemical reaction that involves the breaking of the bonds in the methane and formation of new bonds to produce CO_2 and

---

*There are several different definitions of a calorie in the literature. For example, you might find a reference that states that 1.0 cal = 4.184 J.

FIGURE 7.2

(a) For a spray bottle, the force acting through a distance is an example of mechanical work. (b) For the wind turbine, the pressure of the air causes a torque that acts through an angular displacement. This is also an example of mechanical work.

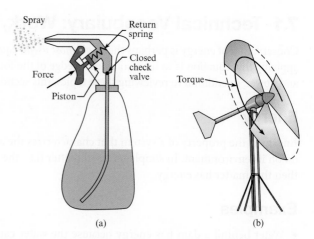

water. This chemical reaction releases heat, which is another way of saying that the chemical energy is converted to thermal energy during combustion.

- **Electrical energy.** This is the energy associated with electrical change. For example, a charged capacitor contains the amount of electrical energy $\Delta E = 1/2\ CV^2$, where C is capacitance and V is voltage.

- **Nuclear energy.** This is energy associated with the binding of the particles in the nucleus of an atom. For example, when the uranium atom divides into two other atoms during fission, energy is released.

Work

Usually, students in college classes are taught the concept of mechanical work first. **Mechanical work** occurs when a force acts through a distance. A better (i.e., more precise) definition is that work is given as the line integral of force **F** and displacement **ds**, as in

$$W = \int_{s_1}^{s_2} \mathbf{F} \cdot \mathbf{ds} \tag{7.1}$$

The units and dimensions of work are the same as the units and dimensions of energy. Fig. 7.2 shows two examples of mechanical work.

To perform energy balances (i.e., apply the law of conservation of energy) for real-world problems, you need a more general definition of work.* The definition that we like best is this:[†] **Work** is any interaction at the boundary of a system that is not heat transfer or the transfer of matter. For example, when electrical power is supplied to a motor, the electric current is classified as a work term.

Power

Power, which expresses a rate of work or energy, is defined by

$$P \equiv \frac{\text{quantity of work (or energy)}}{\text{interval of time}} = \lim_{\Delta t \to 0} \frac{\Delta W}{\Delta t} = \dot{W} \tag{7.2}$$

---

*This generalized kind of work is sometimes called *thermodynamic work* to distinguish it from *mechanical work*. In this text, we use the label *work* to represent all types of work, including mechanical work.

[†]This definition comes from chemical engineering professor and Nobel Prize winner John Fenn in his book *Engines, Energy, and Entropy: A Thermodynamics Primer*, p. 5.

Eq. (7.2) is defined at an instant in time because power can vary with time. To calculate power, engineers use several different equations. For rectilinear motion, such as a car or bicycle, the amount of work is the product of force and displacement: $\Delta W = F\Delta x$. Then, power can be found using

$$P = \lim_{\Delta t \to 0} \frac{F\Delta x}{\Delta t} = FV \qquad (7.3a)$$

where V is the velocity of the moving body.

When a shaft is rotating (Fig. 7.2b), the amount of work is given by the product of torque and angular displacement, $\Delta W = T\Delta\theta$. In this case, the power equation is

$$P = \lim_{\Delta t \to 0} \frac{T\Delta\theta}{\Delta t} = T\omega \qquad (7.3b)$$

where ω is the angular speed. The SI units of angular speed are rad/s.

Because power has units of energy per time, the SI unit is a joule/second, which is called a watt. Common units for power are the watt (W), horsepower (hp), and the ft-lbf/s. Some typical values of power include the following:

- An incandescent lightbulb can use 60 to 100 J/s of energy.
- A well-conditioned athlete can sustain a power output of about 300 J/s for an hour. This is about four-tenths of a horsepower. One horsepower is the approximate power that a draft horse can supply.
- A typical midsize car (a 2011 Toyota Camry) has a rated power of 126 kW (169 hp).
- A large hydroelectric facility (i.e., the Bonneville Dam on the Columbia River 40 miles east of Portland, Oregon) has a rated power of 1080 MW.

Pumps and Turbines

A **turbine** is a machine that is used to extract energy from a flowing fluid.* Examples of turbines include the horizontal axis wind turbine, the gas turbine, the Kaplan turbine, the Francis turbine, and the Pelton wheel.

A **pump** is a machine that is used to provide energy to a flowing fluid. Examples of pumps include the piston pump, the centrifugal pump, the diaphragm pump, and the gear pump.

7.2 Conservation of Energy

When James Prescott Joule died, his obituary in *The Electrical Engineer* (1, p. 311) stated:

> *Very few indeed who read this announcement will realize how great of a man has passed away; and yet it must be admitted by those most competent to judge that his name must be classed among the greatest original workers in science.*

Joule was a brewer who engaged in science as a hobby, yet he formulated one of the most important scientific laws ever developed. However, Joule's theory of conservation of energy was so controversial that he could not get a scientific journal to publish it, so his theory first appeared in a local Manchester newspaper (2). What a fine example of persistence! Nowadays, Joule's ideas about work and energy are foundational to engineering. This section introduces Joule's theory.

---

*The engine on a jet, which is called a gas turbine, is a notable exception. The jet engine adds energy to a flowing fluid, thereby increasing the momentum of a fluid jet and producing thrust.

FIGURE 7.3

The law of conservation of energy for a closed system.

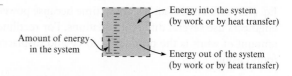

Joule's Theory of Energy Conservation

Joule recognized that the energy of a *closed system* can be changed in only two ways:

- **Work.** The energy of the system can be changed by work interactions at the boundary.
- **Heat transfer.** The energy of the system can change by heat transfer across the boundary. **Heat transfer** can be defined as the transfer of thermal energy from hot to cold by mechanisms of conduction, convection, and radiation.

Joule's idea of energy conservation is illustrated in Fig. 7.3. The system is represented by the blue box. The scale on the left side of the figure represents the quantity of energy in the system. The arrows on the right side illustrate that energy can increase or decrease via work or heat transfer interactions. Note that energy is a property of a system, whereas work and heat transfer are interactions that occur on system boundaries.

The work and energy balance proposed by Joule is captured with an equation called the first law of thermodynamics:

$$\Delta E \qquad = \qquad Q \qquad - \qquad W$$

$$\left\{ \begin{array}{c} \text{increase in} \\ \text{energy stored} \\ \text{in the system} \end{array} \right\} = \left\{ \begin{array}{c} \text{amount of energy} \\ \text{that entered system} \\ \text{by heat transfer} \end{array} \right\} - \left\{ \begin{array}{c} \text{amount of energy} \\ \text{that left system} \\ \text{due to work} \end{array} \right\} \qquad \textbf{(7.4)}$$

The terms in Eq. (7.4) have units of joules, and the equation is applied during a time interval when the system undergoes a process to move from state 1 to state 2. To modify Eq. (7.4) so that it applies at an instant in time, take the derivative to give

$$\frac{dE}{dt} = \dot{Q} - \dot{W} \qquad \textbf{(7.5)}$$

Eq. (7.5) applies at an instant in time and has units of joules per second or watts. The work and heat transfer terms have sign conventions:

- W and $\dot{W}$ are positive if work is done by the system on the surroundings.
- W and $\dot{W}$ are negative if work is done by the surroundings on the system.
- Q and $\dot{Q}$ are positive if heat (i.e., thermal energy) is transferred into the system.
- Q and $\dot{Q}$ are negative if heat (i.e., thermal energy) is transferred out of the system.

Control Volume (Open System)

Eq. (7.5) applies to a *closed system*. To extend it to a CV, apply the Reynolds transport theorem, Eq. (5.23). Let the extensive property be energy ($B_{sys} = E$) and let $b = e$ to obtain

$$\dot{Q} - \dot{W} = \frac{d}{dt} \int_{cv} e\rho \, d\Psi + \int_{cs} e\rho \mathbf{V} \cdot d\mathbf{A} \qquad \textbf{(7.6)}$$

where e is energy per mass in the fluid. Eq. (7.6) is the general form of conservation of energy for a control volume. However, most problems in fluid mechanics can be solved with a simpler form of this equation. This simpler equation will be derived in the next section.

7.3 The Energy Equation

This section shows how to simplify Eq. (7.6) to a form that is convenient for problems that occur in fluid mechanics.

Select Eq. (7.6). Then, let $e = e_k + e_p + u$, where e_k is the kinetic energy per unit mass, e_p is the gravitational potential energy per unit mass, and u is the internal energy* per unit mass:

$$\dot{Q} - \dot{W} = \frac{d}{dt}\int_{cv} (e_k + e_p + u)\rho\, d\forall + \int_{cs} (e_k + e_p + u)\rho\mathbf{V} \cdot d\mathbf{A} \qquad (7.7)$$

Next, let†

$$e_k = \frac{\text{kinetic energy of a fluid particle}}{\text{mass of this fluid particle}} = \frac{mV^2/2}{m} = \frac{V^2}{2} \qquad (7.8)$$

Similarly, let

$$e_p = \frac{\text{gravitational potential energy of a fluid particle}}{\text{mass of this fluid particle}} = \frac{mgz}{m} = gz \qquad (7.9)$$

where z is the elevation measured relative to a datum. When Eqs. (7.8) and (7.9) are substituted into Eq. (7.7), the result is

$$\dot{Q} - \dot{W} = \frac{d}{dt}\int_{cv} \left(\frac{V^2}{2} + gz + u\right)\rho\, d\forall + \int_{cs} \left(\frac{V^2}{2} + gz + u\right)\rho\mathbf{V} \cdot d\mathbf{A} \qquad (7.10)$$

Shaft and Flow Work

To simplify the work term in Eq. (7.10), classify work into two categories:

$$(\text{work}) = (\text{flow work}) + (\text{shaft work})$$

When the work is associated with a pressure force, then the work is called **flow work**. Alternatively, **shaft work** is any work that is not associated with a pressure force. Shaft work is usually done through a shaft (from which the term originates) and is commonly associated with a pump or turbine. According to the sign convention for work, pump work is negative. Similarly, turbine work is positive. Thus,

$$\dot{W}_{\text{shaft}} = \dot{W}_{\text{turbines}} - \dot{W}_{\text{pumps}} = \dot{W}_t - \dot{W}_p \qquad (7.11)$$

To derive an equation for flow work, use the idea that work equals force times distance. Begin the derivation by defining a control volume situated inside a converging pipe (Fig. 7.4). At section 2, the fluid that is inside the control volume will push on the fluid that is outside the control volume. The magnitude of the pushing force is p_2A_2. During a time interval Δt, the displacement of the fluid at section 2 is $\Delta x_2 = V_2\Delta t$. Thus, the amount of work is

$$\Delta W_2 = (F_2)(\Delta x_2) = (p_2 A_2)(V_2 \Delta t) \qquad (7.12)$$

Convert the amount of work given by Eq. (7.12) into a rate of work:

$$\dot{W}_2 = \lim_{\Delta t \to 0} \frac{\Delta W_2}{\Delta t} = p_2 A_2 V_2 = \left(\frac{p_2}{\rho}\right)(\rho A_2 V_2) = \dot{m}\left(\frac{p_2}{\rho}\right) \qquad (7.13)$$

---

*By definition, internal energy contains all forms of energy that are not kinetic energy or gravitational potential energy.

†It is assumed that the control surface is not accelerating, so V, which is referenced to the control surface, is also referenced to an inertial reference frame.

FIGURE 7.4

Sketch for deriving flow work.

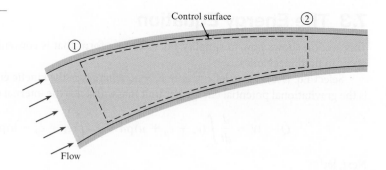

This work is positive because the fluid inside the control volume is doing work on the environment. In a similar manner, the flow work at section 1 is negative and is given by

$$\dot{W}_1 = -\dot{m}\left(\frac{p_1}{\rho}\right)$$

The net flow work for the situation pictured in Fig. 7.4 is

$$\dot{W}_{\text{flow}} = \dot{W}_2 + \dot{W}_1 = \dot{m}\left(\frac{p_2}{\rho}\right) - \dot{m}\left(\frac{p_1}{\rho}\right) \tag{7.14}$$

Eq. (7.14) can be generalized to a situation involving multiple streams of fluid passing across a control surface:

$$\dot{W}_{\text{flow}} = \sum_{\text{outlets}} \dot{m}_{\text{out}}\left(\frac{p_{\text{out}}}{\rho}\right) - \sum_{\text{inlets}} \dot{m}_{\text{in}}\left(\frac{p_{\text{in}}}{\rho}\right) \tag{7.15}$$

To develop a general equation for flow work, use integrals to account for velocity and pressure variations on the control surface. Also, use the dot product to account for flow direction. The general equation for flow work is

$$\dot{W}_{\text{flow}} = \int_{\text{cs}} \left(\frac{p}{\rho}\right)\rho\mathbf{V} \cdot d\mathbf{A} \tag{7.16}$$

In summary, the work term is the sum of flow work, Eq. (7.16), and shaft work, Eq. (7.11):

$$\dot{W} = \dot{W}_{\text{flow}} + \dot{W}_{\text{shaft}} = \left(\int_{\text{cs}} \left(\frac{p}{\rho}\right)\rho\mathbf{V} \cdot d\mathbf{A}\right) + \dot{W}_{\text{shaft}} \tag{7.17}$$

Introduce the work term from Eq. (7.17) into Eq. (7.10), and let $\dot{W}_{\text{shaft}} = \dot{W}_s$:

$$\dot{Q} - \dot{W}_s - \int_{\text{cs}} \frac{p}{\rho}\rho\mathbf{V} \cdot d\mathbf{A}$$
$$= \frac{d}{dt}\int_{\text{cv}} \left(\frac{V^2}{2} + gz + u\right)\rho\,d\Forall + \int_{\text{cs}} \left(\frac{V^2}{2} + gz + u\right)\rho\mathbf{V} \cdot d\mathbf{A} \tag{7.18}$$

In Eq. (7.18), combine the last term on the left side with the last term on the right side:

$$\dot{Q} - \dot{W}_s = \frac{d}{dt}\int_{\text{cv}} \left(\frac{V^2}{2} + gz + u\right)\rho\,d\Forall + \int_{\text{cs}} \left(\frac{V^2}{2} + gz + u + \frac{p}{\rho}\right)\rho\mathbf{V} \cdot d\mathbf{A} \tag{7.19}$$

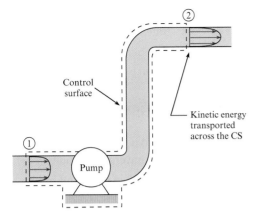

FIGURE 7.5

Flow carries kinetic energy into and out of a control volume.

Replace $p/\rho + u$ with the specific enthalpy, h. The integral form of the energy principle is

$$\dot{Q} - \dot{W}_s = \frac{d}{dt}\int_{cv}\left(\frac{V^2}{2} + gz + u\right)\rho\, d\Psi + \int_{cs}\left(\frac{V^2}{2} + gz + h\right)\rho\mathbf{V}\cdot d\mathbf{A} \qquad (7.20)$$

Kinetic Energy Correction Factor

The next simplification is to extract the velocity terms out of the integrals on the right side of Eq. (7.20). This is done by introducing the kinetic energy correction factor.

Fig. 7.5 shows fluid that is pumped through a pipe. At sections 1 and 2, kinetic energy is transported across the control surface by the flowing fluid. To derive an equation for this kinetic energy, start with the mass flow rate equation:

$$\dot{m} = \rho A\overline{V} = \int_A \rho V dA$$

This integral can be conceptualized as adding up the mass of each fluid particle that is crossing the section area and then dividing by the time interval associated with this crossing. To convert this integral to kinetic energy (KE), multiply the mass of each fluid particle by ($V^2/2$):

$$\left\{\begin{array}{c} \text{rate of KE} \\ \text{transported} \\ \text{across a section} \end{array}\right\} = \int_A \rho V\left(\frac{V^2}{2}\right)dA = \int_A \frac{\rho V^3 dA}{2}$$

The **kinetic energy correction factor** is defined as

$$\alpha = \frac{\text{actual KE/time that crosses a section}}{\text{KE/time by assuming a uniform velocity distribution}} = \frac{\displaystyle\int_A \frac{\rho V^3 dA}{2}}{\displaystyle\frac{\overline{V}^3}{2}\int_A \rho dA}$$

For a constant density fluid, this equation simplifies to

$$\alpha = \frac{1}{A}\int_A \left(\frac{V}{\overline{V}}\right)^3 dA \qquad (7.21)$$

For theoretical development, α is found by integrating the velocity profile using Eq. (7.21). This approach, illustrated in Example 7.1, is a lot of work. Thus in application, engineers commonly estimate a value of α. Some guidelines are as follows:

- For fully developed laminar flow in a pipe, the velocity distribution is parabolic. Use $\alpha = 2.0$ because this is the correct value as shown by Example 7.1.

- For fully developed turbulent flow in a pipe, $\alpha \approx 1.05$ because the velocity profile is plug-like. Use $\alpha = 1.0$ for this case.

- For flow at the exit of a nozzle or converging section, use $\alpha = 1.0$ because converging flow leads to a uniform velocity profile. This is why wind tunnels use converging sections.

- For a uniform flow (such as air flow in a wind tunnel or air flow incident on a wind turbine), use $\alpha = 1.0$.

EXAMPLE 7.1

Calculating the Kinetic Energy Correction Factor for Laminar Flow

Problem Statement

The velocity distribution for laminar flow in a pipe is given by the equation

$$V(r) = V_{\max}\left[1 - \left(\frac{r}{r_0}\right)^2\right]$$

where $V_{\max}$ is the velocity in the center of the pipe, r_0 is the radius of the pipe, and r is the radial distance from the center. Find the KE correction factor α.

Define the Situation

There is laminar flow in a round pipe.

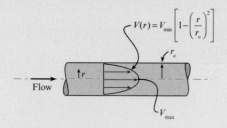

State the Goal

$\alpha \leftarrow$ find the KE correction factor (no units)

Generate Ideas and Make a Plan

Because the goal is α, apply the definition given by Eq. (7.21):

$$\alpha = \frac{1}{A}\int_A \left(\frac{V(r)}{\overline{V}}\right)^3 dA \qquad \textbf{(a)}$$

Eq. (a) has one known (A) and two unknowns (dA, $\overline{V}$). To find dA, see Chapter 5, Fig. 5.3.

$$dA = 2\pi r\,dr \qquad \textbf{(b)}$$

To find $\overline{V}$, apply the *flow rate equation*:

$$\overline{V} = \frac{1}{A}\int_A V(r)\,dA = \frac{1}{\pi r_o^2}\int_{r=0}^{r=r_o} V(r)\,2\pi r\,dr \qquad \textbf{(c)}$$

Now, the problem is cracked. There are three equations and three unknowns. The plan is as follows:

1. Find the mean velocity $\overline{V}$ using Eq. (c).
2. Plug $\overline{V}$ into Eq. (a) and integrate.

Take Action (Execute the Plan)

1. Flow rate equation (find mean velocity):

$$\overline{V} = \frac{1}{\pi r_0^2}\left[\int_0^{r_0} V_{\max}\left(1 - \frac{r^2}{r_0^2}\right)2\pi r\,dr\right]$$

$$= \frac{2V_{\max}}{r_0^2}\left[\int_0^{r_0}\left(1 - \frac{r^2}{r_0^2}\right)r\,dr\right] = \frac{2V_{\max}}{r_0^2}\left[\int_0^{r_0}\left(r - \frac{r^3}{r_0^2}\right)dr\right]$$

$$= \frac{2V_{\max}}{r_0^2}\left[\left(\frac{r^2}{2} - \frac{r^4}{4r_0^2}\right)\Big|_0^{r_0}\right] = \frac{2V_{\max}}{r_0^2}\left[\frac{r_0^2}{2} - \frac{r_0^2}{4}\right] = V_{\max}/2$$

2. Definition of α:

$$\alpha = \frac{1}{A}\left[\int_A\left(\frac{V(r)}{\overline{V}}\right)^3 dA\right] = \frac{1}{\pi r_0^2 \overline{V}^3}\left[\int_0^{r_0} V(r)^3\,2\pi r\,dr\right]$$

$$= \frac{1}{\pi r_0^2 (V_{\max}/2)^3}\left[\int_0^{r_0}\left[V_{\max}\left(1 - \frac{r^2}{r_0^2}\right)\right]^3 2\pi r\,dr\right]$$

$$= \frac{16}{r_0^2}\left[\int_0^{r_0}\left(1 - \frac{r^2}{r_0^2}\right)^3 r\,dr\right]$$

To evaluate the integral, make a change of variable by letting $u = (1 - r^2/r_0^2)$. The integral becomes

$$\alpha = \left(\frac{16}{r_0^2}\right)\left(-\frac{r_0^2}{2}\right)\left(\int_1^0 u^3\,du\right) = 8\left(\int_0^1 u^3\,du\right)$$

$$= 8\left(\frac{u^4}{4}\Big|_0^1\right) = 8\left(\frac{1}{4}\right)$$

$$\boxed{\alpha = 2}$$

Review the Solution and the Process

1. *Knowledge.* Laminar fully developed flow in a round pipe is called Poiseuille flow. Useful facts:
   - The velocity profile is parabolic.
   - The mean velocity is one-half of the maximum (centerline) velocity: $\overline{V} = V_{max}/2$.
   - The kinetic energy correction factor is $\alpha = 2$.
2. *Knowledge.* In practice, engineers commonly estimate α. The purpose of this example is to illustrate how to calculate α.

Last Steps of the Derivation

Now that the KE correction factor is available, the derivation of the energy equation may be completed. Begin by applying Eq. (7.20) to the control volume shown in Fig. 7.5. Assume steady flow and that velocity is normal to the control surfaces. Then, Eq. (7.20) simplifies to

$$\dot{Q} - \dot{W}_s + \int_{A_1}\left(\frac{p_1}{\rho} + gz_1 + u_1\right)\rho V_1\,dA_1 + \int_{A_1}\frac{\rho V_1^3}{2}\,dA_1$$
$$= \int_{A_2}\left(\frac{p_2}{\rho} + gz_2 + u_2\right)\rho V_2\,dA_2 + \int_{A_2}\frac{\rho V_2^3}{2}\,dA_2 \tag{7.22}$$

Assume that piezometric head $p/\gamma + z$ is constant across sections 1 and 2.* If temperature is also assumed constant across each section, then $p/\rho + gz + u$ can be taken outside the integral to yield

$$\dot{Q} - \dot{W}_s + \left(\frac{p_1}{\rho} + gz_1 + u_1\right)\int_{A_1}\rho V_1\,dA_1 + \int_{A_1}\rho\frac{V_1^3}{2}\,dA_1$$
$$= \left(\frac{p_2}{\rho} + gz_2 + u_2\right)\int_{A_2}\rho V_2\,dA_2 + \int_{A_2}\rho\frac{V_2^3}{2}\,dA_2 \tag{7.23}$$

Next, factor out $\int\rho V\,dA = \rho\overline{V}A = \dot{m}$ from each term in Eq. (7.23). Because $\dot{m}$ does not appear as a factor of $\int(\rho V^3/2)dA$, express $\int(\rho V^3/2)dA$ as $\alpha(\rho\overline{V}^3/2)A$, where α is the kinetic energy correction factor:

$$\dot{Q} - \dot{W}_s + \left(\frac{p_1}{\rho} + gz_1 + u_1 + \alpha_1\frac{\overline{V}_1^2}{2}\right)\dot{m} = \left(\frac{p_2}{\rho} + gz_2 + u_2 + \alpha_2\frac{\overline{V}_2^2}{2}\right)\dot{m} \tag{7.24}$$

Divide through by $\dot{m}$:

$$\frac{1}{\dot{m}}(\dot{Q} - \dot{W}_s) + \frac{p_1}{\rho} + gz_1 + u_1 + \alpha_1\frac{\overline{V}_1^2}{2} = \frac{p_2}{\rho} + gz_2 + u_2 + \alpha_2\frac{\overline{V}_2^2}{2} \tag{7.25}$$

Introduce Eq. (7.11) into Eq. (7.25):

$$\frac{\dot{W}_p}{\dot{m}g} + \frac{p_1}{\gamma} + z_1 + \alpha_1\frac{\overline{V}_1^2}{2g} = \frac{\dot{W}_t}{\dot{m}g} + \frac{p_2}{\gamma} + z_2 + \alpha_2\frac{\overline{V}_2^2}{2g} + \frac{u_2 - u_1}{g} - \frac{\dot{Q}}{\dot{m}g} \tag{7.26}$$

*Euler's equation can be used to show that pressure variation normal to rectilinear streamlines is hydrostatic.

Introduce **pump head** and **turbine head**:

$$pump\ head = h_p = \frac{\dot{W}_p}{\dot{m}g} = \frac{\text{work/time done by pump on flow}}{\text{weight/time of flowing fluid}}$$

$$turbine\ head = h_t = \frac{\dot{W}_t}{\dot{m}g} = \frac{\text{work/time done by flow on turbine}}{\text{weight/time of flowing fluid}}$$

(7.27)

Equation (7.26) becomes

$$\frac{p_1}{\gamma} + \alpha_1 \frac{\overline{V}_1^2}{2g} + z_1 + h_p = \frac{p_2}{\gamma} + \alpha_2 \frac{\overline{V}_2^2}{2g} + z_2 + h_t + \left[\frac{1}{g}(u_2 - u_1) - \frac{\dot{Q}}{\dot{m}g} \right]$$

(7.28)

Eq. (7.28) is separated into terms that represent mechanical energy (nonbracketed terms) and terms that represent thermal energy (bracketed terms). A bracketed term is always positive because of the second law of thermodynamics. This term is called head loss and is represented by h_L. **Head loss** is the conversion of useful mechanical energy to waste thermal energy through viscous action. Head loss is analogous to thermal energy (heat) that is produced by Coulomb friction. When the bracketed term is replaced by head loss h_L, Eq. (7.28) becomes the *energy equation*:

$$\left(\frac{p_1}{\gamma} + \alpha_1 \frac{\overline{V}_1^2}{2g} + z_1 \right) + h_p = \left(\frac{p_2}{\gamma} + \alpha_2 \frac{\overline{V}_2^2}{2g} + z_2 \right) + h_t + h_L$$

(7.29)

Physical Interpretation of the Energy Equation

The energy equation describes an energy balance for a control volume (Fig. 7.6). The inflows of energy are balanced with the outflows of energy.* Regarding inflows, energy can be transported across the control surface by the flowing fluid, or a pump can do work on the fluid and thereby add energy to the fluid. Regarding outflows, energy within the flow can be used to do work on a turbine, energy can be transported across the control surface by the flowing fluid, or mechanical energy can be converted to waste thermal heat via head loss.

The energy balance can also be expressed using head:

$$\left(\frac{p_1}{\gamma} + \alpha_1 \frac{\overline{V}_1^2}{2g} + z_1 \right) + h_p = \left(\frac{p_2}{\gamma} + \alpha_2 \frac{\overline{V}_2^2}{2g} + z_2 \right) + h_t + h_L$$

$$\begin{pmatrix} \text{pressure head} \\ \text{velocity head} \\ \text{elevation head} \end{pmatrix}_1 + \begin{pmatrix} \text{pump} \\ \text{head} \end{pmatrix} = \begin{pmatrix} \text{pressure head} \\ \text{velocity head} \\ \text{elevation head} \end{pmatrix}_2 + \begin{pmatrix} \text{turbine} \\ \text{head} \end{pmatrix} + \begin{pmatrix} \text{head} \\ \text{loss} \end{pmatrix}$$

Head can be thought of as the *ratio of energy to weight for a fluid particle*, or head can describe the *energy per time that is passing across a section* because head and power are related by $P = \dot{m}gh$.

Working Equations

Table 7.1 summarizes the energy equation, its variables, and the main assumptions.

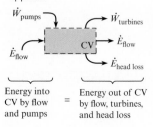

FIGURE 7.6

The energy balance for a CV when the energy equation is applied.

$\dot{W}_{\text{pumps}}$ $\dot{W}_{\text{turbines}}$

$\dot{E}_{\text{flow}}$ CV $\dot{E}_{\text{flow}}$

$\dot{E}_{\text{head loss}}$

Energy into CV by flow and pumps = Energy out of CV by flow, turbines, and head loss

---

*The term $\dot{E}_{\text{flow}}$ includes a work term—namely, flow work. Remember that energy is a property of a system, whereas work and heat transfer are interactions that occur on system boundaries. Here, we are using the term "energy balance" to describe (energy terms) + (work terms) + (heat transfer terms).

TABLE 7.1 Summary of the Energy Equation

Description	Equation	Terms

The energy equation has only one form.
 	
Major assumptions:
 	
• Steady state; no energy accumulation in CV.
• The CV has one inlet and one outlet.
• Constant density flow.
• All thermal energy terms (except for head loss) can be neglected.
• Streamlines are straight and parallel at each section.
• Temperature is constant across each section.
	$\left(\dfrac{p_1}{\gamma} + \alpha_1 \dfrac{\overline{V}_1^{\,2}}{2g} + z_1\right) + h_p$
	
	$= \left(\dfrac{p_2}{\gamma} + \alpha_2 \dfrac{\overline{V}_2^{\,2}}{2g} + z_2\right) + h_t + h_L$
	
	Eq. (7.29)
		$\left(\dfrac{p}{\gamma} + \alpha\dfrac{\overline{V}^2}{2g} + z\right) = \begin{pmatrix}\text{energy/weight transported}\\ \text{into or out of CV}\\ \text{by fluid flow}\end{pmatrix} =$ total head
		
		p/γ = pressure head at CS (m)
		$\alpha\dfrac{\overline{V}^2}{2g}$ = velocity head at CS (m)
		  (α = kinetic energy (KE) correction factor at CS)
		  ($\alpha \approx 1.0$ for turbulent flow)
		  ($\alpha \approx 1.0$ for nozzles)
		  ($\alpha \approx 2.0$ for fully-developed laminar flow in round pipe)
		z = elevation head at CS (m)
		
		h_p = head added by a pump (m)
		h_t = head removed by a turbine (m)
		h_L = head loss (m)
		  (to predict head loss, apply Eq. (10.45))

The process for applying the energy equation is as follows:

Step 1: Selection. Select the energy equation when the problem involves a pump, a turbine, or head loss. Check to ensure that the assumptions used to derive the energy equation are satisfied. The assumptions are steady flow, one inlet port and one outlet port, constant density, and negligible thermal energy terms (except for head loss).

Step 2: CV selection. Select and label section 1 (inlet port) and section 2 (outlet port). Locate sections 1 and 2 where (a) you know information or (b) where you want information. By convention, engineers usually do not sketch a CV when applying the energy equation.

Step 3: Analysis. Write the general form of the energy equation. Conduct a term-by-term analysis. Simplify the general equation to the reduced equation.

Step 4: Validation. Check units. Check the physics: (head in via fluid flow and pump) = (head out via fluid flow, turbine, and head loss).

EXAMPLE 7.2

Applying the Energy Equation to Predict the Speed of Water in a Pipe Connected to a Reservoir

Problem Statement

A horizontal pipe carries cooling water at 10°C for a thermal power plant. The head loss in the pipe is

$$h_L = \frac{0.02(L/D)V^2}{2g}$$

where L is the length of the pipe from the reservoir to the point in question, V is the mean velocity in the pipe, and D is the diameter of the pipe. If the pipe diameter is 20 cm and the rate of flow is 0.06 m³/s, what is the pressure in the pipe at $L = 2000$ m? Assume $\alpha_2 = 1$.

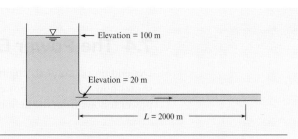

Define the Situation

Water flows in a system.

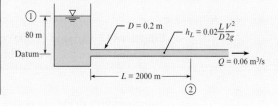

Assumptions:

- $\alpha_2 = 1.0$
- Steady flow

Properties: Water (10°C, 1 atm, Table A.5): $\gamma = 9810 \text{ N/m}^3$

State the Goal

$p_2(\text{kPa})$ ← pressure at section 2

Generate Ideas and Make a Plan

Select the energy equation because (a) the situation involves water flowing through a pipe and (b) the energy equation contains the goal (p_2). Locate section 1 at the surface and section 2 at the location where we want to know pressure. The plan is as follows:

1. Write the general form of the energy equation (7.29).
2. Analyze each term in the energy equation.
3. Solve for p_2.

Take Action (Execute the Plan)

1. Energy equation (general form):

$$\frac{p_1}{\gamma} + \alpha_1 \frac{\overline{V}_1^2}{2g} + z_1 + h_p = \frac{p_2}{\gamma} + \alpha_2 \frac{\overline{V}_2^2}{2g} + z_2 + h_t + h_L$$

2. Term-by-term analysis:

- $p_1 = 0$ because the pressure at top of a reservoir is $p_{\text{atm}} = 0$ gage.
- $V_1 \approx 0$ because the level of the reservoir is constant or changing very slowly.
- $z_1 = 100 \text{ m}$; $z_2 = 20 \text{ m}$.
- $h_p = h_t = 0$ because there are no pumps or turbines in the system.

- Find V_2 using the flow rate equation (5.3):

$$V_2 = \frac{Q}{A} = \frac{0.06 \text{ m}^3/\text{s}}{(\pi/4)(0.2 \text{ m})^2} = 1.910 \text{ m/s}$$

- Head loss is

$$h_L = \frac{0.02(L/D)V^2}{2g} = \frac{0.02(2000 \text{ m}/0.2 \text{ m})(1.910 \text{ m/s})^2}{2(9.81 \text{ m/s}^2)}$$
$$= 37.2 \text{ m}$$

3. Combine steps 1 and 2:

$$(z_1 - z_2) = \frac{p_2}{\gamma} + \alpha_2 \frac{\overline{V}_2^2}{2g} + h_L$$

$$80 \text{ m} = \frac{p_2}{\gamma} + 1.0 \frac{(1.910 \text{ m/s})^2}{2(9.81 \text{ m/s}^2)} + 37.2 \text{ m}$$

$$80 \text{ m} = \frac{p_2}{\gamma} + (0.186 \text{ m}) + (37.2 \text{ m})$$

$$p_2 = \gamma(42.6 \text{ m}) = (9810 \text{ N/m}^3)(42.6 \text{ m}) = \boxed{418 \text{ kPa}}$$

Review the Solution and the Process

1. *Skill.* Notice that section 1 was set at the free surface because properties are known there. Section 2 was set where we want to find information.

2. *Knowledge.* Regarding selection of an equation, we could have chosen the Bernoulli equation. However, it would have been an unwise choice because the Bernoulli equation assumes inviscid flow.

- *Key idea.* Select the Bernoulli equation if viscous effects can be neglected; select the energy equation if viscous effects are significant.
- *Rule of thumb.* When fluid is flowing through a pipe that is more than about five diameters long—that is, ($L/D > 5$)—viscous effects are significant.

7.4 The Power Equation

Depending on context, engineers use various equations for calculating power. This section shows how to calculate power associated with pumps and turbines. An equation for pump power follows from the definition of pump head given in Eq. (7.27):

$$\dot{W}_p = \gamma Q h_p = \dot{m} g h_p \tag{7.30a}$$

Similarly, the power delivered from a flow to a turbine is

$$\dot{W}_t = \gamma Q h_t = \dot{m} g h_t \tag{7.30b}$$

Equations (7.30a) and (7.30b) can be generalized to give an equation for calculating power associated with a pump or turbine:

$$P = \dot{m} g h = \gamma Q h \tag{7.31}$$

Equations for calculating power are summarized in Table 7.2.

TABLE 7.2 Summary of the Power Equation

Description	Equation		Terms
Rectilinear motion of an object such as an airplane, a submarine, or a car	$P = FV$	(7.3a)	P = power (W) F = force doing work (N) V = speed of object (m/s)
Rotational motion, such as a shaft driving a pump or an output shaft from a turbine	$P = T\omega$	(7.3b)	T = torque (N·m) ω = angular speed (rad/s)
Power supplied from a pump to a flowing fluid Power supplied from a flowing fluid to a turbine	$P = \dot{m}gh = \gamma Qh$	(7.31)	$\dot{m}$ = mass flow rate through machine (kg/s) g = gravitational constant = 9.81 (m/s²) h = head of pump or head of turbine (m) γ = specific weight (N/m³) Q = volume flow rate (m³/s)

EXAMPLE 7.3

Applying the Energy Equation to Calculate the Power Required by a Pump

Problem Statement

A pipe 50 cm in diameter carries water (10°C) at a rate of 0.5 m³/s. A pump in the pipe is used to move the water from an elevation of 30 m to 40 m. The pressure at section 1 is 70 kPa gage, and the pressure at section 2 is 350 kPa gage. What power in kilowatts and in horsepower must be supplied to the flow by the pump? Assume $h_L = 3$ m of water and $\alpha_1 = \alpha_2 = 1$.

Define the Situation

Water is being pumped through a system.

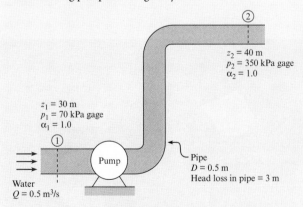

$z_2 = 40$ m
$p_2 = 350$ kPa gage
$\alpha_2 = 1.0$

$z_1 = 30$ m
$p_1 = 70$ kPa gage
$\alpha_1 = 1.0$

Pipe
$D = 0.5$ m
Head loss in pipe = 3 m

Water
$Q = 0.5$ m³/s

Properties: Water (10°C, 1 atm, Table A.5): $\gamma = 9810$ N/m³

State the Goal

P(W and hp) ⬅ power the pump is supplying to the water in units of watts and horsepower

Generate Ideas and Make a Plan

Because this problem involves water being pumped through a system, it is an energy equation problem. However, the goal is to find power, so the power equation will also be needed. The steps are as follows:

1. Write the energy equation between section 1 and section 2.
2. Analyze each term in the energy equation.
3. Calculate the head of the pump h_p.
4. Find the power by applying the power equation (7.30a).

Take Action (Execute the Plan)

1. Energy equation (general form):

$$\frac{p_1}{\gamma} + \alpha_1\frac{\overline{V}_1^2}{2g} + z_1 + h_p = \frac{p_2}{\gamma} + \alpha_2\frac{\overline{V}_2^2}{2g} + z_2 + h_t + h_L$$

2. Term-by-term analysis:
   - Velocity head cancels because $V_1 = V_2$.
   - $h_t = 0$ because there are no turbines in the system.
   - All other head terms are given.
   - Inserting terms into the general equation gives

$$\frac{p_1}{\gamma} + z_1 + h_p = \frac{p_2}{\gamma} + z_2 + h_L$$

3. Pump head (from step 2):

$$h_p = \left(\frac{p_2 - p_1}{\gamma}\right) + (z_2 - z_1) + h_L$$

$$= \left(\frac{(350,000 - 70,000)\ \text{N/m}^2}{9810\ \text{N/m}^3}\right) + (10\ \text{m}) + (3\ \text{m})$$

$$= (28.5\ \text{m}) + (10\ \text{m}) + (3\ \text{m}) = 41.5\ \text{m}$$

Physics: The head provided by the pump (41.5 m) is balanced by the increase in pressure head (28.5 m) plus the increase in elevation head (10 m) plus the head loss (3 m).

4. Power equation:

$$P = \gamma Q h_p$$

$$= (9810 \text{ N/m}^3)(0.5 \text{ m}^3/\text{s})(41.5 \text{ m})$$

$$= \boxed{204 \text{ kW}} = (204 \text{ kW}) \left(\frac{1.0 \text{ hp}}{0.746 \text{ kW}} \right) = \boxed{273 \text{ hp}}$$

Review the Solution and the Process

Discussion. The calculated power represents the work/time being done by the pump impeller on the water. The electrical power supplied to the pump would need to be larger than this because of energy losses in the electrical motor and because the pump itself is not 100% efficient. These factors can be accounted for using pump efficiency (η_{pump}) and motor efficiency (η_{motor}), respectively.

7.5 Mechanical Efficiency

Fig. 7.7 shows an electric motor connected to a centrifugal pump. Motors, pumps, turbines, and similar devices have energy losses. In pumps and turbines, energy losses are due to factors such as mechanical friction, viscous dissipation, and leakage. Energy losses are accounted for by using efficiency.

Mechanical efficiency is defined as the ratio of power output to power input:

$$\eta \equiv \frac{\text{power output from a machine or system}}{\text{power input to a machine or system}} = \frac{P_{\text{output}}}{P_{\text{input}}} \tag{7.32}$$

The symbol for mechanical efficiency is the Greek letter η, which is pronounced as "eta." In addition to mechanical efficiency, engineers also use *thermal efficiency,* which is defined using thermal energy input into a system. In this text, only mechanical efficiency is used, and we sometimes use the label "efficiency" instead of "mechanical efficiency."

EXAMPLE. Suppose an electric motor like the one shown in Fig. 7.7 is drawing 1000 W of electrical power from a wall circuit. As shown in Fig. 7.8, the motor provides 750 J/s of power to its output shaft. This power drives the pump, and the pump supplies 450 J/s to the fluid.

FIGURE 7.7

CAD drawing of a centrifugal pump and electric motor. (Image courtesy of Ted Kyte; www.ted-kyte.com.)

FIGURE 7.8

The energy flow through a pump that is powered by an electric motor.

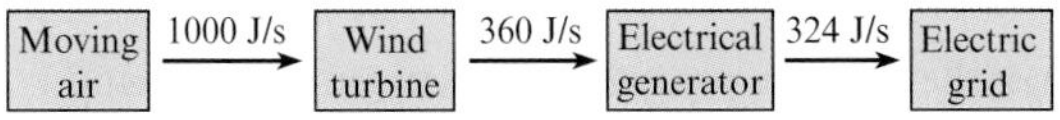

Moving air → 1000 J/s → Wind turbine → 360 J/s → Electrical generator → 324 J/s → Electric grid

FIGURE 7.9

The energy flow associated with generating electrical power from a wind turbine.

In this example, the efficiency of the electric motor is

$$\eta_{motor} = (750 \text{ J/s})/(1000 \text{ J/s}) = 0.75 = 75\%$$

Similarly, the efficiency of the pump is

$$\eta_{pump} = (450 \text{ J/s})/(750 \text{ J/s}) = 0.60 = 60\%$$

and the combined efficiency is

$$\eta_{combined} = (450 \text{ J/s})/(1000 \text{ J/s}) = 0.45 = 45\%$$

EXAMPLE. Suppose that wind incident on a wind turbine contains 1000 J/s of energy, as shown in Fig. 7.9. Because a wind turbine cannot extract all the energy and because of losses, the work that the wind turbine does on its output shaft is 360 J/s. This power drives an electric generator, and the generator produces 324 J/s of electrical power, which is supplied to the power grid. Calculate the system efficiency and the efficiency of the components.

The efficiency of the wind turbine is

$$\eta_{wind\ turbine} = (360 \text{ J/s})/(1000 \text{ J/s}) = 0.36 = 36\%$$

The efficiency of the electric generator is

$$\eta_{electric\ generator} = (324 \text{ J/s})/(360 \text{ J/s}) = 0.90 = 90\%$$

The combined efficiency is

$$\eta_{combined} = (324 \text{ J/s})/(1000 \text{ J/s}) = 0.324 = 32.4\%$$

We can generalize the results of the last two examples to summarize the efficiency equations (Table 7.3). Example 7.4 shows how efficiency enters into a calculation of power.

TABLE 7.3 Summary of the Efficiency Equation

Description	Equation	Terms
Pump	$P_{pump} = \eta_{pump} P_{shaft}$ (7.33a)	P_{pump} = power that the pump supplies to the fluid (W) $[P_{pump} = \dot{m}gh_p = \gamma Qh_p]$ η_{pump} = efficiency of pump () P_{shaft} = power that is supplied to the pump shaft (W)
Turbine	$P_{shaft} = \eta_{turbine} P_{turbine}$ (7.33b)	$P_{turbine}$ = power that the fluid supplies to a turbine (W) $[P_{turbine} = \dot{m}gh_t = \gamma Qh_t]$ $\eta_{turbine}$ = efficiency of turbine () P_{shaft} = power that is supplied by the turbine shaft (W)

EXAMPLE 7.4

Applying the Energy Equation to Predict the Power Produced by a Turbine

Problem Statement

At the maximum rate of power generation, a small hydro-electric power plant takes a discharge of 14.1 m³/s through an elevation drop of 61 m. The head loss through the intakes, penstock, and outlet works is 1.5 m. The combined efficiency of the turbine and electrical generator is 87%. What is the rate of power generation?

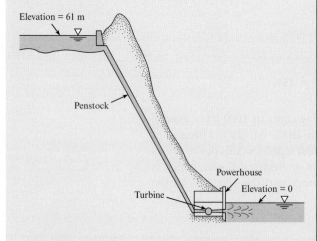

Define the Situation

A small hydroelectric plant is producing electrical power:

- Combined head loss: $h_L = 1.5$ m
- Combined efficiency (turbine/generator): $\eta = 0.87$

Properties: Water (10°C, 1 atm, Table A.5): $\gamma = 9810$ N/m³

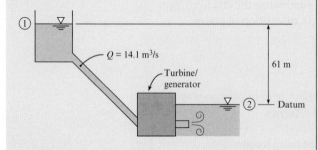

State the Goal

$P_{\text{output from generator}}$ (MW) ⬅ power produced by generator

Generate Ideas and Make a Plan

Because this problem involves a fluid system for producing power, select the energy equation. Because power is the goal, also select the power equation. The plan is as follows:

1. Write the energy equation (7.29) between section 1 and section 2.
2. Analyze each term in the energy equation.
3. Solve for the head of the turbine h_t.
4. Find the input power to the turbine using the power equation (7.30b).
5. Find the output power from the generator by using the efficiency equation (7.33b).

Take Action (Execute the Plan)

1. Energy equation (general form):

$$\frac{p_1}{\gamma} + \alpha_1 \frac{\overline{V}_1^2}{2g} + z_1 + h_p = \frac{p_2}{\gamma} + \alpha_2 \frac{\overline{V}_2^2}{2g} + z_2 + h_t + h_L$$

2. Term-by-term analysis:
   - Velocity heads are negligible because $V_1 \approx 0$ and $V_2 \approx 0$.
   - Pressure heads are zero because $p_1 = p_2 = 0$ gage.
   - $h_p = 0$ because there is no pump in the system.
   - Elevation head terms are given.

3. Combine steps 1 and 2:

$$h_1 = (z_1 - z_2) - h_L$$
$$= (61 \text{ m}) - (1.5 \text{ m}) = 59.5 \text{ m}$$

Physics: Head supplied to the turbine (59.5 m) is equal to the net elevation change of the dam (61 m) minus the head loss (1.5 m).

4. Power equation:

$$P_{\text{input to turbine}} = \gamma Q h_t = (9810 \text{ N/m}^3)(14.1 \text{ m}^3/\text{s})(59.5 \text{ m})$$
$$= 8.23 \text{ MW}$$

5. Efficiency equation:

$$P_{\text{output from generator}} = \eta P_{\text{input to turbine}} = 0.87(8.23 \text{ MW})$$
$$= \boxed{7.16 \text{ MW}}$$

Review the Solution and the Process

1. *Knowledge.* Notice that sections 1 and 2 were located on the free surfaces. This is because information is known at these locations.

2. *Discussion.* The maximum power that can be generated is a function of the elevation head and the flow rate. This maximum power is decreased by head loss and by energy losses in the turbine and the generator.

7.6 Contrasting the Bernoulli Equation and the Energy Equation

Although the Bernoulli equation (Eq. 4.21b) and the energy equation (Eq. 7.29) have a similar form and several terms in common, they are not the same equation. This section explains the differences between these two equations. This information is important for conceptual understanding of these two important equations.

The Bernoulli equation and the energy equation are derived in different ways. The Bernoulli equation was derived by applying Newton's second law to a particle and then integrating the resulting equation along a streamline. The energy equation was derived by starting with the first law of thermodynamics and then using the Reynolds transport theorem. Consequently, the Bernoulli equation involves only mechanical energy, whereas the energy equation includes both mechanical and thermal energy.

The two equations have different methods of application. The Bernoulli equation is applied by selecting two points on a streamline and then equating terms at these points:

$$\frac{p_1}{\gamma} + \frac{V_1^2}{2g} + z_1 = \frac{p_2}{\gamma} + \frac{V_2^2}{2g} + z_2$$

In addition, these two points can be anywhere in the flow field for the special case of irrotational flow. The energy equation is applied by selecting an inlet section and an outlet section and then equating terms as they apply to a control volume located between the inlet and outlet:

$$\left(\frac{p_1}{\gamma} + \alpha_1 \frac{\overline{V}_1^2}{2g} + z_1 \right) + h_p = \left(\frac{p_2}{\gamma} + \alpha_2 \frac{\overline{V}_2^2}{2g} + z_2 \right) + h_t + h_L$$

The two equations have different assumptions. The Bernoulli equation applies to steady, incompressible, and inviscid flow. The energy equation applies to steady, viscous, incompressible flow in a pipe, with additional energy being added through a pump or extracted through a turbine.

Under special circumstances, the energy equation can be reduced to the Bernoulli equation. If the flow is inviscid, there is no head loss; that is, $h_L = 0$. If the "pipe" is regarded as a small stream tube enclosing a streamline, then $\alpha = 1$. There is no pump or turbine along a streamline, so $h_p = h_t = 0$. In this case, the energy equation is identical to the Bernoulli equation. Note that the energy equation cannot be derived by starting with the Bernoulli equation.

Summary. The energy equation is *not* the Bernoulli equation. However, both equations can be related to the law of conservation of energy. Thus, similar terms appear in each equation.

7.7 Transitions

The purpose of this section is to illustrate how the energy, momentum, and continuity equations can be used together to analyze (a) head loss for an abrupt expansion and (b) forces on transitions. These results are useful for designing systems, especially those with large pipes, such as the penstock in a dam.

Abrupt Expansion

An **abrupt or sudden expansion** in a pipe or duct is a change from a smaller section area to a larger section area, as shown in Fig. 7.10. Notice that a confined jet of fluid from the smaller pipe discharges into the larger pipe and creates a zone of separated flow. Because the streamlines in the jet are initially straight and parallel, the piezometric pressure distribution across the jet at section 1 will be uniform.

FIGURE 7.10

Flow through an abrupt expansion.

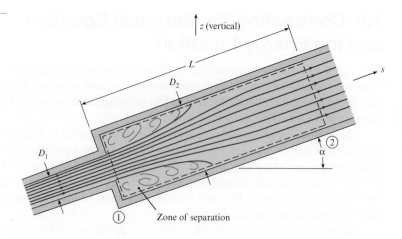

To analyze the transition, apply the energy equation between sections 1 and 2:

$$\frac{p_1}{\gamma} + \alpha_1 \frac{V_1^2}{2g} + z_1 = \frac{p_2}{\gamma} + \alpha_2 \frac{V_2^2}{2g} + z_2 + h_L \tag{7.34}$$

Assume turbulent flow conditions, so $\alpha_1 = \alpha_2 \approx 1$. The momentum equation is

$$\sum F_s = \dot{m}V_2 - \dot{m}V_1$$

Next, let $\dot{m} = \rho A V$ and then identify the forces. Note that the shear force can be neglected because it is small relative to the pressure force. The momentum equation becomes

$$p_1 A_2 - p_2 A_2 - \gamma A_2 L \sin \alpha = \rho V_2^2 A_2 - \rho V_1^2 A_1$$

or

$$\frac{p_1}{\gamma} - \frac{p_2}{\gamma} - (z_2 - z_1) = \frac{V_2^2}{g} - \frac{V_1^2}{g} \frac{A_1}{A_2} \tag{7.35}$$

The continuity equation simplifies to

$$V_1 A_1 = V_2 A_2 \tag{7.36}$$

Combining Eqs. (7.34) to (7.36) gives an equation for the head loss h_L caused by a sudden expansion:

$$h_L = \frac{(V_1 - V_2)^2}{2g} \tag{7.37}$$

If a pipe discharges fluid into a reservoir, then $V_2 = 0$, and the head loss simplifies to

$$h_L = \frac{V^2}{2g}$$

which is the velocity head in the pipe. This energy is dissipated by the viscous action of the fluid in the reservoir.

Forces on Transitions

To find forces on transitions in pipes, apply the momentum equation in combination with the energy equation, the flow rate equation, and the head loss equation. This approach is illustrated by Example 7.5.

EXAMPLE 7.5

Applying the Energy and Momentum Equations to Find Force on a Pipe Contraction

Problem Statement

A pipe 30 cm in diameter carries water (10°C, 250 kPa) at a rate of 0.707 m³/s. The pipe contracts to a diameter of 20 cm. The head loss through the contraction is given by

$$h_L = 0.1 \frac{V_2^2}{2g}$$

where V_2 is the velocity in the 20 cm pipe. What horizontal force is required to hold the transition in place? Assume the kinetic energy correction factor is 1.0 at both the inlet and exit.

Define the Situation

Water flows through a contraction.

- $\alpha_1 = \alpha_2 = 1.0$
- $h_L = 0.1 \ (V_2^2/(2g))$

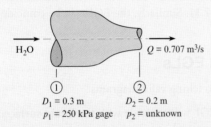

①
$D_1 = 0.3$ m
$p_1 = 250$ kPa gage

②
$D_2 = 0.2$ m
$p_2 =$ unknown

Properties: Water (10°C, 1 atm, Table A.5): $\gamma = 9810$ N/m³

State the Goal

F_x(N) ← horizontal force acting on the contraction

Generate Ideas and Make a Plan

Because force is the goal, start with the momentum equation. To solve the momentum equation, we need p_2. Find this with the energy equation. The step-by-step plan is as follows:

1. Derive an equation for F_x by applying the momentum equation.
2. Derive an equation for p_2 by applying the energy equation.
3. Calculate p_2.
4. Calculate F_x.

Take Action (Execute the Plan)

1. Momentum equation:
   - Sketch a force diagram and a momentum diagram:

- Write the x direction momentum equation:

$$p_1 A_1 - p_2 A_2 + F_x = \dot{m} V_2 - \dot{m} V_1$$

- Rearrange to give

$$F_x = \rho Q (V_2 - V_1) + p_2 A_2 - p_1 A_1$$

2. Energy equation (from section 1 to section 2):
   - Let $\alpha_1 = \alpha_2 = 1$, $z_1 = z_2$, and $h_p = h_t = 0$.
   - Eq. (7.29) simplifies to

$$\frac{p_1}{\gamma} + \frac{V_1^2}{2g} = \frac{p_2}{\gamma} + \frac{V_2^2}{2g} + h_L$$

   - Rearrange to give

$$p_2 = p_1 - \gamma \left(\frac{V_2^2}{2g} - \frac{V_1^2}{2g} + h_L \right)$$

3. Pressure at section 2:
   - Find velocities using the flow rate equation:

$$V_1 = \frac{Q}{A_1} = \frac{0.707 \text{ m}^3/\text{s}}{(\pi/4) \times (0.3 \text{ m})^2} = 10 \text{ m/s}$$

$$V_2 = \frac{Q}{A_2} = \frac{0.707 \text{ m}^3/\text{s}}{(\pi/4) \times (0.2 \text{ m})^2} = 22.5 \text{ m/s}$$

   - Calculate head loss:

$$h_L = \frac{0.1 \ V_2^2}{2g} = \frac{0.1 \times (22.5 \text{ m/s})^2}{2 \times (9.81 \text{ m/s}^2)} = 2.58 \text{ m}$$

   - Calculate pressure:

$$p_2 = p_1 - \gamma \left(\frac{V_2^2}{2g} - \frac{V_1^2}{2g} + h_L \right)$$

$$= 250 \text{ kPa} - 9.81 \text{ kN/m}^3$$

$$\times \left(\frac{(22.5 \text{ m/s})^2}{2(9.81 \text{ m/s}^2)} - \frac{(10 \text{ m/s})^2}{2(9.81 \text{ m/s}^2)} + 2.58 \text{ m} \right)$$

$$= 21.6 \text{ kPa}$$

4. Calculate F_x:

$$F_x = \rho Q (V_2 - V_1) + p_2 A_2 - p_1 A_1$$

$$= (1000 \text{ kg/m}^3)(0.707 \text{ m}^3/\text{s})(22.5 - 10)(\text{m/s})$$

$$+ (21,600 \text{ Pa}) \left(\frac{\pi (0.2 \text{ m})^2}{4} \right) - (250,000 \text{ Pa})$$

$$\times \left(\frac{\pi (0.3 \text{ m})^2}{4} \right)$$

$$= (8837 + 677 - 17,670)\text{N} = -8.16 \text{ kN}$$

$$\boxed{F_x = 8.16 \text{ kN acting to the left}}$$

7.8 The Hydraulic and Energy Grade Lines

This section introduces the hydraulic grade line (HGL) and the energy grade line (EGL), which are graphical representations that show head in a system. This visual approach provides insights and helps one locate and correct trouble spots in the system (usually points of low pressure).

The **EGL**, shown in Fig. 7.11, is a line that indicates the total head at each location in a system. The EGL is related to terms in the energy equation by

$$\text{EGL} = \left(\begin{array}{c}\text{velocity} \\ \text{head}\end{array}\right) + \left(\begin{array}{c}\text{pressure} \\ \text{head}\end{array}\right) + \left(\begin{array}{c}\text{elevation} \\ \text{head}\end{array}\right) = \alpha\frac{V^2}{2g} + \frac{p}{\gamma} + z = \left(\begin{array}{c}\text{total} \\ \text{head}\end{array}\right) \quad \textbf{(7.38)}$$

Notice that **total head**, which characterizes the energy that is carried by a flowing fluid, is the sum of velocity head, pressure head, and elevation head.

The **HGL**, shown in Fig. 7.11, is a line that indicates the piezometric head at each location in a system:

$$\text{HGL} = \left(\begin{array}{c}\text{pressure} \\ \text{head}\end{array}\right) + \left(\begin{array}{c}\text{elevation} \\ \text{head}\end{array}\right) = \frac{p}{\gamma} + z = \left(\begin{array}{c}\text{piezometric} \\ \text{head}\end{array}\right) \quad \textbf{(7.39)}$$

Because the HGL gives piezometric head, the HGL will be coincident with the liquid surface in a piezometer, as shown in Fig. 7.11. Similarly, the EGL will be coincident with the liquid surface in a stagnation tube.

Tips for Drawing HGLs and EGLs

This section presents 10 useful ideas for sketching valid diagrams.

1. In a lake or reservoir, the HGL and EGL will coincide with the liquid surface. Also, both the HGL and EGL will indicate piezometric head.

2. A pump causes an abrupt rise in the EGL and HGL by adding energy to the flow. For example, see Fig. 7.12.

3. For steady flow in a pipe of constant diameter and wall roughness, the slope ($\Delta h_L/\Delta_L$) of the EGL and the HGL will be constant. For examples, see Figs. 7.11 to 7.13.

4. Locate the HGL below the EGL by a distance of the velocity head ($\alpha V^2/2g$).

5. The height of the EGL decreases in the flow direction unless a pump is present.

FIGURE 7.11

EGL and HGL in a straight pipe.

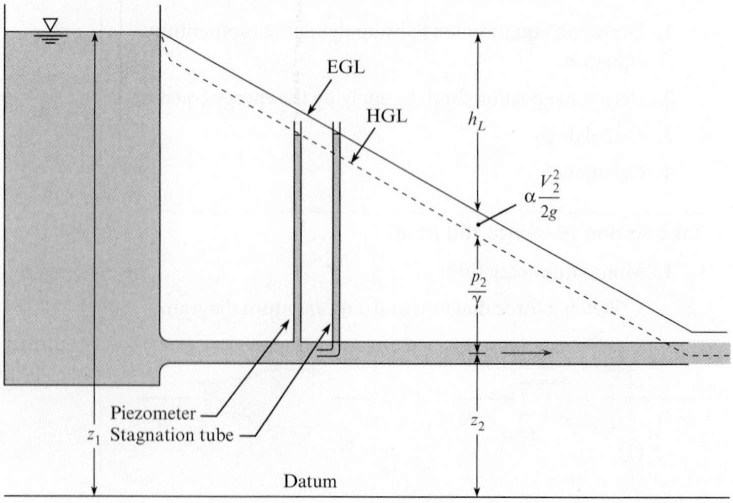

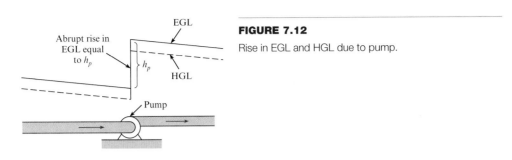

FIGURE 7.12

Rise in EGL and HGL due to pump.

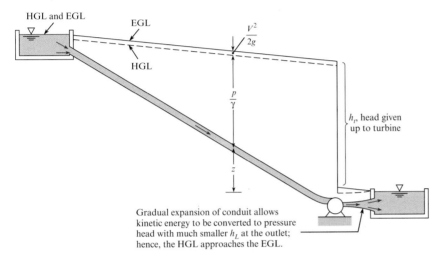

FIGURE 7.13

Drop in EGL and HGL due to turbine.

6. A turbine causes an abrupt drop in the EGL and HGL by removing energy from the flow. For example, see Fig. 7.13.

7. Power generated by a turbine can be increased by using a gradual expansion at the turbine outlet. As shown in Fig. 7.13, the expansion converts kinetic energy to pressure. If the outlet to a reservoir is an abrupt expansion, as in Fig. 7.14, then this kinetic energy is lost.

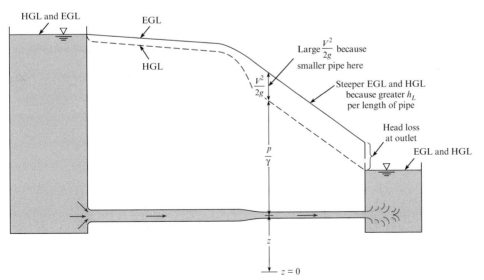

FIGURE 7.14

Change in EGL and HGL due to change in diameter of pipe.

FIGURE 7.15

Change in HGL and EGL due to flow through a nozzle.

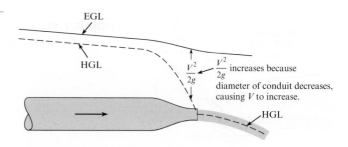

EGL

HGL

$\frac{V^2}{2g}$ $\frac{V^2}{2g}$ increases because diameter of conduit decreases, causing V to increase.

HGL

FIGURE 7.16

Subatmospheric pressure when pipe is above HGL.

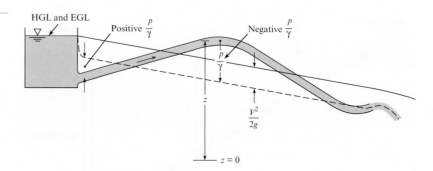

HGL and EGL

Positive $\frac{p}{\gamma}$

Negative $\frac{p}{\gamma}$

$\frac{p}{\gamma}$

z

$\frac{V^2}{2g}$

$z = 0$

8. When a pipe discharges into the atmosphere, the HGL is coincident with the system because $p/\gamma = 0$ at these points. For example, in Figs. 7.15 and 7.16, the HGL in the liquid jet is drawn through the centerline of the jet.

9. When a flow passage changes diameter, the distance between the EGL and the HGL will change (see Figs. 7.14 and 7.15) because velocity changes. In addition, the slope on the EGL will change because the head loss per length will be larger in the conduit with the larger velocity (see Fig. 7.14).

10. If the HGL falls below the pipe, then p/γ is negative, indicating subatmospheric pressure (see Fig. 7.16) and a potential location of cavitation.

The recommended procedure for drawing an EGL and HGL is shown in Example 7.6. Notice how the tips from the preceding section are applied.

EXAMPLE 7.6

Sketching the EGL and HGL for a Piping System

Problem Statement

A pump draws water (50°F) from a reservoir, where the water surface elevation is 520 ft, and forces the water through a pipe 5000 ft long and 1 ft in diameter. This pipe then discharges the water into a reservoir with water surface elevation of 620 ft. The flow rate is 7.85 cfs, and the head loss in the pipe is given by

$$h_L = 0.01\left(\frac{L}{D}\right)\left(\frac{V^2}{2g}\right)$$

Determine the head supplied by the pump, h_p, and the power supplied to the flow, and draw the HGL and EGL for the system. Assume that the pipe is horizontal and is 510 ft in elevation.

Define the Situation

Water is pumped from a lower reservoir to a higher reservoir.

• $h_L = 0.01\left(\frac{L}{D}\right)\left(\frac{V^2}{2g}\right)$

• **Properties:** Water (50°F, 1 atm, Table A.5): $\gamma = 62.4 \text{ lbf/ft}^3$

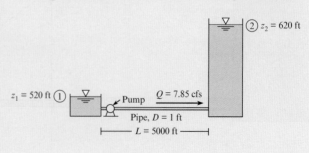

② $z_2 = 620$ ft

$z_1 = 520$ ft ①

Pump $Q = 7.85$ cfs

Pipe, $D = 1$ ft

$L = 5000$ ft

State the Goals

1. h_p(ft) ← pump head
2. P(hp) ← power supplied by the pump
3. Draw the HGL and the EGL.

Generate Ideas and Make a Plan

Because pump head and power are goals, apply the energy equation and the power equation, respectively. The step-by-step plan is as follows:

1. Locate section 1 and section 2 at the top of the reservoirs (see sketch). Then, apply the energy equation (7.29).
2. Calculate terms in the energy equation.
3. Calculate power using the power equation (7.30a).
4. Draw the HGL and EGL.

Take Action (Execute the Plan)

1. Energy equation (general form):

$$\frac{p_1}{\gamma} + \alpha_1 \frac{\overline{V}_1^2}{2g} + z_1 + h_p = \frac{p_2}{\gamma} + \alpha_2 \frac{\overline{V}_2^2}{2g} + z_2 + h_t + h_L$$

- Velocity heads are negligible because $V_1 \approx 0$ and $V_2 \approx 0$.
- Pressure heads are zero because $p_1 = p_2 = 0$ gage.
- $h_t = 0$ because there are no turbines in the system.

$$h_p = (z_2 - z_1) + h_L$$

Interpretation: Head supplied by the pump provides the energy to lift the fluid to a higher elevation plus the energy to overcome head loss.

2. Calculations:

- Calculate V using the flow rate equation:

$$V = \frac{Q}{A} = \frac{7.85 \text{ ft}^3/\text{s}}{(\pi/4)(1 \text{ ft})^2} = 10 \text{ ft/s}$$

- Calculate head loss:

$$h_L = 0.01\left(\frac{L}{D}\right)\left(\frac{V^2}{2g}\right) = 0.01\left(\frac{5000 \text{ ft}}{1.0 \text{ ft}}\right)\left(\frac{(10 \text{ ft/s})^2}{2 \times (32.2 \text{ ft/s}^2)}\right)$$
$$= 77.6 \text{ ft}$$

- Calculate h_p:

$$h_p = (z_2 - z_1) + h_L = (620 \text{ ft} - 520 \text{ ft}) + 77.6 \text{ ft} = \boxed{178 \text{ ft}}$$

3. Power:

$$\dot{W}_p = \gamma Q h_p = \left(\frac{62.4 \text{ lbf}}{\text{ft}^3}\right)\left(\frac{7.85 \text{ ft}^3}{s}\right)(178 \text{ ft})\left(\frac{\text{hp} \cdot s}{550 \text{ ft} \cdot \text{lbf}}\right)$$
$$= \boxed{159 \text{ hp}}$$

4. HGL and EGL:

- From Tip 1 in the preceding section, locate the HGL and EGL along the reservoir surfaces.
- From Tip 2, sketch in a head rise of 178 ft corresponding to the pump.
- From Tip 3, sketch the EGL from the pump outlet to the reservoir surface. Use the fact that the head loss is 77.6 ft. Also, sketch EGL from the reservoir on the left to the pump inlet. Show a small head loss.
- From Tip 4, sketch the HGL below the EGL by a distance of $V^2/2g \approx 1.6$ ft.
- From Tip 5, check the sketches to ensure that EGL and HGL are decreasing in the direction of flow (except at the pump).

The sketch follows. The HGL is shown as a dashed black line. The EGL is shown as a solid blue line.

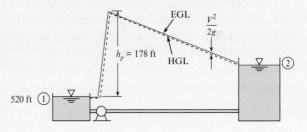

7.9 Summarizing Key Knowledge

Foundational Concepts

- *Energy* is a property of a system that allows the system to do work on its surroundings. Energy can be classified into five categories: mechanical energy, thermal energy, chemical energy, electrical energy, and nuclear energy.
- *Mechanical work* is done by a force that acts through a distance. A more general definition of work is that *work* is any interaction at a system boundary that is not heat transfer or the transfer of matter.

- *Power* is the ratio of work to time or energy to time at an instant in time. Note the key difference between energy and power:
- Energy (and work) describes an *amount* (e.g., how many joules).
- Power describes an *amount/time* or *rate* (e.g., how many joules/second or watts).

- Machines can be classified into two categories:
  - A *pump* is any machine that adds energy to a flowing fluid.
  - A *turbine* is any machine that extracts energy from a flowing fluid.

Conservation of Energy and Derivation of the Energy Equation

- The law of conservation of energy asserts that work and energy balance.
  - The balance for a closed system is (energy changes of the system) = (energy increases due to heat transfer) − (energy decreases due to the system doing work).
  - The balance for a CV is (energy changes in the CV) = (energy increases in the CV due to heat transfer) − (energy out of CV via work done on the surroundings) + (energy transported into the CV by fluid flow).
- Work can be classified into two categories:
  - *Flow work* is work that is done by the pressure force in a flowing fluid.
  - *Shaft work* is any work that is not flow work.

The Energy Equation

- The energy equation is the law of conservation of energy simplified so that it applies to common situations that occur in fluid mechanics. Some of the most important assumptions are steady state, one inflow and one outflow port to the CV, constant density, and that all thermal energy terms (except for head loss) are neglected.
- The energy equation describes an energy balance for a control volume (CV):

$$(\text{energy into CV}) = (\text{energy out of CV})$$

(energy into CV by flow and pumps)
   = (energy out by flow, turbines, and head loss)

- The energy equation, using math symbols, is

$$\left(\frac{p_1}{\gamma} + \alpha_1\frac{V_1^2}{2g} + z_1\right) + h_p = \left(\frac{p_2}{\gamma} + \alpha_2\frac{V_2^2}{2g} + z_2\right) + h_t + h_L$$

$$\begin{pmatrix}\text{pressure head} \\ \text{velocity head} \\ \text{elevation head}\end{pmatrix}_1 + \begin{pmatrix}\text{pump} \\ \text{head}\end{pmatrix} = \begin{pmatrix}\text{pressure head} \\ \text{velocity head} \\ \text{elevation head}\end{pmatrix}_2$$

$$+ \begin{pmatrix}\text{turbine} \\ \text{head}\end{pmatrix} + \begin{pmatrix}\text{head} \\ \text{loss}\end{pmatrix}$$

- Regarding head:
  - Head can be thought of as the ratio of energy to weight for a fluid particle.

- Head can also describe the energy per time that is passing across a section because head and power are related by $P = \dot{m}gh$.
- Regarding head loss (h_L):
  - Head loss represents an irreversible conversion of mechanical energy to thermal energy through the action of viscosity.
  - Head loss is always positive and is analogous to frictional heating.
  - Head loss for a sudden expansion is given by

$$h_L = \frac{(V_1 - V_2)^2}{2g}$$

- Regarding the kinetic energy correction factor α:
  - This factor accounts for the distribution of kinetic energy in a flowing fluid. It is defined as the ratio of the actual KE/time that crosses a surface to the KE/time that would cross if the velocity were uniform.
  - For most situations, engineers set $\alpha = 1$. If the flow is known to be fully developed and laminar, then engineers use $\alpha = 2$. In other cases, one can go back to the mathematical definition and calculate a value of α.

Power and Mechanical Efficiency

- Mechanical efficiency is the ratio of (power output) to (power input) for a machine or system.
- There are several equations that engineers use to calculate power.
  - For translational motion such as a car or an airplane, $P = FV$
  - For rotational motion such as the shaft on a pump, $P = T\omega$
  - For the pump, the power added to the flow is $P = \gamma Q h_p$.
  - For a turbine, the power extracted from the flow is $P = \gamma Q h_t$.

The HGL and EGL

- The hydraulic grade line (HGL) is a profile of the piezometric head, $p/\gamma + z$, along a pipe.
- The energy grade line (EGL) is a profile of the total head, $V^2/2g + p/\gamma + z$, along a pipe.
- If the hydraulic grade line falls below the elevation of a pipe, subatmospheric pressure exists in the pipe at that location, giving rise to the possibility of cavitation.

REFERENCES

1. *Electrical Engineer,* October 18, 1889.

2. Winhoven, S. H., and N. K. Gibbs, "James Prescott Joule (1818–1889): A Manchester Son and the Father of the International Unit of Energy," British Association of Dermatologists. Accessed January 23, 2011, http://www.bad.org.uk/Portals/_Bad/History/Historical%20poster%2006.pdf.

3. Cengel, Y. A., and M. A. Boles. *Thermodynamics: An Engineering Approach.* New York: McGraw-Hill, 1998.

4. Moran, M. J., and H. N. Shapiro. *Fundamentals of Engineering Thermodynamics.* New York: John Wiley, 1992.

PROBLEMS

Work, Energy and Power (§7.1)

7.1 Fill in the blanks. Show your work.

  a. 2000 J = _____ Cal.

  b. _____ ft·lbf = energy to lift a 20 N weight through an elevation difference of 115 m.

  c. 90000 Btu = _____ kWh.

  d. 65 ft·lbf/s = _____ hp.

  e. $[E]$ = [energy] = _____.

7.2 From the list below, select one topic that is interesting to you. Then, use references such as the Internet to research your topic and prepare one page of documentation that you could use to present your topic to your peers.

  a. Explain how hydroelectric power is produced.

  b. Explain how a Kaplan turbine works, how a Francis turbine works, and the differences between these two types of turbines.

  c. Explain how a horizontal-axis wind turbine is used to produce electrical power.

  d. Explain how a steam turbine is used to produce electrical power.

7.3 Apply the grid method to each situation.

  a. Calculate the energy in joules used by a 10 hp pump that is operating for 410 hours. Also, calculate the cost of electricity for this time period. Assume that electricity costs $0.12 per kWh.

  b. A motor is being used to turn the shaft of a centrifugal pump. Apply Eq. (7.3b) to calculate the power in watts corresponding to a torque of 355 lbf·in. and a rotation speed of 1100 rpm.

  c. A turbine produces a power of 5000 ft·lbf/s. Calculate the power in hp and in watts.

7.4 Energy (select all that are correct):

  a. has same units as power

  b. has same units as work

  c. has same units work/time

  d. can have units of Watt

  e. can have units of Joule

  f. can have units of calories

  g. can have units of ft·lbf

7.5 Power (select all that are correct)

  a. has same units as energy

  b. has same units as work/time

  c. has same units as energy/time

  d. can have units of watts

  e. can have units of joules

  f. can have units of horsepower

  g. can have units of ft·lbf

Conservation of Energy (§7.2)

7.6 The first law of thermodynamics for a closed system can be characterized in words as

  a. (change in energy in a system) = (thermal energy out) − (work done by surroundings)

  b. (change in energy in a system) = (thermal energy in) − (work done on surroundings)

  c. either of the above

7.7 The application of the Reynolds transport theorem to the first law of thermodynamics (select all that are correct)

  a. refers only to heat transfer, and not to work

  b. refers to the increase of energy stored in a closed system

  c. extends the applicability of the first law from a closed system to an open system (control volume)

The Kinetic Energy Correction Factor (§7.3)

7.8 An approximate equation for the velocity distribution in a pipe with turbulent flow is

$$\frac{V}{V_{max}} = \left(\frac{y}{r_0}\right)^n$$

where V_{max} is the centerline velocity, y is the distance from the wall of the pipe, r_0 is the radius of the pipe, and n is an exponent that depends on the Reynolds number and varies between 1/6 and 1/8 for most applications. Derive a formula for α as a function of n. What is α if $n = 1/7$?

7.9 An approximate equation for the velocity distribution in a rectangular channel with turbulent flow is

$$\frac{u}{u_{max}} = \left(\frac{y}{d}\right)^n$$

where u_{max} is the velocity at the surface, y is the distance from the floor of the channel, d is the depth of flow, and n is an exponent that varies from about 1/6 to 1/8, depending on the Reynolds number. Derive a formula for α as a function of n. What is the value of α for $n = 1/7$?

7.10 The following data were taken for turbulent flow in a circular pipe with a radius of 3.5 cm. Evaluate the kinetic energy correction factor. The velocity at the pipe wall is zero.

r (cm)	V (m/s)	r (cm)	V (m/s)
0.0	32.50	2.8	22.03
0.5	32.44	2.9	21.24
1.0	32.27	3.0	20.49
1.5	31.22	3.1	19.60
2.0	28.21	3.2	18.69
2.25	26.51	3.25	18.16
2.5	24.38	3.3	17.54
2.6	23.70	3.35	17.02
2.7	22.88	3.4	16.14

The Energy Equation (§7.3)

7.11 Using Section 7.3 and other resources, answer the questions below. Strive for depth, clarity, and accuracy. Also, strive for effective use of sketches, words, and equations.

   a. What is the conceptual meaning of the first law of thermodynamics for a system?

   b. What is flow work? How is the equation for flow work (Eq. 7.16) derived?

   c. What is shaft work? How is shaft work different than flow work?

7.12 Using Section 7.3 and other resources, answer the questions below. Strive for depth, clarity, and accuracy. Also, strive for effective use of sketches, words, and equations.

   a. What is head? How is head related to energy? To power?

   b. What is head of a turbine?

   c. How is head of a pump related to power? To energy?

   d. What is head loss?

7.13 Using Sections 7.3 and 7.7 and using other resources, answer the following questions. Strive for depth, clarity, and accuracy. Also, strive for effective use of sketches, words and equations.

   a. What are the five main terms in the energy equation (7.29)? What does each term mean?

   b. How are terms in the energy equation related to energy? To power?

   c. What assumptions are required for using the energy equation (7.29)?

7.14 Using the energy equation (7.29), prove that fluid in a pipe will flow from a location with high piezometric head to a location with low piezometric head. Assume there are no pumps or turbines and that the pipe has a constant diameter.

7.15 It is necessary to find the head loss for the pipe reducer shown, installed in a system with 10°C water flowing at 0.070 m³/s. The diameter reduces from 20 cm to 12 cm across this fitting (flow direction arrow shown), and the centerline pressure is measured to drop from 490 kPa to 465 kPa for the given flow rate. Assume the kinetic energy correction factor is 1.05 at the reducer inlet and exit and that the reducer is horizontal.

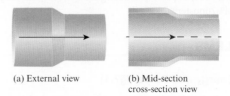

(a) External view   (b) Mid-section cross-section view

Problem 7.15

7.16 A pipe drains a tank as shown. If $x = 10$ ft, $y = 3$ ft, and head losses are neglected, what is the pressure at point A and what is the velocity at the exit? Assume $\alpha = 1.0$ at all locations.

7.17 A pipe drains a tank as shown. If $x = 5$ m, $y = 1$ m, and head losses are neglected, what is the pressure at point A and what is the velocity at the exit? Assume $\alpha = 1.0$ at all locations.

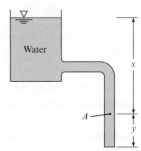

Problems 7.16, 7.17

7.18 For this system, the discharge of water is 0.3 m³/s, $x = 1.0$ m, $y = 2.0$ m, $z = 7.0$ m, and the pipe diameter is 60 cm. Assuming a head loss of 0.5 m, what is the pressure head at point 2 if the jet from the nozzle is 10 cm in diameter? Assume $\alpha = 1.0$ at all locations.

7.19 For this diagram of an industrial pressure washer system, $x = 2$ ft, $y = 4$ ft, $z = 9$ ft, $Q = 3.0$ ft³/s, and the hose diameter is 3 in. Assuming a head loss of 4 ft is derived over the distance from point 2 to the jet, what is the pressure at point 2 if the jet from the nozzle is 1 in. in diameter? Assume $\alpha = 1.0$ throughout.

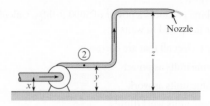

Problems 7.18, 7.19

7.20 Gasoline having a specific gravity of 0.74 is flowing in the pipe shown at a rate of 4 cfs. What is the pressure at section 2 when the pressure at section 1 is 18 psig and the head loss is 5 ft between the two sections? Assume $\alpha = 1.0$ at all locations.

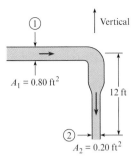

Problem 7.20

7.21 An artery in the human arm, diameter $D = 0.7$ cm, tapers gradually over a distance of 10 cm to a diameter of $d = 0.5$ cm. The blood pressure (gage) at diameter D is 110 mm Hg and at d is 85 mm Hg. What is the head loss (m) that occurs over this distance if the blood ($SG = 1.06$) is moving with a flow rate of 20 cm³/s and the arm is being held horizontally? Idealize the flow in the artery as steady, $\alpha = 1$, the fluid as Newtonian, and the walls of the artery as rigid.

7.22 As shown, a microchannel is being designed to transfer fluid in a MEMS (microelectrical mechanical system) application. The channel is 300 micrometers in diameter and is 8 cm long. Ethyl alcohol is driven through the system at the rate of 0.1 microliters/s (μL/s) with a syringe pump, which is essentially a moving piston. The pressure at the exit of the channel is atmospheric. The flow is laminar, so $\alpha = 2$. The head loss in the channel is given by

$$h_L = \frac{32\mu LV}{\gamma D^2}$$

where L is the channel length, D the diameter, V the mean velocity, μ the viscosity of the fluid, and γ the specific weight of the fluid. Find the pressure in the syringe pump. The velocity head associated with the motion of the piston in the syringe pump is negligible.

Problem 7.22

7.23 The discharge in the siphon is 3.5 cfs, $D = 6$ in., $L_1 = 4$ ft, and $L_2 = 6$ ft. Determine the head loss between the reservoir surface and point C. Determine the pressure at point B if three-quarters of the head loss (as found above) occurs between the reservoir surface and point B. Assume $\alpha = 1.0$ at all locations.

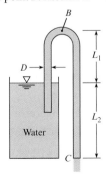

Problem 7.23

7.24 A pump is used to pressurize air in a tank to 300 kPa abs. The tank has a diameter of 2 m and a height of 4 m. The initial level of water in the tank is 1 m, and the initial pressure at the tank water surface is 0 kPa gage. The atmospheric pressure is 100 kPa. The pump provides a constant head of 50 m. The water is drawn from a reservoir that is 4 m below the tank bottom. The pipe connecting the reservoir and the tank is 4 cm in diameter and the head loss, including the expansion loss at the tank, is 10 $V^2/2g$. The flow is turbulent.

Assume the compression of the air in the tank takes place isothermally, so the tank pressure is given by

$$p_T = \frac{3}{4 - z_t} p_0$$

where z_t is the depth of water in the tank in meters. Write a computer program that will show how the air pressure varies in the tank with time, and find the time to pressurize the tank to 300 kPa abs.

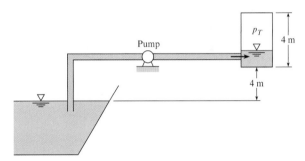

Problem 7.24

The Power Equation (§7.4)

7.25 The pump shown in the figure supplies energy to the flow such that the upstream pressure (15 in. pipe) is 3 psi and the downstream pressure (6 in. pipe) is 59 psi when the flow of water is 7 cfs. What horsepower is delivered by the pump to the flow? Assume $\alpha = 1.0$ at all locations.

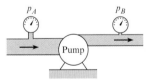

Problem 7.25

7.26 Neglecting head losses, determine what horsepower the pump must deliver to produce the flow as shown. Here, the elevations at points A, B, C, and D are 120 ft, 180 ft, 110 ft, and 90 ft, respectively. The nozzle area is 0.10 ft².

7.27 Neglecting head losses, determine what power the pump must deliver to produce the flow as shown. Here, the elevations at points A, B, C, and D are 42 m, 70 m, 35 m, and 30 m, respectively. The nozzle area is 11 cm².

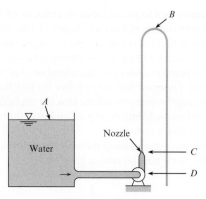

Problems 7.26, 7.27

7.28 If the discharge is 520 cfs, what power output may be expected from the turbine? Assume that the turbine efficiency is 88% and that the overall head loss is 1.3 $V^2/2g$, where V is the velocity in the 7 ft penstock. Assume $\alpha = 1.0$ at all locations.

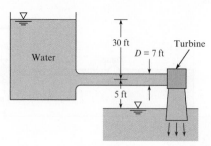

Problem 7.28

Mechanical Efficiency (§7.5)

7.29 The pump shown draws water through a 6 in. suction pipe and discharges it through a 4 in. pipe in which the velocity is 10 ft/s. The 6 in. pipe discharges horizontally into air at C. To what height h above the water surface at A can the water be raised if 14 hp is used by the pump? The pump operates at 60% efficiency and that the head loss in the pipe between A and C is equal to 2 $V_C^2/2g$. Assume $\alpha = 1.0$ throughout.

7.30 The pump shown draws water (20°C) through a 18 cm suction pipe and discharges it through a 10 cm pipe in which the velocity is 3 m/s. The 10 cm pipe discharges horizontally into air at point C. To what height h above the water surface at A can the water be raised if 28 kW is delivered to the pump? Assume that the pump operates at 60% efficiency and that the head loss in the pipe between A and C is equal to 2 $V_C^2/2g$. Assume $\alpha = 1.0$ throughout.

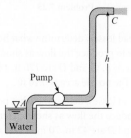

Problems 7.29, 7.30

7.31 A pumping system is to be designed to pump crude oil a distance of 0.85 mile in a 1-ft-diameter pipe at a rate of 3900 gpm. The pressures at the entrance and exit of the pipe are atmospheric, and the exit of the pipe is 250 feet higher than the entrance. The pressure loss in the system due to pipe friction is 50 psi. The specific weight of the oil is 53 lbf/ft³. Find the power, in horsepower, required for the pump.

Contrasting Bernoulli Eqn. to Energy Eqn. (§7.6)

7.32 How is the energy equation (7.29) in §7.3 similar to the Bernoulli equation? How is it different? Give two important similarities and three important differences. Prepare a concept map or other type of diagram suitable for explaining the differences to your peers.

Transitions (§7.7)

7.33 What is the head loss at the outlet of the pipe that discharges water into the reservoir at a rate of 36 cfs if the diameter of the pipe is 18 in.?

7.34 What is the head loss at the outlet of the pipe that discharges water into the reservoir at a rate of 1.2 m³/s if the diameter of the pipe is 53 cm?

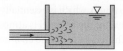

Problems 7.33, 7.34

7.35 Fluid flowing along a pipe of diameter D accelerates around a disk of diameter d as shown in the figure. The velocity far upstream of the disk is U, and the fluid density is ρ. Assuming incompressible flow and that the pressure downstream of the disk is the same as that at the plane of separation, develop an expression for the force required to hold the disk in place in terms of U, D, d, and ρ. Using the expression you developed, determine the force when $U = 10$ m/s, $D = 5$ cm, $d = 4$ cm, and $\rho = 1.2$ kg/m³. Assume $\alpha = 1.0$ at all locations.

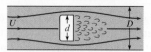

Problem 7.35

Hydraulic and Energy Grade Lines (§7.8)

7.36 The HGL and the EGL are as shown for a certain flow system.

    a. Is flow from A to E or from E to A?

    b. Is there a pump in the system?

    c. Does the pipe at E have a uniform or a variable diameter?

    d. Does a reservoir exists in the system?

    e. Sketch the physical setup that could yield the conditions shown between C and D.

    f. Is anything else revealed by the sketch?

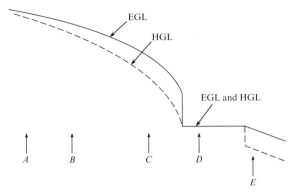

Problem 7.36

7.37 In the figure shown, the magnitude of the EGL changes from 10 m to 37 m. What is the pump head, h_p?

7.38 The pump shown is supplied with 2 kW from the shaft of a motor, to provide a mass flow rate of 35 kg/s. If the pump operates at 70% efficiency, what is the increase in the EGL?

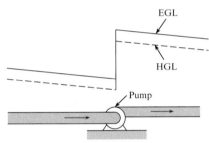

Problems 7.37, 7.38

7.39 Refer to Fig. 7.14. Assume that the head loss in the pipes is given by $h_L = 0.03(L/D)(V^2/2g)$, where V is the mean velocity in the pipe, D is the pipe diameter, and L is the pipe length. The water surface elevations of the upper and lower reservoirs are 150 m and 60 m, respectively. The respective dimensions for upstream and downstream pipes are $D_u = 32$ cm, $L_u = 190$ m, $D_d = 12$ cm, and $L_d = 110$ m. Determine the discharge of water in the system.

7.40 Water flows from reservoir A to reservoir B in a desert community. The water temperature in the system is 100°F, the pipe diameter D is 2.5 ft, and the pipe length L is 200 ft. If $H = 40$ ft, $h = 10$ ft, and the pipe head loss is given by $h_L = 0.01(L/D)(V^2/2g)$, where V is the velocity in the pipe, what will be the discharge in the pipe? In your solution, include the head loss at the pipe outlet. What will be the pressure at point P halfway between the two reservoirs? Assume $\alpha = 1.0$ at all locations.

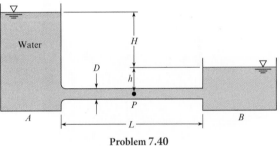

Problem 7.40

CHAPTER EIGHT

Dimensional Analysis and Similitude

CHAPTER ROAD MAP Because of the complexity of flows, designs are often based on experimental results, which are commonly done using scale models. The theoretical basis of experimental testing is called dimensional analysis, the topic of this chapter. This topic is also used to simplify analysis and to present results.

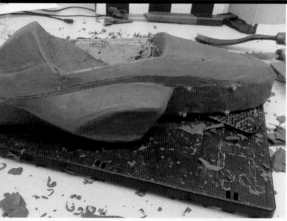

FIGURE 8.1
The photo shows a model of a formula racing car that was built out of clay for testing in a small wind tunnel. The purpose of the testing was to assess the drag characteristics. The work was done by Josh Hartung while he was an undergraduate engineering student. (Photo courtesy of Josh Hartung.)

LEARNING OUTCOMES

DIMENSIONAL ANALYSIS (§8.1, §8.2).

- Explain why dimensional analysis is useful to engineers.
- Define a π-group.
- Explain or apply the Buckingham Π theorem.

METHODS (§8.3).

- Apply the step-by-step method.
- Apply the exponent method.

COMMON π-GROUPS (§8.4).

- Define and describe the common fluids π-groups.
- Explain how a π-group can be understood as a ratio of physically significant terms.

EXPERIMENTS (§8.5).

- Define model and prototype.
- Explain what similitude means and how to achieve similitude.
- Relate physical variables between a model and a prototype by matching the π-groups.

8.1 The Need for Dimensional Analysis

Fluid mechanics is more heavily involved with experimental testing than other disciplines because the analytical tools currently available to solve the momentum and energy equations are not capable of providing accurate results. This is particularly evident in turbulent, separating flows. The solutions obtained by utilizing techniques from computational fluid dynamics with the largest computers available yield only fair approximations for turbulent flow problems—hence the need for experimental evaluation and verification.

For analyzing model studies and for correlating the results of experimental research, it is essential that researchers employ dimensionless groups. To appreciate the advantages of using dimensionless groups, consider the flow of water through the unusual orifice illustrated in Fig. 8.2. Actually, this is much like a nozzle used for flow metering, except that the flow is in

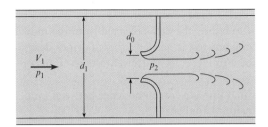

FIGURE 8.2

Flow through inverted flow nozzle.

the opposite direction. An orifice operating in this flow condition will have a much different performance than one operating in the normal mode. However, it is not unlikely that a firm or city water department might have such a situation in which the flow may occur the "right way" most of the time and the "wrong way" part of the time—hence the need for such knowledge.

Because of size and expense, it is not always feasible to carry out tests on a full-scale prototype. Thus, engineers will test a subscale model and measure the pressure drop across the model. The test procedure may involve testing several orifices, each with a different throat diameter d_0. For purposes of discussion, assume that three nozzles are to be tested. The Bernoulli equation, introduced in Chapter 4, suggests that the pressure drop will depend on flow velocity and fluid density. It may also depend on the fluid viscosity.

The test program may be carried out with a range of velocities and possibly with fluids of different density (and viscosity). The pressure drop, $p_1 - p_2$, is a function of the velocity V_1, density ρ, and diameter d_0. By carrying out numerous measurements at different values of V_1 and ρ for the three different nozzles, the data could be plotted as shown in Fig. 8.3a for tests using water. In addition, further tests could be planned with different fluids at considerably more expense.

The material introduced in this chapter leads to a much better approach. Through dimensional analysis, it can be shown that the pressure drop can be expressed as

$$\frac{p_1 - p_2}{(\rho V^2)/2} = f\left(\frac{d_0}{d_1}, \frac{\rho V_1 d_0}{\mu}\right) \tag{8.1}$$

which means that the dimensionless group for pressure, $(p_1 - p_2)/(\rho V^2/2)$, is a function of the dimensionless throat/pipe diameter ratio d_0/d_1 and the dimensionless group $(\rho V_1 d_0)/\mu$, which will be identified later as the Reynolds number. The purpose of the experimental program is to establish the functional relationship. As will be shown later, if the Reynolds number is sufficiently large, the results are independent of the Reynolds number. Then

$$\frac{p_1 - p_2}{(\rho V^2)/2} = f\left(\frac{d_0}{d_1}\right) \tag{8.2}$$

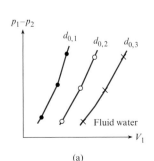

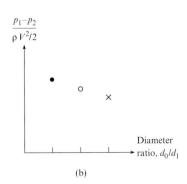

(a)   (b)

FIGURE 8.3

Relations for pressure, velocity, and diameter. (a) Using dimensional variables. (b) Using dimensionless groups.

Thus, for any specific orifice design (the same d_0/d_1) the pressure drop, $p_1 - p_2$, divided by $\rho V_1^2/2$ for the model is same for the prototype. Therefore, the data collected from the model tests can be applied directly to the prototype. Only one test is needed for each orifice design. Consequently, only three tests are needed, as shown in Fig. 8.2b. The fewer tests result in considerable savings in effort and expense.

The identification of dimensionless groups that provide correspondence between model and prototype data is carried out through **dimensional analysis**.

8.2 Buckingham Π Theorem

In 1915, Buckingham (1) showed that the number of independent dimensionless groups of variables (dimensionless parameters) needed to correlate the variables in a given process is equal to $n - m$, where n is the number of variables involved and m is the number of basic dimensions included in the variables.

Buckingham referred to the dimensionless groups as Π, which is the reason the theorem is called the Π theorem. Henceforth, dimensionless groups will be referred to as **π-groups**. If the equation describing a physical system has n dimensional variables and is expressed as

$$y_1 = f(y_2, y_3, \dots y_n)$$

then it can be rearranged and expressed in terms of $(n - m)$ π-groups as

$$\pi_1 = \varphi(\pi_2, \pi_3, \dots \pi_{n-m})$$

Thus, if the drag force F of a fluid flowing past a sphere is known to be a function of the velocity V, mass density ρ, viscosity μ, and diameter D, then five variables (F, V, ρ, μ, and D) and three basic dimensions (L, M, and T) are involved.* By the Buckingham Π theorem, there will be $5 - 3 = 2$ π-groups that can be used to correlate experimental results in the form

$$\pi_1 = \varphi(\pi_2)$$

8.3 Dimensional Analysis

Dimensional analysis is the process for applying π-groups to analysis, experiment design, and the presentation of results. This section presents two methods for finding π-groups: the step-by-step method and the exponent method.

The Step-by-Step Method

Several methods may be used to carry out the process of finding the π-groups, but the step-by-step approach, very clearly presented by Ipsen (2), is one of the easiest and reveals much about the process. The process for the step-by-step method is laid out in Table 8.1.

The final result can be expressed as a functional relationship of the form

$$\pi_1 = f(\pi_2, \pi_2, \dots \pi_n) \tag{8.3}$$

The selection of the dependent and independent π-groups depends on the application. Also, the selection of variables used to eliminate dimensions is arbitrary.

---

*Note that only three basic dimensions will be considered here. Temperature will not be included.

TABLE 8.1 The Step-by-Step Approach

Step	Action Taken during This Step
1	Identify the significant dimensional variables and write out the primary dimensions of each.
2	Apply the Buckingham Π theorem to find the number of π-groups.*
3	Set up a table with the number of rows equal to the number of dimensional variables and the number of columns equal to the number of basic dimensions plus one ($m + 1$).
4	List all the dimensional variables in the first column with primary dimensions.
5	Select a dimension to be eliminated, choose a variable with that dimension in the first column, and combine with remaining variables to eliminate the dimension. List combined variables in the second column with the remaining primary dimensions.
6	Select another dimension to be eliminated, choose from variables in the second column that have that dimension, and combine with the remaining variables. List the new combinations with the remaining primary dimensions in the third column.
7	Repeat Step 6 until all dimensions are eliminated. The remaining dimensionless groups are the π-groups. List the π-groups in the last column.

*Note that, in rare instances, the number of π-groups may be one more than predicted by the Buckingham Π theorem. This anomaly can occur because it is possible that two dimensional categories can be eliminated when dividing (or multiplying) by a given variable. See Ipsen (2) for an example of this.

Example 8.1 shows how to use the step-by-step method to find the π-groups for a body falling in a vacuum.

EXAMPLE 8.1

Finding the π-Group for a Body Falling in a Vacuum

Problem Statement

There are three significant dimensional variables for a body falling in a vacuum (no viscous effects): the velocity, V; the acceleration due to gravity, g; and the distance through which the body falls, h. Find the π-groups using the step-by-step method.

Define the Situation

A body is falling in a vacuum, $V = f(g, h)$.

State the Goal

Find the π-groups.

Generate Ideas and Make a Plan

Apply the step-by-step method laid out in Table 8.1.

Take Action (Execute the Plan)

1. Significant variables and dimensions:

$$[V] = L/T$$
$$[g] = L/T^2$$
$$[h] = L$$

There are only two dimensions, L and T.

2. From the Buckingham Π theorem, there is only one (three variables; two dimensions) π-group.

3. Set up a table with three rows (number of variables) and three (dimensions + 1) columns.

4. List variables and primary dimensions in the first column.

Variable	[ ]	Variable	[ ]	Variable	[ ]
V	$\dfrac{L}{T}$	$\dfrac{V}{h}$	$\dfrac{1}{T}$	$\dfrac{V}{\sqrt{gh}}$	0
g	$\dfrac{L}{T^2}$	$\dfrac{g}{h}$	$\dfrac{1}{T^2}$		
h	L				

5. Select h to eliminate L. Divide g by h and enter in the second column with dimension $1/T^2$. Divide V by h and enter in the second column with dimension $1/T$.

6. Select g/h to eliminate T. Divide V/h by $\sqrt{g/h}$ and enter in the third column.

As expected, there is only one π-group:

$$\pi = \frac{V}{\sqrt{gh}}$$

The final functional form of the equation is

$$\frac{V}{\sqrt{gh}} = C$$

Review the Solution and the Process

1. *Knowledge.* From physics, one can show that $C = \sqrt{2}$.

2. *Knowledge.* The proper relationship between V, h, and g was found with dimensionless analysis. If the value of C was not known, then it could be determined from experiment.

Example 8.2 illustrates the application of the step-by-step method for finding π-groups for a problem with five variables and three primary dimensions.

EXAMPLE 8.2

Finding π-Groups for Drag on a Sphere Using the Step-by-Step Method

Problem Statement

The drag F_D of a sphere in a fluid flowing past the sphere is a function of the viscosity, μ, the mass density, ρ, the velocity of flow, V, and the diameter of the sphere, D. Use the step-by-step method to find the π-groups.

Define the Situation

The functional relationship is $F_D = f(V, \rho, \mu, D)$.

State the Goal

Find the π-groups using the step-by-step method.

Generate Ideas and Make a Plan

Apply the step-by-step procedure from Table 8.1.

Take Action (Execute the Plan)

1. Dimensions of significant variables:

$$F = \frac{ML}{T^2}, \quad V = \frac{L}{T}, \quad \rho = \frac{M}{L^3}, \quad \mu = \frac{M}{LT}, \quad D = L$$

2. Number of π-groups: $5 - 3 = 2$.

3. Set up a table with five rows and four columns.

4. Write variables and dimensions in the first column.

Variable []		Variable []		Variable []		Variable []	
F_D	$\frac{ML}{T^2}$	F_D	$\frac{M}{T^2}$	F_D	$\frac{1}{\rho D^4}\frac{1}{T^2}$	$\frac{F_D}{\rho V^2 D^2}$	0
V	$\frac{L}{T}$	V	$\frac{1}{D}\frac{1}{T}$	V	$\frac{1}{D}\frac{1}{T}$		
ρ	$\frac{M}{L^3}$	ρD^3	M				
μ	$\frac{M}{LT}$	μD	$\frac{M}{T}$	$\frac{\mu}{\rho D^2}\frac{1}{T}$		$\frac{\mu}{\rho V D}$	0
D	L						

5. Eliminate L using D and write new variable combinations with corresponding dimensions in the second column.

6. Eliminate M using ρD^3 and write new variable combinations with dimensions in the third column.

7. Eliminate T using V/D and write new combinations in the fourth column.

The final two π-groups are

$$\pi_1 = \frac{F_D}{\rho V^2 D^2} \quad \text{and} \quad \pi_2 = \frac{\mu}{\rho V D}$$

The functional equation can be written as

$$\frac{F_D}{\rho V^2 D^2} = f\left(\frac{\mu}{\rho V D}\right)$$

The form of the π-groups obtained will depend on the variables selected to eliminate dimensions. For example, if in Example 8.2 $\mu/\rho D^2$ had been used to eliminate the time dimension, then the two π-groups would have been

$$\pi_1 = \frac{\rho F_D}{\mu^2} \quad \text{and} \quad \pi_2 = \frac{\mu}{\rho V D}$$

TABLE 8.2 The Exponent Method

Step	Action Taken during This Step
1	Identify the significant dimensional variables, y_i, and write out the primary dimensions of each, $[y_i]$.
2	Apply the Buckingham Π theorem to find the number of π-groups.
3	Write out the product of the primary dimensions in the form $$[y_1] = [y_2]^a \times [y_3]^b \times \cdots \times [y_n]^k$$ where n is the number of dimensional variables and a, b, and so on are exponents.
4	Find the algebraic equations for the exponents that satisfy dimensional homogeneity (same power for dimensions on each side of equation).
5	Solve the equations for the exponents.
6	Express the dimensional equation in the form $y_1 = y_2^a y_3^b \ldots y_n^k$ and identify the π-groups.

The result is still valid but may not be convenient to use. The form of any π-group can be altered by multiplying or dividing by another π-group. Multiplying π_1 by the square of π_2 yields the original π_1 in Example 8.2:

$$\frac{\rho F_D}{\mu^2} \times \left(\frac{\mu}{\rho V D}\right)^2 = \frac{F_D}{\rho V^2 D^2}$$

By so doing, the two π-groups would be the same as in Example 8.2.

The Exponent Method

An alternative method for finding the π-groups is the exponent method. This method involves solving a set of algebraic equations to satisfy dimensional homogeneity. The process for the exponent method is listed in Table 8.2.

Example 8.3 illustrates how to apply the exponent method to find the π-groups of the same problem addressed in Example 8.2.

EXAMPLE 8.3

Finding π-Groups for Drag on a Sphere Using the Exponent Method

Problem Statement

The drag of a sphere, F_D, in a flowing fluid is a function of the velocity, V, the fluid density, ρ, the fluid viscosity, μ, and the sphere diameter, D. Find the π-groups using the exponent method.

Define the Situation

The functional equation is $F_D = f(V, \rho, \mu, D)$.

State the Goal

Find the π-groups using the exponent method.

Generate Ideas and Make a Plan

Apply the process for the exponent method from Table 8.2.

Take Action (Execute the Plan)

1. Dimensions of significant variables are

$$[F] = \frac{ML}{T^2}, [V] = \frac{L}{T}, [\rho] = \frac{M}{L^3}, [\mu] = \frac{M}{LT}, [D] = L$$

2. Number of π-groups: $5 - 3 = 2$.

3. Form product with dimensions:

$$\frac{ML}{T^2} = \left[\frac{L}{T}\right]^a \times \left[\frac{M}{L^3}\right]^b \times \left[\frac{M}{LT}\right]^c \times [L]^d$$

$$= \frac{L^{a-3b-c+d} M^{b+c}}{T^{a+c}}$$

4. *Dimensional homogeneity*. Equate powers of dimensions on each side:

$$L: a - 3b - c + d = 1$$
$$M: b + c = 1$$
$$T: a + c = 2$$

5. Solve for exponents a, b, and c in terms of d:

$$\begin{pmatrix} 1 & -3 & -1 \\ 0 & 1 & 1 \\ 1 & 0 & 1 \end{pmatrix} \begin{pmatrix} a \\ b \\ c \end{pmatrix} = \begin{pmatrix} 1-d \\ 1 \\ 2 \end{pmatrix}$$

The value of the determinant is -1, so a unique solution is achievable. The solution is $a = d, b = d-1, c = 2-d$.

6. Write the dimensional equation with exponents.

$$F = V^d \rho^{d-1} \mu^{2-d} D^d$$
$$F = \frac{\mu^2}{\rho} \left(\frac{\rho V D}{\mu} \right)^d$$
$$\frac{F\rho}{\mu^2} = \left(\frac{\rho V D}{\mu} \right)^d$$

There are two π-groups:

$$\pi_1 = \frac{F\rho}{\mu^2} \quad \text{and} \quad \pi_2 = \frac{\rho V D}{\mu}$$

By dividing π_1 by the square of π_2, the π_1 group can be written as $F_D/(\rho V^2 D^2)$, so the functional form of the equation can be written as

$$\frac{F}{\rho V^2 D^2} = f \left(\frac{\rho V D}{\mu} \right)$$

Review the Solution and the Process

Discussion. The functional relationship between the two π-groups can be obtained from experiments.

Selection of Significant Variables

All the foregoing procedures deal with straightforward situations. However, some problems do occur. To apply dimensional analysis, one must first decide which variables are significant. If the problem is not sufficiently well understood to make a good choice of the significant variables, then dimensional analysis seldom provides clarification.

A serious shortcoming might be the omission of a significant variable. If this is done, one of the significant π-groups will likewise be missing. In this regard, it is often best to identify a list of variables that one regards as significant to a problem and to determine if only one dimensional category (such as M or L or T) occurs. When this happens, it is likely that there is an error in choice of significant variables because it is not possible to combine two variables to eliminate the lone dimension. Either the variable with the lone dimension should not have been included in the first place (it is not significant), or another variable should have been included.

How does one know if a variable is significant for a given problem? Probably the truest answer is by experience. After working in the field of fluid mechanics for several years, one develops a feel for the significance of variables to certain kinds of applications. However, even the inexperienced engineer will appreciate the fact that free-surface effects have no significance in closed-conduit flow; consequently, surface tension, σ, would not be included as a variable. In closed-conduit flow, if the velocity is less than approximately one-third the speed of sound, compressibility effects are usually negligible. Such guidelines, which have been observed by previous experimenters, help the novice engineer develop confidence in her or his application of dimensional analysis and similitude.

8.4 Common π-Groups

The most common π-groups can be found by applying dimensional analysis to the variables that might be significant in a general flow situation. The purpose of this section is to develop these common π-groups and discuss their significance.

Variables that have significance in a general flow field are the velocity, V, the density, ρ, the viscosity, μ, and the acceleration due to gravity, g. In addition, if fluid compressibility were likely, then the bulk modulus of elasticity, E_v, should be included. If there is a liquid-gas interface, then the surface tension effects may also be significant. Finally, the flow field will be affected by a general length, L, such as the width of a building or the diameter of a pipe. These variables will be regarded as the independent variables. The primary dimensions of the significant independent variables are

$$[V] = L/T \quad [\rho] = M/L^3 \quad [\mu] = M/LT$$

$$[g] = L/T^2 \quad [E_v] = M/LT^2 \quad [\sigma] = M/T^2 \quad [L] = L$$

There are several other independent variables that could be identified for thermal effects, such as temperature, specific heat, and thermal conductivity. Inclusion of these variables is beyond the scope of this text.

Products that result from a flowing fluid are pressure distributions (p), shear stress distributions (τ), and forces on surfaces and objects (F) in the flow field. These will be identified as the dependent variables. The primary dimensions of the dependent variables are

$$[p] = M/LT^2 \quad [\tau] = [\Delta p] = M/LT^2 \quad [F] = (ML)/T^2$$

There are other dependent variables not included here, but they will be encountered and introduced for specific applications.

Altogether, there are 10 significant variables, which, by application of the Buckingham Π theorem, means there are seven π-groups. Utilizing either the step-by-step method or the exponent method yields

$$\frac{p}{\rho V^2} \quad \frac{\tau}{\rho V^2} \quad \frac{F}{\rho V^2 L^2}$$

$$\frac{\rho V L}{\mu} \quad \frac{V}{\sqrt{E_v/\rho}} \quad \frac{\rho L V^2}{\sigma} \quad \frac{V^2}{gL}$$

The first three groups, the dependent π-groups, are identified by specific names. For these groups, it is common practice to use the kinetic pressure, $\rho V^2/2$, instead of ρV^2. In most applications, one is concerned with a pressure difference, so the pressure π-group is expressed as

$$C_p = \frac{p - p_0}{\frac{1}{2}\rho V^2}$$

where C_p is called the pressure coefficient and p_0 is a reference pressure. The pressure coefficient was introduced earlier in Chapter 4 and discussed in Section 8.1. The π-group associated with shear stress is called the shear-stress coefficient and is defined as

$$c_f = \frac{\tau}{\frac{1}{2}\rho V^2}$$

where the subscript f denotes "friction." The π-group associated with force is referred to, here, as a force coefficient and is defined as

$$C_F = \frac{F}{\frac{1}{2}\rho V^2 L^2}$$

TABLE 8.3 Common Π-Groups

π-Group	Symbol	Name	Ratio
$\dfrac{p - p_0}{(\rho V^2)/2}$	C_p	Pressure coefficient	$\dfrac{\text{Pressure difference}}{\text{Kinetic pressure}}$
$\dfrac{\tau}{(\rho V^2)/2}$	c_f	Shear-stress coefficient	$\dfrac{\text{Shear stress}}{\text{Kinetic pressure}}$
$\dfrac{F}{(\rho V^2 L^2)/2}$	C_F	Force coefficient	$\dfrac{\text{Force}}{\text{Kinetic force}}$
$\dfrac{\rho L V}{\mu}$	Re	Reynolds number	$\dfrac{\text{Kinetic force}}{\text{Viscous force}}$
$\dfrac{V}{c}$	M	Mach number	$\sqrt{\dfrac{\text{Kinetic force}}{\text{Compressive force}}}$
$\dfrac{\rho L V^2}{\sigma}$	We	Weber number	$\dfrac{\text{Kinetic force}}{\text{Surface-tension force}}$
$\dfrac{V}{\sqrt{gL}}$	Fr	Froude number	$\sqrt{\dfrac{\text{Kinetic force}}{\text{Gravitational force}}}$

This coefficient will be used extensively in Chapter 11 for lift and drag forces on airfoils and hydrofoils.

The independent π-groups are named after earlier contributors to fluid mechanics. The π-group $VL\rho/\mu$ is called the Reynolds number, after Osborne Reynolds, and is designated by Re. The group $V/(\sqrt{E_v/\rho})$ is rewritten as (V/c) because $\sqrt{E_v/\rho}$ is the speed of sound, c. This π-group is called the Mach number and is designated by M. The π-group $\rho L V^2/\sigma$ is called the Weber number and is designated by We. The remaining π-group is usually expressed as $V/\sqrt{gL}$ and identified as the Froude (rhymes with "food") number* and is written as Fr.

The general functional form for all the π-groups is

$$C_p, c_f, C_F = f(\text{Re, M, We, Fr}) \tag{8.4}$$

which means that either of the three dependent π-groups are functions of the four independent π-groups; that is, the pressure coefficient, the shear-stress coefficient, or the force coefficient are functions of the Reynolds number, Mach number, Weber number, and Froude number.

The π-groups, their symbols, and their names are summarized in Table 8.3. Each independent π-group has an important physical interpretation, as indicated by the Ratio column. The Reynolds number can be viewed as the ratio of kinetic to viscous forces. The kinetic forces are the forces associated with fluid motion. The Bernoulli equation indicates that the pressure difference required to bring a moving fluid to rest is the kinetic pressure, $\rho V^2/2$, so the kinetic forces,[†] F_k, should be proportional to

$$F_k \propto \rho V^2 L^2$$

The shear force due to viscous effects, F_v, is proportional to the shear stress and area

$$F_v \propto \tau A \propto \tau L^2$$

---

*Sometimes, the Froude number is written as $V/\sqrt{(\Delta \gamma g L)/\gamma}$ and called the densimetric Froude number. It has application in studying the motion of fluids in which there is density stratification, such as between saltwater and freshwater in an estuary or heated-water effluents associated with thermal power plants.

[†]Traditionally, the kinetic force has been identified as the "inertial" force.

and the shear stress is proportional to

$$\tau \propto \mu \frac{dV}{dy} \propto \frac{\mu V}{L}$$

so $F_v \propto \mu VL$. Taking the ratio of the kinetic to the viscous forces

$$\frac{F_k}{F_v} \propto \frac{\rho VL}{\mu} = \text{Re}$$

yields the Reynolds number. The magnitude of the Reynolds number provides important information about the flow. A low Reynolds number implies that viscous effects are important; a high Reynolds number implies that kinetic forces predominate. The Reynolds number is one of the most widely used π-groups in fluid mechanics. It is also often written using kinematic viscosity, $\text{Re} = \rho VL/\mu = VL/\nu$.

The ratios of the other independent π-groups have similar significance. The Mach number is an indicator of how important compressibility effects are in a fluid flow. If the Mach number is small, then the kinetic force associated with the fluid motion does not cause a significant density change, and the flow can be treated as incompressible (constant density). On the other hand, if the Mach number is large, there are often appreciable density changes that must be considered in model studies.

The Weber number is an important parameter in liquid atomization. The surface tension of the liquid at the surface of a droplet is responsible for maintaining the droplet's shape. If a droplet is subjected to an air jet and there is a relative velocity between the droplet and the gas, kinetic forces due to this relative velocity cause the droplet to deform. If the Weber number is too large, the kinetic force overcomes the surface-tension force to the point that the droplet shatters into even smaller droplets. Thus, a Weber number criterion can be useful in predicting the droplet size to be expected in liquid atomization. The size of the droplets resulting from liquid atomization is a very significant parameter in gas turbine and rocket combustion.

The Froude number is unimportant when gravity causes only a hydrostatic pressure distribution, such as in a closed conduit. However, if the gravitational force influences the pattern of flow, such as in flow over a spillway or in the formation of waves created by a ship as it cruises over the sea, then the Froude number is a most significant parameter.

8.5 Similitude

Scope of Similitude

Similitude is the theory and art of predicting prototype performance from model observations. Whenever it is necessary to perform tests on a model to obtain information that cannot be obtained by analytical means alone, the rules of similitude must be applied. The theory of similitude involves the application of π-groups, such as the Reynolds number or the Froude number, to predict prototype performance from model tests. The art of similitude enters the problem when the engineer must make decisions about model design, model construction, performance of tests, or analysis of results that are not included in the basic theory.

Present engineering practice makes use of model tests more frequently than most people realize. For example, whenever a new airplane is being designed, tests are made not only on the general scale model of the prototype airplane but also on various components of the plane. Numerous tests are made on individual wing sections as well as on the engine pods and tail sections.

Models of automobiles and high-speed trains are also tested in wind tunnels to predict the drag and flow patterns for the prototype. Information derived from these model studies often

indicates potential problems that can be corrected before the prototype is built, thereby saving considerable time and expense in development of the prototype.

In civil engineering, model tests are always used to predict flow conditions for the spillways of large dams. In addition, river models assist the engineer in the design of flood-control structures as well as in the analysis of sediment movement in the river. Marine engineers make extensive tests on model ship hulls to predict the drag of the ships. Much of this type of testing is done at the David Taylor Model Basin, Naval Surface Warfare Center, Carderock Division, near Washington, D.C. (see Fig. 8.4). Tests are also regularly performed on models of tall buildings to help predict the wind loads on the buildings, the stability characteristics of the buildings, and the airflow patterns in their vicinity. The latter information is used by the architects to design walkways and passageways that are safer and more comfortable for pedestrians to use.

Geometric Similitude

Geometric similitude means that the model is an exact geometric replica of the prototype.* Consequently, if a 1:10 scale model is specified, all linear dimensions of the model must be 1/10 of those of the prototype. In Fig. 8.5, if the model and prototype are geometrically similar, the following equalities hold:

$$\frac{\ell_m}{\ell_p} = \frac{w_m}{w_p} = \frac{c_m}{c_p} = L_r \tag{8.5}$$

Here, ℓ, w, and c are specific linear dimensions associated with the model and prototype, and L_r is the scale ratio between model and prototype. It follows that the ratio of corresponding areas between model and prototype will be the square of the length ratio: $A_r = L_r^2$. The ratio of corresponding volumes will be given by $\Psi_m/\Psi_p = L_r^3$.

---

*For most model studies, this is a basic requirement. However, for certain types of problems, such as river models, distortion of the vertical scale is often necessary to obtain meaningful results.

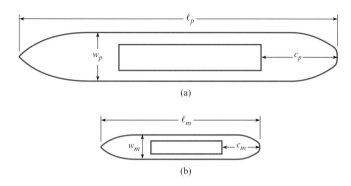

FIGURE 8.5

(a) Prototype. (b) Model.

Dynamic Similitude

Dynamic similitude means that the forces that act on corresponding masses in the model and prototype are in the same ratio (F_m/F_p = constant) throughout the entire flow field. For example, the ratio of the kinetic to viscous forces must be the same for the model and the prototype. Because the forces acting on the fluid elements control the motion of those elements, it follows that dynamic similarity will yield similarity of flow patterns. Consequently, the flow patterns for the model and the prototype will be the same if geometric similitude is satisfied and if the relative forces acting on the fluid are the same in the model as in the prototype. This latter condition requires that the appropriate π-groups introduced in Section 8.4 be the same for the model and prototype because these π-groups are indicators of relative forces within the fluid.

A more physical interpretation of the force ratios can be illustrated by considering the flow over the spillway shown in Fig. 8.6a. Here, corresponding masses of fluid in the model and prototype are acted on by corresponding forces. These forces are the force of gravity F_g, the pressure force F_p, and the viscous resistance force F_v. These forces add vectorially, as shown in Fig. 8.6, to yield a resultant force F_R, which will in turn produce an acceleration of the volume of fluid in accordance with Newton's second law of motion. Hence, because the force polygons in the prototype and model are similar, the magnitudes of the forces in the prototype and model will be in the same ratio as the magnitude of the vectors representing mass times acceleration:

$$\frac{m_m a_m}{m_p a_p} = \frac{F_{gm}}{F_{gp}}$$

or

$$\frac{\rho_m L_m^3 (V_m/t_m)}{\rho_p L_p^3 (V_p/t_p)} = \frac{\gamma_m L_m^3}{\gamma_p L_p^3}$$

which reduces to

$$\frac{V_m}{g_m t_m} = \frac{V_p}{g_p t_p}$$

But

$$\frac{t_m}{t_p} = \frac{L_m/V_m}{L_p/V_p}$$

so

$$\frac{V_m^2}{g_m L_m} = \frac{V_p^2}{g_p L_p} \tag{8.6}$$

FIGURE 8.6

Model-prototype relations: prototype view (a) and
model view (b).

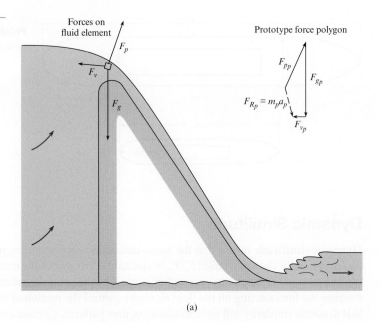

(a)

(b)

Taking the square root of each side of Eq. (8.6) gives

$$\frac{V_m}{\sqrt{g_m L_m}} = \frac{V_p}{\sqrt{g_p L_p}} \qquad \text{or} \qquad \text{Fr}_m = \text{Fr}_p \tag{8.7}$$

Thus, the Froude number for the model must be equal to the Froude number for the prototype to have the same ratio of forces on the model and the prototype.

Equating the ratio of the forces producing acceleration to the ratio of viscous forces,

$$\frac{m_m a_m}{m_p a_p} = \frac{F_{vm}}{F_{vp}} \tag{8.8}$$

where $F_v \propto \mu V L$, leads to

$$\text{Re}_m = \text{Re}_p$$

The same analysis can be carried out for the Mach number and the Weber number. To summarize, if the independent π-groups for the model and prototype are equal, then the condition for dynamic similitude is satisfied.

Referring back to Eq. (8.4) for the general functional relationship,

$$C_p, c_f, C_F = f(\text{Re}, \text{M}, \text{We}, \text{Fr})$$

If the independent π-groups are the same for the model and the prototype, then dependent π-groups must also be equal, so

$$C_{p,m} = C_{p,p} \qquad c_{f,m} = c_{f,p} \qquad C_{F,m} = C_{F,p} \qquad \text{(8.9)}$$

To have complete similitude between the model and the prototype, it is necessary to have both geometric and dynamic similitude.

In many situations, it may not be possible nor necessary to have all the independent π-groups the same for the model and the prototype to carry out useful model studies. For the flow of a liquid in a horizontal pipe, for example, in which the fluid completely fills the pipe (no free surface), there would be no surface tension effects, so the Weber number would be inappropriate. Compressibility effects would not be important, so the Mach number would not be needed. In addition, gravity would not be responsible for the flow, so the Froude number would not have to be considered. The only significant π-group would be the Reynolds number; thus dynamic similitude would be achieved by matching the Reynolds number between the model and the prototype.

On the other hand, if a model test were to be done for the flow over a spillway, then the Froude number would be a significant π-group because gravity is responsible for the motion of the fluid. Also, the action of viscous stresses due to the spillway surface could possibly affect the flow pattern, so the Reynolds number may be a significant π-group. In this situation, dynamic similitude may require that both the Froude number and the Reynolds number be the same for the model and prototype.

The choice of significant π-groups for dynamic similitude and their actual use in predicting prototype performance are considered in the next two sections.

8.6 Model Studies for Flows without Free-Surface Effects

Free-surface effects are absent in the flow of liquids or gases in closed conduits, including control devices such as valves, or in the flow about bodies (e.g., aircraft) that travel through air or are deeply submerged in a liquid such as water (submarines). Free-surface effects are also absent where a structure such as a building is stationary and wind flows past it. In all these cases, given relatively low Mach numbers, the Reynolds-number criterion is the most significant for dynamic similarity. That is, the Reynolds number for the model must equal the Reynolds number for the prototype.

Example 8.4 illustrates the application of Reynolds-number similitude for the flow over a blimp.

EXAMPLE 8.4
Reynolds-Number Similitude

Problem Statement

The drag characteristics of a blimp 5 m in diameter and 60 m long are to be studied in a wind tunnel. If the speed of the blimp through still air is 10 m/s, and if a 1/10 scale model is to be tested, what airspeed in the wind tunnel is needed for dynamically similar conditions? Assume the same air pressure and temperature for both model and prototype.

Define the Situation

A one-tenth-scale model blimp is being tested in a wind tunnel. Prototype speed is 10 m/s.

Assumptions: Same air pressure and temperature for model and prototype, therefore $v_m = v_p$

State the Goal

Find the air speed (m/s) in the wind tunnel for dynamic similitude.

Generate Ideas and Make a Plan

The only π-group that is appropriate is the Reynolds number (there are no compressibility effects, free-surface effects, or gravitation effects). Thus, equating the model and prototype Reynolds number satisfies dynamic similitude.

1. Equate the Reynolds number of the model and the prototype.
2. Calculate model speed.

Take Action (Execute the Plan)

1. Reynolds number similitude:

$$Re_m = Re_p$$

$$\frac{V_m L_m}{\nu_m} = \frac{V_p L_p}{\nu_p}$$

2. Model velocity:

$$V_m = V_p \frac{L_p}{L_m} \frac{\nu_m}{\nu_p} = 10 \text{ m/s} \times 10 \times 1 = \boxed{100 \text{ m/s}}$$

Example 8.4 shows that the airspeed in the wind tunnel must be 100 m/s for true Reynolds number similitude. This speed is quite large, and in fact Mach number effects may start to become important at such a speed. However, it will be shown in Section 8.8 that it is not always necessary to operate models at true Reynolds number similitude to obtain useful results.

If the engineer feels that it is essential to maintain Reynolds number similitude, then only a few alternatives are available. One way to produce high Reynolds numbers at nominal airspeeds is to increase the density of the air. A NASA wind tunnel at the Ames Research Center at Moffett Field in California is one such facility. It has a 12 ft diameter test section, it can be pressurized up to 90 psia (620 kPa), it can be operated to yield a Reynolds number per foot up to 1.2×10^7, and the maximum Mach number at which a model can be tested in this wind tunnel is 0.6. The airflow in this wind tunnel is produced by a single-stage, 20-blade, axial flow fan, which is powered by a 15,000-horsepower, variable-speed, synchronous electric motor (3). Several problems are peculiar to a pressurized tunnel. First, a shell (essentially a pressurized bottle) must surround the entire tunnel and its components, adding to the cost of the tunnel. Second, it takes a long time to pressurize the tunnel in preparation for operation, increasing the time from the start to the finish of runs. In this regard, it should be noted that the original pressurized wind tunnel at the Ames Research Center was built in 1946; however, because of extensive use, the tunnel's pressure shell began to deteriorate, so a new facility (the one previously described) was built and put in operation in 1995. Improvements over the old facility include a better data collection system, very low turbulence, and capability of depressurizing only the test section instead of the entire 620,000 ft^3 wind tunnel circuit when installing and removing models. The original pressurized wind tunnel was used to test most models of U.S. commercial aircraft over the past half-century, including the Boeing 737, 757, and 767; Lockheed L-1011; and McDonnell Douglas DC-9 and DC-10.

The Boeing 777 was tested in the low-speed, pressurized, 5 m by 5 m tunnel in Farnborough, England. This tunnel, operated by the Defence Evaluation and Research Agency (DERA) of Great Britain, can operate at three atmospheres with Mach numbers up to 0.2. Approximately 15,000 hours of total testing time was required for the Boeing 777 (4).

Another method of obtaining high Reynolds numbers is to build a tunnel in which the test medium (gas) is at a very low temperature, thus producing a relatively high-density, low-viscosity fluid. NASA has built such a tunnel and operates it at the Langley Research Center. This tunnel, called the National Transonic Facility, can be pressurized up to 9 atmospheres. The test medium is nitrogen, which is cooled by injecting liquid nitrogen into the system. In this wind tunnel, it is possible to reach Reynolds numbers of 10^8 based on a model size of 0.25 m (5). Because of its sophisticated design, its initial cost was approximately $100,000,000 (6), and its operating expenses are high.

Another modern approach in wind tunnel technology is the development of magnetic or electrostatic suspension of models. The use of the magnetic suspension with model airplanes has been studied (6), and the electrostatic suspension for the study of single-particle aerodynamics has been reported (7).

The use of wind tunnels for aircraft design has grown significantly as the size and sophistication of aircraft have increased. For example, in the 1930s the DC-3 and B-17 each had about 100 hours of wind tunnel tests at a rate of $100 per hour of run time. By contrast, the F-15 fighter required about 20,000 hours of tests at a cost of $20,000 per hour (6). The latter test time is even more staggering when one realizes that a much greater volume of data per hour at higher accuracy is obtained from the modern wind tunnels because of the high-speed data acquisition made possible by computers.

Example 8.5 illustrates the use of Reynolds number similitude to design a test for a valve.

EXAMPLE 8.5

Reynolds Number Similitude of a Valve

Problem Statement

The valve shown is the type used in the control of water in large conduits. Model tests are to be done, using water as the fluid, to determine how the valve will operate under wide-open conditions. The prototype size is 6 ft in diameter at the inlet. What flow rate is required for the model if the prototype flow is 700 cfs? Assume that the temperature for model and prototype is 60°F and that the model inlet diameter is 1 ft.

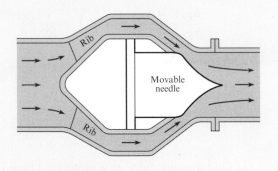

Define the Situations

A one-sixth-scale model of a valve will be tested in a water tunnel. Prototype flow rate is 700 cfs.

Assumptions:

1. No compressibility, free-surface, or gravitational effects.
2. The temperature of water in the model and prototype is the same. Therefore, kinematic viscosity for the model and prototype are equal.

State the Goal

Find the flow rate through the model in cfs.

Generate Ideas and Make a Plan

Dynamic similitude is obtained by equating the model and prototype Reynolds number. The model/prototype area ratio is the square of the scale ratio.

1. Equate the Reynolds numbers of the model and prototype.
2. Calculate the velocity ratio.
3. Calculate the discharge ratio using the model/prototype area ratio.

Take Action (Execute the Plan)

1. Reynolds number similitude:

$$\text{Re}_m = \text{Re}_p$$

$$\frac{V_m L_m}{\nu_m} = \frac{V_p L_p}{\nu_p}$$

2. Velocity ratio:

$$\frac{V_m}{V_p} = \frac{L_p}{L_m} \frac{\nu_m}{\nu_p}$$

Since $\nu_p = \nu_m$,

$$\frac{V_m}{V_p} = \frac{L_p}{L_m}$$

3. Discharge:

$$\frac{Q_m}{Q_p} = \frac{V_m}{V_p} \frac{A_m}{A_p} = \frac{L_p}{L_m} \left(\frac{L_m}{L_p} \right)^2 = \frac{L_m}{L_p}$$

$$Q_m = 700 \text{ cfs} \times \frac{1}{6} = \boxed{117 \text{ cfs}}$$

Review the Solution and the Process

Discussion. This discharge is very large and serves to emphasize that very few model studies are made that completely satisfy the Reynolds number criterion. This subject will be discussed further in the next sections.

8.7 Model-Prototype Performance

Geometric (scale model) and dynamic (same π-groups) similitude mean that the dependent π-groups are the same for both the model and the prototype. For this reason, measurements made with the model can be applied directly to the prototype. Such correspondence is illustrated in this section.

Example 8.6 shows how the pressure difference measured in a model test can be used to find the pressure difference between the corresponding two points on the prototype.

EXAMPLE 8.6

Application of Pressure Coefficient

Problem Statement

A one-tenth-scale model of a blimp is tested in a wind tunnel under dynamically similar conditions. The speed of the blimp through still air is 10 m/s. A 17.8 kPa pressure difference is measured between two points on the model. What will be the pressure difference between the two corresponding points on the prototype? The temperature and pressure in the wind tunnel is the same as the prototype.

Define the Situation

A one-tenth-scale model of a blimp is tested in a wind tunnel under dynamically similar conditions. A pressure difference of 17.8 kPa is measured on the model.

Properties: Pressure and temperature are the same for wind tunnel test and prototype, so $v_m = v_p$.

State the Goal

Find the corresponding pressure difference (Pa) on the prototype.

Generate Ideas and Make a Plan

Eq. (8.4) reduces to

$$C_p = f(\text{Re})$$

1. Equate the Reynolds numbers to find the velocity ratio.
2. Equate the coefficient of pressure to find the pressure difference.

Take Action (Execute the Plan)

1. Reynolds number similitude:

$$\text{Re}_m = \text{Re}_p$$

$$\frac{V_m L_m}{v_m} = \frac{V_p L_p}{v_p}$$

$$\frac{V_p}{V_m} = \frac{L_m}{L_p} = \frac{1}{10}$$

2. Pressure coefficient correspondence:

$$\frac{\Delta p_m}{\frac{1}{2}\rho_m V_m^2} = \frac{\Delta p_p}{\frac{1}{2}\rho_p V_p^2}$$

$$\frac{\Delta p_p}{\Delta p_m} = \left(\frac{V_p}{V_m}\right)^2 = \left(\frac{L_m}{L_p}\right)^2 = \frac{1}{100}$$

Pressure difference on the prototype:

$$\Delta p_p = \frac{\Delta p_m}{100} = \frac{17.8 \text{ kPa}}{100} = \boxed{178 \text{ Pa}}$$

Example 8.7 illustrates calculating the fluid dynamic force on a prototype blimp from wind tunnel data using similitude.

EXAMPLE 8.7

Drag Force from Wind Tunnel Testing

Problem Statement

A one-tenth-scale model of a blimp is tested in a wind tunnel under dynamically similar conditions. If the drag force on the model blimp is measured to be 1530 N, what corresponding force could be expected on the prototype? The air pressure and temperature are the same for both model and prototype.

Define the Situation

A one-tenth-scale model of blimp is tested in a wind tunnel, and a drag force of 1530 N is measured.

Properties: Pressure and temperature are the same, $v_m = v_p$.

State the Goal

Find the drag force (in newtons) on the prototype.

Generate Ideas and Make a Plan

The Reynolds number is the only significant π-group, so Eq. (8.4) reduces to $C_F = f(\text{Re})$.

1. Find the velocity ratio by equating Reynolds numbers.
2. Find the force by equating the force coefficients.

Take Action (Execute the Plan)

1. Reynolds number similitude:

$$\text{Re}_m = \text{Re}_p$$

$$\frac{V_m L_m}{\nu_m} = \frac{V_p L_p}{\nu_p}$$

$$\frac{V_p}{V_m} = \frac{V_m}{L_p} = \frac{1}{10}$$

2. Force coefficient correspondence:

$$\frac{F_p}{\frac{1}{2}\rho_p V_p^2 L_p^2} = \frac{F_m}{\frac{1}{2}\rho_m V_m^2 L_m^2}$$

$$\frac{F_p}{F_m} = \frac{V_p^2 L_p^2}{V_m^2 L_m^2} = \frac{L_m^2 L_p^2}{L_p^2 L_m^2} = 1$$

Therefore,

$$F_p = 1530 \text{ N}$$

Review the Solution and the Process

Discussion. The result that the model force is the same as the prototype force is interesting. When Reynolds number similitude is used and the fluid properties are the same, the forces on the model will always be the same as the forces on the prototype.

8.8 Approximate Similitude at High Reynolds Numbers

The primary justification for model tests is that it is more economical to get answers needed for engineering design by such tests than by any other means. However, as revealed by Examples 8.3, 8.4, and 8.6, Reynolds number similitude requires expensive model tests (high-pressure facilities, large test sections, or using different fluids). This section shows that approximate similitude is achievable even though high Reynolds numbers cannot be reached in model tests.

Consider the size and power required for wind tunnel tests of the blimp in Example 8.4. The wind tunnel would probably require a section at least 2 m by 2 m to accommodate the model blimp. With a 100 m/s airspeed in the tunnel, the power required for producing continuously a stream of air of this size and velocity is in the order of 4 MW. Such a test is not prohibitive, but it is very expensive. It is also conceivable that the 100 m/s airspeed would introduce Mach-number effects not encountered with the prototype, thus generating concern over the validity of the model data. Furthermore, a force of 1530 N is generally larger than that usually associated with model tests. Therefore, especially in the study of problems involving non-free-surface flows, it is desirable to perform model tests in such a way that large magnitudes of forces or pressures are not encountered.

For many cases, it is possible to obtain all the needed information from abbreviated tests. Often, the Reynolds number effect (relative viscous effect) either becomes insignificant at high Reynolds numbers or becomes independent of the Reynolds number. The point at which testing can be stopped often can be detected by inspection of a graph of the pressure coefficient C_p versus the Reynolds number Re. Such a graph for a venturi meter in a pipe is shown in Fig. 8.7. In this meter, Δp is the pressure difference between the points shown, and V is the velocity in the restricted section of the venturi meter. Here it is seen that viscous forces affect the value of C_p below a Reynolds number of approximately 50,000. However, for higher Reynolds numbers, C_p is virtually constant. Physically, this means that at low Reynolds numbers (relatively high viscous forces), a significant part of the change in pressure comes from viscous resistance, and the remainder comes from the acceleration (change in kinetic energy) of the fluid as it passes through the venturi meter. However, with high Reynolds numbers (resulting from either small viscosity or a large product of V, D, and ρ), the viscous resistance is negligible compared with the force required to accelerate the fluid. Because the ratio of Δp to the kinetic

FIGURE 8.7

C_p for a venturi meter as a function of the Reynolds numbers.

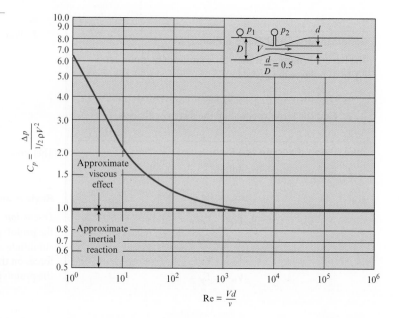

pressure does not change (constant C_p) for high Reynolds numbers, there is no need to carry out tests at higher Reynolds numbers. This is true in general, so long as the flow pattern does not change with the Reynolds number.

In a practical sense, whoever is in charge of the model test will try to predict from previous works approximately what maximum Reynolds number will be needed to reach the point of insignificant Reynolds number effect and then will design the model accordingly. After a series of tests has been made on the model, C_p versus Re will be plotted to see whether the range of constant C_p has indeed been reached. If so, then no more data are needed to predict the prototype performance. However, if C_p has not reached a constant value, then the test program has to be expanded or results extrapolated. Thus, the results of some model tests can be used to predict prototype performance, even though the Reynolds numbers are not the same for the model and the prototype. This is especially valid for angular-shaped bodies, such as model buildings, tested in wind tunnels.

In addition, the results of model testing can be combined with analytic results. Computational fluid dynamics (CFD) may predict the change in performance with the Reynolds number, but may not be reliable to predict the performance level. In this case, the model testing would be used to establish the level of performance, and the trends predicted by CFD would be used to extrapolate the results to other conditions.

Example 8.8 is an illustration of the approximate similitude at a high Reynolds number for flow through a constriction.

EXAMPLE 8.8

Measuring Head Loss in a Nozzle in Reverse Flow

Problem Statement

Tests are to be performed to determine the head loss in a nozzle under a reverse-flow situation. The prototype operates with water at 50°F and with a nominal reverse-flow velocity of 5 ft/s. The diameter of the prototype is 3 ft. The tests are done in a 1/12-scale model facility with water at 60°F. A head loss (pressure drop) of 1 psid is measured with

a velocity of 20 ft/s. What will be the head loss in the actual nozzle?

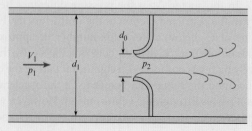

Define the Situation

A 1/12-scale model tests for head loss in a reverse-flow nozzle. A pressure difference of 1 psid is measured with the model at 20 ft/s.

Properties:

- Water (50°F, Table F.5): $\rho = 1.94$ slugs/ft^3, $\nu = 1.41 \times 10^{-5}$ ft^2/s
- Water (60°F, Table F.5): $\rho = 1.94$ slugs/ft^3, and $\nu = 1.22 \times 10^{-5}$ ft^2/s

State the Goal

Find the pressure drop (psid) for the prototype nozzle.

Generate Ideas and Make a Plan

The only significant π-group is the Reynolds number, so Eq. (8.4) reduces to $C_p = f(\mathrm{Re})$. Dynamic similitude is achieved if $\mathrm{Re}_m = \mathrm{Re}_p$, then $C_{p,m} = C_{p,p}$. From Fig. 8.7, if $\mathrm{Re}_m, \mathrm{Re}_p > 10^3$, then $C_{p,m} = C_{p,p}$.

1. Calculate the Reynolds number for the model and prototype.
2. Check if both exceed 10^3. If not, model tests need to be reevaluated.
3. Calculate pressure coefficient.
4. Evaluate pressure drop in the prototype.

Take Action (Execute the Plan)

1. Reynolds numbers:

$$\mathrm{Re}_m = \frac{VD}{\nu} = \frac{20\ \mathrm{ft/s} \times (3/12\ \mathrm{ft})}{1.22 \times 10^{-5}\ \mathrm{ft}^2/\mathrm{s}} = 4.10 \times 10^5$$

$$\mathrm{Re}_p = \frac{5\ \mathrm{ft/s} \times 3\ \mathrm{ft}}{1.41 \times 10^{-5}\ \mathrm{ft}^2/\mathrm{s}} = 1.06 \times 10^6$$

2. Both Reynolds numbers exceed 10^3. Therefore $C_{p,m} = C_{p,p}$. The test is valid.

3. Pressure coefficient from model tests:

$$C_{p,m} = \frac{\Delta p}{\frac{1}{2}\rho V^2} = \frac{1\ \mathrm{lbf/in}^2 \times 144\ \mathrm{in}^2/\mathrm{ft}^2}{\frac{1}{2} \times 1.94\ \mathrm{slug/ft}^3 \times (20\ \mathrm{ft/s})^2} = 0.371$$

4. Pressure drop in prototype:

$$\Delta p_p = 0.371 \times \frac{1}{2}\rho V^2 = 0.371 \times 0.5 \times 1.94\ \mathrm{slug/ft}^3 \times (5\ \mathrm{ft/s})^2$$

$$= 9.0\ \mathrm{lbf/ft}^2 = \boxed{0.0625\ \mathrm{psid}}$$

Review the Solution and the Process

1. *Knowledge.* Because the Reynolds numbers are so much greater than 10^3, the equation for pressure drop is valid over a wide range of velocities.

2. *Discussion.* This example justifies the independence of the Reynolds number referred to in Section 8.1.

In some situations, viscous and compressibility effects may both be important, but it is not possible to have dynamic similitude with both π-groups. Which π-group is chosen for similitude depends a great deal on what information the engineer is seeking. If the engineer is interested in the viscous motion of fluid near a wall in shock-free supersonic flow, then the Reynolds number should be selected as the significant π-group. However, if the shockwave pattern over a body is of interest, then the Mach number should be selected for similitude. A useful rule of thumb is that compressibility effects are unimportant for M < 0.3.

Example 8.9 shows the difficulty in having Reynolds number similitude and avoiding Mach number effects in wind tunnel tests of an automobile.

EXAMPLE 8.9

Model Tests for Drag Force on an Automobile

Problem Statement

A one-tenth-scale model of an automobile is tested in a wind tunnel with air at atmospheric pressure and 20°C. The automobile is 4 m long and travels at a velocity of 100 km/hr in air at the same conditions. What should the wind tunnel speed be such that the measured drag can be related to the drag of the prototype? Experience shows that the dependent π-groups are independent of Reynolds numbers for values exceeding 10^5. The speed of sound is 1235 km/hr.

Define the Situation

A one-tenth-scale model of a 4-m-long automobile moving at 100 km/hr is tested in wind tunnel.

Properties: Air (20°C), Table A.3, $\rho = 1.2$ kg/m^3, $\nu = 1.51 \times 10^{-5}$ N·s/m^2

State the Goal

Find the wind tunnel speed to achieve similitude.

Generate Ideas and Make a Plan

Mach number of the prototype is about 0.08 (100/1235), so Mach number effects are unimportant. Dynamic similitude is achieved with Reynolds numbers, $Re_m = Re_p$. With dynamic similitude, $C_{F,m} = C_{F,p}$, and model measurements can be applied to prototype.

1. Determine the model speed for dynamic similitude.
2. Evaluate the model speed. If it is not feasible, continue to next step.
3. Calculate the prototype Reynolds number. If $Re_p > 10^5$, then $Re_m \geq 10^5$, for $C_{F,m} = C_{F,p}$.
4. Find the speed for which $Re_m \geq 10^5$.

Take Action (Execute the Plan)

1. Velocity from Reynolds number similitude:

$$\left(\frac{VL}{\nu}\right)_m = \left(\frac{VL}{\nu}\right)_p$$

$$\frac{V_m}{V_p} = \frac{L_p}{L_m} = 10$$

$$V_m = 10 \times 100 \text{ km/hr} = 1000 \text{ km/hr}$$

2. With this velocity, M = 1000/1235 = 0.81. This is too high for model tests because it would introduce unwanted compressibility effects.

3. Reynolds number of prototype:

$$Re_p = \frac{VL\rho}{\mu} = \frac{100 \text{ km/hr} \times 0.278 \text{ (m/s)(km/hr)} \times 4 \text{ m}}{1.51 \times 10^{-5} \text{ m}^2/\text{s}}$$

$$= 7.4 \times 10^6$$

Therefore, $C_{F,m} = C_{F,p}$ if $Re_m \geq 10^5$.

4. Wind tunnel speed:

$$V_m \geq Re_m \frac{\nu_m}{L_m} = 10^5 \times \frac{1.51 \times 10^{-5} \text{ m}^2/\text{s}}{0.4 \text{ m}}$$

$$\geq \boxed{3.8 \text{ m/s}}$$

Review the Solution and the Process

Discussion. The wind tunnel speed must exceed 3.8 m/s. From a practical point of view, the speed will be chosen to provide sufficiently large forces for reliable and accurate measurements.

8.9 Free-Surface Model Studies

Spillway Models

The flow over a spillway is a classic case of a free-surface flow. The major influence, besides the spillway geometry itself, on the flow of water over a spillway is the action of gravity. Hence, the Froude number similarity criterion is used for such model studies. It can be appreciated for large spillways with depths of water on the order of 3 m or 4 m and velocities on the order of 10 m/s or more that the Reynolds number is very large. At high values of the Reynolds number, the relative viscous forces are often independent of the Reynolds number, as noted in the foregoing section (§8.8). However, if the reduced-scale model is made too small, then the viscous forces as well as the surface tension forces would have a larger relative effect on the flow in the model than in the prototype. Therefore, in practice, spillway models are made large enough so that the viscous effects have about the same relative effect in the model as in the prototype (i.e., the viscous effects are nearly independent of the Reynolds number). Then, the Froude number is the significant π-group. Most model spillways are made at least 1 m high, and for precise studies, such as calibration of individual spillway bays, it is not uncommon to design and construct model spillway sections that are 2 m or 3 m high. Figs. 8.8 and 8.9 show a comprehensive model and spillway model for Hell's Canyon Dam in Idaho.

Example 8.10 is an application of Froude number similitude in modeling discharge over a spillway.

FIGURE 8.8

Comprehensive model for Hell's Canyon Dam. Tests were made at the Albrook Hydraulic Laboratory, Washington State University. (Photo courtesy of Albrook Hydraulic Laboratory, Washington State University.)

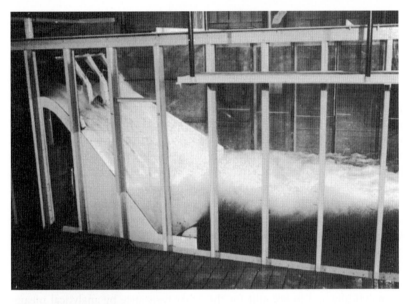

FIGURE 8.9

Spillway model for Hell's Canyon Dam. Tests were made at the Albrook Hydraulic Laboratory, Washington State University. (Photo courtesy of Albrook Hydraulic Laboratory, Washington State University.)

EXAMPLE 8.10

Modeling Flood Discharge over a Spillway

Problem Statement

A 1/49-scale model of a proposed dam is used to predict prototype flow conditions. If the design flood discharge over the spillway is 15,000 m^3/s, what water flow rate should be established in the model to simulate this flow? If a velocity of 1.2 m/s is measured at a point in the model, what is the velocity at a corresponding point in the prototype?

Define the Situation

A 1/49-scale model of a spillway will be tested.
Prototype discharge is 15,000 m^3/s.

State the Goal

Find:

1. The flow rate over the model

2. Velocity on the prototype at the point where velocity is 1.2 m/s on the model

Generate Ideas and Make a Plan

Gravity is responsible for the flow, so the significant π-group is the Froude number. For dynamic similitude, $\mathrm{Fr}_m = \mathrm{Fr}_p$.

1. Calculate velocity ratio from Froude number similitude.

2. Calculate the discharge ratio using the scale ratio and calculate the model discharge.

3. Use the velocity ratio from step 1 to find the velocity at the corresponding point in the prototype.

Take Action (Execute the Plan)

1. Froude number similitude:

$$Fr_m = Fr_p$$

$$\frac{V_m}{\sqrt{g_m L_m}} = \frac{V_p}{\sqrt{g_p L_p}}$$

The acceleration due to gravity is the same, so

$$\frac{V_m}{V_p} = \sqrt{\frac{L_m}{L_p}}$$

2. Discharge ratio:

$$\frac{Q_m}{Q_p} = \frac{A_m}{A_p}\frac{V_m}{V_p} = \frac{L_m^2}{L_p^2}\sqrt{\frac{L_m}{L_p}} = \left(\frac{L_m}{L_p}\right)^{5/2}$$

Discharge for model:

$$Q_m = Q_p\left(\frac{1}{49}\right)^{5/2} = 15{,}000\,\frac{m^3}{s} \times \frac{1}{16{,}800} = \boxed{0.89\ m^3/s}$$

3. Velocity on prototype:

$$\frac{V_p}{V_m} = \sqrt{\frac{L_p}{L_m}}$$

$$V_p = \sqrt{49} \times 1.2\ m/s = \boxed{8.4\ m/s}$$

Ship Model Tests

The largest facility for ship testing in the United States is the David Taylor Model Basin, Naval Surface Warfare Center, Carderock Division, near Washington, D.C. Two of the core facilities are the towing basins and the rotating arm facility. In the rotating arm facility, models are suspended from the end of a rotating arm in a larger circular basin. Forces and moments can be measured on ship models up to 9 m in length at steady-state speeds as high as 15.4 m/s (30 knots). In the high-speed towing basin, models 1.2 m to 6.1 m can be towed at speeds up to 16.5 m/s (32 knots).

The aim of the ship model testing is to determine the resistance that the propulsion system of the ship must overcome. This resistance is the sum of the wave resistance and the surface resistance of the hull. The wave resistance is a free-surface, or Froude number, phenomenon, and the hull resistance is a viscous, or Reynolds number, phenomenon. Because both wave and viscous effects contribute significantly to the overall resistance, it would appear that both the Froude and Reynolds criteria should be used. However, it is impossible to satisfy both if the model liquid is water (the only practical test liquid) because the Reynolds number similitude dictates a higher velocity for the model than for the prototype [equal to $V_p(L_p/L_m)$], whereas the Froude number similitude dictates a lower velocity for the model [equal to $V_p(\sqrt{L_m}/\sqrt{L_p})$]. To circumvent such a dilemma, the procedure is to model for the phenomenon that is the most difficult to predict analytically and to account for the other resistance by analytical means. Because the wave resistance is the most difficult problem, the model is operated according to the Froude number similitude, and the hull resistance is accounted for analytically.

To illustrate how the test results and the analytical solutions for surface resistance are merged to yield design data, the following necessary sequential steps are indicated:

1. Make model tests according to Froude number similitude, and the total model resistance is measured. This total model resistance will be equal to the wave resistance plus the surface resistance of the hull of the model.

2. Estimate the surface resistance of the model by analytical calculations.

3. Subtract the surface resistance calculated in step 2 from the total model resistance of step 1 to yield the wave resistance of the model.

4. Using the Froude number similitude, scale the wave resistance of the model up to yield the wave resistance of the prototype.

5. Estimate the surface resistance of the hull of the prototype by analytical means.

6. The sum of the wave resistance of the prototype from step 4 and the surface resistance of the prototype from step 5 yields the total prototype resistance, or drag.

8.10 Summarizing Key Knowledge

Rationale and Description of Dimensional Analysis

- Dimensional analysis involves combining dimensional variables to form dimensionless groups. These groups, called π-groups, can be regarded as the scaling parameters for fluid flow. Dimensional analysis is applied to analysis, experiment design, and to the presentation of results.

- The *Buckingham Π theorem* states that the number of independent π-groups is $n - m$, where n is the number of dimensional variables and m is the number of basic dimensions included in the variables.

Rationale and Description of Dimensional Analysis

- The π-groups can be found by either the *step-by-step method* or the *exponent method*:

  - In the *step-by-step method,* each dimension is removed by successively using a dimensional variable until the π-groups are obtained.

  - In the *exponent method,* each variable is raised to a power, they are multiplied together, and three simultaneous algebraic equations formulated for dimensional homogeneity are solved to yield the π-groups.

Common π-Groups

- Four common *independent π-groups* are

$$\text{Reynolds number, Re} = \frac{\rho V L}{\mu} \qquad \text{Mach number, M} = \frac{V}{c}$$

$$\text{Weber number, We} = \frac{\rho V^2 L}{\sigma} \qquad \text{Froude number, Fr} = \frac{V}{\sqrt{gL}}$$

- Three common *dependent π-groups* are

$$\text{Pressure coefficient, } C_p = \frac{\Delta p}{(\rho V^2)/2}$$

$$\text{Shear stress coefficient, } c_f = \frac{\tau}{(\rho V^2)/2}$$

$$\text{Force coefficient, } C_F = \frac{F}{(\rho V^2 L^2)/2}$$

- The general functional form of the common π-groups is

$$C_F, c_f, C_p = f(\text{Re}, \text{M}, \text{We}, \text{Fr})$$

Dimensional Analysis in Experimental Testing

- Experimental testing is often performed with a small-scale replica (*model*) of the full-scale structure (*prototype*).

- *Similitude* is the art and theory of predicting prototype performance from model observations. To achieve exact similitude:

  - The model must be a scale model of the prototype (*geometric similitude*).

    - Values of the π-groups must be the same for the model and the prototype (*dynamic similitude*).

- In practice, it is not always possible to have complete dynamic similitude, so *only the most important π-groups are matched.*

REFERENCES

1. Buckingham, E. "Model Experiments and the Forms of Empirical Equations." *Trans. ASME,* vol. 37, 263, 1915.

2. Ipsen, D. C. *Units, Dimensions and Dimensionless Numbers.* New York: McGraw-Hill, 1960.

3. NASA publication available from the U.S. Government Printing Office: No. 1995-685-893.

4. Personal communication. Mark Goldhammer, Manager, Aerodynamic Design of the 777.

5. Kilgore, R.A., and D.A. Dress. "The Application of Cryogenics to High Reynolds-Number Testing in Wind Tunnels, Part 2. Development and Application of the Cryogenic Wind Tunnel Concept." *Cryogenics,* Vol. 24, no. 9, 1984.

6. Baals, D.D., and W.R. Corliss. *Wind Tunnels of NASA.* Washington, DC: U.S. Govt. Printing Office, 1981.

7. Kale, S., et al. "An Experimental Study of Single-Particle Aerodynamics." *Proc. of First Nat. Congress on Fluid Dynamics,* Cincinnati, Ohio, July 1988.

PROBLEMS

Dimensional Analysis (§8.3)

8.1 Determine which of the following equations are dimensionally homogeneous:

a. $V = \dfrac{1.49}{n} R^{2/3} S^{1/2}$

where V is velocity, n is here assumed to have dimensions of length to the one-sixth power, R is length, and S is slope.

b. $Q = \frac{2}{3} CL \sqrt{2g} H^{3/2}$

where Q is discharge, C is a number without units, L is length, g is acceleration due to gravity, and H is head.

c. $h_f = f \dfrac{L}{D} \dfrac{V^2}{2g}$

where h_f is head loss, f is a dimensionless resistance coefficient, L is length, D is diameter, V is velocity, and g is acceleration due to gravity.

d. $D = \dfrac{0.074}{Re^{0.2}} \dfrac{Bx\rho V^2}{2}$

where D is drag force, Re is Vx/ν, B is width, x is length, ρ is mass density, ν is the kinematic viscosity, and V is velocity.

8.2 Determine the primary dimensions of the following terms.

a. $\rho V^2/2$, where V is velocity and ρ is mass density

b. Q/ND^3, where Q is discharge, D is diameter, and N is angular speed of a pump

c. T (torque)

d. $\sqrt{\tau/\rho}$, where τ is shear stress

8.3 A flow-metering device, called a vortex meter, consists of a square element mounted inside a pipe. Vortices are generated by the element, which gives rise to an oscillatory pressure measured on the leeward side of the element. The fluctuation frequency is related to the flow velocity. The discharge in the pipe is a function of the frequency of the oscillating pressure ω, the pipe diameter D, the length of the element l, the density ρ, and the viscosity μ. Thus

$$Q = f(\omega, D, l, \rho, \mu)$$

Find the π-groups in the form

$$\dfrac{Q}{\omega D^3} = f(\pi_1, \pi_2)$$

8.4 An experimental test program is being set up to calibrate a new flow meter. The flow meter is to measure the mass flow rate of liquid flowing through a pipe. It is assumed that the mass flow rate is a function of the following variables:

$$\dot{m} = f(\Delta p, D, \mu, \rho)$$

where Δp is the pressure difference across the meter, D is the pipe diameter, μ is the liquid viscosity, and ρ is the liquid density.

Using dimensional analysis, find the π-groups. Express your answer in the form

$$\dfrac{\dot{m}}{\sqrt{\rho \Delta p D^4}} = f(\pi)$$

8.5 Engineers are measuring the drag forces on an oscillating fin in a water tunnel. It is assumed that the drag force F_D is a function of the liquid density ρ, the fluid velocity V, the surface area of the fin S, and the frequency of oscillation ω:

$$F_D = f(\rho, V, S, \omega)$$

By dimensional analysis, find the dimensionless parameters for this problem. Express your answer in the form

$$\dfrac{F_D}{\rho V^2 S} = f(\pi)$$

Common π-Groups (§8.4)

8.6 For each item listed, which π-group (Re, We, M, or Fr) would best match the given description?

a. (kinetic force)/(gravitational force)

b. (kinetic force)/(surface-tension force)

c. (kinetic force)/(viscous force)

d. (kinetic force)/(compressive force)

e. Used for analyzing the drag on a car in a wind tunnel

f. Used for modeling water flowing over a spillway on a dam

g. Used for designing laser jet printers

h. Used to analyze the flight of supersonic jets

Similitude (§8.5)

8.7 The drag on a submarine moving completely submerged is to be determined by a test on a 1/19-scale model in a water tunnel. The velocity of the prototype in seawater ($\rho = 1030$ kg/m³, $\nu = 1.4 \times 10^{-6}$ m²/s) is 3 m/s. The test is done in fresh water at 20°C. Determine the speed of the water in the water tunnel for dynamic similitude and the ratio of the drag force on the model to the drag force on the prototype.

8.8 In a study of the power required to overcome drag, an engineer is using a π-group given by $\dfrac{P}{\rho A V^3}$, where P is the power lost, ρ is the fluid density, A is area, and V is the fluid velocity. In laboratory tests with a one-eighth-scale model, the power lost was measured as 6 W when the air velocity was 0.4 m/s. Calculate the power lost in the prototype (kW) when the air velocity is 2 m/s. Temperature is the same in both cases.

8.9 Water with a kinematic viscosity of 10^{-6} m²/s flows through a 30 cm pipe. What would the velocity of water have to be for the water flow to be dynamically similar to oil ($\nu = 10^{-5}$ m²/s) flowing through the same pipe at a velocity of 30 m/s?

8.10 Oil with a kinematic viscosity of 3×10^{-6} m²/s flows through a smooth pipe 20 cm in diameter at 4 m/s. What velocity should water have at 20°C in a smooth pipe 5 cm in diameter to be dynamically similar?

8.11 A one-tenth-scale model of an experimental deep sea submersible that will operate at great depths is to be tested to determine its drag characteristic by towing it behind a submarine. For true similitude, what should be the towing speed relative to the speed of the prototype?

8.12 An engineer needs a value of lift force for an airplane that has a coefficient of lift (C_L) of 0.5. The π-group is defined as

$$C_L = 2\,\frac{F_L}{\rho V^2 S}$$

where F_L is the lift force, ρ is the density of ambient air, V is the speed of the air relative to the airplane, and S is the area of the wings from a top view. Estimate the lift force in newtons for a speed of 90 m/s, an air density of 1.1 kg/m³, and a wing area (planform area) of 18 m².

8.13 A smooth pipe designed to carry crude oil (D = 72 in., $\rho = 1.75$ slugs/ft³, and $\mu = 4 \times 10^{-4}$ lbf-s/ft²) is to be modeled with a smooth pipe 4 in. in diameter carrying water ($T = 60°F$). If the mean velocity in the prototype is 5.5 ft/s, what should be the mean velocity of water in the model to ensure dynamically similar conditions?

8.14 The Airbus A380-300 has a wing span of 79.8 m. The cruise altitude is 10,000 m in a standard atmosphere. Assume you are designing a wind tunnel to operate with air at 20°C. The span of the scale model A380 in the wind tunnel is 1 m. Assume Mach-number correspondence between model and prototype. Both the speed of sound and the dynamic viscosity vary linearly with the square root of the absolute temperature. What would the pressure of the air in the wind tunnel have to be to have Reynolds-number similitude? Assume that $T = 237.3$K, $p = 28$ kPa abs and $\rho = 0.411$ kg/m³.

8.15 Water at 10°C flowing through a rough pipe 10 cm in diameter is to be simulated by air (20°C) flowing through the same pipe. If the velocity of the water is 1.5 m/s, what will the air velocity have to be to achieve dynamic similarity? Assume the absolute air pressure in the pipe to be 150 kPa. If the pressure difference between two sections of the pipe during air flow was measured as 780 Pa, what pressure difference occurs between these two sections when water is flowing under dynamically similar conditions?

8.16 Colonization of the moon will require an improved understanding of fluid flow under reduced gravitational forces. The gravitational force on the moon is one-fifth that on the surface of the earth. An engineer is designing a model experiment for flow in a conduit on the moon. The important scaling parameters are the Froude number and the Reynolds number. The model will be full scale. The kinematic viscosity of the fluid to be used on the moon is 3×10^{-5} m²/s. What should be the kinematic viscosity of the fluid to be used for the model on earth?

8.17 A drying tower at an industrial site is 5 m in diameter. The air inside the tower has a kinematic viscosity of 4×10^{-5} m²/s and enters at 30 m/s. A 1/20-scale model of this tower is fabricated

to operate with water that has a kinematic viscosity of 10^{-6} m²/s. What should be the entry velocity of the water be to achieve Reynolds-number scaling?

8.18 A flow meter to be used in a 50 cm pipeline carrying oil ($v = 10^{-5}$ m²/s, $\rho = 890$ kg/m³) is to be calibrated by means of a model (1/12 scale) carrying water ($T = 20°C$ and standard atmospheric pressure). If the model is operated with a velocity of 7 m/s, find the velocity for the prototype based on Reynolds-number scaling. For the given conditions, if the pressure difference in the model was measured as 4 kPa, what pressure difference would you expect for the discharge meter in the oil pipeline?

8.19 Water is sprayed from a nozzle at 30 m/s into air at atmospheric pressure and 20°C. Estimate the size of the droplets produced if the Weber number for breakup is 5.2 based on the droplet diameter.

8.20 To study flow over a spillway in a new dam, a 1/30-scale model is constructed. The maximum design flow rate in the actual spillway will be 500 m³/s. Calculate the corresponding flow rate in the model. The π-groups that you should match are

$$\pi_1 = \frac{Q}{AV} \text{ and } \pi_2 = \sqrt{\frac{V}{gy}}$$

8.21 A 1/10 scale model of an automobile is tested in a pressurized wind tunnel. The test is to simulate the automobile traveling at 100 km/h in air at atmospheric pressure and 25°C. The wind tunnel operates with air at 25°C. At what pressure in the test section must the tunnel operate to have the same Mach and Reynolds numbers? The speed of sound in air at 25°C is 345 m/s.

8.22 If the tunnel in Prob. 8.21 were to operate at atmospheric pressure and 25°C, what speed would be needed to achieve the same Reynolds number for the prototype? At this speed, would you conclude that Mach-number effects were important?

8.23 The scale ratio between a model dam and its prototype is 1/25. In the model test, the velocity of flow near the crest of the spillway was measured to be 2.5 m/s. What is the corresponding prototype velocity? If the model discharge is 0.10 m³/s, what is the prototype discharge?

8.24 The depth and velocity at a point in a river are measured to be 20 ft and 15 ft/s, respectively. If a 1/64 scale model of this river is constructed and the model is operated under dynamically similar conditions to simulate the free-surface conditions, then what velocity and depth can be expected in the model at the corresponding point?

8.25 A ship model 4 ft long is tested in a towing tank at a speed that will produce waves that are dynamically similar to those observed around the prototype. The test speed is 5 ft/s. What should the prototype speed be, given that the prototype length is 100 ft? Assume both the model and the prototype are to operate in freshwater.

8.26 A 1/20 scale model building is tested in a wind tunnel. If the drag of the model in the wind tunnel is measured to be 200 N for a wind speed of 20 m/s, then the prototype building in a 40 m/s wind (same temperature) should have a drag of about (a) 40 kN, (b) 80 kN, (c) 230 kN, or (d) 320 kN.

8.27 Experiments were carried out in a water tunnel and a wind tunnel to measure the drag force on an object. The water tunnel was operated with freshwater at 20°C, and the wind tunnel was operated at 20°C and atmospheric pressure. Three models were used with dimensions of 5 cm, 8 cm, and 15 cm. The drag force on each model was measured at different velocities. The following data were obtained.

Data for the water tunnel

Model Size, cm	Velocity, m/s	Force, N
5	1.0	0.064
5	4.0	0.69
5	8.0	2.20
8	1.0	0.135
8	4.0	1.52
8	8.0	4.52

Data for the wind tunnel

Model Size, cm	Velocity, m/s	Force, N
8	10	0.025
8	40	0.21
8	80	0.64
15	10	0.06
15	40	0.59
15	80	1.82

The drag force is a function of the density, viscosity, velocity, and model size,

$$F_D = f(\rho, \mu, V, D)$$

Using dimensional analysis, express this equation using π-groups and then write a computer program or use a spreadsheet to analyze the data and then plot the data using π-groups.

8.28 Experiments are performed to measure the pressure drop in a pipe with water at 20°C and crude oil at the same temperature. Data are gathered with pipes of two diameters, 5 cm and 10 cm. The following data were obtained for pressure drop per unit length.

For water

Pipe Diameter, cm	Velocity, m/s	Pressure Drop, N/m^3
5	1	210
5	2	730
5	5	3750
10	1	86
10	2	320
10	5	1650

For crude oil

Pipe Diameter, cm	Velocity, m/s	Pressure Drop, N/m^3
5	1	310
5	2	1040
5	5	5300
10	1	130
10	2	450
10	5	2210

The pressure drop per unit length is assumed to be a function of the pipe diameter, liquid density and viscosity, and the velocity,

$$\frac{\Delta p}{L} = f(\rho, \mu, V, D)$$

Perform a dimensional analysis to obtain the π-groups and then write a computer program or use a spreadsheet to analyze the data and then plot the results using π-groups.

Viscous Flow Over a Flat Surface

CHAPTER ROAD MAP Knowledge of viscous flow will equip you to solve many problems in engineering. Much of what we know about viscous flow has come from the study of flow over a flat surface. This chapter introduces both viscous flow and the Navier-Stokes equation.

FIGURE 9.1
The design of boats can involve the application of viscous flow theory. (Foucras G./StockImage/Getty Images.)

LEARNING OUTCOMES

NAVIER-STOKES EQUATIONS (§9.1, §9.2, §9.3).

- List the steps to derive the Navier-Stokes equation for steady and uniform flow.
- For Couette flow, describe the flow and apply the working equations.
- For Poiseuille flow in a channel, describe the flow and apply the working equations.

BOUNDARY LAYER (QUALITATIVE) (§9.4, §9.5).

- Explain the boundary layer concept.
- Sketch the laminar and turbulent boundary layers and describe the main features.
- Sketch the velocity profile in the laminar and turbulent boundary layers.

BOUNDARY LAYER (CALCULATIONS) (§9.4, §9.6).

- Define and calculate Re_x and Re_L.
- Calculate the boundary layer thickness δ.
- Calculate the local shear stress coefficient c_f and the wall shear stress.
- Calculate the average shear stress coefficient C_f and the drag force F_D.
- For a moving body, calculate the power needed to overcome the drag force.

9.1 The Navier-Stokes Equation for Uniform Flow

Many solutions to viscous flow problems come from solving the Navier-Stokes equation. In this section, we derive this equation for uniform and steady flows of a Newtonian fluid.

Derivation

Step 1: **Sketch a fluid particle.** Select a viscous flow that is uniform and steady, and then sketch a rectangular fluid particle with dimensions of Δs by Δy by 1 unit (Fig. 9.2).

Step 2: **Apply Newton's second law.** Apply $\Sigma F = ma$ in the s direction. The acceleration is zero because the flow is uniform and steady. Thus,

$$\Sigma F_s = 0 \tag{9.1}$$

The forces are the weight, the pressure force, and the shear force. Thus,

$$\text{(weight)} + \text{(pressure force)} + \text{(shear force)} = 0$$
$$W_s + F_{ps} + F_{\tau s} = 0 \tag{9.2}$$

Each variable in Eq. (9.2) represents a component of force in the s direction.

Step 3: **Analyze the weight.** The weight of the fluid particle is

$$W = mg = \rho(\Delta s)(\Delta y)(1)g \tag{9.3}$$

The component of the weight in the s direction is $\rho(\Delta s)(\Delta y)g \sin(\theta)$. By using the triangle that is sketched in the upper part of Fig. 9.2, it can be shown that $\sin(\theta) = -dz/ds$. Thus, the component of the weight in the s direction is

$$W_s = -\gamma \Delta y \Delta s \frac{dz}{ds} \tag{9.4}$$

Step 4: **Analyze the pressure force.** The pressure terms in Fig. 9.2 come from a Taylor series expansion of the pressure field. The net pressure force in the s direction is

$$F_{ps} = -\frac{dp}{ds} \Delta s \Delta y \tag{9.5}$$

FIGURE 9.2

This sketch shows the forces acting on a fluid particle that is situated in a viscous flow. The particle is rectangular in shape and extends 1.0 unit into the paper.

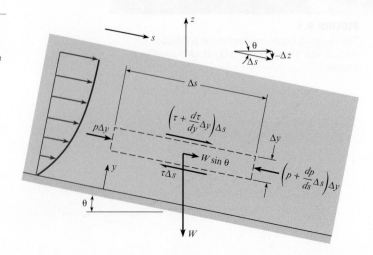

Step 5: **Analyze the shear force.** The shear stress terms in Fig. 9.2 come from a Taylor series expansion of the shear stress field. The net shear force in the s direction is

$$F_{\tau s} = \frac{d\tau}{dy} \Delta s \Delta y \qquad (9.6)$$

If the fluid is assumed to be a Newtonian fluid, then $\tau = \mu du/dy$. Here, μ is the viscosity of the fluid and u is the component of the velocity vector in the s direction. Substitute $\tau = \mu du/dy$ into Eq. (9.6) to give

$$F_{\tau s} = \mu \frac{d^2 u}{dy^2} \Delta s \Delta y \qquad (9.7)$$

Step 6: **Combine terms.** Substitute the terms from steps 2 to 4 into Eq. (9.2). Then, divide each term by $\Delta s \Delta y(1)$, which is the volume of the fluid particle. The result is the Navier-Stokes equation for uniform and steady flow of a Newtonian fluid:

$$\frac{d^2 u}{dy^2} = \frac{1}{\mu} \frac{d}{ds}(p + \gamma z) \qquad (9.8)$$

Eq. (9.8) describes, for a fluid particle, the balance between the pressure force, the shear force, and the weight. A more general form of the Navier-Stokes equation, derived in §16.4, includes terms that account for the acceleration of a fluid particle.

Solving the Navier-Stokes Equation

Couette flow and Poiseuille flow, which are analyzed in the next two sections, are classified as exact solutions. There are only a few exact solutions in existence; all other solutions involve an approximation of one form or another.

When engineers solve the Navier-Stokes equation, the goal is usually the velocity field. After the velocity field has been found, other parameters of engineering interest—for example, shear stress and mean flow rate—can be derived. This approach will be illustrated in the next two sections.

9.2 Couette Flow

Couette flow (see Fig. 9.3 and §2.5) is used to idealize flows such as the flow of oil in a journal bearing.

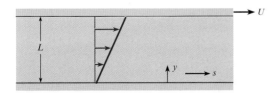

FIGURE 9.3

Couette flow involves two flat plates, each large enough that the dimensions of the plates can be idealized as infinite. The lower plate is stationary and the upper plate is moving with a constant velocity U. The gap L is small—for example, a fraction of a millimeter. The equation for the pictured velocity profile is derived in the current section.

Solving for the Velocity Field

To solve for the velocity field in Couette flow, take the following steps.

Step 1:   **Apply the Navier-Stokes equation.** In Couette flow, the pressure gradient in the stream-wise direction is zero ($dp/ds = 0$) and the streamlines are in the horizontal direction, which means that $dz/ds = 0$. Therefore, the right side of Eq. (9.8) is zero, and this equation reduces to

$$\frac{d^2u}{dy^2} = 0 \tag{9.9}$$

To solve this differential equation, integrate twice using the method known as "separation of variables":

$$u = C_1 y + C_2 \tag{9.10}$$

Step 2:   **Write the boundary conditions.** To solve for the two constants in Eq. (9.10), apply the following two boundary conditions:

$$u = 0 \text{ at } y = 0 \tag{9.11}$$
$$u = U \text{ at } y = L \tag{9.12}$$

Step 3:   **Apply the boundary conditions.** Combine Eq. (9.10) with Eqs. (9.11) and (9.12) to give an equation for the velocity field:

$$u = Uy/L \tag{9.13}$$

The velocity profile (Fig. 9.3) reveals a linear velocity profile.

Deriving Working Equations

To develop an equation for the shear stress field, apply $\tau = \mu(du/dy)$ to Eq. (9.13):

$$\tau = \mu U/L \tag{9.14}$$

Eq. (9.14) reveals that the shear stress is a constant at every point. To develop an equation for mean velocity, substitute Eq. (9.13) into $V = (1/A) \int u dA$. After integration, the result is $\overline{V} = U/2$.

9.3 Poiseuille Flow in a Channel

The conduit considered in this section is a rectangular channel (Fig. 9.4). A **channel** is a flow passage between two parallel plates, when each plate is wide enough so that the end effects caused by the side walls of the channel can be neglected.

FIGURE 9.4

Poiseuille flow refers to a laminar flow in a conduit. This sketch shows a rectangular channel and the associated velocity profile.

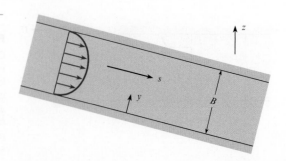

Experiments have revealed that the flow in a rectangular channel will be laminar for a Reynolds number below 1000. Thus, channel Poiseuille flow applies to steady, laminar, and fully developed flow of a Newtonian fluid with $Re_B < 1000$, where $Re_B = UB/\nu$.

Solving for the Velocity Field

To derive an equation for the velocity field, take the following steps.

Step 1: **Apply the Navier-Stokes equation**. Prior to solving Eq. (9.8), recognize that the right side of the equation must be a constant. *Rationale*. The independent variable on the left side of the equation is y. The independent variable on the right side of the equation is s. Because each of these variables can change their value independently, both sides of the equation must equal a constant to preserve the equality. Thus, the right side of the equation must be a constant.

Integrate Eq. (9.8) twice to give the general solution:

$$u = \frac{y^2}{2\mu} \frac{d}{ds}(p + \gamma z) + C_1 y + C_2 \tag{9.15}$$

Step 2: **Apply the boundary conditions**. To develop boundary conditions, apply the no-slip condition at each wall:

$$u = 0 \text{ at } y = 0 \tag{9.16}$$
$$u = 0 \text{ at } y = B \tag{9.17}$$

To satisfy Eq. (9.16), set $C_2 = 0$. To satisfy Eq. (9.17), solve for C_1 as follows:

$$C_1 = -\frac{B}{2\mu} \frac{d}{ds}(p + \gamma z) \tag{9.18}$$

Step 3: **Build the particular solution**. Combine Eqs. (9.15) and (9.18) to give an equation for the velocity field:

$$u = -\frac{(By - y^2)}{2\mu} \frac{d}{ds}(p + \gamma z) \tag{9.19}$$

Notice that Eq. (9.19) is the equation of a parabola (Fig. 9.4). To check any solution of a differential equation—for example, to validate Eq. (9.19)—we recommend three practices: (a) Verify that the solution satisfies the original differential equation, (b) verify that the solution is dimensionally homogeneous (DH), and (c) verify that the solution satisfies the boundary conditions.

Deriving Working Equations

To develop an equation for the maximum velocity, recognize that u_{max} occurs at $y = B/2$. Then, substitute this value into Eq. (9.19) to give

$$u_{max} = -\left(\frac{B^2}{8\mu}\right) \frac{d}{ds}(p + \gamma z) \tag{9.20}$$

Next, simplify the right side of Eq. (9.20) by introducing piezometric head ($p + \gamma z = \gamma h$). The result is

$$u_{max} = -\left(\frac{B^2 \gamma}{8\mu}\right) \frac{dh}{ds} \tag{9.21}$$

To develop an equation for discharge per unit width q, substitute Eq. (9.19) into $Q = qw = \int u dA = \int_0^B u w dy$. The result is

$$q = -\left(\frac{B^3}{12\mu}\right)\frac{d}{ds}(p + \gamma z) = -\left(\frac{B^3\gamma}{12\mu}\right)\frac{dh}{ds} \tag{9.22}$$

Eq. (9.22) reveals that Poiseuille flow is driven by a gradient in piezometric head (dh/ds). The negative sign on the right side of Eq. (9.22) means that a fluid will flow from high to low piezometric head. Sometimes, we hear someone say that a fluid flows from high to low pressure or down the pressure gradient; be aware that this idea about pressure is not true in general.

To develop an equation for mean velocity $\overline{V}$, apply the equation $\overline{V} = Q/A = q/B$ and introduce u_{max} as given in Eq. (9.21). The result is

$$\overline{V} = 2u_{max}/3 \tag{9.23}$$

Eq. (9.23) reveals that the mean velocity is two-thirds of the maximum velocity.

9.4 The Boundary Layer (Description)

The Boundary Layer Concept

When he was 29 years old, Ludwig Prandtl wrote a paper (1) about the boundary layer concept. The impact of this idea, as described by John Anderson (2, p. 42), is that "the modern world of aerodynamics and fluid dynamics is still dominated by Prandtl's idea. By every right, his boundary layer concept was worthy of the Nobel Prize."

The central idea of the boundary layer concept (Fig. 9.5) is to idealize the flow near the surface of a body as a thin layer that is distinct from the surrounding flow.

The boundary layer concept is applied to idealize many real-world flows; some examples follow:

- The water flow past the pier of a bridge
- The air flow past the surface of the earth
- The air flow past an automobile
- The air or water flow in the entrance of a pipe

The Boundary Layer on a Flat Plate

Because flow over a flat plate (Fig. 9.6) is simple, this flow is the foundation for understanding most boundary layer flows. Notice that the flow outside the boundary layer is called the **free stream**.

The boundary layer starts from the leading edge of the plate. The thickness of the boundary layer δ increases with x. The edge of the boundary layer is defined as the point at which the local velocity equals 99% of the free stream velocity. Therefore, $y = \delta$ when Eq. (9.24) is satisfied:

$$\frac{u}{U_o} = 0.99 \tag{9.24}$$

FIGURE 9.5

The boundary layer is a thin layer near the body that is analyzed separately from the surrounding flow.

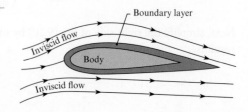

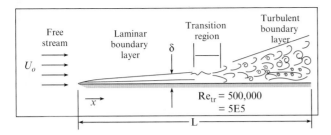

FIGURE 9.6

The boundary layer on a flat surface. Here, U_o is the free stream velocity, x is distance, δ is the boundary layer thickness, L is the length of the plate, and Re_{tr} is the Reynolds number associated with transition.

The boundary layer on a flat plate starts out laminar (Fig. 9.6). This means the flow is smooth, layer-like, and steady. If plate length L is large enough, the laminar boundary layer will be followed by a transition boundary layer and then a turbulent boundary layer. A **transition region** is the zone in which the boundary layer changes from laminar to turbulent. The turbulent boundary layer has several distinctive characteristics: (a) At each point, the velocity varies with time; (b) at each point, the pressure varies with time; (c) the flow is always unsteady; and (d) the flow is dominated by eddies. An **eddy** is a circular movement of fluid. The largest eddy in the boundary layer has a diameter that is approximately equal to the boundary layer thickness. The smallest eddy in the boundary layer has a diameter that is given by a length scale called the Kolmogorov length scale. Due to the eddies in a turbulent flow, there is strong mixing in any such flow. This mixing tends to even out the velocity profile.

The Reynolds Number

The nature of the boundary layer (laminar versus turbulent) is correlated with the Reynolds number. The variable Re_x, called the "Reynolds number based on x," or the "local Reynolds number," is defined as follows:

$$Re_x = \frac{U_o x}{\nu} \tag{9.25}$$

The variable U_o represents the free stream velocity, x is the distance from the leading edge of the plate, and ν is the kinematic viscosity of the fluid. The variable Re_L, called the "Reynolds number based on L," is defined as

$$Re_L = \frac{U_o L}{\nu} \tag{9.26}$$

A variable like x or L in the Reynolds number is called a length scale.

The Transition Reynolds Number

The value of Re_x at which transition starts and ends varies from experiment to experiment. Thus, there is not a unique criterion for transition. This book follows a common engineering approach that involves three assumptions:

- If $Re_x \leq 500,000$, the boundary layer is assumed to be laminar.
- If $Re_x > 500,000$, the boundary layer is assumed to be turbulent.
- The length of the transition zone is neglected.

The transition Reynolds number is $Re_{tr} = 500,000$.

9.5 Velocity Profiles in the Boundary Layer

In fluid mechanics, engineers want to find the velocity profile. Once we know the velocity profile, we can determine other parameters (e.g., drag force and shear stress) that we want to know.

Boundary Layer Theory

To analyze a boundary layer, Prandtl created an approximate (not exact) form of the governing equations called the boundary layer equations. Blasius (3), a graduate student working with Prandtl, solved the boundary layer equations for a laminar boundary layer over a flat plate; engineers call this solution the **Blasius solution**.

The Laminar Velocity Profile

For a laminar boundary layer on a flat plate, the Blasius solution (Fig. 9.7) gives a plot of the velocity profile. Notice that $u = 0$ at $y = 0$ because of the no-slip condition. Also, notice that the velocity profile merges smoothly with the free stream.

The Turbulent Velocity Profile

Researchers have developed multiple solutions for the turbulent velocity profile. The research method involves fitting curves to experimental data with the addition of an ad hoc theory. The results can be confusing because there are many different velocity profiles. In this textbook, we only present some of the most common profiles.

In the turbulent boundary layer, the velocity profile (Fig. 9.8) is nearly uniform away from the wall. This is because of eddies that mix the flow. Near the wall, the velocity profile exhibits a steep velocity gradient. Thus, τ_o for the turbulent boundary layer is generally larger than τ_o for the laminar boundary layer.

Logarithmic velocity distribution. A common equation for describing the turbulent boundary layer is the logarithmic velocity distribution, which is

$$\frac{u}{u_\star} = 2.44 \ln\left(\frac{yu_\star}{v}\right) + 5.56 \tag{9.27}$$

The **friction velocity** $u_\star$ is defined by $u_\star = \sqrt{\tau_o/\rho}$. This term is not really a velocity; instead, it is a term that has the same units as velocity. To calculate a value for $u_\star$, engineers calculate a value of τ_o using the methods described in §9.6.

Eq. (9.27) applies to the part of the velocity profile where the value of $yu_\star/v$ is between 30 and 500.

FIGURE 9.7

The Blasius solution (3) for the velocity profile in a laminar boundary layer on a flat plate. Here, u is the velocity at a height y above the surface. The velocity U_0 is the free stream velocity, and Re_x is the local Reynolds number.

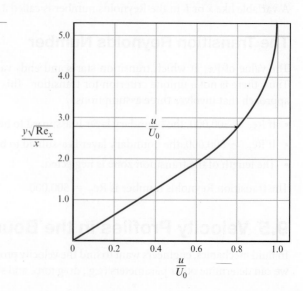

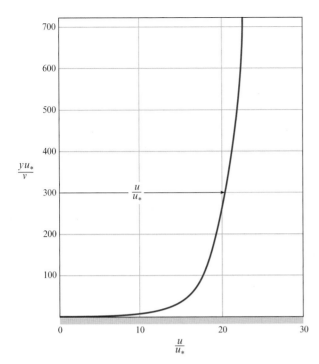

FIGURE 9.8

An example of a turbulent velocity profile. Here, u is the time-averaged velocity and u_* is the friction velocity.

Power law formula. For Reynolds numbers between 10^5 and 10^7 (i.e., $10^5 < \mathrm{Re}_x < 10^7$), the velocity profile in the turbulent boundary layer on a flat plate is well approximated by the power law equation, which is

$$\frac{u}{U_o} = \left(\frac{y}{\delta}\right)^{1/7} \tag{9.28}$$

Eq. (9.28) matches experimental results for the range $(0.1 < y/\delta < 1.0)$. At the wall, Eq. (9.28) cannot be valid because $du/dy \to \infty$ at $y = 0$, which means that $\tau_o \to \infty$.

9.6 The Boundary Layer (Calculations)

Equations in this section are based on four assumptions: (1) uniform flow over a smooth and flat surface, (2) steady flow with the free stream parallel to the flat surface, (3) a Newtonian fluid, and (4) a large enough Reynolds number to ensure that the boundary layer concept is valid.

Boundary Layer Thickness

For the laminar boundary layer, Schlichting (4, p. 140) shows that the thickness δ of the boundary layer is given by

$$\frac{\delta}{x} = \frac{5.0}{\mathrm{Re}_x^{1/2}} \tag{9.29}$$

For the turbulent boundary layer, White (5, p. 430) shows that δ is given by

$$\frac{\delta}{x} = \frac{0.16}{\mathrm{Re}_x^{1/7}} \tag{9.30}$$

EXAMPLE. Water ($\nu = 1.2$E-6 m^2/s) flows over the top of a flat plate. The plate length is 1.5 m. The free stream velocity is 5 m/s. Calculate the boundary layer thickness at $x/L = 0.5$.

Reasoning.

1. To determine if the boundary layer is laminar or turbulent, calculate the Reynolds number. $\text{Re}_x = (5 \text{ m/s})(0.75 \text{ m})/(1.2\text{E-6 m}^2/\text{s}) = 3.125\text{E6}$.
2. At the point in question, the boundary layer is turbulent.
3. Thus, $\delta = (0.16)(0.75 \text{ m})/(3.125\text{E6})^{1/7} = 1.42$ cm.

Shear Stress at the Wall

Engineers incorporate the wall shear stress τ_o into a π-group as follows:

$$c_f = \frac{\text{(shear stress at the wall)}}{\text{(kinetic pressure)}} = \frac{\tau_o}{(\rho U_o^2/2)} \tag{9.31}$$

The name of c_f is the **local shear stress coefficient**. For the laminar boundary layer, Schlichting (**4**, p. 138) shows that

$$c_f = \frac{0.664}{\text{Re}_x^{1/2}} \tag{9.32}$$

For the turbulent boundary layer, researchers have proposed many different formulas. White (**5**, p. 432) recommends

$$c_f = \frac{0.455}{\ln^2(0.06 \text{ Re}_x)} \tag{9.33}$$

After a value of c_f has been calculated, τ_o can be found by using

$$\tau_o = c_f(\rho U_o^2/2) \tag{9.34}$$

EXAMPLE. By applying Eqs. (9.32), (9.33), and (9.34), we calculated τ_o for water flowing over a plate (Fig. 9.9). Notice that τ_o for the turbulent boundary layer is usually greater than τ_o for the laminar boundary layer and that τ_o decreases with x. For the laminar boundary layer, this decrease is proportional to $x^{-0.5}$, a fact that we deduced from Eq. (9.32).

FIGURE 9.9

This figure shows the wall shear stress τ_o calculated for water flowing over a flat plate. The problem variables are $L = 1.0$ m, $U_o = 1.1$ m/s, $\nu = 1.1$E-6 m^2/s, and $\rho = 1000$ kg/m^3.

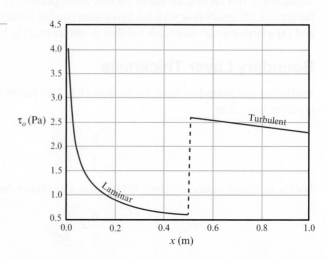

Drag Force (Shear Force, Skin Friction Drag)

Engineers incorporate the drag force into a π-group as follows:

$$C_f \equiv \frac{(\text{drag force})}{(\text{kinetic pressure})(\text{reference area})} = \frac{F_D}{(\rho U_o^2 / 2)(A_{\text{Ref}})} \qquad (9.35)$$

The term C_f is called the "coefficient of drag" or the "average shear stress coefficient." Note that the free stream velocity U_o is measured relative to an observer situated on the plate. **Example.** If a plate is moving at a speed of 6 m/s in still air, then $U_o = 6$ m/s.

The reference area A_{Ref} is either (a) the surface area of one side of the plate or (b) the surface area of both sides of the plate. The drag force F_D can also be represented as a shear force F_s because the drag force on a flat plate is due to shear stress only. Also, the drag force associated with shear stress is sometimes called the shear force, the skin friction drag, or the surface drag.

To calculate the drag force, rearrange Eq. (9.35) to give

$$F_D = C_f(\rho U_o^2 / 2)(A_{\text{Ref}}) \qquad (9.36)$$

To calculate C_f, determine the nature of the boundary layer and then select the appropriate equation from the three options that follow:

- **Laminar boundary layer.** If $\text{Re}_L \leq 500{,}000$, assume the boundary layer is laminar and select Eq. (9.37).

$$C_f = \frac{1.33}{\text{Re}_L^{1/2}} \qquad (9.37)$$

- **Mixed boundary layer.** If $\text{Re}_L > 500{,}000$, assume the boundary layer is mixed and select Eq. (9.38). A **mixed boundary layer** is a boundary layer that starts as laminar and then transitions to turbulent.

$$C_f = \frac{0.523}{\ln^2(0.06\text{Re}_L)} - \frac{1520}{\text{Re}_L} \qquad (9.38)$$

- **Tripped boundary layer.** For this case, select Eq. (9.39). A **tripped boundary layer** is a boundary layer that is turbulent over the entire length of the plate. Engineers idealize the boundary layer as tripped when there are roughness elements that cause all of the boundary layer to be turbulent. Examples of roughness elements include sand glued to the plate, electrical components on a printed circuit board, and wires situated near the leading edge of the plate. If $\text{Re}_L \gg 500{,}000$, the boundary layer can also be idealized as tripped because the length of the laminar boundary layer is short relative to the plate length.

$$C_f = \frac{0.032}{\text{Re}_L^{1/7}} \qquad (9.39)$$

The correlations for drag force are plotted in Fig. 9.10. If $\text{Re}_L \gtrsim 10^7$, then Eqs. (9.38) and (9.39) give the same value of C_f. For this condition, the boundary layer can be modeled as a "tripped boundary layer."

FIGURE 9.10

This figure shows C_f on a flat plate as a function of the Reynolds number.

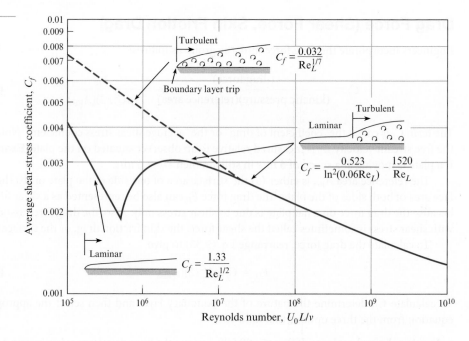

The drag force on a flat plate can be used to idealize a more complex flow. **Example.** The drag force on an airplane can be estimated by idealizing the airplane as a flat plate with an equivalent surface area. Thus, an airplane with an exterior surface area of 6.2 m² could be idealized as a flat plate with a surface area of 3.1 m² on the top of the plate and 3.1 m² on the bottom. To achieve scaling, the plate length and other variables should be set so that Re_L for the real system (i.e., the airplane) will match Re_L for the idealization (i.e., the flat plate).

EXAMPLE PROBLEM. Oil (ρ = 900 kg/m³, ν = 8.0E-5 m²/s) flows over both sides of a flat plate. The plate dimensions are W = 250 mm and L = 750 mm. The free stream velocity is U_o = 4.5 m/s. Calculate the drag force on the plate.

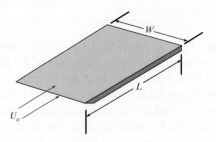

Reasoning.

1. The drag force can be calculated using $F_D = C_f(\rho U_o^2/2)(A_{Ref})$.
2. The Reynolds number is Re_L = (4.5m/s)(0.75m)/(8.0E-5) = 42,187.
3. Because $Re_L \leq$ 500,000, the boundary layer is laminar.
4. Thus, the drag coefficient is C_f = 1.33/ $\sqrt{42,187}$ = 6.4752E-3.
5. The reference area is A_{Ref} = (2)(0.75 m)(0.25 m) = 0.375 m².
6. From step 1, F_D = (6.4752E-3)(900 kg/m³)(4.5 m/s)²(0.375 m²)/2 = 22.1 N.

The Power Equation

When a body (e.g., a car, boat, or airplane) moves through a fluid, the body must do work on the fluid to overcome the drag force. When the body is moving in a straight line at a constant speed (i.e., rectilinear motion), this rate of work is given by the power equation:

$$P = F_D V \qquad\qquad (9.40)$$

Here, P is the power needed to overcome the drag force, F_D is the drag force acting on the body, and V is the speed of the body relative to a ground-based reference frame.

> **EXAMPLE PROBLEM.** A submarine (length = 150 m, average diameter = 12 m) moves through seawater (ν = 1.4E-6 m^2/s, ρ = 1030 kg/m^3) at a speed of 9 m/s. Estimate the power needed to overcome the drag force. Idealize the submarine as a flat plate that has the same surface area as the submarine.
>
> **Solution.**
> 1. The power needed to overcome drag is $P = F_D U$.
> 2. The drag force is given by $F_D = C_f(\rho V^2/2)(A_{Ref})$.
> 3. The reference area is $A_{Ref} = (150\text{ m})(\pi)(12\text{ m}) = 5655\text{ m}^2$.
> 4. The Reynolds number is $\text{Re}_L = (150\text{ m})(9\text{ m/s})/(1.4\text{E-6 m}^2/\text{s}) = 9.643\text{E8}$.
> 5. From either Fig. 9.10 or Eq. (9.38), $C_f = 0.00164$.
> 6. The drag force (see step 2) is $F_D = 3.858\text{E5 N}$.
> 7. The power (see step 1) is $P = (3.858\text{E5 N})(9\text{ m/s}) = 3.47\text{ MW}$.

9.7 Summarizing Key Knowledge

The Navier-Stokes Equation (Uniform Flow)

- The Navier-Stokes equation is derived by applying Newton's second law of motion to a fluid particle.
- The Navier-Stokes equation only has a few exact solutions. Two of these exact solutions are Couette flow and Poiseuille flow.
- For Couette flow, the following facts apply: (a) The pressure is uniform, (b) the shear stress is equal everywhere to $\mu U/L$, and (c) the velocity profile is linear.
- For Poiseuille flow in a channel, the following facts apply: (a) The criterion for laminar flow is $\text{Re}_B = VB/\nu < 1000$, (b) the mean velocity is two-thirds of the maximum velocity ($\overline{V} = 2u_{max}/3$), (c) the discharge per length is $q = -(B^3\gamma/12\mu)dh/ds$, and (d) the velocity profile is parabolic.

The Boundary Layer

- The boundary layer is the thin region of fluid near a solid body. In the boundary layer, viscous stresses cause a velocity profile.

- The boundary layer thickness δ is the distance from the wall to the location where the velocity is 99% of the free stream velocity.
- If Re_L is large enough, the boundary layer will have three regions: (a) the laminar boundary layer, (b) the transition boundary layer, and (c) the turbulent boundary layer.
- For engineering purposes, transition to a turbulent boundary layer occurs at $\text{Re}_{tr} = 500,000$.

Predicting Boundary Layer Parameters

- Table 9.1 lists equations that are commonly applied.
- Wall shear stress τ_o is calculated using $\tau_o = c_f(\rho U_o^2/2)$.
- The drag force F_D is calculated from $F_D = C_f(\rho U_o^2/2)(A_{Ref})$.

TABLE 9.1 Summary of Equations for a Boundary Layer on a Flat Plate

Parameter	Laminar Flow Re_x, $Re_L < 5 \times 10^5$	Turbulent Flow Re_x, $Re_L \geq 5 \times 10^5$
Boundary Layer Thickness, δ	$\delta = \dfrac{5x}{Re_x^{1/2}}$	$\delta = \dfrac{0.16x}{Re_x^{1/7}}$
Local Shear Stress Coefficient, c_f	$c_f = \dfrac{0.664}{Re_x^{1/2}}$	$c_f = \dfrac{0.445}{\ln^2(0.06 Re_x)}$
Average Shear Stress Coefficient, C_f (mixed boundary layer)	$C_f = \dfrac{1.33}{Re_L^{1/2}}$	$C_f = \dfrac{0.523}{\ln^2(0.06 Re_L)} - \dfrac{1520}{Re_L}$
Average Shear Stress Coefficient, C_f (tripped boundary layer)		$C_f = \dfrac{0.032}{Re_L^{1/7}}$

REFERENCES

1. Prandtl, L. "Über Flussigkeitsbewegung bei sehr kleiner Reibung." Verhandlungen des III. Internationalen Mathematiker–Kongresses. Leipzig, 1905.

2. Anderson, John D., Jr. "Ludwig Prandtl's Boundary Layer." *Physics Today* 58, no. 12, 42-48, 2005.

3. Blasius, H. "Grenzschichten in Flüssigkeiten mit kleiner Reibung." *Z. Mat. Physik*, 1908, English translation in NACA TM 1256.

4. Schlichting, Herman. *Boundary-Layer Theory*. 7th ed. New York: McGraw-Hill, 1979.

5. White, Frank M. *Viscous Fluid Flow*. 2nd ed. New York: McGraw-Hill, 1991.

PROBLEMS

Couette Flow (§9.2)

9.1 A board 2 ft by 2 ft that weighs 32 lbf slides down an inclined ramp with a velocity of 0.6 ft/s. The board is separated from the ramp by a layer of oil 0.02 in. thick. Calculate the dynamic viscosity μ of the oil.

9.2 A board 0.5 m by 0.5 m that weighs 35 N slides down an inclined ramp with a velocity of 40 cm/s. The board is separated from the ramp by a layer of oil 0.6 mm thick. Neglecting the edge effects of the board, calculate the dynamic viscosity μ of the oil.

9.3 Information is needed about the thickness of oil necessary to lubricate metal parts sliding down an inclined plane. A square metal part with 0.6 m sides that weighs 25 N is required to slide down the inclined plane shown, at a velocity of 20 cm/s. Neglecting edge effects, calculate the necessary thickness of oil, if $\mu = 5.43 \times 10^{-2}$ N·s/m².

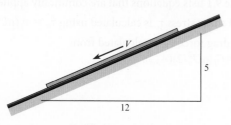

Problems 9.1, 9.2, 9.3

9.4 A plate 2 mm thick and 2.5 m wide (normal to the page) is pulled between the walls shown in the figure at a speed of 0.7 m/s. Note that the space that is not occupied by the plate is filled with glycerine at a temperature of 20°C. Also, the plate is positioned midway between the walls. Sketch the velocity distribution of the glycerine at section *A-A*. Neglecting the weight of the plate, estimate the force required to pull the plate at the speed given.

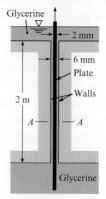

Problem 9.4

9.5 An important application of viscous flow is found in lubrication theory. Consider a shaft that turns inside a stationary cylinder, with a lubricating fluid in the annular region. By considering a system consisting of an annulus of fluid of radius r and width

Δr, and realizing that under steady-state operation the net torque on this ring is zero, show that $d(r^2\tau)/dr = 0$, where τ is the viscous shear stress. For a flow that has a tangential component of velocity only, the shear stress is related to the velocity by $\tau = \mu r\, d(V/r)/dr$. Show that the torque per unit length acting on the inner cylinder is given by $T = 4\pi\mu\omega r_s^2/(1 - r_s^2/r_0^2)$, where ω is the angular velocity of the shaft.

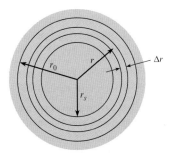

Problem 9.5

9.6 Using the equation for torque developed in Prob. 9.5, find the power necessary to rotate a 2 cm shaft at 60 rad/s if the inside diameter of the casing is 2.2 cm, the bearing is 3 cm long, and SAE 30 oil at 38°C is the lubricating fluid.

9.7 The analysis developed in Prob. 9.5 applies to a device used to measure the viscosity of a fluid. By applying a known torque to the inner cylinder and measuring the angular velocity achieved, you can calculate the viscosity of the fluid. Assume you have a 4 cm inner cylinder and a 4.5 cm outer cylinder. The cylinders are 10 cm long. When a force of 0.6 N is applied to the tangent of the inner cylinder, it rotates at 20 rpm. Calculate the viscosity of the fluid.

Poiseuille Flow (§9.3)

9.8 Glycerin at 20°C flows downward in the annular region between two cylinders. The internal diameter of the outer cylinder is 3 cm, and the external diameter of the inner cylinder is 2.8 cm. The pressure is constant along the flow direction. The flow is laminar. Calculate the discharge. (*Hint:* The flow between the two cylinders can be treated as the flow between two flat plates.)

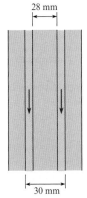

28 mm

30 mm

Problem 9.8

9.9 Uniform, steady flow is occurring between horizontal parallel plates as shown.

- **a.** Where is the maximum velocity located?
- **b.** Where is the minimum shear stress located?
- **c.** Where is the maximum shear stress located?
- **d.** The flow is Poiseuille; therefore, what is causing the fluid to move?

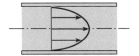

Problem 9.9

9.10 Under certain conditions (pressure decreasing in the x direction, the upper plate fixed, and the lower plate moving to the right in the positive x direction), the laminar velocity distribution will be as shown. For such conditions, indicate whether each of the following statements is true or false:

- **a.** The minimum shear stress occurs next to the moving plate.
- **b.** The minimum shear stress occurs where the velocity is the greatest.
- **c.** The shear stress midway between the plates is zero.
- **d.** The maximum shear stress occurs where the velocity is the greatest.

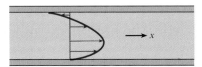

Problem 9.10

9.11 The velocity distribution that is shown represents laminar flow. Indicate which of the following statements are true:

- **a.** The maximum shear stress in the liquid occurs midway between the walls.
- **b.** The maximum shear stress in the liquid occurs next to the boundary.
- **c.** The velocity gradient at the boundary is infinitely large.
- **d.** The flow is rotational.
- **e.** The flow is irrotational.

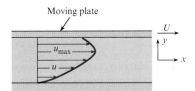

Problem 9.11

9.12 Glycerine flows downward between two vertical parallel plates separated by a distance of 1.5 cm. $T = 20°C$. The ends are open, so there is no pressure gradient. Calculate the discharge per unit width, q, in m^2/s.

9.13 Two vertical parallel plates are spaced 0.02 ft apart. If the pressure decreases at a rate of 75 psf/ft in the vertical z direction in the fluid between the plates, what is the maximum fluid velocity in the z direction? The fluid has a viscosity of 10^{-3} lbf·s/ft^2 and a specific gravity of 0.80.

9.14 An engineer is designing a very thin, horizontal channel for cooling electronic circuitry. The channel is 2 cm wide and 7 cm long. The distance between the plates is 0.25 mm. The average fluid velocity is 7 cm/s. The fluid used has a viscosity of 1.2 cp and a density of 800 kg/m^3. Assuming no change in viscosity or density, find the pressure drop in the channel and the power required to move the flow through the channel.

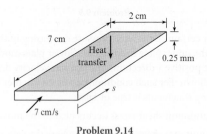

Problem 9.14

The Boundary Layer (Description) (§9.4)

9.15 Which of the following are features of a laminar boundary layer? (Select all that are correct.)

a. The boundary layer thickness increases in the downstream direction.

b. A decreasing boundary layer thickness correlates with decreasing shear stress.

c. An increasing boundary layer thickness correlates with decreasing shear stress.

d. Flow is smooth.

Velocity Profiles in the Boundary Layer (§9.5)

9.16 A liquid ($\rho = 998$ kg/m^3; $\mu = 2 \times 10^{-2}$ N·s/m^2; $v = 2 \times 10^{-5}$ m^2/s) flows tangentially past a flat plate. If the approach velocity is 1 m/s, what is the liquid velocity 2 m downstream from the leading edge of the plate, at 0.8 mm away from the plate?

9.17 Oil ($v = 2 \times 10^{-4}$ m^2/s) flows tangentially past a thin plate. If the free-stream velocity is 2 m/s, what is the velocity 5 m downstream from the leading edge and 15 mm away from the plate?

9.18 Oil ($v = 10^{-4}$ m^2/s; $SG = 0.8$) flows past a plate in a tangential direction so that a boundary layer develops. If the velocity of approach is 0.3 m/s, what is the oil velocity 0.2 m downstream from the leading edge, 0.1 cm away from the plate?

9.19 A turbulent boundary layer exists in the flow of water at 20°C over a flat plate. The local shear stress measured at the surface of the plate is 0.3 N/m^2. What is the velocity at a point 0.2 cm from the plate surface?

The Boundary Layer (Calculations) (§9.6)

9.20 Classify each of the following features into one of two categories: laminar boundary layer (L) or turbulent boundary layer (T).

a. The boundary layer contains eddies that mix the flow.

b. The flow occurs in smooth layers.

c. The velocity profile is a function of $\sqrt{\text{Re}}$.

d. The velocity profile can be written with a power law equation.

e. The velocity profile can be described with the logarithmic velocity distribution.

f. The boundary layer thickness δ varies as $x^{6/7}$.

g. The boundary layer thickness δ varies as $x^{1/2}$.

h. Even when the mean flow is steady, the velocity in the boundary layer will be unsteady.

i. Shear stress is a function of $\sqrt{\text{Re}}$.

j. Shear stress is a function of the natural log.

9.21 A model airplane has a wing span of 4 ft and a chord (leading edge–trailing edge distance) of 10 in. The model flies in air at 60°F and atmospheric pressure. The wing can be regarded as a flat plate so far as drag is concerned. (a) At what speed will a turbulent boundary layer start to develop on the wing? (b) What will be the total drag force on the wing just before turbulence appears?

9.22 A liquid ($\rho = 998$ kg/m^3; $\mu = 4 \times 10^{-2}$ N·s/m^2; $v = 2 \times 10^{-5}$ m^2/s) flows tangentially past a flat plate with total length of 4 m (parallel to the flow direction), a velocity of 1 m/s, and a width of 1.5 m. What is the skin friction drag (shear force) on one side of the plate?

9.23 A transducer for measuring shear stress is positioned on a flat plate 0.8 meter from the leading edge. The element consists of a small plate, 1 cm × 1 cm, mounted flush with the wall, and the shear force is measured on the plate. The fluid flowing by the plate is air with a free-stream velocity of $V = 75$ m/s, a density of 1.2 kg/m^3, and a kinematic viscosity of 1.5×10^{-5} m^2/s. The boundary layer is tripped at the leading edge. What is the magnitude of the force due to shear stress acting on the element?

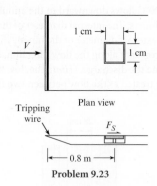

Problem 9.23

9.24 A liquid flows tangentially past a flat plate. The fluid properties are $\mu = 10^{-5}$ N · s/m^2 and $\rho = 1.5$ kg/m^3. Find the skin friction drag on the plate per unit width if the plate is 2.5 m long and the approach velocity is 16 m/s. Also, what is the velocity gradient at a point that is 1 m downstream of the leading edge and just next to the plate ($y = 0$)?

9.25 For the hypothetical boundary layer on the flat plate shown, what is the shear stress on the plate at the downstream end (point A)? Here $\rho = 1.2$ kg/m^3 and $\mu = 7.0 \times 10^{-5}$ N·s/m^2.

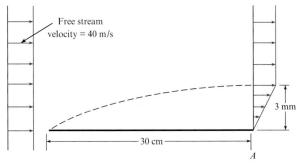

Problem 9.25

9.26 Calculate the power required to pull the sign shown if it is towed at 55 m/s and if it is assumed that the sign has the same drag force as an equivalent flat plate. Assume standard atmospheric pressure and a temperature of 10°C.

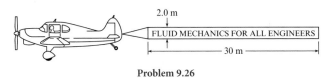

Problem 9.26

9.27 A flat plate is oriented parallel to a 42 m/s airflow at 20°C and atmospheric pressure. The plate is $L = 1$ m in the flow direction

and 0.5 m wide. On one side of the plate, the boundary layer is tripped at the leading edge, and on the other side there is no tripping device. Find the total drag force on the plate.

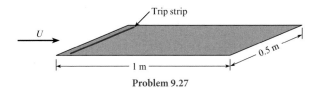

Problem 9.27

9.28 An outboard racing boat "planes" at 70 mph over water at 60°F. The part of the hull in contact with the water has an average width of 3 ft and a length of 8 ft. Estimate the power required to overcome its shear force.

9.29 Consider the boundary layer next to the smooth hull of a ship. The ship is cruising at a speed of 45 ft/s in 60°F freshwater. Assuming that the boundary layer on the ship hull develops the same as on a flat plate, determine

  a. The thickness of the boundary layer at a distance $x = 100$ ft downstream from the bow.

  b. The velocity of the water at a point in the boundary layer at $x = 100$ ft and $y/\delta = 0.50$.

  c. The shear stress, τ_0, adjacent to the hull at $x = 100$ ft.

CHAPTER**TEN**

Flow in Conduits

CHAPTER ROAD MAP This chapter explains how to analyze flow in conduits. The primary tool, the energy equation, was presented in Chapter 7. This chapter expands on this knowledge by describing how to calculate head loss. In addition, this chapter explains how to analyze pumps and how to analyze a network of pipes.

FIGURE 10.1
The Alaskan pipeline, a significant accomplishment of the engineering profession, transports oil 1286 km across the state of Alaska. The pipe diameter is 1.2 m, and 44 pumps are used to drive the flow. This chapter presents information for designing systems involving pipes, pumps, and turbines. (Photo © Eastcott/Momatiuk/The Image Works.)

LEARNING OUTCOMES

FLOW IN CONDUITS (§10.1, §10.2).
- Define a conduit.
- Know the main ideas about flow classification and Reynolds number.
- Specify a pipe size using the NPS standard.

HEAD LOSS (§10.3).
- Describe total head loss, pipe head loss, and component head loss.
- Define the friction factor f.
- For the Darcy-Weisbach equation, list the steps of the derivation, describe the physics, explain the meaning of the variables, and apply this equation.

FRICTION FACTOR (§10.5, §10.6).
- Calculate h_f or f using the relevant equations.
- Describe the Moody diagram and apply this diagram to find f.

EQUATION SOLVING (§10.7).
- Solve turbulent flow problems when the equations cannot be solved by algebra alone.

COMBINED HEAD LOSS (§10.8).
- Define the minor loss coefficient.
- Describe and apply the combined head loss equation.

HYDRAULIC DIAMETER (§10.9).
- Define and calculate hydraulic diameter and hydraulic radius.
- Solve relevant problems.

CENTRIFUGAL PUMPS (§10.9).
- Sketch and explain the system curve and the pump curve.
- Solve relevant problems.

10.1 Classifying Flow

This section describes how to classify flow in a conduit by considering (a) whether the flow is laminar or turbulent and (b) whether the flow is developing or fully developed. Classifying flow is essential for selecting the proper equation for calculating head loss.

A **conduit** is any pipe, tube, or duct that is completely filled with a flowing fluid. Examples include a pipeline transporting liquefied natural gas, a microchannel transporting hydrogen in a fuel cell, and a duct transporting air for heating a building. A pipe that is partially filled with a flowing fluid (e.g., a drainage pipe) is classified as an open-channel flow and is analyzed using ideas from Chapter 15.

Laminar Flow and Turbulent Flow

Flow in a conduit is classified as being either laminar or turbulent, depending on the magnitude of the Reynolds number. The original research involved visualizing flow in a glass tube, as shown in Fig. 10.2a. In the 1880s, Reynolds (1) injected dye into the center of the tube and observed the following:

- When the velocity was low, the streak of dye flowed down the tube with little expansion, as shown in Fig. 10.2b. However, if the water in the tank was disturbed, the streak would shift about in the tube.

- If velocity was increased, at some point in the tube the dye would all at once mix with the water, as shown in Fig. 10.2c.

- When the dye exhibited rapid mixing (Fig. 10.2c), illumination with an electric spark revealed eddies in the mixed fluid, as shown in Fig. 10.2d.

The flow regimes shown in Fig. 10.2 are laminar flow (Fig. 10.2b) and turbulent flow (Figs. 10.2c and 10.2d). Reynolds showed that the onset of turbulence was related to a π-group that is now called the Reynolds number (Re $= \rho VD/\mu$) in honor of Reynolds' pioneering work.

The Reynolds number is often written as Re_D, where the subscript "D" denotes that diameter is used in the formula. This subscript is called a *length scale*. Indicating the length scale for the Reynolds number is good practice because multiple values are used. For example, Chapter 9 introduced Re_x and Re_L.

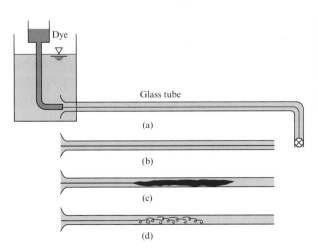

FIGURE 10.2

Reynolds' experiment:
(a) apparatus,
(b) laminar flow of dye in tube,
(c) turbulent flow of dye in tube,
(d) eddies in turbulent flow.

Reynolds number can be calculated with four different equations. These equations are equivalent because one can start with one formula and derive the others. The formulas are

$$\mathrm{Re}_D = \frac{VD}{\nu} = \frac{\rho VD}{\mu} = \frac{4Q}{\pi D \nu} = \frac{4\dot{m}}{\pi D \mu} \tag{10.1}$$

Reynolds discovered that if the fluid in the upstream reservoir was not completely still or if the pipe had some vibrations, then the change from laminar to turbulent flow occurred at $\mathrm{Re}_D \sim 2100$. However, if conditions were ideal, then it was possible to reach a much higher Reynolds number before the flow became turbulent. Reynolds also found that, when going from high velocity to low velocity, the change back to laminar flow occurred at $\mathrm{Re}_D \sim 2000$. Based on Reynolds' experiments, engineers use guidelines to establish whether or not flow in a conduit will be laminar or turbulent. The guidelines used in this text are as follows:

$$
\begin{array}{ll}
\mathrm{Re}_D \leq 2000 & \text{laminar flow} \\
2000 \leq \mathrm{Re}_D \leq 3000 & \text{unpredictable} \\
\mathrm{Re}_D \geq 3000 & \text{turbulent flow}
\end{array} \tag{10.2}
$$

In Eq. (10.2), the middle range ($2000 \leq \mathrm{Re}_D \leq 3000$) corresponds to a type of flow that is unpredictable because it can change back and forth between laminar and turbulent states. Recognize that precise values of the Reynolds number versus flow regime do not exist. Thus, the guidelines given in Eq. (10.2) are approximate, and other references may give different values. For example, some references use $\mathrm{Re}_D = 2300$ as the criterion for turbulence.

Developing Flow and Fully Developed Flow

Flow in a conduit is classified as either developing flow or fully developed flow. For example, consider laminar fluid entering a pipe from a reservoir as shown in Fig. 10.3. As the fluid moves down the pipe, the velocity profile changes in the streamwise direction as viscous effects cause the plug-type profile to gradually change into a parabolic profile. This region of changing velocity profile is called **developing flow**. After the parabolic distribution is achieved, the flow profile remains unchanged in the streamwise direction, and flow is called **fully developed flow**.

The distance required for flow to develop is called the **entry** or **entrance length** (L_e). In the entry length, the wall shear stress is decreasing in the streamwise (i.e., s) direction. For

FIGURE 10.3

In developing flow, the wall shear stress is changing. In fully developed flow, the wall shear stress is constant.

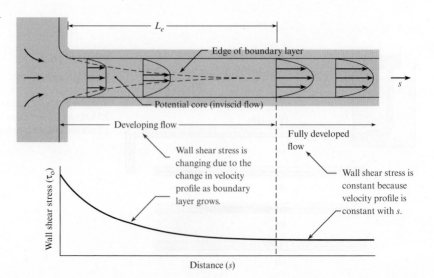

laminar flow, the wall shear stress distribution is shown in Fig. 10.3. Near the pipe entrance, the radial velocity gradient (change in velocity with distance from the wall) is high, so the shear stress is large. As the velocity profile progresses to a parabolic shape, the velocity gradient and the wall shear stress decrease until a constant value is achieved. The entry length is defined as the distance at which the shear stress reaches 2% of the fully developed value. Correlations for entry length are

$$\frac{L_e}{D} = 0.05\,\mathrm{Re}_D \qquad (\text{laminar flow: Re}_D \leq 2000) \qquad \textbf{(10.3a)}$$

$$\frac{L_e}{D} = 50 \qquad (\text{turbulent flow: Re}_D \geq 3000) \qquad \textbf{(10.3b)}$$

Eq. (10.3) is valid for flow entering a circular pipe from a reservoir under quiescent conditions. Other upstream components, such as valves, elbows, and pumps, produce complex flow fields that require different lengths to achieve fully developing flow.

In summary, flow in a conduit is classified into four categories: laminar developing, laminar fully developed, turbulent developing, or turbulent fully developed. The key to classification is to calculate the Reynolds number, as shown by Example 10.1.

EXAMPLE 10.1

Classifying Flow in Conduits

Problem Statement

Consider fluid flowing in a round tube of length 1 m and diameter 5 mm. Classify the flow as laminar or turbulent and calculate the entrance length for (a) air (50°C) with a speed of 12 m/s and (b) water (15°C) with a mass flow rate of $\dot{m} = 8$ g/s.

Define the Situation

Fluid is flowing in a round tube (two cases given).

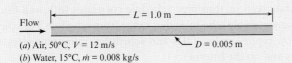

(a) Air, 50°C, $V = 12$ m/s
(b) Water, 15°C, $\dot{m} = 0.008$ kg/s

Properties:

- Air (50°C): Table A.3, $\nu = 1.79 \times 10^{-5}\ \mathrm{m^2/s}$
- Water (15°C): Table A.5, $\mu = 1.14 \times 10^{-3}\ \mathrm{N \cdot s/m^2}$

Assumptions:

- The pipe is connected to a reservoir.
- The entrance is smooth and tapered.

State the Goal

- Determine whether each flow is laminar or turbulent.
- Calculate the entrance length (in meters) for each case.

Generate Ideas and Make a Plan

- Calculate the Reynolds number using Eq. (10.1).
- Establish whether the flow is laminar or turbulent using Eq. (10.2).
- Calculate the entrance length using Eq. (10.3).

Take Action (Execute the Plan)

a. Air:

$$\mathrm{Re}_D = \frac{VD}{\nu} = \frac{(12\ \mathrm{m/s})(0.005\ \mathrm{m})}{1.79 \times 10^{-5}\ \mathrm{m^2/s}} = 3350$$

Because $\mathrm{Re}_D > 3000$, the $\boxed{\text{flow is turbulent.}}$

$$L_e = 50D = 50(0.005\ \mathrm{m}) = \boxed{0.25\ \mathrm{m}}$$

b. Water:

$$\mathrm{Re}_D = \frac{4\dot{m}}{\pi D \mu} = \frac{4(0.008\ \mathrm{kg/s})}{\pi(0.005\ \mathrm{m})(1.14 \times 10^{-3}\ \mathrm{N \cdot s/m^2})}$$
$$= 1787$$

Because $\mathrm{Re}_D < 2000$, the $\boxed{\text{flow is laminar.}}$

$$L_e = 0.05\mathrm{Re}_D D = 0.05(1787)(0.005\ \mathrm{m}) = \boxed{0.447\ \mathrm{m}}$$

10.2 Specifying Pipe Sizes

This section describes how to specify pipes using the Nominal Pipe Size (NPS) standard. This information is useful for specifying a size of pipe that is available commercially.

TABLE 10.1 Nominal Pipe Sizes

NPS (in.)	OD (in.)	Schedule	Wall Thickness (in.)	ID (in.)
1/2	0.840	40	0.109	0.622
		80	0.147	0.546
1	1.315	40	0.133	1.049
		80	0.179	0.957
2	2.375	40	0.154	2.067
		80	0.218	1.939
4	4.500	40	0.237	4.026
		80	0.337	3.826
8	8.625	40	0.322	7.981
		80	0.500	7.625
14	14.000	10	0.250	13.500
		40	0.437	13.126
		80	0.750	12.500
		120	1.093	11.814
24	24.000	10	0.250	23.500
		40	0.687	22.626
		80	1.218	21.564
		120	1.812	20.376

FIGURE 10.4

Section view of a pipe.

A larger schedule indicates thicker walls. A schedule 40 pipe has thicker walls than a schedule 10 pipe.

ID (inside diameter)

OD (outside diameter)

Standard Sizes for Pipes (NPS)

One of the most common standards for pipe sizes is the NPS system. The terms used in the NPS system are introduced in Fig. 10.4. The ID (pronounced "eye dee") indicates the inner pipe diameter, and the OD ("oh dee") indicates the outer pipe diameter. As shown in Table 10.1, an NPS pipe is specified using two values: a nominal pipe size and a schedule. The nominal pipe size determines the outside diameter or OD. For example, pipes with a nominal size of 2 inches have an OD of 2.375 inches. Once the nominal size reaches 14 inches, the nominal size and the OD are equal. That is, a pipe with a nominal size of 24 inches will have an OD of 24 inches.

Pipe schedule is related to the thickness of the wall. The original meaning of "schedule" was the ability of a pipe to withstand pressure; thus pipe schedule correlates with wall thickness. Each nominal pipe size has many possible schedules that range from schedule 5 to schedule 160. The data in Table 10.1 show representative ODs and schedules; more pipe sizes are specified in engineering handbooks and on the Internet.

10.3 Pipe Head Loss

This section presents the Darcy-Weisbach equation, which is used for calculating head loss in a straight run of pipe. This equation is one of the most useful equations in fluid mechanics.

Combined (Total) Head Loss

Pipe head loss is one type of head loss; the other type is called component head loss. All head loss is classified using these two categories:

$$\text{(total head loss)} = \text{(pipe head loss)} + \text{(component head loss)} \tag{10.4}$$

Component head loss is associated with flow through devices such as valves, bends, and tees. **Pipe head loss** is associated with fully developed flow in conduits, and it is caused by shear

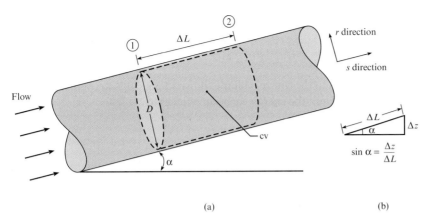

FIGURE 10.5

Initial situation for the derivation of the Darcy-Weisbach equation.

(a)   (b)

stresses that act on the flowing fluid. Note that pipe head loss is sometimes called major head loss, and component head loss is sometimes called minor head loss. Pipe head loss is predicted with the Darcy-Weisbach equation.

Derivation of the Darcy-Weisbach Equation

To derive the Darcy-Weisbach equation, start with the situation shown in Fig. 10.5. Assume fully developed and steady flow in a round tube of constant diameter D. Situate a cylindrical control volume of diameter D and length ΔL inside the pipe. Define a coordinate system with an axial coordinate in the streamwise direction (s direction) and a radial coordinate in the r direction.

Apply the momentum equation to the control volume shown in Fig. 10.5:

$$\sum \mathbf{F} = \frac{d}{dt} \int_{cv} \mathbf{v}\rho\, d V + \int_{cs} \mathbf{v}\rho\mathbf{V} \cdot d\mathbf{A} \tag{10.5}$$

(net forces) = (momentum accumulation rate) + (net efflux of momentum)

Select the streamwise direction and analyze each of the three terms in Eq. (10.5). The net efflux of momentum is zero because the velocity distribution at section 2 is identical to the velocity distribution at section 1. The momentum accumulation term is also zero because the flow is steady. Thus, Eq. (10.5) simplifies to $\sum \mathbf{F} = \mathbf{0}$. Forces are shown in Fig. 10.6. Summing of forces in the streamwise direction gives

$$F_{\text{pressure}} + F_{\text{shear}} + F_{\text{weight}} = 0$$

$$(p_1 - p_2)\left(\frac{\pi D^2}{4}\right) - \tau_0(\pi D \Delta L) - \gamma\left[\left(\frac{\pi D^2}{4}\right)\Delta L\right]\sin \alpha = 0 \tag{10.6}$$

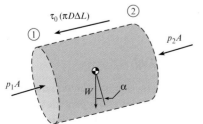

FIGURE 10.6

Force diagram.

Figure 10.5b shows that $\sin \alpha = (\Delta z/\Delta L)$. Equation (10.6) becomes

$$(p_1 + \gamma z_1) - (p_2 + \gamma z_2) = \frac{4\,\Delta L \tau_0}{D} \tag{10.7}$$

Next, apply the energy equation to the control volume shown in Fig. 10.5. Recognize that $h_p = h_t = 0$, $V_1 = V_2$, and $\alpha_1 = \alpha_2$. Thus, the energy equation reduces to

$$\frac{p_1}{\gamma} + z_1 = \frac{p_2}{\gamma} + z_2 + h_L \tag{10.8}$$

$$(p_1 + \gamma z_1) - (p_2 + \gamma z_2) = \gamma h_L$$

Combine Eqs. (10.7) and (10.8) and replace ΔL with L. Also, introduce a new symbol, h_f, to represent head loss in a pipe:

$$h_f = \left(\begin{array}{c}\text{head loss} \\ \text{in a pipe}\end{array}\right) = \frac{4L\tau_0}{D\gamma} \tag{10.9}$$

Rearrange the right side of Eq. (10.9):

$$h_f = \left(\frac{L}{D}\right)\left\{\frac{4\tau_0}{\rho V^2/2}\right\}\left\{\frac{\rho V^2/2}{\gamma}\right\} = \left\{\frac{4\tau_0}{\rho V^2/2}\right\}\left(\frac{L}{D}\right)\left\{\frac{V^2}{2g}\right\} \tag{10.10}$$

Define a new π-group called the **friction factor** (f) that gives the ratio of wall shear stress (τ_0) to kinetic pressure ($\rho V^2/2$):

$$f \equiv \frac{(4 \cdot \tau_0)}{(\rho V^2/2)} \approx \frac{\text{shear stress acting at the wall}}{\text{kinetic pressure}} \tag{10.11}$$

In the technical literature, the friction factor is identified by several different labels that are synonymous: friction factor, Darcy friction factor, Darcy-Weisbach friction factor, and the resistance coefficient. There is also another coefficient called the Fanning friction factor, often used by chemical engineers, which is related to the Darcy-Weisbach friction factor by a factor of 4:

$$f_{\text{Darcy}} = 4f_{\text{Fanning}}$$

This text uses only the Darcy-Weisbach friction factor. Combining Eqs. (10.10) and (10.11) gives the Darcy-Weisbach equation:

$$h_f = f\frac{L}{D}\frac{V^2}{2g} \tag{10.12}$$

To use the Darcy-Weisbach equation, the flow should be fully developed and steady. The Darcy-Weisbach equation is used for either laminar flow or turbulent flow and for either round pipes or nonround conduits, such as a rectangular duct.

The Darcy-Weisbach equation shows that head loss depends on the friction factor, the pipe length-to-diameter ratio, and the mean velocity squared. The key to using the Darcy-Weisbach equation is calculating a value of the friction factor f. This topic is addressed in the next sections of this text.

10.4 Stress Distributions in Pipe Flow

This section derives equations for the stress distributions on a plane that is oriented normal to streamlines. These equations, which apply to both laminar and turbulent flow, provide insights about the nature of the flow. Also, these equations are used for subsequent derivations.

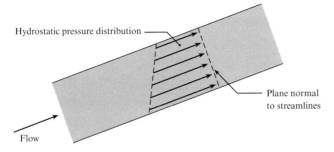

Hydrostatic pressure distribution

Plane normal
to streamlines

Flow

FIGURE 10.7

For fully developed flow in a pipe,
the pressure distribution on an
area normal to streamlines is
hydrostatic.

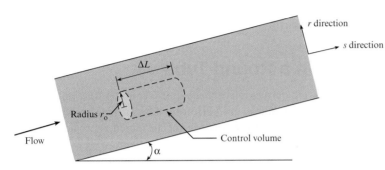

r direction

s direction

ΔL

Radius r_o

Flow

Control volume

α

FIGURE 10.8

Sketch for derivation of an equation for shear stress.

In pipe flow, the pressure acting on a plane that is normal to the direction of flow is hydrostatic. This means that the pressure distribution varies linearly, as shown in Fig. 10.7. The reason that the pressure distribution is hydrostatic can be explained with Euler's equation (see §4.5).

To derive an equation for the shear stress variation, consider flow of a Newtonian fluid in a round tube that is inclined at an angle α with respect to the horizontal, as shown in Fig. 10.8. Assume that the flow is fully developed, steady, and laminar. Define a cylindrical control volume of length ΔL and radius r.

Apply the momentum equation in the s direction. The net momentum efflux is zero because the flow is fully developed; that is, the velocity distribution at the inlet is the same as the velocity distribution at the exit. The momentum accumulation is also zero because the flow is steady. The momentum equation simplifies to force equilibrium:

$$\sum F_s = F_{\text{pressure}} + F_{\text{weight}} + F_{\text{shear}} = 0 \tag{10.13}$$

Analyze each term in Eq. (10.13) using the force diagram shown in Fig. 10.9:

$$pA - \left(p + \frac{dp}{ds}\Delta L \right) A - W\sin\alpha - \tau(2\pi r)\Delta L = 0 \tag{10.14}$$

Let $W = \gamma A \Delta L$, and let $\sin\alpha = \Delta z/\Delta L$, as shown in Fig. 10.5b. Next, divide Eq. (10.14) by $A\Delta L$:

$$\tau = \frac{r}{2}\left[-\frac{d}{ds}(p + \gamma z) \right] \tag{10.15}$$

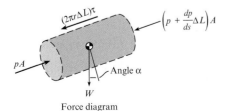

$(2\pi r\Delta L)\tau$

$\left(p + \frac{dp}{ds}\Delta L \right)A$

pA

Angle α

W

Force diagram

FIGURE 10.9

Force diagram corresponding to the control volume
defined in Fig. 10.8.

FIGURE 10.10

In fully developed flow (laminar or turbulent), the shear stress distribution is linear.

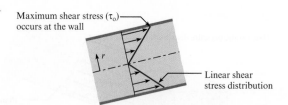

Maximum shear stress (τ_0) occurs at the wall

Linear shear stress distribution

Eq. (10.15) shows that the shear stress distribution varies linearly with r, as shown in Fig. 10.10. Notice that the shear stress is zero at the centerline, that it reaches a maximum value of τ_0 at the wall, and that the variation is linear in between. This linear shear stress variation applies to both laminar and turbulent flow.

10.5 Laminar Flow in a Round Tube

This section describes laminar flow and derives relevant equations. Laminar flow is important for flow in small conduits called microchannels, for lubrication flow, and for analyzing other flows in which viscous forces are dominant. Also, knowledge of laminar flow provides a foundation for the study of advanced topics.

Laminar flow is a flow regime in which fluid motion is smooth, the flow occurs in layers (laminae), and the mixing between layers occurs by molecular diffusion, a process that is much slower than turbulent mixing. According to Eq. (10.2), laminar flow occurs when $\text{Re}_D \leq 2000$. Laminar flow in a round tube is called **Poiseuille flow** or **Hagen-Poiseuille flow** in honor of researchers who studied low-speed flows in the 1840s.

Velocity Profile

To derive an equation for the velocity profile in laminar flow, begin by relating stress to rate of strain using the viscosity equation:

$$\tau = \mu \frac{dV}{dy}$$

where y is the distance from the pipe wall. Change variables by letting $y = r_0 - r$, where r_0 is pipe radius and r is the radial coordinate. Next, use the chain rule of calculus:

$$\tau = \mu \left(\frac{dV}{dy} \right) = \mu \left(\frac{dV}{dr} \right) \left(\frac{dr}{dy} \right) = - \left(\mu \frac{dV}{dr} \right) \tag{10.16}$$

Substitute Eq. (10.16) into Eq. (10.15):

$$\left(\frac{2\mu}{r} \right) \left(\frac{dV}{dr} \right) = \frac{d}{ds} (p + \gamma z) \tag{10.17}$$

In Eq. (10.17), the left side of the equation is a function of radius r, and the right side is a function of axial location s. This can be true if and only if each side of Eq. (10.17) is equal to a constant. Thus,

$$\text{constant} = \frac{d}{ds}(p + \gamma z) = \left(\frac{\Delta(p + \gamma z)}{\Delta L} \right) = \left(\frac{\gamma \Delta h}{\Delta L} \right) \tag{10.18}$$

where Δh is the change in piezometric head over a length ΔL of conduit. Combine Eqs. (10.17) and (10.18):

$$\frac{dV}{dr} = \left(\frac{r}{2\mu} \right) \left(\frac{\gamma \Delta h}{\Delta L} \right) \tag{10.19}$$

Integrate Eq. (10.19):

$$V = \left(\frac{r^2}{4\mu}\right)\left(\frac{\gamma\Delta h}{\Delta L}\right) + C \tag{10.20}$$

To evaluate the constant of integration C in Eq. (10.20), apply the no-slip condition, which states that the velocity of the fluid at the wall is zero. Thus,

$$V(r = r_0) = 0$$

$$0 = \frac{r_0^2}{4\mu}\left(\frac{\gamma\Delta h}{\Delta L}\right) + C$$

Solve for C and substitute the result into Eq. (10.20):

$$V = \frac{r_0^2 - r^2}{4\mu}\left[-\frac{d}{ds}(p + \gamma z)\right] = -\left(\frac{r_0^2 - r^2}{4\mu}\right)\left(\frac{\gamma\Delta h}{\Delta L}\right) \tag{10.21}$$

The maximum velocity occurs at $r = 0$:

$$V_{max} = -\left(\frac{r_0^2}{4\mu}\right)\left(\frac{\gamma\Delta h}{\Delta L}\right) \tag{10.22}$$

Combine Eqs. (10.21) and (10.22):

$$V(r) = -\left(\frac{r_0^2 - r^2}{4\mu}\right)\left(\frac{\gamma\Delta h}{\Delta L}\right) = V_{max}\left(1 - \left(\frac{r}{r_0}\right)^2\right) \tag{10.23}$$

Equation (10.23) shows that velocity varies as radius squared ($V \sim r^2$), meaning that the velocity distribution in laminar flow is parabolic, as plotted in Fig. 10.11.

Discharge and Mean Velocity V

To derive an equation for discharge Q, introduce the velocity profile from Eq. (10.23) into the flow rate equation:

$$Q = \int V \, dA$$
$$= -\int_0^{r_0} \frac{(r_0^2 - r^2)}{4\mu}\left(\frac{\gamma\Delta h}{\Delta L}\right)(2\pi r \, dr) \tag{10.24}$$

Integrate Eq. (10.24):

$$Q = -\left(\frac{\pi}{4\mu}\right)\left(\frac{\gamma\Delta h}{\Delta L}\right)\frac{(r^2 - r_0^2)^2}{2}\Big|_0^{r_0} = -\left(\frac{\pi r_0^4}{8\mu}\right)\left(\frac{\gamma\Delta h}{\Delta L}\right) \tag{10.25}$$

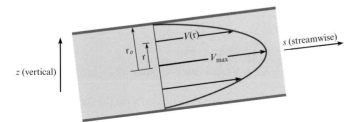

FIGURE 10.11

The velocity profile in Poiseuille flow is parabolic.

To derive an equation for mean velocity, apply $Q = \overline{V}A$ and use Eq. (10.25):

$$\overline{V} = -\left(\frac{r_0^2}{8\mu}\right)\left(\frac{\gamma\Delta h}{\Delta L}\right) \tag{10.26}$$

Comparing Eqs. (10.26) and (10.22) reveals that $\overline{V} = V_{\max}/2$. Next, substitute $D/2$ for r_0 in Eq. (10.26). The final result is an equation for mean velocity in a round tube:

$$\overline{V} = -\left(\frac{D^2}{32\mu}\right)\left(\frac{\gamma\Delta h}{\Delta L}\right) = \frac{V_{\max}}{2} \tag{10.27}$$

Head Loss and Friction Factor *f*

FIGURE 10.12

Flow in a pipe.

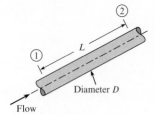

To derive an equation for head loss in a round tube, assume fully developed flow in the pipe shown in Fig. 10.12. Apply the energy equation from sections 1 to 2 and simplify to give

$$\left(\frac{p_1}{\gamma} + z_1\right) = \left(\frac{p_2}{\gamma} + z_2\right) + h_L \tag{10.28}$$

Let $h_L = h_f$, and then Eq. (10.28) becomes

$$\left(\frac{p_1}{\gamma} + z_1\right) = \left(\frac{p_2}{\gamma} + z_2\right) + h_f \tag{10.29}$$

Expand Eq. (10.27):

$$\overline{V} = -\left(\frac{\gamma D^2}{32\mu}\right)\left(\frac{\Delta h}{\Delta L}\right) = -\left(\frac{\gamma D^2}{32\mu}\right)\frac{\left(\frac{p_2}{\gamma} + z_2\right) - \left(\frac{p_1}{\gamma} + z_1\right)}{\Delta L} \tag{10.30}$$

Reorganize Eq. (10.30) and replace ΔL with L:

$$\left(\frac{p_1}{\gamma} + z_1\right) = \left(\frac{p_2}{\gamma} + z_2\right) + \frac{32\mu\overline{V}L}{\gamma D^2} \tag{10.31}$$

Comparing Eqs. (10.29) and (10.31) gives an equation for head loss in a pipe:

$$h_f = \frac{32\mu L\,\overline{V}}{\gamma D^2} \tag{10.32}$$

Key assumptions for Eq. (10.32) are (a) laminar flow, (b) fully developed flow, (c) steady flow, and (d) Newtonian fluid.

Eq. (10.32) shows that head loss in laminar flow varies linearly with velocity. Also, head loss is influenced by viscosity, pipe length, specific weight, and pipe diameter squared.

To derive an equation for the friction factor *f*, combine Eq. (10.32) with the Darcy-Weisbach equation (10.12):

$$h_f = \frac{32\,\mu L V}{\gamma D^2} = f\frac{L}{D}\frac{V^2}{2g} \tag{10.33}$$

$$\text{or } f = \left(\frac{32\mu L V}{\gamma D^2}\right)\left(\frac{D}{L}\right)\left(\frac{2g}{V^2}\right) = \frac{64\mu}{\rho D V} = \frac{64}{\mathrm{Re}_D} \tag{10.34}$$

Eq. (10.34) shows that the friction factor for laminar flow depends only on the Reynolds number. Example 10.2 illustrates how to calculate head loss.

EXAMPLE 10.2

Head Loss for Laminar Flow

Problem Statement

Oil ($SG = 0.85$) with a kinematic viscosity of 6×10^{-4} m^2/s flows in a 15-cm-diameter pipe at a rate of 0.020 m^3/s. What is the head loss for a 100 m length of pipe?

Define the Situation

- Oil is flowing in a pipe at a flow rate of $Q = 0.02$ m^3/s.
- Pipe diameter is $D = 0.15$ m.

Assumptions: Fully developed, steady flow

Properties: Oil: S = 0.85, $v = 6 \times 10^{-4}$ m^2/s

State the Goal

Calculate head loss (in meters) for a pipe length of 100 m.

Generate Ideas and Make a Plan

1. Calculate the mean velocity using the flow rate equation.
2. Calculate the Reynolds number using Eq. (10.1).
3. Check whether the flow is laminar or turbulent using Eq. (10.2).
4. Calculate head loss using Eq. (10.32).

Take Action (Execute the Plan)

1. Mean velocity:

$$V = \frac{Q}{A} = \frac{0.020 \text{ m}^3\text{/s}}{(\pi D^2)/4} = \frac{0.020 \text{ m}^3\text{/s}}{\pi((0.15 \text{ m})^2/4)} = 1.13 \text{ m/s}$$

2. Reynolds number:

$$\text{Re}_D = \frac{VD}{v} = \frac{(1.13 \text{ m/s})(0.15 \text{ m})}{6 \times 10^{-4} \text{ m}^2\text{/s}} = 283$$

3. Because $\text{Re}_D < 2000$, the flow is laminar.
4. Head loss (laminar flow):

$$h_f = \frac{32\mu LV}{\gamma D^2} = \frac{32\rho v LV}{\rho g D^2} = \frac{32v LV}{g D^2}$$

$$= \frac{32(6 \times 10^{-4} \text{ m}^2\text{/s})(100 \text{ m})(1.13 \text{ m/s})}{(9.81 \text{ m/s}^2)(0.15 \text{ m})^2}$$

$$= \boxed{9.83 \text{ m}}$$

Review the Solution and the Process

Knowledge. An alternative way to calculate head loss for laminar flow is to use the Darcy-Weisbach equation (10.12) as follows:

$$f = \frac{64}{\text{Re}_D} = \frac{64}{283} = 0.226$$

$$h_f = f\left(\frac{L}{D}\right)\left(\frac{V^2}{2g}\right) = 0.226\left(\frac{100 \text{ m}}{0.15 \text{ m}}\right)\left(\frac{(1.13 \text{ m/s})}{2 \times 9.81 \text{ m/s}^2}\right)^2$$

$$= 9.83 \text{ m}$$

10.6 Turbulent Flow and the Moody Diagram

This section describes the characteristics of turbulent flow, presents equations for calculating the friction factor f, and presents a famous graph called the Moody diagram. This information is important because most flows in conduits are turbulent.

Qualitative Description of Turbulent Flow

Turbulent flow is a flow regime in which the movement of fluid particles is chaotic, eddying, and unsteady, with significant movement of particles in directions transverse to the flow direction. Because of the chaotic motion of fluid particles, turbulent flow produces high levels of mixing and has a velocity profile that is more uniform or flatter than the corresponding laminar velocity profile. According to Eq. (10.2), turbulent flow occurs when Re ≥ 3000.

Engineers and scientists model turbulent flow by using an empirical approach. This is because the complex nature of turbulent flow has prevented researchers from establishing a mathematical solution of general utility. Still, the empirical information has been used successfully and extensively in system design. Over the years, researchers have proposed many equations for shear stress and head loss in turbulent pipe flow. The empirical equations that have proven to be the most reliable and accurate for engineering use are presented in the next section.

TABLE 10.2 Exponents for Power-Law Equation and Ratio of Mean to Maximum Velocity

Re	4×10^3	2.3×10^4	1.1×10^5	1.1×10^6	3.2×10^6
m	$\dfrac{1}{6.0}$	$\dfrac{1}{6.6}$	$\dfrac{1}{7.0}$	$\dfrac{1}{8.8}$	$\dfrac{1}{10.0}$
u_{max}/V	1.26	1.24	1.22	1.18	1.16

Source of data: Schlichting (2).

Equations for the Velocity Distribution

The time-average velocity distribution is often described using an equation called the power-law formula:

$$\frac{u(r)}{u_{max}} = \left(\frac{r_0 - r}{r_0}\right)^m \tag{10.35}$$

where u_{max} is velocity in the center of the pipe, r_0 is the pipe radius, and m is an empirically determined variable that depends on Re, as shown in Table 10.2. Notice in Table 10.2 that the velocity in the center of the pipe is typically about 20% higher than the mean velocity V. Although Eq. (10.35) provides an accurate representation of the velocity profile, it does not predict an accurate value of wall shear stress.

An alternative approach to Eq. (10.35) is to use the turbulent boundary layer equations presented in Chapter 9. The most significant of these equations, called the logarithmic velocity distribution, is given by Eq. (9.27) and repeated here:

$$\frac{u(r)}{u_*} = 2.44 \ln \frac{u_*(r_0 - r)}{\nu} + 5.56 \tag{10.36}$$

where u_*, the shear velocity, is given by $u_* = \sqrt{\tau_0/\rho}$.

Equations for the Friction Factor, f

To derive an equation for f in turbulent flow, substitute the log law in Eq. (10.36) into the definition of mean velocity given by Eq. (5.10):

$$V = \frac{Q}{A} = \left(\frac{1}{\pi r_0^2}\right) \int_0^{r_0} u(r) 2\pi r \, dr = \left(\frac{1}{\pi r_0^2}\right) \int_0^{r_0} u_* \left[2.44 \ln \frac{u_*(r_0 - r)}{\nu} + 5.56\right] 2\pi r \, dr$$

After integration, algebra, and tweaking the constants to better fit experimental data, the result is

$$\frac{1}{\sqrt{f}} = 2.0 \log_{10}(\mathrm{Re} \sqrt{f}) - 0.8 \tag{10.37}$$

Equation (10.37), first derived by Prandtl in 1935, gives the friction factor for turbulent flow in tubes that have smooth walls. The details of the derivation of Eq. (10.37) are presented by White (20). To determine the influence of roughness on the walls, Nikuradse (4), one of Prandtl's graduate students, glued uniform-sized grains of sand to the inner walls of a tube and then measured pressure drops and flow rates.

Nikuradse's data, Fig. 10.13, shows the friction factor f plotted as function of the Reynolds number for various sizes of sand grains. To characterize the size of sand grains, Nikuradse used a variable called the **sand roughness height** with the symbol k_s. The π-group, k_s/D, is given the name **relative roughness**.

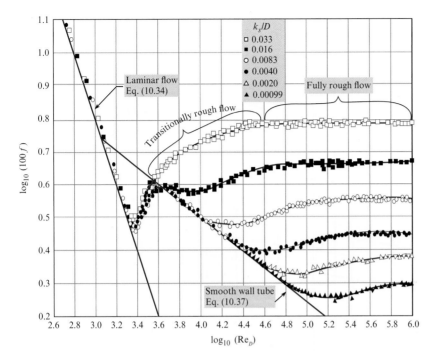

FIGURE 10.13

Resistance coefficient f versus Reynolds number for sand-roughened pipe. [After Nikuradse (4).]

In laminar flow, the data in Fig. 10.13 show that wall roughness does not influence f. In particular, notice how the data corresponding to various values of k_s/D collapse into a single blue line that is labeled "laminar flow."

In turbulent flow, the data in Fig. 10.13 show that wall roughness has a major impact on f. When $k_s/D = 0.033$, then values of f are about 0.04. As the relative roughness drops to 0.002, values of f decrease by a factor of about 3. Eventually, wall roughness does not matter, and the value of f can be predicted by assuming that the tube has a smooth wall. This latter case corresponds to the blue curve in Fig. 10.13 labeled "smooth wall tube." The effects of roughness are summarized by White (5) and presented in Table 10.3. These regions are also labeled in Fig. 10.13.

Moody Diagram

Colebrook (6) advanced Nikuradse's work by acquiring data for commercial pipes and then developing an empirical equation, called the Colebrook-White formula, for the friction factor. Moody (3) used the Colebrook-White formula to generate a design chart similar to that shown in Fig. 10.14. This chart is now known as the **Moody diagram** for commercial pipes.

TABLE 10.3 Effects of Wall Roughness

Type of Flow	Parameter Ranges		Influence of Parameters on f
Laminar flow	$Re_D < 2000$	NA	f depends on Reynolds number. f is independent of wall roughness (k_s/D).
Turbulent flow, smooth tube	$Re_D > 3000$	$\left(\dfrac{k_s}{D}\right) Re_D < 10$	f depends on Reynolds number. f is independent of wall roughness (k_s/D).
Transitionally rough turbulent flow	$Re_D > 3000$	$10 < \left(\dfrac{k_s}{D}\right) Re_D < 1000$	f depends on Reynolds number. f depends on wall roughness (k_s/D).
Fully rough turbulent flow	$Re_D > 3000$	$\left(\dfrac{k_s}{D}\right) Re_D > 1000$	f is independent of Reynolds number. f depends on wall roughness (k_s/D).

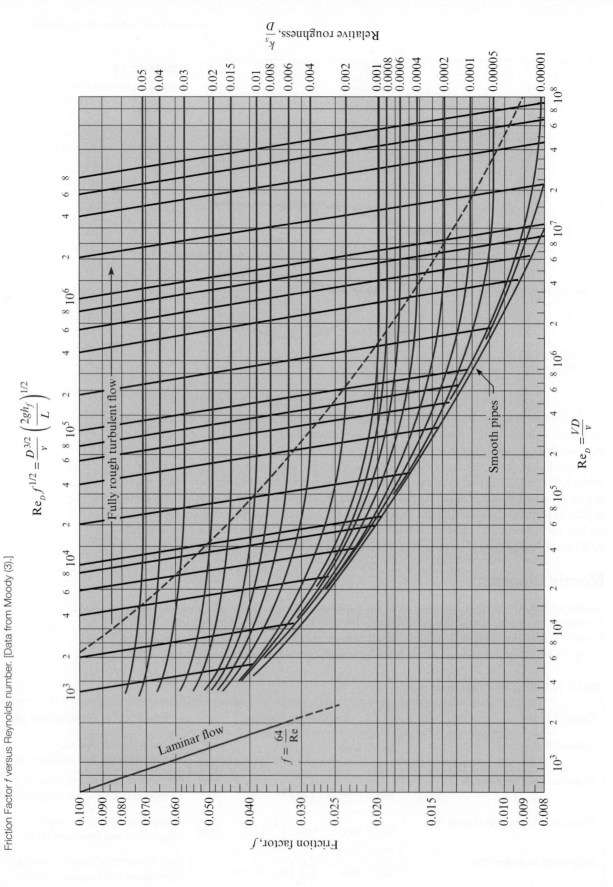

FIGURE 10.14

Friction Factor f versus Reynolds number. [Data from Moody (3).]

Relative roughness, $\dfrac{k_s}{D}$

$\mathrm{Re}_D f^{1/2} = \dfrac{D^{3/2}}{\nu}\left(\dfrac{2gh_f}{L}\right)^{1/2}$

Fully rough turbulent flow

$\mathrm{Re}_D = \dfrac{VD}{\nu}$

Smooth pipes

Laminar flow

$f = \dfrac{64}{\mathrm{Re}}$

Friction factor, f

TABLE 10.4 Equivalent Sand-Grain Roughness, (k_s), for Various Pipe Materials

Boundary Material	k_s, Millimeters	k_s, Inches
Glass, plastic	0.00 (smooth)	0.00 (smooth)
Copper or brass tubing	0.0015	6×10^{-5}
Wrought iron, steel	0.046	0.002
Asphalted cast iron	0.12	0.005
Galvanized iron	0.15	0.006
Cast iron	0.26	0.010
Concrete	0.3 to 3.0	0.012–0.12
Riveted steel	0.9–9	0.035–0.35
Rubber pipe (straight)	0.025	0.001

In the Moody diagram, Fig. 10.14, the variable k_s denotes the **equivalent sand roughness**. That is, a pipe that has the same resistance characteristics at high Re values as a sand-roughened pipe is said to have a roughness equivalent to that of the sand-roughened pipe. Table 10.4 gives the equivalent sand roughness for various kinds of pipes. This table can be used to calculate the relative roughness for a given pipe diameter, which, in turn, is used in Fig. 10.14 to find the friction factor.

In the Moody diagram, Fig. 10.14, the abscissa is the Reynolds number Re, and the ordinate is the resistance coefficient f. Each blue curve is for a constant relative roughness k_s/D, and the values of k_s/D are given on the right at the end of each curve. To find f, given Re and k_s/D, go to the right to find the correct relative roughness curve. Then, look at the bottom of the chart to find the given value of Re and, with this value of Re, move vertically upward until the given k_s/D curve is reached. Finally, from this point, move horizontally to the left scale to read the value of f. If the curve for the given value of k_s/D is not plotted in Fig. 10.14, then simply find the proper position on the graph by interpolation between the k_s/D curves that bracket the given k_s/D.

To provide a more convenient solution to some types of problems, the top of the Moody diagram presents a scale based on the parameter $\mathrm{Re}\, f^{1/2}$. This parameter is useful when h_f and k_s/D are known but the velocity V is not. Using the Darcy-Weisbach equation given in Eq. (10.12) and the definition of Reynolds number, one can show that

$$\mathrm{Re}\, f^{1/2} = \frac{D^{3/2}}{\nu}(2gh_f/L)^{1/2} \tag{10.38}$$

In the Moody diagram, Fig. 10.14, curves of constant $\mathrm{Re}\, f^{1/2}$ are plotted using heavy black lines that slant from the left to right. For example, when $\mathrm{Re}\, f^{1/2} = 10^5$ and $k_s/D = 0.004$, then $f = 0.029$. When using computers to carry out pipe-flow calculations, it is much more convenient to have an equation for the friction factor as a function of the Reynolds number and relative roughness. By using the Colebrook-White formula, Swamee and Jain (7) developed an explicit equation for friction factor, namely,

$$f = \frac{0.25}{\left[\log_{10}\left(\dfrac{k_s}{3.7D} + \dfrac{5.74}{\mathrm{Re}_D^{0.9}}\right)\right]^2} \tag{10.39}$$

It is reported that this equation predicts friction factors that differ by less than 3% from those on the Moody diagram for $4 \times 10^3 < \mathrm{Re}_D < 10^8$ and $10^{-5} < k_s/D < 2 \times 10^{-2}$.

10.7 A Strategy for Solving Problems

Analyzing flow in conduits can be challenging because the equations often cannot be solved with algebra. Thus, this section presents a strategy.

Fig. 10.15 provides a strategy for problem solving. When flow is laminar, solutions are straightforward because head loss is linear with velocity V and the equations are simple enough to solve with algebra. When flow is turbulent, head loss is nonlinear with V and the equations are too complex to solve with algebra. Thus, for turbulent flow, engineers use computer solutions or the traditional approach.

To solve a turbulent flow problem using the *traditional approach*, one classifies the problems into three cases:

Case 1 applies when the goal is to find the *head loss*, given the pipe length, pipe diameter, and flow rate. This problem is straightforward because it can be solved using algebra; see Example 10.3.

Case 2 applies when the goal is to find the *flow rate*, given the head loss (or pressure drop), the pipe length, and the pipe diameter. This problem usually requires an iterative approach. See Examples 10.4 and 10.5.

Case 3 applies when the goal is to find the *pipe diameter*, given the flow rate, length of pipe, and head loss (or pressure drop). This problem usually requires an iterative approach; see Example 10.6.

There are several approaches that sometimes eliminate the need for an iterative approach. For case 2, an iterative approach can sometimes be avoided by using an explicit equation developed by Swamee and Jain (7):

$$Q = -2.22 \, D^{5/2} \sqrt{gh_f/L} \, \log \left(\frac{k_s}{3.7 \, D} + \frac{1.78 \, v}{D^{3/2} \sqrt{gh_f/L}} \right) \tag{10.40}$$

Using Eq. (10.40) is equivalent to using the top of the Moody diagram, which presents a scale for $\mathrm{Re} \, f^{1/2}$. For case 3, one can sometimes use an explicit equation developed by Swamee and Jain (7) and modified by Streeter and Wylie (8):

$$D = 0.66 \left[k_s^{1.25} \left(\frac{LQ^2}{gh_f} \right)^{4.75} + vQ^{9.4} \left(\frac{L}{gh_f} \right)^{5.2} \right]^{0.04} \tag{10.41}$$

Example 10.3 shows an example of a case 1 problem.

FIGURE 10.15

A strategy for solving conduit flow problems.

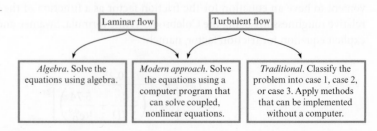

EXAMPLE 10.3

Head Loss in a Pipe (Case 1)

Problem Statement

Water ($T = 20°C$) flows at a rate of 0.05 m³/s in a 20 cm asphalted cast iron pipe. What is the head loss per kilometer of pipe?

Define the Situation

Water is flowing in a pipe.

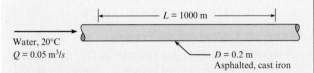

Water, 20°C
$Q = 0.05$ m³/s

$L = 1000$ m

$D = 0.2$ m
Asphalted, cast iron

Assumptions: Fully developed flow

Properties: Water (20°C): Table A.5, $v = 1 \times 10^{-6}$ m²/s

State the Goal

Calculate the head loss (in meters) for $L = 1000$ m.

Generate Ideas and Make a Plan

Because this is a case 1 problem (head loss is the goal), the solution is straightforward.

1. Calculate the mean velocity using the flow rate equation.
2. Calculate the Reynolds number using Eq. (10.1).

3. Calculate the relative roughness and then look up f on the Moody diagram.
4. Find head loss by applying the Darcy-Weisbach equation (10.1).

Take Action (Execute the Plan)

1. Mean velocity:

$$V = \frac{Q}{A} = \frac{0.05 \text{ m}^3/\text{s}}{(\pi/4)(0.2 \text{ m})^2} = 1.59 \text{ m/s}$$

2. Reynolds number:

$$\text{Re}_D = \frac{VD}{v} = \frac{(1.59 \text{ m/s})(0.20 \text{ m})}{10^{-6} \text{ m}^2/\text{s}} = 3.18 \times 10^5$$

3. Resistance coefficient:
   - Equivalent sand roughness (Table 10.4):
   $$k_s = 0.12 \text{ mm}$$
   - Relative roughness:
   $$k_s/D = (0.00012 \text{ m})/(0.2 \text{ m}) = 0.0006$$
   - Look up f on the Moody diagram for Re = 3.18×10^5 and $k_s/D = 0.0006$:
   $$f = 0.019$$

4. Darcy-Weisbach equation:

$$h_f = f\left(\frac{L}{D}\right)\left(\frac{V^2}{2g}\right) = 0.019\left(\frac{1000 \text{ m}}{0.20 \text{ m}}\right)\left(\frac{1.59^2 \text{ m}^2/\text{s}^2}{2(9.81 \text{ m/s}^2)}\right)$$
$$= \boxed{12.2 \text{ m}}$$

Example 10.4 shows an example of a case 2 problem. Notice that the solution involves application of the scale on the top of the Moody diagram, thereby avoiding an iterative solution.

EXAMPLE 10.4

Flow Rate in a Pipe (Case 2)

Problem Statement

The head loss per kilometer of 20 cm asphalted cast iron pipe is 12.2 m. What is the flow rate of water through the pipe?

Define the Situation

This is the same situation as Example 10.3 except that the head loss is now specified and the discharge is unknown.

State the Goal

Calculate the discharge (m³/s) in the pipe.

Generate Ideas and Make a Plan

This is a case 2 problem because flow rate is the goal. However, a direct (i.e., noniterative) solution is possible because head loss is specified. The strategy will be to use the horizontal scale on the top of the Moody diagram.

1. Calculate the parameter on the top of the Moody diagram.
2. Using the Moody diagram, find the friction factor f.
3. Calculate mean velocity using the Darcy-Weisbach equation (10.12).
4. Find discharge using the flow rate equation.

Take Action (Execute the Plan)

1. Compute the parameter $D^{3/2}\sqrt{2gh_f/L}/\nu$:

$$D^{3/2}\frac{\sqrt{2gh_f/L}}{\nu} = (0.20\text{ m})^{3/2}$$

$$\times \frac{[2(9.81\text{ m/s}^2)(12.2\text{ m}/1000\text{ m})]^{1/2}}{1.0\times10^{-6}\text{ m}^2/\text{s}}$$

$$= 4.38\times10^4$$

2. Determine resistance coefficient:

- Relative roughness:

$$k_s/D = (0.00012\text{ m})/(0.2\text{ m}) = 0.0006$$

- Look up f on the Moody diagram for $D^{3/2}\sqrt{2gh_f/L}/\nu = 4.4\times10^4$ and $k_s/D = 0.0006$:

$$f = 0.019$$

3. Find V using the Darcy-Weisbach equation:

$$h_f = f\left(\frac{L}{D}\right)\left(\frac{V^2}{2g}\right)$$

$$12.2\text{ m} = 0.019\left(\frac{1000\text{ m}}{0.20\text{ m}}\right)\left(\frac{V^2}{2(9.81\text{ m/s}^2)}\right)$$

$$V = 1.59\text{ m/s}$$

4. Use flow rate equation to find discharge:

$$Q = VA = (1.59\text{ m/s})(\pi/4)(0.2\text{ m})^2 = \boxed{0.05\text{ m}^3/\text{s}}$$

Review the Solution and the Process

Validation. The calculated flow rate matches the value from Example 10.3. This is expected because the data are the same.

When case 2 problems require iteration, several methods can be used to find a solution. One of the easiest ways is a method called "successive substitution," which is illustrated in Example 10.5.

EXAMPLE 10.5

Flow Rate in a Pipe (Case 2)

Problem Statement

Water ($T = 20°C$) flows from a tank through a 50 cm diameter steel pipe. Determine the discharge of water.

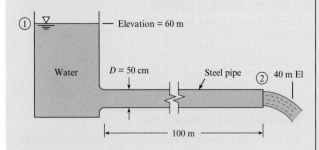

Define the Situation

Water is draining from a tank through a steel pipe.

Assumptions:

- Flow is fully developed.
- Include only the head loss in the pipe.

Properties:

- Water (20°C): Table A.5, $\nu = 1\times10^{-6}\text{ m}^2/\text{s}$.
- Steel pipe: Table 10.4, equivalent sand roughness, $k_s = 0.046$ mm. Relative roughness (k_s/D) is 9.2×10^{-5}.

State the Goal

Find: Discharge (m^3/s) for the system.

Generate Ideas and Make a Plan

This is a case 2 problem because flow rate is the goal. An iterative solution is used because V is unknown, so there is no direct way to use the Moody diagram.

1. Apply the energy equation from section 1 to section 2.
2. First trial: Guess a value of f and then solve for V.
3. Second trial: Using V from the first trial, calculate a new value of f.
4. Convergence: If the value of f is constant within a few percent between trials, then stop. Otherwise, continue with more iterations.
5. Calculate flow rate using the flow rate equation.

Take Action (Execute the Plan)

1. Energy equation (reservoir surface to outlet):

$$\frac{p_1}{\gamma} + \frac{V_1^2}{2g} + z_1 = \frac{p_2}{\gamma} + \frac{V_2^2}{2g} + z_2 + h_L$$

$$0 + 0 + 60 = 0 + \frac{V_2^2}{2g} + 40 + f\frac{L}{D}\frac{V_2^2}{2g}$$

or

$$V = \left(\frac{2g \times 20}{1 + 200f} \right)^{1/2}$$ **(a)**

2. First trial (iteration 1):
   - Guess a value of $f = 0.020$.
   - Use Eq. (a) to calculate $V = 8.86$ m/s.
   - Use $V = 8.86$ m/s to calculate Re $= 4.43 \times 10^6$.
   - Use Re $= 4.43 \times 10^6$ and $k_s/D = 9.2 \times 10^{-5}$ on the Moody diagram to find that $f = 0.012$.
   - Use Eq. (a) with $f = 0.012$ to calculate $V = 10.7$ m/s.

3. Second trial (iteration 2):
   - Use $V = 10.7$ m/s to calculate Re$_D = 5.35 \times 10^6$.
   - Use Re$_D = 5.35 \times 10^6$ and $k_s/D = 9.2 \times 10^{-5}$ on the Moody diagram to find that $f = 0.012$.

4. Convergence: The value of $f = 0.012$ is unchanged between the first and second trials. Therefore, there is no need for more iterations.

5. Flow rate:

   $$Q = VA = (10.7 \text{ m/s}) \times (\pi/4) \times (0.50)^2 \text{ m}^2 = 2.10 \text{ m}^3/\text{s}$$

In a case 3 problem, derive an equation for diameter D and then use the method of successive substitution to find a solution. Iterative approaches, as illustrated in Example 10.6, can employ a spreadsheet program to perform the calculations.

EXAMPLE 10.6

Finding Pipe Diameter (Case 3)

Problem Statement

What size of asphalted cast iron pipe is required to carry water (60°F) at a discharge of 3 cfs and with a head loss of 4 ft per 1000 ft of pipe?

Define the Situation

Water is flowing in a asphalted cast iron pipe. $Q = 3$ ft^3/s.

Assumptions: Fully developed flow

Properties:
- Water (60°F): Table A.5, $\nu = 1.22 \times 10^{-5}$ ft^2/s
- Asphalted cast iron pipe: Table 10.4, equivalent sand roughness, $k_s = 0.005$ in.

State the Goal

Calculate the pipe diameter (in ft) so that head loss is 4 ft per 1000 ft of pipe length.

Generate Ideas and Make a Plan

Because this is a case 3 problem (pipe diameter is the goal), use an iterative approach.

1. Derive an equation for pipe diameter by using the Darcy-Weisbach equation.
2. For iteration 1, guess f, solve for pipe diameter, and then recalculate f.
3. To complete the problem, build a table in a spreadsheet program.

Take Action (Execute the Solution)

1. Develop an equation to use for iteration.
   - Darcy-Weisbach equation:

   $$h_f = f\left(\frac{L}{D}\right)\left(\frac{V^2}{2g}\right) = f\left(\frac{L}{D}\right)\left(\frac{Q^2/A^2}{2g}\right) = \frac{fLQ^2}{2g(\pi/4)^2 D^5}$$

   - Solve for pipe diameter:

   $$D^5 = \frac{fLQ^2}{0.785^2(2gh_f)}$$ **(a)**

2. Iteration 1:
   - Guess $f = 0.015$.
   - Solve for diameter using Eq. (a):

   $$D^5 = \frac{0.015(1000 \text{ ft})(3 \text{ ft}^3/\text{s})^2}{0.785^2(64.4 \text{ ft/s}^2)(4 \text{ ft})} = 0.852 \text{ ft}^5$$

   $$D = 0.968 \text{ ft}$$

   - Find parameters needed for calculating f:

   $$V = \frac{Q}{A} = \frac{3 \text{ ft}^3/\text{s}}{(\pi/4)(0.968^2 \text{ ft}^2)} = 4.08 \text{ ft/s}$$

   $$\text{Re} = \frac{VD}{\nu} = \frac{(4.08 \text{ ft/s})(0.968 \text{ ft})}{1.22 \times 10^{-5} \text{ ft}^2/\text{s}} = 3.26 \times 10^5$$

   $$k_s/D = 0.005/(0.97 \times 12) = 0.00043$$

   - Calculate f using Eq. (10.39): $f = 0.0178$.

3. In the following table, the first row contains the values from iteration 1. The value of $f = 0.0178$ from iteration | 1 is used for the initial value for iteration 2. Notice how the solution has converged by iteration 2.

Iteration #	Initial f	D	V	Re	k_s/D	New f
		(ft)	(ft/s)			
1	0.0150	0.968	4.08	3.26E+05	4.3E−04	0.0178
2	0.0178	1.002	3.81	3.15E+05	4.2E−04	0.0178
3	0.0178	1.001	3.81	3.15E+05	4.2E−04	0.0178
4	0.0178	1.001	3.81	3.15E+05	4.2E−04	0.0178

Specify a pipe with a 12-inch inside diameter.

10.8 Combined Head Loss

Previous sections have described how to calculate head loss in pipes. This section completes the story by describing how to calculate head loss in components. This knowledge is essential for modeling and design of systems.

The Minor Loss Coefficient, K

When fluid flows through a component such as a partially open value or a bend in a pipe, viscous effects cause the flowing fluid to lose mechanical energy. For example, Fig. 10.16 shows flow through a "generic component." At section 2, the head of the flow will be less than at section 1. To characterize component head loss, engineers use a π-group called the **minor loss coefficient** K:

$$K \equiv \frac{(\Delta h)}{(V^2/2g)} = \frac{(\Delta p_z)}{(\rho V^2/2)} \tag{10.42}$$

where Δh is the drop in piezometric head that is caused by a component, Δp_z is the drop in piezometric pressure, and V is mean velocity. As shown in Eq. (10.42), the minor loss coefficient has two useful interpretations:

$$K = \frac{\text{drop in piezometric head across component}}{\text{velocity head}} = \frac{\text{pressure drop due to component}}{\text{kinetic pressure}}$$

Thus, the head loss across a single component or transition is $h_L = K(V^2/(2g))$, where K is the minor loss coefficient for that component or transition.

Most values of K are found by experiment. For example, consider the setup shown in Fig. 10.17. To find K, flow rate is measured and mean velocity is calculated using $V = (Q/A)$. Pressure and elevation measurements are used to calculate the change in piezometric head:

$$\Delta h = h_2 - h_1 = \left(\frac{p_2}{\gamma} + z_2\right) - \left(\frac{p_1}{\gamma} + z_1\right) \tag{10.43}$$

Then, values of V and Δh are used in Eq. (10.42) to calculate K. The next section presents typical data for K.

FIGURE 10.16

Flow through a generic component.

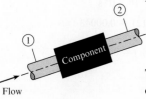

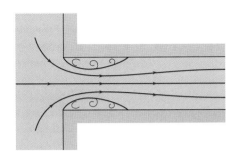

FIGURE 10.17

Flow at a sharp-edged inlet.

Data for the Minor Loss Coefficient This section presents K data and relates these data to flow separation and wall shear stress. This information is useful for system modeling.

Pipe inlet. Near the entrance to a pipe when the entrance is rounded, flow is developing as shown in Fig. 10.3, and the wall shear stress is higher than that found in fully developed flow. Alternatively, if the pipe inlet is abrupt, or sharp-edged, as in Fig. 10.17, separation occurs just downstream of the entrance. Hence, the streamlines converge and then diverge with consequent turbulence and relatively high head loss. The loss coefficient for the abrupt inlet is $K_e = 0.5$. This value is found in Table 10.5 using the row labeled "Pipe entrance" and the criteria of $r/d = 0.0$. Other values of head loss are summarized in Table 10.5.

Flow in an elbow. In an elbow (90° smooth bend), considerable head loss is produced by secondary flows and by separation that occurs near the inside of the bend and downstream of the midsection, as shown in Fig. 10.18.

The loss coefficient for an elbow at high Reynolds numbers depends primarily on the shape of the elbow. For a very short-radius elbow, the loss coefficient is quite high. For larger-radius elbows, the coefficient decreases until a minimum value is found at an r/d value of about 4 (see Table 10.5). However, for still larger values of r/d, an increase in loss coefficient occurs because the elbow itself is significantly longer.

Other components. The loss coefficients for a number of other fittings and flow transitions are given in Table 10.5. This table is representative of engineering practice. For more extensive tables, see references (10–15).

In Table 10.5, the K was found by experiment, so one must be careful to ensure that Reynolds number values in the application correspond to Reynolds number values used to acquire the data.

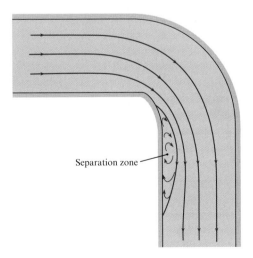

FIGURE 10.18

Flow pattern in an elbow.

Separation zone

TABLE 10.5 Loss Coefficients for Various Transitions and Fittings

Description	Sketch	Additional Data		K	Source
Pipe entrance $h_L = K_e V^2/2g$		r/d 0.0 0.1 >0.2		K_e 0.50 0.12 0.03	(10)
Contraction $h_L = K_C V_2^2/2g$		D_2/D_1 0.00 0.20 0.40 0.60 0.80 0.90	K_C $\theta = 60°$ 0.08 0.08 0.07 0.06 0.06 0.06	K_C $\theta = 180°$ 0.50 0.49 0.42 0.27 0.20 0.10	(10)
Expansion $h_L = K_E V_1^2/2g$		D_1/D_2 0.00 0.20 0.40 0.60 0.80	K_E $\theta = 20°$ 0.30 0.25 0.15 0.10	K_E $\theta = 180°$ 1.00 0.87 0.70 0.41 0.15	(9)
90° miter bend		Without vanes		$K_b = 1.1$	(15)
90° smooth bend		With vanes r/d 1 2 3 4 6 8 10	$K_b = 0.35$ 0.19 0.16 0.21 0.28 0.32	$K_b = 0.2$	(15) (16) and (9)

Threaded pipe fittings	Globe valve—wide open
	Angle valve—wide open
	Gate valve—wide open
	Gate valve—half open
	Return bend
	Tee
	  Straight-through flow
	  Side-outlet flow
	90° elbow
	45° elbow
				$K_v = 10.0$
				$K_v = 5.0$
				$K_v = 0.2$
				$K_v = 5.6$
				$K_b = 2.2$
				
				$K_t = 0.4$
				$K_t = 1.8$
				$K_b = 0.9$
				$K_b = 0.4$
					(15)

Combined Head Loss Equation

The total head loss is given by Eq. (10.4), which is repeated here:

$$\{\text{total head loss}\} = \{\text{pipe head loss}\} + \{\text{component head loss}\} \tag{10.44}$$

To develop an equation for the combined head loss, substitute Eqs. (10.12) and (10.42) in Eq. (10.44):

$$h_L = \sum_{\text{pipes}} f \frac{L}{D} \frac{V^2}{2g} + \sum_{\text{components}} K \frac{V^2}{2g} \qquad \textbf{(10.45)}$$

Equation (10.45) is called the *combined head loss equation*. To apply this equation, follow the same approaches that were used for solving pipe problems. That is, classify the flow as case 1, 2, or 3, and apply the usual equations: the energy, Darcy-Weisbach, and flow rate equations. Example 10.7 illustrates this approach for a case 1 problem.

EXAMPLE 10.7

Pipe System with Combined Head Loss

Problem Statement

If oil ($v = 4 \times 10^{-5}$ m²/s; $SG = 0.9$) flows from the upper to the lower reservoir at a rate of 0.028 m³/s in the 15 cm smooth pipe, then what is the elevation of the oil surface in the upper reservoir?

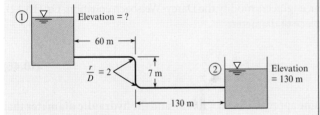

Define the Situation

Oil is flowing from a upper reservoir to a lower reservoir.

Properties:

- Oil: $v = 4 \times 10^{-5}$ m²/s, $SG = 0.9$
- Minor head loss coefficients: Table 10.5, entrance = $K_e = 0.5$; bend = $K_b = 0.19$; outlet = $K_E = 1.0$

State the Goal

Calculate the elevation (in meters) of the free surface of the upper reservoir.

Generate Ideas and Make a Plan

This is a case 1 problem because flow rate and pipe dimensions are known. Thus, the solution is straightforward.

1. Apply the energy equation from 1 to 2.
2. Apply the combined head loss equation (10.45).
3. Develop an equation for z_1 by combining results from steps 1 and 2.
4. Calculate the resistance coefficient f.
5. Solve for z_1 using the equation from step 3.

Take Action (Execute the Plan)

1. Energy equation and term-by-term analysis:

$$\frac{p_1}{\gamma} + \alpha_1 \frac{\overline{V}_1^2}{2g} + z_1 + h_p = \frac{p_2}{\gamma} + \alpha_2 \frac{\overline{V}_2^2}{2g} + z_2 + h_t + h_L$$

$$0 + 0 + z_1 + 0 = 0 + 0 + z_2 + 0 + h_L$$

$$z_1 = z_2 + h_L$$

Interpretation: Change in elevation head is balanced by the total head loss.

2. Combined head loss equation:

$$h_L = \sum_{\text{pipes}} f \frac{L}{D} \frac{V^2}{2g} + \sum_{\text{components}} K \frac{V^2}{2g}$$

$$h_L = f \frac{L}{D} \frac{V^2}{2g} + \left(2K_b \frac{V^2}{2g} + K_e \frac{V^2}{2g} + K_E \frac{V^2}{2g} \right)$$

$$= \frac{V^2}{2g} \left(f \frac{L}{D} + 2K_b + K_e + K_E \right)$$

3. Combine Eqs. (1) and (2):

$$z_1 = z_2 + \frac{V^2}{2g} \left(f \frac{L}{D} + 2K_b + K_e + K_E \right)$$

4. Resistance coefficient:

- Flow rate equation:

$$V = \frac{Q}{A} = \frac{(0.028 \text{ m}^3/\text{s})}{(\pi/4)(0.15 \text{ m})^2} = 1.58 \text{ m/s}$$

- Reynolds number:

$$\text{Re}_D = \frac{VD}{v} = \frac{1.58 \text{ m/s}(0.15 \text{ m})}{4 \times 10^{-5} \text{ m}^2/\text{s}} = 5.93 \times 10^3$$

Thus, flow is turbulent.

- Swamee-Jain equation (10.39):

$$f = \frac{0.25}{\left[\log_{10} \left(\dfrac{k_s}{3.7D} + \dfrac{5.74}{\text{Re}^{0.9}} \right) \right]^2} = \frac{0.25}{\left[\log_{10} \left(0 + \dfrac{5.74}{5930^{0.9}} \right) \right]^2} = 0.036$$

5. Calculate z_1 using the equation from step (3):

$$z_1 = (130 \text{ m}) + \frac{(1.58 \text{ m/s})^2}{2(9.81)\text{m/s}^2}$$

$$\left(0.036\frac{(197 \text{ m})}{(0.15 \text{ m})} + 2(0.19) + 0.5 + 1.0 \right)$$

$$\boxed{z_1 = 136 \text{ m}}$$

Review the Solution and the Process

1. *Discussion.* Notice that the difference is the magnitude of the pipe head loss versus the magnitude of the component head loss:

$$\text{pipe head loss} \sim \Sigma f \frac{L}{D} = 0.036\frac{(197 \text{ m})}{(0.15 \text{ m})} = 47.2$$

$$\text{component head loss} \sim \Sigma K = 2(0.19) + 0.5 + 1.0 = 1.88$$

Thus pipe losses ≫ component losses for this problem.

2. *Skill.* When pipe head loss is dominant, make simple estimates of K because these estimates will not impact the prediction very much.

10.9 Nonround Conduits

Previous sections have considered round pipes. This section extends this information by describing how to account for conduits that are square, triangular, or any other nonround shape. This information is important for applications such as sizing ventilation ducts in buildings and for modeling flow in open channels.

When a conduit is noncircular, engineers modify the Darcy-Weisbach equation, Eq. (10.12), to use hydraulic diameter D_h in place of diameter:

$$h_L = f\frac{L}{D_h} \frac{V^2}{2g} \tag{10.46}$$

Eq. (10.46) is derived using the same approach as Eq. (10.12), and the **hydraulic diameter** that emerges from this derivation is

$$D_h \equiv \frac{4 \times \text{cross-section area}}{\text{wetted perimeter}} \tag{10.47}$$

where the "wetted perimeter" is that portion of the perimeter that is physically touching the fluid. The wetted perimeter of a rectangular duct of dimension $L \times w$ is $2L + 2w$. Thus, the hydraulic diameter of this duct is

$$D_h \equiv \frac{4 \times Lw}{2L + 2w} = \frac{2Lw}{L + w}$$

Using Eq. (10.47), the hydraulic diameter of a round pipe is the pipe's diameter D. When Eq. (10.46) is used to calculate head loss, the resistance coefficient f is found using a Reynolds number based on hydraulic diameter. Use of hydraulic diameter is an approximation. According to White (20), this approximation introduces an uncertainty of 40% for laminar flow and 15% for turbulent flow:

$$f = \left(\frac{64}{\text{Re}_{D_h}} \right) \pm 40\% \text{ (laminar flow)}$$

$$f = \frac{0.25}{\left[\log_{10}\left(\dfrac{k_s}{3.7D_h} + \dfrac{5.74}{\text{Re}_{D_h}^{0.9}} \right) \right]^2} \pm 15\% \text{ (turbulent flow)} \tag{10.48}$$

In addition to hydraulic diameter, engineers also use **hydraulic radius**, which is defined as

$$R_h \equiv \frac{\text{section area}}{\text{wetted perimeter}} = \frac{D_h}{4} \qquad \text{(10.49)}$$

Notice that the ratio of R_h to D_h is 1/4 instead of 1/2. Although this ratio is not logical, it is the convention used in the literature and is useful to remember. Chapter 15, which focuses on open-channel flow, will present examples of hydraulic radius.

   Summary. To model flow in a nonround conduit, the approaches of the previous sections are followed, with the only difference being the use of hydraulic diameter in place of diameter. This is illustrated by Example 10.8.

EXAMPLE 10.8

Pressure Drop in an HVAC Duct

Problem Statement

Air ($T = 20°C$ and $p = 101$ kPa absolute) flows at a rate of 2.5 m³/s in a horizontal, commercial steel, HVAC duct. (HVAC is an acronym for heating, ventilating, and air conditioning.) What is the pressure drop in inches of water per 50 m of duct?

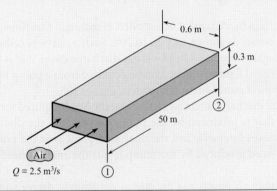

Define the Situation

Air is flowing through a duct.

Assumptions:

- Fully developed flow, meaning that $V_1 = V_2$. Thus, the velocity head terms in the energy equation cancel out.
- No sources of component head loss.

Properties:

- Air (20°C, 1 atm, Table A.2): $\rho = 1.2$ kg/m³, $\nu = 15.1 \times 10^{-6}$ m²/s
- Steel pipe: Table 10.4, $k_s = 0.046$ mm

State the Goal

Find: Pressure drop (inch H_2O) in a length of 50 m.

Generate Ideas and Make a Plan

This is a case 1 problem because flow rate and duct dimensions are known. Thus, the solution is straightforward.

1. Derive an equation for pressure drop by using the energy equation.
2. Calculate parameters needed to find head loss.
3. Calculate head loss by using the Darcy-Weisbach equation (10.12).
4. Calculate pressure drop Δp by combining steps 1, 2, and 3.

Take Action (Execute the Plan)

1. Energy equation (after term-by-term analysis):

$$p_1 - p_2 = \rho g h_L$$

2. Intermediate calculations:

- Flow rate equation:

$$V = \frac{Q}{A} = \frac{2.5 \text{ m}^3/\text{s}}{(0.3 \text{ m})(0.6 \text{ m})} = 13.9 \text{ m/s}$$

- Hydraulic diameter:

$$D_h \equiv \frac{4 \times \text{section area}}{\text{wetted perimeter}} = \frac{4(0.3 \text{ m})(0.6 \text{ m})}{(2 \times 0.3 \text{ m}) + (2 \times 0.6 \text{ m})} = 0.4 \text{ m}$$

- Reynolds number:

$$\text{Re} = \frac{V D_h}{\nu} = \frac{(13.9 \text{ m/s})(0.4 \text{ m})}{(15.1 \times 10^{-6} \text{ m}^2/\text{s})} = 368{,}000$$

Thus, flow is turbulent.

- Relative roughness:

$$k_s/D_h = (0.000046 \text{ m})/(0.4 \text{ m}) = 0.000115$$

- Resistance coefficient (Moody diagram): $f = 0.015$

3. Darcy-Weisbach equation:

$$h_f = f\left(\frac{L}{D_h}\right)\left(\frac{V^2}{2g}\right) = 0.015\left(\frac{50 \text{ m}}{0.4 \text{ m}}\right)\left\{\frac{(13.9 \text{ m/s})^2}{2(9.81 \text{ m/s}^2)}\right\}$$

$$= 18.6 \text{ m}$$

4. Pressure drop (from step 1):

$$p_1 - p_2 = \rho g h_L = (1.2 \text{ kg/m}^3)(9.81 \text{ m/s}^2)(18.6 \text{ m}) = 220 \text{ Pa}$$

$$\boxed{p_1 - p_2 = 0.883 \text{ inch } H_2O}$$

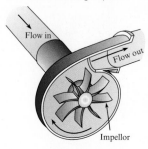

Flow in

Flow out

Impellor

10.10 Pumps and Systems of Pipes

This section explains how to model flow in a network of pipes and how to incorporate performance data for a centrifugal pump. These topics are important because pumps and pipe networks are common.

Modeling a Centrifugal Pump

As shown in Fig. 10.19, a **centrifugal pump** is a machine that uses a rotating set of blades situated within a housing to add energy to a flowing fluid. The amount of energy that is added is represented by the head of the pump h_p, and the rate at which work is done on the flowing fluid is $P = \dot{m}gh_p$.

To model a pump in a system, engineers commonly use a graphical solution involving the energy equation and a pump curve. To illustrate this approach, consider flow of water in the system of Fig. 10.20a. The energy equation applied from the reservoir water surface to the outlet stream is

$$\frac{p_1}{\gamma} + \frac{V_1^2}{2g} + z_1 + h_p = \frac{p_2}{\gamma} + \frac{V_2^2}{2g} + z_2 + \sum K_L \frac{V^2}{2g} + \sum \frac{fL}{D}\frac{V^2}{2g}$$

For a system with one size of pipe, this simplifies to

$$h_p = (z_2 - z_1) + \frac{V^2}{2g}\left(1 + \sum K_L + \frac{fL}{D}\right) \tag{10.50}$$

Hence, for any given discharge, a certain head h_p must be supplied to maintain that flow. Thus, construct a head-versus-discharge curve, as shown in Fig. 10.20b. Such a curve is called the **system curve**. Now, a given centrifugal pump has a head-versus-discharge curve that is characteristic of that pump. This curve, called a **pump curve**, can be acquired from a pump manufacturer, or it can be measured. A typical pump curve is shown in Fig. 10.20b.

Figure 10.20b reveals that as the discharge increases in a pipe, the head required for flow also increases. However, the head that is produced by the pump decreases as the discharge increases. Consequently, the two curves intersect, and the operating point is at the point of intersection—that point where the head produced by the pump is just the amount needed to overcome the head loss in the pipe.

To incorporate performance data for a pump, use the energy equation to derive a system curve. Then, acquire a pump curve from a manufacturer or other source and plot the two curves together. The point of intersection shows where the pump will operate. This process is illustrated in Example 10.9.

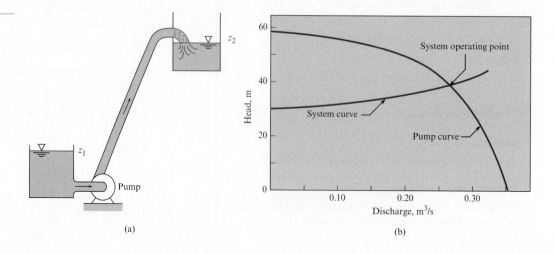

(a)

(b)

EXAMPLE 10.9

Finding a System Operating Point

Define the Situation

- The system diagram is sketched ahead.
- The pump curve is given in Fig. 10.20b.
- The friction factor is $f = 0.015$.

State the Goal

Calculate the discharge (m³/s) in the system.

Generate Ideas and Make a Plan

1. Develop an equation for the system curve by applying the energy equation.
2. Plot the given pump curve and the system curve on the same graph.
3. Find discharge Q by finding the intersection of the system and pump curve.

Take Action (Execute the Plan)

Energy equation:

$$\frac{p_1}{g} + \frac{V_1^2}{2g} + z_1 + h_p = \frac{p_2}{g} + \frac{V_2^2}{2g} + z_2 + \sum h_L$$

$$0 + 0 + 200 + h_p = 0 + 0 + 230 + \left(\frac{fL}{D} + K_e + K_b + K_E\right)\frac{V^2}{2g}$$

Here, $K_e = 0.5$, $K_b = 0.35$ and $K_E = 1.0$. Hence

$$h_p = 30 + \frac{Q^2}{2gA^2}\left[\frac{0.015(1000)}{0.40} + 0.5 + 0.35 + 1\right]$$

$$= 30 + \frac{Q^2}{2 \times 9.81 \times [(\pi/4) \times 0.4^2]^2}(39.3)$$

$$= 30\ m + 127Q^2\ m$$

Now, make a table of Q versus h_p (as follows) to give values to produce a system curve that will be plotted with the pump curve. When the system curve is plotted on the same graph as the pump curve, it is seen (Fig. 10.20b) that the operating condition occurs at $\boxed{Q = 0.27\ m^3/s.}$

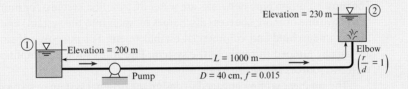

Q(m³/s)	$h_p = (30\ m + 127Q^2)\ m$
0	30
0.1	31.3
0.2	35.1
0.3	41.4

Pipes in Parallel

Consider a pipe that branches into two parallel pipes and then rejoins, as shown in Fig. 10.21. A problem involving this configuration might be to determine the division of flow in each pipe, given the total flow rate.

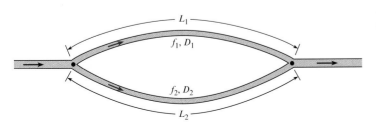

FIGURE 10.21

Flow in parallel pipes.

No matter which pipe is involved, the pressure difference between the two junction points is the same. Also, the elevation difference between the two junction points is the same. Because $h_L = (p_1/\gamma + z_1) - (p_2/\gamma + z_2)$, it follows that h_L between the two junction points is the same in both of the pipes of the parallel pipe system. Thus,

$$h_{L_1} = h_{L_2}$$

$$f_1 \frac{L_1}{D_1} \frac{V_1^2}{2g} = f_2 \frac{L_2}{D_2} \frac{V_2^2}{2g}$$

Then,

$$\left(\frac{V_1}{V_2}\right)^2 = \frac{f_2 L_2 D_1}{f_1 L_1 D_2} \qquad \text{or} \qquad \frac{V_1}{V_2} = \left(\frac{f_2 L_2 D_1}{f_1 L_1 D_2}\right)^{1/2}$$

If f_1 and f_2 are known, the division of flow can be easily determined. However, some trial-and-error analysis may be required if f_1 and f_2 are in the range in which they are functions of the Reynolds number.

Pipe Networks

The most common pipe networks are water distribution systems for municipalities. These systems have one or more sources (discharges of water into the system) and numerous loads: one for each household and commercial establishment. For purposes of simplification, the loads are usually lumped throughout the system. Fig. 10.22 shows a simplified distribution system with two sources and seven loads.

The engineer is often engaged to design the original system or to recommend an economical expansion to the network. An expansion may involve additional housing or commercial developments, or it may be designed to handle increased loads within the existing area.

In the design of such a system, the engineer will have to estimate the future loads for the system and will need to have sources (wells or direct pumping from streams or lakes) to satisfy the loads. Also, the layout of the pipe network must be made (usually parallel to streets), and pipe sizes will have to be determined. The object of the design is to arrive at a network of pipes that will deliver the design flow at the design pressure for minimum cost. The cost will include first costs (materials and construction) as well as maintenance and operating costs. The design process usually involves a number of iterations on pipe sizes and layouts before the optimum design (minimum cost) is achieved.

So far as the fluid mechanics of the problem are concerned, the engineer must determine pressures throughout the network for various conditions—that is, for various combinations of

FIGURE 10.22

Pipe network.

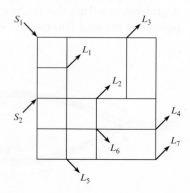

pipe sizes, sources, and loads. The solution of a problem for a given layout and a given set of sources and loads requires that two conditions be satisfied:

1. The continuity equation must be satisfied. That is, the flow into a junction of the network must equal the flow out of the junction. This must be satisfied for all junctions.

2. The head loss between any two junctions must be the same regardless of the path in the series of pipes taken to get from one junction point to the other. This requirement results because pressure must be continuous throughout the network (pressure cannot have two values at a given point). This condition leads to the conclusion that the algebraic sum of head losses around a given loop must be equal to zero. Here the sign (positive or negative) for the head loss in a given pipe is given by the sense of the flow with respect to the loop—that is, whether the flow has a clockwise or counterclockwise direction.

At one time, these solutions were made by trial-and-error hand computation, but computers have made the older methods obsolete. Even with these advances, however, the engineer charged with the design or analysis of such a system must understand the basic fluid mechanics of the system to be able to interpret the results properly and to make good engineering decisions based on the results. Therefore, an understanding of the original method of solution by Hardy Cross (17) may help you to gain this basic insight. The Hardy Cross method is as follows.

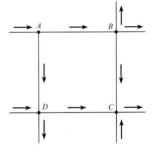

The engineer first distributes the flow throughout the network so that loads at various nodes are satisfied. In the process of distributing the flow through the pipes of the network, the engineer must be certain that continuity is satisfied at all junctions (flow into a junction equals flow out of the junction), thus satisfying requirement 1. The first guess at the flow distribution obviously will not satisfy requirement 2 regarding head loss; therefore, corrections are applied. For each loop of the network, a discharge correction is applied to yield a zero net head loss around the loop. For example, consider the isolated loop in Fig. 10.23. In this loop, the loss of head in the clockwise direction will be given by

$$\sum h_{L_c} = h_{L_{AB}} + h_{L_{BC}}$$
$$= \sum k Q_c^n \tag{10.51}$$

The loss of head for the loop in the counterclockwise direction is

$$\sum h_{L_{cc}} = \sum_{cc} k Q_{cc}^n \tag{10.52}$$

For a solution, the clockwise and counterclockwise head losses have to be equal, or

$$\sum h_{L_c} = \sum h_{L_{cc}}$$
$$\sum k Q_c^n = \sum k Q_{cc}^n$$

As noted, the first guess for flow in the network will undoubtedly be in error; therefore, a correction in discharge, ΔQ, will have to be applied to satisfy the head loss requirement. If the clockwise head loss is greater than the counterclockwise head loss, ΔQ will have to be applied in the counterclockwise direction. That is, subtract ΔQ from the clockwise flows and add it to the counterclockwise flows:

$$\sum k(Q_c - \Delta Q)^n = \sum k(Q_{cc} + \Delta Q)^n \tag{10.53}$$

Expand the summation on either side of Eq. (10.53) and include only two terms of the expansion:

$$\sum k(Q_c^n - n Q_c^{n-1} \Delta Q) = \sum k(Q_{cc}^n + n Q_{cc}^{n-1} \Delta Q)$$

Solve for ΔQ:

$$\Delta Q = \frac{\sum kQ_c^n - \sum kQ_{cc}^n}{\sum nkQ_c^{n-1} + \sum nkQ_{cc}^{n-1}} \qquad (10.54)$$

Thus, if ΔQ as computed from Eq. (10.54) is positive, the correction is applied in a counterclockwise sense (add ΔQ to counterclockwise flows and subtract it from clockwise flows).

A different ΔQ is computed for each loop of the network and applied to the pipes. Some pipes will have two ΔQs applied because they will be common to two loops. The first set of corrections usually will not yield the final desired result because the solution is approached only by successive approximations. Thus, the corrections are applied successively until the corrections are negligible. Experience has shown that for most loop configurations, applying ΔQ as computed by Eq. (10.54) produces too large a correction. Fewer trials are required to solve for Qs if approximately 0.6 of the computed ΔQ is used.

More information on methods of solution of pipe networks is available in references 18 and 19. Searching the Internet for "pipe networks" yields information on software available from various sources.

EXAMPLE 10.10

Discharge in a Piping Network

Problem Statement

A simple pipe network with water flow consists of three valves and a junction, as shown in the figure. The piezometric head at points 1 and 2 is 1 ft and reduces to zero at point 4. There is a wide-open globe valve in line A, a gate valve half open in line B, and a wide-open angle valve in line C. The pipe diameter in all lines is 2 inches. Find the flow rate in each line. Assume that the head loss in each line is due only to the valves.

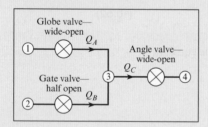

Define the Situation

Water flows through a network of pipes.

- $h_1 = h_2 = 1$ ft.
- $h_4 = 0$ ft.
- Pipe diameter (all pipes) is 0.167 ft.

Assumptions: Head loss is due to valves only.

State the Goal

Find the flow rate (in cfs) in each pipe.

Generate Ideas and Make a Plan

1. Let $h_{L,1\to3} = h_{L,2\to3}$.
2. Let $h_{L,2\to4} = 1$ ft.
3. Solve equations using the Hardy Cross approach.

Take Action (Execute the Plan)

The piezometric heads at points 1 and 2 are equal, so

$$h_{L,1\to3} + h_{L,3\to2} = 0$$

The head loss between points 2 and 4 is 1 ft, so

$$h_{L,2\to3} + h_{L,3\to4} = 0$$

Continuity must be satisfied at point 3, so

$$Q_A + Q_B = Q_C$$

The head loss through a valve is given by

$$h_L = K_V \frac{V^2}{2g}$$

$$= K_V \frac{1}{2g}\left(\frac{Q}{A}\right)^2$$

where K_V is the loss coefficient. For a 2-inch pipe, the head loss becomes

$$h_L = 32.6 K_V Q^2$$

where h_L is in feet and Q is in cfs.

The head loss equation between points 1 and 2 expressed in term of discharge is

$$32.6 K_A Q_A^2 - 32.6 K_B Q_B^2 = 0$$

or

$$K_A Q_A^2 - K_B Q_B^2 = 0$$

where K_A is the loss coefficient for the wide-open globe valve ($K_A = 10$) and K_B is the loss coefficient for the half-open gate valve ($K_B = 5.6$). The head loss equation between points 2 and 4 is

$$32.6 K_B Q_B^2 + 32.6 K_C Q_C^2 = 1$$

or

$$K_B Q_B^2 + K_C Q_C^2 = 0.0307$$

where K_C is the loss coefficient for the wide-open angle valve ($K_C = 5$). The two head loss equations and the continuity equation comprise three equations for Q_A, Q_B, and Q_C. However, the equations are nonlinear and require linearization and solution by iteration (Hardy Cross approach). The discharge is written as

$$Q = Q_0 + \Delta Q$$

where Q_0 is the starting value and ΔQ is the change. Then,

$$Q^2 \cdot Q_0^2 + 2 Q_0 \Delta Q$$

where the $(\Delta Q)^2$ term is neglected. The equations in terms of ΔQ become

$$2 K_A Q_{0,A} \Delta Q_A - 2 K_B Q_{0,B} \Delta Q_B = K_B Q_{0,B}^2 - K_A Q_{0,A}^2$$

$$2 K_C Q_{0,C} \Delta Q_C - 2 K_B Q_{0,B} \Delta Q_B = 0.0307 - K_B Q_{0,B}^2 - K_C Q_{0,C}^2$$

$$\Delta Q_A + \Delta Q_B = \Delta Q_C$$

which can be expressed in matrix form as

$$\begin{bmatrix} 2 K_A Q_{0,A} & -2 K_B Q_{0,B} & 0 \\ 0 & 2 K_B Q_{0,B} & 2 K_C Q_{0,C} \\ 1 & 1 & -1 \end{bmatrix} \begin{Bmatrix} \Delta Q_A \\ \Delta Q_B \\ \Delta Q_C \end{Bmatrix}$$

$$= \begin{bmatrix} K_B Q_{0,B}^2 - K_A Q_{0,A}^2 \\ 0.0307 - K_B Q_{0,B}^2 - K_C Q_{0,C}^2 \\ 0 \end{bmatrix}$$

The procedure begins by selecting values for $Q_{0,A}$, $Q_{0,B}$, and $Q_{0,C}$. Assume $Q_{0,A} = Q_{0,B}$ and $Q_{0,C} = 2 Q_{0,A}$. Then, from the head loss equation from points 2 to 4,

$$K_B Q_{0,B}^2 + K_C Q_{0,C}^2 = 0.0307$$
$$(K_B + 4 K_C) Q_{0,B}^2 = 0.0307$$
$$(5.6 + 4 \times 5) Q_{0,B}^2 = 0.0307$$
$$Q_{0,B} = 0.0346$$

and $Q_{0,A} = 0.0346$ and $Q_{0,C} = 0.0693$. These values are substituted into the matrix equation to solve for the ΔQs. The discharges are corrected by $Q_0^{new} = Q_0^{old} + \Delta Q$ and substituted into the matrix equation again to yield new ΔQs. The iterations are continued until sufficient accuracy is obtained. The accuracy is judged by how close the column matrix on the right approaches zero. A table with the results of iterations for this example is as follows:

	Initial	Iteration			
		1	2	3	4
Q_A	0.0346	0.0328	0.0305	0.0293	0.0287
Q_B	0.0346	0.0393	0.0384	0.0394	0.0384
Q_C	0.0693	0.0721	0.0689	0.0687	0.0671

Review the Solution and the Process

Knowledge. This solution technique is called the Newton-Raphson method. This method is useful for nonlinear systems of algebraic equations. It can be implemented easily on a computer. The solution procedure for more complex systems is the same.

10.11 Summarizing Key Knowledge

Classifying Flow in Conduits

- A *conduit* is any pipe, tube, or duct that is filled with a flowing fluid.
- Flow in a conduit is characterized using the Reynolds number based on pipe diameter. This π-group is given by several equivalent formulas:

$$\text{Re}_D = \frac{VD}{\nu} = \frac{\rho VD}{\mu} = \frac{4Q}{\pi D \nu} = \frac{4\dot{m}}{\pi D \mu}$$

- To classify a flow as *laminar* or *turbulent*, calculate the Reynolds number:

$$\text{Re}_D \leq 2000 \quad \text{laminar flow}$$
$$\text{Re}_D \geq 3000 \quad \text{turbulent flow}$$

- Flow in a conduit can be developing or fully developed:
  - *Developing flow* occurs near an entrance or after the flow is disrupted (i.e., downstream of a valve, a bend, or an orifice). *Developing flow* means that the velocity profile and wall shear stress are changing with axial location.

- *Fully developed flow* occurs in straight runs of pipe that are long enough to allow the flow to develop. Fully developed flow means that the velocity profile and the shear stress are constant with axial location x. In fully developed flow, the flow is uniform, and the pressure gradient (dp/dx) is constant.

- To classify a flow at a pipe inlet as developing or fully developed, calculate the *entrance length* (L_e). At any axial location greater than L_e, the flow will be fully developed. The equations for entrance length are

$$\frac{L_e}{D} = 0.05 \text{Re}_D \qquad \text{(laminar flow: Re}_D \leq 2000)$$

$$\frac{L_e}{D} = 50 \qquad \text{(turbulent flow: Re}_D \geq 3000)$$

- To describe commercial pipe in the NPS system, specify a nominal diameter in inches and a schedule number. The schedule number characterizes the wall thickness. Actual dimensions need to be looked up.

Head Loss (Pipe Head Loss)

- The sum of head losses in a piping system is called the *total head loss*. Sources of head loss classify into two categories:
  - *Pipe head loss.* Head loss in straight runs of pipe with fully developed flow.
  - *Component head loss.* Head loss in components and transitions such as valves, elbows, and bends.

- To characterize *pipe head loss*, engineers use a π-group called the *friction factor*. The friction factor f gives the ratio of wall shear stress ($4\tau_0$) to kinetic pressure ($\rho V^2/2$).

- Pipe head loss has two symbols that are used: h_L and h_f. To predict pipe head loss, apply the Darcy-Weisbach equation (DWE):

$$h_L = h_f = f \frac{L}{D} \frac{V^2}{2g}$$

There are three methods for using the DWE:

- *Method 1 (laminar flow).* Apply the DWE in this form:

$$h_f = \frac{32\mu LV}{\gamma D^2}$$

- *Method 2 (laminar or turbulent flow).* Apply the DWE and use a formula for f:

$$f = \frac{64}{\text{Re}} \qquad \text{laminar flow}$$

$$f = \frac{0.25}{\left[\log_{10}\left(\dfrac{k_s}{3.7D} + \dfrac{5.74}{\text{Re}_D^{0.9}} \right) \right]^2} \qquad \text{turbulent flow}$$

- *Method 3 (laminar or turbulent flow).* Apply the DWE, and look up f on the Moody diagram.

- The *roughness* of a pipe wall sometimes affects the friction factor:
  - *Laminar flow.* The roughness does not matter; the friction factor f is independent of roughness.
  - *Turbulent flow.* The roughness is characterized by looking up an equivalent sand roughness height k_s and then finding f as a function of the Reynolds number and k_s/D. When the flow is *fully turbulent,* then f is independent of the Reynolds number.

Head Loss (Component Head Loss)

- To characterize the head loss in a component, engineers use a π-group called the *minor loss coefficient, K,* which gives the ratio of head loss to velocity head. Values of K, which come from experimental studies, are tabulated in engineering references. Each component has a specific value of K, which is looked up. The head loss for a component is

$$h_L = K_{\text{component}} \frac{V^2}{2g}$$

- The *total head loss* in a pipe is given by

$$\text{overall (total) head loss} = \sum (\text{pipe head losses}) + \sum (\text{component head losses})$$

$$h_L = \sum_{\text{pipes}} f \frac{L}{D} \frac{V^2}{2g} + \sum_{\text{components}} K \frac{V^2}{2g}$$

Additional Useful Results

- Noncircular conduits can be analyzed using the hydraulic diameter D_h or the hydraulic radius (R_h). To analyze a noncircular conduit, apply the same equations that are used for round conduits and replace D with D_h in the formulas. The equations for D_h and R_h are

$$D_h = 4R_h = \frac{4 \times \text{section area}}{\text{wetted perimeter}}$$

- To find the operating point of a centrifugal pump in a system, the traditional approach is a graphical solution. Plot a system curve that is derived using the energy equation, and plot the head versus flow rate curve of the centrifugal pump. The intersection of these two curves gives the operating point of the system.

- The analysis of pipe networks is based on the continuity equation being satisfied at each junction and the head loss between any two junctions being independent of pipe path between the two junctions. A series of equations based on these principles are solved iteratively to obtain the flow rate in each pipe and the pressure at each junction in the network.

REFERENCES

1. Reynolds, O. "An Experimental Investigation of the Circumstances Which Determine Whether the Motion of Water Shall Be Direct or Sinuous and of the Law of Resistance in Parallel Channels." *Phil. Trans. Roy. Soc. London,* 174, part III (1883).

2. Schlichting, Hermann. *Boundary Layer Theory,* 7th ed. New York: McGraw-Hill, 1979.

3. Moody, Lewis F. "Friction Factors for Pipe Flow." *Trans. ASME,* 671 (November 1944).

4. Nikuradse, J. "Strömungsgesetze in rauhen Rohren." *VDI-Forschungsh.,* no. 361 (1933). Also translated in *NACA Tech. Memo,* 1292.

5. White, F. M. *Viscous Fluid Flow.* New York: McGraw-Hill, 1991.

6. Colebrook, F. "Turbulent Flow in Pipes with Particular Reference to the Transition Region between the Smooth and Rough Pipe Laws." *J. Inst. Civ. Eng.,* vol. 11, 133–156 (1939).

7. Swamee, P. K., and A. K. Jain. "Explicit Equations for Pipe-Flow Problems." *J. Hydraulic Division of the ASCE,* vol. 102, no. HY5 (May 1976).

8. Streeter, V. L., and E. B. Wylie. *Fluid Mechanics,* 7th ed. New York: McGraw-Hill, 1979.

9. Barbin, A. R., and J. B. Jones. "Turbulent Flow in the Inlet Region of a Smooth Pipe." *Trans. ASME, Ser. D: J. Basic Eng.,* vol. 85, no. 1 (March 1963).

10. ASHRAE. *ASHRAE Handbook—1977 Fundamentals.* New York: Am. Soc. of Heating, Refrigerating and Air Conditioning Engineers, Inc., 1977.

11. Crane Co. "Flow of Fluids Through Valves, Fittings and Pipe." Technical Paper No. 410, Crane Co. (1988), 104 N. Chicago St., Joliet, IL 60434.

12. Fried, Irwin, and I. E. Idelchik. *Flow Resistance: A Design Guide for Engineers.* New York: Hemisphere, 1989.

13. Hydraulic Institute. *Engineering Data Book,* 2nd ed., Hydraulic Institute, 30200 Detroit Road, Cleveland, OH 44145.

14. Miller, D. S. *Internal Flow—A Guide to Losses in Pipe and Duct Systems.* British Hydrodynamic and Research Association (BHRA), Cranfield, England (1971).

15. Streeter, V.L. (ed.). *Handbook of Fluid Dynamics.* New York: McGraw-Hill, 1961.

16. Beij, K.H. "Pressure Losses for Fluid Flow in 90% Pipe Bends." *J. Res. Nat. Bur. Std.,* 21 (1938). Information cited in Streeter (20).

17. Cross, Hardy. "Analysis of Flow in Networks of Conduits or Conductors." *Univ. Illinois Bull.,* 286 (November 1936).

18. Hoag, Lyle N., and Gerald Weinberg. "Pipeline Network Analysis by Digital Computers." *J. Am. Water Works Assoc.,* 49 (1957).

19. Jeppson, Roland W. *Analysis of Flow in Pipe Networks.* Ann Arbor, MI: Ann Arbor Science Publishers, 1976.

20. White, F. M. *Fluid Mechanics,* 5th ed. New York: McGraw-Hill, 2003.

PROBLEMS

Notes on Pipe Diameter for Chapter 10 Problems

When a pipe diameter is given using the label "NPS" or "nominal," find the dimensions using Table 10.1 of §10.2. Otherwise, assume the specified diameter is an inside diameter (ID).

Classifying Flow (§10.1)

10.1 Kerosene (20°C) flows at a rate of 0.06 m³/s in a 13 cm diameter pipe. Would you expect the flow to be laminar or turbulent? Calculate the entrance length.

10.2 A compressor draws 0.02 m³/s of ambient air (20°C) in from the outside through a round duct that is 10 m long and 80 mm in diameter. Determine the entrance length and establish whether the flow is laminar or turbulent.

10.3 Design a lab demo for laminar flow. Specify the diameter and length for a tube that carries SAE 10W-30 oil at 38°C so that the flow is laminar and fully developed with a discharge of $Q = 0.1$ L/s.

Darcy-Weisbach Equation for Head Loss (§10.3)

10.4 For the 2 cases that follow, apply the Darcy-Weisbach equation from Eq. (10.12) in §10.3 to calculate the head loss in a pipe. Apply the grid method to carry and cancel units.

    a. Water flows at a rate of 15 gpm and a mean velocity of 210 ft/min in a pipe of length 180 feet. For a resistance coefficient of $f = 0.02$, find the head loss in feet.

    b. The head loss in a section of PVC pipe is 0.08 m, the resistance coefficient is $f = 0.012$, the length is 15 m, and the flow rate is 2 cfs. Find the pipe diameter in meters.

10.5 Air ($\rho = 1.4$ kg/m³) flows in a straight round tube. The mean velocity is 17 m/s. The friction factor is 0.03. Flow is fully developed. Calculate the wall shear stress in units of Pa. Choose the closest answer (Pa): (a) 1.0, (b) 1.5, (c) 2.0, (d) 2.5, (e) 3.5.

10.6 The head loss from section 1 to 2 is 0.8 m. The Darcy friction factor is 0.01. $D = 1.0$ m. $L = 50$ m. The flow is steady and fully developed. Calculate the mean velocity in m/s. Choose the closest answer (m/s): (a) 1.2, (b) 2.4, (c) 3.2, (d) 4.4, (e) 5.6.

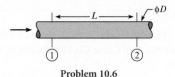

Problem 10.6

10.7 In Case A, water flows through an 8-inch schedule 40 pipe with a discharge of 83 liters per second. In Case B, the schedule is changed to 80. The mean velocity is the same in both cases. Calculate the discharge in units of L/s for case B. Choose the closest answer (L/s): (a) 68, (b) 74, (c) 75, (d) 79, (e) 82.

10.8 Water (15°C) flows through a garden hose (ID = 32 mm) with a mean velocity of 1.5 m/s. Find the pressure drop for a section of hose that is 20 meters long and situated horizontally. Assume that $f = 0.012$.

10.9 Water flows in the pipe shown, and the manometer deflects 111 cm. What is f for the pipe if $V = 15$ m/s?

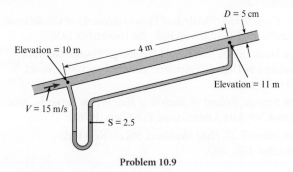

Problem 10.9

Laminar Flow in Pipes (§10.5)

10.10 Using §10.5 and other resources, answer the questions that follow. Strive for depth, clarity, and accuracy while also combining sketches, words, and equations in ways that enhance the effectiveness of your communication.

a. What are the main characteristics of laminar flow?

b. What is the meaning of each variable that appears in Eq. (10.27) in §10.5?

c. In Eq. (10.33) of §10.5, what is the meaning of h_f?

10.11 A Newtonian fluid is flowing in a round conduit. The flow is laminar, steady, and fully developed. The Darcy friction factor is 7.7. Calculate Re_D. Choose the closest answer: (a) 4.0, (b) 6.1, (c) 8.3, (d) 16.6, (e) 32.2.

10.12 Flow of a liquid in a smooth 5 cm pipe yields a head loss of 2 m per meter of pipe length when the mean velocity is 2 m/s. Calculate f and the Reynolds number. Prove that doubling the flow rate will double the head loss. Assume fully developed flow.

10.13 Liquid ($\gamma = 9.8$ kN/m³) is flowing in a pipe at a steady rate, but the direction of flow is unknown. Is the liquid moving

upward or moving downward in the pipe? If the pipe diameter is 12 mm and the liquid viscosity is 3.0×10^{-3} N·s/m², what is the magnitude of the mean velocity in the pipe?

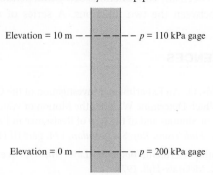

Problem 10.13

10.14 Oil ($SG = 0.85$, $\mu = 10^{-2}$ lbf-s/ft²) is pumped through a nominal 1 in., schedule 80 pipe at the rate of 0.005 cfs. What is the head loss per 100 ft of level pipe?

10.15 A liquid ($\rho = 1000$ kg/m³; $\mu = 10^{-1}$ N·s/2 m²; $v = 10^{-4}$ m²/s) flows uniformly with a mean velocity of 1.3 m/s in a pipe with a diameter of 175 mm. Show that the flow is laminar. Also, find the friction factor f and the head loss per meter of pipe length.

10.16 Oil ($SG = 0.82$; $\mu = 0.048$ N·s/m²) is pumped through a horizontal 40 cm pipe. Mean velocity is 0.2 m/s. What is the pressure drop per 10 m of pipe?

10.17 Oil ($S = 0.80$; $\mu = 10^{-2}$ lbf-s/ft²; $v = 0.0057$ ft²/s) flows downward in the pipe, which is 0.10 ft in diameter and has a slope of 30° with the horizontal. Mean velocity is 3 ft/s. What is the pressure gradient (dp/ds) along the pipe?

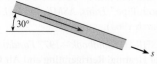

Problem 10.17

10.18 Glycerine ($T = 68°F$) flows in a pipe with a 6-in. diameter at a mean velocity of 1.5 ft/s. Is the flow laminar or turbulent? Plot the velocity distribution across the flow section, in 0.5-in. increments of radius.

10.19 Glycerine at 20°C flows at 0.6 m/s in the 2-cm commercial steel pipe. Two piezometers are used as shown to measure the piezometric head. The distance along the pipe between the standpipes is 1 m. The inclination of the pipe is 20°. What is the height difference Δh between the glycerine in the two standpipes?

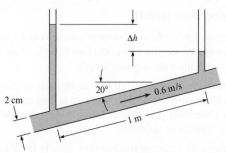

Problem 10.19

10.20 Glycerine ($T = 20°C$) flows through a funnel with $D = 0.9$ cm as shown. Calculate the mean velocity of the glycerine exiting the tube. Assume the only head loss is due to friction in the tube.

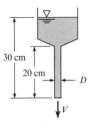

Problem 10.20

10.21 Velocity measurements are made in a 20-cm pipe. The velocity at the center is found to be 2 m/s, and the velocity distribution is observed to be parabolic. If the pressure drop is found to be 2 kPa per 100 m of pipe, what is the kinematic viscosity ν of the fluid? Assume that the fluid's $SG = 0.8$.

10.22 Water is pumped through a heat exchanger consisting of tubes 7 mm in diameter and 6 m long. The velocity in each tube is 18 cm/s. The water temperature increases from 20°C at the entrance to 30°C at the exit. Calculate the pressure difference across the heat exchanger, neglecting entrance losses but accounting for the effect of temperature change by using properties at average temperatures.

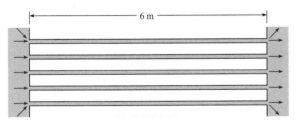

Problem 10.22

Turbulent Flow in Pipes (§10.6)

10.23 A liquid flows in a round pipeline. The mass flow rate is 8500 kg/s. The Reynolds number is 8 million. Kinematic viscosity is 1.4E-6 m²/s and $SG = 0.9$. Calculate the pipe diameter in meters. Choose the closest answer (m): (a) 0.8, (b) 1.1, (c) 1.4, (d) 1.7, (e) 2.0.

10.24 Water (70°F) flows through a nominal 4-in., schedule 40, PVC pipe at the rate of 10 cfs. What is the resistance coefficient f? Use the Swamee-Jain equation (10.39), given in §10.6.

10.25 Water (10°C) flows through a 85-cm diameter smooth pipe at a rate of 0.05 m³/s. What is the resistance coefficient f?

10.26 What is f for the flow of water at 10°C through a 80-cm diameter cast iron pipe with a mean velocity of 24 m/s?

10.27 Water (20°C) flows in a 20-cm cast iron pipe at a rate of 0.5 m³/s. For these conditions, determine or estimate the following:

   a. The Reynolds number

   b. Friction factor f (use Swamee-Jain Eq. (10.39) in §10.6.)

   c. Shear stress at the wall, τ_0

10.28 In a 4.2 in. uncoated cast iron pipe, 10.0 cfs of water flows at 60°F. Determine f from Fig. 10.14.

10.29 Determine the head loss in a 750 ft length of a concrete pipe with a 6 in. diameter ($k_s = 0.0002$ ft) carrying 2.5 cfs of fluid. The properties of the fluid are $\nu = 3.33 \times 10^{-3}$ ft²/s and $\rho = 1.5$ slug/ft³.

10.30 A fluid flows through a pipe. Calculate the drop in piezometer level (Δh) in units of cm. The flow is steady and fully developed. The fluid is Newtonian. The mean velocity is 0.4 m/s. The Reynolds number is 100,000. The relative roughness is 0.002. The length L between piezometers is 21 m and the pipe ID is 3.0 cm. Choose the closest answer (cm): (a) 8, (b) 14, (c) 22, (d) 28, (e) 34.

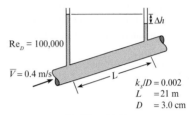

Problem 10.30

10.31 Points A and B are 3 mi apart along a 24-in. new cast-iron pipe carrying water ($T = 50°F$). Point A is 30 ft higher than B. The pressure at B is 20 psi greater than that at A. Determine the direction and rate of flow.

10.32 A train travels through a tunnel as shown. The train and tunnel are circular in cross section. Clearance is small, causing all air (60°F) to be pushed from the front of the train and discharged from the tunnel. The tunnel is 10 ft in diameter and is concrete. The train speed is 50 fps. Assume the concrete is very rough ($k_s = 0.05$ ft).

   a. Determine the change in pressure between the front and rear of the train.

   b. Sketch the energy and hydraulic grade lines for the train position shown.

   c. What power is required to produce the air flow in the tunnel?

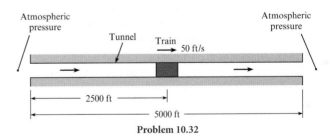

Problem 10.32

10.33 The house shown is flooded by a broken waterline. The owners siphon water out of the basement window and down the hill, with one hose, of length L, and thus an elevation difference of h to drive the siphon. Water drains from the siphon, but too slowly for the desperate home owners. They reason that with a larger head difference, they can generate more flow. So they get another hose, same length as the first, and connect the 2 hoses for total length $2L$. The backyard has a constant slope, so that a hose length of $2L$ correlates to an elevation difference of $2h$.

a. Assume no head loss, and calculate whether the flow rate doubles when the hose length is doubled from Case 1 (length L and height h) to Case 2 (length $2L$ and height $2h$).

b. Assume $h_L = 0.025(L/D)(V^2/2g)$, and calculate the flow rate for Cases 1 and 2, where $D = 1$ in., $L = 50$ ft., and $h = 20$ ft. How much of an improvement in flow rate is accomplished in Case 2 as compared to Case 1?

c. Both the husband and wife of this couple took fluid mechanics in college. They review with new appreciation the energy equation and the form of the head loss term and realize that they should use a larger diameter hose. Calculate the flow rate for Case 3, where $L = 50$ ft, $h = 20$ ft, and $D = 2.5$ in. Use the same expression for h_L as in part (b). How much of an improvement in flow rate is accomplished in Case 3 as compared to Case 1 in part (b)?

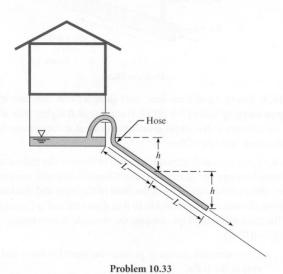

Problem 10.33

Solving Turbulent Flow Problems (§10.7)

10.34 Air (40°C, 1 atm) will be transported in a straight horizontal copper tube over a distance of 150 m at a rate of 0.1 m³/s. If the pressure drop in the tube should not exceed 6 in H₂O, what is the minimum pipe diameter?

10.35 Determine the diameter of commercial steel pipe required to convey 300 cfs of water at 60°F with a head loss of 1 ft per 1000 ft of pipe. Assume pipes are available in the even sizes when the diameters are expressed in inches (that is, 10 in., 12 in., etc.).

Combined Head Loss in Systems (§10.8)

10.36 Use Table 10.5 (§10.8) to select loss coefficients, K, for the following transitions and fittings.

a. A 90° smooth bend with $r/d = 2$

b. A threaded pipe 90° elbow

c. A gate valve, wide open

d. A pipe entrance with r/d of 0.3

e. An abrupt contraction, with $\theta = 180°$, and $D_2/D_1 = 0.60$

10.37 The sketch shows a test of an electrostatic air filter. The pressure drop for the filter is 4 inches of water when the airspeed is 9 m/s. What is the minor loss coefficient for the filter? Assume air properties at 20°C.

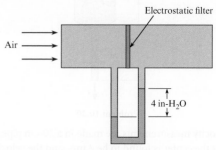

Problem 10.37

10.38 Water flows out of a reservoir, through a penstock, and then through a turbine. Calculate the total head loss in units of meters. The mean velocity is 4.3 m/s. The friction factor is 0.02. The total penstock length is 30 m and the diameter is 0.3 m. There are three minor loss coefficients: 0.5 for the penstock entrance, 0.5 for the bends in the penstock, and 1.0 for the exit. Choose the closest answer (m): (a) 1.2, (b) 2.8, (c) 3.8, (d) 4.8, (e) 5.7.

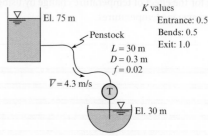

Problem 10.38

10.39 A liquid flows upward through a valve situated in a vertical pipe. Calculate the differential pressure (kPa) between points A and B. The mean velocity of the flow is 1.0 m/s. The specific gravity of the liquid is 1.2. The pipe has constant diameter. The valve has a minor loss coefficient of 4.0. Assume that major losses (i.e., head losses in the pipe itself) can be neglected. Point A is located 3.2 meters below point B. Choose the closest answer (kPa): (a) 3.4, (b) 6.6, (c) 40, (d) 65, (e) 78.

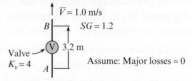

Problem 10.39

10.40 A cast-iron pipe 1.0 ft in diameter and 200 ft long joins two water (60°F) reservoirs. The upper reservoir has a water-surface elevation of 150 ft, and the lower on has a water-surface elevation of 40 ft. The pipe exits from the side of the upper reservoir at an elevation of 120 ft and enters the lower reservoir at an elevation of 30 ft. There are two wide-open gate valves in the pipe. (a) List all sources of h_L and the quantitative factors associated with each.

(b) Draw the EGL and the HGL for the system, and (c) determine the discharge in the pipe.

10.41 The water-surface elevation in a reservoir is 150 ft. A straight pipe 100 ft long and 6 in. in diameter conveys water from the reservoir to an open drain. The pipe entrance (it is abrupt) is at elevation 100 ft, and the pipe outlet is at elevation 60 ft. At the outlet the water discharges freely into the air. The water temperature is 50°F. If the pipe is asphalted cast iron, what will be the discharge rate in the pipe? Consider all head losses. Also draw the HGL and the EGL for this system.

10.42 The heat exchanger shown consists of 10 m of drawn tubing 2 cm in diameter with 19 return bends. The flow rate is 3×10^{-4} m³/s. Water enters at 20°C and exits at 80°C. The elevation difference between the entrance and the exit is 0.8 m. Calculate the pump power required to operate the heat exchanger if the pressure at 1 equals the pressure at 2. Use the viscosity corresponding to the average temperature in the heat exchanger.

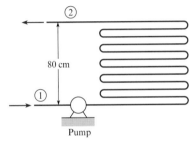

Problem 10.42

10.43 The pressure at a water main is 350 kPa gage. What size of pipe is needed to carry water from the main at a rate of 0.025 m³/s to a factory that is 160 m from the main? Assume that galvanized-steel pipe is to be used and that the pressure required at the factory is 70 kPa gage at a point 8 m above the main connection.

10.44 A tank and piping system are shown. The galvanized pipe diameter is 2 cm, and the total length of pipe is 10 m. The two 90° elbows are threaded fittings. The vertical distance from the water surface to the pipe outlet is 5 m. Find (a) the exit velocity of the water and (b) the height (h) the water jet would rise on exiting the pipe. The water temperature is 20°C.

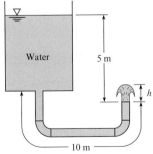

Problem 10.44

10.45 A water turbine is connected to a reservoir as shown. The flow rate in this system is 10 cfs. What power can be delivered by the turbine if its efficiency is 90%? Assume a temperature of 70°F.

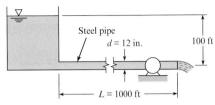

Problem 10.45

10.46 What power must the pump supply to the oil to pump the oil from the lower reservoir to the upper reservoir at a rate of 1.0 m³/s? Sketch the HGL and the EGL for the system.

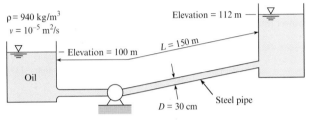

Problem 10.46

10.47 Water is flowing through a gate valve ($K_v = 0.2$). Calculate the value of b in units of mm. The pipe is horizontal. The flow is steady and fully developed. Over a 2.1-meter pipe length upstream of the valve, the EGL drops by $a = 430$ mm. The pipe ID is 0.25 m and the friction factor in the pipe is 0.03. Choose the closest answer (mm): (a) 180, (b) 240, (c) 320, (d) 340, (e) 360.

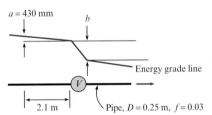

Problem 10.47

10.48 Water flows through a turbine. Calculate the power (MW) transmitted by the output shaft of the turbine. The water density is 1000 kg/m³. The elevation of the upper reservoir surface is 720 m and that of the lower reservoir is 695 m. The pipeline diameter is 2.2 m, the total length is 50 m, and the mean velocity is 3.6 m/s. The friction factor is 0.25. The sum of minor loss coefficients is 4.7. The turbine efficiency is 80%. Choose the closest answer (MW): (a) 1.6, (b) 1.7, (c) 2.0, (d) 2.1, (e) 2.5.

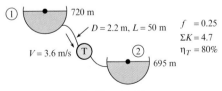

Problem 10.48

10.49 A heat exchanger consists of a closed system with a series of parallel tubes connected by 180° elbows as shown in the figure. There are a total of 14 return elbows. The pipe diameter is 1.5 cm, and the total pipe length is 10 m. The head loss coefficient for each return elbow is 2.2. The tube is copper. Water with an average temperature of 40°C flows through the system with a mean velocity of 10 m/s. Find the power required to operate the pump if the pump is 85% efficient.

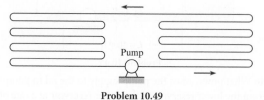

Problem 10.49

10.50 Water is pumped at a rate of 15 m³/s from the lower reservoir and out through the pipe, which has a diameter of 1.50 m. What power must be supplied to the water to produce this discharge?

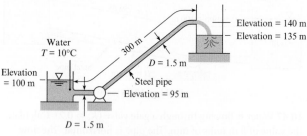

Problem 10.50

10.51 The steel pipe shown carries water from the main pipe A to the reservoir and is 2 in. in diameter and 275 ft long. What must be the pressure in pipe A to provide a flow of 70 gpm?

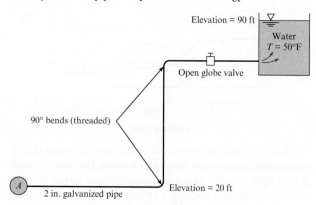

Problem 10.51

Nonround Conduits (§10.9)

10.52 The cross section of an air duct has the dimensions shown in the sketch. Find the hydraulic diameter in units of cm. Choose the closest answer (cm): (a) 0.9, (b) 2.4, (c) 3.2, (d) 6.4, (e) 7.1.

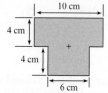

Problem 10.52

10.53 A cold-air duct 100 cm by 15 cm in cross section is 30 m long and made of galvanized iron. This duct is to carry air at a

rate of 6 m³/s at a temperature of 15°C and atmospheric pressure. What is the power loss in the duct?

10.54 An air-conditioning system is designed to have a duct with a rectangular cross section 1 ft by 2 ft, as shown. During construction, a truck driver backed into the duct and made it a trapezoidal section, as shown. The contractor, behind schedule, installed it anyway. For the same pressure drop along the pipe, what will be the ratio of the velocity in the trapezoidal duct to that in the rectangular duct? Assume the Darcy-Weisbach resistance coefficient is the same for both ducts.

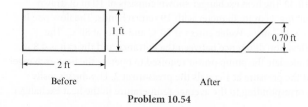

Problem 10.54

Modeling Pumps in Systems (§10.10)

10.55 What power must be supplied by the pump to the flow if water (T = 20°C) is pumped through the 300-mm steel pipe from the lower tank to the upper one at a rate of 0.55 m³/s?

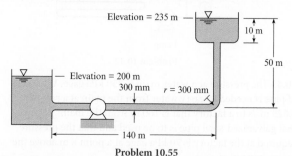

Problem 10.55

10.56 A pump with the characteristic curve shown in the accompanying graph is to be installed in the system shown. If head loss is neglected, what will be the flow rate?

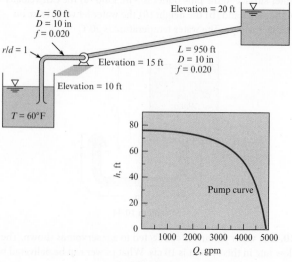

Problem 10.56

Pipes in Parallel and in Networks (§10.10)

10.57 Pipes 1 and 2 are the same material (cast-iron), and pipe 2 is three times as long as pipe 1. They are the same diameter (1 ft). If the discharge of water in pipe 2 is 2.0 cfs, then what will be the discharge in pipe 1? Assume the same value of f in both pipes.

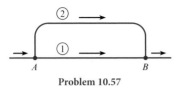

Problem 10.57

10.58 Water flows from left to right in this parallel pipe system. The water having the greatest velocity is in (a) pipe A, (b) pipe B, or (c) pipe C.

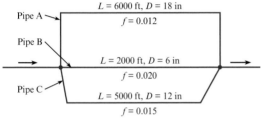

Problem 10.58

10.59 With a total flow of 14 cfs, determine the division of flow and the head loss from A to B.

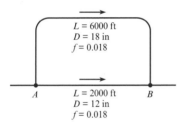

Problem 10.59

10.60 A parallel pipe system is set up as shown. Flow occurs from A to B. To augment the flow, a pump having the characteristics shown in Fig. 10.20b is installed at point C. For a total discharge of 0.60 m³/s, what will be the division of flow between the pipes and what will be the head loss between A and B? Assume commercial steel pipe.

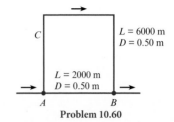

Problem 10.60

10.61 For the given source and loads shown, how will the flow be distributed in the simple network, and what will be the pressures at the load points if the pressure at the source is 60 psi? Assume horizontal pipes and $f = 0.012$ for all pipes.

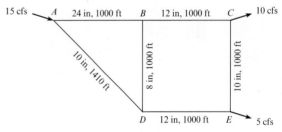

Problem 10.61

CHAPTER**ELEVEN**
Drag and Lift

CHAPTER ROAD MAP Previous chapters have described the hydrostatic force on a panel, the buoyant force on a submerged object, and the shear force on a flat plate. This chapter expands this list by introducing the lift and drag forces.

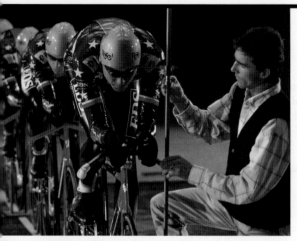

FIGURE 11.1
This photo shows the USA Olympic pursuit team being tested so that aerodynamic drag can be reduced. This wind tunnel is located at the General Motors Tech Center in Warren, Michigan. (Andy Sacks/Photodisc/ Getty Images.)

LEARNING OUTCOMES

UNDERSTANDING DRAG FORCE (§11.1, §11.2).
● Define drag.
● Explain how drag is related to the shear stress and pressure distributions.
● Define form drag and friction drag.
● For flow over a circular cylinder, describe the three drag regimes and the drag crisis.

CALCULATING DRAG FORCE (§11.2 to §11.4).
● Define the coefficient of drag.
● Find C_D values.
● Calculate the drag force.
● Calculate the power required to overcome drag.
● Solve terminal velocity problems.

UNDERSTANDING AND CALCULATING LIFT FORCE (§11.1, §11.8).
● Define lift and the coefficient of lift.
● Calculate the lift force.

When a body moves through a stationary fluid or when a fluid flows past a body, the fluid exerts a resultant force. The component of this resultant force that is parallel to the free-stream velocity is called the **drag force**. Similarly, the **lift force** is the component of the resultant force that is perpendicular to the free stream. For example, as air flows over a kite, it creates a resultant force that can be resolved in lift and drag components, as shown in Fig. 11.2. By definition, lift and drag forces are limited to those forces produced by a flowing fluid.

11.1 Relating Lift and Drag to Stress Distributions

This section explains how lift and drag forces are related to stress distributions. This section also introduces the concepts of form and friction drag. These ideas are fundamental to understanding lift and drag.

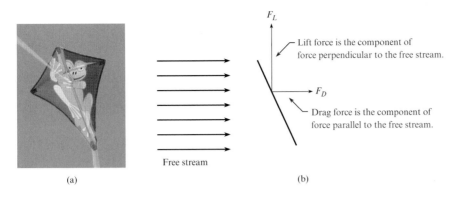

FIGURE 11.2

(a) A kite. (Photo by Donald Elger.)
(b) Forces acting on the kite due to the air flowing over the kite.

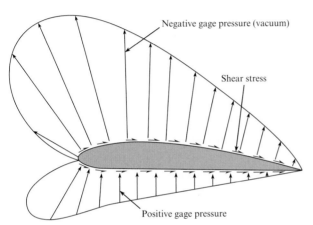

FIGURE 11.3

Pressure and shear stress acting on an airfoil.

Integrating a Stress Distribution to Yield Force

This section applies the ideas of §2.4 to develop equations for the lift and drag forces. Lift and drag forces are related to the stress distribution on a body through integration. For example, consider the stress acting on the airfoil shown in Fig. 11.3. As shown, there is a pressure distribution and a shear stress distribution. To relate stress to force, select a differential area as shown in Fig. 11.4. The magnitude of the pressure force is $dF_p = p\,dA$, and the magnitude of the viscous force is $dF_v = \tau\,dA$.* The differential lift force is normal to the free stream direction,

$$dF_L = -p\,dA\sin\theta - \tau\,dA\cos\theta$$

and the differential drag is parallel to the free stream direction,

$$dF_D = -p\,dA\cos\theta + \tau\,dA\sin\theta$$

Integration over the surface of the airfoil gives the lift force (F_L) and the drag force (F_D) in terms of the stress distribution:

$$F_L = \int (-p\sin\theta - \tau\cos\theta)\,dA \qquad \textbf{(11.1)}$$

$$F_D = \int (-p\cos\theta + \tau\sin\theta)\,dA \qquad \textbf{(11.2)}$$

*The sign convention on τ is such that a clockwise sense of $\tau\,dA$ on the surface of the foil signifies a positive sign for τ.

FIGURE 11.4

Pressure and viscous forces acting on a differential element of area.

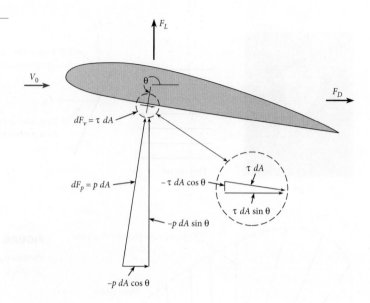

Form Drag and Friction Drag

Notice that Eq. (11.2) can be written as the sum of two integrals:

$$F_D = \underbrace{\int (-p \cos \theta)\, dA}_{\text{form drag}} + \underbrace{\int (\tau \sin \theta)\, dA}_{\text{friction drag}} \tag{11.3}$$

Form drag is the portion of the total drag force that is associated with the pressure distribution. **Friction drag** is the portion of the total drag force that is associated with the viscous shear stress distribution. The drag force on any body is the sum of form drag and friction drag. In words, Eq. (11.3) can be written as

$$(\text{total drag force}) = (\text{form drag}) + (\text{friction drag}) \tag{11.4}$$

11.2 Calculating the Drag Force

This section introduces the drag force equation, the coefficient of drag, and presents data for two-dimensional bodies. This information is used to calculate drag force on objects.

Drag Force Equation

The drag force F_D on a body is found by using the drag force equation:

$$F_D = C_D A_{\text{Ref}} \left(\frac{\rho V_0^2}{2} \right) \tag{11.5}$$

where C_D is called the coefficient of drag, A is a reference area of the body, ρ is the fluid density, and V_0 is the free stream velocity measured relative to the body.

The reference area A depends on the type of body. One common reference area, called **projected area** and given the symbol A_p, is the silhouetted area that would be seen by a person looking at the body from the direction of flow. For example, the projected area of a plate normal to the flow is $b\ell$, and the projected area of a cylinder with its axis normal to the flow is $d\ell$. Other geometries use different reference areas; for example, the reference area for an airplane wing is the planform area, which is the area observed when the wing is viewed from above.

The **coefficient of drag** C_D is a parameter that characterizes the drag force associated with a given body shape. For example, an airplane might have $C_D = 0.03$, and a baseball might have $C_D = 0.4$. The coefficient of drag is a π-group that is defined by

$$C_D \equiv \frac{F_D}{A_{\text{Ref}}(\rho V_0^2/2)} = \frac{\text{(drag force)}}{\text{(reference area)}\,\text{(kinetic pressure)}} \quad \textbf{(11.6)}$$

Values of the coefficient of drag C_D are usually found by experiment. For example, drag force F_D can be measured using a force balance in a wind tunnel. Then C_D can be calculated using Eq. (11.6). For this calculation, the speed of the air in the wind tunnel V_0 can be measured using a Pitot-static tube or similar device, and air density can be calculated by applying the ideal gas law using measured values of temperature and pressure.

Equation (11.5) shows that drag force is related to four variables. Drag is related to the shape of an object because shape is characterized by the value of C_D. Drag is related to the size of the object because size is characterized by the reference area. Drag is related to the density of ambient fluid. Finally, drag is related to the speed of the fluid squared. This means that if the wind velocity doubles and C_D is constant, then the wind load on a building goes up by a factor of four.

Coefficient of Drag (Two-Dimensional Bodies)

This section presents C_D data and describes how C_D varies with the Reynolds number for objects that can be classified as two dimensional. A **two-dimensional body** is a body with a uniform section area and a flow pattern that is independent of the ends of the body. Examples of two-dimensional bodies are shown in Fig. 11.5. In the aerodynamics literature, C_D values for

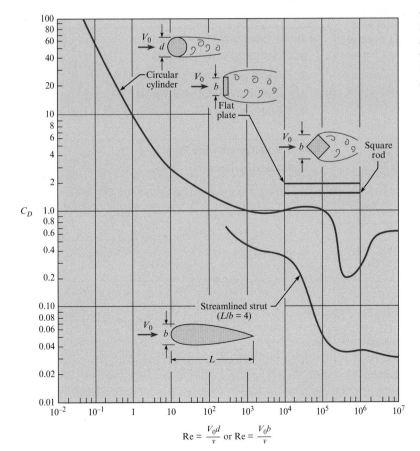

FIGURE 11.5

Coefficient of drag versus Reynolds number for two-dimensional bodies. [Data sources: Bullivant (1), DeFoe (2), Goett and Bullivant (3), Jacobs (4), Jones (5), and Lindsey (6).]

two-dimensional bodies are called **sectional drag coefficients**. Two-dimensional bodies can be visualized as objects that are infinitely long in the direction normal to the flow.

The sectional drag coefficient can be used to estimate C_D for real objects. For example, C_D for a cylinder with a length to diameter ratio of 20 (e.g., $L/D \geq 20$) approaches the sectional drag coefficient because the end effects have an insignificant contribution to the total drag force. Alternatively, the sectional drag coefficient would be inaccurate for a cylinder with a small L/D ratio (e.g., $L/D \approx 1$) because the end effects would be important.

As shown in Fig. 11.5, the Reynolds number sometimes, but not always, influences the sectional drag coefficient. The values of C_D for the flat plate and the square rod are independent of Re. The sharp edges of these bodies produce flow separation, and the drag force is due to the pressure distribution (form drag) and not the shear stress distribution (friction drag, which depends on Re). Alternatively, C_D values for the cylinder and the streamlined strut show strong Re dependence because both form and friction drag are significant.

To calculate drag force on an object, find a suitable coefficient of drag, and then apply the drag force equation. This approach is illustrated by Example 11.1.

EXAMPLE 11.1

Drag Force on a Cylinder

Problem Statement

A vertical cylinder that is 30 m high and 30 cm in diameter is being used to support a television-transmitting antenna. Find the drag force acting on the cylinder and the bending moment at its base. The wind speed is 35 m/s, the air pressure is 1 atm, and temperature is 20°C.

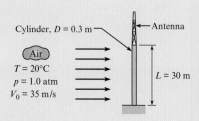

Define the Situation

Wind is blowing across a tall cylinder.

Assumptions:
- Wind speed is steady.
- Effects associated with the ends of the cylinder are negligible because $L/D = 100$.
- Neglect drag force on the antenna because the frontal area is much less than the frontal area of the cylinder.
- The line of action of the drag force is at an elevation of 15 m.

Properties: Air (20°C): Table A.5, $\rho = 1.2$ kg/m³, and $\mu = 1.81 \times 10^{-5}$ N·s/m²

State the Goals

Calculate:
- Drag force (in N) on the cylinder
- Bending moment (in N·m) at the base of the cylinder

Generate Ideas and Make a Plan

1. Calculate the Reynolds number.
2. Find coefficient of drag using Fig. 11.5.
3. Calculate drag force using Eq. (11.5).
4. Calculate bending moment using $M = F_D \cdot L/2$.

Take Action (Execute the Plan)

1. Reynolds number:

$$\text{Re}_D = \frac{V_0 D \rho}{\mu} = \frac{35 \text{ m/s} \times 0.30 \text{ m} \times 1.20 \text{ kg/m}^3}{1.81 \times 10^{-5} \text{ N} \cdot \text{s/m}^2} = 7.0 \times 10^5$$

2. From Fig. 11.5, the coefficient of drag is $C_D = 0.20$.

3. Drag force:

$$F_D = \frac{C_D A_p \rho V_0^2}{2}$$

$$= \frac{(0.2)(30 \text{ m})(0.3 \text{ m})(1.20 \text{ kg/m}^3)(35^2 \text{ m}^2/\text{s}^2)}{2}$$

$$= \boxed{1323 \text{ N}}$$

4. Moment at the base:

$$M = F_D \left(\frac{L}{2}\right) = (1323 \text{ N})\left(\frac{30}{2} \text{ m}\right) = \boxed{19,800 \text{ N} \cdot \text{m}}$$

FIGURE 11.6

Flow pattern around a cylinder for $10^3 < \mathrm{Re} < 10^5$.

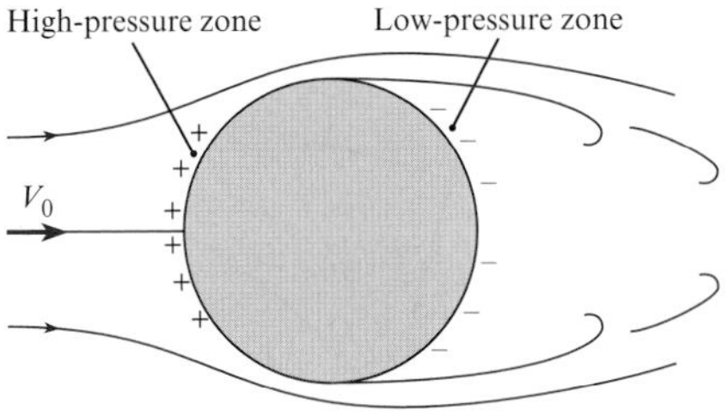

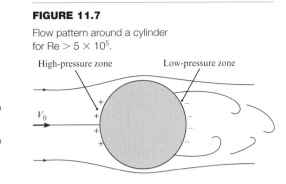

High-pressure zone   Low-pressure zone

V_0

FIGURE 11.7

Flow pattern around a cylinder for $\mathrm{Re} > 5 \times 10^5$.

High-pressure zone   Low-pressure zone

V_0

Discussion of C_D for a Circular Cylinder

Drag Regimes The coefficient of drag C_D, as shown in Fig. 11.5, can be described in terms of three regimes.

- **Regime I (Re < 10^3).** In this regime, C_D depends on both form drag and friction drag. As shown, C_D decreases with increasing Re.
- **Regime II ($10^3 < $ Re $ < 10^5$).** In this regime, C_D has a nearly constant value. The reason is that form drag, which is associated with the pressure distribution, is the dominant cause of drag. Over this range of Reynolds numbers, the flow pattern around the cylinder remains virtually unchanged, thereby producing very similar pressure distributions. This characteristic, the constancy of C_D at high values of Re, is representative of most bodies that have angular form.
- **Regime III ($10^5 < $ Re $ < 5 \times 10^5$).** In this regime, C_D decreases by about 80%, a remarkable change that is called the **drag crisis**. The drag crisis occurs because the boundary layer on the circular cylinder changes. For Reynolds numbers less than 10^5, the boundary layer is laminar, and separation occurs about midway between the upstream side and downstream side of the cylinder (Fig. 11.6). Hence, the entire downstream half of the cylinder is exposed to a relatively low pressure, which in turn produces a relatively high value for C_D. When the Reynolds number is increased to about 10^5, the boundary layer becomes turbulent, which causes higher-velocity fluid to be mixed into the region close to the wall of the cylinder. As a consequence of the presence of this high-velocity, high-momentum fluid in the boundary layer, the flow proceeds farther downstream along the surface of the cylinder against the adverse pressure before separation occurs (Fig. 11.7). This change in separation produces a much smaller zone of low pressure and the lower value of C_D.

Surface Roughness

Surface roughness has a major influence on drag. For example, if the surface of the cylinder is slightly roughened upstream of the midsection, then the boundary layer will be forced to become turbulent at lower Reynolds numbers than those for a smooth cylinder surface. The same trend can also be produced by creating abnormal turbulence in the approach flow. The effects of roughness are shown in Fig. 11.8 for cylinders that were roughened with sand grains of size k. A small to medium size of roughness ($10^{-3} < k/d < 10^{-2}$) on a cylinder triggers an early onset of reduction of C_D. However, when the relative roughness is quite large ($10^{-2} < k/d$), the characteristic dip in C_D is absent.

11.3 Drag of Axisymmetric and 3-D Bodies

Section 11.2 described drag for two-dimensional bodies. Drag on other body shapes is presented in this section. This section also describes power and rolling resistance.

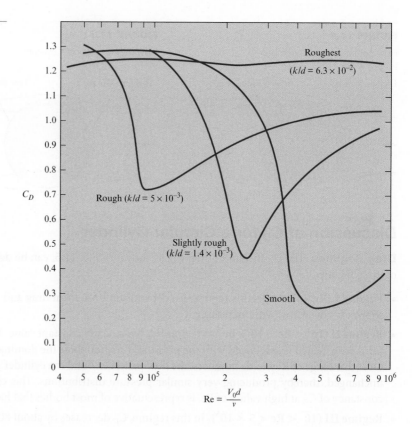

Drag Data

An object is classified as an **axisymmetric body** when the flow direction is parallel to an axis of symmetry of the body and the resulting flow is also symmetric about its axis. Examples of axisymmetric bodies include a sphere, a bullet, and a javelin. When flow is not aligned with an axis of symmetry, the flow field is three-dimensional (3-D), and the body is classified as a **3-D body**. Examples of 3-D bodies include a tree, a building, and an automobile.

The principles that apply to two-dimensional flow over a body also apply to axisymmetric flows. For example, at very low values of the Reynolds number, the coefficient of drag is given by exact equations relating C_D and Re. At high values of Re, the coefficient of drag becomes constant for angular bodies, whereas rather abrupt changes in C_D occur for rounded bodies. All these characteristics can be seen in Fig. 11.9, where C_D is plotted against Re for several axisymmetric bodies.

The drag coefficient of a sphere is of special interest because many applications involve the drag of spherical or near-spherical objects, such as particles and droplets. Also, the drag of a sphere is often used as a standard of comparison for other shapes. For Reynolds numbers less than 0.5, the flow around the sphere is laminar and amenable to analytical solutions. An exact solution by Stokes yielded the following equation, which is called Stokes's equation, for the drag of a sphere:

$$F_D = 3\pi\mu V_0 d \tag{11.7}$$

Note that the drag for this laminar flow condition varies directly with the first power of V_0. This is characteristic of all laminar flow processes. For completely turbulent flow, the drag is a function of the velocity to the second power. When the drag force given by Eq. (11.7) is substituted into Eq. (11.6), the result is the drag coefficient corresponding to Stokes's equation:

$$C_D = \frac{24}{\text{Re}} \tag{11.8}$$

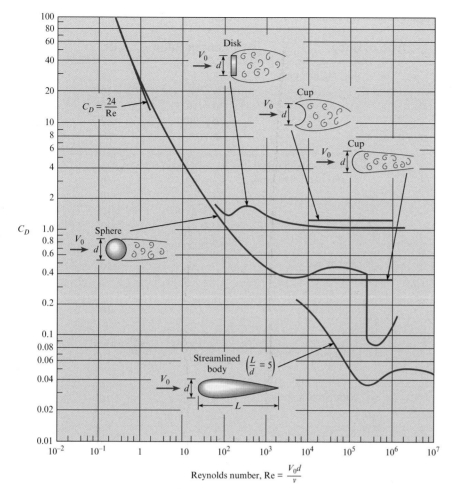

FIGURE 11.9

Coefficient of drag versus Reynolds number for axisymmetric bodies. [Data sources: Abbott (9), Brevoort and Joyner (10), Freeman (11), and Rouse (12).]

Thus for flow past a sphere, when $\text{Re} \leq 0.5$, one may use the direct relation for C_D given in Eq. (11.8).

Several correlations for the drag coefficient of a sphere are available (13). One such correlation has been proposed by Clift and Gauvin (14),

$$C_D = \frac{24}{\text{Re}}(1 + 0.15\text{Re}^{0.687}) + \frac{0.42}{1 + 4.25 \times 10^4 \text{Re}^{-1.16}} \qquad \textbf{(11.9)}$$

which deviates from the *standard drag curve*[*] by -4% to 6% for Reynolds numbers up to 3×10^5. Note that as the Reynolds number approaches zero, this correlation reduces to the equation for Stokes flow.

Values for C_D for other axisymmetric and 3-D bodies at high Reynolds numbers ($\text{Re} > 10^4$) are given in Table 11.1. Extensive data on the drag of various shapes are available in Hoerner (15).

To find the drag force on an object, find or estimate the coefficient of drag and then apply the drag force equation. This approach is illustrated by Example 11.2.

---

[*]The *standard drag curve* represents the best fit of the cumulative data that have been obtained for drag coefficient of a sphere.

TABLE 11.1 Approximate C_D Values for Various Bodies

Type of Body		Length Ratio	Re	C_D
	Rectangular plate	$l/b = 1$	$>10^4$	1.18
		$l/b = 5$	$>10^4$	1.20
		$l/b = 10$	$>10^4$	1.30
		$l/b = 20$	$>10^4$	1.50
		$l/b = \infty$	$>10^4$	1.98
	Circular cylinder with axis parallel to flow	$l/d = 0$ (disk)	$>10^4$	1.17
		$l/d = 0.5$	$>10^4$	1.15
		$l/d = 1$	$>10^4$	0.90
		$l/d = 2$	$>10^4$	0.85
		$l/d = 4$	$>10^4$	0.87
		$l/d = 8$	$>10^4$	0.99
	Square rod	∞	$>10^4$	2.00
	Square rod	∞	$>10^4$	1.50
	Triangular cylinder	∞	$>10^4$	1.39
	Semicircular shell	∞	$>10^4$	1.20
	Semicircular shell	∞	$>10^4$	2.30
	Hemispherical shell		$>10^4$	0.39
	Hemispherical shell		$>10^4$	1.40
	Cube		$>10^4$	1.10
	Cube		$>10^4$	0.81
	Cone—60° vertex		$>10^4$	0.49
	Parachute		$\approx 3 \times 10^7$	1.20

Sources: Brevoort and Joyner (10), Lindsey (6), Morrison (16), Roberson et al. (17), Rouse (12), and Scher and Gale (18).

EXAMPLE 11.2

Drag on a Sphere

Problem Statement

What is the drag of a 12 mm sphere that drops at a rate of 8 cm/s in oil ($\mu = 10^{-1}$ N·s/m², $SG = 0.85$)?

Define the Situation

A sphere ($d = 0.012$ m) is falling in oil.
Speed of the sphere is $V = 0.08$ m/s.

Assumptions: The sphere is moving at a steady speed (terminal velocity).

Properties:
Oil: $\mu = 10^{-1}$ N·s/m², $S = 0.85$, $\rho = 850$ kg/m³

State the Goal

Find: Drag force (in newtons) on the sphere.

Generate Ideas and Make a Plan

1. Calculate the Reynolds number.
2. Find the coefficient of drag using Fig. 11.9.
3. Calculate drag force using Eq. (11.5).

Take Action (Execute the Plan)

1. Reynolds number:

$$\text{Re} = \frac{Vd\rho}{\mu} = \frac{(0.08 \text{ m/s})(0.012 \text{ m})(850 \text{ kg/m}^3)}{10^{-1} \text{ N} \cdot \text{s/m}^2} = 8.16$$

2. Coefficient of drag (from Fig. 11.9) is $C_D = 5.3$.
3. Drag force:

$$F_D = \frac{C_D A_p \rho V_0^2}{2}$$

$$F_D = \frac{(5.3)(\pi/4)(0.012^2 \text{ m}^2)(850 \text{ kg/m}^3)(0.08 \text{ m/s})^2}{2}$$

$$= \boxed{1.63 \times 10^{-3} \text{ N}}$$

Power and Rolling Resistance

When power is involved in a problem, the power equation from Chapter 7 is applied. For example, consider a car moving at a steady speed on a level road. Because the car is not accelerating, the horizontal forces are balanced as shown in Fig. 11.10. Force equilibrium gives

$$F_{\text{Drive}} = F_{\text{Drag}} + F_{\text{Rolling resistance}}$$

The driving force (F_{Drive}) is the frictional force between the driving wheels and the road. The drag force is the resistance of the air on the car. The rolling resistance is the frictional force that occurs when an object such as a ball or tire rolls. It is related to the deformation and types of the materials that are in contact. For example, a rubber tire on asphalt will have a larger rolling resistance than a steel train wheel on a steel rail. The rolling resistance is calculated using

$$F_{\text{Rolling resistance}} = F_r = C_r N \qquad \textbf{(11.10)}$$

where C_r is the coefficient of rolling resistance and N is the normal force.

The power required to move the car shown in Fig. 11.10 at a constant speed is given by Eq. (7.2a):

$$P = FV = F_{\text{Drive}} V_{\text{Car}} = (F_{\text{Drag}} + F_{\text{Rolling resistance}}) V_{\text{Car}} \qquad \textbf{(11.11)}$$

Thus, when power is involved in a problem, apply the equation $P = FV$ while concurrently using a free body diagram to determine the appropriate force. This approach is illustrated in Example 11.3.

F_{Drag}

F_{Drive}

$F_{\text{Rolling resistance}}$

FIGURE 11.10

Horizontal forces acting on car that is moving at a steady speed.

EXAMPLE 11.3

Speed of a Bicycle Rider

Problem Statement

A bicyclist of mass 70 kg supplies 300 watts of power while riding into a 3 m/s headwind. The frontal area of the cyclist and bicycle together is 3.9 ft^2 = 0.362 m^2, the drag coefficient is 0.88, and the coefficient of rolling resistance is 0.007. Determine the speed V_c of the cyclist. Express your answer in mph and in m/s.

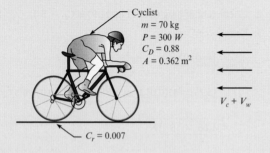

Cyclist
$m = 70$ kg
$P = 300\ W$
$C_D = 0.88$
$A = 0.362$ m^2

$V_c + V_w$

$C_r = 0.007$

Define the Situation

A bicyclist is pedaling into a headwind of magnitude $V_w = 3$ m/s.

Assumptions:

1. The path is level, with no hills.
2. Mechanical losses in the bike gear train are zero.

Properties: Air (20°C, 1 atm): Table A.2, $\rho = 1.2$ kg/m^3

State the Goal

Find the speed (m/s and mph) of the rider.

Generate Ideas and Make a Plan

1. Relate bike speed (V_c) to power using Eq. (11.11).
2. Calculate rolling resistance.
3. Develop an equation for drag force using Eq. (11.5).
4. Combine steps 1 to 3.
5. Solve for V_c.

Take Action (Execute the Plan)

1. Power equation:
   - The power from the bike rider is being used to overcome drag and rolling resistance. Thus,

   $$P = (F_D + F_r)V_c$$

2. Rolling resistance:

   $$F_r = C_r N = C_r mg = 0.007(70\ \text{kg})(9.81\ \text{m/s}^2) = 4.81\ \text{N}$$

3. Drag force:
   - V_0 = speed of the air relative to the bike rider

   $$V_0 = V_c + 3\ \text{m/s}$$

   - Drag force:

   $$F_D = C_D A\left(\frac{\rho V_0^2}{2}\right) = \frac{0.88(0.362\ \text{m}^2)(1.2\ \text{kg/m}^3)}{2}$$
   $$\times (V_c + 3\ \text{m/s})^2$$
   $$= 0.1911(V_c + 3\ \text{m/s})^2$$

4. Combine results:

   $$P = (F_D + F_r)V_c$$
   $$300\ \text{W} = (0.1911(V_c + 3)^2 + 4.81)V_c$$

5. Because the equation is cubic, use a spreadsheet program as shown. In this spreadsheet, let V_c vary, and then search for the value of V_c that causes the right side of the equation to equal 300. The result is

   $$V_c = \boxed{9.12\ \text{m/s} = 20.4\ \text{mph}}$$

V_c	RHS
(m/s)	(W)
0	0.0
5	85.2
8	223.5
9	291.0
9.1	298.4
9.11	299.1
9.12	299.9
9.13	300.6

11.4 Terminal Velocity

Another common application of the drag force equation is finding the steady state speed of a body that is falling through a fluid. When a body is dropped, it accelerates under the action of gravity. As the speed of the falling body increases, the drag increases until the upward force (drag) equals the net downward force (weight minus buoyant force). Once the forces are balanced, the body moves at a constant speed called the **terminal velocity**, which is identified as the maximum velocity attained by a falling body.

To find terminal velocity, balance the forces acting on the object, and then solve the resulting equation. In general, this process is iterative, as illustrated by Example 11.4.

EXAMPLE 11.4

Terminal Velocity of a Sphere in Water

Problem Statement

A 20 mm plastic sphere ($SG = 1.3$) is dropped in water. Determine its terminal velocity. Assume $T = 20°C$.

Define the Situation

A smooth sphere ($D = 0.02$ m, $SG = 1.3$) is falling in water.

Properties: Water (20°C): Table A.5, $\nu = 1 \times 10^{-6}$ m²/s, $\rho = 998$ kg/m³, and $\gamma = 9790$ N/m³

State the Goal

Find the terminal velocity (m/s) of the sphere.

Generate Ideas and Make a Plan

This problem requires an iterative solution because the terminal velocity equation is implicit. The plan steps are as follows:

1. Apply force equilibrium.
2. Develop an equation for terminal velocity.
3. To solve the terminal velocity equation, set up a procedure for iteration.
4. To implement the iterative solution, build a table in a spreadsheet program.

Take Action (Execute the Plan)

1. Force equilibrium:
   - Sketch a free body diagram.

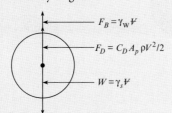

   - Apply force equilibrium (vertical direction):
$$F_{\text{Drag}} + F_{\text{Buoyancy}} = W$$

2. Terminal velocity equation:
   - Analyze terms in the equilibrium equation:
$$C_D A\left(\frac{\rho V_0^2}{2}\right) + \gamma_w \forall = \gamma_s \forall$$
$$C_D\left(\frac{\pi d^2}{4}\right)\left(\frac{\rho V_0^2}{2}\right) + \gamma_w\left(\frac{\pi d^3}{6}\right) = \gamma_s\left(\frac{\pi d^3}{6}\right)$$

- Solve for V_0:
$$V_0 = \left[\frac{(\gamma_s - \gamma_w)(4/3)d}{C_D \rho_w}\right]^{1/2}$$
$$= \left[\frac{(12.7 - 9.79)(10^3 \text{ N/m}^3)(4/3)(0.02 \text{ m})}{C_D \times 998 \text{ kg/m}^3}\right]^{1/2}$$
$$V_0 = \left(\frac{0.0778}{C_D}\right)^{1/2} = \frac{0.279}{C_D^{1/2}} \text{ m/s}$$

3. Iteration 1
   - Initial guess: $V_0 = 1.0$ m/s
   - Calculate Re:
$$\text{Re} = \frac{Vd}{\nu} = \frac{(1.0 \text{ m/s})(0.02 \text{ m})}{1 \times 10^{-6} \text{ m}^2/\text{s}} = 20000$$
   - Calculate C_D using Eq. (11.9):
$$C_D = \frac{24}{20000}(1 + 0.15(20000^{0.687}))$$
$$+ \frac{0.42}{1 + 4.25 \times 10^4 (20000)^{-1.16}} = 0.456$$
   - Find new value of V_0 (use equation from step 2):
$$V_0 = \left(\frac{0.0778}{C_D}\right)^{1/2} = \frac{0.279}{0.456^{0.5}} = 0.413 \text{ m/s}$$

4. Iterative solution
   - As shown, use a spreadsheet program to build a table. The first row shows the results of iteration 1.
   - The terminal velocity from iteration 1 $V_0 = 0.413$ m/s is used as the initial velocity for iteration 2.
   - The iteration process is repeated until the terminal velocity reaches a constant value of $V_0 = 0.44$ m/s. Notice that convergence is reached in two iterations.

Iteration #	Initial V_0	Re	C_D	New V_0
	(m/s)			(m/s)
1	1.000	20000	0.456	0.413
2	0.413	8264	0.406	0.438
3	0.438	8752	0.409	0.436
4	0.436	8721	0.409	0.436
5	0.436	8723	0.409	0.436
6	0.436	8722	0.409	0.436

$$\boxed{V_0 = 0.44 \text{ m/s}}$$

11.5 Vortex Shedding

This section introduces vortex shedding, which is important for two reasons: It can be used to enhance heat transfer and mixing, and it can cause unwanted vibrations and failures of structures.

Flow past a bluff body generally produces a series of vortices that are shed alternatively from each side, thereby producing a series of alternating vortices in the wake. This phenomenon is call **vortex shedding**. Vortex shedding for a cylinder occurs for Re $\gtrsim$ 50 and gives the flow pattern sketched in Fig. 11.11. In this figure, a vortex is in the process of formation near the top of the cylinder. Below and to the right of the first vortex is another vortex, which was formed and shed a short time before. Thus the flow process in the wake of a cylinder involves the formation and shedding of vortices alternately from one side and then the other. This alternate formation and shedding of vortices creates a cyclic change in pressure with consequent periodicity in side thrust on the cylinder. Vortex shedding was the primary cause of failure of the Tacoma Narrows suspension bridge in the state of Washington in 1940.

Experiments reveal that the frequency of shedding can be represented by plotting Strouhal number (St) as a function of Reynolds number. The Strouhal number is a π-group defined as

$$\text{St} = \frac{nd}{V_0} \tag{11.12}$$

where n is the frequency of shedding of vortices from one side of the cylinder, in Hz, d is the diameter of the cylinder, and V_0 is the free stream velocity. The Strouhal number for vortex shedding from a circular cylinder is given in Fig. 11.12. Other cylindrical and two-dimensional bodies also shed vortices. Consequently, the engineer should always be alert to vibration problems when designing structures that are exposed to wind or water flow.

FIGURE 11.11

Formation of a vortex behind a cylinder.

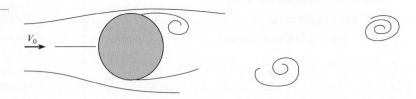

FIGURE 11.12

Strouhal number versus Reynolds number for flow past a circular cylinder. [After Jones (5) and Roshko (8).]

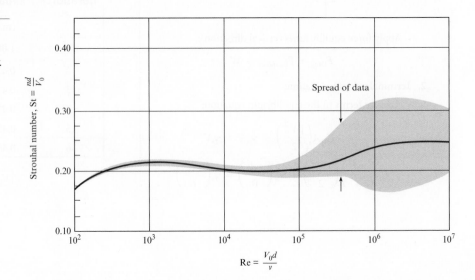

11.6 Reducing Drag by Streamlining

An engineer can design a body shape to minimize the drag force. This process is called **streamlining** and is often focused on reducing form drag. The reason for focusing on form drag is that drag on most bluff objects (e.g., a cylindrical body at Re > 1000) is predominantly due to the pressure variation associated with flow separation. In this case, streamlining involves modifying the body shape to reduce or eliminate separation. The impacts of streamlining can be dramatic. For example, Fig. 11.5 shows that C_D for the streamlined shape is about 1/6 of C_D for the circular cylinder when Re $\approx 5 \times 10^5$.

While streamlining reduces form drag, friction drag is typically increased. This is because there is more surface area on a streamlined body as compared to a nonstreamlined body. Consequently, when a body is streamlined, the optimum condition results when the sum of form drag and friction drag is minimum.

Streamlining to produce minimum drag at high Reynolds numbers will probably not produce minimum drag at very low Reynolds numbers. For example, at Re < 1, the majority of the drag of a cylinder is friction drag. Hence, if the cylinder is streamlined, the friction drag will likely be magnified, and C_D will increase.

Another advantage of streamlining at high Reynolds numbers is that vortex shedding is eliminated. Example 11.5 shows how to estimate the impact of streamlining by using a ratio of C_D values.

EXAMPLE 11.5

Comparing Drag on Bluff and Streamlined Shapes

Problem Statement

Compare the drag of the cylinder of Example 11.1 with the drag of the streamlined shape shown in Fig. 11.5. Assume that both shapes have the same projected area.

Define the Situation

The cylinder from Example 11.1 is being compared to a streamlined shape.

Assumptions:
1. The cylinder and the streamlined body have the same projected area.
2. Both objects are two-dimensional bodies (neglect end effects).

State the Goal

Find the ratio of drag force on the streamlined body to drag force on the cylinder.

Generate Ideas and Make a Plan

1. Retrieve Re and C_D from Example 11.1.

2. Find the coefficient of drag for the streamlined shape using Fig. 11.5.
3. Calculate the ratio of drag forces using Eq. (11.5).

Take Action (Execute the Plan)

1. From Example 11.1, Re = 7×10^5 and C_D(cylinder) = 0.2.
2. Using this Re and Fig. 11.5 gives C_D(streamlined shape) = 0.034.
3. Drag force ratio (derived from Eq. 11.5) is

$$\frac{F_D(\text{streamlined shape})}{F_D(\text{cylinder})} = \frac{C_D(\text{streamlined shape})}{C_D(\text{cylinder})}$$

$$\times \left(\frac{A_p(\rho V_0^2/2)}{A_p(\rho V_0^2/2)}\right)$$

$$\frac{F_D(\text{streamlined shape})}{F_D(\text{cylinder})} = \frac{0.034}{0.2} = \boxed{0.17}$$

Review the Results and the Process

Discussion. The streamlining provided nearly a sixfold reduction in drag!

11.7 Drag in Compressible Flow

So far, this chapter has described drag for flows with constant density. This section describes drag when the density of a gas is changing due to pressure variations. These types of flow are called *compressible flows*. This information is important for modeling of projectiles such as bullets and rockets.

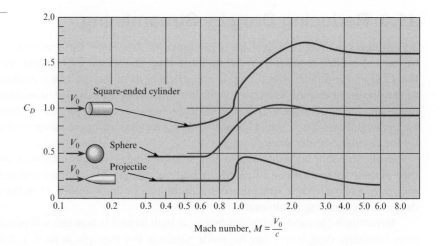

In steady flow, the influence of compressibility depends on the ratio of fluid velocity to the speed of sound. This ratio is a π-group called the Mach number.

The variation of drag coefficient with Mach number for three axisymmetric bodies is shown in Fig. 11.13. In each case, the drag coefficient increases only slightly with the Mach number at low Mach numbers and then increases sharply as transonic flow ($M \approx 1$) is approached. Note that the rapid increase in drag coefficient occurs at a higher Mach number (closer to unity) if the body is slender with a pointed nose. The drag coefficient reaches a maximum at a Mach number somewhat larger than unity and then decreases as the Mach number is further increased.

The slight increase in drag coefficient with low Mach numbers is attributed to an increase in form drag due to compressibility effects on the pressure distribution. However, as the flow velocity is increased, the maximum velocity on the body finally becomes sonic. The Mach number of the free stream flow at which sonic flow first appears on the body is called the *critical Mach number*. Further increases in flow velocity result in local regions of supersonic flow ($M > 1$), which lead to wave drag due to shock wave formation and an appreciable increase in drag coefficient.

The critical Mach number for a sphere is approximately 0.6. Note in Fig. 11.13 that the drag coefficient begins to rise sharply at about this Mach number. The critical Mach number for the pointed body is larger; correspondingly, the rise in drag coefficient occurs at a Mach number closer to unity.

The drag coefficient data for the sphere shown in Fig. 11.13 are for a Reynolds number of the order of 10^4. The data for the sphere shown in Fig. 11.9, on the other hand, are for very low Mach numbers. The question then arises about the general variation of the drag coefficient of a sphere with both Mach number and Reynolds number. Information of this nature is often needed to predict the trajectory of a body through the upper atmosphere or to model the motion of a nanoparticle.

A contour plot of the drag coefficient of a sphere versus both Reynolds and Mach numbers based on available data (19) is shown in Fig. 11.14. Notice the C_D-versus-Re curve from Fig. 11.9 in the $M = 0$ plane. Correspondingly, notice the C_D-versus-M curve from Fig. 11.13 in the $Re = 10^4$ plane. At low Reynolds numbers C_D decreases with an increasing Mach number, whereas at high Reynolds numbers the opposite trend is observed. Using this figure, the engineer can determine the drag coefficient of a sphere at any combination of Re and M. Of course, corresponding C_D contour plots can be generated for any body, provided the data are available.

11.8 The Theory of Lift

This section introduces circulation, the basic cause of lift, as well as the coefficient of lift.

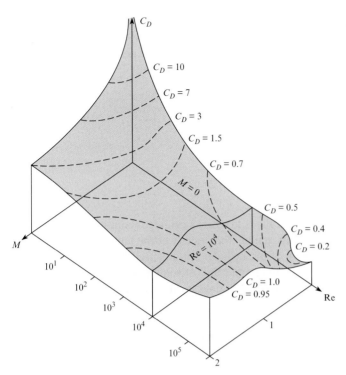

FIGURE 11.14

Contour plot of the drag coefficient of the sphere versus Reynolds and Mach numbers.

Circulation

Circulation, a characteristic of a flow field, gives a measure of the average rate of rotation of fluid particles that are situated in an area that is bounded by a closed curve. Circulation is defined by the path integral, as shown in Fig. 11.15. Along any differential segment of the path, the velocity can be resolved into components that are tangent and normal to the path. Signify the tangential component of velocity as V_L. Integrate $V_L \, dL$ around the curve. The resulting quantity is called circulation, which is represented by the Greek letter Γ (capital gamma). Hence,

$$\Gamma = \oint V_L \, dL \tag{11.13}$$

Sign convention dictates that in applying Eq. (11.13), one takes tangential velocity vectors that have a counterclockwise sense around the curve as negative and those that have

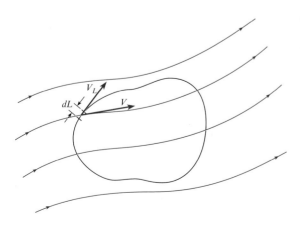

FIGURE 11.15

Concept of circulation.

a clockwise direction as having a positive contribution.* For example, consider finding the circulation for an irrotational vortex. The tangential velocity at any radius is C/r, where a positive C means a clockwise rotation. Therefore, if circulation is evaluated about a curve with radius r, the differential circulation is

$$d\Gamma = V_L \, dL = \frac{C}{r_1} r_1 \, d\theta = C \, d\theta \tag{11.14}$$

Integrate this around the entire circle:

$$\Gamma = \int_0^{2\pi} C \, d\theta = 2\pi C \tag{11.15}$$

One way to induce circulation physically is to rotate a cylinder about its axis. Fig. 11.16a shows the flow pattern produced by such action. The velocity of the fluid next to the surface of the cylinder is equal to the velocity of the cylinder surface itself because of the no-slip condition that must prevail between the fluid and solid. At some distance from the cylinder, however, the velocity decreases with r, much like it does for the irrotational vortex. The next section shows how circulation produces lift.

Combination of Circulation and Uniform Flow around a Cylinder

Superpose the velocity field produced for uniform flow around a cylinder, Fig. 11.16b, onto a velocity field with circulation around a cylinder, Fig. 11.16a. Observe that the velocity is reinforced on the top side of the cylinder and reduced on the other side (Fig. 11.16c). Also observe that the stagnation points have both moved toward the low-velocity side of the cylinder. Consistent with the Bernoulli equation (assuming irrotational flow throughout), the pressure on the high-velocity side is lower than the pressure on the low-velocity side. Hence, a pressure differential exists that

FIGURE 11.16

Ideal flow around a cylinder:
(a) circulation, (b) uniform flow, (c) combination of circulation and uniform flow.

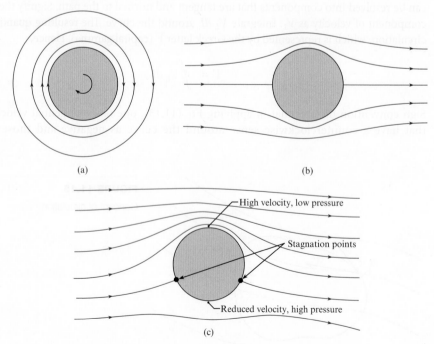

(a)　　　　　　　　　　　　　(b)

(c)

*The sign convention is the opposite of that for the mathematical definition of a line integral.

FIGURE 11.17

Coefficients of lift and drag for a rotating cylinder. [Data from Rouse (12).]

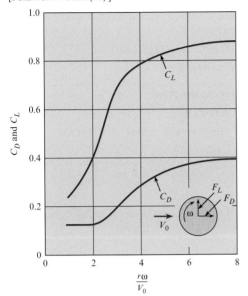

FIGURE 11.18

Coefficients of lift and drag for a rotating sphere. [Data from Barkla et al. (20).]

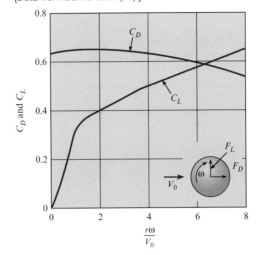

causes a side thrust, or lift, on the cylinder. According to ideal flow theory, the lift per unit length of an infinitely long cylinder is given by $F_L/\ell = \rho V_0 \Gamma$, where F_L is the lift on the segment of length ℓ. For this ideal irrotational flow, there is no drag on the cylinder. For the real flow case, separation and viscous stresses do produce drag, and the same viscous effects will reduce the lift somewhat. Even so, the lift is significant when flow occurs past a rotating body or when a body is translating and rotating through a fluid. Hence, the reason for the "curve" on a pitched baseball or the "drop" on a Ping-Pong ball is a fore spin. This phenomenon of lift produced by rotation of a solid body is called the **Magnus effect** after a nineteenth-century German scientist who made early studies of the lift on rotating bodies. A paper by Mehta (28) offers an interesting account of the motion of rotating sports balls.

Coefficients of lift and drag for the rotating cylinder with end plates are shown in Fig. 11.17. In this figure, the parameter $r\omega/V_0$ is the ratio of cylinder surface speed to the free stream velocity, where r is the radius of the cylinder and ω is the angular speed in radians per second. The corresponding curves for the rotating sphere are given in Fig. 11.18.

Coefficient of Lift

The **coefficient of lift** is a parameter that characterizes the lift that is associated with a body. For example, a wing at a high angle of attack will have a high coefficient of lift, and a wing that has a zero angle of attack will have a low or zero coefficient of lift. The coefficient of lift is defined using a π-group:

$$C_L \equiv \frac{F_L}{A(\rho V_0^2/2)} = \frac{\text{lift force}}{(\text{reference area})(\text{dynamic pressure})} \tag{11.16}$$

To calculate lift force, engineers use the lift equation:

$$F_L = C_L A \left(\frac{\rho V_0^2}{2} \right) \tag{11.17}$$

where the reference area for a rotating cylinder or sphere is the projected area A_p.

EXAMPLE 11.6

Lift on a Rotating Sphere

Problem Statement

A Ping-Pong ball is moving at 10 m/s in air and is spinning at 100 revolutions per second in the clockwise direction. The diameter of the ball is 3 cm. Calculate the lift and drag force and indicate the direction of the lift (up or down). The density of air is 1.2 kg/m^3.

Define the Situation

A Ping-Pong ball is moving horizontally and rotating.

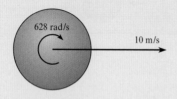

628 rad/s   10 m/s

Properties: Air: $\rho = 1.2$ kg/m^3

State the Goal

Find:

1. Drag force (in newtons) on the ball
2. Lift force (in newtons) on the ball
3. The direction of lift (up or down?)

Generate Ideas and Make a Plan

1. Calculate the value of $r\omega/V_0$.
2. Use the value of $r\omega/V_0$ to look up the coefficients of lift and drag on Fig. 11.7.
3. Calculate lift force using Eq. (11.8).
4. Calculate drag force using Eq. (11.5).

Take Action (Execute the Plan)

The rotation rate in rad/s is

$$\omega = (100 \text{ rev/s})(2\pi \text{ rad/rev}) = 628 \text{ rad/s}$$

The rotational parameter is

$$\frac{\omega r}{V_0} = \frac{(628 \text{ rad/s})(0.015 \text{ m})}{10 \text{ m/s}} = 0.942$$

From Fig. 11.18, the lift coefficient is approximately 0.26, and the drag coefficient is 0.64. The lift force is

$$F_L = \frac{1}{2}\rho V_0^2 C_L A_p$$

$$= \frac{1}{2}(1.2 \text{ kg/m}^3)(10 \text{ m/s})^2 (0.26)\frac{\pi}{4}(0.03 \text{ m})^2$$

$$= \boxed{1.10 \times 10^{-2} \text{ N}}$$

$\boxed{\text{The lift force is downward.}}$ The drag force is

$$F_D = \frac{1}{2}\rho V_0^2 C_D A_p$$

$$= \boxed{27.1 \times 10^{-3} \text{ N}}$$

11.9 Lift and Drag on Airfoils

This section presents information on how to calculate lift and drag on winglike objects. Some typical applications include calculating the takeoff weight of an airplane, determining the size of wings needed, and estimating power requirements to overcome drag force.

Lift of an Airfoil

An **airfoil** is a body designed to produce lift from the movement of fluid around it. Specifically, lift is a result of circulation in the flow produced by the airfoil. To see this, consider the flow of an ideal flow (nonviscous and incompressible) past an airfoil as shown in Fig. 11.19a. Here, as for irrotational flow past a cylinder, the lift and drag are zero. There is a stagnation point on the bottom side near the leading edge, and another on the top side near the trailing edge of the foil. In the real flow (viscous fluid) case, the flow pattern around the upstream half of the foil is plausible. However, the flow pattern in the region of the trailing edge, as shown in Fig. 11.19a, cannot occur. A stagnation point on the upper side of the foil indicates that fluid must flow from the lower side around the trailing edge and then toward the stagnation point. Such a flow pattern implies an infinite acceleration of the fluid particles as they turn the corner around the trailing edge of the wing. This is a physical impossibility, and separation occurs at the sharp edge. As a consequence of the separation, the downstream stagnation point moves

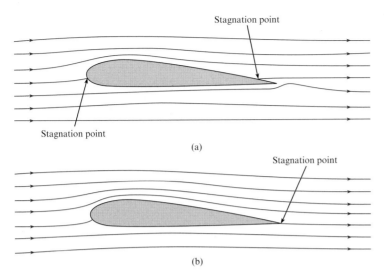

Stagnation point

Stagnation point

(a)

Stagnation point

(b)

FIGURE 11.19

Patterns of flow around an airfoil: (a) ideal flow—no circulation; (b) real flow—circulation.

to the trailing edge. Flow from both the top and bottom sides of the airfoil in the vicinity of the trailing edge then leaves the airfoil smoothly and essentially parallel to these surfaces at the trailing edge (Fig. 11.19b).

To bring theory into line with the physically observed phenomenon, it was hypothesized that a circulation around the airfoil must be induced in just the right amount so that the downstream stagnation point is moved all the way back to the trailing edge of the airfoil, thus allowing the flow to leave the airfoil smoothly at the trailing edge. This is called the *Kutta condition* (21), named after a pioneer in aerodynamic theory. When analyses are made with this simple assumption concerning the magnitude of the circulation, very good agreement occurs between theory and experiment for the flow pattern and the pressure distribution, as well as for the lift on a two-dimensional airfoil section (no end effects). Ideal flow theory then shows that the magnitude of the circulation required to maintain the rear stagnation point at the trailing edge (the Kutta condition) of a symmetric airfoil with a small angle of attack is given by

$$\Gamma = \pi c V_0 \alpha \qquad (11.18)$$

where Γ is the circulation, c is the chord length of the airfoil, and α is the angle of attack of the chord of the airfoil with the free stream direction (see Fig. 11.20 for a definition sketch).

Like that for the cylinder, the lift per unit length for an infinitely long wing is

$$F_L/\ell = \rho V_0 \Gamma$$

The planform area for the length segment ℓ is ℓc. Hence, the lift on segment ℓ is

$$F_L = \rho V_0^2 \pi c \ell \alpha \qquad (11.19)$$

For an airfoil, the coefficient of lift is

$$C_L = \frac{F_L}{S\rho V_0^2/2} \qquad (11.20)$$

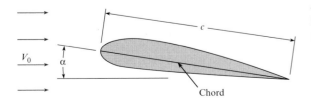

V_0

α

c

Chord

FIGURE 11.20

Definition sketch for an airfoil section.

FIGURE 11.21

Values of C_L for two NACA airfoil
sections. [After Abbott and Van
Doenhoff (22).]

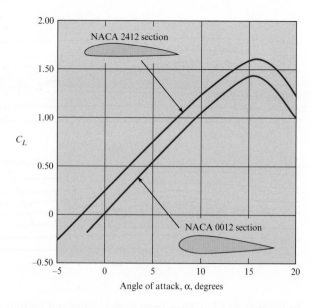

FIGURE 11.21

Values of C_L for two NACA airfoil
sections. [After Abbott and Van
Doenhoff (22).]

where the reference area S is the planform area of the wing—that is, the area seen from the
plan view. On combining Eqs. (11.18) and (11.19) and identifying S as the area associated
with length segment ℓ, one finds that C_L for irrotational flow past a two-dimensional airfoil
is given by

$$C_L = 2\pi\alpha \qquad \qquad \text{(11.21)}$$

Equations (11.19) and (11.21) are the theoretical lift equations for an infinitely long
airfoil at a small angle of attack. Flow separation near the leading edge of the airfoil produces
deviations (high drag and low lift) from the ideal flow predictions at high angles of attack. Therefore, experimental wind-tunnel tests are always made to evaluate the performance of a given
type of airfoil section. For example, the experimentally determined values of lift coefficient
versus α for two NACA (National Advisory Committee for Aeronautics) airfoils are shown in
Fig. 11.21. Note in this figure that the coefficient of lift increases with the angle of attack, α, to
a maximum value and then decreases with further increase in α. This condition, where C_L starts
to decrease with a further increase in α, is called **stall**. Stall occurs because of the onset of
separation over the top of the airfoil, which changes the pressure distribution so that it not only
decreases lift but also increases drag. Data for many other airfoil sections are given by Abbott
and Von Doenhoff (22).

Airfoils of Finite Length—Effect on Drag and Lift

The drag of a two-dimensional foil at a low angle of attack (no end effects) is primarily viscous
drag. However, wings of finite length also have an added drag and a reduced lift associated
with vortices generated at the wing tips. These vortices occur because the high pressure below
the wing and the low pressure on top cause fluid to circulate around the end of the wing from
the high-pressure zone to the low-pressure zone, as shown in Fig. 11.22. This induced flow
has the effect of adding a downward component of velocity, w, to the approach velocity V_0.
Hence, the "effective" free stream velocity is now at an angle ($\phi \approx w/V_0$) to the direction of
the original free stream velocity, and the resultant force is tilted back as shown in Fig. 11.23.
Thus, the effective lift is smaller than the lift for the infinitely long wing because the effective
angle of incidence is smaller. This resultant force has a component parallel to V_0 that is called

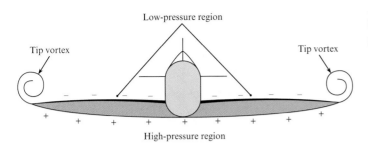

Low-pressure region

Tip vortex

Tip vortex

High-pressure region

FIGURE 11.22

Formation of tip vortices.

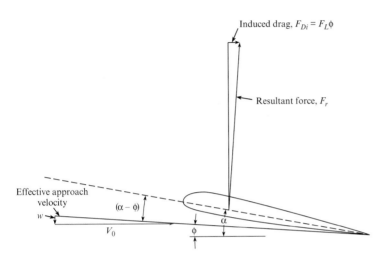

Induced drag, $F_{Di} = F_L \phi$

Resultant force, F_r

Effective approach velocity

w

$(\alpha - \phi)$

V_0

ϕ

α

FIGURE 11.23

Definition sketch for induced-drag relations.

the *induced drag* and is given by $F_L\phi$. Prandtl (23) showed that the induced velocity w for an elliptical spanwise lift distribution is given by the following equation:

$$w = \frac{2F_L}{\pi \rho V_0 b^2}$$

(11.22)

where b is the total length (or span) of the finite wing. Hence,

$$F_{Di} = F_L\phi = \frac{2F_L^2}{\pi \rho V_0^2 b^2} = \frac{C_L^2}{\pi} \frac{S^2}{b^2} \frac{\rho V_0^2}{2}$$

(11.23)

From Eq. (11.23), it can be easily shown that the coefficient of induced drag, C_{Di}, is given by

$$C_{Di} = \frac{C_L^2}{\pi (b^2/S)} = \frac{C_L^2}{\pi \Lambda}$$

(11.24)

which happens to represent the minimum induced drag for any wing planform. Here, the ratio b^2/S is called the aspect ratio Λ of the wing, and S is the planform area of the wing. Thus, for a given wing section (constant C_L and constant chord c), longer wings (larger aspect ratios) have smaller induced-drag coefficients. The induced drag is a significant portion of the total drag of an airplane at low velocities and must be given careful consideration in airplane design. Aircraft (such as gliders) and even birds (such as the albatross and gull) that are required to be airborne for long periods of time with minimum energy expenditure are noted for their long, slender wings. Such a wing is more efficient because the induced drag is small. To illustrate the effect of finite span, look at Fig. 11.24, which shows C_L and C_D versus α for wings with several aspect ratios.

FIGURE 11.24

Coefficients of lift and drag for three wings with aspect ratios of 3, 5, and 7. [After Prandtl (23).]

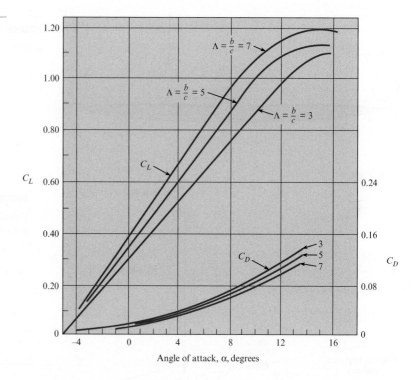

The total drag of a rectangular wing is computed by

$$F_D = (C_{D0} + C_{Di})\frac{bc\rho V_0^2}{2}$$

(11.25)

where C_{D0} is the coefficient of form drag of the wing section and C_{Di} is the coefficient of induced drag.

EXAMPLE 11.7

Wing Area for an Airplane

Problem Statement

An airplane with a weight of 10,000 lbf is flying at 600 ft/s at 36,000 ft, where the pressure is 3.3 psia and the temperature is −67°F. The lift coefficient is 0.2. The span of the wing is 54 ft. Calculate the wing area (in ft^2) and the minimum induced drag.

Define the Situation

An airplane (W = 10,000 lbf) is traveling at V_0 = 600 ft/s.

Coefficient of lift is C_L = 0.2.

Wing span is b = 54 ft.

Properties: Atmosphere (36,000 ft): $T = -67°F$, p = 3.3 psia

State the Goal

- Calculate the required wing area (in ft^2).
- Find the minimum value of induced drag (in N).

Generate Ideas and Make a Plan

1. Apply the idea gas law to calculate density of air.
2. Apply force equilibrium to derive an equation for the required wing area.
3. Calculate the coefficient of induced drag with Eq. (11.24).
4. Calculate the drag using Eq. (11.25) with $C_{D0} = 0$.

Take Action (Execute the Plan)

1. Ideal gas law:

$$\rho = \frac{p}{RT}$$

$$= \frac{(3.3 \text{ lbf/in}^2)(144 \text{ in}^2/\text{ft}^2)}{(1716 \text{ ft-lbf/slug-}°R)(-67 + 460°R)}$$

$$= 0.000705 \text{ slug/ft}^3$$

2. Force equilibrium:

$$W = F_L = \frac{1}{2}\rho V_0^2 C_L S$$

so

$$S = \frac{2W}{\rho V_0^2 C_L}$$

$$= \frac{2 \times 10{,}000 \text{ lbf}}{(0.000705 \text{ slug/ft}^3)(600^2 \text{ ft}^2/\text{s}^2)(0.2)}$$

$$= \boxed{394 \text{ ft}^2}$$

3. Coefficient of induced drag:

$$C_{Di} = \frac{C_L^2}{\pi \left(\dfrac{b^2}{S} \right)} = \frac{0.2^2}{\pi \left(\dfrac{54^2}{394} \right)} = 0.00172$$

4. The induced drag is

$$D_i = \frac{1}{2} \rho V_0^2 C_{Di} S$$

$$= \frac{1}{2} (0.000705 \text{ slug/ft}^3)(600 \text{ ft/s})^2 (0.00172)(394 \text{ ft}^2)$$

$$= \boxed{86.0 \text{ lbf}}$$

A graph showing C_L and C_D versus α is given in Fig. 11.25. Note in this graph that C_D is separated into the induced-drag coefficient C_{Di} and the form-drag coefficient C_{D0}.

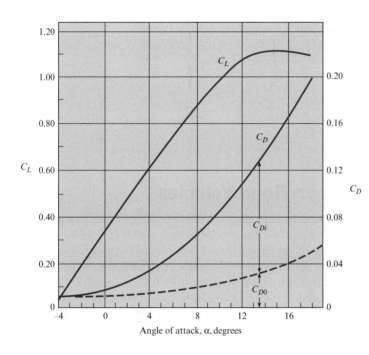

FIGURE 11.25

Coefficients of lift and drag for a wing with an aspect ratio of 5. [After Prandtl (23).]

EXAMPLE 11.8

Takeoff Characteristics of an Airplane

Problem Statement

A light plane (weight = 10 kN) has a wingspan of 10 m and a chord length of 1.5 m. If the lift characteristics of the wing are like those given in Fig. 11.24, what must be the angle of attack for a takeoff speed of 140 km/h? What is the stall speed? Assume two passengers at 800 N each and standard atmospheric conditions.

Define the Situation

- An airplane ($W = 10$ kN) with two passengers $W = 1.6$ kN is taking off.
- Wing span is $b = 10$ m, and chord length is $c = 1.5$ m.
- Lift coefficient information is given by Fig. 11.24.
- Takeoff speed is $V_0 = 140$ km/h.

Assumptions:

1. Ground effects can be neglected.
2. Standard atmospheric conditions prevail.

Properties: Air: $\rho = 1.2 \text{ kg/m}^3$

State the Goal

Find:

1. Angle of attack (in degrees)
2. Stall speed (in km/h)

Generate Ideas and Make a Plan

1. Find the lift by applying force equilibrium.
2. Calculate the coefficient of lift using Eq. (11.20).
3. Find the angle of attack α from Fig. 11.24.
4. Read the maximum angle of attack from Fig. 11.24, and then calculate the corresponding stall speed using the lift force equation (11.17).

Take Action (Execute the Plan)

1. Force equilibrium (y direction), so lift = weight = 11.6 kN.
2. Coefficient of lift:

$$C_L = \frac{F_L}{S\rho V_0^2/2}$$

$$= \frac{11{,}600 \text{ N}}{(15 \text{ m}^2)(1.2 \text{ kg/m}^3)[(140{,}000/3600)^2\text{m}^2/s^2]/2}$$

$$= 0.852$$

3. The aspect ratio is

$$\Lambda = \frac{b}{c} = \frac{10}{1.5} = 6.67$$

4. From Fig. 11.24, the angle of attack is

$$\boxed{\alpha = 7°}$$

From Fig. 11.24, stall will occur when

$$C_L = 1.18$$

Applying the lift force equation gives

$$F_L = C_L A \left(\frac{\rho V_0^2}{2} \right)$$

$$11{,}600 = 1.18(15)\left(\frac{1.2}{2} \right)(V_{\text{stall}})^2$$

$$V_{\text{stall}} = 33.0 \text{ m/s} = \boxed{119 \text{ km/h}}$$

Review the Solution and the Process

Discussion. Notice that the stall speed (119 km/h) is less than the takeoff speed (140 km/h).

11.10 Lift and Drag on Road Vehicles

Early in the development of cars, aerodynamic drag was a minor factor in performance because normal highway speeds were quite low. Thus in the 1920s, coefficients of drag for cars were around 0.80. As highway speeds increased and the science of metal forming became more advanced, cars took on a less angular shape, so that by the 1940s drag coefficients were 0.70 and lower. In the 1970s, the average C_D for U.S. cars was approximately 0.55. In the early 1980s, the average C_D for American cars dropped to 0.45, and currently auto manufacturers are giving even more attention to reducing drag in designing their cars. All major U.S., Japanese, and European automobile companies now have models with C_Ds of about 0.33, and some companies even report C_Ds as low as 0.29 on new models. European manufacturers were the leaders in streamlining cars because European gasoline prices (including tax) have been, for a number of years, about three times those in the United States. Table 11.2 shows the C_D for a 1932 Fiat and for other more contemporary car models.

Great strides have been made in reducing the drag coefficients for passenger cars. However, significant future progress will be very hard to achieve. One of the most streamlined cars was the "Bluebird," which set a world land speed record in 1938. Its C_D was 0.16. The minimum C_D of well-streamlined racing cars is about 0.20. Thus, lowering the C_D for passenger cars below 0.30 will require exceptional design and workmanship. For example, the underside of most cars is aerodynamically very rough (axles, wheels, muffler, fuel tank, shock absorbers, and so on). One way to smooth the underside is to add a panel to the bottom of the car, but then clearance may become a problem, and adequate dissipation of heat from the muffler may be hard to achieve. Other basic features of the automobile that contribute to drag but are not

TABLE 11.2 Coefficients of Drag for Cars

Make and Model	Profile	C_D
1932 Fiat Balillo		0.60
Volkswagen "Bug"		0.46
Plymouth Voyager		0.36
Toyota Paseo		0.31
Dodge Intrepid		0.31
Ford Taurus		0.30
Mercedes-Benz E320		0.29
Ford Probe V (concept car)		0.14
GM Sunraycer (experimental solar vehicle)		0.12

very amenable to drag-reduction modifications are interior airflow systems for engine cooling, wheels, exterior features such as rearview mirrors and antennas, and other surface protrusions. The reader is directed to references (24) and (25), which address the drag and lift of road vehicles in more detail than is possible here.

To produce low-drag vehicles, the basic teardrop shape is an idealized starting point. This shape can be altered to accommodate the necessary functional features of the vehicle. For example, the rear end of the teardrop shape must be lopped off to yield an overall vehicle length that will be manageable in traffic and will fit in our garages. Also, the shape should be wider than its height. Wind-tunnel tests are always helpful in producing the most efficient design. One such test was done on a three-eighths-scale model of a typical notchback sedan. Wind-tunnel test results for such a sedan are shown in Fig. 11.26. Here, the centerline pressure distribution (distribution of C_p) for the conventional sedan is shown by a solid line, and that for a sedan with a 68 mm rear-deck lip is shown by a dashed line. Clearly, the rear-deck lip causes

FIGURE 11.26

Effect of rear-deck lip on model surface. [The data are from Schenkel (25).]

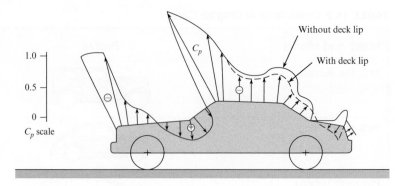

FIGURE 11.26

Effect of rear-deck lip on model surface. [The data are from Schenkel (25).]

the pressure on the rear of the car to increase (C_p is less negative), thereby reducing the drag on the car itself. It also decreases the lift, thereby improving traction. Of course, the lip itself produces some drag, and these tests show that the optimum lip height for greatest overall drag reduction is about 20 mm.

Research and development programs to reduce the drag of automobiles continue. As an entry in the PNGV (Partnership for a New Generation of Vehicles), General Motors (26) has exhibited a vehicle with a drag coefficient as low as 0.163, which is approximately one-half that of the typical midsize sedan. These automobiles will have a rear engine to eliminate the exhaust system underneath the vehicle and allow a flat underbody. Cooling air for the engine is drawn in through inlets on the rear fenders and exhausted out the rear, reducing the drag due to the wake. The protruding rearview mirrors are also removed to reduce the drag. The cumulative effect of these design modifications is a sizable reduction in aerodynamic drag.

The drag of trucks can be reduced by installing vanes near the corners of the truck body to deflect the flow of air more sharply around the corner, thereby reducing the degree of separation. This in turn creates a higher pressure on the rear surfaces of the truck, which reduces the drag of the truck.

One of the desired features in racing cars is the generation of negative lift to improve the stability and traction at high speeds. An idea from Smith (27) is to generate negative gage pressure underneath the car by installing a *ground-effect pod*. This is an airfoil section mounted across the bottom of the car that produces a venturi effect in the channel between the airfoil section and the road surface. The design of ground-effect vehicles involves optimizing design parameters to avoid separation and possible increase in drag. Another scheme to generate negative lift is the use of vanes, as shown in Fig. 11.27. Sometimes "gurneys" are mounted on these vanes to reduce separation effects. Gurneys are small ribs mounted on the upper surface of the vanes near the trailing edge to induce local separation, reduce the separation on the lower surface of the vane, and increase the magnitude of the negative lift. As the speed of racing cars continues to increase, automobile aerodynamics will play an ever-increasing role in traction, stability, and control.

FIGURE 11.27

Racing car with negative-lift devices.

EXAMPLE 11.9

Calculating Negative Lift on a Race Car

Problem Statement

The rear vane installed on the racing car of Fig. 11.27 is at an angle of attack of 8° and has characteristics like those given in Fig. 11.24. Estimate the downward thrust (negative lift) and drag from the vane that is 1.5 m long and has a chord length of 250 mm. Assume the racing car travels at a speed of 270 km/h on a track where normal atmospheric pressure and a temperature of 30°C prevail.

Define the Situation

- A racing car experiences downward lift from a rear-mounted vane.
- Vane overall length is $\ell = 1.5$ m, and chord length is $c = 0.25$ m.
- Car speed is $V_0 = 270$ km/h = 75 m/s.

Properties: Air: $\rho = 1.17$ kg/m^3

State the Goal

Find:

- Downward lift force from vane (in newtons)
- Drag force from vane (in newtons)

Generate Ideas and Make a Plan

1. Find the coefficient of lift C_L and the coefficient of drag C_D from Fig. 11.24.

2. Calculate the downward force using the lift force equation (11.17).

3. Calculate the drag using the drag force equation (11.5).

Take Action (Execute the Plan)

1. The aspect ratio is

$$\Lambda = \frac{\ell}{c} = \frac{1.5}{0.25} = 6$$

From Fig. 11.24, the lift and drag coefficients are

$$C_L = 0.93 \text{ and } C_D = 0.070$$

2. Lift force equation:

$$F_L = C_L A \left(\frac{\rho V_0^2}{2} \right)$$

$$F_L = 0.93 \times 1.5 \times 0.25 \times 1.17 \times (75)^2/2$$

$$= \boxed{1148 \text{ N}}$$

3. Drag force equation:

$$F_D = C_D A \left(\frac{\rho V_0^2}{2} \right) = \left(\frac{C_D}{C_L} \right) F_L$$

$$F_D = (0.070/0.93) \times 1148$$

$$= \boxed{86.4 \text{ N}}$$

11.11 Summarizing Key Knowledge

Relating Lift and Drag to Stress Distributions

- When a body moves relative to a fluid
  - The *drag force* is the component of force that is parallel to the free stream.
  - The *lift force* is the component of force that is perpendicular to the free stream.
- The lift and drag forces are caused by the stress distributions (pressure and shear stress) acting on the body. Integrating the stress distributions over area gives the lift and drag forces.
- The drag force has two parts:
  - *Form drag* is due to pressure stresses acting on the body.
  - *Friction drag* (also called skin friction) is due to shear stresses acting on the body.

Calculating and Understanding the Drag Force

- Drag force depends on four factors: shape of the body, size, fluid density, and fluid speed squared. These four factors are related through the drag force equation:

$$F_D = C_D A \left(\frac{\rho V_0^2}{2} \right)$$

- The *coefficient of drag* (C_D), which characterizes the shape of a body, is a π-group defined by

$$C_D \equiv \frac{F_D}{A_{\text{Ref}}(\rho V_0^2/2)} = \frac{\text{(drag force)}}{\text{(reference area)(kinetic pressure)}}$$

- (C_D) is typically found by experiment and tabulated in engineering references. Objects are classified into three categories: (a) 2-D bodies, (b) axisymmetric bodies, and (c) 3-D bodies.

- For a sphere, two useful equations follow.
  - Stokes flow ($\text{Re}_D < 0.5$):

$$C_D = \frac{24}{\text{Re}}$$

  - Clift and Gauvin correlation ($\text{Re}_D < 3 \times 10^5$):

$$C_D = \frac{24}{\text{Re}}(1 + 0.15\,\text{Re}^{0.687}) + \frac{0.42}{1 + 4.25 \times 10^4\,\text{Re}^{-1.16}}$$

- Drag of bluff bodies and streamlined bodies differs:
  - A *bluff body* is a body with flow separation when the Reynolds number is high enough. When flow separation occurs, the drag is mostly form drag.
  - A *streamlined body* does not have separated flow. Consequently, the drag force is mostly friction drag.
- (C_D) for cylinders and spheres drops dramatically at Reynolds numbers near 10^5 because the boundary layer changes from laminar to turbulent, moving the separation point downstream, reducing the wake region, and decreasing the form drag. This effect is called the *drag crisis*.

Rolling Resistance and Power

- To calculate the power to move a body such as a car or an airplane at a steady speed through a fluid, the usual approach is as follows:
  - *Step 1.* Draw a free body diagram.
  - *Step 2.* Apply the power equation in the form $P = FV$, where F, the force in the direction of motion, is evaluated from the free body diagram.
- The rolling resistance is the frictional force that occurs when an object such as a ball or tire rolls. The rolling resistance is calculated using

$$F_{\text{Rolling resistance}} = F_r = C_r N \qquad \textbf{(11.26)}$$

where C_r is the coefficient of rolling resistance and N is the normal force.

Finding Terminal Velocity

- *Terminal velocity* is the steady state speed of a body that is falling through a fluid.
- When a body has reached terminal velocity, the forces are balanced. These forces typically are weight, drag, and buoyancy.
- To find terminal velocity, sum the forces in the direction of motion and solve the resulting equation. The solution process often needs to be done using iteration (traditional method) or using a computer program (modern method).

Vortex Shedding, Streamlining, and Compressible Flow

- Vortex shedding can cause beneficial effects (better mixing, better heat transfer) and detrimental effects (unwanted structural vibrations, noise).
  - *Vortex shedding* is when cylinders and bluff bodies in a cross-flow produce vortices that are released alternately from each side of the body.
  - The frequency of vortex shedding depends on a π-group called the *Strouhal number.*
- *Streamlining* involves designing a body to minimize the drag force. Usually, streamlining involves designing to reduce or minimize flow separation for a bluff body.
- In high-speed air flows, compressibility effects increase the drag.

The Lift Force

- The lift force on a body depends on four factors: shape, size, density of the flowing fluid, and speed squared. The working equation is

$$F_L = C_L A\left(\frac{\rho V_0^2}{2}\right)$$

- The *coefficient of lift* (C_L) is a π-group defined by

$$C_L \equiv \frac{F_L}{A_{\text{Ref}}(\rho V_0^2/2)} = \frac{(\text{drag force})}{(\text{reference area})(\text{kinetic pressure})}$$

- *Circulation theory of lift.* The lift on an airfoil is due to the circulation produced by the airfoil on the surrounding fluid. This circulatory motion causes a change in the momentum of the fluid and a lift on the airfoil.
- The lift coefficient for a symmetric two-dimensional wing (no tip effect) is

$$C_L = 2\pi\alpha$$

where α is the angle of attack (expressed in radians) and the reference area is the product of the chord and a unit length of wing.
- As the angle of attack increases, the flow separates, the airfoil stalls, and the lift coefficient decreases.
- A wing of finite span produces trailing vortices that reduce the angle of attack and produce an induced drag.
- The drag coefficient corresponding to the minimum induced drag is

$$C_{Di} = \frac{C_L^2}{\pi(b^2/S)} = \frac{C_L^2}{\pi\Lambda}$$

where b is the wing span and S is the planform area of the wing.

REFERENCES

1. Bullivant, W. K. "Tests of the NACA 0025 and 0035 Airfoils in the Full Scale Wind Tunnel." *NACA Rept.,* 708 (1941).

2. DeFoe, G. L. "Resistance of Streamline Wires." *NACA Tech. Note,* 279 (March 1928).

3. Goett, H. J., and W. K. Bullivant. "Tests of NACA 0009, 0012, and 0018 Airfoils in the Full Scale Tunnel." *NACA Rept.,* 647 (1938).

4. Jacobs, E. N. "The Drag of Streamline Wires." *NACA Tech. Note,* 480 (December 1933).

5. Jones, G. W., Jr. "Unsteady Lift Forces Generated by Vortex Shedding about a Large, Stationary, and Oscillating Cylinder at High Reynolds Numbers." *Symp. Unsteady Flow, ASME* (1968).

6. Lindsey, W. F. "Drag of Cylinders of Simple Shapes." *NACA Rept.,* 619 (1938).

7. Miller, B. L., J. F. Mayberry, and I. J. Salter. "The Drag of Roughened Cylinders at High Reynolds Numbers." *NPL Rept. MAR Sci.,* R132 (April 1975).

8. Roshko, A. "Turbulent Wakes from Vortex Streets." *NACA Rept.,* 1191 (1954).

9. Abbott, I. H. "The Drag of Two Streamline Bodies as Affected by Protuberances and Appendages." *NACA Rept.,* 451 (1932).

10. Brevoort, M. J., and U. T. Joyner. "Experimental Investigation of the Robinson-Type Cup Anemometer." *NACA Rept.,* 513 (1935).

11. Freeman, H. B. "Force Measurements on a 1/40-Scale Model of the U.S. Airship 'Akron.'" *NACA Rept.,* 432 (1932).

12. Rouse, H. *Elementary Mechanics of Fluids.* New York: John Wiley, 1946.

13. Clift, R., J. R. Grace, and M. E. Weber. *Bubbles, Drops and Particles.* San Diego, CA: Academic Press, 1978.

14. Clift, R., and W. H. Gauvin. "The Motion of Particles in Turbulent Gas Streams." *Proc. Chemeca '70,* vol. 1, pp. 14–28 (1970).

15. Hoerner, S. F. *Fluid Dynamic Drag.* Published by the author, 1958.

16. Morrison, R. B. (ed). *Design Data for Aeronautics and Astronautics.* New York: John Wiley, 1962.

17. Roberson, J. A., et al. "Turbulence Effects on Drag of Sharp-Edged Bodies." *J. Hydraulics Div., Am. Soc. Civil Eng.* (July 1972).

18. Scher, S. H., and L. J. Gale. "Wind Tunnel Investigation of the Opening Characteristics, Drag, and Stability of Several Hemispherical Parachutes." *NACA Tech. Note,* 1869 (1949).

19. Crowe, C. T., et al. "Drag Coefficient for Particles in Rarefied, Low Mach-Number Flows." In *Progress in Heat and Mass Transfer,* vol. 6, pp. 419–431. New York: Pergamon Press, 1972.

20. Barkla, H. M., et al. "The Magnus or Robins Effect on Rotating Spheres." *J. Fluid Mech.,* vol. 47, part 3 (1971).

21. Kuethe, A. M., and J. D. Schetzer. *Foundations of Aerodynamics.* New York: John Wiley, 1967.

22. Abbott, H., and A. E. Von Doenhoff. *Theory of Wing Sections.* New York: Dover, 1949.

23. Prandtl, L. "Applications of Modern Hydrodynamics to Aeronautics." *NACA Rept.,* 116 (1921).

24. Hucho, Wolf-Heinrich, ed. *Aerodynamics of Road Vehicles.* London: Butterworth, 1987.

25. Schenkel, Franz K. "The Origins of Drag and Lift Reductions on Automobiles with Front and Rear Spoilers." *SAE Paper,* no. 770389 (February 1977).

26. Sharke, P. "Smooth Body." *Mechanical Engineer,* vol. 121, pp. 74–77 (1999).

27. Smith, C. *Engineer to Win.* Osceola, WI: Motorbooks International, 1984.

28. Mehta, R. D. "Aerodynamics of Sports Balls." *Annual Review of Fluid Mechanics,* 17, p. 151 (March 1985).

PROBLEMS

Lift, Drag, and Stress Distribution (§11.1)

11.1 The hypothetical pressure distribution on a rod of triangular (equilateral) cross section is shown, where flow is from left to right. That is, C_p is maximum and equal to $+1.0$ at the leading edge and decreases linearly to zero at the trailing edges. The pressure coefficient on the downstream face is constant with a value of -0.5. Neglecting skin friction drag, find C_D for the rod.

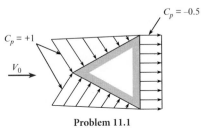

Problem 11.1

11.2 The pressure distribution on a rod having a triangular (equilateral) cross section is shown, where flow is from left to right. What is C_D for the rod?

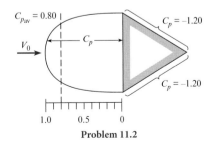

Problem 11.2

11.3 Flow is occurring past the square rod. The pressure coefficient values are as shown. From which direction do you think the flow is coming? (a) NW direction, (b) NE direction, (c) SW direction, or (d) SE direction.

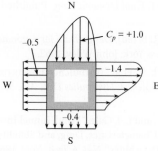

Problem 11.3

Calculating the Drag Force (§11.2)

11.4 The coefficient of drag for a body (select all that apply):

  a. is dimensionless

  b. is usually determined by experiment

  c. depends on thrust

  d. depends on the body's shape

  e. requires an updraft

11.5 Determine the drag of a 2 ft × 4 ft sheet of plywood held at a right angle to a stream of air (60°F, 1 atm) having a velocity of 35 mph.

11.6 Estimate the wind force that would act on *you* if you were standing on top of a tower in a 30 m/s (115 ft/s) wind on a day when the temperature was 20°C (68°F) and the atmospheric pressure was 96 kPa (14 psia).

11.7 Flow over a rectangular plate produces an average wall shear stress of 1.2 pascals. Calculate the friction drag in units of N. The plate dimensions are 1.5 m by 1.5 m. The plate is inclined at 20 degrees with respect to the free stream. Choose the closest answer (N): (a) 2.5, (b) 3.2, (c) 5.1, (d) 6.3, or (e) 6.9.

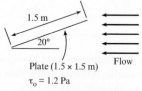

Problem 11.7

11.8 Water is flowing over a sphere. The Reynolds number based on sphere diameter is 1.0. The flow is steady. Calculate the coefficient of drag. Choose the closest answer: (a) 0.01 or less, (b) 0.1, (c) 1, (d) 10, or (e) 20 or greater.

11.9 Compute the overturning moment exerted by a 35 m/s wind on a smokestack that has a diameter of 2.5 m and a height of 75 m. Assume that the air temperature is 20°C and that p_a is 99 kPa absolute.

11.10 A cylindrical anchor (vertical axis) made of concrete ($\gamma = 15$ kN/m³) is reeled in at a rate of 1.0 m/s by a man in a boat. If the anchor is 30 cm in diameter and 30 cm long, what tension must be applied to the rope to pull it up at this rate? Neglect the weight of the rope.

11.11 Estimate the moment at ground level on a signpost supporting a sign measuring 3 m by 2 m if the wind is normal to the surface and has a speed of 35 m/s and the center of the sign is 4 m above the ground. Neglect the wind load on the post itself. Assume $T = 10°C$ and $p = 1$ atm.

11.12 Hoerner (15) presents data that show that fluttering flags of moderate-weight fabric have a drag coefficient (based on the flag area) of about 0.14. Thus the total drag is about 14 times the skin friction drag alone. Design a flagpole that is 100 ft high and is to fly an American flag 6 ft high. Make your own assumptions regarding other required data.

11.13 A circular billboard with a diameter of 5 m is exposed to the wind. Estimate the total force exerted on the structure by a wind that has a direction normal to the structure and a speed of 25 m/s. Assume $T = 10°C$ and $p = 101$ kPa absolute. Use Table 11.1 in §11.3.

11.14 What is the moment at the bottom of a flagpole 18 m high and 20 cm in diameter in a 20 m/s wind? The atmospheric pressure is 100 kPa, and the temperature is 20°C.

11.15 Windstorms sometimes blow empty boxcars off their tracks. The dimensions of one type of boxcar are shown. What minimum wind velocity normal to the side of the car would be required to blow the car over? Assume $C_D = 1.4$.

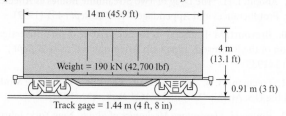

Problem 11.15

Drag on Axisymmetric and 3-D Bodies (§11.3)

11.16 Consider trends from Table 11.1 and Fig. 11.9 in order to classify these statements as true or false:

  a. A value of $C_D = 10$ for a sports car would be a reasonable estimate.

  b. A value of $C_D = 0.15$ for a swimming dolphin would be a reasonable estimate.

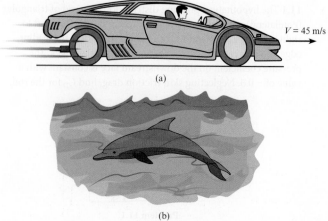

Problem 11.16

Power, Energy, and Rolling Resistance (§11.3)

11.17 A car is driving into a headwind. Calculate the power (in kW) that is required to overcome drag. A car is traveling due east at a steady speed of 22 m/s. The coefficient of drag for the car is 0.4 and the reference area is 1.5 m². The wind is coming from the northeast at a steady speed of 10 m/s. The density of air is 1.1 kg/m³. Choose the closest answer (kW): (a) 8.2 or less, (b) 11, (c) 14, (d) 17, or (e) 27 or greater.

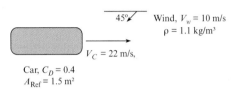

Car, $C_D = 0.4$
$A_{Ref} = 1.5$ m²

Problem 11.17

11.18 A blimp flies at 30 ft/s at an altitude where the specific weight of the air is 0.07 lbf/ft³ and the kinematic viscosity is 1.3×10^{-4} ft²/s. The blimp has a length-to-diameter ratio of 5 and has a drag coefficient corresponding to the streamlined body in Fig. 11.9 in §11.3. The diameter of the blimp is 80 ft. What is the power required to propel the blimp at this speed?

11.19 A cylindrical rod of diameter d and length L is rotated in still air about its midpoint in a horizontal plane. Assume the drag force at each section of the rod can be calculated assuming a two-dimensional flow with an oncoming velocity equal to the relative velocity component normal to the rod. Assume C_D is constant along the rod.

  a. Derive an expression for the average power needed to rotate the rod.

  b. Calculate the power for $\omega = 50$ rad/s, $d = 2$ cm, $L = 1.5$ m, $\rho = 1.2$ kg/m³, and $C_D = 1.2$.

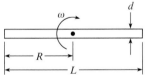

Problem 11.19

11.20 Assume that the horsepower of the engine in the original 1932 Fiat Balillo (see Table 11.2 of §11.10) was 40 bhp (brake horsepower) and that the maximum speed at sea level was 60 mph. Also assume that the projected area of the automobile is 30 ft². Assume that the automobile is now fitted with a modern 220 bhp motor with a weight equal to the weight of the original motor; thus the rolling resistance is unchanged. What is the maximum speed of the "souped-up" Balillo at sea level?

Terminal Velocity (§11.4)

11.21 A sphere is falling in a liquid. Calculate the terminal velocity in units of m/s. The projected area (i.e., reference area) is 7.5 cm². The coefficient of drag is 0.4. The mass of the sphere is 70 grams. The specific gravity of the liquid is 1.2. Choose the closest answer (m/s): (a)1.3, (b) 1.7, (c) 1.8, (d) 1.9, or (e) above 2.1.

$A_R = 7.5$ cm²
$C_D = 0.4$
$m = 70$ g
Liq: $SG = 1.2$
V_T

Problem 11.21

11.22 A grain of pollen is falling at terminal velocity. The fluid is air. Calculate the coefficient of drag. Idealize the pollen grain as a smooth sphere with a diameter of 70 microns. The terminal velocity is 6.0 cm/s. The air has a kinematic viscosity of 15.0×10^6 m²/s. Choose the closest answer: (a) 0.8, or less, (b) 1.8, (c) 18, (d) 86, or (e) 120 or greater.

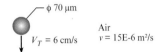

ϕ 70 μm

Air
$V_T = 6$ cm/s $v = 15E\text{-}6$ m²/s

Problem 11.22

11.23 A spherical balloon 2 m in diameter that is used for meteorological observations is filled with helium at standard conditions. The empty weight of the balloon is 3 N. What terminal velocity of ascent will it attain under standard atmospheric conditions?

11.24 A 120-lbf (534 N) skydiver is free-falling at an altitude of 6500 ft (1980 m). Estimate the terminal velocity in mph for minimum and maximum drag conditions. At maximum drag conditions, the product of frontal area and coefficient of drag is $C_D A = 8$ ft² (0.743 m²). At minimum drag conditions, $C_D A = 1$ ft² (0.0929 m²). Assume the pressure and temperature at 6500 ft (1980 m) are 79.5 kPa abs and 277K respectively.

Problem 11.24

11.25 A paratrooper and parachute weigh 900 N. What rate of descent will they have if the parachute is 7 m in diameter and the air has a density of 1.20 kg/m³?

The Theory of Lift (§11.8)

11.26 From the following list, select one topic that is interesting to you. Then, use references such as the Internet to research your topic and prepare one page of written documentation that you could use to present your topic to your peers.

  a. Explain how an airplane works.

  b. Describe the aerodynamics of a flying bird.

  c. Explain how a propeller produces thrust.

  d. Explain how a kite flies.

11.27 Analyses of pitched baseballs indicate that C_L of a rotating baseball is approximately three times that shown in Fig. 11.18 in §11.8. This greater C_L is due to the added circulation caused by the seams of the ball. What is the lift of a ball pitched at a speed of 85 mph and with a spin rate of 35 rps? Also, how much will the ball be deflected from its original path by the time it gets to the plate as a result of the lift force? *Note:* The mound-to-plate distance is 60 ft, the weight of the baseball is 5 oz, and the circumference is 9 in. Assume standard atmospheric conditions, and assume that the axis of rotation is vertical.

Lift and Drag on Airfoils (§11.9)

11.28 A glider at 500 m altitude has a mass of 180 kg and a wing area of 20 m². The glide angle is 1.7°, and the air density is 1.2 kg/m³. If the lift coefficient of the glider is 0.83, how many minutes will it take to reach sea level on a calm day?

11.29 As shown, a glider traveling at a constant velocity will move along a straight glide path that has an angle θ with respect to the horizontal. The angle θ, also called the glide ratio, is given by θ = (C_D/C_L). Use basic principles to prove the preceding statement.

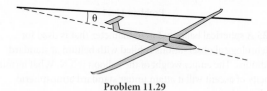

Problem 11.29

11.30 The airstream affected by the wing of an airplane can be considered to be a cylinder (stream tube) with a diameter equal to the wingspan, b. Far downstream from the wing, the tube is deflected through an angle θ from the original direction. Apply the momentum equation to the stream tube between sections 1 and 2 and find the lift of the wing as a function of b, ρ, V, and θ. Relating the lift to the lift coefficient, find θ as a function of b, C_L, and wing area, S. Using the relation for induced drag, $F_{Di} = F_L\theta/2$, show that $C_{Di} = C_L^2/\pi\Lambda$, where Λ is the wing aspect ratio.

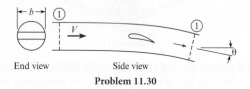

End view   Side view

Problem 11.30

11.31 An airplane has a rectangular-planform wing that has an elliptical spanwise lift distribution. The airplane has a mass of 1000 kg, a wing area of 16 m², and a wingspan of 10 m, and it is flying at 50 m/s at 3000 m altitude in a standard atmosphere. If the form drag coefficient is 0.01, calculate the total drag on the wing and the power ($P = F_D V$) necessary to overcome the drag.

11.32 The total drag coefficient for a wing with an elliptical lift distribution is $C_D = C_{D_0} + C_L^2/\pi\Lambda$, where Λ is the aspect ratio. Derive an expression for C_L that corresponds to minimum C_D/C_L (maximum C_L/C_D) and the corresponding C_L/C_D.

11.33 Your objective is to design a human-powered aircraft using the characteristics of the wing in Fig. 11.24 in §11.9. The pilot weighs 130 pounds and is capable of outputting 1/2 horsepower (225 ft-lbf/s) of continuous power. The aircraft without the wing has a weight of 40 lbf, and the wing can be designed with a weight of 0.12 lbf per square foot of wing area. The drag consists of the drag of the structure plus the drag of the wing. The drag coefficient of the structure, C_{D0} is 0.05, so that the total drag on the craft will be

$$F_D = (C_{D_0} + C_D)\frac{1}{2}\rho V_0^2 S$$

where C_D is the drag coefficient from Fig. 11.24 in §11.9. The power required is equal to $F_D V_0$. The air density is 0.00238 slugs/ft³. Assess whether the airfoil is adequate, and if it is, find the optimum design (wing area and aspect ratio).

CHAPTER TWELVE

Compressible Flow

CHAPTER ROAD MAP The compressibility of a gas that is flowing in a steady state becomes significant when the Mach number exceeds 0.3. For example, the performance of high-speed aircraft, the flow in rocket nozzles, and the reentry mechanics of spacecraft require inclusion of compressible flow effects. This chapter introduces topics in compressible flow.

FIGURE 12.1

The de Laval nozzle is used to accelerate a gas to supersonic speeds. This nozzle is used in turbines, rocket engines, and supersonic jet engines.

This particular nozzle was designed by Andrew Donelick under the guidance of Dr. John Crepeau, Professor of Mechanical Engineering at the University of Idaho. The nozzle was built by Russ Porter, also at the University of Idaho. (Photo by Donald Elger.)

LEARNING OUTCOMES

SOUND WAVES (§12.1).

- Describe the propagation of a sound wave.
- Explain the significance of the Mach number.
- Calculate the speed of sound and the Mach number.

COMPRESSIBLE FLOW (§12.2).

- Explain how properties vary in compressible flow.
- Do relevant calculations.

SHOCK WAVES (§12.3).

- Describe a normal shock wave.
- Calculate the property changes across a normal shock wave.

FLOW IN DUCTS (§12.4).

- Describe how properties vary in a duct when the section area is changing.
- Solve problems involving nozzles.

12.1 Wave Propagation in Compressible Fluids

Wave propagation in a fluid is the mechanism through which the presence of boundaries is communicated to the flowing fluid. In a liquid, the propagation speed of the pressure wave is much higher than the flow velocities, so the flow has adequate time to adjust to a change in boundary shape. Gas flows, on the other hand, can achieve speeds that are comparable to and even exceed the speed at which pressure disturbances are propagated. In this situation, with compressible fluids, the propagation speed is an important parameter and must be incorporated into the flow analysis. In this section, it will be shown how the speed of an infinitesimal pressure disturbance can be evaluated and what its significance is to the flow of a compressible fluid.

Speed of Sound

Everyone has had the experience during a thunderstorm of seeing lightning flash and hearing the accompanying thunder an instant later. Obviously, the sound was produced by the lightning, so the sound wave must have traveled at a finite speed. If the air were totally incompressible (if that were possible), then the sound of thunder and the lightning flash would be simultaneous because all disturbances propagate at infinite speed through incompressible media.* It is analogous to striking one end of a bar of incompressible material and recording instantaneously the response at the other end. Actually, all materials are compressible to some degree and propagate disturbances at finite speeds.

The **speed of sound** is defined as the rate at which an infinitesimal disturbance (pressure pulse) propagates in a medium with respect to the frame of reference of that medium. Actual sound waves, comprised of pressure disturbances of finite amplitude, such that the ear can detect them, travel only slightly faster than the "speed of sound."

To derive an equation for the speed of sound, consider a small section of a pressure wave as it propagates at velocity c through a medium, as depicted in Fig. 12.2. As the wave travels through the gas at pressure p and density ρ, it produces infinitesimal changes of Δp, $\Delta\rho$, and ΔV. These changes must be related through the laws of conservation of mass and momentum. Select a control surface around the wave and let the control volume travel with the wave. The velocities, pressures, and densities relative to the control volume (which is assumed to be very thin) are shown in Fig. 12.3. Conservation of mass in a steady flow requires that the net mass flux across the control surface be zero. Thus,

FIGURE 12.2

Section view of a sound wave.

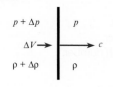

$$-\rho c A + (\rho + \Delta\rho)(c - \Delta V)A = 0 \tag{12.1}$$

where A is the cross-sectional area of the control volume. Neglecting products of higher-order terms ($\Delta\rho\Delta V$) and dividing by the area reduces the conservation-of-mass equation to

$$-\rho\Delta V + c\Delta\rho = 0 \tag{12.2}$$

FIGURE 12.3

Flow relative to the sound wave.

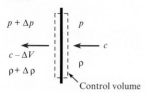

Control volume

The momentum equation for a nonaccelerating steady flow,

$$\sum \mathbf{F} = \dot{m}_o V_o - \dot{m}V_i \tag{12.3}$$

applied to the control volume containing the pressure wave gives

$$(p + \Delta p)A - pA = (-c)(-\rho A c) + (-c + \Delta V)\rho A c \tag{12.4}$$

where the direction to the right is defined as positive. The momentum equation reduces to

$$\Delta p = \rho c \Delta V \tag{12.5}$$

Substituting the expression for ΔV obtained from Eq. (12.2) into Eq. (12.5) gives

$$c^2 = \frac{\Delta p}{\Delta\rho} \tag{12.6}$$

which shows how the speed of propagation is related to the pressure and density change across the wave. It is immediately obvious from this equation that if the flow were ideally incompressible, $\Delta\rho = 0$, the propagation speed would be infinite, which confirms the argument presented earlier.

Equation (12.6) provides an expression for the speed of a general pressure wave. The sound wave is a special type of pressure wave. By definition, a sound wave produces only infinitesimal changes in pressure and density, so it can be regarded as a reversible process. There is also negligibly small heat transfer, so one can assume the process is *adiabatic*. A reversible, adiabatic process is an *isentropic* process; thus the resulting expression for the speed of sound is

$$c^2 = \left.\frac{\partial p}{\partial\rho}\right|_s \tag{12.7}$$

---

*Actually, the thunder would be heard before the lightning was seen, because light also travels at a finite, though very high, speed! However, this would violate one of the basic tenets of relativity theory. No medium can be completely incompressible and propagate disturbances exceeding the speed of light.

This equation is valid for the speed of sound in any substance. However, for many substances the relationship between p and ρ at constant entropy is not very well known.

To reiterate, the speed of sound is the speed at which an infinitesimal pressure disturbance travels through a fluid. Waves of finite strength (finite pressure change across the wave) travel faster than sound waves. Sound speed is the *minimum* speed at which a pressure wave can propagate through a fluid.

For an isentropic process in an ideal gas, the following relationship exists between pressure and density (1):

$$\frac{p}{\rho^k} = \text{constant} \tag{12.8}$$

where k is the ratio of specific heats; that is, the ratio of specific heat at constant pressure to that at constant volume. Thus,

$$k = \frac{c_p}{c_v} \tag{12.9}$$

The values of k for some commonly used gases are given in Table A.2. Taking the derivative of Eq. (12.8) to obtain $\partial p / \partial \rho|_s$ results in

$$\left.\frac{\partial p}{\partial \rho}\right|_s = \frac{kp}{\rho} \tag{12.10}$$

However, from the ideal gas law,

$$\frac{p}{\rho} = RT$$

so the speed of sound is given by

$$c = \sqrt{kRT} \tag{12.11}$$

Thus, the speed of sound in an ideal gas varies with the square root of the temperature. Using this equation to predict sound speeds in real gases at standard conditions gives results very near the measured values. Of course, if the state of the gas is far removed from ideal conditions (high pressures, low temperatures), then using Eq. (12.11) is not valid.

Example 12.1 illustrates the calculation of sound speed for a given temperature.

EXAMPLE 12.1

Speed of Sound Calculation

Define the Situation

Air is at 15°C.

Assume: Air is an ideal gas.

Properties: Air: Table A.2, $R = 287$ J/kg K, and $k = 1.4$

State the Goal

Calculate the speed of sound.

Generate Ideas and Make a Plan

Apply the speed of sound equation, Eq. (12.11), with $T = 288$ K.

Take Action (Execute the Plan)

$$c = \sqrt{kRT}$$
$$c = [(1.4)(287 \text{ J/kg K})(288 \text{ K})]^{1/2} = \boxed{340 \text{ m/s}}$$

Review the Solution and the Process

Knowledge. The absolute temperature must always be used in speed of sound equation.

FIGURE 12.4

Propagation of a sound wave by an airfoil.

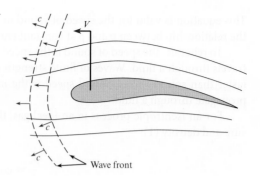

It is possible to demonstrate, in a very simple way, the significance of sound in a compressible flow. Consider the airfoil traveling at speed V in Fig. 12.4. As this airfoil travels through the fluid, the pressure disturbance generated by the airfoil's motion propagates as a wave at sonic speed ahead of the airfoil. These pressure disturbances travel a considerable distance ahead of the airfoil before being attenuated by the viscosity of the fluid, and they "warn" the upstream fluid that the airfoil is coming. In turn, the fluid particles begin to move apart in such a way that there is a smooth flow over the airfoil by the time it arrives. If a pressure disturbance created by the airfoil is essentially attenuated in time Δt, then the fluid at a distance $\Delta t(c - V)$ ahead is alerted to prepare for the airfoil's impending arrival.

What happens as the speed of the airfoil is increased? Obviously, the relative velocity $c - V$ is reduced, and the upstream fluid has less time to prepare for the airfoil's arrival. The flow field is modified by smaller streamline curvatures, and the form drag on the airfoil is increased. If the airfoil speed increases to the speed of sound or greater, then the fluid has no warning whatsoever that the airfoil is coming and cannot prepare for its arrival. At this point, nature resolves the problem by creating a shock wave that stands off the leading edge, as shown in Fig. 12.5. As the fluid passes through the shock wave near the leading edge, it is decelerated to a speed less than sonic speed and therefore has time to divide and flow around the airfoil. Shock waves will be treated in more detail in Section 12.3.

Another approach to appreciating the significance of sound propagation in a compressible fluid is to consider a point source of sound moving in a quiescent fluid, as shown in Fig. 12.6. The sound source is moving at a speed less than the local sound speed in Fig. 12.6a and faster than the local sound speed in Fig. 12.6b. At time $t = 0$, a sound pulse is generated and propagates radially outward at the local speed of sound. At time t_1, the sound source has moved a distance Vt_1, and the circle representing the sound wave emitted at $t = 0$ has a radius of ct_1. The sound source emits a new sound wave at t_1 that propagates radially outward. At time t_2, the sound source has moved to Vt_2, and the sound waves have moved outward as shown.

When the sound source moves at a speed less than the speed of sound, the sound waves form a family of nonintersecting eccentric circles, as shown in Fig. 12.6a. For an observer stationed at A, the frequency of the sound pulses would appear higher than the emitted frequency

FIGURE 12.5

A standing shock wave in front of an airfoil.

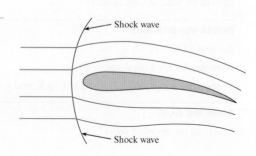

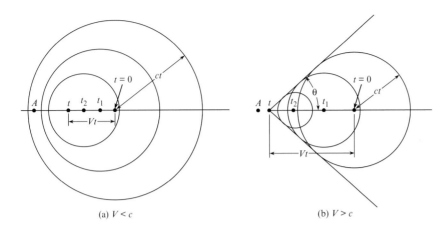

FIGURE 12.6

A sound field generated by a moving point source of sound: (a) the source is moving slower than the speed of sound, (b) the source is moving faster than the speed of sound.

(a) $V < c$ (b) $V > c$

because the sound source is moving toward the observer. In fact, the observer at A will detect a frequency of

$$f = f_0/(1 - V/c)$$

where f_0 is the emitting frequency of the moving sound source. This change in frequency is known as the **Doppler effect**.

When the sound source moves faster than the local sound speed, the sound waves intersect and form the locus of a cone with a half-angle of

$$\theta = \sin^{-1}(c/V)$$

The observer at A will not detect the sound source until it has passed. In fact, only an observer within the cone is aware of the moving sound source.

In view of the physical arguments given, it is apparent that an important parameter relating to sound propagation and compressibility effects is the ratio V/c. This π-group was first proposed by Ernst Mach, an Austrian scientist, and bears his name. The Mach number is defined as

$$M = \frac{V}{c} \tag{12.12}$$

The conical wave surface depicted in Fig. 12.6b is known as a **Mach wave** and the conical half-angle as the **Mach angle**.

Compressible flows are characterized by their Mach number regimes as follows:

$$M < 1 \quad \text{subsonic flow}$$
$$M \approx 1 \quad \text{transonic flow}$$
$$M > 1 \quad \text{supersonic flow}$$

Flows with Mach numbers exceeding 5 are sometimes referred to as **hypersonic**. Airplanes designed to travel at near-sonic speeds and faster are equipped with Mach meters because of the significance of the Mach number with respect to aircraft performance.

Evaluation of the Mach number of an airplane flying at altitude is demonstrated in Example 12.2.

EXAMPLE 12.2

Calculating the Mach Number of an Aircraft

Problem Statement

An F-16 fighter is flying at an altitude of 13 km with a speed of 470 m/s. Assume a U.S. standard atmosphere, and calculate the Mach number of the aircraft.

Define the Situation

A fighter jet is flying at 470 m/s at an altitude of 13 km.

Assumptions: The temperature variation is described by the U.S. standard atmosphere.

Properties: From Table A.2, $R_{air} = 287$ J/kg K, and $k = 1.4$.

State the Goal

Calculate the Mach number of the aircraft.

Generate Ideas and Make a Plan

1. Find the temperature at 13 km by using the 1976 standard atmosphere model (e.g., see http://www.digitaldutch.com/atmoscalc/).

2. Calculate the speed of sound.
3. Calculate the Mach number.

Take Action (Execute the Plan)

1. Temperature at 13 km:

$$T = 217 \text{ K.}$$

2. Speed of sound:

$$c = \sqrt{kRT} = \sqrt{1.4 \times 287 \times 217} = 295 \text{ m/s}$$

3. Mach number:

$$M = \frac{V}{c} = \frac{470 \text{ m/s}}{295 \text{ m/s}} = \boxed{1.59}$$

Review the Solution and the Process

Discussion. The aircraft is flying at supersonic speed.

12.2 Mach Number Relationships

This section will show how fluid properties vary with the Mach number in a compressible flow. Consider a control volume bounded by two streamlines in a steady compressible flow, as shown in Fig. 12.7. Applying the energy equation to this control volume gives

$$-\dot{m}_1\left(h_1 + \frac{V_1^2}{2} + gz_1\right) + \dot{m}_2\left(h_2 + \frac{V_2^2}{2} + gz_2\right) = \dot{Q} \tag{12.13}$$

The elevation terms (z_1 and z_2) can usually be neglected for gaseous flows. If the flow is adiabatic ($\dot{Q} = 0$), the energy equation reduces to

$$\dot{m}_1\left(h_1 + \frac{V_1^2}{2}\right) = \dot{m}_2\left(h_2 + \frac{V_2^2}{2}\right) \tag{12.14}$$

From the principle of continuity, the mass flow rate is constant, $\dot{m}_1 = \dot{m}_2$, so

$$h_1 + \frac{V_1^2}{2} = h_2 + \frac{V_2^2}{2} \tag{12.15}$$

FIGURE 12.7

Control volume enclosed by streamlines.

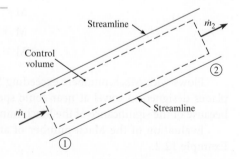

Because positions 1 and 2 are arbitrary points on the same streamline, one can say that

$$h + \frac{V^2}{2} = \text{constant along a streamline in an adiabatic flow} \qquad \textbf{(12.16)}$$

The constant in this expression is called the **total enthalpy,** h_t. It is the enthalpy that would arise if the flow velocity were brought to zero in an adiabatic process. Thus, the energy equation along a streamline under adiabatic conditions is

$$h + \frac{V^2}{2} = h_t \qquad \textbf{(12.17)}$$

If h_t is the same for all streamlines, the flow is **homenergic.**

It is instructive at this point to compare Eq. (12.17) with the Bernoulli equation. Expressing the specific enthalpy as the sum of the specific internal energy and p/ρ, Eq. (12.17) becomes

$$u + \frac{p}{\rho} + \frac{V^2}{2} = \text{constant}$$

If the fluid is incompressible and there is no heat transfer, the specific internal energy is constant and the equation reduces to the Bernoulli equation (excluding the pressure change due to elevation change).

Temperature

The enthalpy of an ideal gas can be written as

$$h = c_p T \qquad \textbf{(12.18)}$$

where c_p is the specific heat at constant pressure. Substituting this relation into Eq. (12.17) and dividing by $c_p T$, results in

$$1 + \frac{V^2}{2c_p T} = \frac{T_t}{T} \qquad \textbf{(12.19)}$$

where T_t is the **total temperature.** From thermodynamics (1), it is known for an ideal gas that

$$c_p - c_v = R \qquad \textbf{(12.20)}$$

or

$$k - 1 = \frac{R}{c_v} = \frac{kR}{c_p}$$

Therefore,

$$c_p = \frac{kR}{k-1} \qquad \textbf{(12.21)}$$

Substituting this expression for c_p back into Eq. (12.19) and realizing that kRT is the speed of sound squared results in the *total temperature* equation:

$$T_t = T\left(1 + \frac{k-1}{2}M^2\right) \qquad \textbf{(12.22)}$$

The temperature T is called the **static temperature**—the temperature that would be registered by a thermometer moving with the flowing fluid. **Total temperature** is analogous to total enthalpy in that it is the temperature that would arise if the velocity were brought to zero adiabatically. If the flow is adiabatic, the total temperature is constant along a streamline. If not, the total temperature varies according to the amount of thermal energy transferred.

Example 12.3 illustrates the evaluation of the total temperature on an aircraft's surface.

EXAMPLE 12.3

Total Temperature Calculation

Problem Statement

An aircraft is flying at M = 1.6 at an altitude where the atmospheric temperature is −50°C. The temperature on the aircraft's surface is approximately the total temperature. Estimate the surface temperature, taking $k = 1.4$.

Define the Situation

An aircraft is flying at M = 1.6. The static temperature is −50°C.

State the Goal

Calculate the total temperature.

Generate Ideas and Make a Plan

This problem can be visualized as the aircraft being stationary and an airstream with a static temperature of −50°C flowing past the aircraft at a Mach number of 1.6.

1. Convert the local static temperature to degrees K.
2. Use total temperature equation, Eq. (12.22).

Take Action (Execute the Plan)

1. Static temperature in absolute temperature units:

$$T = 273 - 50 = 223 \text{ K}$$

2. Total temperature:

$$T_t = 223[1 + 0.2(1.6)^2] = \boxed{337 \text{ K or } 64°\text{C}}$$

If the flow is isentropic, thermodynamics shows that the following relationship for pressure and temperature of an ideal gas between two points on a streamline is valid (1):

$$\frac{p_1}{p_2} = \left(\frac{T_1}{T_2}\right)^{k/(k-1)} \tag{12.23}$$

Isentropic flow means that there is no heat transfer, so the total temperature is constant along the streamline. Therefore,

$$T_t = T_1\left(1 + \frac{k-1}{2}M_1^2\right) = T_2\left(1 + \frac{k-1}{2}M_2^2\right) \tag{12.24}$$

Solving for the ratio T_1/T_2 and substituting into Eq. (12.23) shows that the pressure variation with the Mach number is given by

$$\frac{p_1}{p_2} = \left\{\frac{1 + [(k-1)/2]M_2^2}{1 + [(k-1)/2]M_1^2}\right\}^{k/(k-1)} \tag{12.25}$$

In the ideal gas law used to derive Eq. (12.23), absolute pressures must always be used in calculations with these equations.

The **total pressure** in a compressible flow is given by

$$p_t = p\left(1 + \frac{k-1}{2}M^2\right)^{k/(k-1)} \tag{12.26}$$

which is the pressure that would result if the flow were decelerated to zero speed reversibly and adiabatically. Unlike total temperature, total pressure may not be constant along streamlines in adiabatic flows. For example, it will be shown that flow through a shock wave, although adiabatic, is not reversible and therefore not isentropic. The total pressure variation along a streamline in an adiabatic flow can be obtained by substituting Eqs. (12.26) and (12.24) into Eq. (12.25) to give

$$\frac{p_{t_1}}{p_{t_2}} = \frac{p_1}{p_2}\left\{\frac{1 + [(k-1)/2]M_1^2}{1 + [(k-1)/2]M_2^2}\right\}^{k/(k-1)} = \frac{p_1}{p_2}\left(\frac{T_2}{T_1}\right)^{k/(k-1)} \tag{12.27}$$

Unless the flow is also reversible and Eq. (12.23) is applicable, the total pressures at points 1 and 2 will not be equal. However, if the flow is isentropic, total pressure is constant along streamlines.

Density

Analogous to the total pressure, the **total density** in a compressible flow is given by

$$\rho_t = \rho\left(1 + \frac{k-1}{2}M^2\right)^{1/(k-1)} \tag{12.28}$$

where ρ is the local or static density. If the flow is isentropic, then ρ_t is a constant along streamlines and Eq. (12.28) can be used to determine the variation of gas density with the Mach number.

In literature dealing with compressible flows, one often finds reference to "stagnation" conditions—that is, **stagnation temperature** and **stagnation pressure**. By definition, *stagnation* refers to the conditions that exist at a point in the flow where the velocity is zero, regardless of whether the zero velocity has been achieved by an adiabatic, or reversible, process. For example, if one were to insert a Pitot-static tube into a compressible flow, strictly speaking, one would measure stagnation pressure, not total pressure, because the deceleration of the flow would not be reversible. In practice, however, the difference between stagnation and total pressure is insignificant.

Kinetic Pressure

The kinetic pressure, $q = \rho V^2/2$, is often used to calculate aerodynamic forces with the use of appropriate coefficients. It can also be related to the Mach number. Using the ideal gas law to replace ρ gives

$$q = \frac{1}{2}\frac{pV^2}{RT} \tag{12.29}$$

Then using the equation for the speed of sound, Eq. (12.11), results in

$$q = \frac{k}{2}pM^2 \tag{12.30}$$

where p must always be an absolute pressure because it derives from the ideal gas law.

The use of the equation for kinetic pressure to evaluate the drag force is shown in Example 12.4.

EXAMPLE 12.4

Calculating the Drag Force on a Sphere

Problem Statement

The drag coefficient for a sphere at a Mach number of 0.7 is 0.95. Determine the drag force on a sphere 10 mm in diameter in air if $p = 101$ kPa.

Define the Situation

A sphere is moving at a Mach number of 0.7 in air.

Properties: From Table A.2, $k_{air} = 1.4$.

State the Goal

Find the drag force (in newtons) on the sphere.

Generate Ideas and Make a Plan

The drag force on a sphere is $F_D = qC_D A$.

1. Calculate the kinetic pressure q from Eq. (12.30).
2. Calculate the drag force.

Take Action (Execute the Plan)

1. Kinetic pressure:

$$q = \frac{k}{2}pM^2 = \frac{1.4}{2}(101 \text{ kPa})(0.7)^2 = 34.6 \text{ kPa}$$

2. Drag force:

$$F_D = C_D q\left(\frac{\pi}{4}\right)D^2 = 0.95\left(34.6 \times 10^3 \frac{\text{N}}{\text{m}^2}\right)\left(\frac{\pi}{4}\right)(0.01 \text{ m})^2$$

$$= \boxed{2.58 \text{ N}}$$

The Bernoulli equation is not valid for compressible flows. Consider what would happen if one decided to measure the Mach number of a high-speed air flow with a Pitot-static tube, assuming that the Bernoulli equation was valid. Assume a total pressure of 180 kPa and a static pressure of 100 kPa were measured. By the Bernoulli equation, the kinetic pressure is equal to the difference between the total and static pressures, so

$$\frac{1}{2}\rho V^2 = p_t - p \qquad \text{or} \qquad \frac{k}{2}pM^2 = p_t - p$$

Solving for the Mach number,

$$M = \sqrt{\frac{2}{k}\left(\frac{p_t}{p} - 1\right)}$$

and substituting in the measured values, one obtains

$$M = 1.07$$

The correct approach is to relate the total and static pressures in a compressible flow using Eq. (12.26). Solving that equation for the Mach number gives

$$M = \left\{\frac{2}{k-1}\left[\left(\frac{p_t}{p}\right)^{(k-1)/k} - 1\right]\right\}^{1/2} \tag{12.31}$$

and substituting in the measured values yields

$$M = 0.96$$

Thus, applying the Bernoulli equation would have led one to say that the flow was supersonic, whereas the flow was actually subsonic. In the limit of low velocities ($p_t/p \to 1$), Eq. (12.31) reduces to the expression derived using the Bernoulli equation, which is indeed valid for very low ($M \ll 1$) Mach numbers.

It is instructive to see how the pressure coefficient at the stagnation (total pressure) condition varies with Mach number. The pressure coefficient is defined by

$$C_p = \frac{p_t - p}{\frac{1}{2}\rho V^2}$$

Using Eq. (12.30) for the kinetic pressure enables one to express C_p as a function of the Mach number and the ratio of specific heats:

$$C_p = \frac{2}{kM^2}\left[\left(1 + \frac{k-1}{2}M^2\right)^{k/(k-1)} - 1\right]$$

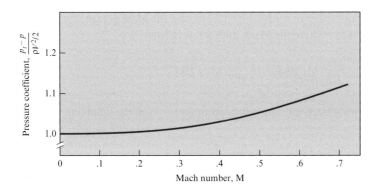

FIGURE 12.8

Variation of the pressure coefficient with Mach number.

The variation of C_p with the Mach number is shown in Fig. 12.8. At a Mach number of zero, the pressure coefficient is unity, which corresponds to incompressible flow. The pressure coefficient begins to depart significantly from unity at a number of about 0.3. From this observation, it is inferred that compressibility effects in the flow field are unimportant for Mach numbers less than 0.3.

12.3 Normal Shock Waves

Normal shock waves are wave fronts normal to the flow across which a supersonic flow is decelerated to a subsonic flow with an attendant increase in static temperature, pressure, and density. The purpose of this section is to develop relations for property changes across normal shock waves.

Change in Flow Properties across a Normal Shock Wave

The most straightforward way to analyze a normal shock wave is to draw a control surface around the wave, as shown in Fig. 12.9, and write down the continuity, momentum, and energy equations.

The net mass flux into the control volume is zero because the flow is steady. Therefore,

$$-\rho_1 V_1 A + \rho_2 V_2 A = 0 \qquad (12.32)$$

where A is the cross-sectional area of the control volume. Equating the net pressure forces acting on the control surface to the net efflux of momentum from the control volume gives

$$\rho_1 V_1 A(-V_1 + V_2) = (p_1 - p_2)A \qquad (12.33)$$

The energy equation can be expressed simply as

$$T_{t_1} = T_{t_2} \qquad (12.34)$$

because the temperature gradients on the control surface are assumed negligible, and thus heat transfer is neglected (adiabatic).

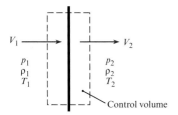

FIGURE 12.9

Control volume enclosing a normal shock wave.

Using the equation for the speed of sound, Eq. (12.11), and the ideal gas law, the continuity equation can be rewritten to include the Mach number as follows:

$$\frac{p_1}{RT_1} M_1 \sqrt{kRT_1} = \frac{p_2}{RT_2} M_2 \sqrt{kRT_2} \qquad (12.35)$$

The Mach number can be introduced into the momentum equation in the following way:

$$\rho_2 V_2^2 - \rho_1 V_1^2 = p_1 - p_2$$

$$p_1 + \frac{p_1}{RT_1} V_1^2 = p_2 + \frac{p_2}{RT_2} V_2^2 \qquad (12.36)$$

$$p_1(1 + kM_1^2) = p_2(1 + kM_2^2)$$

Rearranging Eq. (12.36) for the static pressure ratio across the shock wave results in

$$\frac{p_2}{p_1} = \frac{(1 + kM_1^2)}{(1 + kM_2^2)} \qquad (12.37)$$

As will be shown later, the Mach number of a normal shock wave is always greater than unity upstream and less than unity downstream, so the static pressure always increases across a shock wave.

Rewriting the energy equation in terms of the temperature and Mach number, as done in Eq. (12.22), by utilizing the fact that $T_{t_2}/T_{t_1} = 1$, yields the static temperature ratio across the shock wave:

$$\frac{T_2}{T_1} = \frac{\{1 + [(k-1)/2]M_1^2\}}{\{1 + [(k-1)/2]M_2^2\}} \qquad (12.38)$$

Substituting Eqs. (12.37) and (12.38) into Eq. (12.35) gives the following relationship for the Mach numbers upstream and downstream of a normal shock wave:

$$\frac{M_1}{1 + kM_1^2}\left(1 + \frac{k-1}{2}M_1^2\right)^{1/2} = \frac{M_2}{1 + kM_2^2}\left(1 + \frac{k-1}{2}M_2^2\right)^{1/2} \qquad (12.39)$$

Solving this equation for M_2 as a function of M_1 results in two solutions. One solution is trivial: $M_1 = M_2$, which corresponds to no shock wave in the control volume. The other solution gives the Mach number downstream of the shock wave:

$$M_2^2 = \frac{(k-1)M_1^2 + 2}{2kM_1^2 - (k-1)} \qquad (12.40)$$

Note: Because of the symmetry of Eq. (12.39), one can also use Eq. (12.40) to solve for M_1 given M_2 by simply interchanging the subscripts on the Mach numbers.

Setting $M_1 = 1$ in Eq. (12.40) results in M_2 also being equal to unity. Equations (12.38) and (12.39) also show that there would be no pressure or temperature increase across such a wave. In fact, the wave corresponding to $M_1 = 1$ is the sound wave across which, by definition, pressure and temperature changes are infinitesimal. Thus, the sound wave represents a degenerate normal shock wave.

Example 12.5 demonstrates how to calculate properties downstream of a normal shock wave given the upstream Mach number.

EXAMPLE 12.5

Property Changes across a Normal Shock Wave

Problem Statement

A normal shock wave occurs in air flowing at a Mach number of 1.6. The static pressure and temperature of the air upstream of the shock wave are 100 kPa absolute and 15°C. Determine the Mach number, pressure, and temperature downstream of the shock wave.

Define the Situation

The Mach number upstream of a normal shock wave in air is 1.6.

$$M_1 = 1.6 \longrightarrow \quad \longrightarrow M_2$$

$p_1 = 100$ kPa abs $\qquad p_2$
$T_1 = 15°C$ $\qquad T_2$

Properties: From Table A.2, $k = 1.4$.

State the Goal

Calculate the downstream Mach number, pressure, and temperature.

Generate Ideas and Make a Plan

1. Use Eq. (12.40) to calculate M_2.
2. Use Eq. (12.37) to calculate p_2.
3. Convert upstream temperature to degrees Kelvin and use Eq. (12.38) to find T_2.

Take Action (Execute the Plan)

1. Downstream Mach number:

$$M_2^2 = \frac{(k-1)M_1^2 + 2}{2kM_1^2 - (k-1)} = \frac{(0.4)(1.6)^2 + 2}{(2.8)(1.6)^2 - 0.4} = 0.447$$

$$M_2 = \boxed{0.668}$$

2. Downstream pressure:

$$p_2 = p_1\left(\frac{1 + kM_1^2}{1 + kM_2^2}\right)$$

$$= (100 \text{ kPa})\left[\frac{1 + (1.4)(1.6)^2}{1 + (1.4)(0.668)^2}\right] = \boxed{282 \text{ kPa, absolute}}$$

3. Downstream temperature:

$$T_2 = T_1\left\{\frac{1 + [(k-1)/2]M_1^2}{1 + [(k-1)/2]M_2^2}\right\}$$

$$= (288 \text{ K})\left[\frac{1 + (0.2)(2.56)}{1 + (0.2)(0.447)}\right] = \boxed{400 \text{ K or } 127°C}$$

Review the Solution and the Process

Knowledge. Note that absolute values for the pressure and temperature have to be used in the equations for property changes across shock waves.

The changes in flow properties across a shock wave are presented in Table A.1 for a gas, such as air, for which $k = 1.4$.

A shock wave is an adiabatic process in which no shaft work is done. Thus, for ideal gases the total temperature (and total enthalpy) is unchanged across the wave. The total pressure, however, does change across a shock wave. The total pressure upstream of the wave in Example 12.5 is

$$p_{t_1} = p_1\left(1 + \frac{k-1}{2}M_1^2\right)^{k/(k-1)}$$

$$= 100 \text{ kPa}[1 + (0.2)(1.6^2)]^{3.5} = 425 \text{ kPa}$$

The total pressure downstream of the same wave is

$$p_{t_2} = p_2\left(1 + \frac{k-1}{2}M_2^2\right)^{k/(k-1)}$$

$$= 282 \text{ kPa}[1 + (0.2)(0.668^2)]^{3.5} = 380 \text{ kPa}$$

Thus, the total pressure decreases through the wave, which occurs because the flow through the shock wave is not an isentropic process. Total pressure remains constant along streamlines only in isentropic flow. Values for the ratio of total pressure across a normal shock wave are also provided in Table A.1.

Existence of Shock Waves Only in Supersonic Flows

Refer back to Eq. (12.40), which gives the Mach number downstream of a normal shock wave. If one were to substitute a value for M_1 less than unity, it is easy to see that a value for M_2 would be larger than unity. For example, if $M_1 = 0.5$ in air, then

$$M_2^2 = \frac{(0.4)(0.5)^2 + 2}{(2.8)(0.5)^2 - 0.4}$$

$$M_2 = 2.65$$

Is it possible to have a shock wave in a subsonic flow across which the Mach number becomes supersonic? In this case the total pressure would also increase across the wave; that is,

$$\frac{p_{t_2}}{p_{t_1}} > 1$$

The only way to determine whether such a solution is possible is to invoke the second law of thermodynamics, which states that for any process the entropy of the universe must remain unchanged or increase:

$$\Delta s_{univ} \geq 0 \tag{12.41}$$

Because the shock wave is an adiabatic process, there is no change in the entropy of the surroundings; thus the entropy of the system must remain unchanged or increase:

$$\Delta s_{sys} \geq 0 \tag{12.42}$$

The entropy change of an ideal gas between pressures p_1 and p_2 and temperatures T_1 and T_2 is given by (1)

$$\Delta s_{1 \to 2} = c_p \ln \frac{T_2}{T_1} - R \ln \frac{p_2}{p_1} \tag{12.43}$$

Using the relationship between c_p and R, Eq. (12.21), one can express the entropy change as

$$\Delta s_{1 \to 2} = R \ln \left[\frac{p_1}{p_2} \left(\frac{T_2}{T_1} \right)^{k/(k-1)} \right] \tag{12.44}$$

Note that the quantity in the square brackets is simply the total pressure ratio as given by Eq. (12.27). Therefore, the entropy change across a shock wave can be rewritten as

$$\Delta s = R \ln \frac{p_{t_1}}{p_{t_2}} \tag{12.45}$$

A shock wave across which the Mach number changes from subsonic to supersonic would give rise to a total pressure ratio less than unity and a corresponding decrease in entropy,

$$\Delta s_{sys} < 0$$

which violates the second law of thermodynamics. Therefore, shock waves can exist only in supersonic flow.

The total pressure ratio approaches unity for $M_1 \to 1$, which conforms with the definition that sound waves are isentropic ($\ln 1 = 0$). Example 12.6 demonstrates the increase in entropy across a normal shock wave.

EXAMPLE 12.6

Entropy Increase across Shock Wave

Problem Statement

A normal shock wave occurs in air flowing at a Mach number of 1.5. Find the change in entropy across the wave.

Define the Situation

A normal shock wave in air has an upstream Mach number of 1.5.

Properties: From Table A.2, $R_{air} = 287$ J/kg K, and $k = 1.4$.

State the Goal

Find the change in entropy (in J/kg K) across the wave.

Generate Ideas and Make a Plan

1. Calculate downstream Mach number using Eq. (12.40).
2. Calculate pressure ratio across wave using Eq. (12.37).
3. Calculate temperature across the wave using Eq. (12.38).
4. Calculate entropy change using Eq. (12.44).

Take Action (Execute the Plan)

1. Downstream Mach number:

$$M_2^2 = \frac{(k-1)M_1^2 + 2}{2kM_1^2 - (k-1)} = \frac{(0.4)(1.5)^2 + 2}{(2.8)(1.5)^2 - 0.4} = 0.492$$

$$M_2 = 0.701$$

2. Pressure ratio:

$$\frac{p_2}{p_1} = \left(\frac{1 + kM_1^2}{1 + kM_2^2}\right) = \left[\frac{1 + (1.4)(1.5)^2}{1 + (1.4)(0.701)^2}\right] = 2.46$$

3. Temperature ratio:

$$\frac{T_2}{T_1} = \left\{\frac{1 + [(k-1)/2]M_1^2}{1 + [(k-1)/2]M_2^2}\right\}$$

$$= \left[\frac{1 + (0.2)(2.25)}{1 + (0.2)(0.492)}\right] = 1.32$$

4. Entropy change:

$$\Delta s = R \ln\left[\left(\frac{p_1}{p_2}\right)\left(\frac{T_2}{T_1}\right)^{k/(k-1)}\right]$$

$$= 287 \,(\text{J/kg K}) \ln\left[\left(\frac{1}{2.46}\right)(1.32)^{3.5}\right]$$

$$= \boxed{20.5 \text{ J/kg K}}$$

More examples of shock waves will be given in the next section. This section is concluded by qualitatively discussing other features of shock waves.

Besides the normal shock waves studied here, there are oblique shock waves that are inclined with respect to the flow direction. Look once again at the shock wave structure in front of a blunt body, as depicted qualitatively in Fig. 12.10. The portion of the shock wave immediately in front of the body behaves like a normal shock wave. As the shock wave bends in the free stream direction, oblique shock waves result. The same relationships derived earlier for the normal shock waves are valid for the velocity components normal to oblique waves. The oblique shock waves continue to bend in the downstream direction until the Mach number of the velocity component normal to the wave is unity. Then the oblique shock has degenerated into a so-called Mach wave, across which changes in flow properties are infinitesimal.

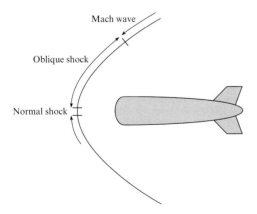

FIGURE 12.10

Shock wave structure in front of a blunt body.

The familiar sonic booms are the result of weak oblique shock waves that reach ground level. One can appreciate the damage that would ensue from stronger oblique shock waves if aircraft were permitted to travel at supersonic speeds near ground level.

12.4 Isentropic Compressible Flow through a Duct with Varying Area

With the flow of incompressible fluids through a venturi configuration, as the flow approaches the throat (smallest area), the velocity increases and the pressure decreases; then as the area again increases, the velocity decreases. The same velocity-area relationship is not always found for compressible flows. The purpose of this section is to show the dependence of flow properties on changes in the cross-sectional area with compressible flow in variable area ducts.

Dependence of the Mach Number on Area Variation

Consider the duct of varying area shown in Fig. 12.11. It is assumed that the flow is isentropic and that the flow properties at each section are uniform. This type of analysis, in which the flow properties are assumed to be uniform at each section yet in which the cross-sectional area is allowed to vary (nonuniform), is classified as "quasi one-dimensional."

The mass flow through the duct is given by

$$\dot{m} = \rho A V \tag{12.46}$$

where A is the duct's cross-sectional area. Because the mass flow is constant along the duct,

$$\frac{d\dot{m}}{dx} = \frac{d(\rho A V)}{dx} = 0 \tag{12.47}$$

which can be written as,*

$$\frac{1}{\rho}\frac{d\rho}{dx} + \frac{1}{A}\frac{dA}{dx} + \frac{1}{V}\frac{dV}{dx} = 0 \tag{12.48}$$

the flow is assumed to be inviscid, so Euler's equation is valid. For steady flow,

$$\rho V \frac{dV}{dx} + \frac{dp}{dx} = 0$$

FIGURE 12.11

Duct with variable area.

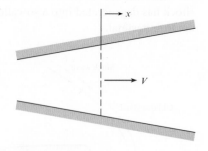

---

*This step can easily be seen by first taking the logarithm of Eq. (12.46),

$$\ln(\rho A V) = \ln\rho + \ln A + \ln V$$

and then taking the derivative of each term:

$$\frac{d}{dx}[\ln(\rho A V)] = 0 = \frac{1}{\rho}\frac{d\rho}{dx} + \frac{1}{A}\frac{dA}{dx} + \frac{1}{V}\frac{dV}{dx}$$

Making use of Eq. (12.7), which relates $dp/d\rho$ to the speed of sound in an isentropic flow, gives

$$\frac{-V}{c^2}\frac{dV}{dx} = \frac{1}{\rho}\frac{d\rho}{dx} \tag{12.49}$$

Using this relationship to eliminate ρ in Eq. (12.48) results in

$$\frac{1}{V}\frac{dV}{dx} = \frac{1}{M^2-1}\frac{1}{A}\frac{dA}{dx} \tag{12.50a}$$

which can be written in an alternate form as

$$\frac{dV}{dA} = \frac{V}{A}\frac{1}{M^2-1} \tag{12.50b}$$

This equation, although simple, leads to the following important, far-reaching conclusions.

Subsonic Flow

For subsonic flow, M^2-1 is negative, so $dV/dA < 0$, which means that a decreasing area leads to an increasing velocity, and correspondingly an increasing area leads to a decreasing velocity. This velocity-area relationship parallels the trend for incompressible flows.

Supersonic Flow

For supersonic flow, M^2-1 is positive, so $dV/dA > 0$, which means that a decreasing area leads to a decreasing velocity, and an increasing area leads to an increasing velocity. Thus, the velocity at the minimum area of a duct with supersonic compressible flow is a minimum. This is the principle underlying the operation of diffusers on jet engines for supersonic aircraft, as shown in Fig. 12.12. The purpose of the diffuser is to decelerate the flow so that there is sufficient time for combustion in the chamber. Then the diverging nozzle accelerates the flow again to achieve a larger kinetic energy of the exhaust gases and an increased engine thrust.

Transonic Flow ($M \approx 1$)

Stations along a duct corresponding to $dA/dx = 0$ represent either a local minimum or a local maximum in the duct's cross-sectional area, as illustrated in Fig. 12.13. If at these stations the flow was either subsonic ($M < 1$) or supersonic ($M > 1$), then by Eq. (12.50a), $dV/dx = 0$, so the flow velocity would have either a maximum or a minimum value. In particular, if the flow were supersonic through the duct of Fig. 12.13a, then the velocity would be a minimum at the throat; if subsonic, a maximum.

What happens if the Mach number is unity? Equation (12.50a) states that if the Mach number is unity and dA/dx is not equal to zero, then the velocity gradient dV/dx is infinite—

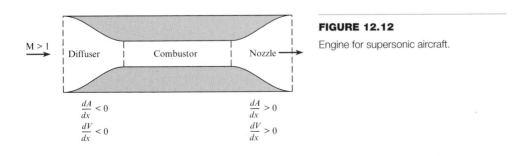

FIGURE 12.12

Engine for supersonic aircraft.

FIGURE 12.13

Duct contours for which dA/dx is zero.

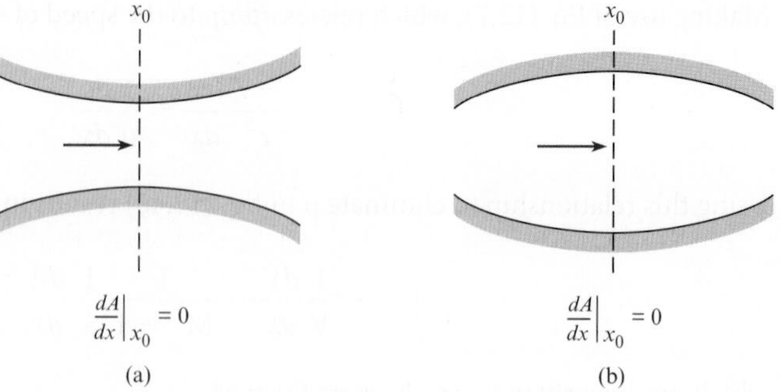

$$\left.\frac{dA}{dx}\right|_{x_0} = 0$$

(a)

$$\left.\frac{dA}{dx}\right|_{x_0} = 0$$

(b)

a physically impossible situation. Therefore, dA/dx must be zero, where the Mach number is unity for a finite, physically reasonable velocity gradient to exist.*

The argument can be taken one step further here to show that sonic flow can occur only at a minimum area. Consider Fig. 12.13a. If the flow is initially subsonic, the converging duct accelerates the flow toward a sonic velocity. If the flow is initially supersonic, the converging duct decelerates the flow toward a sonic velocity. Using this same reasoning, one can prove that sonic flow is impossible in the duct depicted in Fig. 12.13b. If the flow is initially supersonic, the diverging duct increases the Mach number even more. If the flow is initially subsonic, the diverging duct decreases the Mach number; thus sonic flow cannot be achieved at a maximum area. Hence, the Mach number in a duct of varying cross-sectional area can be unity only at a local area minimum (throat). This does not imply, however, that the Mach number must always be unity at a local area minimum.

de Laval Nozzle

The de Laval nozzle is a duct of varying area that produces supersonic flow. The nozzle is named after its inventor, de Laval (1845–1913), a Swedish engineer. According to the foregoing discussion, the nozzle must consist of a converging section to accelerate the subsonic flow, a throat section for transonic flow, and a diverging section to further accelerate the supersonic flow. Thus, the shape of the de Laval nozzle is as shown in Fig. 12.14.

One very important application of the de Laval nozzle is the supersonic wind tunnel, which has been an indispensable tool in the development of supersonic aircraft. Basically, the supersonic wind tunnel (as illustrated in Fig. 12.15) consists of a high-pressure source of gas, a de Laval nozzle to produce supersonic flow, and a test section. The high-pressure source may be from a large pressure tank, which is connected to the de Laval nozzle through a regulator valve to maintain a constant upstream pressure, or from a pumping system that provides a continuous high-pressure supply of gas.

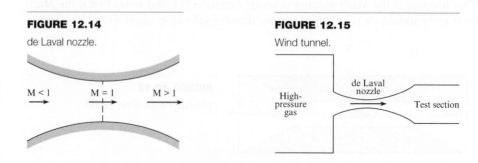

FIGURE 12.14

de Laval nozzle.

M < 1   M = 1   M > 1

FIGURE 12.15

Wind tunnel.

High-pressure gas

de Laval nozzle

Test section

*Actually, the velocity gradient is indeterminate because the numerator and denominator are both zero. However, it can be shown by application of L'Hôpital's rule that the velocity gradient is finite.

The equations relating to the compressible flow through a de Laval nozzle have already been developed. Because the mass flow rate is the same at every cross section,

$$\rho V A = \text{constant}$$

and the constant is usually evaluated corresponding to those conditions that exist when the Mach number is unity. Thus,

$$\rho V A = \rho_* V_* A_* \qquad \textbf{(12.51)}$$

where the asterisk signifies conditions wherein the Mach number is equal to unity. Rearranging Eq. (12.51) gives

$$\frac{A}{A_*} = \frac{\rho_* V_*}{\rho V}$$

However, the velocity is the product of the Mach number and the local speed of sound. Therefore,

$$\frac{A}{A_*} = \frac{\rho_*}{\rho} \, \frac{M_* \sqrt{kRT_*}}{M \sqrt{kRT}} \qquad \textbf{(12.52)}$$

By definition, $M_* = 1$, so

$$\frac{A}{A_*} = \frac{\rho_*}{\rho} \left(\frac{T_*}{T}\right)^{1/2} \frac{1}{M} \qquad \textbf{(12.53)}$$

Because the flow in a de Laval nozzle is assumed to be isentropic, the total temperature and total pressure (and total density) are constant throughout the nozzle. From Eq. (12.28),

$$\frac{\rho_*}{\rho} = \left\{ \frac{1 + [(k-1)/2]M^2}{(k+1)/2} \right\}^{1/(k-1)}$$

and from Eq. (12.24),

$$\frac{T_*}{T} = \frac{1 + [(k-1)/2]M^2}{(k+1)/2}$$

Substituting these expressions into Eq. (12.53) yields the following relationship for area ratio as a function of Mach number in a variable area duct:

$$\frac{A}{A_*} = \frac{1}{M} \left\{ \frac{1 + [(k-1)/2]M^2}{(k+1)/2} \right\}^{(k+1)/2(k-1)} \qquad \textbf{(12.54)}$$

This equation is valid, of course, for all Mach numbers: subsonic, transonic, and supersonic. The area ratio A/A_* is the ratio of the area at the station where the Mach number is M to the area where M is equal to unity. Many supersonic wind tunnels are designed to maintain the same test section area and to vary the Mach number by varying the throat area.

Example 12.7 illustrates the use of the Mach number–area ratio expression to size the test section of a supersonic wind tunnel.

Example 12.7 demonstrates that it is a straightforward task to calculate the area ratio given the Mach number and ratio of specific heats. However, in practice, one usually knows the area ratio and wishes to determine the Mach number. It is not possible to solve Eq. (12.54) for the Mach number as an explicit function of the area ratio. For this reason, compressible flow tables have been developed that allow one to obtain the Mach number easily given the area ratio (as shown in Table A.1).

EXAMPLE 12.7

Finding the Test Section Size in a Supersonic Wind Tunnel

Problem Statement

Suppose a supersonic wind tunnel is being designed to operate with air at a Mach number of 3. If the throat area is 10 cm², then what must the cross-sectional area of the test section be?

Define the Situation

Design a supersonic wind tunnel with a Mach number of 3.0 in the test section.

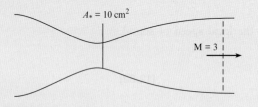

$A_* = 10 \text{ cm}^2$

$M = 3$

Properties: From Table A.2, $k_{air} = 1.4$.

State the Goal

Find the cross-sectional area (in cm²) of the test section.

Generate Ideas and Make a Plan

1. Use Eq. (12.54), which gives area ratio with respect to the throat section.
2. Calculate the area of the test section.

Take Action (Execute the Plan)

1. Area ratio:

$$\frac{A}{A_*} = \frac{1}{M}\left\{\frac{1 + [(k-1)/2]M^2}{(k+1)/2}\right\}^{(k+1)/2(k-1)}$$

$$= \frac{1}{3}\left[\frac{1 + (0.2)3^2}{1.2}\right]^3 = 4.23$$

2. Cross-sectional area of test section:

$$A = 4.23 \times 10 \text{ cm}^2 = \boxed{42.3 \text{ cm}^2}$$

Consider again Table A.1. This table has been developed for a gas, such as air, for which $k = 1.4$. The symbols that head each column are defined at the beginning of the table. Tables for both subsonic and supersonic flow are provided. Example 12.8 shows how to use the tables to find flow properties at a given area ratio.

EXAMPLE 12.8

Flow Properties in a Supersonic Wind Tunnel

Problem Statement

The test section of a supersonic wind tunnel using air has an area ratio of 10. The absolute total pressure and temperature are 4 MPa and 350 K. Find the Mach number, pressure, temperature, and velocity at the test section.

Define the Situation

Situation. A supersonic wind tunnel has an area ratio of 10.

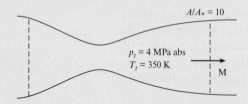

$A/A_* = 10$

$p_t = 4 \text{ MPa abs}$
$T_t = 350 \text{ K}$

M

Properties: From Table A.2, $k_{air} = 1.4$, $R_{air} = 287$ J/kg K.

State the Goal

Find the Mach number, pressure, temperature, and velocity at the test section.

Generate Ideas and Make a Plan

1. Use Table A.1 and interpolate to find the Mach number at the test section.
2. Use Table A.1 to find the pressure and temperature ratios at the test section.
3. Evaluate the pressure and temperature in the test section.
4. Calculate the speed of sound using Eq. (12.11).
5. Find the velocity using $V = MC$.

Take Action (Execute the Plan)

1. From Table A.1:

M	A/A⋆
3.5	6.79
4.0	10.72

Interpolating between the two points gives $\boxed{M = 3.91}$ at $A/A_* = 10.0$.

2. Interpolation using Table A.1 to find the pressure and temperature ratios:

$$\frac{p}{p_t} = 0.00743 \quad \text{and} \quad \frac{T}{T_t} = 0.246$$

3. In the test section,

$$p = 0.00743 \times 4 \text{ MPa} = \boxed{29.7 \text{ kPa}}$$
$$T = 0.246 \times 350 \text{ K} = \boxed{86 \text{ K}}$$

4. Speed of sound:

$$c = \sqrt{kRT} = \sqrt{1.4 \times 287 \times 86} = 186 \text{ m/s}$$

5. Velocity:

$$V = 3.91 \times 186 \text{ m/s} = \boxed{727 \text{ m/s}}$$

Review the Solution and the Process

Knowledge. Low temperatures can cause problems. Notice that the temperature of air in the test section is only 86 K, or $-187°C$. At this temperature, the water vapor in the air can condense out, creating fog in the tunnel and compromising tunnel utility.

Mass Flow Rate through a de Laval Nozzle

An important consideration in the design of a supersonic wind tunnel is size. A large wind tunnel requires a large mass flow rate, which, in turn, requires a large pumping system for a continuous-flow tunnel or a large tank for sufficient run time in an intermittent tunnel. The purpose of this section is to develop an equation for the mass flow rate.

The easiest station at which to calculate the mass flow rate is the throat because the Mach number is unity there:

$$\dot{m} = \rho_* A_* V_* = \rho_* A_* \sqrt{kRT_*}$$

It is more convenient, however, to express the mass flow in terms of total conditions. The local density and static temperature at sonic velocity are related to the total density and temperature by

$$\frac{T_*}{T_t} = \left(\frac{2}{k+1}\right)$$
$$\frac{\rho_*}{\rho_t} = \left(\frac{2}{k+1}\right)^{1/(k-1)}$$

which, when substituted into the foregoing equation, give

$$\dot{m} = \rho_t \sqrt{kRT_t}\, A_* \left(\frac{2}{k+1}\right)^{(k+1)/2(k-1)} \tag{12.55}$$

Usually, the total pressure and temperature are known. Using the ideal gas law to eliminate ρ_t yields the expression for *critical mass flow rate*

$$\dot{m} = \frac{p_t A_*}{\sqrt{RT_t}}\, k^{1/2} \left(\frac{2}{k+1}\right)^{(k+1)/2(k-1)} \tag{12.56}$$

For gases with a ratio of specific heats of 1.4,

$$\dot{m} = 0.685 \frac{p_t A_*}{\sqrt{RT_t}} \tag{12.57}$$

For gases with $k = 1.67$,

$$\dot{m} = 0.727 \frac{p_t A_*}{\sqrt{RT_t}} \tag{12.58}$$

Example 12.9 illustrates how to calculate mass flow rate in a supersonic wind tunnel given the conditions in the test section.

EXAMPLE 12.9

Mass Flow Rate in Supersonic Wind Tunnel

Problem Statement

A supersonic wind tunnel with a square test section 15 cm by 15 cm is being designed to operate at a Mach number of 3 using air. The static temperature and pressure in the test section are −20°C and 50 kPa abs, respectively. Calculate the mass flow rate.

Define the Situation

A Mach 3 supersonic wind tunnel has a 15 cm by 15 cm test section.

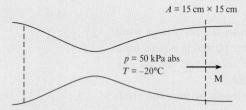

$A = 15 \text{ cm} \times 15 \text{ cm}$

$p = 50$ kPa abs
$T = -20°C$

M

Properties: From Table A.2, $k_{air} = 1.4$ and $R_{air} = 287$ J/kg K.

State the Goal

Calculate the mass flow rate (kg/s) in the tunnel.

Generate Ideas and Make a Plan

1. Use Eq. (12.54) to find area ratio and calculate throat area.
2. Use Eq. (12.22) to find total temperature.
3. Use Eq. (12.26) to find total pressure.
4. Use Eq. (12.56) to find the mass flow rate.

Take Action (Execute the Plan)

1. Area ratio:

$$\frac{A}{A_*} = \frac{1}{M}\left\{\frac{1 + [(k-1)/2]M^2}{(k+1)/2}\right\}^{(k+1)/2(k-1)}$$

$$= \frac{1}{3}\left[\frac{1 + 0.2 \times 3^2}{1.2}\right]^3 = 4.23$$

Throat area:

$$A_* = \frac{225 \text{ cm}^2}{4.23} = 53.2 \text{ cm}^2 = 0.00532 \text{ m}^2$$

2. Total temperature:

$$T_t = T\left(1 + \frac{k-1}{2}M^2\right) = 253 \text{ K} (2.8) = 708 \text{ K}$$

3. Total pressure:

$$p_t = p\left(1 + \frac{k-1}{2}M^2\right)^{k/(k-1)} = (50 \text{ kPa})(36.7)$$

$$= 1840 \text{ kPa} = 1.84 \text{ MPa}$$

4. Mass flow rate:

$$\dot{m} = 0.685\frac{p_t A_*}{\sqrt{RT_t}} = \frac{(0.685)[1.840(10^6 \text{ N/m}^2)](0.00532 \text{ m}^2)}{[(287 \text{ J/kg K})(708 \text{ K})]^{1/2}}$$

$$= \boxed{14.9 \text{ kg/s}}$$

Review the Solution and the Process

1. *Discussion.* An alternate way to solve this problem is to calculate the density in the test section using the ideal gas law, calculate the speed of sound with the speed of sound equation, find the air speed using the Mach number, and finally determine the mass flow rate with $\dot{m} = \rho VA$.

2. *Discussion.* A pump capable of moving air at this rate against a 1.8 MPa pressure would require over 6000 kW of power input. Such a system would be large and costly to build and to operate.

Classification of Nozzle Flow by Exit Conditions

Nozzles are classified by the conditions at the nozzle exit. Consider the de Laval nozzle depicted in Fig. 12.16 with the corresponding pressure and Mach number distributions plotted beneath it. The pressure at the nozzle entrance is very near the total pressure because the Mach number is small. As the area decreases toward the throat, the Mach number increases and the pressure decreases. The static to total pressure ratio at the throat, where conditions are sonic, is called the **critical pressure ratio**. It has a value of

$$\frac{p_*}{p_t} = \left(\frac{2}{k+1}\right)^{k/(k-1)}$$

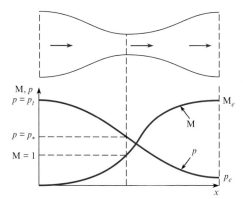

FIGURE 12.16

Distribution of static pressure and Mach number in a de Laval nozzle.

which for air with $k = 1.4$ is

$$\frac{p_*}{p_t} = 0.528$$

It is called a critical pressure ratio because to achieve sonic flow with air in a nozzle, it is necessary that the exit pressure be equal to or less than 0.528 times the total pressure. The pressure continues to decrease until it reaches the exit pressure corresponding to the nozzle-exit area ratio. Similarly, the Mach number monotonically increases with distance down the nozzle.

The nature of the exit flow from the nozzle depends on the difference between the exit pressure, p_e, and the back pressure (the pressure to which the nozzle exhausts). If the exit pressure is higher than the back pressure, then an expansion wave exists at the nozzle exit, as shown in Fig. 12.17a. These waves, which will not be studied here, effect a turning and further acceleration of the flow to achieve the back pressure. As one watches the exhaust of a rocket motor as it rises through the ever-decreasing pressure of higher altitudes, one can see the plume fan out as the flow turns more in response to the lower back pressure. A nozzle for which the exit pressure is larger than the back pressure is called an **underexpanded nozzle** because the flow could have expanded further.

If the exit pressure is less than the back pressure, shock waves occur. If the exit pressure is only slightly less than the back pressure, then pressure equalization can be obtained by oblique shock waves at the nozzle exit, as shown in Fig. 12.17b.

If, however, the difference between back pressure and exit pressure is larger than can be accommodated by oblique shock waves, then a normal shock wave will occur in the nozzle, as shown in Fig. 12.17c. A pressure jump occurs across the normal shock wave. The flow becomes subsonic and decelerates in the remaining portion of the diverging section in such a way that

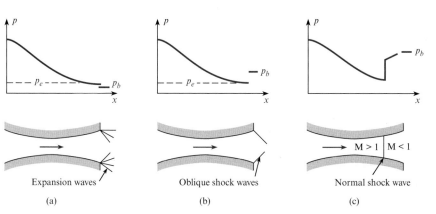

FIGURE 12.17

Conditions at a nozzle exit: (a) expansion waves, (b) oblique shock waves, (c) normal shock wave.

the exit pressure is equal to the back pressure. As the back pressure is further increased, the shock wave moves toward the throat region until, finally, there is no region of supersonic flow. A nozzle in which the exit pressure corresponding to the exit area ratio of the nozzle is less than the back pressure is called an **overexpanded nozzle**. Any flow that exits from a duct (or pipe) subsonically must always exit at the local back pressure.

A nozzle with supersonic flow in which the exit pressure is equal to the back pressure is **ideally expanded**.

The assessment of the nozzle exit conditions is provided by Example 12.10.

EXAMPLE 12.10

Finding a Nozzle Exit Condition

Problem Statement

The total pressure in a nozzle with an area ratio (A/A_*) of 4 is 1.3 MPa. Air is flowing through the nozzle. If the back pressure is 100 kPa, is the nozzle overexpanded, ideally expanded, or underexpanded?

Define the Situation

Air flows through a nozzle with exit area ratio of 4.

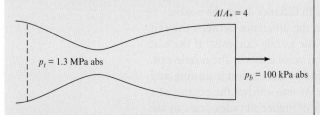

$A/A_* = 4$

$p_t = 1.3$ MPa abs

$p_b = 100$ kPa abs

State the Goal

Determine the state of the exit condition (ideally expanded, overexpanded, or underexpanded).

Generate Ideas and Make a Plan

1. Interpolate Table A.1 to find Mach number corresponding to exit area ratio.

2. Calculate exit pressure using Eq. (12.26).

3. Compare exit pressure with back pressure to determine exit condition.

Take Action (Execute the Plan)

1. Interpolation for Mach number from Table A.1:

M	A/A_*
2.90	3.850
3.00	4.235

M = 2.94 at $A/A_* = 4.0$.

2. Exit pressure:

$$\frac{p_t}{p_e} = \left(1 + \frac{k-1}{2}M^2\right)^{k/(k-1)}$$

$$p_e = \frac{1300 \text{ kPa}}{(1 + 0.2 \times 2.94^2)^{3.5}} = 38.7 \text{ kPa}.$$

3. Because $p_e < p_b$, the nozzle is overexpanded.

Review the Solution and the Process

Knowledge. Because the nozzle is overexpanded, there will be a shock wave structure inside the nozzle to achieve pressure equilibration at the nozzle exit.

Example 12.11 illustrates how to calculate the static pressure at the exit of a de Laval nozzle with overexpanded flow.

EXAMPLE 12.11

A Shock Wave in a de Laval Nozzle

Problem Statement

The de Laval nozzle shown in the figure has an expansion ratio of 4 (exit area/throat area). Air flows through the nozzle, and a normal shock wave occurs where the area ratio is 2. The total pressure upstream of the shock is 1 MPa. Determine the static pressure at the exit.

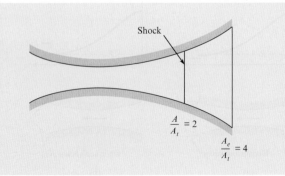

Shock

$\frac{A}{A_t} = 2$

$\frac{A_e}{A_t} = 4$

Define the Situation

Air flows in de Laval nozzle with an area ratio (A_e/A_*) of 4 and a normal shock at $A/A_* = 2$.

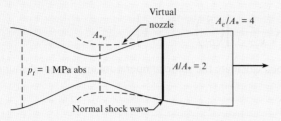

Properties: $k_{air} = 1.4$.

State the Goal

Calculate the static pressure (in kPa) at the exit.

Generate Ideas and Make a Plan

This problem will require the identification of a "virtual nozzle" shown in the sketch. The virtual nozzle is an expanding nozzle with subsonic flow and with a Mach number equal to the downstream Mach number behind the normal shock wave.

1. From Table A.1, interpolate to find the Mach number for $A/A_* = 2$.

2. Using the same table, find the Mach number downstream of shock and total pressure ratio across shock.

3. Calculate total pressure downstream of the shock wave.

4. Treat the problem as flow in a virtual subsonic nozzle with a Mach number equal to the Mach number behind the wave with new total pressure. Calculate the exit area ratio of the virtual nozzle.

5. Use the subsonic flow table to find the subsonic Mach number at exit.

6. Use the total pressure equation to calculate static pressure at exit.

Take Action (Execute the Plan)

1. From interpolation of the supersonic-flow part of Table A.1,

$$\text{at } A/A_* = 2, \text{ and } M = 2.2.$$

2. From the same entry in the table,

$$M_2 = 0.547$$
$$\frac{p_{t_2}}{p_{t_1}} = 0.6281$$

3. Total pressure downstream of the shock wave:

$$p_{t_2} = 0.6281 \times 1 \text{ MPa} = 6.28 \text{ kPa}$$

4. From the subsonic part of Table A.1,

$$\text{at } M = 0.547, \text{ and } A/A_{*_v} = 1.26.$$

5. Exit area ratio of virtual nozzle:

$$\frac{A_e}{A_{*_v}} = \frac{A_e}{A_*} \times \frac{A_*}{A_s} \times \frac{A_s}{A_{*_v}}$$

$$= 4 \times \frac{1}{2} \times 1.26 = 2.52$$

where A_s is the cross-sectional area at the shock wave.

6. From the subsonic part of Table A.1,

$$\text{at } A/A_* = 2.52, \text{ M} = 0.24$$

Exit pressure from Eq. (12.26):

$$\frac{p_t}{p_e} = \left(1 + \frac{k-1}{2}M^2\right)^{k/(k-1)}$$

$$p_e = \frac{628 \text{ kPa}}{[1 + (0.2)(0.24)^2]^{3.5}} = \boxed{603 \text{ kpa}}$$

Mass Flow through a Truncated Nozzle

The **truncated nozzle** is a de Laval nozzle cut off at the throat, as shown in Fig. 12.18. The nozzle exits to a back pressure p_b. This type of nozzle is important to engineers because of its frequent use as a flow-metering device for compressible flows. The purpose of this section is to develop an equation for mass flow through a truncated nozzle.

To calculate the mass flow, one must first determine whether the flow at the exit is sonic or subsonic. Of course, the flow at the exit could never be supersonic because the nozzle area does not diverge. First, calculate the value of the critical pressure ratio,

$$\frac{p_*}{p_t} = \left(\frac{2}{k+1}\right)^{k/(k-1)}$$

which, for air, is 0.528. Then evaluate the ratio of back pressure to total pressure, p_b/p_t, and compare it with the critical pressure ratio:

1. If $p_b/p_t \le p_*/p_t$, the exit pressure is higher than or equal to the back pressure, so the exit flow must be sonic. Pressure equilibration is achieved after exit by a series of expansion

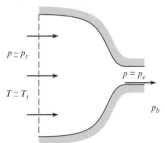

FIGURE 12.18

Truncated nozzle.

waves. The mass flow is calculated using Eq. (12.56), where A_* is the area at the truncated station.

2. If $p_b/p_t > p_*/p_t$, the flow exits subsonically. In this case the exit pressure is equal to the back pressure. One must first determine the Mach number at the exit by using Eq. (12.31):

$$M_e = \sqrt{\frac{2}{k-1}\left[\left(\frac{p_t}{p_b}\right)^{(k-1)/k} - 1\right]}$$

Using this value for the Mach number, calculate the static temperature and speed of sound at the exit:

$$T_e = \frac{T_t}{\{1 + [(k-1)/2]M_e^2\}}$$
$$c_e = \sqrt{kRT_e}$$

The gas density at the nozzle exit is determined by using the ideal gas law with the exit temperature and back pressure:

$$\rho_e = \frac{p_b}{RT_e}$$

Finally, the mass flow is given by

$$\dot{m} = \rho_e A_e M_e c_e$$

where A_e is the area at the truncated section.

Example 12.12 shows how to calculate mass flow in a truncated nozzle.

EXAMPLE 12.12

Mass Flow in a Truncated Nozzle

Problem Statement

Air exhausts through a truncated nozzle 3 cm in diameter from a reservoir at a pressure of 160 kPa and a temperature of 80°C. Calculate the mass flow rate if the back pressure is 100 kPa.

Define the Situation

Air flows through a 3 cm diameter truncated nozzle.

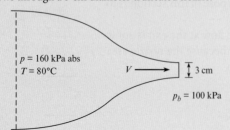

p = 160 kPa abs
T = 80°C
V → 3 cm
p_b = 100 kPa

Properties: From Table A.2, $k_{air} = 1.4$.

State the Goal

Calculate the mass flow rate (in kg/s) through the nozzle.

Generate Ideas and Make a Plan

1. Determine the exit condition by comparing exit pressure with back pressure. If $p_b/p_t < p_*/p_t$, exit flow is sonic. If $p_b/p_t > p_*/p_t$, exit flow is subsonic.

2. Calculate mass flow according to the exit condition.

Take Action (Execute the Plan)

1. Ratio of exit pressure to total pressure:

$$p_b/p_t = 100/160 = 0.625$$

Because 0.625 is larger than the critical pressure ratio for air (0.528), the flow at the nozzle exit must be subsonic.

2. Mach number at exit, from the total pressure equation, Eq. (12.26):

$$M_e^2 = \frac{2}{k-1}\left[\left(\frac{p_t}{p_b}\right)^{(k-1)/k} - 1\right]$$
$$M_e = 0.85$$

Static temperature at exit, from the total temperature equation, Eq. (12.22):

$$T_e = \frac{T_t}{\{1 + [(k-1)/2]M_e^2\}} = 308 \text{ K}$$

Static density at exit, from the ideal gas law:

$$\rho_e = \frac{p_b}{RT_e} = \frac{100 \times 10^3 \text{ N/m}^2}{(287 \text{ J/kg K})(309 \text{ K})} = 1.13 \text{ kg/m}^3$$

Speed of sound at the exit from the speed-of-sound equation, Eq. (12.11):

$$c_e = \sqrt{kRT_e} = [(1.4)(287 \text{ J/kg K})(309 \text{ K})]^{1/2}$$
$$= 352 \text{ m/s}$$

Mass flow rate:

$$\dot{m} = \rho_e A_e M_e c_e$$
$$\dot{m} = (1.13 \text{ kg/m}^3)(\pi/4)(0.03^2 \text{ m}^2)(0.85)(352 \text{ m/s})$$
$$= \boxed{0.239 \text{ kg/s}}$$

Review the Solution and the Process

Had p_b/p_t been less than 0.528, then Eq. (12.56) would have been used to calculate the mass flow rate.

Further information and other topic areas in compressible flow can be found in other sources, such as Anderson (2) and Shapiro (3).

12.5 Summarizing Key Knowledge

Speed of Sound and Compressible Flow

- The speed of sound is the speed at which an infinitesimal pressure disturbance travels through a fluid.
- The speed of sound in an ideal gas is

$$c = \sqrt{kRT}$$

where k is the ratio of specific heats, R is the gas constant, and T is the absolute temperature.

- The Mach number is defined as

$$M = \frac{V}{c}$$

- Compressible flows are classified as

$$\begin{array}{ll} M < 1 & \text{subsonic} \\ M \approx 1 & \text{transonic} \\ M > 1 & \text{supersonic} \end{array}$$

- In general, if the Mach number is less than 0.3, then a steady flow can be regarded as incompressible.

Property Variations along a Streamline

- For an adiabatic flow (no heat transfer), the temperature varies along a streamline according to

$$T = T_t\left(1 + \frac{k-1}{2}M^2\right)^{-1}$$

where T_t, the total temperature, is the temperature attained if the flow is decelerated to zero velocity.

- If the flow is isentropic, the pressure varies along a streamline as

$$p = p_t\left(1 + \frac{k-1}{2}M^2\right)^{-k/(k-1)}$$

where p_t is the total pressure, the pressure achieved if the flow is decelerated to zero velocity isentropically.

The Normal Shock Wave

- A normal shock wave is a narrow region in which a supersonic flow is decelerated to a subsonic flow with an attendant rise in pressure, temperature, and density. The total temperature does not change through a shock wave, but the total pressure decreases. The shock wave is a nonisentropic process and can only occur in supersonic flows.

The de Laval Nozzle

- A de Laval nozzle is a duct with a converging and expanding area that is used to accelerate a compressible fluid to supersonic speeds. Sonic flow can occur only at the nozzle throat (minimum area).
- The ratio of the area at a location in the nozzle to the throat area, A/A_*, is a function of the local Mach number and the ratio of specific heats.
- The flow rate through a de Laval nozzle is given by

$$\dot{m} = 0.685\frac{p_t A_*}{\sqrt{RT_t}}$$

• A de Laval nozzle is classified by comparing the pressure at the exit, p_e, for supersonic flow in the nozzle with the back (ambient) pressure, p_b:

$$p_e/p_b > 1 \quad \text{underexpanded}$$
$$p_e/p_b = 1 \quad \text{ideally expanded}$$
$$p_e/p_b < 1 \quad \text{overexpanded}$$

• Shock waves occur in overexpanded nozzles, yielding a subsonic flow at the exit.

• A truncated nozzle is a de Laval nozzle terminated at the throat. The truncated nozzle is typically used for mass flow measurement.

REFERENCES

1. Cengel, Y. A., and M. A. Boles, *Thermodynamics*. New York: McGraw-Hill, 1994.

2. Anderson, J. D., Jr. *Modern Compressible Flow with Historical Perspective*. New York: McGraw-Hill, 1991.

3. Shapiro, A. H. *The Dynamics and Thermodynamics of Compressible Fluid Flow*. New York: Ronald Press, 1953.

PROBLEMS

Speed of Sound and Mach Number (§12.1)

12.1 Find from available sources the Mach number at which modern-day airliners fly at altitude. Discuss whether it is possible to have regions of supersonic flow on the aircraft.

12.2 From available sources, find the orbital velocity of a satellite circling the earth. If the satellite entered the earth's atmosphere at this speed and the air temperature were $-60°C$, what would the Mach number be? Classify the flow.

12.3 Determine what the equation for the speed of sound in an ideal gas would be if the sound wave were an isothermal process.

12.4 The relationship between pressure and density for the propagation of a sound wave through a fluid is

$$p - p_0 = E_v \ln (\rho/\rho_0)$$

where p_0 and ρ_0 are the reference pressure and density (constants) and E_v is the bulk modulus of elasticity. Determine the equation for the speed of a sound wave in terms of E_v and ρ. Calculate the sound speed for water with $\rho = 1000$ kg/m^3 and $E_v = 2.20$ GN/m^2.

Mach-Number Relationships (§12.2)

12.5 A supersonic aircraft is flying at Mach 1.6 through air at $30°C$. What temperature could be expected on exposed aircraft surfaces?

12.6 A total heat tube inserted in the flow of a compressible fluid measures the stagnation pressure. Explain the difference between the total and stagnation pressure.

12.7 An object is immersed in an airflow with a static pressure of 200 kPa absolute, a static temperature of $20°C$, and a velocity of 250 m/s. What are the pressure and temperature at the stagnation point?

12.8 One problem in creating high-Mach-number flows is condensation of the oxygen component in the air when the temperature reaches 50 K. If the temperature of the reservoir is 300 K and the flow is isentropic, at what Mach number will condensation of oxygen occur?

12.9 Using Eq. (12.26) in §12.2, develop an expression for the pressure coefficient at stagnation conditions—that is, $C_p = (p_t - p)/[(1/2)\rho V^2]$—in terms of Mach number and ratio of specific heats, $C_p = f(k, M)$. Evaluate C_p at $M = 0$, 2, and 4 for $k = 1.4$. What would its value be for incompressible flow?

12.10 For low velocities, the total pressure is only slightly larger than the static pressure. Thus one can write $p_t/p = 1 + \epsilon$, where ϵ is a small positive number ($\epsilon \ll 1$). Using this approximation, show that as $\epsilon \to 0$ ($M \to 0$), Eq. (12.31) in §12.2 reduces to

$$M = \left[\frac{2(p_t/p - 1)}{k} \right]^{1/2}$$

Normal Shock Waves (§12.3)

12.11 Show that the lowest Mach number possible downstream of a normal shock wave is

$$M_2 = \sqrt{\frac{k-1}{2k}}$$

and that the largest density ratio possible is

$$\frac{\rho_2}{\rho_1} = \frac{k+1}{k-1}$$

What are the limiting values of M_2 and ρ_2/ρ_1 for air?

12.12 Show that the Mach number downstream of a weak wave ($M \approx 1$) is approximated by

$$M_2^2 = 2 - M_1^2$$

[*Hint:* Let $M_1^2 = 1 + \epsilon$, where $\epsilon \ll 1$, and expand Eq. (12.40) in §12.3 in terms of ϵ.] Compare values for M_2 obtained using this equation with values for M_2 from Table A.1 for $M_1 = 1$, 1.05, 1.1, and 1.2.

Flow in Truncated Nozzles (§12.4)

12.13 Develop a computer program for calculating the mass flow in a truncated nozzle. The input to the program would be total pressure, total temperature, back pressure, ratio of specific heats, gas constant, and nozzle diameter. Run the program and compare the results with Example 12.12 in §12.4. Run the program for back pressures of 80, 90, 110, 120, and 130 kPa and make a table for the variation of mass flow rate with back pressure. What trends do you observe?

This program will be useful for Prob. 12.15.

12.14 A truncated nozzle with an exit area of 8 cm^2 is used to measure a mass flow of air of 0.40 kg/s. The static temperature of the air at the exit is 0°C, and the back pressure is 100 kPa. Determine the total pressure.

12.15 A truncated nozzle with a 10 cm^2 exit area is supplied from a helium reservoir in which the absolute pressure is first 130 kPa and then 350 kPa. The temperature in the reservoir is 28°C, and the back pressure is 100 kPa. Calculate the mass flow rate of helium for the two reservoir pressures.

12.16 Truncated nozzles are often used for flow-metering devices. Assume that you want to design a truncated nozzle, or a series of truncated nozzles, to measure the performance of an air compressor. The compressor is rated at 100 scfm (standard cubic feet per minute) at 120 psig. (A standard cubic foot is the volume the air would occupy at atmospheric conditions.) A performance curve for the compressor would be a plot of flow rate versus supply pressure. Explain how you would carry out the test program.

Flow in de Laval Nozzles (§12.4)

12.17 Develop a computer program that requires the Mach number and ratio of specific heats as input and prints out the area ratio, the ratio of static to total pressure, the ratio of static to total temperature, the ratio of density to total density, and the ratio of pressure before and after a shock wave. Run the program for a Mach number of 2 and a ratio of specific heats of 1.4, and compare with results in Table A.1. Then run the program for the same Mach number with ratios of specific heats of 1.3 (carbon dioxide) and 1.67 (helium).

This program will be useful for Prob. 12.22.

12.18 Develop a computer program that, given the area ratio, ratio of specific heats, and flow condition (subsonic or supersonic) as input, provides the Mach number. This will require a numerical root-finding algorithm. Run the program for an area ratio of 5 and ratio of specific heats of 1.4. Compare the results with those in Table A.1. Then run the program for the same area ratio but with the ratios of specific heats of 1.67 (helium) and 1.31 (methane).

This program will be useful for Probs. 12.20–12.23.

12.19 Determine the Mach number and area ratio at which the dynamic pressure is maximized in a de Laval nozzle with air. [*Hint:* Express q in terms of p and M, and use Eq. (12.26) in §12.2 for p. Differentiate with respect to M and equate to zero.]

12.20 A rocket motor operates at an altitude where the atmospheric pressure is 30 kPa. The expansion ratio of the nozzle is 4 (exit area/throat area). The chamber pressure of the motor

(total pressure) is 1.2 MPa, and the chamber temperature (total temperature) is 3000°C. The ratio of the specific heats of the exhaust gas is 1.2, and the gas constant is 400 J/kg K. The throat area of the rocket nozzle is 100 cm^2.

    a. Determine the Mach number, density, pressure, and velocity at the nozzle exit.

    b. Determine the mass flow rate.

    c. Calculate the thrust of the rocket using

$$T = \dot{m}V_e + (p_e - p_0)A_e$$

    d. What would the chamber pressure of the rocket have to be to have an ideally expanded nozzle? Calculate the rocket thrust under this condition.

12.21 A rocket motor is being designed to operate at sea level, where the pressure is 100 kPa absolute. The chamber pressure (total pressure) is 2.0 MPa, and the chamber temperature (total temperature) is 3300 K. The throat area of the nozzle is 10 cm^2. The ratio of the specific heats (k) of the exhaust gas is 1.2, and the gas constant is 400 J/kg K.

    a. Determine the nozzle expansion ratio that is required to achieve an ideally expanded nozzle, and determine the nozzle thrust under these conditions (see Prob. 12.20c for the thrust equation).

    b. Determine the thrust that would be obtained if the expansion ratio were reduced by 10% to achieve an underexpanded nozzle.

12.22 Air flows through a de Laval nozzle with an expansion ratio of 4. The total pressure of the air entering the nozzle is 200 kPa, and the back pressure is 100 kPa. Determine the area ratio at which the shock wave occurs in the expansion section of the nozzle. (*Hint:* This problem can be solved graphically by calculating the exit pressure corresponding to different shock wave locations and finding the location where the exit pressure is equal to the back pressure.)

12.23 A normal shock wave occurs in a nozzle at an area ratio of 5. Determine the entropy increase if the gas is hydrogen.

12.24 Design a supersonic wind tunnel that achieves a Mach number of 1.5 in a test section 5 cm by 5 cm. The tunnel is to be attached to a vacuum tank as shown in the figure. After the tank is evacuated, the valve is opened and atmospheric air is drawn through the tunnel into the tank. The tunnel should operate for 30 seconds before the pressure rises to the point in the tank that supersonic flow is no longer achievable. Do a preliminary design of this system including details such as nozzle dimensions, configuration, and tank size.

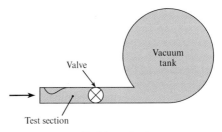

Valve

Vacuum tank

Test section

Problem 12.24

CHAPTER**THIRTEEN**

Flow Measurements

CHAPTER ROAD MAP Measurement techniques are important because fluid mechanics relies heavily on experiments. Thus, this chapter describes ways to measure flow rate, pressure, and velocity. Also, this chapter describes how to estimate the uncertainty of a measurement.

FIGURE 13.1
This photograph shows a laminar flow element being used to measure the volume flow rate of air for testing of fans. (Photo by Donald Elger.)

LEARNING OUTCOMES

VELOCITY AND PRESSURE (§13.1).

 • Describe common instruments for measuring velocity and pressure.

FLOW RATE (§13.2).

 • Calculate flow rate by integrating velocity distribution data.
 • Calculate flow rate for an obstruction flowmeter (i.e., an orifice, venturi, flow nozzle).
 • Calculate flow rate for a rectangular or triangular weir.

13.1 Measuring Velocity and Pressure

Stagnation (Pitot) Tube

The stagnation tube, also called the Pitot tube, is shown in Fig. 13.2a. A Pitot tube measures stagnation pressure with an open tube that is aligned parallel with the velocity direction and then senses pressure in the tube using a pressure gage or transducer.

When the stagnation tube was introduced in Chapter 4, viscous effects were not discussed. Viscous effects are notable because they can influence the accuracy of a measurement. The effects of viscosity, from reference (1), are shown in Fig. 13.3. This shows the pressure coefficient C_p plotted as a function of the Reynolds number. Viscous effects are important when $C_p > 1.0$. This guideline can be used to establish a Reynolds number range.

In Fig. 13.3, it is seen that when the Reynolds number for the circular stagnation tube is greater than 60, the error in measured velocity is less than 1%. For boundary layer measurements, a stagnation tube with a flattened end can be used. By flattening the end of the tube, the velocity measurement can be taken nearer the boundary than if a circular tube were used. For these flattened tubes, the pressure coefficient remains near unity for a Reynolds number as low as 30.

FIGURE 13.2

Section views of (a) Pitot tube, (b) static tube, and (c) Pitot-static tube.

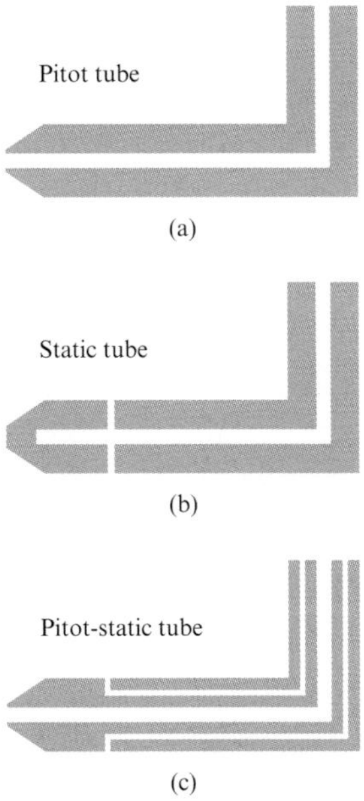

FIGURE 13.3

Viscous effects on C_p. [Data are from Hurd, Chesky, and Shapiro (1).]

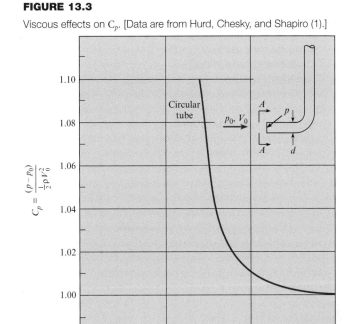

Static Tube

A **static tube**, as shown in Fig. 13.2b, is an instrument for measuring static pressure. Static pressure is the pressure in a fluid that is stationary or in a fluid that is flowing. When the fluid is flowing, the static pressure must be measured in a way that does not disturb the pressure. Thus, in the design of the static tube, as shown in Fig. 13.4, the placement of the holes along the probe is critical because the rounded nose on the tube causes some decrease of pressure along the tube, and the downstream stem causes an increase in pressure in front of it. Hence, the location for sensing the static pressure must be at the point where these two effects cancel each other. Experiments reveal that the optimum location is at a point approximately 6 diameters downstream of the front of the tube and 8 diameters upstream from the stem.

Pitot-Static Tube

The Pitot-static tube, Fig. 13.2c, measures velocity by using concentric tubes to measure static pressure and dynamic pressure. Application of the Pitot-static tube is presented in Chapter 4.

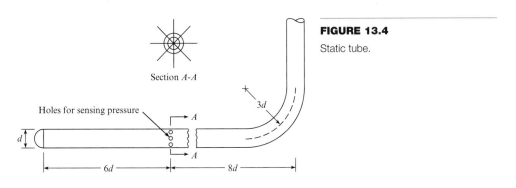

FIGURE 13.4

Static tube.

FIGURE 13.5

Various types of yaw meters:
(a) cylindrical-tube yaw meter,
(b) two-tube yaw meter,
(c) three-dimensional yaw meter.

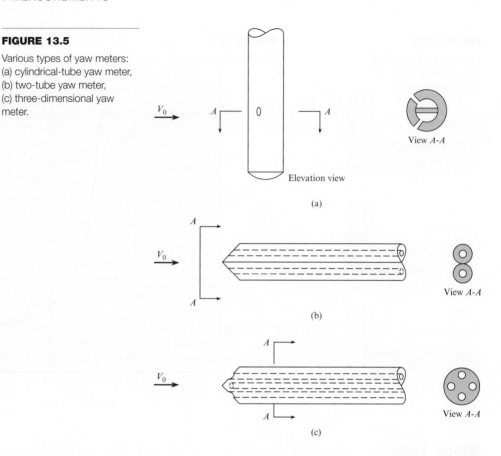

Yaw Meters

A **yaw meter**, Fig. 13.5, is an instrument for measuring velocity by using multiple pressure ports to determine the magnitude and direction of fluid velocity. The first two yaw meters in Fig. 13.5 can be used for two-dimensional flow, where flow direction in only one plane needs to be found. The third yaw meter in Fig. 13.5 is used for determining flow direction in three dimensions. In all these devices, the tube is turned until the pressure on symmetrically opposite openings is equal. This pressure is sensed by a differential pressure gage or manometer connected to the openings in the yaw meter. The flow direction is sensed when a null reading is indicated on the differential gage. The velocity magnitude is found by using equations that depend on the type of yaw meter that is used.

The Vane or Propeller Anemometer

The term **anemometer** originally meant an instrument that was used to measure the velocity of the wind. However, anemometer now means an instrument that is used to measure fluid velocity because anemometers are used in water, air, nitrogen, blood, and many other fluids.

The **vane anemometer** (Fig. 13.6a) and the **propeller anemometer** (Fig. 13.6b) measure velocity by using vanes typical of a fan or propeller, respectively. These blades rotate with a speed of rotation that depends on the wind speed. Typically, an electronic circuit converts the rotational speed into a velocity reading. On some older instruments, the rotor drives a low-friction gear train that, in turn, drives a pointer that indicates feet on a dial. Thus, if the anemometer is held in an airstream for 1 min and the pointer indicates a 300 ft change on the scale, the average airspeed is 300 ft/min.

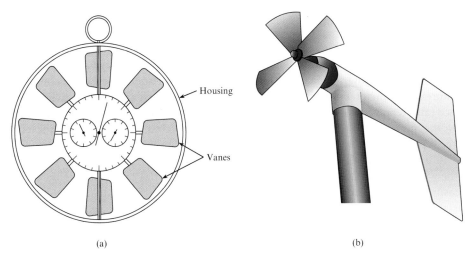

FIGURE 13.6

(a) Vane anemometer,
(b) propeller anemometer.

(a)   (b)

Cup Anemometer

Instead of using vanes, the **cup anemometer** (Fig. 13.7) is a device that uses the drag on cup-shaped objects to spin a rotor around a central axis. Because the rotational speed of the rotor is related to drag force, the frequency of rotation is related to the fluid velocity by appropriate calibration data. A typical rotor comprises three to five hemispherical or conical cups. In addition to applications in air, engineers use a cup anemometer to measure the velocity in streams and rivers.

FIGURE 13.7

Cup anemometer.

Hot-Wire and Hot-Film Anemometers

The **hot-wire anemometer** (HWA; Fig. 13.8) is an instrument for measuring velocity by sensing the heat transfer from a heated wire. As velocity increases, more energy is needed to keep the wire hot, and the corresponding changes in electrical characteristics can be used to determine the velocity of the fluid that is passing by the wire.

The HWA has advantages over other instruments. The HWA is well suited for measuring velocity fluctuations that occur in turbulent flow, whereas instruments such as the Pitot-static tube are only suitable for measuring velocity that either is steady or changes slowly with time. The sensing element of the HWA is quite small, allowing the HWA to be used in locations such as the boundary layer, where the velocity is varying in a region that is small in size. Many other instruments are too large for recording velocity in a region that is geometrically small. Another advantage of the HWA is that it is sensitive to low-velocity flows, a characteristic lacking in the Pitot tube and other instruments. The main disadvantages of the HWA are its delicate nature (the sensor wire is easily broken), its relatively high cost, and its need for an experienced user.

The basic principle of the hot-wire anemometer is described as follows: A wire of very small diameter—the sensing element of the hot-wire anemometer—is welded to supports as shown in Fig. 13.8. In operation, the wire either is heated by a fixed flow of electric current (the constant-current anemometer) or is maintained at a constant temperature by adjusting the current (the constant-temperature anemometer).

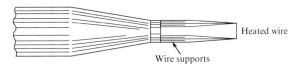

Heated wire

Wire supports

FIGURE 13.8

Probe for hot-wire anemometer (enlarged).

A flow of fluid past the hot wire causes the wire to cool because of convective heat transfer. In the constant-current anemometer, the cooling of the wire causes its resistance to change, and a corresponding voltage change occurs across the wire. Because the rate of cooling is a function of the speed of flow past the heated wire, the voltage across the wire is correlated with the flow velocity. The more popular type of anemometer, the constant-temperature anemometer, operates by varying the current in such a manner as to keep the resistance (and temperature) constant. The flow of current is correlated with the speed of the flow: the higher the speed, the greater the current needed to maintain a constant temperature. Typically, the wires are 1 mm to 2 mm in length and heated to 150°C. The wires may be 10 μm or less in diameter; the time response improves with the smaller wire. The lag of the wire's response to a change in velocity (thermal inertia) can be compensated for more easily, using modern electronic circuitry, in constant-temperature anemometers than in constant-current anemometers. The signal from the hot wire is processed electronically to give the desired information, such as mean velocity or the root mean square of the velocity fluctuation.

To illustrate the versatility of these instruments, note that the hot-wire anemometer can accurately measure gas flow velocities from 30 cm/s to 150 m/s; it also can measure fluctuating velocities with frequencies up to 100,000 Hz, and it has been used satisfactorily for both gases and liquids.

The single hot wire mounted normal to the mean flow direction measures the fluctuating component of velocity in the mean flow direction. Other probe configurations and electronic circuitry can be used to measure other components of velocity.

For velocity measurements in liquids or dusty gases, where wire breakage is a problem, the hot-film anemometer is more suitable. This anemometer consists of a thin conducting metal film (less than 0.1 μm thick) mounted on a ceramic support, which may be 50 μm in diameter. The hot film operates in the same fashion as the hot wire. Recently, the split film has been introduced. It consists of two semicylindrical films mounted on the same cylindrical support and electrically insulated from each other. The split film provides both speed and directional information.

For more detailed information on the hot-wire and hot-film anemometers, see King and Brater (2) and Lomas (3).

Laser-Doppler Anemometer

The **laser-Doppler anemometer** (LDA) is an instrument for measuring velocity by using the Doppler shift that occurs when a particle in a flow scatters light from crossed laser beams. Advantages of the LDA are that the flow field is not disturbed by the presence of a probe and it provides excellent spatial resolution. Disadvantages of the LDA include cost, complexity, the need for a transparent fluid, and requirements for particle seeding.

There are several different configurations for the LDA. The dual-beam mode (Fig. 13.9) splits a laser beam into two parallel beams and then uses a converging lens to cause the two beams to cross. The point where beams cross is called the measuring volume, which might best be described as an ellipsoid that is typically 0.3 mm in diameter and 2 mm long, illustrating the excellent spatial resolution achievable. The interference of the two beams generates a series of light and dark fringes in the measuring volume perpendicular to the plane of the two beams. As a particle passes through the fringe pattern, light is scattered, and a portion of the scattered light passes through the collecting lens toward the photodetector. A typical signal obtained from the photodetector is shown in the figure.

It can be shown from optics theory that the spacing between the fringes is given by

$$\Delta x = \frac{\lambda}{2 \sin \phi} \tag{13.1}$$

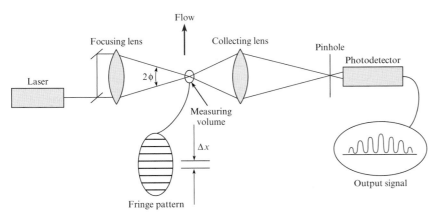

FIGURE 13.9

Dual-beam laser-Doppler anemometer.

where λ is the wavelength of the laser beam and ϕ is the half-angle between the crossing beams. By suitable electronic circuitry, the frequency of the signal (f) is measured, so the velocity is given by

$$U = \frac{\Delta x}{\Delta t} = \frac{\lambda f}{2 \sin \phi} \qquad (13.2)$$

The operation of the laser-Doppler anemometer depends on the presence of particles to scatter the light. These particles need to move at the same velocity as the fluid. Thus, the particles need to be small relative to the size of flow patterns, and they need to have a density near that of the ambient fluid. In liquid flows, impurities of the fluid can serve as scattering centers. In water flows, adding a few drops of milk is common. In gaseous flows, it is common to "seed" the flow with small particles. Smoke is often used for this seeding.

Laser-Doppler anemometers that provide two or three velocity components of a particle traveling through the measuring volume are now available. This is accomplished by using laser beam pairs of different colors (wavelengths). The measuring volumes for each color are positioned at the same physical location but oriented differently to measure a different component. The signal-processing system can discriminate the signals from each color and thereby provide component velocities.

Another recent technological advance in laser-Doppler anemometry is the use of fiber optics, which transmit the laser beams from the laser to a probe that contains optical elements to cross the beams and generate a measuring volume. Thus, measurements at different locations can be made by moving the probe and without moving the laser. For more applications of the laser-Doppler technique, see Durst (4).

Marker Methods

The marker method for determining velocity involves particles that are placed in the stream. By analyzing the motion of these particles, one can deduce the velocity of the flow itself. Of course, this requires that the markers follow virtually the same path as the surrounding fluid elements. Therefore, the marker must have nearly the same density as the fluid, or it must be so small that its motion relative to the fluid is negligible. Thus, for water flow it is common to use colored droplets from a liquid mixture that has nearly the same density as the water. For example, Macagno (6) used a mixture of *n*-butyl phthalate and xylene with a bit of white paint to yield a mixture that had the same density as water and could be photographed effectively. Solid particles (such as plastic beads) that have densities near that of the liquid being studied can also be used as markers.

FIGURE 13.10

When the hydrogen bubble visualization method is applied, hydrogen bubbles are produced by the electrolysis of water.

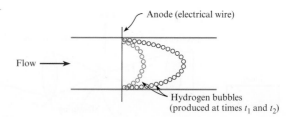

For a flow of water, hydrogen bubble visualization (Fig. 13.10) is a useful method. The technique involves the generation of small hydrogen bubbles from a tiny electrical wire (25 to 50 μm in diameter) using electrolysis. The electrical wire acts as the negative electrode (i.e., the anode). The cathode (i.e., the positive electrode) is situated where it will not disturb the flow. The wire is pulsed with a current, and hydrogen bubbles form on the wire and then are transported downstream by the flowing water. By repeating the electrical signal at various times, you can produce lines of bubbles. Other details concerning the marker methods of flow visualization are described by Macagno (6).

A relatively new marker method is particle image velocimetry (PIV), which provides a measurement of the velocity field. In PIV, the marker or seeding particles may be minuscule spheres of aluminum, glass, or polystyrene, or they may be oil droplets, oxygen bubbles (liquids only), or smoke particles (gases only). The seeding particles are illuminated to produce a photographic record of their motion. In particular, a sheet of light passing through a cross section of the flow is pulsed on twice, and the scattered light from the particles is recorded by a camera. The first pulse of light records the position of each particle at time t, and the second pulse of light records the position at time $t + \Delta t$. Thus, the displacement $\Delta\mathbf{r}$ of each particle is recorded on the photograph. Dividing $\Delta\mathbf{r}$ by Δt yields the velocity of each particle. Because PIV uses a sheet of light, the method provides a simultaneous measurement of velocity at locations throughout a cross section of the flow. Hence, PIV is identified as a whole-field technique. Other velocity measurements (e.g., the LDA method) are limited to measurements at one location.

PIV measurement of the velocity field for flow over a backward-facing step is shown in Fig. 13.11. This experiment was carried out in water using 15-μm-diameter, silver-coated

FIGURE 13.11

Velocity vectors from PIV measurements. (Courtesy of TSI Incorporated and Florida State University.)

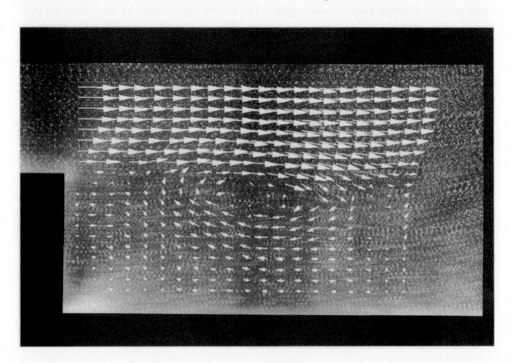

FIGURE 13.12

Flow pattern produced by a model truck in a wind tunnel. (Photo by Stephen Lyda.)

hollow spheres as seeding particles. Notice that the PIV method provided data over the cross section of the flow. Although the data shown in Fig. 13.11 are qualitative, numerical values of the velocity at each location are also available.

The PIV method is typically performed using digital hardware and computers. For example, images may be recorded with a digital camera. Each resulting digital image is evaluated with software that calculates the velocity at points throughout the image. This evaluation proceeds by dividing the image into small subareas called "interrogation areas." Within a given interrogation area, the displacement vector ($\Delta\mathbf{r}$) of each particle is found by using statistical techniques (auto- and cross-correlation). After processing, the PIV data are typically available on a computer screen. Additional information on PIV systems is provided by Raffel et al. (7).

Smoke is often used as a marker in flow measurement. One technique is to suspend a wire vertically across the flow field and allow oil to flow down the wire. The oil tends to accumulate in droplets along the wire. Applying a voltage to the wire vaporizes the oil, creating streaks from the droplets. Fig. 13.12 is an example of a flow pattern revealed by such a method. Smoke generators that provide smoke by heating oils are also commercially available. It is also possible to position a thin sheet of laser light through the smoke field to obtain an improved spatial definition of the flow field indicated by the smoke. Another technique is to introduce titanium tetrachloride ($TiCl_4$) in a dried-air flow, which reacts with the water vapor in the ambient air to produce micron-sized titanium oxide particles, which serve as tracers.

13.2 Measuring Flow Rate (Discharge)

Measuring flow rate is important in research, design, and testing and in many commercial applications.

Direct Measurement of Volume or Weight

For liquids, a simple and accurate method is to collect a sample of the flowing fluid over a given period of time Δt. The sample is weighed, and the average weight rate of flow is $\Delta W/\Delta t$, where ΔW is the weight of the sample. The volume of a sample can also be measured (usually in a calibrated tank), and from this the average volume rate of flow is calculated as $\Delta V\!\!\!/\Delta t$, where $\Delta V\!\!\!/$ is the volume of the sample. This method has several disadvantages; for example, it cannot be used for an unsteady flow, and it is not always possible to collect a sample.

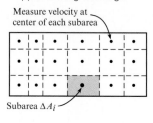

Measure velocity at center of each subarea

Subarea ΔA_i

Integrating a Measured Velocity Distribution

Flow rate can be found by measuring a velocity distribution and then integrating using the volume flow rate equation:

$$Q = \int_A V\, dA$$

For example, one can divide a rectangular conduit into subareas and then measure velocity at the center of each subarea, as shown in Fig. 13.13. Then, flow rate is determined by

$$Q = \int_A V\, dA \approx \sum_{i=1}^{N} V_i(\Delta A)_i \tag{13.3}$$

where N is the number of subareas. When the flow area occurs in a round pipe, then the subarea is a ring, as shown by Example 13.1.

EXAMPLE 13.1

Calculating Flow Rate from Velocity Data

Problem Statement

The data given in the table are for a velocity traverse of air flow in a pipe 100 cm in diameter. What is the volume rate of flow in cubic meters per second?

r (cm)	V (m/s)
0.00	50.0
5.00	49.5
10.00	49.0
15.00	48.0
20.00	46.5
25.00	45.0
30.00	43.0
35.00	40.5
40.00	37.5
45.00	34.0
47.50	25.0
50.00	0.0

Define the Situation

Air is flowing in a round pipe ($D = 1.0$ m).

Velocity in m/s is known as a function of radius (see table).

Assumptions: The velocity distribution is symmetric around the centerline of the pipe.

State the Goal

Calculate the volume flow rate (m³/s) in the pipe.

Generate Ideas and Make a Plan

1. Develop an equation for a round pipe by applying Eq. (13.3).

2. Find discharge by using a spreadsheet program.

Take Action (Execute the Plan)

The flow rate is given by

$$Q = \sum_{i=1}^{N} V_i(\Delta A)_i$$

The area ΔA_i is shown in the sketch. Visualize this area as a strip of length $2\pi r_i$ and width Δr_i. Then $\Delta A_i \approx (2\pi r_i)\Delta r_i$. The flow rate equation becomes

$$Q = \sum_{i=1}^{N} V_i(\Delta A)_i = \sum_{i=1}^{N} V_i(2\pi r_i)\Delta r_i$$

$\Delta A_i = (2\pi r_i)\Delta r_i$ Δr_i

i	r_i (cm)	V_i (m/s)	$2*\pi*r_i$ (m)	Δr_i (m)	ΔA_i (m²)	$V_i*\Delta A_i$ (m³/s)
1	0.0	50.0	0.0000	0.0250	0.0000	0.000
2	5.0	49.5	0.3142	0.0500	0.0157	0.778
3	10.0	49.0	0.6283	0.0500	0.0314	1.539
4	15.0	48.0	0.9425	0.0500	0.0471	2.262
5	20.0	46.5	1.2566	0.0500	0.0628	2.922
6	25.0	45.0	1.5708	0.0500	0.0785	3.534
7	30.0	43.0	1.8850	0.0500	0.0942	4.053
8	35.0	40.5	2.1991	0.0500	0.1100	4.453
9	40.0	37.5	2.5133	0.0500	0.1257	4.712
10	45.0	34.0	2.8274	0.0375	0.1060	3.605
11	47.5	25.0	2.9845	0.0250	0.0746	1.865
12	50.0	0.0	3.1416	0.0125	0.0393	0.000
			SUM⇒	0.50	0.79	29.72

To perform the sum, use a spreadsheet as shown. To see how the table is set up, consider the row $i = 2$. The area is

$$\Delta A_2 = (2\pi r_2)\Delta r_2 = (2\pi(0.05 \text{ m}))(0.05 \text{ m}) = 0.0157 \text{ m}^2$$

which is given in the sixth column. The last column gives

$$V_2(\Delta A)_2 = (49.5 \text{ m/s})(0.0157 \text{ m}^2) = 0.778 \text{ m}^3/\text{s}$$

Discharge is found by summing the last column. As shown,

$$Q = \sum_{i=1}^{12} V_i(\Delta A)_i = \boxed{29.7 \frac{\text{m}^3}{\text{s}}}$$

To check the validity of the calculation, sum the column labeled Δr_i and check to ensure that this value equals the radius of the pipe. As shown, this sum is 0.5 m. Similarly, the pipe area of

$$A = \pi r^2 = \pi(0.5 \text{ m})^2 = 0.785 \text{ m}^2$$

should be produced by summing the column labeled ΔA_i. As shown, this is the case.

Calibrated Orifice Meter

An **orifice meter** is an instrument for measuring flow rate by using a carefully designed plate with a round opening and situating this device in a pipe, as shown in Fig. 13.14. Flow rate is found by measuring the pressure drop across the orifice and then using an equation to calculate the appropriate flow rate. One common application of the orifice meter is metering of natural gas in pipelines. Because large quantities of natural gas are measured and the associated costs are high, accuracy is very important. This section describes the main ideas associated with orifice meters. Details about using orifice meters are presented in standards such as reference (10).

Flow through a sharp-edged orifice is shown in Fig. 13.14. Note that the streamlines continue to converge a short distance downstream of the plane of the orifice. Hence, the minimum-flow area is actually smaller than the area of the orifice. To relate the minimum-flow area, often called the contracted area of the jet, or **vena contracta**, to the area of the orifice A_o, one uses the contraction coefficient, which is defined as

$$A_j = C_c A_o$$

$$C_c = \frac{A_j}{A_o}$$

Then, for a circular orifice,

$$C_c = \frac{(\pi/4)d_j^2}{(\pi/4)d^2} = \left(\frac{d_j}{d}\right)^2$$

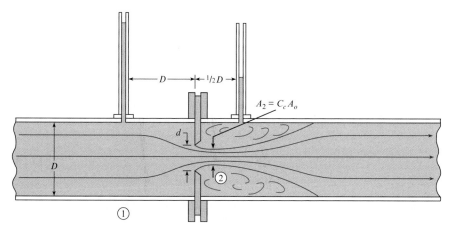

FIGURE 13.14

Flow through a sharp-edged pipe orifice.

Because d_j and d_2 are identical, $C_c = (d_2/d)^2$. At low values of the Reynolds number, C_c is a function of the Reynolds number. However, at high values of the Reynolds number, C_c is only a function of the geometry of the orifice. For d/D ratios less than 0.3, C_c has a value of approximately 0.62. However, as d/D is increased to 0.8, C_c increases to a value of 0.72.

To derive the orifice equation, consider the situation shown in Fig. 13.14. Apply the Bernoulli equation between section 1 and section 2:

$$\frac{p_1}{\gamma} + \frac{V_1^2}{2g} + z_1 = \frac{p_2}{\gamma} + \frac{V_2^2}{2g} + z_2$$

V_1 is eliminated by means of the continuity equation $V_1 A_1 = V_2 A_2$. Solving for V_2 gives

$$V_2 = \left\{ \frac{2g\left[(p_1/\gamma + z_1) - (p_2/\gamma + z_2) \right]}{1 - (A_2/A_1)^2} \right\}^{1/2} \tag{13.4a}$$

However, $A_2 = C_c A_o$ and $h = p/\gamma + z$, so Eq. (13.4a) reduces to

$$V_2 = \sqrt{\frac{2g(h_1 - h_2)}{1 - C_c^2 A_o^2/A_1^2}} \tag{13.4b}$$

Our primary objective is to obtain an expression for discharge in terms of h_1, h_2, and the geometric characteristics of the orifice. The discharge is given by $V_2 A_2$. Hence, multiply both sides of Eq. (13.4b) by $A_2 = C_c A_o$ to give the desired result:

$$Q = \frac{C_c A_o}{\sqrt{1 - C_c^2 A_o^2/A_1^2}} \sqrt{2g(h_1 - h_2)} \tag{13.5}$$

Equation (13.5) is the discharge equation for the flow of an incompressible inviscid fluid through an orifice. However, it is valid only at relatively high Reynolds numbers. For low and moderate values of the Reynolds number, viscous effects are significant, and an additional coefficient called the *coefficient of velocity*, C_v, must be applied to the discharge equation to relate the ideal to the actual flow.[*] Thus, for viscous flow through an orifice, we have the following discharge equation:

$$Q = \frac{C_v C_c A_o}{\sqrt{1 - C_c^2 A_o^2/A_1^2}} \sqrt{2g(h_1 - h_2)}$$

The product $C_v C_c$ is called the **discharge coefficient**, C_d, and the combination $C_v C_c/(1 - C_c^2 A_o^2/A_1^2)^{1/2}$ is called the **flow coefficient**, K. Thus, $Q = KA_o \sqrt{2g(h_1 - h_2)}$, where

$$K = \frac{C_d}{\sqrt{1 - C_c^2 A_o^2/A_1^2}} \tag{13.6}$$

If Δh is defined as $h_1 - h_2$, then the final form of the orifice equation reduces to

$$Q = KA_o \sqrt{2g\Delta h} \tag{13.7a}$$

If a differential pressure transducer is connected across the orifice, it will sense a piezometric pressure change that is equivalent to $\gamma \Delta h$, so the orifice equation becomes

$$Q = KA_o \sqrt{2 \frac{\Delta p_z}{\rho}} \tag{13.7b}$$

---

[*]At low Reynolds numbers the coefficient of velocity may be quite small; however, at Reynolds numbers above 10^5, C_v typically has a value close to 0.98. See Lienhard (8) for C_v analyses.

Experimentally determined values of K as a function of d/D and Reynolds number based on orifice size are given in Fig. 13.15. If Q is given, Re_d is equal to $4Q/\pi\, dv$. K is obtained from Fig. 13.15 (using the vertical lines and the bottom scale), and Δh is computed from Eq. (13.7a), or Δp_z can be computed from Eq. (13.7b). However, the problem of determining the discharge Q when a certain value of Δh or a certain value of Δp_z is given often arises. When Q is to be determined, there is no direct way to obtain K by entering Fig. 13.15 with Re, because Re is a function of the flow rate, which is still unknown. Hence, another scale, which does not involve Q, is constructed on the graph of Fig. 13.15. The variables for this scale are obtained in the following manner: Because $\mathrm{Re}_d = 4Q/\pi\, dv$ and $Q = K(\pi d^2/4)\sqrt{2g\Delta h}$, write Re_d in terms of Δh,

$$\mathrm{Re}_d = K\frac{d}{v}\sqrt{2g\Delta h}$$

or

$$\frac{\mathrm{Re}_d}{K} = \frac{d}{v}\sqrt{2g\Delta h} = \frac{d}{v}\sqrt{\frac{2\Delta p_z}{\rho}}$$

Thus, the slanted dashed lines and the top scale are used in Fig. 13.15 when Δh is known and the flow rate is to be determined. If a certain value of Δp is given, then apply Fig. 13.15 by using $\Delta p_z/\rho$ in place of $g\Delta h$ in the parameter at the top of Fig. 13.15.

The literature on orifice flow contains numerous discussions concerning the optimum placement of pressure taps on both the upstream side and the downstream side of an orifice.

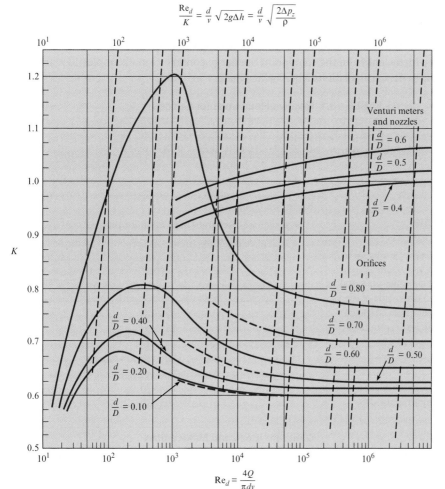

FIGURE 13.15

Flow coefficient K and Re_d/K versus the Reynolds number for orifices, nozzles, and venturi meters. [Data source: Tuve and Sprenkle (9) and ASME (10).]

The data given in Fig. 13.15 are for "corner taps." That is, on the upstream side the pressure readings were taken immediately upstream of the orifice plate (at the corner of the orifice plate and the pipe wall), and the downstream tap was at a similar downstream location. However, pressure data from flange taps (1 in. upstream and 1 in. downstream) and from the taps shown in Fig. 13.14 all yield virtually the same values for K; the differences are no greater than the deviations involved in reading Fig. 13.15. For more precise values of K with specific types of taps, see the ASME report on fluid meters (10).

Head Loss for Orifices

Some head loss occurs between the upstream side of the orifice and the vena contracta. However, this head loss is very small compared with the head loss that occurs downstream of the vena contracta. This downstream portion of the head loss is like that for an abrupt expansion. Neglecting all head loss except that due to the expansion of the flow gives

$$h_L = \frac{(V_2 - V_1)^2}{2g} \tag{13.8}$$

where V_2 is the velocity at the vena contracta and V_1 is the velocity in the pipe. It can be shown that the ratio of this expansion loss, h_L, to the change in head across the orifice, Δh, is given as

$$\frac{h_L}{\Delta h} = \frac{\dfrac{V_2}{V_1} - 1}{\dfrac{V_2}{V_1} + 1} \tag{13.9}$$

Table 13.1 shows how the ratio increases with increasing values of V_2/V_1. It is obvious that an orifice is very inefficient from the standpoint of energy conservation. Examples 13.2 and 13.3 illustrate how to make calculations when orifice meters are used.

TABLE 13.1 Relative Head Loss for Orifices

$V_2/V_1 \rightarrow$	1	2	4	6	8	10
$h_L/\Delta h \rightarrow$	0	0.33	0.60	0.71	0.78	0.82

EXAMPLE 13.3

Applying an Orifice Meter to Measure the Flow Rate of Water

Problem Statement

A 15 cm orifice is located in a horizontal 24 cm water pipe, and a water-mercury manometer is connected to either side of the orifice. When the deflection on the manometer is 25 cm, what is the discharge in the system, and what head loss is produced by the orifice? Assume the water temperature is 20°C.

Define the Situation

Water flows through an orifice ($d = 0.15$ m) in a pipe ($D = 0.24$ m). A mercury-water manometer is used to measure pressure drop.

Properties:

- Water (20°C): Table A.5, $\nu = 1 \times 10^{-6}$ m²/s
- Mercury (20°C): Table A.4, $SG = 13.6$

State the Goal

- Calculate discharge (in m³/s) in the pipe.
- Calculate head loss (in meters) produced by the orifice.

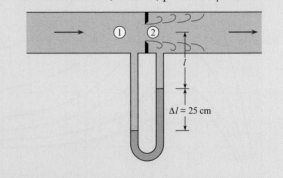

Generate Ideas and Make a Plan

1. Calculate $\Delta h = h_1 - h_2$ using the manometer equation.
2. Find the flow coefficient K using Fig. 13.15.

3. Find discharge Q using Eq. (13.7a).

4. Calculate the coefficient of contraction C_c using Eq. (13.6).

5. Solve for the velocity V_2 at the vena contracta.

6. Calculate head loss using Eq. (13.8).

Take Action (Execute the Plan)

1. Change in piezometric head:
   - Apply manometer equation from 1 to 2:
   $$p_1 + \gamma_w(l + \Delta l) - \gamma_{Hg}\Delta l - \gamma_w l = p_2$$
   - Solve for Δh:
   $$\Delta h = \frac{p_1 - p_2}{\gamma_w} = \Delta l \frac{\gamma_{Hg} - \gamma_w}{\gamma_w} = \Delta l \left(\frac{\gamma_{Hg}}{\gamma_w} - 1\right)$$
   $$\Delta h = (0.25\ \text{m})(13.6 - 1) = 3.15\ \text{m of water}$$

2. Flow coefficient:
   - Calculate (Re_d/K):
   $$\frac{\text{Re}_d}{K} = \frac{d\sqrt{2g\Delta h}}{v} = \frac{0.15\ \text{m}\ \sqrt{2(9.81\ \text{m/s}^2)(3.15\ \text{m})}}{1.0 \times 10^{-6}\ \text{m}^2/\text{s}}$$
   $$= 1.2 \times 10^6$$
   - From Fig. 13.15 with $d/D = 0.625$, $K = 0.66$ (interpolated).

3. Discharge:
   $$Q = 0.66 A_o \sqrt{2g\Delta h}$$
   $$= 0.66\ \frac{\pi}{4}d^2\ \sqrt{2(9.81\ \text{m/s}^2)(3.15\ \text{m})}$$
   $$= 0.66\ (0.785)(0.15^2\ \text{m}^2)(7.86\ \text{m/s}) = \boxed{0.092\ \text{m}^3/\text{s}}$$

4. Coefficient of contraction C_c:
   $$K = \frac{C_d}{\sqrt{1 - C_c^2 A_o^2/A_1^2}}$$
   Let $K = 0.66$. The ratio $(A_o/A_1)^2 = (0.625)^4 = 0.1526$ and $C_d = C_v C_c$. Assuming $C_v = 0.98$ (see the discussion of C_v in §13.2.) and solving for C_c gives $C_c = 0.633$.

5. Velocity at the vena contracta:
   $$V_2 = Q/(C_c A_o)$$
   $$(0.092\ \text{m}^3/\text{s})/[(0.633)(\pi/4)(0.15^2\ \text{m}^2)] = 8.23\ \text{m/s}$$
   $$V_1 = Q/A_{\text{pipe}}$$
   $$(0.092\ \text{m}^3/\text{s})/[(\pi/4)(0.24^2\ \text{m}^2)] = 2.03\ \text{m/s}$$

6. Head loss:
   $$h_L = (V_2 - V_1)^2/2g = (8.23 - 2.03)^2/(2 \times 9.81)$$
   $$= \boxed{1.96\ \text{m}}$$

EXAMPLE 13.4

Applying an Orifice Meter

Problem Statement

An air-water manometer is connected to either side of an 8 in. orifice in a 12 in. water pipe. If the maximum flow rate is 2 cfs, what is the deflection on the manometer? The water temperature is 60°F.

Define the Situation

- Water flows ($Q = 2$ cfs) through an orifice ($d = 8$ in.) in a pipe ($D = 2$ in.)
- An air-water manometer is used to measure pressure drop.

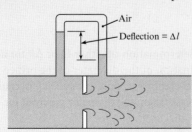

Properties: Water (60°F): Table A.5, $v = 1.22 \times 10^{-5}\ \text{ft}^2/\text{s}$.

State the Goal

Calculate the deflection (in ft) of water in the manometer.

Generate Ideas and Make a Plan

1. Calculate Reynolds number.

2. Find the flow coefficient K from Fig. 13.15.

3. Solve for Δh by using Eq. (13.7a).

4. Solve for Δl by using the manometer equation.

Take Action (Execute the Plan)

1. Reynolds number:
   $$\text{Re} = \frac{4Q}{\pi d v} = \frac{(4)(2\ \text{ft}^3/\text{s})}{\pi((8/12)\ \text{ft})(1.22 \times 10^{-5}\ \text{ft}^2/\text{s})} = \boxed{3.1 \times 10^5}$$

2. Flow coefficient:
   - Use Fig. 13.15. Interpolate for $d/D = 8/12 = 0.667$ to find $K \approx 0.68$.

3. Change in piezometric head:
   - From $Q = KA_o\sqrt{2g\Delta h}$, solve for Δh:
   $$\Delta h = \frac{Q^2}{2gK^2A_o^2} = \frac{4}{64.4(0.68^2)[((\pi)/4)(8/12)^2]^2} = 1.1\ \text{ft}$$

4. Manometer deflection:
   - The deflection is related to Δh by
   $$\Delta h = \Delta l \left(\frac{\gamma_w - \gamma_{air}}{\gamma_w}\right)$$
   - Because $\gamma_w \gg \gamma_{air}$, $\Delta l = \Delta h = 1.1$ ft. $\boxed{\Delta l = 1.1\ \text{ft}}$

The sharp-edged orifice can also be used to measure the mass flow rate of gases. The discharge equation [Eq. (13.7b)] is multiplied by the upstream gas density and an empirical factor to account for compressibility effects (10). The resulting equation is

$$\dot{m} = YA_oK\sqrt{2\rho_1(p_1 - p_2)} \tag{13.10}$$

where K, the flow coefficient, is found using Fig. 13.15 and Y is the compressibility factor given by the empirical equation

$$Y = 1 - \left\{\frac{1}{k}\left(1 - \frac{p_2}{p_1}\right)\left[0.41 + 0.35\left(\frac{A_o}{A_1}\right)^2\right]\right\} \tag{13.11}$$

In this case, both the pressure difference across the orifice and the absolute pressure of the gas are needed. When using the equation for the compressibility factor, remember that the absolute pressure must be used.

EXAMPLE 13.5

Applying an Orifice Meter to Measure the Flow Rate of Natural Gas

Problem Statement

The mass flow rate of natural gas is to be measured using a sharp-edged orifice. The upstream pressure of the gas is 101 kPa absolute, and the pressure difference across the orifice is 10 kPa. The upstream temperature of the methane is 15°C. The pipe diameter is 10 cm, and the orifice diameter is 7 cm. What is the mass flow rate?

Define the Situation

- Natural gas (methane) is flowing through a sharp-edged orifice.
- Pipe diameter is $D = 0.1$ m. Orifice diameter is $d = 0.07$ m.
- Pressure difference across orifice is 10 kPa.

Properties: Natural gas (15°C, 1 atm): Table A.2,
$\rho = 0.678$ kg/m^3, $v = 1.59 \times 10^{-5}$ m^2/s, $K = 1.31$.

State the Goal

Find the mass flow rate (in kg/s).

Generate Ideas (Make a Plan)

1. Find the flow coefficient K from Fig. 13.15.
2. Calculate the compressibility factor Y using Eq. (13.11).
3. Calculate the mass flow rate using Eq. (13.10).

Take Action (Execute the Plan)

1. Flow coefficient:
   - Calculate (Re_d/K):
   $$\frac{Re_d}{K} = \frac{d}{v}\sqrt{2\frac{\Delta p}{\rho_1}} = \frac{0.07}{1.59(10^{-5})}\sqrt{2\frac{10^4}{0.678}} = 7.56 \times 10^5$$
   - Using Fig. 13.15, $K = 0.7$.

2. Compressibility factor:
   $$Y = 1 - \left\{\frac{1}{1.31}\left(1 - \frac{91}{101}\right)(0.41 + 0.35 \times 0.7^4)\right\} = 0.962$$

3. Mass flow rate of methane:
   $$\dot{m} = YA_oK\sqrt{2\rho_1(p_1 - p_2)}$$
   $$= 0.962\left(\frac{\pi}{4}0.07^2\right)(0.7)\sqrt{2(0.678)(10^4)}$$
   $$= \boxed{0.302 \text{ kg/s}}$$

The foregoing examples involved the determination of either Q or Δh for a given size of orifice. Another type of problem is determination of the diameter of the orifice for a given Q and Δh. For this type of problem, a trial-and-error procedure is required. Because one knows the approximate value of K, that is guessed first. Then, the diameter is solved for, after which a better value of K can be determined, and so on.

Venturi Meter

The **venturi meter** (Fig. 13.16) is an instrument for measuring flow rate by using measurements of pressure across a converging-diverging flow passage. The main advantage of the venturi meter as compared to the orifice meter is that the head loss for a venturi meter is much smaller.

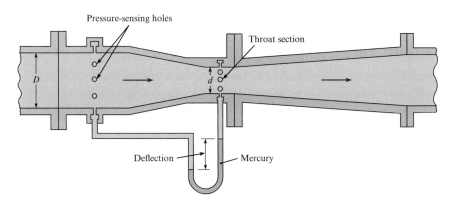

FIGURE 13.16

Typical venturi meter.

The lower head loss results from streamlining the flow passage, as shown in Fig. 13.16. Such streamlining eliminates any jet contraction beyond the smallest flow section. Consequently, the coefficient of contraction has a value of unity, and the venturi equation is

$$Q = \frac{A_t C_d}{\sqrt{1 - (A_t/A_p)^2}} \sqrt{2g(h_p - h_t)} \qquad (13.12)$$

$$Q = KA_t \sqrt{2g\Delta h} \qquad (13.13)$$

where A_t is the throat area and Δh is the difference in piezometric head between the venturi entrance (pipe) and the throat. Note that the venturi equation is the same as the orifice equation. However, K for the venturi meter approaches unity at high values of the Reynolds number and small d/D ratios. This trend can be seen in Fig. 13.15, where values of K for the venturi meter are plotted along with similar data for the orifice.

Flow Nozzles

The **flow nozzle** (Fig. 13.17) is an instrument for measuring flow rate by using the pressure drop across a nozzle that is typically placed inside a conduit. Similar to an orifice meter, design and application of the flow nozzle are described by engineering standards (10). As compared to an orifice meter, the flow nozzle is better in flows that cause wear (e.g., particle-laden flow). The reason is that erosion of an orifice will produce more change in the pressure-drop versus flow-rate relationship. Both the flow nozzle and orifice meter will produce about the same overall head loss.

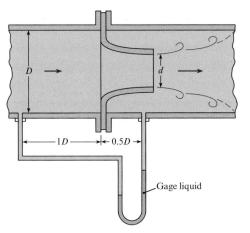

FIGURE 13.17

Typical flow nozzle.

EXAMPLE 13.6

Applying a Venturi Meter to Measure the Flow Rate of Water

Problem Statement

The pressure difference between the taps of a horizontal venturi meter carrying water is 35 kPa. If $d = 20$ cm and $D = 40$ cm, what is the discharge of water at 10°C?

Define the Situation

- Water flows through a horizontal venturi meter.
- Pipe diameter is $D = 0.40$ m. Venturi throat diameter is $d = 0.2$ m.

Properties: Water (10°C): Table A.5, $v = 1.31 \times 10^{-6}$ m^2/s, and $\gamma = 9810$ N/m^3.

State the Goal

Find the discharge (m^3/s).

Generate Ideas and Make a Plan

1. Compute $\Delta h = h_1 - h_2$.
2. Find the flow coefficient K from Fig. 13.15.
3. Find discharge Q using Eq. (13.7a).

Take Action (Execute the Plan)

1. Change in piezometric head:

$$\Delta h = \frac{\Delta p}{\gamma} + \Delta z = \frac{\Delta p}{\gamma} + 0 = \frac{35,000 \text{ N/m}^2}{9810 \text{ N/m}^3} = 3.57 \text{ m of water}$$

2. Flow coefficient:
   - Calculate (Re$_d$/K):

$$\frac{\text{Re}_d}{K} = \frac{d\sqrt{2g\Delta h}}{v} = \frac{0.20\sqrt{2(9.81)(3.57)}}{1.31(10^{-6})} = 1.28 \times 10^6$$

   - From Fig. 13.15, find that $K = 1.02$.

3. Discharge:

$$Q = 1.02 A_2 \sqrt{2g\Delta h}$$

$$= 1.02(0.785)(0.20^2)\sqrt{2(9.81)(3.57)} = \boxed{0.268 \text{ m}^3/\text{s}}$$

Electromagnetic Flowmeter

All the flowmeters described so far require that some sort of obstruction be placed in the flow. The obstruction may be the rotor of a vane anemometer or the reduced cross section of an orifice or venturi meter. A meter that neither obstructs the flow nor requires pressure taps (which are subject to clogging) is the **electromagnetic flowmeter**. Its basic principle is that a conductor that moves in a magnetic field produces an electromotive force. Hence, liquids with a degree of conductivity generate a voltage between the electrodes, as in Fig. 13.18, and this voltage is proportional to the velocity of flow in the conduit. It is interesting to note that the basic principle of the electromagnetic flowmeter was investigated by Faraday in 1832. However, practical application of the principle was not made until approximately a century later, when it was used to measure blood flow. Recently, with the need for a meter to measure the flow of liquid metal in nuclear reactors and with the advent of sophisticated electronic signal detection, this type of meter has found extensive commercial use.

FIGURE 13.18

Electromagnetic flowmeter.

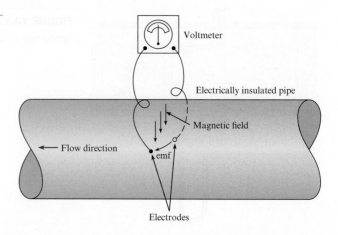

The main advantages of the electromagnetic flowmeter are that the output signal varies linearly with the flow rate and that the meter causes no resistance to the flow. The major disadvantages are its high cost and its unsuitability for measuring gas flow.

For a summary of the theory and application of the electromagnetic flowmeter, the reader is referred to Shercliff (11). This reference also includes a comprehensive bibliography on the subject.

Ultrasonic Flowmeter

Another form of nonintrusive flowmeter that is used in diverse applications ranging from blood flow measurement to open-channel flow is the **ultrasonic flowmeter**. Basically, there are two different modes of operation for ultrasonic flowmeters. One mode involves measuring the difference in travel time for a sound wave traveling upstream and downstream between two measuring stations. The difference in travel time is proportional to flow velocity. The second mode of operation is based on the Doppler effect. When an ultrasonic beam is projected into an inhomogeneous fluid, some acoustic energy is scattered back to the transmitter at a different frequency (Doppler shift). The measured frequency difference is related directly to the flow velocity.

Turbine Flowmeter

The **turbine flowmeter** consists of a wheel with a set of curved vanes (blades) mounted inside a duct. The volume rate of flow through the meter is related to the rotational speed of the wheel. This rotational rate is generally measured by a blade passing an electromagnetic pickup mounted in the casing. The meter must be calibrated for the given flow conditions. The turbine meter is versatile in that it can be used for either liquids or gases. It has an accuracy of better than 1% over a wide range of flow rates, and it operates with small head loss. The turbine flowmeter is used extensively in monitoring flow rates in fuel-supply systems.

Vortex Flowmeter

The **vortex flowmeter** (Fig. 13.19) measures flow rate by relating vortex shedding frequency to flow rate. The vortices are shed from a sensor tube that is situated in the center of a pipe. These vortices cause vibrations, which are sensed by piezoelectric crystals that are located inside the sensor tube and are converted to an electronic signal that is directly proportional to flow rate. This vortex meter gives accurate and repeatable measurements with no moving parts. However, the corresponding head loss is comparable to that from other obstruction-type meters.

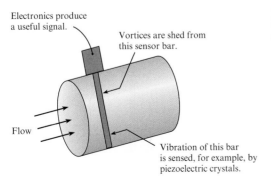

FIGURE 13.19

Vortex flowmeter.

FIGURE 13.20

Rotameter.

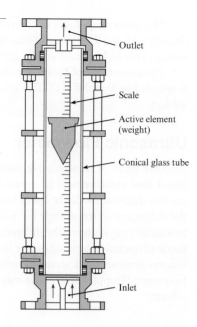

FIGURE 13.20

Rotameter.

Rotameter

The **rotameter** (Fig. 13.20) is an instrument for measuring flow rate by sensing the position of an active element (weight) that is situated in a tapered tube. The equilibrium position of the weight is related to the flow rate. Because the velocity is lower at the top of the tube (greater flow section there) than at the bottom, the rotor seeks a neutral position where the drag on it just balances its weight. Thus the rotor "rides" higher or lower in the tube depending on the rate of flow. The weight is designed so that it spins and thus stays in the center of the tube. A calibrated scale on the side of the tube indicates the rate of flow. Although venturi and orifice meters have better accuracy (approximately 1% of full scale) than the rotameter (approximately 5% of full scale), the rotameter offers other advantages, such as ease of use and low cost.

Rectangular Weir

A **weir** (Fig. 13.21) is an instrument for determining flow rate in liquids by measuring the height of water relative to an obstruction in an open channel. The discharge over the weir

FIGURE 13.21

Definition sketch for sharp-crested weir: (a) plan view, (b) elevation view.

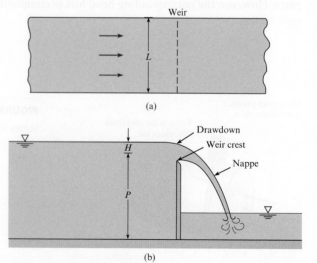

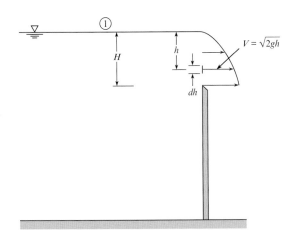

FIGURE 13.22

Theoretical velocity distribution over a weir.

is a function of the weir geometry and of the head on the weir, H, which is defined as the vertical distance between the weir crest and the liquid surface taken far enough upstream of the weir to avoid local free-surface curvature.

The discharge equation for the weir is derived by integrating $V\,dA = VL\,dh$ over the total head on the weir. Here, L is the length of the weir and V is the velocity at any given distance h below the free surface. Neglecting streamline curvature and assuming negligible velocity of approach upstream of the weir, one obtains an expression for V by writing the Bernoulli equation between a point upstream of the weir and a point in the plane of the weir (see Fig. 13.22). Assuming the pressure in the plane of the weir is atmospheric, this equation is

$$\frac{p_1}{\gamma} + H = (H - h) + \frac{V^2}{2g} \tag{13.14}$$

Here, the reference elevation is the elevation of the crest of the weir, and the reference pressure is atmospheric pressure. Therefore, $p_1 = 0$, and Eq. (13.14) reduces to

$$V = \sqrt{2gh}$$

Then $dQ = \sqrt{2gh}\,Ldh$, and the discharge equation becomes

$$Q = \int_0^H \sqrt{2gh}\,Ldh$$
$$= \frac{2}{3}L\sqrt{2g}\,H^{3/2} \tag{13.15}$$

In the case of actual flow over a weir, the streamlines converge downstream of the plane of the weir, and viscous effects are not entirely absent. Consequently, a discharge coefficient C_d must be applied to the basic expression on the right-hand side of Eq. (13.15) to bring the theory in line with the actual flow rate. Thus, the rectangular weir equation is

$$Q = \frac{2}{3}C_d\sqrt{2g}\,LH^{3/2}$$
$$= K\sqrt{2g}\,LH^{3/2} \tag{13.16}$$

FIGURE 13.23

Rectangular weir with end contractions: (a) plan view, (b) elevation view.

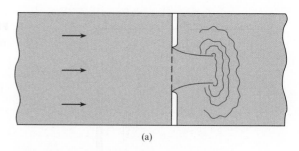

(a)

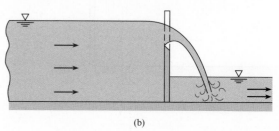

(b)

For low-viscosity liquids, the flow coefficient K is primarily a function of the relative head on the weir, H/P. An empirically determined equation for K adapted from Kindsvater and Carter (12) is

$$K = 0.40 + 0.05\frac{H}{P} \tag{13.17}$$

This is valid up to an H/P value of 10 as long as the weir is well ventilated so that atmospheric pressure prevails on both the top and the bottom of the weir nappe.

When the rectangular weir does not extend the entire distance across the channel, as in Fig. 13.23, additional end contractions occur. Therefore, K will be smaller than for the weir without end contractions. Refer to King (13) for additional information on flow coefficients for weirs.

EXAMPLE 13.7

Applying a Rectangular Weir to Measure the Flow Rate of Water

Problem Statement

The head on a rectangular weir that is 60 cm high in a rectangular channel that is 1.3 m wide is measured to be 21 cm. What is the discharge of water over the weir?

Define the Situation

- Water flows over a rectangular weir.
- The weir has a height of $P = 0.6$ m and a width of $L = 1.3$ m.
- Head on the weir is $H = 0.21$ m.

State the Goal

Find the discharge (m³/s).

Generate Ideas and Make a Plan

1. Calculate the flow coefficient K using Eq. (13.17).
2. Calculate flow rate using the rectangular weir equation (13.16).

Take Action (Execute the Plan)

1. Flow coefficient:

$$K = 0.40 + 0.05\frac{H}{P} = 0.40 + 0.05\left(\frac{21}{60}\right) = 0.417$$

2. Discharge:

$$Q = K\sqrt{2g}\,LH^{3/2} = 0.417\,\sqrt{2(9.81)}\,(1.3)(0.21^{3/2})$$

$$= \boxed{0.23 \text{ m}^3/\text{s}}$$

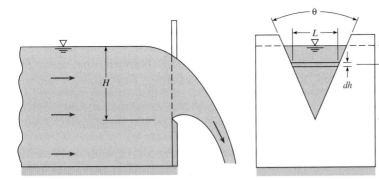

FIGURE 13.24

Definition sketch for the triangular weir.

Triangular Weir

A definition sketch for the **triangular weir** is shown in Fig. 13.24. The primary advantage of the triangular weir is that it has a higher degree of accuracy over a much wider range of flow than does the rectangular weir because the average width of the flow section increases as the head increases.

The discharge equation for the triangular weir is derived in the same manner as that for the rectangular weir. The differential discharge $dQ = V\,dA = VL\,dh$ is integrated over the total head on the weir to give

$$Q = \int_0^H \sqrt{2gh}\,(H - h)\,2\tan\left(\frac{\theta}{2}\right)dh$$

which integrates to

$$Q = \frac{8}{15}\sqrt{2g}\tan\left(\frac{\theta}{2}\right)H^{5/2}$$

However, a coefficient of discharge must still be used with the basic equation. Hence,

$$Q = \frac{8}{15}C_d\sqrt{2g}\tan\left(\frac{\theta}{2}\right)H^{5/2} \tag{13.18}$$

Experimental results with water flow over weirs with $\theta = 60°$ and $H > 2$ cm indicate that C_d has a value of 0.58. Hence, the triangular weir equation with these limitations is

$$Q = 0.179\sqrt{2g}\,H^{5/2} \tag{13.19}$$

EXAMPLE 13.8

Flow Rate for a Triangular Weir

Problem Statement

The head on a 60° triangular weir is measured to be 43 cm. What is the flow of water over the weir?

Define the Situation

- Water flows over a 60° triangular weir.
- Head on the weir is $H = 0.43$ m.

State the Goal

Calculate the discharge (m^3/s).

Generate Ideas and Make a Plan

Apply the triangular weir equation (13.19).

Take Action (Execute the Plan)

$$Q = 0.179\sqrt{2g}\,H^{5/2} = 0.179 \times \sqrt{2 \times 9.81} \times (0.43)^{5/2}$$

$$= \boxed{0.096\ m^3/s}$$

More details about flow-measuring devices for incompressible flow can be found in references (14) and (15).

13.3 Summarizing Key Knowledge

Measuring Velocity and Pressure

- Instruments for *velocity measurement* include the stagnation tube, Pitot-static tube, yaw meter, vane and cup anemometers, hot-wire and hot-film anemometers, laser-Doppler anemometer, and particle image velocimeter.

- Instruments for *pressure measurement* include the static tube, piezometer, differential manometer, Bourdon-tube gage, and several types of pressure transducers.

Measuring Flow Rate (Discharge)

- To measure flow rate, there are several direct methods, including the following:
  - Measure volume (or weight) and divide by time.
  - Measure velocities at points on a cross section and integrate using $Q = \int V\, dA$.

- Common instruments for *flow measurement* include the orifice meter, flow nozzle, venturi meter, electromagnetic flow meter, ultrasonic flow meter, turbine flow meter, vortex flow meter, rotameter, and weir.

- Flow rate or discharge for a flowmeter that uses a restricted opening (i.e., an orifice, flow nozzle, or venturi) is calculated using

$$Q = KA_o\sqrt{2g\Delta h} = KA_o\sqrt{2\Delta p_z/\rho}$$

where K is a flow coefficient that depends on Reynolds number and the type of flowmeter, A_o is the area of the opening, Δh is the change in piezometric head across the flowmeter, and Δp_z is drop in piezometric pressure across the flowmeter.

- Discharge for a rectangular weir of length L is given by

$$Q = K\sqrt{2g}\,LH^{3/2}$$

where K is the flow coefficient that depends on H/P. The term H is the height of the water above the crest of the weir, as measured upstream of the weir, and P is the height of the weir.

- Discharge for a 60° triangular weir with $H > 2$ cm is given by

$$Q = 0.179\sqrt{2g}\,H^{5/2}$$

REFERENCES

1. Hurd, C.W., K.P. Chesky, and A.H. Shapiro. "Influence of Viscous Effects on Impact Tubes." *Trans. ASME J. Applied Mechanics,* vol. 75 (June 1953).

2. King, H.W., and E.F. Brater. *Handbook of Hydraulics.* New York: McGraw-Hill, 1963.

3. Lomas, Charles C. *Fundamentals of Hot Wire Anemometry.* New York: Cambridge University Press, 1986.

4. Durst, Franz. *Principles and Practice of Laser-Doppler Anemometry.* New York: Academic Press, 1981.

5. Kline, J.J. "Flow Visualization." In *Illustrated Experiments in Fluid Mechanics: The NCFMF Book of Film Notes.* Cambridge, MA: MIT Press, 1972.

6. Macagno, Enzo O. "Flow Visualization in Liquids." *Iowa Inst. Hydraulic Res. Rept.,* 114 (1969).

7. Raffel, M., C. Wilbert, and J. Kompenhans. *Particle Image Velocimetry.* New York: Springer, 1998.

8. Lienhard, J.H., V, and J.H. Lienhard, IV. "Velocity Coefficients for Free Jets from Sharp-Edged Orifices." *Trans. ASME J. Fluids Engineering,* 106 (March 1984).

9. Tuve, G.L., and R.E. Sprenkle. "Orifice Discharge Coefficients for Viscous Liquids." *Instruments,* vol. 8 (1935).

10. ASME. *Fluid Meters,* 6th ed. New York: ASME, 1971.

11. Shercliff, J.A. *Electromagnetic Flow-Measurement.* New York: Cambridge University Press, 1962.

12. Kindsvater, Carl E., and R.W. Carter. "Discharge Characteristics of Rectangular Thin-Plate Weirs." *Trans. Am. Soc. Civil Eng.,* 124 (1959), 772–822.

13. King, L.V. *Phil. Trans. Roy. Soc. London, Ser. A,* 14 (1914), 214.

14. Miller, R.W. *Flow Measurement Engineering Handbook.* New York: McGraw-Hill, 1983.

15. Scott, R.W.W., ed. *Developments in Flow Measurement–1.* Englewood Cliffs, NJ: Applied Science, 1982.

PROBLEMS

Measuring Velocity and Pressure (§13.1)

13.1 List three different instruments or approaches that engineers use to measure fluid velocity, and three more that are used to measure pressure. For each instrument or approach, list two advantages and two disadvantages. *Hint:* Use this text or sources on the Internet.

Flow Velocity: Stagnation Tubes (§13.1)

13.2 A stagnation tube 6 mm in diameter is used to measure the velocity in a stream of air as shown. What is the air velocity if the deflection of the water column is 2.3 mm? Air temperature = 10°C, and p = 1 atm.

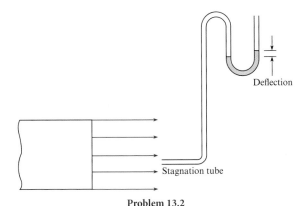
Deflection

Stagnation tube

Problem 13.2

13.3 Consider measuring the speed of an automobile by building a stagnation tube from a drinking straw and then using this device with a water-filled *U*-tube manometer.

a. Make a sketch that illustrates how you would propose making this measurement.

b. Determine the lowest velocity that could be measured. Assume that the lower limit is based on the resolution of the manometer.

Flow Velocity: Laser-Doppler Anemometers (§13.1)

13.4 On the Internet, locate technically sound resources relevant to the LDA. Skim these resources, and then

a. write down five details about the LDA that are relevant to engineering practice and interesting to you, and

b. write down two questions about LDAs that are interesting and insightful.

Measuring Volume Flow Rate or Discharge (§13.2)

13.5 Classify the following devices as to whether they are used to measure velocity (V), pressure (p), or discharge (Q).

a. venturi meter

b. hot-wire anemometer

c. differential manometer

d. stagnation tube

e. orifice meter

f. rotameter

g. ultrasonic flow meter

h. weir

i. Bourdon-tube gage

j. laser-Doppler anemometer

13.6 Water from a pipe is diverted into a tank for 3 min. If the weight of diverted water is measured to be 20 kN, what is the discharge in cubic meters per second? Assume the water temperature is 20°C.

13.7 A velocity traverse inside a 16-in.-circular air duct yields the data in the table. What is the rate of flow in cubic feet per second and cubic feet per minute? What is the ratio of V_{max} to V_{mean}? Does it appear that the flow is laminar or turbulent? If p = 14.3 psia and T = 70°F, what is the mass flow rate?

y^*	V (ft/s)	y^*	V (ft/s)
0.0	0	2.0	110
0.1	72	3.0	117
0.2	79	4.0	122
0.4	88	5.0	126
0.6	93	6.0	129
1.0	100	7.0	132
1.5	106	8.0	135

*Distance from pipe wall, in.

13.8 The asymmetry of the flow in stacks means that flow velocity must be measured at several locations on the cross-flow plane. Consider the cross section of the cylindrical stack shown. The two access holes through which probes can be inserted are separated by 90°. Velocities can be measured at the five points shown (five-point method).

a. Determine the ratio r_m/D such that the areas of the five measuring segments are equal.

b. Determine the ratio r/D (probe location) that corresponds to the centroid of the segment.

c. The data in the table are taken for a stack 2 m in diameter in which the gas temperature is 300°C, the pressure is 110 kPa absolute, and the gas constant is 400 J/kg K. The data represent the deflection within a water manometer connected to a Pitot-static tube located at the measuring stations. Calculate the mass flow rate.

Station	Δh (cm)
1	1.2
2	1.1
3	1.1
4	0.9
5	1.05

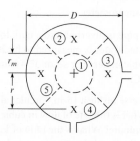

Problem 13.8

13.9 Repeat Prob. 13.8 for the case in which three access holes are separated by 60° and seven measuring points are used. The diameter of the stack is 1.5 m, the gas temperature is 250°C, the pressure is 115 kPa absolute, and the gas constant is 420 J/kg K. The data in the table shown represent the deflection of a water manometer connected to a Pitot-static tube at the measuring stations. Calculate the mass flow rate.

Station	Δh (cm)
1	8.2
2	8.6
3	8.2
4	8.9
5	8.0
6	8.5
7	8.4

Discharge: Orifice Meters (§13.2)

13.10 A fluid jet discharging from a 8.0 cm orifice has a diameter of 7.6 cm at its vena contracta. What is the coefficient of contraction?

13.11 On the Internet, locate quality knowledge resources relevant to orifice meters. Skim these resources, and then

  a. Write down five findings that are relevant to engineering practice and interesting to you.

  b. Write down two questions that are interesting, insightful, and relevant to orifice meters.

13.12 New orifices such as that shown in Fig. 13.14 will have definite flow coefficients as given in Fig. 13.15. With age,

however, physical changes could occur to the orifice. Explain what changes these might be and how (if at all) these physical changes might affect the flow coefficients.

13.13 The flow coefficient values for orifices given in Fig. 13.15 were obtained by testing orifices in relatively smooth pipes. If an orifice were used in a pipe that was very rough, do you think you would get a valid indication of discharge by using the flow coefficient of Fig. 13.15? Justify your conclusion.

13.14 What size orifice is required to produce a change in head of 6 m for a discharge of 2 m³/s of water in a pipe 1 m in diameter?

13.15 Semicircular orifices such as the one shown are sometimes used to measure the flow rate of liquids that also transport sediments. The opening at the bottom of the pipe allows free passage of the sediment. Derive a formula for Q as a function of Δp, D, and other relevant variables associated with the problem. Then, using that formula and guessing any unknown data, estimate the water discharge through such an orifice when Δp is read as 80 kPa and flow is in a 30-cm pipe.

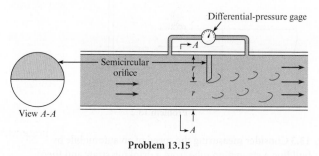

Problem 13.15

13.16 What is the discharge of gasoline ($SG = 0.68$) in a 22 cm horizontal pipe if the differential pressure across a 11 cm orifice in the pipe is 60 kPa?

Discharge: Venturi Meters (§13.2)

13.17 Engineers are calibrating a poorly designed venturi meter for the flow of petroleum by relating the pressure difference between taps 1 and 2 to the discharge. By applying the Bernoulli equation and assuming a quasi-one-dimensional flow (velocity uniform across every cross section), the engineers find that

$$Q_0 = A_2[2(p_1 - p_2)/\rho]^{0.5}[1 - (d/D)^4]^{-0.5}$$

where D and d are the duct diameters at stations 1 and 2. However, they realize that the flow is not quasi-one-dimensional and that the pressure at tap 2 is not equal to the average pressure in the throat because of streamline curvature. Thus the engineers introduce a correction factor K into the foregoing equation to yield

$$Q = KQ_0$$

Use your knowledge of pressure variation across curved streamlines to decide whether K is larger or smaller than unity, and support your conclusion by presenting a rational argument.

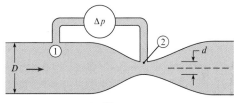

Problem 13.17

Weirs (§13.2)

13.18 Water flows over a rectangular weir that is 4 m wide and 35 cm high. If the head on the weir is 10 cm, what is the discharge in cubic meters per second?

13.19 What is the water discharge over a rectangular weir 4.5 ft high and 20 ft long in a rectangular channel 20 ft wide if the head on the weir is 2.2 ft?

13.20 A basin is 50 ft long, 2 ft wide, (L) and 4 ft deep. A sharp-crested rectangular weir is located at one end of the basin, and it spans the width of the basin (the weir is 2 ft long). The crest of the weir is 2 ft above the bottom of the basin. At a given instant water in the basin is 3 ft deep; thus water is flowing over the weir and out of the basin. Estimate the time it will take for the water in the basin to go from the 3 ft depth to a depth of 2 ft 2 in.

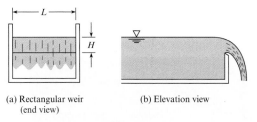

(a) Rectangular weir   (b) Elevation view
(end view)

Problem 13.20

CHAPTER FOURTEEN

Turbomachinery

CHAPTER ROAD MAP Machines to move fluids or to extract power from moving fluids have been designed since the beginning of recorded history. Fluid machines are everywhere. They are the essential components of the automobiles we drive, the supply systems for the water we drink, the power generation plants for the electricity we use, and the air-conditioning and heating systems that provide the comfort we enjoy. This chapter introduces the concepts underlying various types of machines.

FIGURE 14.1

This figure shows the impeller from the blower inside a vacuum cleaner. This impeller rotates inside a housing. This rotational motion creates a suction pressure that draws air into the center hole. The air is flung outward by the spinning blades of the impeller.

This impeller was "liberated" from the vacuum cleaner by Jason Stirpe while he was an engineering student. Jason used this impeller with a DC motor and a homemade housing to fabricate a blower for a design that he was creating. Being resourceful is at the heart of technology innovation. (Photo by Donald Elger.)

LEARNING OUTCOMES

PROPELLER THEORY (§14.1).

- Describe the factors that influence the thrust and efficiency of a propeller.
- Calculate the thrust and efficiency of a propeller.

CENTRIFUGAL PUMPS (§14.2 to §14.4).

- Describe axial flow and radial flow pumps.
- Define the head coefficient and the discharge coefficient.
- Sketch a pump performance curve and describe the relevant π-groups that appear.
- Explain how specific speed is used to select an appropriate type of pump for an application.
- Explain how to use NPSH to avoid cavitation.

TURBINES (§14.8).

- Describe an impulse turbine and a reaction turbine.
- Describe the maximum power that can be produced by a wind turbine.

Fluid machines are separated into two broad categories: positive-displacement machines and turbomachines. **Positive-displacement machines** operate by forcing fluid into or out of a chamber. Examples include the bicycle tire pump, the gear pump, the peristaltic pump, and the human heart. **Turbomachines** involve the flow of fluid through rotating blades or rotors that remove or add energy to the fluid. Examples include propellers, fans, water pumps, windmills, and compressors.

Axial-flow turbomachines operate with the flow entering and leaving the machine in the direction that is parallel to the axis of rotation of blades. A radial-flow machine can have the flow either entering or leaving the machine in the radial direction that is normal to the axis of rotation.

TABLE 14.1 Categories of Turbomachinery

	Power Absorbing	**Power Producing**
Axial machines	Axial pumps Axial fans Propellers Axial compressors	Axial turbine (Kaplan) Wind turbine Gas turbine
Radial machines	Centrifugal pump Centrifugal fan Centrifugal compressor	Impulse turbine (Pelton wheel) Reaction turbine (Francis turbine)

Table 14.1 provides a classification for turbomachinery. Power-absorbing machines require power to increase head (or pressure). A power-producing machine provides shaft power at the expense of head (or pressure) loss. Pumps are associated with liquids, whereas fans (blowers) and compressors are associated with gases. Both gases and liquids produce power through turbines. Often, "gas turbine" refers to an engine that has both a compressor and turbine and produces power.

14.1 Propellers

A **propeller** is a fan that converts rotational motion into thrust. The design of a propeller is based on the fundamental principles of airfoil theory (1). For example, consider a section of the propeller in Fig. 14.2; notice the analogy between the lifting vane and the propeller. This propeller is rotating at an angular speed ω, and the speed of advance of the airplane and

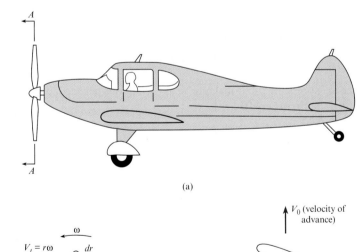

FIGURE 14.2

Propeller motion: (a) airplane motion, (b) view *A-A*, (c) view *B-B*, (d) velocity relative to blade element.

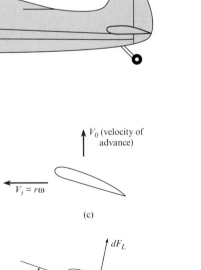

propeller is V_0. Focusing on an elemental section of the propeller, Fig. 14.2c, note that the given section has a velocity with components V_0 and V_t. Here, V_t is tangential velocity, and $V_t = r\omega$, resulting from the rotation of the propeller. Reversing and adding the velocity vectors V_0 and V_t yield the velocity of the air relative to the particular propeller section (Fig. 14.2d).

The angle θ is given by

$$\theta = \arctan\left(\frac{V_0}{r\omega}\right) \tag{14.1}$$

For a given forward speed and rotational rate, this angle is a minimum at the propeller tip ($r = r_0$) and increases toward the hub as the radius decreases. The angle β is known as the **pitch angle**. The local angle of attack of the elemental section is

$$\alpha = \beta - \theta \tag{14.2}$$

The propeller can be analyzed as a series of elemental sections (of width dr) producing lift and drag, which provide the propeller thrust and create resistive torque. This torque multiplied by the rotational speed is the power input to the propeller.

The propeller is designed to produce thrust; because the greatest contribution to thrust comes from the lift force F_L, the goal is to maximize lift and minimize drag, F_D. For a given shape of propeller section, the optimum angle of attack can be determined from data such as are given in Fig. 11.24. Because the angle θ increases with decreasing radius, the local pitch angle has to change to achieve the optimum angle of attack. This is done by twisting the blade.

A dimensional analysis can be performed to determine the π-groups that characterize the performance of a propeller. For a given propeller shape and pitch distribution, the thrust of a propeller T will depend on the propeller diameter D, the rotational speed n, the forward speed V_0, the fluid density ρ, and the fluid viscosity μ:

$$T = f(D, \omega, V_0, \rho, \mu) \tag{14.3}$$

Performing a dimensional analysis results in

$$\frac{T}{\rho n^2 D^4} = f\left(\frac{V_0}{nD}, \frac{\rho D^2 n}{\mu}\right) \tag{14.4}$$

It is conventional practice to express the rotational rate, n, as revolutions per second (rps). The π-group on the left is called the **thrust coefficient**,

$$C_T = \frac{T}{\rho n^2 D^4} \tag{14.5}$$

The first π-group on the right is the **advance ratio**. The second group is a Reynolds number based on the tip speed and diameter of the propeller. For most applications, the Reynolds number is high, and experience shows that the thrust coefficient is unaffected by the Reynolds number, so

$$C_T = f\left(\frac{V_0}{nD}\right) \tag{14.6}$$

The angle θ at the propeller tip is related to the advance ratio by

$$\theta = \arctan\left(\frac{V_0}{\omega r_0}\right) = \arctan\left(\frac{1}{\pi}\frac{V_0}{nD}\right) \tag{14.7}$$

As the advance ratio increases and θ increases, the local angle of attack at the blade element decreases, the lift increases, and the thrust coefficient goes down. This trend is illustrated in

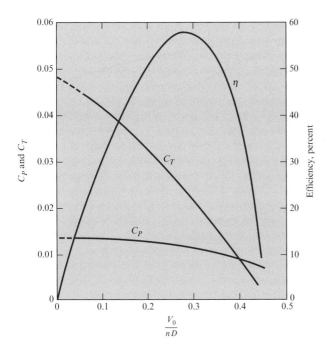

FIGURE 14.3

Dimensionless performance curves for a typical propeller; $D = 2.90$ m, $n = 1400$ rpm. [After Weick (2).]

Fig. 14.3, which shows the dimensionless performance curves for a typical propeller. Ultimately, an advance ratio is reached where the thrust coefficient goes to zero.

Performing a dimensional analysis for the power, P, shows

$$\frac{P}{\rho n^3 D^5} = f\left(\frac{V_0}{nD}, \frac{\rho D^2 n}{\mu}\right) \tag{14.8}$$

The π-group on the left is the **power coefficient**,

$$C_P = \frac{P}{\rho n^3 D^5} \tag{14.9}$$

As with the thrust coefficient, the power coefficient is not significantly influenced by the Reynolds number at high Reynolds numbers, so C_P reduces to a function of the advance ratio only:

$$C_P = f\left(\frac{V_0}{nD}\right) \tag{14.10}$$

The functional relationship between C_P and V_0/nD for an actual propeller is also shown in Fig. 14.3. Even though the thrust coefficient approaches zero for a given advance ratio, the power coefficient shows little decrease because it still takes power to overcome the torque on the propeller blade.

The curves for C_T and C_P are evaluated from performance characteristics of a given propeller operating at different values of V_0, as shown in Fig. 14.4. Although the data for the curves are obtained for a given propeller, the values for C_T and C_P, as a function of advance ratio, can be applied to geometrically similar propellers of different sizes and angular speeds.[*] Example 14.1 illustrates such an application.

---

[*]The speed of sound was not included in the dimensional analysis. However, the propeller performance is reduced because the Mach number based on the propeller tip speed leads to shock waves and other compressible-flow effects.

FIGURE 14.4

Power and thrust of a propeller 2.90 m in diameter at a rotational speed of 1400 rpm. [After Weick (2).]

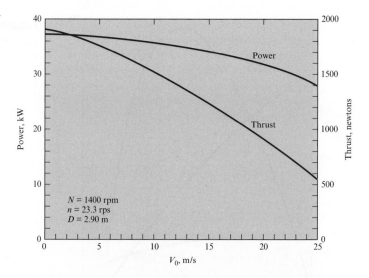

EXAMPLE 14.1

Propeller Application

Problem Statement

A propeller with the characteristics shown in Fig. 14.3 is to be used to drive a swamp boat. If the propeller is to have a diameter of 2 m and a rotational speed of $N = 1200$ rpm, what should be the thrust starting from rest? If the boat resistance (air and water) is given by the empirical equation $F_D = 0.003\rho V_0^2/2$, where V_0 is the boat speed in meters per second, F_D is the drag, and ρ is the mass density of the water, what will be the maximum speed of the boat and what power will be required to drive the propeller? Assume $\rho_{air} = 1.20$ kg/m^3 and $\rho_{water} = 1000$ kg/m^3.

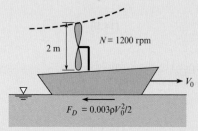

Define the Situation

A propeller is being used to drive a swamp boat.

Properties: $\rho = 1.2$ kg/m^3, $\rho_w = 1000$ kg/m^3.

State the Goals

- Calculate thrust (in N) starting from rest.
- Find maximum speed (in m/s) of swamp boat.
- Calculate power required (in kW) to operate propeller.

Generate Ideas and Make a Plan

1. From Fig. 14.3, find thrust coefficient for zero advance ratio.
2. Calculate thrust using Eq. (14.5).

3. To calculate maximum speed, plot propeller thrust versus boat speed and on same graph plot resistance of swamp boat versus boat speed. The maximum speed is where the curves intersect.

4. The maximum power will be when the boat speed is zero, so use Eq. (14.9) with C_P for zero advance ratio from Fig. 14.3.

Take Action (Execute the Plan)

1. From Fig. 14.3, $C_T = 0.048$ for $V_0/nD = 0$.

2. Thrust:
$$F_T = C_T \rho_a D^4 n^2 = 0.048(1.20 \text{ kg/m}^3)(2 \text{ m})^4(20 \text{ rps})^2$$
$$= \boxed{369 \text{ N}}$$

3. Table of thrust versus speed of swamp boat:

V_0	V_0/nD	C_T	$F_T = C_T\rho_a D^4 n^2$	$F_D = 0.003\rho_w V_0^2/2$
5 m/s	0.125	0.040	307 N	37.5 N
10 m/s	0.250	0.027	207 N	150 N
15 m/s	0.375	0.012	90 N	337 N

Graph of propeller thrust and swamp boat drag versus speed:

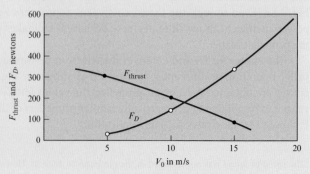

Curves intersect at $V_0 = 11$ m/s. Hence, the maximum speed of the swamp boat is 11 m/s.

4. At $V_0/nD = 0$, $C_P = 0.014$:

$$P = 0.014(1.20 \text{ kg/m}^3)(2 \text{ m})^5(20 \text{ rps})^3$$
$$= 4300 \text{ m} \cdot \text{N/s} = \boxed{4.30 \text{ kW}}$$

Review the Solution and the Process

Discussion. In an actual application, the starting rotational rate of the propeller need not be 1200 rpm but can be a lower value. After the boat is gaining speed, the rotational rate can be increased to achieve maximum speed.

The efficiency of a propeller is defined as the ratio of the power output—that is, thrust times velocity of advance—to the power input. Hence, the efficiency η is given as

$$\eta = \frac{F_T V_0}{P} = \frac{C_T \rho D^4 n^2 V_0}{C_P \rho D^5 n^3}$$

or

$$\eta = \frac{C_T}{C_P}\left(\frac{V_0}{nD}\right) \tag{14.11}$$

The variation of efficiency with advance ratio for a typical propeller is also shown in Fig. 14.3. The efficiency can be calculated directly from C_T and C_P performance curves. Note that at low advance ratios the efficiency increases with advance ratio and then reaches a maximum value before the decreasing thrust coefficient causes the efficiency to drop toward zero. The maximum efficiency represents the best operating point for fuel efficiency.

Many propeller systems are designed to have variable pitch; that is, pitch angles can be changed during propeller operation. Different efficiency curves corresponding to varying pitch angles are shown in Fig. 14.5. The envelope for the maximum efficiency is also shown in the figure. During operation of the aircraft, the pitch angle can be controlled to achieve maximum efficiency corresponding to the propeller rpm and forward speed.

The best source for propeller performance information is from propeller manufacturers. There are many speciality manufacturers for everything from marine to aircraft applications.

14.2 Axial-Flow Pumps

The axial-flow pump acts much like a propeller enclosed in a housing, as shown in Fig. 14.6. The rotating element, the impeller, causes a pressure change between the upstream and downstream sections of the pump. In practical applications, axial-flow machines are best suited to

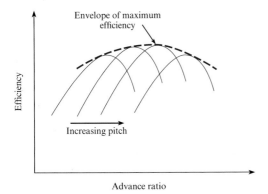

FIGURE 14.5

Efficiency curves for variable-pitch propeller.

FIGURE 14.6

Axial-flow blower in a duct.

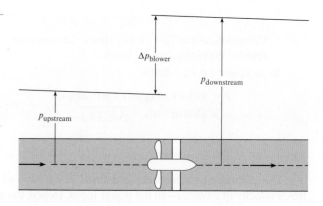

deliver relatively low heads and high flow rates. Hence, pumps used for dewatering lowlands, such as those behind dikes, are almost always of the axial-flow type. Water turbines in low-head dams (less than 30 m) where the flow rate and power production are large are also generally of the axial type.

Head and Discharge Coefficients for Pumps

The thrust coefficient is defined as $F_T/\rho D^4 n^2$ for use with propellers, and if the same variables are applied to flow in an axial pump, the thrust can be expressed as $F_T = \Delta p A = \gamma \Delta H A$ or

$$C_T = \frac{\gamma \Delta H A}{\rho D^4 n^2} = \frac{\pi}{4} \frac{\gamma \Delta H D^2}{\rho D^4 n^2} = \frac{\pi}{4} \frac{g \Delta H}{D^2 n^2} \qquad (14.12)$$

A new parameter, called the **head coefficient**, C_H, is defined using the variables of Eq. (14.12), as

$$C_H = \frac{4}{\pi} C_T = \frac{\Delta H}{D^2 n^2/g} \qquad (14.13)$$

which is a π-group that relates head delivered to fan diameter and rotational speed.

The independent π-group relating to propeller operation is V_0/nD; however, multiplying the numerator and denominator by the diameter squared gives $V_0 D^2/nD^3$, and $V_0 D^2$ is proportional to the discharge, Q. Thus, the π-group for pump similarity studies is Q/nD^3 and is identified as the **discharge coefficient** C_Q. The power coefficient used for pumps is the same as the power coefficient used for propellers. Summarizing, the π-groups used in the similarity analyses of pumps are

$$C_H = \frac{\Delta H}{D^2 n^2/g} \qquad (14.14)$$

$$C_P = \frac{P}{\rho D^5 n^3} \qquad (14.15)$$

$$C_Q = \frac{Q}{n D^3} \qquad (14.16)$$

where C_H and C_P are functions of C_Q for a given type of pump.

Fig. 14.7 is a set of curves of C_H and C_P versus C_Q for a typical axial-flow pump. Also plotted on this graph is the efficiency of the pump as a function of C_Q. The dimensional curves (head and power versus Q for a constant speed of rotation) from which Fig. 14.7 was developed

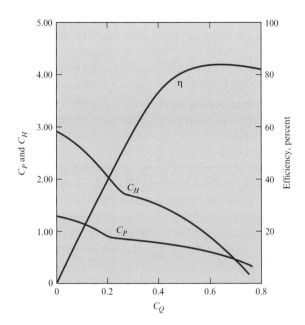

FIGURE 14.7

Dimensionless performance curves for a typical axial-flow pump. [After Stepanoff (3).]

are shown in Fig. 14.8. Because curves like those shown in Fig. 14.7 or Fig. 14.8 characterize pump performance, they are often called **characteristic curves** or **performance curves**. These curves are obtained by experiment.

There can be a problem with overload when operating axial-flow pumps. As seen in Fig. 14.7, when the pump flow is throttled below maximum-efficiency conditions, the required power increases with decreasing flow, thus leading to the possibility of overloading at low-flow conditions. For very large installations, special operating procedures are followed to avoid such overloading. For instance, the valve in the bypass from the pump discharge back to the pump inlet can be adjusted to maintain a constant flow through the pump. However, for small-scale applications, it is often desirable to have complete flexibility in flow control without the complexity of special operating procedures.

Performance curves are used to predict prototype operation from model tests or the effect of changing the speed of the pump. Example 14.2 shows how to use pump curves to calculate discharge and power.

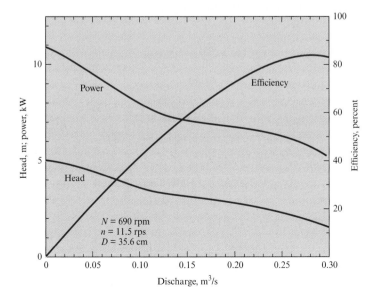

FIGURE 14.8

Performance curves for a typical axial-flow pump. [After Stepanoff (3).]

EXAMPLE 14.2

Discharge and Power for an Axial-Flow Pump

Define the Situation

For the pump represented by Figs. 14.7 and 14.8, what discharge of water in cubic meters per second will occur when the pump is operating against a 2 m head and at a speed of 600 rpm? What power in kilowatts is required for these conditions?

Define the Situation

This problem involves an axial-flow pump with water.

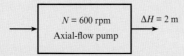

$N = 600$ rpm
Axial-flow pump $\Delta H = 2$ m

Properties: Assume $\rho = 1000$ kg/m³.

State the Goal

- Calculate discharge (in m³/s).
- Calculate power (in kW).

Generate Ideas and Make a Plan

1. Calculate C_H.

2. From Fig. 14.7 find C_Q and C_P.
3. Use C_Q to calculate discharge.
4. Use C_P to calculate power.

Take Action (Execute the Plan)

1. Rotational rate is (600 rev/min)/(60 s/min) = 10 rps. $D = 35.6$ cm.

$$C_H = \frac{2 \text{ m}}{(0.356 \text{ m})^2(10^2 \text{ s}^{-2})/(9.81 \text{ m/s}^2)} = 1.55$$

2. From Fig. 14.7, $C_Q = 0.40$ and $C_P = 0.72$.
3. Discharge is

$$Q = C_Q n D^3$$
$$Q = 0.40(10 \text{ s}^{-1})(0.356 \text{ m})^3 = \boxed{0.180 \text{ m}^3/\text{s}}$$

4. Power is

$$P = 0.72\rho D^5 n^3$$
$$= 0.72(10^3 \text{ kg/m}^3)(0.356 \text{ m})^5(10 \text{ s}^{-1})^3$$
$$= 4.12 \text{ km} \cdot \text{N/s} = 4.12 \text{ kJ/s} = \boxed{4.12 \text{ kW}}$$

Example 14.3 illustrates how to calculate head and power for an axial flow pump.

EXAMPLE 14.3

Head and Power for an Axial-Flow Pump

Problem Statement

If a 30 cm axial-flow pump with the characteristics shown in Fig. 14.7 is operated at a speed of 800 rpm, what head ΔH will be developed when the water-pumping rate is 0.127 m³/s? What power is required for this operation?

Define the Situation

This problem involves a 30 cm axial-flow pump with water.

$Q = 0.127$ m³/s $N = 800$ rpm
Axial-flow pump $\Delta H = ?$
$P = ?$

Properties: Water: $\rho = 10^3$ kg/m³.

State the Goals

- Calculate H = head (in meters) developed.
- Calculate power (in kW) required.

Generate Ideas and Make a Plan

1. Calculate the discharge coefficient, C_Q.

2. From Fig. 14.7, read C_H, and C_P.
3. Use Eq. (14.14) to calculate head produced.
4. Use Eq. (14.15) to calculate power required.

Take Action (Execute the Plan)

1. Discharge coefficient is

$$Q = 0.127 \text{ m}^3/\text{s}$$
$$n = \frac{800}{60} = 13.3 \text{ rps}$$
$$D = 30 \text{ cm}$$
$$C_Q = \frac{0.127 \text{ m}^3/\text{s}}{(13.3 \text{ s}^{-1})(0.30 \text{ m})^3} = 0.354$$

2. From Fig. 14.7, $C_H = 1.70$ and $C_P = 0.80$.
3. Head produced is

$$\Delta H = \frac{C_H D^2 n^2}{g} = \frac{1.70(0.30 \text{ m})^2(13.3 \text{ s}^{-1})^2}{(9.81 \text{ m/s}^2)} = \boxed{2.76 \text{ m}}$$

4. Power required is

$$P = C_P \rho D^5 n^3$$
$$= 0.80(10^3 \text{ kg/m}^3)(0.30 \text{ m})^5(13.3 \text{ s}^{-1})^3 = \boxed{4.57 \text{ kW}}$$

Fan Laws

The **fan laws** are used extensively by designers and practitioners involved with axial fans and blowers. The fan laws are equations that provide the discharge, pressure rise, and power requirements for a fan that operates at different speeds. The laws are based on the discharge, head, and power coefficients being the same at any other state as at the reference state, o; namely, $C_Q = C_{Qo}$, $= C_{Ho}$, and $C_P = C_{Po}$. Because the size and design of fan is unchanged, the discharge at speed n is

$$Q = Q_o \frac{n}{n_o} \qquad \text{(14.17a)}$$

and the pressure rise is

$$\Delta p = \Delta p_o \left(\frac{n}{n_o} \right)^2 \qquad \text{(14.17b)}$$

and finally the power required is

$$P = P_o \left(\frac{n}{n_o} \right)^3 \qquad \text{(14.17c)}$$

These fan laws cannot be applied between fans of different size and design. Of course, the fan laws do not provide exact values because of design considerations and manufacturing tolerances, but they are very useful in estimating fan performance.

14.3 Radial-Flow Machines

Radial-flow machines are characterized by the radial flow of the fluid through the machine. Radial-flow pumps and fans are better suited for larger heads at lower flow rates than axial machines.

Centrifugal Pumps

A sketch of the **centrifugal pump** is shown in Fig. 14.9. Fluid from the inlet pipe enters the pump through the eye of the impeller and then travels outward between the vanes of the impeller to its edge, where the fluid enters the casing of the pump and is then conducted to the discharge pipe. The principle of the radial-flow pump is different from that of the axial-flow

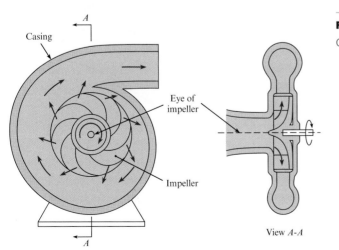

Casing

Eye of impeller

Impeller

View A-A

FIGURE 14.9

Centrifugal pump.

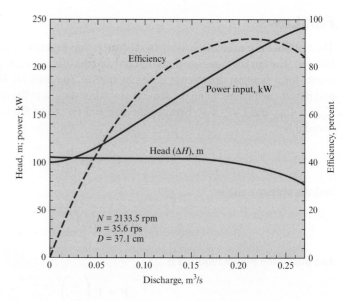

pump in that the change in pressure results in large part from rotary action (pressure increases outward like that in the rotating tank in §4.11) produced by the rotating impeller. Additional pressure increase is produced in the radial-flow pump when the high velocity of the flow leaving the impeller is reduced in the expanding section of the casing.

Although the basic designs are different for radial- and axial-flow pumps, it can be shown that the same similarity parameters (C_Q, C_P, and C_H) apply for both types. Thus, the methods that have already been discussed for relating size, speed, and discharge in axial-flow machines also apply to radial-flow machines.

The major practical difference between axial- and radial-flow pumps so far as the user is concerned is the difference in the performance characteristics of the two designs. The dimensional performance curves for a typical radial-flow pump operating at a constant speed of rotation are shown in Fig. 14.10. The corresponding dimensionless performance curves for the same pump are shown in Fig. 14.11. Note that the power required at shutoff flow is less than that required for flow at maximum efficiency. Normally, the motor used to drive the pump is chosen for conditions corresponding to maximum pump efficiency. Hence, the flow can be

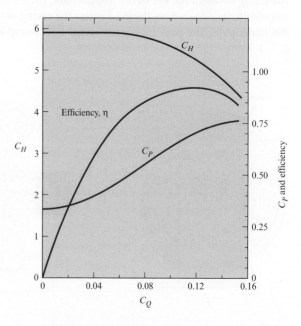

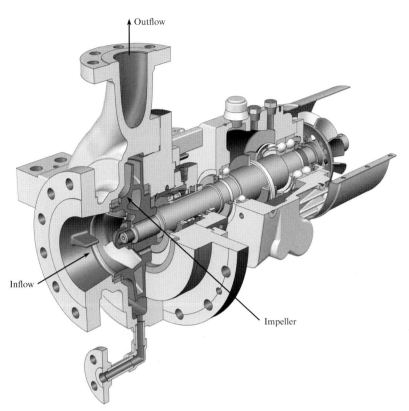

Outflow

Inflow

Impeller

FIGURE 14.12

Cutaway view of a single-suction, single-stage, horizontal-shaft radial pump. Pump inlet, outlet, and impeller are marked. (Copyright Sulzer Pumps.)

throttled between the limits of shutoff condition and normal operating conditions with no chance of overloading the pump motor. In this latter case, a radial-flow pump offers a distinct advantage over axial-flow pumps.

Radial-flow pumps are manufactured in sizes from 1 hp or less and heads of 50 or 60 ft to thousands of horsepower and heads of several hundred feet. Fig. 14.12 shows a cutaway view of a single-suction, single-stage, horizontal-shaft radial pump. Fluid enters in the direction of the rotating shaft and is accelerated outward by the rotating impeller. There are many other configurations designed for specific applications.

Example 14.4 shows how to find the speed and discharge for a centrifugal pump needed to provide a given head.

EXAMPLE 14.4

Speed and Discharge of Centrifugal Pump

Problem Statement

A pump that has the characteristics given in Fig. 14.10 when operated at 2133.5 rpm is to be used to pump water at maximum efficiency under a head of 76 m. At what speed should the pump be operated, and what will the discharge be for these conditions?

Define the Situation

A centrifugal pump operated at 2133.5 rpm pumps water to head of 76 m at maximum efficiency.

Assumptions: Assume the pump is the same size as that corresponding to Fig. 14.10 and the water properties are the same.

State the Goal

1. Find the operational speed of the pump (rpm).
2. Calculate discharge (m³/s).

Generate Ideas and Make a Plan

The C_H, C_P, C_Q, and η are the same for any pump with the same characteristics operating at maximum efficiency. Thus,

$$(C_H)_N = (C_H)_{2133.5 \text{ rpm}}$$

where N represents the unknown speed. Also, $(C_Q)_N = (C_Q)_{2133.5 \text{ rpm}}$.

1. Calculate speed using the same head coefficient.
2. Calculate discharge using the same discharge coefficient.

Take Action (Execute the Plan)

1. Speed calculation: From Fig. 14.10, at maximum efficiency $\Delta H = 90$ m.

$$\left(\frac{g\Delta H}{n^2 D^2}\right)_N = \left(\frac{g\Delta H}{n^2 D^2}\right)_{2133.5}$$

$$\frac{76 \text{ m}}{N^2} = \frac{90 \text{ m}}{2133.5^2 \text{ rpm}^2}$$

$$N = 2133.5 \times \left(\frac{76}{90}\right)^{1/2} = \boxed{1960 \text{ rpm}}$$

2. Discharge calculation: From Fig. 14.10, at maximum efficiency $Q = 0.255$ m³/s.

$$\left(\frac{Q}{nD^3}\right)_N = \left(\frac{Q}{nD^3}\right)_{2133.5}$$

$$\frac{Q_{1960}}{Q_{2133.5}} = \frac{1960}{2133.5} = 0.919$$

$$Q_{1960} = \boxed{0.234 \text{ m}^3/\text{s}}$$

Example 14.5 shows how to scale up data for a specific centrifugal pump to predict performance.

EXAMPLE 14.5

Head, Discharge, and Power of a Centrifugal Pump

Problem Statement

The pump with the characteristics shown in Figs. 14.10 and 14.11 is a model of a pump that was actually used in one of the pumping plants of the Colorado River Aqueduct [see Daugherty and Franzini (4)]. For a prototype that is 5.33 times larger than the model and operates at a speed of 400 rpm, what head, discharge, and power are to be expected at maximum efficiency?

Define the Situation

A prototype pump is 5.33 times larger than the corresponding model. The prototype operates at 400 rpm.

Assumptions: Pumping water with $\rho = 10^3$ kg/m³.

State the Goal

Find (at maximum efficiency):

1. Head (in meters)
2. Discharge (in m³/s)
3. Power (in kW)

Generate Ideas and Make a Plan

1. Find C_Q, C_H, and C_P at maximum efficiency from Fig. 14.11.

2. Evaluate speed in rps and calculate the new diameter.
3. Use Eqs. (14.14) through (14.16) to calculate head, discharge, and power.

Take Action (Execute the Plan)

1. From Fig. 14.11 at maximum efficiency, $C_Q = 0.12$, $C_H = 5.2$, and $C_P = 0.69$.

2. Speed in rps: $n = (400/60)$ rps $= 6.67$ rps
   $D = 0.371 \times 5.33 = 1.98$ m.

3. Pump performance:
   - Head:
   $$\Delta H = \frac{C_H D^2 n^2}{g} = \frac{5.2(1.98 \text{ m})^2 (6.67 \text{ s}^{-1})^2}{(9.81 \text{ m/s}^2)} = \boxed{92.4 \text{ m}}$$

   - Discharge:
   $$Q = C_Q n D^3 = 0.12(6.67 \text{ s}^{-1})(1.98 \text{ m})^3 = \boxed{6.21 \text{ m}^3/\text{s}}$$

   - Power:
   $$P = C_P \rho D^5 n^3 = 0.69 \left((10^3 \text{ kg})/\text{m}^3\right)(1.98 \text{ m})^5 (6.67 \text{ s}^{-1})^3$$
   $$= \boxed{6230 \text{ kW}}$$

14.4 Specific Speed

The preceding sections pointed out that axial-flow pumps are best suited for high discharge and low head, whereas radial machines perform better for low discharge and high head. A tool for selecting the best pump is the value of a π-group called the specific speed, n_s. The

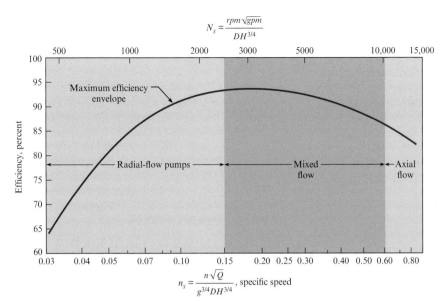

FIGURE 14.13

Optimum efficiency and impeller design versus specific speed.

(a) Optimum efficiency and impeller designs versus specific speed, n_s.

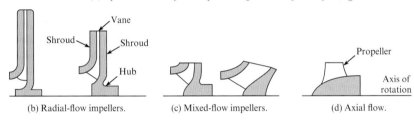

(b) Radial-flow impellers.   (c) Mixed-flow impellers.   (d) Axial flow.

specific speed is obtained by combining both C_H and C_Q in such a manner that the diameter D is eliminated:

$$n_s = \frac{C_Q^{1/2}}{C_H^{3/4}} = \frac{(Q/nD^3)^{1/2}}{[\Delta H/(D^2 n^2/g)]^{3/4}} = \frac{nQ^{1/2}}{g^{3/4}\,\Delta H^{3/4}}$$

Thus, specific speed relates different types of pumps without reference to their sizes.

As shown in Fig. 14.13, when efficiencies of different types of pumps are plotted against n_s, it is seen that certain types of pumps have higher efficiencies for certain ranges of n_s. For low specific speeds, the radial-flow pump is more efficient, whereas high specific speeds favor axial-flow machines. In the range between the completely axial-flow machine and the completely radial-flow machine, there is a gradual change in impeller shape to accommodate the particular flow conditions with maximum efficiency. The boundaries between axial, mixed, and radial machines are somewhat vague, but the value of the specific speed provides some guidance on which machine would be most suitable. The final choice would depend on which pumps were commercially available as well as their purchase price, operating cost, and reliability.

It should be noted that the specific speed traditionally used for pumps in the United States is defined as $N_s = NQ^{1/2}/\Delta H^{3/4}$. Here, the speed N is in revolutions per minute, Q is in gallons per minute, and ΔH is in feet. This form is not dimensionless. Therefore, its values are much larger than those found for n_s (the conversion factor is 17,200). Most texts and references published before the introduction of the SI system of units use this traditional definition for specific speed.

Example 14.6 illustrates the use of specific speed to select a pump type.

EXAMPLE 14.6

Using Specific Speed to Select a Pump

Problem Statement

What type of pump should be used to pump water at the rate of 10 cfs and under a head of 600 ft? Assume $N = 1100$ rpm.

Define the Situation

A pump will be pumping water at 10 cfs for a head of 600 ft.

State the Goal

Find the best type of pump for this application.

Generate Ideas and Make a Plan

1. Calculate specific speed.
2. Use Fig. 14.13 to select a pump type.

Take Action (Execute the Plan)

1. Rotational rate in rps:

$$n = \frac{1100}{60} = 18.33 \text{ rps}$$

Specific speed:

$$n_s = \frac{n\sqrt{Q}}{(g\Delta H)^{3/4}}$$

$$= \frac{18.33 \text{ rps} \times (10 \text{ cfs})^{1/2}}{(32.2 \text{ ft/s}^2 \times 600 \text{ ft})^{3/4}} = 0.035$$

2. From Fig. 14.13, a radial-flow pump is the best choice.

14.5 Suction Limitations of Pumps

The pressure at the suction side of a pump is important because of the possibility that cavitation may occur. As water flows past the impeller blades of a pump, local high-velocity flow zones produce low relative pressures (Bernoulli effect), and if these pressures reach the vapor pressure of the liquid, then cavitation will occur. For a given type of pump operating at a given speed and a given discharge, there will be certain pressure at the suction side of the pump below which cavitation will occur. In their testing procedures, pump manufacturers always determine this limiting pressure and include it with their pump performance curves.

More specifically, the pressure that is significant is the difference in pressure between the suction side of the pump and the vapor pressure of the liquid being pumped. In practice, engineers express this difference in terms of pressure head, called the **net positive suction head** (NPSH). To calculate NPSH for a pump that is delivering a given discharge, first apply the energy equation from the reservoir from which water is being pumped to the section of the intake pipe at the suction side of the pump. Then, subtract the vapor pressure head of the water to determine NPSH.

In Fig. 14.14, points 1 and 2 are the points between which the energy equation would be written to evaluate NPSH.

A more general parameter for indicating susceptibility to cavitation is specific speed. However, instead of using head produced (ΔH), one uses NPSH for the variable to the 3/4 power. This is

$$n_{ss} = \frac{nQ^{1/2}}{g^{3/4}(\text{NPSH})^{3/4}}$$

FIGURE 14.14

Locations used to evaluate NPSH for a pump.

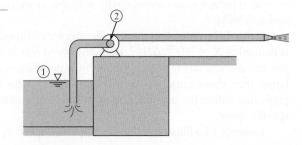

Here, n_{ss} is called the suction specific speed. The more traditional suction specific speed used in the United States is $N_{ss} = NQ^{1/2}/(NPSH)^{3/4}$, where N is in rpm, Q is in gallons per minute (gpm), and NPSH is in feet. Analyses of data from pump tests show that the value of the suction specific speed is a good indicator of whether cavitation may be expected. For example, the Hydraulic Institute (5) indicates that the critical value of N_{ss} is 8500. The reader is directed to manufacturer's data or the Hydraulic Institute for more details about critical NPSH or N_{ss}.

An analysis to find NPSH for a pump system is illustrated in Example 14.7.

EXAMPLE 14.7

Calculating Net Positive Suction Head

Problem Statement

In Fig. 14.14, the pump delivers 2 cfs flow of 80°F water, and the intake pipe diameter is 8 in. The pump intake is located 6 ft above the water surface level in the reservoir. The pump operates at 1750 rpm. What are the net positive suction head and the traditional suction specific speed for these conditions?

Define the Situation

A pump delivers 2 cfs flow of 80°F water.

Assumptions:

1. Entrance loss coefficient = 0.10.
2. Bend loss coefficient = 0.20.

Properties: Water at 80°F: Table A.5, $\gamma = 62.2$ lbf/ft^3, and $p_{vap} = 0.506$ psi.

State the Goal

- Calculate the positive suction head (NPSH).
- Calculate the traditional suction specific speed (N_{ss}).

Generate Ideas and Make a Plan

The net positive suction head is the difference between pressure at pump inlet and the vapor pressure.

1. Determine the atmospheric pressure in head of water for reservoir surface.
2. Determine velocity in 8 in. pipe.
3. Apply the energy equation between the reservoir and pump entrance.
4. Calculate NPSH.
5. Calculate N_{ss} with $N_{ss} = (NQ^{1/2})/(NPSH)^{3/4}$.

Take Action (Execute the Plan)

1. Pressure head at reservoir:

$$\frac{p_1}{\gamma} = \frac{14.7 \text{ lbf/in}^2 \times 144 \text{ (in}^2/\text{ft}^2)}{62.2 \text{ lbf/ft}^3} = 34 \text{ ft}$$

2. Velocity in pipe:

$$V_2 = \frac{Q}{A} = \frac{2 \text{ cfs}}{\pi \times ((4 \text{ in})/12)^2} = 5.73 \text{ ft/s}$$

3. Energy equation between points 1 and 2:

$$\frac{p_1}{\gamma} + \frac{V_1^2}{2g} + z_1 = \frac{p_2}{\gamma} + \frac{V_2^2}{2g} + z_2 + \sum h_L$$

- Input values:

$$V_1 = 0, \quad z_1 = 0, \quad z_2 = 6$$

- Head loss:

$$\sum h_L = (0.1 + 0.2)\frac{V_2^2}{2g}$$

- Head at pump entrance:

$$\frac{p_2}{\gamma} = \frac{p_1}{\gamma} - z_2 - \frac{V_2^2}{2g}(1 + 0.3)$$

$$= 34 - 6 - 1.3 \times \frac{5.73^2}{2 \times 32.2} = 27.3 \text{ ft}$$

4. Vapor pressure in feet of head:

$$0.506 \times 144/62.2 = 1.17 \text{ ft.}$$

Net positive suction head:

$$\text{NPSH} = 27.3 - 1.17 = 26.1 \text{ ft}$$

5. Traditional suction specific speed:

$$Q = 2 \text{ cfs} = 898 \text{ gpm}$$

$$N_{ss} = (1750)(898)^{1/2}/(26.1)^{3/4} = \boxed{4540}$$

Review the Solution and the Process

1. *Discussion.* For a typical single-stage centrifugal pump with an intake diameter of 8 in. and pumping 2 cfs, the critical NPSH is normally about 10 ft; therefore, the pump of this example is operating well within the safe range with respect to cavitation susceptibility.

2. *Discussion.* This value of N_{ss} is much below the critical limit of 8500; therefore, it is in a safe operating range so far as cavitation is concerned.

FIGURE 14.15

Centrifugal pump performance curve. [After McQuiston and Parker (6). Used with permission of John Wiley and Sons.]

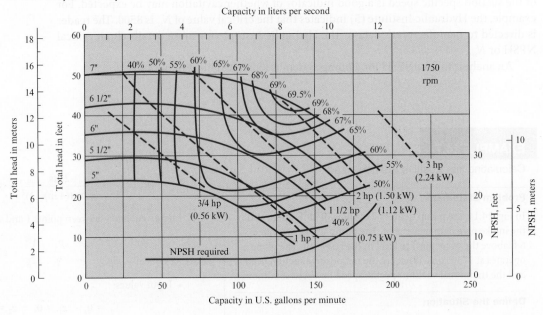

A typical pump performance curve for a centrifugal pump that would be supplied by a pump manufacturer is shown in Fig. 14.15. The solid lines labeled from 5 in. to 7 in. represent different impeller sizes that can be accommodated by the pump housing. These curves give the head delivered as a function of discharge. The dashed lines represent the power required by the pump for a given head and discharge. Lines of constant efficiency are also shown. Obviously, when selecting an impeller, one would like to have the operating point as close as possible to the point of maximum efficiency. The NPSH value gives the minimum head (absolute head) at the pump intake for which the pump will operate without cavitation.

14.6 Viscous Effects

In the foregoing sections, similarity parameters were developed to predict prototype results from model tests, neglecting viscous effects. The latter assumption is not necessarily valid, especially if the model is quite small. To minimize the viscous effects in modeling pumps, the Hydraulic Institute standards (5) recommend that the size of the model be such that the model impeller is not less than 30 cm in diameter. These same standards state that the model should have geometric similarity with the prototype.

Even with geometric similarity, one can expect the model to be less efficient than the prototype. An empirical formula proposed by Moody (7) is used for estimating prototype efficiencies of radial- and mixed-flow pumps and turbines from model efficiencies. That formula is

$$\frac{1 - e_1}{1 - e} = \left(\frac{D}{D_1}\right)^{1/5} \tag{14.18}$$

Here, e_1 is the efficiency of the model and e is the efficiency of the prototype.

Example 14.8 shows how to estimate the efficiency due to viscous effects.

EXAMPLE 14.8

Calculating Viscous Effects on Pump Efficiency

Problem Statement

A model with an impeller diameter of 45 cm is tested and found to have an efficiency of 85%. If a geometrically similar prototype has an impeller diameter of 1.80 m, estimate its efficiency when it is operating under conditions that are dynamically similar to those in the model test ($C_{Q,\,\text{model}} = C_{Q,\,\text{prototype}}$).

Define the Situation

A pump with a 45-cm-diameter impeller has 85% efficiency.

Assumptions: The efficiency differences are due to viscous effects.

State the Goal

Find the efficiency of a pump with a 1.6 m impeller.

Generate Ideas and Make a Plan

Use Eq. (14.18) to determine viscous effects.

Take Action (Execute the Plan)

Efficiency:

$$e = 1 - \frac{1 - e_1}{(D/D_1)^{1/5}} = 1 - \frac{0.15}{1.32} = 1 - 0.11 = 0.89$$

or

$$\boxed{e = 89\%}$$

14.7 Centrifugal Compressors

Centrifugal compressors are similar in design to centrifugal pumps. Because the density of the air or gases used is much less than the density of a liquid, the compressor must rotate at much higher speeds than the pump to effect a sizable pressure increase. If the compression process were isentropic and the gases ideal, the power necessary to compress the gas from p_1 to p_2 would be

$$P_{\text{theo}} = \frac{k}{k-1} Q_1 p_1 \left[\left(\frac{p_2}{p_1} \right)^{(k-1)/k} - 1 \right] \tag{14.19}$$

where Q_1 is the volume flow rate into the compressor and k is the ratio of specific heats. The power calculated using Eq. (14.19) is referred to as the **theoretical adiabatic power**. The efficiency of a compressor with no water cooling is defined as the ratio of the theoretical adiabatic power to the actual power required at the shaft. Ordinarily, the efficiency improves with higher inlet-volume flow rates, increasing from a typical value of 0.60 at 0.6 m³/s to 0.74 at 40 m³/s. Higher efficiencies are obtainable with more expensive design refinements.

Example 14.9 shows how to calculate the shaft power required to operate a compressor.

EXAMPLE 14.9

Calculating Shaft Power for a Centrifugal Compressor

Problem Statement

Determine the shaft power required to operate a compressor that compresses air at the rate of 1 m³/s from 100 kPa to 200 kPa. The efficiency of the compressor is 65%.

Define the Situation

The inlet flow rate to a compressor is 1.0 m³/s. The pressure change is from 100 kPa to 200 kPa.

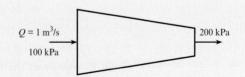

$Q = 1$ m³/s
100 kPa
200 kPa

From Table A.2, $k = 1.4$.

State the Goal

$P_{\text{shaft}}(\text{kW}) \leftarrow$ required shaft power (in kW)

Generate Ideas and Make a Plan

1. Use Eq. (14.19) to calculate theoretical power.
2. Divide theoretical power by efficiency to find shaft (required) power.

Take Action (Execute the Plan)

1. Theoretical power:

$$P_{theo} = \frac{k}{k-1}Q_1 p_1 \left[\left(\frac{p_2}{p_1} \right)^{(k-1)/k} - 1 \right]$$

$$= (3.5)(1 \text{ m}^3/\text{s})(10^5 \text{ N/m}^2)[(2)^{0.286} - 1]$$

$$= 0.767 \times 10^5 \text{ N} \cdot \text{m/s} = 76.7 \text{ kW}$$

2. Shaft power:

$$P_{shaft} = \frac{76.7}{0.65} \text{kW} = \boxed{118 \text{ kW}}$$

Cooling is necessary for high-pressure compressors because of the high gas temperatures resulting from the compression process. Cooling can be achieved through the use of water jackets or intercoolers that cool the gases between stages. The efficiency of water-cooled compressors is based on the power required to compress ideal gases isothermally, or

$$P_{theo} = p_1 Q_1 \ln \frac{p_2}{p_1} \tag{14.20}$$

which is usually called the **theoretical isothermal power**. The efficiencies of water-cooled compressors are generally lower than those of noncooled compressors. If a compressor is cooled by water jackets, its efficiency characteristically ranges between 55% and 60%. The use of intercoolers results in efficiencies from 60% to 65%.

Application to Fluid Systems

The selection of a pump, fan, or compressor for a specific application depends on the desired flow rate. This process requires the acquisition or generation of a system curve for the flow system of interest and a performance curve for the fluid machine. The intersection of these two curves provides the operating point, as discussed in Chapter 10.

For example, consider using the centrifugal pump with the characteristics shown in Fig. 14.15 to pump water at 60°F from a wall into a tank, as shown in Fig. 14.16. A pumping capacity of at least 80 gpm is required. Two hundred feet of standard schedule-40 2 in. galvanized iron pipe are to be used. There is a check valve in the system as well as an open gate valve. There is a 20 ft elevation between the well and the top of the fluid in the tank. Applying the energy equation, the head required by the pump is

$$h_p = \Delta z + \frac{V^2}{2g} \left(\frac{fL}{D} + \sum K_L \right)$$

where K_L represents the head loss coefficients for the entrance, check valve, gate valve, and sudden-expansion loss entering the tank. Using representative values for the loss coefficients and evaluating the friction factor from the Moody diagram in Chapter 10 leads to

$$h_p = 20 + 0.00305 \, Q^2$$

where Q is the flow rate in gpm. This is the system curve.

The result of plotting the system curve on the pump-performance curves is shown in Fig. 14.17. The locations where the lines cross are the operating points. Note that a discharge

FIGURE 14.16

System for pumping water from a well into a tank.

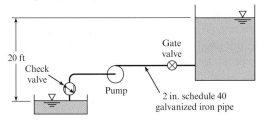

FIGURE 14.17

System and pump performance curves for pumping application.

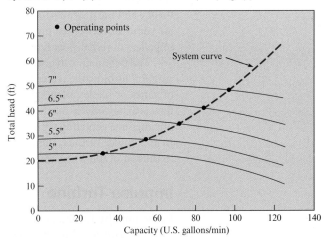

of just over 80 gpm is achieved with the 6.5 in. impeller. Also, referring back to Fig. 14.15, the efficiency at this point is about 62%. To ensure that the design requirements are satisfied, the engineer may select the larger impeller, which has an operating point of 95 gpm. If the pump is to be used in continuous operation and the efficiency is important to operating costs, the engineer may choose to consider another pump that would have a higher efficiency at the operation point. An engineer experienced in the design of pump systems would be very familiar with the trade-offs for economy and performance and could make a design decision relatively quickly.

In some systems, it may be advantageous to use two pumps in series or in parallel. If two pumps are used in series, the performance curve is the sum of the pump heads of the two machines at the same flow rate, as shown in Fig. 14.18a. This configuration would be desirable for a flow system with a steep system curve, as shown in the figure. If two pumps are connected in parallel, the performance curve is obtained by adding the flow rates of the two pumps at the same pump heads, as shown in Fig. 14.18b. This configuration would be advisable for flow systems with shallow system curves, as shown in the figure. The concepts presented here for pumps also apply to fans and compressors.

FIGURE 14.18

Pump performance curves for pumps connected in series (a) and in parallel (b).

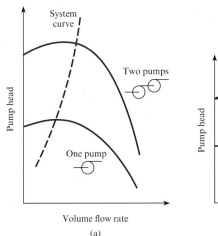

(a)

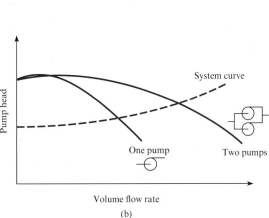

(b)

14.8 Turbines

A **turbine** is defined as a machine that extracts energy from a moving fluid. Much of the basic theory and most similarity parameters used for pumps also apply to turbines. However, there are some differences in physical features and terminology. The details of the flow through the impellers of radial-flow machines will now be addressed.

The two main categories of hydraulic machines are **impulse** and **reaction** turbines. In a reaction turbine, the water flow is used to rotate a turbine wheel or runner through the action of vanes or blades attached to the wheel. When the blades are oriented like a propeller, the flow is axial and the machine is called a **Kaplan turbine**. When the vanes are oriented like an impeller in a centrifugal pump, the flow is radial, and the machine is called a **Francis turbine**. In an impulse turbine, the water accelerates through a nozzle and impinges on vanes attached to the rim of the wheel. This machine is called a **Pelton wheel**.

Impulse Turbine

In the impulse turbine, a jet of fluid issuing from a nozzle impinges on vanes of the turbine wheel, or **runner**, thus producing power as the runner rotates (see Fig. 14.19 and Fig. 14.20). The primary feature of the impulse turbine with respect to fluid mechanics is the power production as the jet is deflected by the moving vanes. When the momentum equation is applied to this deflected jet, it can be shown [see Daugherty and Franzini (4)] for idealized conditions that the maximum power will be developed when the vane speed is one-half of the initial jet speed. Under such conditions, the exiting jet speed will be zero; all the kinetic energy of the jet will have been expended in driving the vane. Thus, if one applies the energy equation, between the incoming jet and the exiting fluid (assuming negligible head loss and negligible kinetic energy at exit), it is found that the head given up to the turbine is $h_t = (V_j^2/2g)$, and the power thus developed is

$$P = Q\gamma h_t \tag{14.21}$$

where Q is the discharge of the incoming jet, γ is the specific weight of jet fluid, and $h_t = V_j^2/2g$, or the velocity head of the jet. Thus, Eq. (14.21) reduces to

$$P = \rho Q \frac{V_j^2}{2} \tag{14.22}$$

FIGURE 14.19

Impulse turbine.

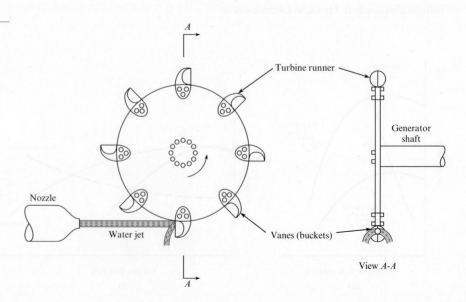

FIGURE 14.20

A photograph of the runner from a Pelton wheel turbine. (Alberto Pomares/Getty Images, Inc.)

To obtain the torque on the turbine shaft, the angular momentum equation is applied to a control volume, as shown in Fig. 14.21. For steady flow,

$$\sum M = \sum_{cs} \mathbf{r}_o \times (\dot{m}_o \mathbf{v}_o) - \sum_{cs} \mathbf{r}_i \times (\dot{m}_i \mathbf{v}_i)$$

Generally, it is assumed that the exiting fluid has negligible angular momentum. The moment acting on the system is the torque T acting on the shaft. Thus, the angular momentum equation reduces to

$$T = -\dot{m} r V_j \qquad\qquad \textbf{(14.23)}$$

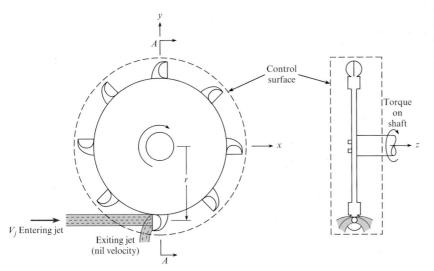

FIGURE 14.21

Control-volume approach for the impulse turbine using the angular momentum principle.

The mass flow rate across the control surface is ρQ, so the torque is

$$T = -\rho Q V_j r$$

The minus sign indicates that the torque applied to the system (to keep it rotating at constant angular velocity) is in the clockwise direction. However, the torque applied by the system to the shaft is in the counterclockwise direction, which is the direction of wheel rotation, so

$$T = \rho Q V_j r \qquad (14.24)$$

The power developed by the turbine is $T\omega$, or

$$P = \rho Q V_j r \omega \qquad (14.25)$$

Furthermore, if the velocity of the turbine vanes is $(1/2)V_j$ for maximum power, as noted earlier, then $P = \rho Q V_j^2/2$, which is the same as Eq. (14.22).

The calculation of torque for an impulse turbine is illustrated in Example 14.10.

EXAMPLE 14.10

Analyzing an Impulse Turbine

Problem Statement

What power in kilowatts can be developed by the impulse turbine shown if the turbine efficiency is 85%? Assume that the resistance coefficient f of the penstock is 0.015 and the head loss in the nozzle itself is negligible. What will be the angular speed of the wheel, assuming ideal conditions ($V_j = 2V_{\text{bucket}}$), and what torque will be exerted on the turbine shaft?

Define the Situation

This problem involves an impulse turbine with an efficiency of 85%.

Assumptions:

1. There is no entrance loss.
2. Head loss in nozzle is negligible.
3. Water density is 1000 kg/m³.

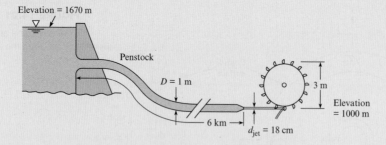

State the Goal

Find:

- Power (kW) developed by turbine
- Angular speed (rpm) of wheel for maximum efficiency
- Torque (kN·m) on turbine shaft

Generate Ideas and Make a Plan

1. Apply the energy equation to find nozzle velocity.
2. Use Eq. (14.22) for power.
3. For maximum efficiency, $\omega r = (V_j/2)$.
4. Calculate torque from $P = T\omega$.

Take Action (Execute the Plan)

1. Energy equation:

$$\frac{p_1}{\gamma} + \frac{V_1^2}{2g} + z_1 = \frac{p_j}{\gamma} + \frac{V_j^2}{2g} + z_j + h_L$$

- Values in energy equation:

$$p_1 = 0, z_1 = 1670 \text{ m}, V_1 = 0, p_j = 0, z_j = 1000 \text{ m}$$

- Penstock-supply pipe velocity ratio:

$$V_{\text{penstock}} = \frac{V_j A_j}{A_{\text{penstock}}} = V_j \left(\frac{0.18 \text{ m}}{1 \text{ m}}\right)^2 = 0.0324\, V_j$$

- Head loss:

$$h_L = f \frac{L}{D} \frac{1}{2g} V_{\text{penstock}}^2$$

$$= \frac{0.015 \times 6000}{1}(0.0324)^2 \frac{V_j^2}{2g} = 0.094 \frac{V_j^2}{2g}$$

- Jet velocity:

$$z_1 - z_2 = 1.094 \frac{V_j^2}{2g}$$

$$V_j = \left(\frac{2 \times 9.81 \text{ m/s}^2 \times 670 \text{ m}}{1.094} \right)^{1/2} = 109.6 \text{ m/s}$$

2. Gross power:

$$P = Q\gamma \frac{V_j^2}{2g} = \frac{\gamma A_j V_j^3}{2g}$$

$$= \frac{9810(\pi/4)(0.18)^2(109.6)^3}{2 \times 9.81} = 16{,}750 \text{ kW}$$

Power delivered:

$$P = 16{,}750 \times \text{efficiency} = \boxed{14{,}240 \text{ kW}}$$

3. Angular speed of wheel:

$$V_{\text{bucket}} = \frac{1}{2}(109.6 \text{ m/s}) = 54.8 \text{ m/s}$$

$$r\omega = 54.8 \text{ m/s}$$

$$\omega = \frac{54.8 \text{ m/s}}{1.5 \text{ m}} = 36.5 \text{ rad/s}$$

Wheel speed:

$$N = (36.5 \text{ rad/s}) \frac{1 \text{ rev}}{2\pi \text{ rad}} (60 \text{ s/min}) = \boxed{349 \text{ rpm}}$$

4. Torque:

$$T = \frac{\text{power}}{\omega} = \frac{14{,}240 \text{ kW}}{36.5 \text{ rad/s}} = \boxed{390 \text{ kN} \cdot \text{m}}$$

Reaction Turbine

In contrast to the impulse turbine, in which a jet under atmospheric pressure impinges on only one or two vanes at a time, flow in a reaction turbine is under pressure and reacts on all vanes of the impeller turbine simultaneously. Also, this flow completely fills the chamber in which the impeller is located (see Fig. 14.22). There is a drop in pressure from the outer radius of the impeller, r_1, to the inner radius, r_2. This is another point of difference with the impulse turbine, in which the pressure is the same for the entering and exiting flows. The original form of the reaction turbine, first extensively tested by J. B. Francis, had a completely radial-flow impeller (Fig. 14.23). That is, the flow passing through the impeller had velocity components only in a plane normal to the axis of the runner. However, more recent impeller designs, such as the mixed-flow and axial-flow types, are still called reaction turbines.

Torque and Power Relations for the Reaction Turbine

As for the impulse turbine, the angular momentum equation is used to develop formulas for the torque and power for the reaction turbine. The segment of turbine runner shown in Fig. 14.23 depicts the flow conditions that occur for the entire runner. The guide vanes outside the runner itself cause the fluid to have a tangential component of velocity around the entire circumference of the runner. Thus, the fluid has an initial amount of angular momentum with respect to the turbine axis when it approaches the turbine runner. As the fluid passes through the passages of the runner, the runner vanes effect a change in the magnitude and direction of its velocity. Thus, the angular momentum of the fluid is changed, which produces a torque on the runner. This torque drives the runner, which, in turn, generates power.

To quantify the preceding, let V_1 and α_1 represent the incoming velocity and the angle of the velocity vector with respect to a tangent to the runner, respectively. Similar terms at the inner-runner radius are V_2 and α_2. Applying the angular momentum equation for steady flow, Eq. (6.27), to the control volume shown in Fig. 14.23 yields

$$\begin{aligned} T &= \dot{m}(-r_2 V_2 \cos \alpha_2) - \dot{m}(-r_1 V_1 \cos \alpha_1) \\ &= \dot{m}(r_1 V_1 \cos \alpha_1 - r_2 V_2 \cos \alpha_2) \end{aligned} \tag{14.26}$$

The power from this turbine will be $T\omega$, or

$$P = \rho Q\omega(r_1 V_1 \cos \alpha_1 - r_2 V_2 \cos \alpha_2) \tag{14.27}$$

FIGURE 14.22

Schematic view of a reaction turbine installation: (a) elevation view, (b) plan view, section *A-A*.

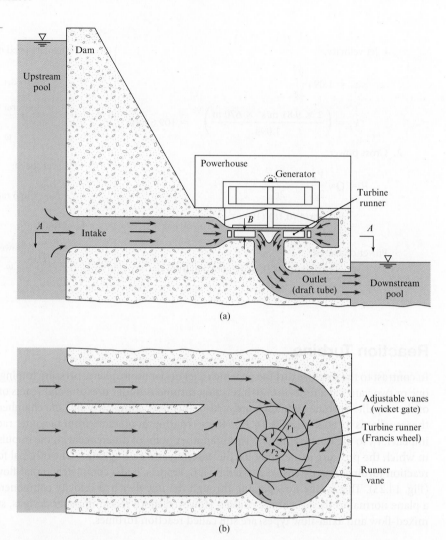

(a)

(b)

Equation (14.27) shows that the power production is a function of the directions of the flow velocities entering and leaving the impeller—that is, α_1 and α_2.

It is interesting to note that even though the pressure varies within the flow in a reaction turbine, it does not enter into the expressions derived using the angular momentum equation. The reason it does not appear is that the chosen outer and inner control surfaces are concentric with the axis about which the moments and angular momentum are evaluated. The pressure forces acting on these surfaces all pass through the given axis; therefore, they do not produce moments about the given axis.

Vane Angles

It should be apparent that the head loss in a turbine will be less if the flow enters the runner with a direction tangent to the runner vanes than if the flow approaches the vane with an angle of attack. In the latter case, separation will occur with consequent head loss. Thus, vanes of an impeller designed for a given speed and discharge and with fixed guide vanes will have a particular optimum blade angle β_1. However, if the discharge is changed from the condition of the original design, the guide vanes and impeller vane angles will not "match" the new flow condition. Most turbines for hydroelectric installations are made with movable guide vanes on the inlet side to effect a better match at all flows. Thus, α_1 is increased or decreased automatically through governor action to accommodate fluctuating power demands on the turbine.

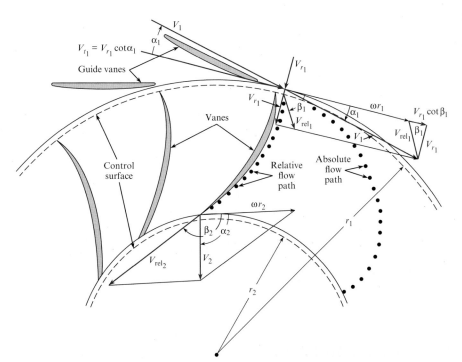

FIGURE 14.23

Velocity diagrams for the impeller for a Francis turbine.

To relate the incoming-flow angle α_1 and the vane angle β_1, first assume that the flow entering the impeller is tangent to the blades at the periphery of the impeller. Likewise, the flow leaving the stationary guide vane is assumed to be tangent to the guide vane. To develop the desired equations, consider both the radial and the tangential components of velocity at the outer periphery of the wheel ($r = r_1$). It is easy to compute the radial velocity, given Q and the geometry of the wheel, by the continuity equation:

$$V_{r_1} = \frac{Q}{2\pi r_1 B} \tag{14.28}$$

where B is the height of the turbine blades. The tangential (tangent to the outer surface of the runner) velocity of the incoming flow is

$$V_{t_1} = V_{r_1} \cot \alpha_1 \tag{14.29}$$

However, this tangential velocity is equal to the tangential component of the relative velocity in the runner, $V_{r_1} \cot \beta_1$, plus the velocity of the runner itself, ωr_1. Thus, the tangential velocity, when viewed with respect to the runner motion, is

$$V_{t_1} = r_1 \omega + V_{r_1} \cot \beta_1 \tag{14.30}$$

Now, eliminating V_{t_1} between Eqs. (14.29) and (14.30) results in

$$V_{r_1} \cot \alpha_1 = r_1 \omega + V_{r_1} \cot \beta_1 \tag{14.31}$$

Equation (14.31) can be rearranged to yield

$$\alpha_1 = \operatorname{arccot}\left(\frac{r_1 \omega}{V_{r_1}} + \cot \beta_1\right) \tag{14.32}$$

Example 14.11 illustrates how to calculate the inlet blade angle to avoid separation.

EXAMPLE 14.11

Analyzing a Francis Turbine

Problem Statement

A Francis turbine is to be operated at a speed of 600 rpm and with a discharge of 4.0 m³/s. If $r_1 = 0.60$ m, $\beta_1 = 110°$, and the blade height B is 10 cm, what should be the guide vane angle α_1 for a nonseparating flow condition at the runner entrance?

Define the Situation

A Francis turbine is operating with an angular speed of 600 rpm and a discharge of 4.0 m³/s.

State the Goal

Find the inlet guide vane angle, α_1.

Generate Ideas and Make a Plan

Use Eq. (14.32) for inlet guide angle.

Take Action (Execute the Plan)

Radial velocity at inlet:

$$\alpha_1 = \text{arccot}\left(\frac{r_1\omega}{V_{r_1}} + \cot\beta_1\right)$$

$$r_1\omega = 0.6 \times 600 \text{ rpm} \times 2\pi \text{ rad/rev} \times 1/60 \text{ min/s}$$
$$= 37.7 \text{ m/s}$$

Inlet guide vane angle:

$$V_{r_1} = \frac{Q}{2\pi r_1 B} = \frac{4.00 \text{ m}^3/\text{s}}{2\pi \times 0.6 \text{ m} \times 0.10 \text{ m}} = 10.61 \text{ m/s}$$

$$\cot\beta_1 = \cot(110°) = -0.364$$

$$\alpha_1 = \text{arccot}\left(\frac{37.7}{10.61} - 0.364\right) = \boxed{17.4°}$$

Specific Speed for Turbines

Because of the attention focused on the production of power by turbines, the specific speed for turbines is defined in terms of power:

$$n_s = \frac{nP^{1/2}}{g^{3/4}\gamma^{1/2}h_t^{5/4}}$$

It should also be noted that large water turbines are innately more efficient than pumps. The reason for this is that as the fluid leaves the impeller of a pump, it decelerates appreciably over a relatively short distance. Also, because guide vanes are generally not used in the flow passages with pumps, large local velocity gradients develop, which in turn cause intense mixing and turbulence, thereby producing large head losses. In most turbine installations, the flow that exits the turbine runner is gradually reduced in velocity through a gradually expanding **draft tube**, thus producing a much smoother flow situation and less head loss than for the pump. For additional details of hydropower turbines, see Daugherty and Franzini (4).

Gas Turbines

The conventional gas turbine consists of a compressor that pressurizes the air entering the turbine and delivers it to a combustion chamber. The high-temperature, high-pressure gases resulting from combustion in the combustion chamber expand through a turbine, which both drives the compressor and delivers power. The theoretical efficiency (power delivered/rate of energy input) of a gas turbine depends on the pressure ratio between the combustion chamber and the intake; the higher the pressure ratio, the higher the efficiency. The reader is directed to Cohen et al. (8) for more detail.

Wind Turbines

Wind energy is discussed frequently as an alternative energy source. The application of wind turbines* as potential sources for power becomes more attractive as utility power rates increase

---

*The phrase "wind turbine" is used to convey the idea of conversion of wind to electrical energy. A windmill converts wind energy to mechanical energy.

FIGURE 14.24

Horizontal-axis wind turbine showing capture area.

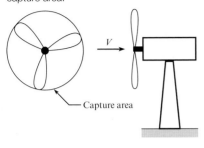

FIGURE 14.25

Typical wind turbine power curve.

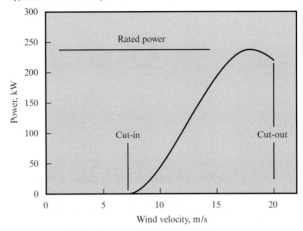

and the concern over greenhouse gases grows. In many European countries, especially northern Europe, the wind turbine is playing an ever-increasing role in power generation.

In essence, the wind turbine is just a reverse application of the process of introducing energy into an airstream to derive a propulsive force. The wind turbine extracts energy from the wind to produce power. There is one significant difference, however. The theoretical upper limit of efficiency of a propeller supplying energy to an airstream is 100%; that is, it is theoretically possible, neglecting viscous and other effects, to convert all the energy supplied to a propeller into energy of the airstream. This is not the case for a wind turbine.

A sketch of a horizontal-axis wind turbine is shown in Fig. 14.24. The wind blows along the axis of the turbine. The area of the circle traced out by the rotating blades is the **capture area**. The power associated with the wind passing through the capture area is

$$P = \rho Q \frac{V^2}{2} = \rho A \frac{V^3}{2} \tag{14.33}$$

where ρ is the air density and V is the wind speed. In an analysis attributed to Glauert/Betz (9), the theoretical maximum power attainable from a wind turbine is 16/27 or 59.3% of this power, or

$$P_{max} = \frac{16}{27}\left(\frac{1}{2}\rho V^3 A\right) \tag{14.34}$$

Other factors, such as swirl of the airstream and viscous effects, further reduce the power achievable from a wind turbine.

The power output of any wind turbine is related to the wind speed through the wind turbine power curve. A typical curve is shown in Fig. 14.25. This curve can usually be obtained from the manufacturer. The wind turbine is inoperative below the cut-in speed. After cut-in, the power increases with wind speed reaching a maximum value, which is the rated power output for the turbine. Engineering design and safety constraints impose an upper limit on the rotational velocity and establish the cutout speed. A braking system is used to prevent operation of the wind turbine beyond this velocity.

The conventional horizontal-axis wind turbine has been the focus of most research and design. Considerable effort has also been devoted to assessment of the Savonius rotor and the Darrieus turbine, both of which are vertical-axis turbines, as shown in Fig. 14.26. The Savonius rotor consists of two curved blades forming an S-shaped passage for the air flow. The Darrieus

FIGURE 14.26

Wind turbine configurations:
(a) Savonius turbine, (b) Darrieus turbine.

(a) Savonius rotor   (b) Darrieus rotor

turbine consists of two or three airfoils attached to a vertical shaft; the unit resembles an egg beater. The advantage of vertical-axis turbines is that their operation is independent of wind direction. The Darrieus wind turbine is considered superior in performance but has a disadvantage in that it is not self-starting. Frequently, a Savonius rotor is mounted on the axis of a Darrieus turbine to provide the starting torque.

For more information on wind turbines and wind turbine systems, refer to *Wind Energy Explained* (10).

EXAMPLE 14.12

Calculating the Capture Area of a Wind Turbine

Problem Statement

Calculate the minimum capture area necessary for a windmill that has to operate five 100-watt bulbs if the wind velocity is 20 km/h and the air density is 1.2 kg/m³.

Define the Situation

A wind turbine needs to generate 500 W of electrical power.

State the Goal

Find the minimum capture area of the windmill.

Generate Ideas and Make a Plan

Use the equation for maximum power of a windmill.

Take Action (Execute the Plan)

Capture area for maximum power:

$$A = P_{max} \frac{54}{16} \frac{1}{\rho V^3}$$

Wind velocity in m/s:

$$20 \text{ km/h} = \frac{20 \times 1000}{3600} = 5.56 \text{ m/s}$$

Minimum capture area:

$$A = 500 \text{ W} \times \frac{54}{16} \times \frac{1}{1.2 \text{ kg/m}^3 \times (5.56 \text{ m/s})^3}$$

$$= \boxed{8.18 \text{ m}^2}$$

Review the Solution and the Process

Discussion. This area corresponds to a windmill diameter of 3.23 m, or about 10.6 ft.

14.9 Summarizing Key Knowledge

The Propeller

• The thrust of a propeller is calculated using

$$F_T = C_T \rho n^2 D^4$$

where ρ is the fluid density, n is the rotational rate of the propeller, and D is the propeller diameter. The thrust coefficient C_T is a function of the advance ratio V_0/nD.

• The efficiency of a propeller is the ratio of the power delivered by the propeller to the power provided to the propeller:

$$\eta = \frac{F_T V_0}{P}$$

Pumps

- Pumps can be *axial flow* or *radial flow*:
  - An axial-flow pump consists of an impeller, much like a propeller, mounted in a housing.
  - In a radial-flow pump, fluid enters near the eye of the impeller, passes through the vanes, and exits at the edge of the vanes.
- The head provided by a pump is quantified by the *head coefficient*, C_H, defined as

$$C_H = \frac{g\Delta H}{n^2 D^2}$$

where ΔH is the head across the pump.
- The head coefficient is a function of the *discharge coefficient*, which is

$$C_Q = \frac{Q}{nD^3}$$

where Q is the discharge.
- Pump performance curves show head delivered, power required, and efficiency as a function of discharge.
- The specific speed of a pump can be used to select an appropriate type of pump for a given application:
  - Axial-flow pumps are best suited for high-discharge, low-head applications.
  - Radial-flow pumps are best suited for low-discharge, high-head applications.

Water Turbines

- Turbines convert the energy associated with a moving fluid to shaft work.
- Turbines are classified into two categories:
  - The *impulse turbine* consists of a liquid jet impinging on vanes of a turbine wheel or runner.
  - A *reaction turbine* consists of a series of rotating vanes immersed in a flowing fluid. The pressure on the vanes provides the torque for the power.

Wind Turbines

- Wind turbines are classified based on the axis of the rotor:
  - The rotor of a turbine can revolve around a *horizontal axis*. Most commercial wind turbines use this design.
  - The rotor of a turbine can revolve around a *vertical axis*. Two types of turbine in this category are the Darrieus turbine and the Savonius turbine.
- The maximum power derivable from a wind turbine is

$$P_{\max} = \frac{16}{27}\left(\frac{1}{2}\rho V_0^3 A\right)$$

where A is the capture area of the wind turbine (projected area from direction of wind) and V_0 is the wind speed.

REFERENCES

1. Weick, F.E. *Aircraft Propeller Design*. New York: McGraw-Hill, 1930.

2. Weick, Fred E. "Full Scale Tests on a Thin Metal Propeller at Various Pit Speeds." *NACA Report*, 302 (January 1929).

3. Stepanoff, A.J. *Centrifugal and Axial Flow Pumps*, 2nd ed. New York: John Wiley, 1957.

4. Daugherty, Robert L., and Joseph B. Franzini. *Fluid Mechanics with Engineering Applications*. New York: McGraw-Hill, 1957.

5. Hydraulic Institute. *Centrifugal Pumps*. Parsippany, NJ: Hydraulic Institute, 1994.

6. McQuiston, F.C., and J.D. Parker. *Heating, Ventilating and Air Conditioning*. New York: John Wiley, 1994.

7. Moody, L.F. "Hydraulic Machinery." In *Handbook of Applied Hydraulics*, ed. C. V. Davis. New York: McGraw-Hill, 1942.

8. Cohen, H., G.F.C. Rogers, and H.I.H. Saravanamuttoo. *Gas Turbine Theory*. New York: John Wiley, 1972.

9. Glauert, H. "Airplane Propellers." *Aerodynamic Theory*, vol. IV, ed. W. F. Durand. New York: Dover, 1963.

10. Manwell, J. F., J. G. McGowan, and A. L. Rogers. *Wind Energy Explained: Theory, Design and Application*. Chichester, UK: John Wiley, 2002.

PROBLEMS

Propellers (§14.1)

14.1 Explain why the thrust of a fixed-pitch propeller decreases with increasing forward speed.

14.2 What thrust is obtained from a propeller 3 m in diameter that has the characteristics given in Fig. 14.3 when the propeller is operated at an angular speed of 1100 rpm and an advance velocity of zero? Assume $\rho = 1.05$ kg/m^3.

14.3 What thrust is obtained from a propeller 3 m in diameter that has the characteristics given in Fig. 14.3 when the propeller is operated at an angular speed of 1400 rpm and an advance velocity of 80 km/h? What power is required to operate the propeller under these conditions? Assume $\rho = 1.05$ kg/m³.

14.4 A propeller 8 ft in diameter, like the one for which characteristics are given in Fig. 14.3, is to be used on a swamp boat and is to operate at maximum efficiency when cruising. If the cruising speed is to be 30 mph, what should the angular speed of the propeller be? Assume $\rho = 0.0024$ slugs/ft³.

14.5 A propeller is being selected for an airplane that will cruise at 2000 m altitude, where the pressure is 60 kPa absolute and the temperature is 10°C. The mass of the airplane is 1200 kg, and the planform area of the wing is 10 m². The lift-to-drag ratio is 30:1. The lift coefficient is 0.4. The engine speed at cruise conditions is 3000 rpm. The propeller is to operate at maximum efficiency, which corresponds to a thrust coefficient of 0.025. Calculate the diameter of the propeller and the speed of the aircraft.

14.6 A propeller 2 m in diameter, like the one for which characteristics are given in Fig. 14.3, is to be used on a swamp boat and is to operate at maximum efficiency when cruising. If the cruising speed is to be 40 km/h, what should the angular speed of the propeller be?

14.7 A propeller 2 m in diameter and like the one for which characteristics are given in Fig. 14.3 is used on a swamp boat. If the angular speed is 1000 rpm and if the boat and passengers have a combined mass of 300 kg, estimate the initial acceleration of the boat when starting from rest. Assume $\rho = 1.1$ kg/m³.

Axial Flow Pumps and Fans (§14.2)

14.8 Answer the following questions about axial-flow pumps.

  a. Axial-flow pumps are best suited for what conditions of head produced and discharge?

  b. For an axial-flow pump, how does the head produced by the pump and the power required to operate a pump vary with flow rate through the pump?

14.9 The pump used in the system shown has the characteristics given in Fig. 14.8. What discharge will occur under the conditions shown, and what power is required?

14.10 If the conditions are the same as in Problem 14.9 except that the speed is increased to 900 rpm, what discharge will occur, and what power is required for the operation?

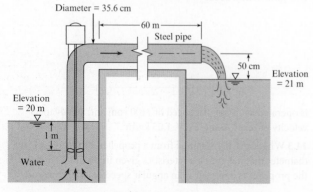

Problems 14.9, 14.10

14.11 A pump has the characteristics given by Fig. 14.7. What discharge and head will be produced at maximum efficiency if the pump size is 50 cm and the angular speed is 45 rps? What power is required when pumping water at 10°C under these conditions?

14.12 For a pump having the characteristics of Fig. 14.7, plot the head-discharge curve if the pump diameter is 60 cm and the speed is 690 rpm.

14.13 An axial-flow blower is used to air-condition an office building that has a volume of 10^5 m³. It is decided that the air at 60°F in the building must be completely changed every 15 min. Assume that the blower operates at 600 rpm at maximum efficiency and has the characteristics shown in Fig. 14.7. Calculate the diameter and power requirements for two blowers operating in parallel.

14.14 An axial fan 2 m in diameter is used in a wind tunnel as shown (test section 1.2 m in diameter; test section velocity of 60 m/s). The rotational speed of the fan is 1800 rpm. Assume the density of the air is constant at 1.2 kg/m³. There are negligible losses in the tunnel. The performance curve of the fan is identical to that shown in Fig. 14.7. Calculate the power needed to operate the fan.

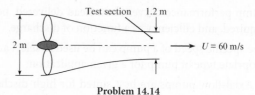

Problem 14.14

Radial Flow Pumps (§14.3)

14.15 The radial flow pump is best suited for what conditions of head produced and discharge?

14.16 If a pump with the characteristics given in Fig. 14.10 is doubled in size but halved in speed, what will be the head and discharge at maximum efficiency?

14.17 If a pump with the characteristics given in Fig. 14.10 or 14.11 is operated at a speed of 1600 rpm, what will be the discharge when the head is 135 ft?

14.18 If a pump with the performance curve shown is operated at a speed of 1600 rpm, what will be the maximum possible head developed?

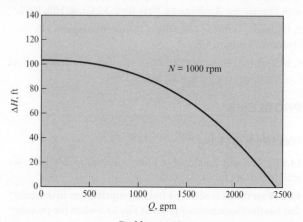

Problem 14.18

14.19 If a pump with the characteristics given in Fig. 14.10 is operated at a speed of 30 rps, what will be the shutoff head?

14.20 A centrifugal pump 20 cm in diameter is used to pump kerosene at a speed of 5000 rpm. Assume that the pump has the characteristics shown in Fig. 14.11. Calculate the flow rate, the pressure rise across the pump, and the power required if the pump operates at maximum efficiency.

Specific Speed and Pump Selection (§14.4)

14.21 The value of the specific speed suggests the type of pump to be used for a given application. A high specific speed suggests the use of what kind of pump?

14.22 The pump curve for a given pump is represented by

$$h_{p,\text{pump}} = 20\left[1 - \left(\frac{Q}{100}\right)^2\right]$$

where $h_{p,\text{pump}}$ is the head provided by the pump in feet and Q is the discharge in gpm. The system curve for a pumping application is

$$h_{p,\text{sys}} = 5 + 0.002Q^2$$

where $h_{p,\text{sys}}$ is the head in feet required to operate the system and Q is the discharge in gpm. Find the operating point (Q) for (a) one pump, (b) two identical pumps connected in series, and (c) two identical pumps connected in parallel.

14.23 What is the suction specific speed for the pump that is operating under the conditions given in Problem 14.9? Is this a safe operation with respect to susceptibility to cavitation?

14.24 What type of pump should be used to pump water at a rate of 10 cfs and under a head of 30 ft? Assume $N = 1500$ rpm.

14.25 An axial-flow pump is to be used to lift water against a head (friction and static) of 15 ft. If the discharge is to be 4000 gpm, what maximum speed in revolutions per minute is allowed if the suction head is 5 ft?

14.26 A pump is needed to pump water at a rate of 0.2 m³/s from the lower to the upper reservoir shown in the figure. What type of pump would be best for this operation if the impeller speed is to be 600 rpm? Assume $f = 0.02$ and $K_e = 0.5$.

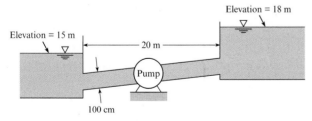

Problem 14.26

Compressors (§14.7)

14.27 Methane flowing at the rate of 1 kg/s is to be compressed by a noncooled centrifugal compressor from 100 kPa to 165 kPa. The temperature of the methane entering the compressor is 27°C. The efficiency of the compressor is 70%. Calculate the shaft power necessary to run the compressor.

14.28 A 36 kW (shaft output) motor is available to run a noncooled compressor for carbon dioxide. The pressure is to be increased from 100 kPa to 150 kPa. If the compressor is 60% efficient, calculate the volume flow rate into the compressor.

Impulse Turbines (§14.8)

14.29 A penstock 1 m in diameter and 10 km long carries water at 10°C from a reservoir to an impulse turbine. If the turbine is 85% efficient, what power can be produced by the system if the upstream reservoir elevation is 650 m above the turbine jet and the jet diameter is 16.0 cm? Assume that $f = 0.016$ and neglect head losses in the nozzle. What should the diameter of the turbine wheel be if it is to have an angular speed of 360 rpm? Assume ideal conditions for the bucket design [$V_{\text{bucket}} = (1/2)V_j$].

14.30 Consider a single jet of water striking the buckets of the impulse wheel as shown. Assume ideal conditions for power generation [$V_{\text{bucket}} = (1/2)V_j$ and the jet is turned through 180° of arc]. With the foregoing conditions, solve for the jet force on the bucket and then solve for the power developed. Note that this power is not the same as that given by Eq. (14.24)! Study the figure to resolve the discrepancy.

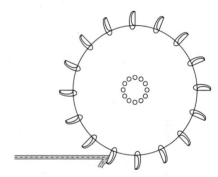

Problem 14.30

Reaction Turbines (§14.8)

14.31 Answer the following questions about reaction turbines.

    a. How does a reaction turbine differ from a centrifugal pump?

    b. What is meant by the "runner" in a reaction turbine?

14.32 For a given Francis turbine, $\beta_1 = 60°$, $\beta_2 = 90°$, $r_1 = 5$ m, $r_2 = 3$ m, and $B = 1$ m. The discharge is 126 m³/s, and the rotational speed is 60 rpm. Assume $T = 10°C$.

    a. What should α_1 be for a nonseparating flow condition at the entrance to the runner?

    b. What is the maximum attainable power with the conditions noted?

    c. If you were to redesign the turbine blades of the runner, what changes would you suggest to increase the power production if the discharge and overall dimensions are to be kept the same?

Wind Turbines (§14.8)

14.33 Calculate the minimum capture area necessary for a wind turbine that will be required to power the 2 kW demands of an energy-efficient home. Assume a wind velocity of 10 mph and an air density of 1.2 kg/m^3.

14.34 Calculate the maximum power derivable from a conventional horizontal-axis wind turbine with a propeller 2.3 m in diameter in a 47 km/h wind with density 1.2 kg/m^3.

14.35 A wind farm consists of 20 Darrieus turbines, each 15 m high. The total output from the turbines is to be 2 MW in a wind of 20 m/s and an air density of 1.2 kg/m^3. The Darrieus turbine shown has the shape of an arc of a circle. Find the minimum width, W, of the turbine needed to provide this power output.

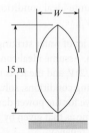

Problem 14.35

14.36 A windmill is connected directly to a mechanical pump that is to pump water from a well 10 ft deep as shown. The windmill is a conventional horizontal-axis type with a fan diameter of 10 ft. The efficiency of the mechanical pump is

80%. The density of the air is 0.07 lbm/ft^3. Assume the windmill delivers the maximum power available. There is 20 ft of 2 in. galvanized pipe in the system. What would the discharge of the pump be (in gallons per minute) for a 30 mph wind? (1 cfm = 7.48 gpm)

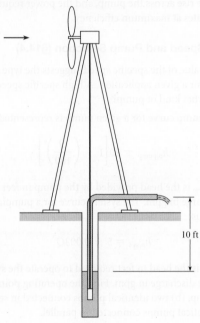

Problem 14.36

FLOW IN OPEN CHANNELS

CHAPTER ROAD MAP   The flow of water in open channels can be observed in aqueducts, rivers, flumes, irrigation ditches, and other contexts. Although these contexts are quite different, a small set of concepts and a few equations generalize to most applications of open-channel flow. These ideas are introduced in this chapter.

FIGURE 15.1
Aerial view of the California Aqueduct at the southwest end of the Tehachapi Mountains. (Macduff Everton/The Image Bank/Getty Images.)

LEARNING OUTCOMES

DESCRIBING FLOW (§15.1).

- Define an open channel.
- Define uniform flow and nonuniform flow.
- Define the Froude number.
- Calculate the hydraulic radius and the Reynolds number.
- List the criteria for laminar and turbulent flow.

UNIFORM FLOW (§15.2, §15.3).

- Explain the physics of the energy equation and also explain the corresponding HGL and EGL.
- Calculate flow rate with the Darcy-Weisbach approach or the Manning equation.
- Define and explain the best hydraulic section.

NONUNIFORM FLOW (§15.4 to §15.7).

- Describe and compare rapidly varied flow and gradually varied flow.
- Describe critical depth, specific energy, supercritical flow, and subcritical flow.
- Describe a hydraulic jump and perform calculations.
- Describe the factors that are used to classify surface profiles that occur in gradually varied flow.

15.1 Describing Open-Channel Flow

An **open channel** is one in which a liquid flows with a free surface. A **free surface** means that the liquid surface is exposed to the atmosphere. Examples of open channels are natural creeks and rivers, artificial channels such as irrigation ditches and canals, and pipelines or sewers flowing less than full. In most cases, water or wastewater is the flowing liquid.

Flow in an open channel is described as uniform or nonuniform, as distinguished in Fig. 15.2. As defined in Chapter 4, **uniform flow** means that the velocity is constant along a streamline, which in open-channel flow means that depth and cross section are constant along the length of a channel. The depth for uniform-flow conditions is called **normal depth** and is designated by y_n. For **nonuniform flow**, the velocity changes from section to section along the channel, and thus one observes changes in depth. The velocity change may be due to a change in channel configuration, such as a bend, change in cross-sectional shape, or change in channel slope. For example, Fig. 15.2 shows steady flow over a spillway of constant width, where the water must flow progressively faster as it goes over the brink of the spillway (from A to B), caused by the suddenly steeper slope. The faster velocity requires a smaller depth, in accordance with conservation of mass (continuity). From reach B to C, the flow is uniform because the velocity (and thus depth) are constant. After reach C, the abrupt flattening of channel slope requires the velocity to suddenly and turbulently slow down. Thus, there is a deeper depth downstream of C than in reach B to C.

The most complicated open-channel flow is unsteady nonuniform flow. An example of this is a breaking wave on a sloping beach. Theory and analysis of unsteady nonuniform flow are reserved for more advanced courses.

Dimensional Analysis in Open-Channel Flow

Open-channel flow results from gravity moving water from higher to lower elevations and is impeded by friction forces caused by the roughness of the channel. Thus, the functional equation $Q = f(\mu, \rho, V, L)$ and dimensional analysis lead to two important π-groups: the Froude number and the Reynolds number. The Froude number squared is the ratio of kinetic force to gravity force:

$$\text{Fr}^2 = \frac{\text{kinetic force}}{\text{gravity force}} = \frac{\rho L^2 V^2}{\gamma L^3} = \frac{V^2}{L \gamma / \rho} \qquad (15.1)$$

$$\text{Fr} = \frac{V}{\sqrt{gL}} \qquad (15.2)$$

FIGURE 15.2

Distinguishing uniform and nonuniform flow: This example shows steady flow over a spillway, such as the emergency overflow channel of a dam.

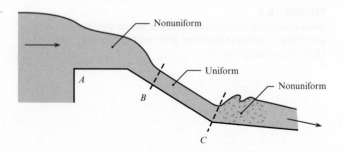

The Froude number is important if the gravitational force influences the direction of flow, such as in flow over a spillway, or the formation of surface waves. However, it is unimportant when gravity causes only a hydrostatic pressure distribution, such as in a closed conduit.

The use of Reynolds number for determining whether the flow in open channels will be laminar or turbulent depends on the **hydraulic radius**, given by

$$R_h = \frac{A}{P} \tag{15.3}$$

where A is the cross-sectional area of flow and P is the wetted perimeter. The characteristic length R_h is analogous to diameter D in pipe flow. Recall that for pipe flow (Chapter 10), if the Reynolds number ($VD\rho/\mu = VD/\nu$) is less than 2000, the flow will be laminar, and if it is greater than about 3000, one can expect the flow to be turbulent. The Reynolds number criterion for open-channel flow would be 2000 if one were to replace D in the Reynolds number with $4R_h$, where R_h is the hydraulic radius. For this definition of Reynolds number, laminar flow would occur in open channels if $V(4R_h)/\nu < 2000$.

However, the standard convention in open-channel flow analysis is to define the Reynolds number as

$$\mathrm{Re} = \frac{VR_h}{\nu} \tag{15.4}$$

Therefore, in open channels, if the Reynolds number is less than 500, the flow is laminar, and if Re is greater than about 750, one can expect to have turbulent flow. A brief analysis of this turbulent criterion (see Example 15.1) will show that water flow in channels will usually be turbulent, unless the velocity and/or the depth is very small.

It should be noted that for rectangular channels (see Fig. 15.3), the hydraulic radius is

$$R_h = \frac{A}{P} = \frac{By}{B + 2y} \tag{15.5}$$

For a wide and shallow channel, $B \gg y$ and Eq. (15.5) reduces to $R_h \approx y$, which means that the hydraulic radius approaches the depth of the channel.

Most open-channel flow problems involve turbulent flow. If one calculates the conditions needed to maintain laminar flow, as in Example 15.1, one sees that laminar flow is uncommon.

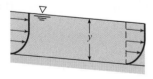

Side view

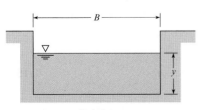

End view

FIGURE 15.3

Open-channel relations.

EXAMPLE 15.1

Calculating Reynolds Number and Classifying Flow for a Rectangular Open Channel

Problem Statement

Water (60 °F) flows in a 10-ft-wide rectangular channel at a depth of 6 ft. What is the Reynolds number if the mean velocity is 0.1 ft/s? With this velocity, at what maximum depth can one be assured of having laminar flow?

Define the Situation

Water flows in a rectangular channel.

$$B = 10 \text{ ft}, y = 6 \text{ ft}, V = 0.1 \text{ ft/s}.$$

Properties:

Water (60 °F, 1 atm, Table A.5): $\nu = 1.22 \times 10^{-5} \text{ ft}^2/\text{s}$.

State the Goal

1. Re ← Reynolds number
2. y_m(ft) ← maximum depth for laminar flow

Generate Ideas and Make a Plan

To find Re, apply Eq. (15.4). To find y_m, apply the criterion that laminar flow occurs for Re < 500. The plan is as follows:

1. Calculate the hydraulic radius using Eq. (15.5).
2. Calculate the Reynolds number using Eq. (15.4).
3. Let Re = 500, solve for R_h, and then solve for y_m.

Take Action (Execute the Plan)

1. Hydraulic radius:

$$R_h = \frac{By}{B + 2y} = \frac{(10 \text{ ft})(6 \text{ ft})}{(10 \text{ ft}) + 2(6 \text{ ft})} = 2.727 \text{ ft}$$

2. Reynolds number:

$$\text{Re} = \frac{VR_h}{\nu} = \frac{(0.1 \text{ ft/s})(2.727 \text{ ft})}{(1.22 \times 10^{-5} \text{ ft}^2/\text{s})} = \boxed{22,400}$$

3. Laminar flow criterion (Re < 500):

$$\text{Re} = VR_h/\nu = (0.10 \text{ ft/s})R_h/(1.22 \times 10^{-5} \text{ ft}^2/\text{s}) = 500$$
$$R_h = (500)(1.22 \times 10^{-5} \text{ ft}^2/\text{s})/(0.10 \text{ ft/s}) = 0.061 \text{ ft}$$

For a rectangular channel,

$$R_h = (By)/(B + 2y)$$
$$(By)/(B + 2y) = (10y)/(10 + 2y) = 0.061 \text{ ft}$$
$$y_m = \boxed{0.062 \text{ ft}}$$

Review the Solution and the Process

1. *Knowledge.* Velocity or depth must be very small to yield laminar flow of water in an open channel.
2. *Knowledge.* Depth and hydraulic radius are virtually the same when depth is very small relative to width.

15.2 Energy Equation for Steady Open-Channel Flow

To derive the energy equation for flow in an open channel, begin with Eq. (7.29) and let the pump head and turbine head equal zero: $h_p = h_t = 0$. Equation (7.29) becomes

$$\frac{p_1}{\gamma} + \alpha_1 \frac{V_1^2}{2g} + z_1 = \frac{p_2}{\gamma} + \alpha_2 \frac{V_2^2}{2g} + z_2 + h_L \tag{15.6}$$

Use Fig. 15.4 to show that

$$\frac{p_1}{\gamma} + z_1 = y_1 + S_0\Delta x \quad \text{and} \quad \frac{p_2}{\gamma} + z_2 = y_2$$

where S_0 is the slope of the channel bottom and y is the depth of flow. Assume the flow in the channel is turbulent, so $\alpha_1 = \alpha_2 \approx 1.0$. Equation (15.6) becomes

$$y_1 + \frac{V_1^2}{2g} + S_0\Delta x = y_2 + \frac{V_2^2}{2g} + h_L \tag{15.7}$$

In addition to the foregoing assumptions, Eq. (15.7) also requires that the channel have a uniform cross section and that the flow be steady.

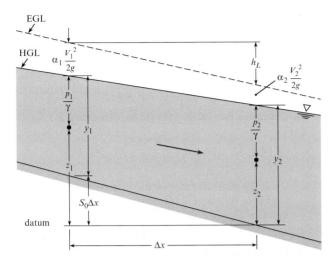

EGL

HGL

$\alpha_1 \dfrac{V_1^2}{2g}$

$\dfrac{p_1}{\gamma}$

y_1

z_1

$S_0 \Delta x$

datum

Δx

h_L

$\alpha_2 \dfrac{V_2^2}{2g}$

$\dfrac{p_2}{\gamma}$

y_2

z_2

FIGURE 15.4

Definition sketch for flow in open channels.

15.3 Steady Uniform Flow

Uniform flow requires that velocity be constant in the flow direction, so the shape of the channel and the depth of fluid are the same from section to section. Consideration of the foregoing slope equations shows that for uniform flow the slope of the HGL will be the same as the channel slope because the velocity and depth are the same in both sections. The HGL, and thus the slope of the water surface, is controlled by head loss. If one restates the Darcy-Weisbach equation introduced in Chapter 10 with D replaced by $4R_h$, the head loss is

$$h_f = \frac{fL}{4R_h}\frac{V^2}{2g} \quad \text{or} \quad \frac{h_f}{L} = \frac{f}{4R_h}\frac{V^2}{2g} \tag{15.8}$$

From Fig. 15.4, S_0 is equal to the slope of the EGL, which is a function of the head loss, so $S_0 = (h_f/L)$, yielding the following equation for velocity:

$$V = \sqrt{\frac{8g}{f}R_h S_0} \tag{15.9}$$

To solve Eq. (15.9) for velocity, the friction factor f can be found from the Moody diagram (Fig. 10.14) and can then be used to solve iteratively for the velocity for a given uniform-flow condition. This is demonstrated in Example 15.2.

EXAMPLE 15.2

Applying the Darcy-Weisbach Equation to Find the Flow Rate in a Rectangular Open Channel

Problem Statement

Estimate the discharge of water that a concrete channel 10 ft wide can carry if the depth of flow is 6 ft and the slope of the channel is 0.0016.

Define the Situation

- Water flows in a rectangular channel.
- $B = 10$ ft, $y = 6$ ft, $S_0 = 0.0016$.

Assumptions: Uniform flow

Properties:

- Water (60 °F, 1 atm, Table A.5): $\nu = 1.22 \times 10^{-5}$ ft²/s
- Concrete (Table 10.4): $k_s \approx 0.005$ ft

State the Goal

$Q(\text{ft}^3/\text{s}) \leftarrow$ discharge in the channel

Generate Ideas and Make a Plan

Because the goal is Q, apply the flow rate equation:

$$Q = VA \tag{a}$$

To find V in Eq. (a), apply Eq. (15.9):

$$V = \sqrt{\frac{8g}{f}R_h S_0} \tag{b}$$

To find R_h in Eq. (b), apply Eq. (15.5):

$$R_h = \frac{By}{B + 2y} = \frac{(10 \text{ ft})(6 \text{ ft})}{(10 \text{ ft}) + 2(6 \text{ ft})} = 2.727 \text{ ft} \qquad \textbf{(c)}$$

To find f in Eq. (b), use an iterative approach with the Moody diagram. This is a Case 2 problem from Chapter 10. Therefore:

1. Calculate relative roughness. Then, guess a value of f.
2. Calculate V using Eq. (b).
3. Calculate Reynolds number, then look up f on the Moody diagram and compare to the guess in step 1. If needed, go back to step 2.
4. Calculate Q using Eq. (a).

Take Action (Execute the Plan)

1. Calculate relative roughness:

$$\frac{k_s}{4R_h} = \frac{0.005 \text{ ft}}{4(60 \text{ ft}^2/22 \text{ ft})} = \frac{0.005 \text{ ft}}{4(2.73 \text{ ft})} = 0.00046$$

Use the value of $k_s/4R_h = 0.00046$ as a guide to estimate $f = 0.016$.

2. Calculate V based on guess of f:

$$V = \sqrt{\frac{8(32.2 \text{ ft/s}^2)(2.73 \text{ ft})(0.0016)}{0.016}}$$

$$= \sqrt{70.6 \text{ ft}^2/\text{s}^2} = 8.39 \text{ ft/s}$$

3. Calculate a new value of f based on V from step 2:

$$Re = V\frac{4R_h}{\nu} = \frac{8.39 \text{ ft/s}(10.9 \text{ ft})}{1.2(10^{-5} \text{ ft}^2/\text{s})} = 7.62 \times 10^6$$

Using this new value of Re and $k_s/4R_h = 0.00046$, read f as 0.016. This value of f is the same as the previous estimate. Thus, we conclude that

$$V = 8.39 \text{ ft/s}$$

4. Flow rate equation:

$$Q = VA = 8.39 \text{ ft/s}(60 \text{ ft}^2) = \boxed{503 \text{ cfs}}$$

Review the Solution and the Process

1. *Notice.* The approach to solving this problem parallels the approach presented in Chapter 10 for solving problems that involve flow in conduits.

2. *Knowledge.* Hydraulic diameter is four times the hydraulic radius. This is why the relative roughness formula in step 1 is $k_s/(4R_h)$.

Rock-Bedded Channels

For rock-bedded channels, such as those in some natural streams or unlined canals, the larger rocks produce most of the resistance to flow; essentially none of this resistance is due to viscous effects. Thus, the friction factor is independent of the Reynolds number. This is analogous to the fully rough region of the Moody diagram for pipe flow. For a rock-bedded channel, Limerinos (1) has shown that the resistance coefficient f can be given in terms of the size of rock in the streambed as

$$f = \frac{1}{\left[1.2 + 2.03 \log\left(\dfrac{R_h}{d_{84}}\right)\right]^2} \qquad \textbf{(15.10)}$$

where d_{84} is a measure of the rock size.*

The Chezy Equation

Leaders in open-channel research have recommended the use of the methods already presented (involving the Reynolds number and relative roughness k_s) for channel design (2). However, many engineers continue to use two traditional methods: the Chezy equation and the Manning equation.

---

*Most river-worn rocks are somewhat elliptical in shape. Limerinos (1) showed that the intermediate dimension d_{84} correlates best with f. The d_{84} refers to the size of rock (intermediate dimension) for which 84% of the rocks in the random sample are smaller than the d_{84} size. Details for choosing the sample are given by Wolman (3).

EXAMPLE 15.3

Resistance Coefficient for Boulders

Problem Statement

Determine the value of the resistance coefficient, f, for a natural rock-bedded channel that is 100 ft wide and has an average depth of 4.3 ft. The d_{84} size of boulders in the streambed is 0.72 ft.

Define the Situation

A natural channel is lined with boulders.

State the Goal

Find the friction factor, f.

Generate Ideas and Make a Plan

1. Since the channel is wide, approximate R_h as the depth of the channel.
2. Use Eq. (15.10) to find f on the basis of the d_{84} boulder size.

Take Action (Execute the Plan)

1. R_h is 4.3 ft.
2. Evaluate f.

$$f = \cfrac{1}{\left[1.2 + 2.03 \log\left(\cfrac{4.3}{0.72}\right)\right]^2} = \boxed{0.130}$$

As noted earlier, the depth in uniform flow, called normal depth, y_n, is constant. Consequently, h_f/L is the slope S_0 of the channel, and Eq. (15.8) can be written as

$$R_h S_0 = \frac{f}{8g} V^2$$

or

$$V = C\sqrt{R_h S_0} \qquad\qquad (15.11)$$

where

$$C = \sqrt{8g/f} \qquad\qquad (15.12)$$

Because $Q = VA$, the discharge in a channel is given by

$$Q = CA\sqrt{R_h S_0} \qquad\qquad (15.13)$$

This equation is known as the **Chezy equation** after a French engineer of that name. For practical application, the coefficient C must be determined. One way to determine C is by knowing an acceptable value of the friction factor f and using Eq. (15.12).

The Manning Equation

The second, and more common, way to determine C in the SI system of units is given as

$$C = \frac{R_h^{1/6}}{n} \qquad\qquad (15.14)$$

where n is a resistance coefficient called **Manning's n,** which has different values for different types of boundary roughness. When this expression for C is inserted into Eq. (15.13), the result is a common form of the discharge equation for uniform flow in open channels for SI units, referred to as the Manning equation:

$$Q = \frac{1.0}{n} A R_h^{2/3} S_0^{1/2} \qquad\qquad (15.15)$$

Table 15.1 gives values of n for various types of boundary surfaces. The major limitation of this approach is that the viscous or relative-roughness effects are not present in the design formula. Hence, application outside the range of normal-sized channels carrying water is not recommended.

TABLE 15.1 Typical Values of Roughness Coefficient, Manning's *n*

Lined Canals	*n*
Cement plaster	0.011
Untreated gunite	0.016
Wood, planed	0.012
Wood, unplaned	0.013
Concrete, troweled	0.012
Concrete, wood forms, unfinished	0.015
Rubble in cement	0.020
Asphalt, smooth	0.013
Asphalt, rough	0.016
Corrugated metal	0.024
Unlined Canals	
Earth, straight and uniform	0.023
Earth, winding and weedy banks	0.035
Cut in rock, straight and uniform	0.030
Cut in rock, jagged and irregular	0.045
Natural Channels	
Gravel beds, straight	0.025
Gravel beds plus large boulders	0.040
Earth, straight, with some grass	0.026
Earth, winding, no vegetation	0.030
Earth, winding, weedy banks	0.050
Earth, very weedy and overgrown	0.080

Manning Equation: Traditional System of Units

The form of the Manning equation depends on the system of units because Manning's equation is not dimensionally homogeneous. In Eq. (15.15), notice that the primary dimensions on the left side of the equation are L^3/T and the primary dimensions on the right side are $L^{8/3}$.

To convert the Manning equation from SI to traditional units, one must apply a factor equal to 1.49 if the same value of *n* is used in the two systems. Thus, in the traditional system, the discharge equation using Manning's *n* is

$$Q = \frac{1.49}{n} A R_h^{2/3} S_0^{1/2} \qquad (15.16)$$

In Example 15.4, a value for Manning's *n* is calculated from known information about a channel and compared to tabulated values for *n* in Table 15.1.

EXAMPLE 15.4

Apply the Chezy Equation to find Manning's Value of *n* for Flow in a Channel

Problem Statement

If a channel with boulders has a slope of 0.0030, is 100 ft wide, has an average depth of 4.3 ft, and is known to have a friction factor of 0.130, what is the discharge in the channel, and what is the numerical value of Manning's *n* for this channel?

Define the Situation

Water flows in a channel with boulders:

$$S_0 = 0.003, B = 100 \text{ ft}, y = 4.3 \text{ ft}, f = 0.13$$

Assumptions: $R_h \approx y = 4.3$ ft (because the channel is wide).

State the Goal

1. Q(cfs) ← discharge in the channel
2. n ← Manning's *n*

Generate Ideas and Make a Plan

To find Q, apply the flow rate equation:

$$Q = VA \qquad \text{(a)}$$

To find V in Eq. (a), apply Eq. (15.9):

$$V = \sqrt{\frac{8g}{f} R_h S_0} \qquad \text{(b)}$$

To find n, apply Eq. (15.16):

$$Q = \frac{1.49}{n} AR_h^{2/3} S_0^{1/2} \qquad \text{(c)}$$

Because Eqs. (a) to (c) form a set of three equations with three unknowns, they can be solved. The plan is as follows:

1. Calculate V using Eq. (b).
2. Calculate Q using Eq. (a).
3. Calculate n using Eq. (c).

Take Action (Execute the Plan)

1. Velocity:

$$V = \left[\sqrt{\frac{(8)(32.2 \ \text{ft/s}^2)}{0.130}} \right] \left[\sqrt{(4.3 \ \text{ft})(0.0030)} \right] = 5.06 \ \text{ft/s}$$

2. Flow rate equation:

$$Q = VA = (5.06 \ \text{ft/s})(100 \times 4.3 \ \text{ft}^2) = \boxed{2180 \ \text{cfs}}$$

3. Manning's n (traditional units):

$$n = \frac{1.49}{Q} AR_h^{2/3} S_0^{1/2}$$

$$n = \left(\frac{1.49}{2176 \ \text{ft}^3/s} \right) (100 \times 4.3 \ \text{ft}^2)(4.3 \ \text{ft})^{2/3} (0.003)^{1/2}$$

$$n = \boxed{0.0426}$$

Review the Solution and the Process

1. *Validation.* This calculated value of n is within the range of typical values in Table 15.1 under the category of "Unlined Canals: Cut in rock."

2. *Notice.* For uniform flow, f in the Darcy-Weisbach equation can be related to Manning's n (as shown by this example).

In Example 15.5, the Chezy equation for traditional units is used to compute discharge.

EXAMPLE 15.5

Discharge Using Chezy Equation

Problem Statement

Using the Chezy equation with Manning's n, compute the discharge in a concrete channel 10 ft wide if the depth of flow is 6 ft and the slope of the channel is 0.0016.

Define the Situation

Water flows in a concrete channel. Width = 10 ft. Depth = 6 ft. Slope = 0.0016.

Properties: $n = 0.015$ for concrete channel (Table 15.1).

State the Goal

Find the discharge, Q.

Generate Ideas and Make a Plan

Use the Chezy equation for traditional units, Eq. (15.16).

Take Action (Execute the Plan)

$$Q = \frac{1.49}{n} AR_h^{2/3} S_0^{1/2}$$

$$R_h = \frac{60}{22} = 2.73 \ \text{ft} \quad \text{and} \quad R_h^{2/3} = 1.95$$

$$S_0^{1/2} = 0.04 \quad \text{and} \quad A = 60 \ \text{ft}^2$$

$$Q = \frac{1.49}{0.015} (60)(1.96)(0.04) = \boxed{467 \ \text{cfs}}$$

The two results (Examples 15.4 and 15.5) are within expected engineering accuracy for this type of problem. For a more complete discussion of the historical development of Manning's equation and the choice of n values for use in design or analysis, refer to Yen (4) and Chow (5).

FIGURE 15.5

Best hydraulic sections for
different geometries.

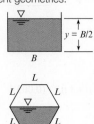

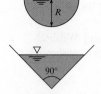

Best Hydraulic Section for Uniform Flow

The **best hydraulic section** is the channel geometry that gives the maximum discharge for a given cross-sectional area. Maximum discharge occurs when a geometry has the minimum wetted perimeter. Therefore, it yields the least viscous energy loss for a given area. Consider the quantity $AR_h^{2/3}$ in Manning's equation given in Eqs. (15.15) and (15.16), which is referred to as the section factor. Because $R_h = A/P$, the section factor relating to uniform flow is given by $A(A/P)^{2/3}$. Thus, for a channel of given resistance and slope, the discharge will increase with increasing cross-sectional area but decrease with increasing wetted perimeter P. For a given area A and a given shape of channel—for example, a rectangular cross section—there will be a certain ratio of depth to width (y/B) for which the section factor will be maximum. This ratio is the best hydraulic section.

Example 15.6 shows that the best hydraulic section for a rectangular channel occurs when $y = \frac{1}{2}B$. It can be shown that the best hydraulic section for a trapezoidal channel is half a hexagon, as shown; for the circular section, it is the half-circle with depth equal to radius; and for the triangular section, it is a triangle with a vertex of 90° (Fig. 15.5). Of all the various shapes, the half circle has the best hydraulic section because it has the smallest perimeter for a given area.

The best hydraulic section can be relevant to the cost of the channel. For example, if a trapezoidal channel were to be excavated and if the water surface were to be at adjacent ground level, the minimum amount of excavation (and excavation cost) would result if the channel of best hydraulic section were used.

EXAMPLE 15.6

Finding the Best Hydraulic Section for a Rectangular Channel

Problem Statement

Determine the best hydraulic section for a rectangular channel with depth y and width B.

Define the Situation

Water flows in a rectangular channel. Depth = y. Width = B.

State the Goal

Find the best hydraulic section (relate B and y).

Generate Ideas and Make a Plan

1. Set $A = By$ and $P = B + 2y$ so that both are a function of y.

2. Let A be constant, and minimize P:
   - Differentiate P with respect to y and set the derivative equal to zero.
   - Express the result of minimizing P as a relation between y and B.

Take Action (Execute the Plan)

1. Relate A and P in terms of y:

$$P = \frac{A}{y} + 2y$$

2a. Minimize P:

$$\frac{dP}{dy} = \frac{-A}{y^2} + 2 = 0$$

$$\frac{A}{y^2} = 2$$

2b. Express result in terms of y and B:

$$A = By, \text{ so}$$

$$\frac{By}{y^2} = 2 \quad \text{or} \quad \boxed{y = \frac{1}{2}B}$$

Review the Solution and the Process

Knowledge. The best hydraulic section for a rectangular channel occurs when the depth is one-half the width of the channel (see Fig. 15.5).

Uniform Flow in Culverts and Sewers

Sewers are conduits that carry sewage (liquid domestic, commercial, or industrial waste) from households, businesses, and factories to sewage disposal sites. These conduits are often circular in cross section, but elliptical and rectangular conduits are also used. The volume rate of sewage varies throughout the day and season, but of course sewers are designed to carry the

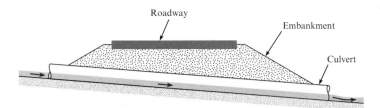

Roadway

Embankment

Culvert

FIGURE 15.6

Culvert under a highway embankment.

maximum design discharge flowing full or nearly full. At discharges less than the maximum, the sewers will operate as open channels.

Sewage usually consists of about 99% water and 1% solid waste. Because most sewage is so dilute, it is assumed that it has the same physical properties as water for purposes of discharge computations. However, if the velocity in the sewer is too small, then the solid particles may settle out and cause blockage of the flow. Therefore, sewers are usually designed to have a minimum velocity of about 2 ft/s (0.60 m/s) at times when the sewer is flowing full. This condition is met by choosing a slope on the sewer line to achieve the desired velocity.

A culvert is a conduit placed under a fill, such as a highway embankment. It is used to convey stream flow from the uphill side of the fill to the downhill side. Fig. 15.6 shows the essential features of a culvert. A culvert should be able to convey runoff from a **design storm** without overtopping the fill and without erosion of the fill at either the upstream or downstream end of the culvert. The design storm, for example, might be the maximum storm that could be expected to occur once in 50 years at the particular site.

The flow in a culvert is a function of many variables, including cross-sectional shape (circular or rectangular), slope, length, roughness, entrance design, and exit design. Flow in a culvert may occur as an open channel throughout its length, it may occur as a completely full pipe, or it may occur as a combination of both. The complete design and analysis of culverts are beyond the scope of this text; therefore, only simple examples are included here (Examples 15.7 and 15.8). For more extensive treatment of culverts, please refer to Chow (5), Henderson (6), and American Concrete Pipe Assoc. (7).

EXAMPLE 15.7

Sizing a Round Concrete Sewer Line

Problem Statement

A sewer line is to be constructed of concrete pipe to be laid on a slope of 0.006. If $n = 0.013$ and if the design discharge is 110 cfs, then what size pipe (commercially available) should be selected for a full-flow condition? What will be the mean velocity in the sewer pipe for these conditions? (It should be noted that concrete pipe is readily available in commercial sizes of 8 in., 10 in., and 12 in. diameter and then in 3 in. increments up to 36 in. diameter. From 36 in. diameter up to 144 in., the sizes are available in 6 in. increments.)

Define the Situation

Sewer line, $S_0 = 0.006$, Q (design) = 110 cfs.

Assumptions: Can only use a standard pipe size.

State the Goal

Find: The pipe diameter large enough to carry design discharge and that allows $V \geq 2$ ft/s at full-flow condition.

Generate Ideas and Make a Plan

1. Use Chezy equation for traditional units, Eq. (15.16).
2. Solve for $AR^{2/3}$.
3. For pipe flowing full, relate A and P to diameter through R_h.
4. Solve for diameter, and use the next commercial size larger.
5. Check that velocity for full flow is greater than 2 ft/s.

Take Action (Execute the Plan)

1. Chezy equation for traditional units is

$$Q = \frac{1.49}{n} AR^{2/3} S_0^{1/2}$$

$Q = 110 \text{ ft}^3/\text{s}$

$n = 0.013$

$S_0 = 0.006$ (assume atmospheric pressure in the pipe)

2. Solve for $AR^{2/3}$. Note that units of $AR^{2/3}$ are $ft^{8/3}$ because A is in ft^2 and R_h is in $ft^{2/3}$.

$$AR^{2/3} = \frac{(110\ ft^3/s)(0.013)}{(1.49)(0.006)^{1/2}} = 12.39\ ft^{8/3}$$

3. Relate A and P to diameter by relating to R_h:

$$R_h = \frac{A}{P} \quad \text{and} \quad R_h^{2/3} = \left(\frac{A}{P}\right)^{2/3}$$

$$AR_h^{2/3} = \frac{A^{5/3}}{P^{2/3}} = 12.39\ ft^{8/3}$$

For a pipe flowing full, $A = \pi D^2/4$ and $P = \pi D$, or

$$\frac{(\pi D^2/4)^{5/3}}{(\pi D)^{2/3}} = 12.39\ ft^{8/3}$$

4. Solving for diameter yields $D = 3.98\ ft = 47.8\ in$. Use the next commercial size larger, which is $\boxed{D = 48\ in.}$

$$A = \frac{\pi D^2}{4} = 50.3\ ft^2 \text{ (for pipe flowing full)}$$

5. Verify that velocity of full flow is greater than 2 ft/s:

$$V = \frac{Q}{A} = \frac{(110\ ft^3/s)}{(50.3\ ft^2)} = \boxed{2.19\ ft/s}$$

Example 15.8 demonstrates the calculation of necessary slope given all sources of head loss and a required discharge.

EXAMPLE 15.8

Culvert Design

Problem Statement

A 54-in.-diameter culvert laid under a highway embankment has a length of 200 ft and a slope of 0.01. This was designed to pass a 50-year flood flow of 225 cfs under full-flow conditions (see figure). For these conditions, what head H is required? When the discharge is only 50 cfs, what will be the uniform flow depth in the culvert? Assume $n = 0.012$.

Define the Situation

A culvert has been designed to carry 225 cfs with the given dimensions.

Assumptions: Uniform flow; pipe head loss h_f can be related to S_0.

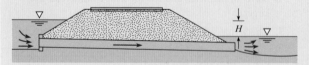

State the Goal

Find:

1. The height H required between the two free surfaces when flowing full.

2. The uniform flow depth in the culvert when $Q = 50$ cfs.

Generate Ideas and Make a Plan

1. Use the energy equation between the two end sections, accounting for head loss.

2. Document all sources of head loss.

3. Find pipe head loss h_f using Eq. (15.17) and the fact that

$$S_0 = \frac{h_f}{L}$$

4. Use continuity equation to find V, the uniform flow velocity, needed to calculate head loss.

5. Solve for H.

6. Solve for depth of flow, for $Q = 50$ cfs, using Eq. (15.16) and pipe geometry relations for pipe flowing partly full.

Take Action (Execute the Plan)

1. Energy equation:

$$\frac{p_1}{g} + \frac{V_1^2}{2g} + z_1 = \frac{p_2}{g} + \frac{V_2^2}{2g} + z_2 + \sum h_L$$

Let points 1 and 2 be at the upstream and downstream water surfaces, respectively.

Thus, $(p_1 = p_2 = 0$ gage and $V_1 = V_2 = 0)$.

Also, $(z_1 - z_2 = H)$.

Therefore, $(H = \sum h_L)$.

2. Head losses occur at culvert entrance and exit, as well as over the length of pipe:

H = pipe head loss + entrance head loss + exit head loss

$$H = \frac{V^2}{2g}(K_e + K_E) + \text{pipe head loss}$$

$$K_e = 0.50 \text{ (from Table 10.5)}$$
$$K_E = 1.00 \text{ (from Table 10.5)}$$

3. Pipe head loss is

$$Q = \frac{1.49}{n} A R_h^{2/3} S_0^{1/2}$$

$$Q = 225 \text{ ft}^3/\text{s}$$

$$A = \frac{\pi D^2}{4} = 15.90 \text{ ft}^2$$

$$R_h = \frac{A}{P} = \frac{\pi D^2/4}{\pi D} = \frac{D}{4} = 1.125 \text{ ft}$$

$$R_h^{2/3} = (1.125 \text{ ft})^{2/3} = 1.0817 \text{ ft}^{2/3}$$

$$S_0 = \frac{h_f}{L}$$

$$225 = \frac{1.49}{0.012}(15.90 \text{ ft}^2)(1.0817 \text{ ft}^{2/3})\left(\frac{h_f}{200}\right)^{1/2}$$

$$h_f = 2.22 \text{ ft}$$

4. Continuity equation yields

$$V = \frac{Q}{A} = \frac{225 \text{ ft}^3/\text{s}}{15.90 \text{ ft}^2} = 14.15 \text{ ft/s}$$

5. Solve for H:

$$H = \frac{14.15^2}{64.4}(0.50 + 1.0) + 2.22$$

$$H = 4.66 \text{ ft} + 2.22 \text{ ft} = \boxed{6.88 \text{ ft}}$$

6. Depth of flow for $Q = 50$ cfs is

$$50 = \frac{1.49}{0.012} A R_h^{2/3}(0.01)^{1/2}$$

Values of A and R_h will depend on the geometry of the partly full pipe, as shown:

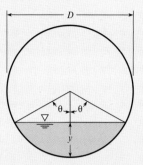

Area A if angle θ is given in degrees:

$$A = \left[\left(\frac{\pi D^2}{4}\right)\left(\frac{20}{360°}\right)\right] - \left(\frac{D}{2}\right)^2 (\sin\theta\cos\theta)$$

Wetted perimeter will be $P = \pi D(\pi/180°)$, so

$$R_h = \frac{A}{P} = \left(\frac{D}{4}\right)\left[1 - \left(\frac{\sin\theta\cos\theta}{(\pi\theta/180°)}\right)\right]$$

Substituting these relations for A and R_h into the discharge equation and solving for θ yields $\theta = 70°$. Therefore, y is

$$y = \frac{D}{2} - \frac{D}{2}\cos\theta = \left(\frac{54 \text{ in}}{2}\right)(1 - 0.342) = \boxed{17.8 \text{ in}}$$

15.4 Steady Nonuniform Flow

As stated in the beginning of this chapter and shown in Fig. 15.2, all open-channel flows are classified as either uniform or nonuniform. Recall that uniform flow has constant velocity along a streamline and thus has constant depth for a constant cross section. In steady nonuniform flow, the depth and velocity change over distance (but not with time). For all such cases, the energy equation as generally introduced in Section 15.2 is invoked to compare two cross sections. However, for analysis of nonuniform flow, it is useful to distinguish whether the depth and velocity change occurs over a short distance, referred to as **rapidly varied flow**, or over a long reach of the channel, referred to as **gradually varied flow** (Fig. 15.7). The head loss term is different for these two cases. For rapidly varied flow, one can neglect the resistance of the channel walls and bottom because it occurs over a short distance. For gradually varied flow, because of the long distances involved, the surface resistance is a significant variable in the energy balance.

15.5 Rapidly Varied Flow

Rapidly varied flow is analyzed with the energy equation presented previously for open-channel flow, Eq. (15.7), with the additional assumptions that the channel bottom is horizontal ($S_0 = 0$) and the head loss is zero ($h_L = 0$). Therefore, Eq. (15.7) becomes

$$y_1 + \frac{V_1^2}{2g} = y_2 + \frac{V_2^2}{2g} \qquad \textbf{(15.17)}$$

FIGURE 15.7

Classifying nonuniform flow.

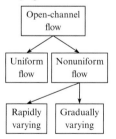

FIGURE 15.8

Relation between depth and specific energy.

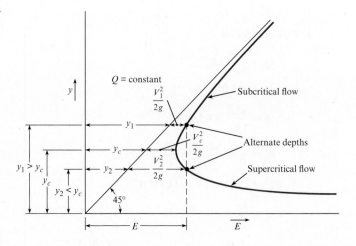

Specific Energy

The sum of the depth of flow and the velocity head is defined as the **specific energy**:

$$E = y + \frac{V^2}{2g} \tag{15.18}$$

Note that specific energy has dimensions of length; that is, it is a head term. Equation (15.17) states that the specific energy at section 1 is equal to the specific energy at section 2, or $E_1 = E_2$. The continuity equation between sections 1 and 2 is

$$A_1 V_1 = A_2 V_2 = Q \tag{15.19}$$

Therefore, Eq. (15.17) can be expressed as

$$y_1 + \frac{Q^2}{2gA_1^2} = y_2 + \frac{Q^2}{2gA_2^2} \tag{15.20}$$

Because A_1 and A_2 are functions of the depths y_1 and y_2, respectively, the magnitude of the specific energy at section 1 or 2 is solely a function of the depth at each section. If, for a given channel and given discharge, one plots depth versus specific energy, then a relationship such as that shown in Fig. 15.8 is obtained. By studying Fig. 15.8 for a given value of specific energy, one can see that the depth may be either large or small. This means that for the small depth, the bulk of the energy of flow is in the form of kinetic energy—that is, $Q^2/(2gA^2) \gg y$—whereas for a larger depth, most of the energy is in the form of potential energy. Flow under a **sluice gate** (Fig. 15.9) is an example of flow in which two depths occur for a given value of specific energy. The large depth and low kinetic energy occur upstream of the gate; the low depth and large kinetic energy occur downstream. The depths as used here are called **alternate depths**. That is, for a given value of E, the large depth is alternate to the low depth, or vice versa. Returning to the flow under the sluice gate, one finds that if the same rate of flow is maintained, but the gate is set with a larger opening, as in Fig. 15.9b, the upstream depth will drop, and the downstream depth will rise. This results in different alternate depths and a smaller value of specific energy than before. This is consistent with the diagram in Fig. 15.8.

Finally, it can be seen in Fig. 15.8 that a point will be reached where the specific energy is minimum and only a single depth occurs. At this point, the flow is termed critical. Thus, one definition of **critical flow** is the flow that occurs when the specific energy is minimum for a given discharge. The flow for which the depth is less than critical (velocity is greater than critical) is termed **supercritical flow**, and the flow for which the depth is greater than critical (velocity is less than critical) is termed **subcritical flow**. Therefore, subcritical flow occurs

FIGURE 15.9

Flow under a sluice gate: (a) smaller gate opening, (b) larger gate opening.

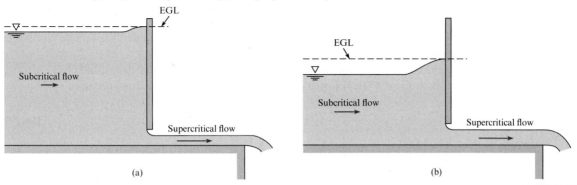

(a)   (b)

upstream and supercritical flow occurs downstream of the sluice gate in Fig. 15.9. Subcritical flows corresponds to a Froude number less than 1 (Fr < 1), and supercritical flow corresponds to Fr > 1. Some engineers refer to subcritical and supercritical flow as **tranquil** and **rapid** flow, respectively. Other aspects of critical flow are shown in the next section.

Characteristics of Critical Flow

Critical flow occurs when the specific energy is minimum for a given discharge. The depth for this condition may be determined by solving for dE/dy from $E = y + Q^2/2gA^2$ and setting dE/dy equal to zero:

$$\frac{dE}{dy} = 1 - \frac{Q^2}{gA^3} \cdot \frac{dA}{dy} \qquad (15.21)$$

However, $dA = T\,dy$, where T is the width of the channel at the water surface, as shown in Fig. 15.10. Then Eq. (15.21), with $dE/dy = 0$, will reduce to

$$\frac{Q^2 T_c}{gA_c^3} = 1 \qquad (15.22)$$

or

$$\frac{A_c}{T_c} = \frac{Q^2}{gA_c^2} \qquad (15.23)$$

If the **hydraulic depth**, D, is defined as

$$D = \frac{A}{T} \qquad (15.24)$$

then Eq. (15.23) will yield a critical hydraulic depth D_c, given by

$$D_c = \frac{Q^2}{gA_c^2} = \frac{V^2}{g} \qquad (15.25)$$

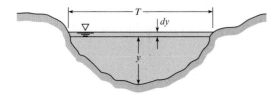

FIGURE 15.10

Open-channel relations.

Dividing Eq. (15.25) by D_c and taking the square root yields

$$1 = \frac{V}{\sqrt{gD_c}} \tag{15.26}$$

Note: $V/\sqrt{gD_c}$ is the Froude number. Therefore, it has been shown that the Froude number is equal to unity when critical flow prevails.

If a channel is of rectangular cross section, then A/T is the actual depth, and $Q^2/A^2 = q^2/y^2$, so the condition for **critical depth** (Eq. 15.23) for a rectangular channel becomes

$$y_c = \left(\frac{q^2}{g}\right)^{1/3} \tag{15.27}$$

where q is the discharge per unit width of channel.

EXAMPLE 15.9

Calculating Critical Depth in a Channel

Problem Statement

Determine the critical depth in this trapezoidal channel for a discharge of 500 cfs. The width of the channel bottom is $B = 20$ ft, and the sides slope upward at an angle of $45°$.

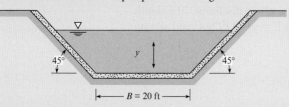

Define the Situation

Water flows in a trapezoidal channel with known geometry.

State the Goal

Calculate the critical depth.

Generate Ideas and Make a Plan

1. For critical flow, Eq. (15.22) must apply.

2. Relate this channel geometry to width T and area A in Eq. (15.22).

3. By iteration (choose y and compute A^3/T), find y that will yield A^3/T equal to 7764 ft^2. This y will be critical depth y_c.

Take Action (Execute the Plan)

1. Apply Eq. (15.22) to show that

$$\frac{Q^2}{g} = \frac{A_c^3}{T_c}$$

2. For $Q = 500$ cfs,

$$\frac{A_c^3}{T_c} = \frac{500^2}{32.2} = 7764 \text{ ft}^2$$

For this channel, $A = y(B + y)$ and $T = B + 2y$.

3. Iterate to find y_c:

$$y_c = \boxed{2.57 \text{ ft}}$$

Critical flow may also be examined in terms of how the discharge in a channel varies with depth for a given specific energy. For example, consider flow in a rectangular channel where

$$E = y + \frac{Q^2}{2gA^2}$$

or

$$E = y + \frac{Q^2}{2gy^2B^2}$$

If one considers a unit width of the channel and lets $q = Q/B$, then the foregoing equation becomes

$$E = y + \frac{q^2}{2gy^2}$$

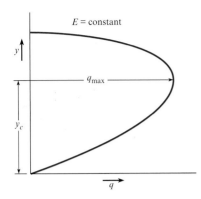

FIGURE 15.11

Variation of q and y with constant specific energy.

If one determines how q varies with y for a constant value of specific energy, one sees that critical flow occurs when the discharge is maximum (see Fig. 15.11).

Originally, the term **critical flow** probably related to the unstable character of the flow for this condition. Referring to Fig. 15.8, one can see that only a slight change in specific energy will cause the depth to increase or decrease a significant amount; this is a very unstable condition. In fact, observations of critical flow in open channels show that the water surface consists of a series of standing waves. Because of the unstable nature of the depth in critical flow, designing canals so that normal depth is either well above or well below critical depth is usually best. The flow in canals and rivers is usually subcritical; however, the flow in steep chutes or over spillways is supercritical.

In this section, various characteristics of critical flow have been explored. The main ones can be summarized as follows:

1. Critical flow occurs when specific energy is minimum for a given discharge (Fig. 15.8).

2. Critical flow occurs when the discharge is maximum for a given specific energy.

3. Critical flow occurs when

$$\frac{A^3}{T} = \frac{Q^2}{g}$$

4. Critical flow occurs when Fr $= 1$. Subcritical flow occurs when Fr < 1. Supercritical flow occurs when Fr > 1.

5. For rectangular channels, critical depth is given as $y_c = (q^2/g)^{1/3}$.

Common Occurrence of Critical Flow

Critical flow occurs when a liquid passes over a broad-crested weir (Fig. 15.12). The principle of the broad-crested weir is illustrated by first considering a closed sluice gate that prevents water from being discharged from the reservoir, as shown in Fig. 15.12a. If the gate is opened a small amount (gate position $a'\text{-}a'$), the flow upstream of the gate will be subcritical, and the flow downstream will be supercritical (as in the condition shown in Fig. 15.9). As the gate is opened further, a point is finally reached where the depths immediately upstream and downstream of the gate are the same. This is the critical condition. At this gate opening and beyond, the gate has no influence on the flow; this is the condition shown in Fig. 15.12b, the broad-crested weir. If the depth of flow over the weir is measured, the rate of flow can easily be computed from Eq. (15.27):

$$q = \sqrt{gy_c^3}$$

or

$$Q = L\sqrt{gy_c^3} \tag{15.28}$$

where L is the length of the weir crest normal to the flow direction.

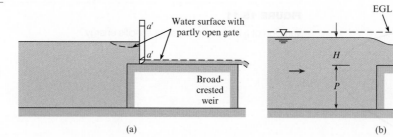

(a)   (b)

Because $y_c/2 = (V_c^2/2g)$, from Eq. (15.25), it can be shown that $y_c = (2/3E)$, where E is the total head above the crest $(H + V_{approach}^2/2g)$; hence Eq. (15.28) can be rewritten as

$$Q = L\sqrt{g}\left(\frac{2}{3}\right)^{3/2} E^{3/2}$$

or

$$Q = 0.385L\sqrt{2g}\,E_c^{3/2} \tag{15.29}$$

For high weirs, the upstream velocity of approach is almost zero. Hence, Eq. (15.29) can be expressed as

$$Q_{theor} = 0.385L\sqrt{2g}\,H^{3/2} \tag{15.30}$$

If the height P of the broad-crested weir is relatively small, then the velocity of approach may be significant, and the discharge produced will be greater than that given by Eq. (15.30). Also, head loss will have some effect. To account for these effects, a discharge coefficient C is defined as

$$C = Q/Q_{theor} \tag{15.31}$$

Then

$$Q = 0.385CL\sqrt{2g}\,H^{3/2} \tag{15.32}$$

where Q is the actual discharge over the weir. An analysis of experimental data by Raju (15) shows that C varies with $H/(H + P)$ as shown in Fig. 15.13. The curve in Fig. 15.13 is for a weir with a vertical upstream face and a sharp corner at the intersection of the upstream face and the weir crest. If the upstream face is sloping at a $45°$ angle, the discharge coefficient should be increased 10% over that given in Fig. 15.13. Rounding of the upstream corner will also produce a coefficient of discharge as much as 3% greater.

Equation (15.32) reveals a definite relationship for Q as a function of the head, H. This type of discharge-measuring device is in the broad class of discharge meters called **critical-flow**

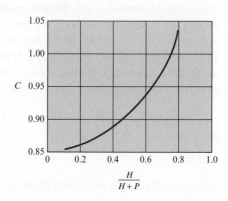

$\dfrac{H}{H+P}$

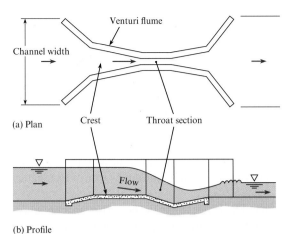

FIGURE 15.14

Flow through a venturi flume.

(a) Plan

Channel width

Venturi flume

Crest   Throat section

Flow

(b) Profile

flumes. Another very common critical-flow flume is the **venturi flume**, which was developed and calibrated by Parshall (8). Fig. 15.14 shows the essential features of the venturi flume. The discharge equation for the venturi flume is in the same form as Eq. (15.32), the only difference being that the experimentally determined coefficient C will have a different value from the C for the broad-crested weir. For more details on the venturi flume, you may refer to Roberson et al. (9), Parshall (8), and Chow (5). The venturi flume is especially useful for discharge measurement in irrigation systems because little head loss is required for its use, and sediment is easily flushed through if the water happens to be silty.

The depth also passes through a critical stage in channel flow where the slope changes from a mild one to a steep one. A **mild slope** is defined as a slope for which the normal depth y_n is greater than y_c. Likewise, a **steep slope** is one for which $y_n < y_c$. This condition is shown in Fig. 15.15. Note that y_c is the same for both slopes in the figure because y_c is a function of the discharge only. However, normal depth (uniform-flow depth) for the mild upstream channel is greater than critical, whereas the normal depth for the steep downstream channel is less than critical; hence it is obvious that the depth must pass through a critical stage. Experiments show that critical depth occurs a very short distance upstream of the intersection of the two channels.

Another place where critical depth occurs is upstream of a free overfall at the end of a channel with a mild slope (Fig. 15.16). Critical depth will occur at a distance of $3y_c$ to $4y_c$ upstream of the

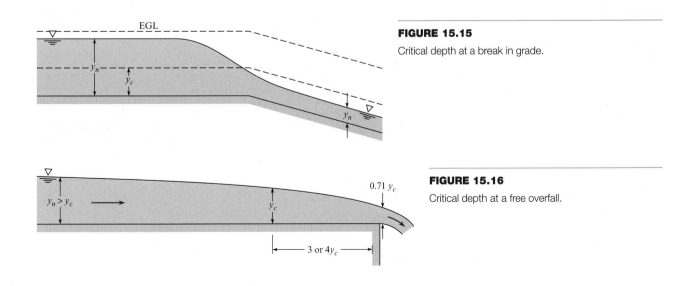

FIGURE 15.15

Critical depth at a break in grade.

EGL

y_n

y_c

y_n

FIGURE 15.16

Critical depth at a free overfall.

$y_n > y_c$

y_c

0.71 y_c

3 or 4y_c

brink. Such occurrences of critical depth (at a break in grade or at a brink) are useful in computing surface profiles because they provide a point for starting surface-profile calculations.*

Channel Transitions

Whenever a channel's cross-sectional configuration (shape or dimension) changes along its length, the change is termed a **transition**. Concepts previously presented are used to show how the flow depth changes when the floor of a rectangular channel is increased in elevation or when the width of the channel is decreased. In these developments, negligible energy losses are assumed. First, the case where the floor of the channel is raised (an upstep) is considered. Later in this section, configurations of transitions used for subcritical flow from a rectangular to a trapezoidal channel are presented.

Consider the rectangular channel shown in Fig. 15.17, where the floor rises an amount Δz. To help in evaluating depth changes, one can use a diagram of specific energy versus depth, which is similar to Fig. 15.8. This diagram is applied both at the section upstream of the transition and at the section just downstream of the transition. Because the discharge, Q, is the same at both sections, the same diagram is valid at both sections. As noted in Fig. 15.17, the depth of flow at section 1 can be either large (subcritical) or small (supercritical) if the specific energy E_1 is greater than that required for critical flow. It can also be seen in Fig. 15.17 that when the upstream flow is subcritical, a decrease in depth occurs in the region of the elevated channel bottom. This occurs because the specific energy at this section, E_2, is less than that at section 1 by the amount Δz. Therefore, the specific energy diagram indicates that y_2 will be less than y_1. In a similar manner, it can be seen that when the upstream flow is supercritical, the depth as well as the actual water-surface elevation increases from section 1 to section 2. A further note should be made about the effect on flow depth of a change in bottom-surface elevation. If the channel bottom at section 2 is at an elevation greater than that just sufficient to establish critical flow at section 2, then there is not enough head at section 1 to cause flow to occur over the rise under steady-flow conditions. Instead, the water level upstream will rise until it is just sufficient to reestablish steady flow.

When the channel bottom is kept at the same elevation but the channel is decreased in width, then the discharge per unit of width between sections 1 and 2 increases, but the specific energy E remains constant. Thus, when utilizing the diagram of q versus depth for the given specific energy E, note that the depth in the restricted section increases if the upstream flow is supercritical and decreases if it is subcritical (see Fig. 15.18).

FIGURE 15.17

Change in depth with change in bottom elevation of a rectangular channel.

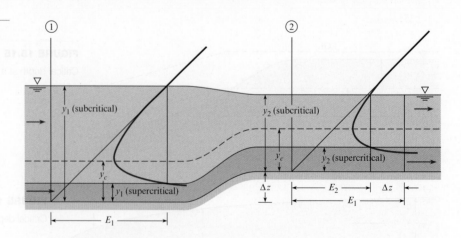

*The procedure for making these computations starts in §15.7 in the subsection titled "Quantitative Evaluation of the Water-Surface Profile."

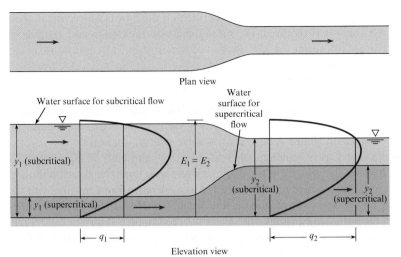

FIGURE 15.18

Change in depth with change in channel width.

The foregoing paragraphs describe gross effects for the simplest transitions. In practice, it is more common to find transitions between a channel of one shape (rectangular cross section, for example) and a channel with a different cross section (trapezoidal, for example). A very simple transition between two such channels consists of two straight vertical walls joining the two channels, as shown by the half-section in Fig. 15.19.

This type of transition can work, but it will produce excessive head loss because of the abrupt change in cross section and the ensuing separation that will occur. To reduce the head losses, a more gradual type of transition is used. Fig. 15.20 is a half-section of a transition similar to that of Fig. 15.19, but with the angle θ much greater than 90°. This is called a **wedge transition**.

The **warped-wall** transition shown in Fig. 15.21 will yield even smoother flow than either of the other two, and it will thus have less head loss. In the practical design and analysis of transitions, engineers usually use the complete energy equation, including the kinetic energy factors α_1 and α_2 as well as a head loss term h_L, to define velocity and water-surface elevation through the transition. Analyses of transitions utilizing the one-dimensional form of the

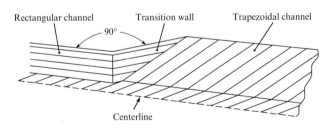

FIGURE 15.19

Simplest type of transition between a rectangular channel and a trapezoidal channel.

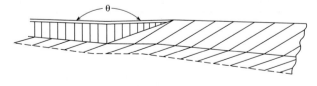

FIGURE 15.20

Half-section of a wedge transition.

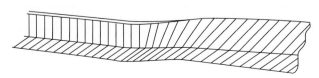

FIGURE 15.21

Half-section of a warped-wall transition.

energy equation are applicable only if the flow is subcritical. If the flow is supercritical, then a much more involved analysis is required. For more details on the design and analysis of transitions, refer to Hinds (10), Chow (5), U.S. Bureau of Reclamation (11), and Rouse (12).

Wave Celerity

Wave celerity is the velocity at which an infinitesimally small wave travels relative to the velocity of the fluid. It can be used to characterize the velocity of waves in the ocean or propagation of a flood wave following a dam failure. A derivation of wave celerity, c, follows.

Consider a small solitary wave moving with velocity c in a calm body of liquid of small depth (Fig. 15.22a). Because the velocity in the liquid changes with time, this is a condition of unsteady flow. However, if one were to refer all velocities to a reference frame moving with the wave, the shape of the wave would be fixed, and the flow would be steady. Then, the flow is amenable to analysis with the Bernoulli equation. The steady-flow condition is shown in Fig. 15.22b. When the Bernoulli equation is written between a point on the surface of the undisturbed fluid and a point at the wave crest, the following equation results:

$$\frac{c^2}{2g} + y = \frac{V^2}{2g} + y + \Delta y \tag{15.33}$$

In Eq. (15.33), V is the velocity of the liquid in the section where the crest of the wave is located. From the continuity equation, $cy = V(y + \Delta y)$. Hence,

$$V = \frac{cy}{y + \Delta y}$$

and

$$V^2 = \frac{c^2 y^2}{(y + \Delta y)^2} \tag{15.34}$$

When Eq. (15.34) is substituted into Eq. (15.33), it yields

$$\frac{c^2}{2g} + y = \frac{c^2 y^2}{2g[y^2 + 2y\Delta y + (\Delta y)^2]} + y + \Delta y \tag{15.35}$$

FIGURE 15.22

Solitary wave (exaggerated vertical scale): (a) unsteady flow, (b) steady flow.

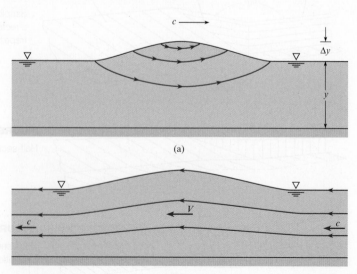

(a)

(b)

Solving Eq. (15.35) for c after discarding terms with $(\Delta y)^2$, assuming an infinitesimally small wave, yields the **wave celerity equation**:

$$c = \sqrt{gy} \qquad\qquad (15.36)$$

It has thus been shown that the speed of a small solitary wave is equal to the square root of the product of the depth and g.

15.6 Hydraulic Jump

Occurrence of the Hydraulic Jump

An interesting and important case of rapidly varied flow is the hydraulic jump. A **hydraulic jump** occurs when the flow is supercritical in an upstream section of a channel and is then forced to become subcritical in a downstream section (the change in depth can be forced by a sill in the downstream part of the channel or just by the prevailing depth in the stream further downstream), resulting in an abrupt increase in depth and considerable energy loss. Hydraulic jumps (Fig. 15.23) are often considered in the design of open channels and spillways of dams. If a channel is designed to carry water at supercritical velocities, the designer must be certain that the flow will not become subcritical prematurely. If it did, overtopping of the channel walls would undoubtedly occur, with consequent failure of the structure. Because the energy loss in the hydraulic jump is initially not known, the energy equation is not a suitable tool for analysis of the velocity-depth relationships. Because there is a significant difference in hydrostatic head on both sides of the equation causing opposing pressure forces, the momentum equation can be applied to the problem, as developed in the following sections.

Derivation of Depth Relationships in Hydraulic Jumps

Consider flow as shown in Fig. 15.23. Here, it is assumed that uniform flow occurs both upstream and downstream of the jump and that the resistance of the channel bottom over the relatively short distance L is negligible. The derivation is for a horizontal channel, but experiments show that the results of the derivation will apply to all channels of moderate slope ($S_0 < 0.02$). The derivation is started by applying the momentum equation in the x direction to the control volume shown in Fig. 15.24:

$$\sum F_x = \dot{m}_2 V_2 - \dot{m}_1 V_1$$

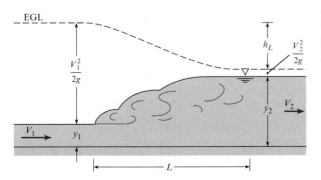

FIGURE 15.23

Definition sketch for the hydraulic jump.

FIGURE 15.24

Control-volume analysis for the hydraulic jump.

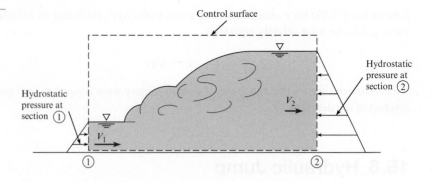

The forces are the hydrostatic forces on each end of the system; thus the following is obtained:

$$\bar{p}_1 A_1 - \bar{p}_2 A_2 = \rho V_2 A_2 V_2 - \rho V_1 A_1 V_1$$

or

$$\bar{p}_1 A_1 + \rho Q V_1 = \bar{p}_2 A_2 + \rho Q V_2 \qquad (15.37)$$

In Eq. (15.37), $\bar{p}_1$ and $\bar{p}_2$ are the pressures at the centroids of the respective areas A_1 and A_2.

A representative problem (e.g., Example 15.10) is to determine the downstream depth y_2 given the discharge and upstream depth. The left-hand side of Eq. (15.37) would be known because V, A, and p are all functions of y and Q, and the right-hand side is a function of y_2; therefore, y_2 can be determined.

EXAMPLE 15.10

Calculating Downstream Depth for a Hydraulic Jump

Problem Statement

Water flows in a trapezoidal channel at a rate of 300 cfs. The channel has a bottom width of 10 ft and side slopes of 1 vertical to 1 horizontal. If a hydraulic jump is forced to occur where the upstream depth is 1.0 ft, what will be the downstream depth and velocity? What are the values of Fr_1 and Fr_2?

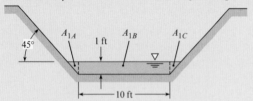

Define the Situation

A hydraulic jump is forced in a trapezoidal channel.

Properties: Water (50°F), Table A.5:

$\gamma = 62.4 \ \mathrm{lbf/ft^3}$, and $\rho = 1.94 \ \mathrm{slugs/ft^3}$.

State the Goal

1. Downstream depth and velocity
2. Values of Fr_1 and Fr_2

Generate Ideas and Make a Plan

1. Find cross section, velocity, and hydraulic depth in the upstream section.
2. Find pressure in the upstream section to use for the left-hand side of Eq. (15.37).
3. Use channel geometry information to solve for y_2 in right-hand side of Eq. (15.37).
4. Use Eq. (15.2) to solve for the Froude number at both sections.

Take Action (Execute the Plan)

1. By inspection, for the upstream section, the cross-sectional flow area is 11 ft².

   Therefore, the mean velocity is $V_1 = Q/A_1 = 27.3$ ft/s. The hydraulic depth is $D_1 = A_1/T_1 = 11 \ \mathrm{ft^2}/12 \ \mathrm{ft} = 0.9167$ ft.

2. The location of the centroid ($\bar{y}$) of the area A_1 can be obtained by taking moments of the subareas about the water surface (see example sketch).

   $A_1 \bar{y}_1 = A_{1A} \times 0.333 \ \mathrm{ft} + A_{1B} \times 0.500 \ \mathrm{ft} + A_{1C} \times 0.333 \ \mathrm{ft}$

   $(11 \ \mathrm{ft^2})\bar{y}_1 = (0.333 \ \mathrm{ft})(0.500 \ \mathrm{ft^2} \times 2) + (0.50 \ \mathrm{ft})(10.00 \ \mathrm{ft^2})$

   $\bar{y} = 0.485 \ \mathrm{ft}$

Pressure $p_1 = 62.4$ lbf/ft$^3 \times 0.485$ ft $= 30.26$ lbf/ft^2. Therefore,

$$30.26 \times 11 + 1.94 \times 300 \times 27.3 = \bar{p}_2 A_2 + \rho Q V_2$$

3. Using the right-hand side of Eq. (15.37), solve for y_2:

$$\bar{p}_2 A_2 + \rho Q V_2 = 16{,}221 \text{ lbf}$$

$$\gamma \bar{y}_2 A_2 + \frac{\rho Q^2}{A_2} = 16{,}221$$

$$\bar{y}_2 = \frac{\sum A_i y_i}{A_2} = \frac{B y_2^2/2 + y_2^3/3}{A_2}$$

Using $B = 10$ ft, $Q = 300$ ft^2/s, and material properties assumed earlier,

$$y_2 = \boxed{5.75 \text{ ft}}$$

4. Froude numbers at both sections are

$$Fr_1 = \frac{V_1}{\sqrt{gD_1}} = \frac{27.3 \text{ ft/s}}{\sqrt{32.2 \text{ ft/s}^2 \times 0.9167 \text{ ft}}} = \boxed{5.02}$$

$$V_2 = \frac{Q}{A_2} = \frac{300}{57.5 + 5.75^2} = 3.31 \text{ ft/s}$$

$$D_2 = \frac{A_2}{T_2} = \frac{90.56}{21.5} = 4.21 \text{ ft}$$

$$Fr_2 = \frac{V}{\sqrt{gD}} = \frac{3.31}{\sqrt{32.2 \times 4.21}} = \boxed{0.284}$$

Hydraulic Jump in Rectangular Channels

If one writes Eq. (15.37) for a unit width of a rectangular channel where $\bar{p}_1 = \gamma y_1/2$, $\bar{p}_2 = \gamma y_2/2$, $Q = q$, $A_1 = y_1$, and $A_2 = y_2$, this will yield

$$\gamma \frac{y_1^2}{2} + \rho q V_1 = \gamma \frac{y_2^2}{2} + \rho q V_2 \qquad \textbf{(15.38a)}$$

but $q = Vy$, so Eq. (15.38a) can be rewritten as

$$\frac{\gamma}{2}(y_1^2 - y_2^2) = \frac{\gamma}{g}(V_2^2 y_2 - V_1^2 y_1) \qquad \textbf{(15.38b)}$$

The preceding equation can be further manipulated to yield

$$\frac{2V_1^2}{g y_1} = \left(\frac{y_2}{y_1}\right)^2 + \frac{y_2}{y_1} \qquad \textbf{(15.39)}$$

The term on the left-hand side of Eq. (15.39) will be recognized as twice Fr_1^2. Hence, Eq. (15.39) is written as

$$\left(\frac{y_2}{y_1}\right)^2 + \frac{y_2}{y_1} - 2Fr_1^2 = 0 \qquad \textbf{(15.40)}$$

By use of the quadratic formula, it is easy to solve for y_2/y_1 in terms of the upstream Froude number. This yields an equation for **depth ratio** across a hydraulic jump:

$$\frac{y_2}{y_1} = \frac{1}{2}\left(\sqrt{1 + 8Fr_1^2} - 1\right) \qquad \textbf{(15.41)}$$

or

$$y_2 = \frac{y_1}{2}\left(\sqrt{1 + 8Fr_1^2} - 1\right) \qquad \textbf{(15.42)}$$

The other solution of Eq. (15.40) gives a negative downstream depth, which is not physically possible. Hence, the downstream depth is expressed in terms of the upstream depth and the upstream Froude number. In Eqs. (15.41) and (15.42), the depths y_1 and y_2 are said to be **conjugate** or **sequent** (both terms are in common use) to each other, in contrast to the alternate depths obtained from the energy equation. Numerous experiments show that the relation represented by Eqs. (15.41) and (15.42) is valid over a wide range of Froude numbers.

Although no theory has been developed to predict the length of a hydraulic jump, experiments [see Chow (5)] show that the relative length of the jump, L/y_2, is approximately 6 for Fr_1 ranging from 4 to 18.

Head Loss in a Hydraulic Jump

In addition to determining the geometric characteristics of the hydraulic jump, it is often desirable to determine the head loss produced by it. This is obtained by comparing the specific energy before the jump to that after the jump, the head loss being the difference between the two specific energies. It can be shown that the head loss for a jump in a rectangular channel is

$$h_L = \frac{(y_2 - y_1)^3}{4y_1 y_2} \tag{15.43}$$

For more information on the hydraulic jump, see Chow (5). The following example shows that Eq. (15.43) yields a magnitude that equals the difference between the specific energies at the two ends of the hydraulic jump.

EXAMPLE 15.11

Calculating Head Loss in a Hydraulic Jump

Problem Statement

Water flows in a rectangular channel at a depth of 30 cm with a velocity of 16 m/s, as shown in the following sketch. If a downstream sill (not shown) forces a hydraulic jump, what will be the depth and velocity downstream of the jump? What head loss is produced by the jump?

Define the Situation

A hydraulic jump is occurring in a rectangular channel.

State the Goal

- Calculate downstream depth and velocity.
- Calculate head loss produced by the jump.

Generate Ideas and Make a Plan

1. To calculate h_L using Eq. (15.43), calculate y_2 from the depth ratio equation (Eq. 15.42). This requires Fr_1.
2. Check validity of head loss by comparing to $E_1 - E_2$.

Take Action (Execute the Plan)

1. Calculate Fr_1, y_2, V_2, and h_L from Eqs. (Eq. 15.42) and (15.43):

$$Fr_1 = \frac{V}{\sqrt{gy_1}} = \frac{16}{\sqrt{9.81\,(0.30)}} = 9.33$$

$$y_2 = \frac{0.30}{2}\left[\sqrt{1 + 8(9.33)^2} - 1\right] = \boxed{3.81 \text{ m}}$$

$$V_2 = \frac{q}{y_2} = \frac{(16 \text{ m/s})(0.30 \text{ m})}{3.81 \text{ m}} = \boxed{1.26 \text{ m/s}}$$

$$h_L = \frac{(3.81 - 0.30)^3}{4(0.30)(3.81)} = \boxed{9.46 \text{ m}}$$

2. Compare the head loss to $E_1 - E_2$:

$$h_L = \left(0.30 + \frac{16^2}{2 \times 9.81}\right) - \left(3.81 + \frac{1.26^2}{2 \times 9.81}\right) = 9.46 \text{ m}$$

The value is the same, so $\boxed{\text{validity of } h_L \text{ equation is verified.}}$

Use of Hydraulic Jump on Downstream End of Dam Spillway

Previously it was shown that the transition from supercritical to subcritical flow produces a hydraulic jump and that the relative height of the jump (y_2/y_1) is a function of Fr_1. Because flow over the spillway of a dam invariably results in supercritical flow at the lower end of the spillway and because flow in the channel downstream of a spillway is usually subcritical, it is obvious that a hydraulic jump must form near the base of the spillway (see Fig. 15.25). The downstream portion of the spillway, called the spillway **apron**, must be designed so that the hydraulic jump always forms on the concrete structure itself. If the hydraulic jump were allowed to form beyond the concrete structure, as in Fig. 15.26, severe erosion of the foundation material as a result of the high-velocity supercritical flow could undermine the dam and cause its complete failure. One way to solve this problem might be to incorporate a long, sloping apron into the design of the spillway, as shown in Fig. 15.27. A design like this would work very satisfactorily from the hydraulics point of view. For all combinations of Fr_1 and

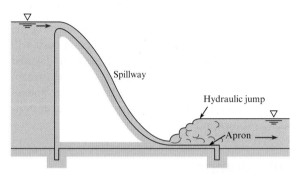

FIGURE 15.25

Spillway of dam and hydraulic jump.

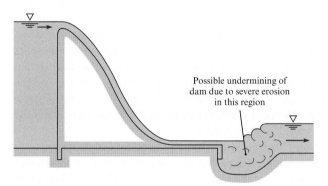

FIGURE 15.26

Hydraulic jump occurring downstream of spillway apron.

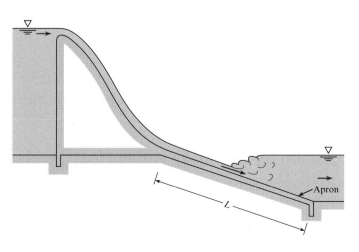

FIGURE 15.27

Long sloping apron.

FIGURE 15.28

Spillway with stilling basin Type III as recommended by the USBR (13).

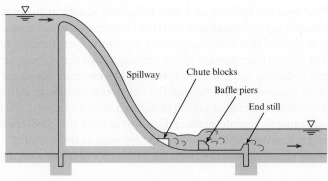

water-surface elevation in the downstream channel, the jump would always form on the sloping apron. However, its main drawback is cost of construction. Construction costs will be reduced as the length, L, of the stilling basin is reduced. Much research has been devoted to the design of stilling basins that will operate properly for all upstream and downstream conditions and yet be relatively short to reduce construction cost. Research by the U.S. Bureau of Reclamation (13) has resulted in sets of standard designs that can be used. These designs include sills, baffle piers, and chute blocks, as shown in Fig. 15.28.

Naturally Occurring Hydraulic Jumps

Hydraulic jumps can occur naturally in creeks and rivers, providing spectacular standing waves, called rollers. Kayakers and white-water rafters must exercise considerable skill when navigating hydraulic jumps because the significant energy loss that occurs over a short distance can be dangerous, potentially engulfing the boat in turbulence. A special case of hydraulic jump, referred to as a submerged hydraulic jump, can be deadly to white-water enthusiasts because it is not easy to see. A **submerged hydraulic jump** occurs when the downstream depth predicted by conservation of momentum is exceeded by the tailwater elevation, and the jump cannot move upstream in response to this disequilibrium because of a buried obstacle [see Valle and Pasternak (14)]. Thus, the visual markers of a hydraulic jump, particularly the rolling waves depicted in Figs. 15.23 and 15.24, are hidden.

A **surge**, or **tidal bore**, is a moving hydraulic jump that may occur for a high tide entering a bay or river mouth. Tides are generally low enough that the waves they produce are smooth and nondestructive. However, in some parts of the world the tides are so high that their entry into shallow bays or mouths of rivers causes a surge to form. Surges may be very hazardous to small boats. The same analytical methods used for the jump can be used to solve for the speed of the surge.

15.7 Gradually Varied Flow

For gradually varied flow, channel resistance is a significant factor in the flow process. Therefore, the energy equation is invoked by comparing S_0 and S_f.

Basic Differential Equation for Gradually Varied Flow

There are a number of cases of open-channel flow in which the change in water-surface profile is so gradual that it is possible to integrate the relevant differential equation from one section to another to obtain the desired change in depth. This may be either an analytical integration or, more commonly, a numerical integration. In Section 15.2, the energy equation was written between two sections of a channel Δx distance apart. Because the only head loss here is the channel resistance, the h_L is given by Δh_f, and Eq. (15.7) becomes

$$y_1 + \frac{V_1^2}{2g} + S_0\,\Delta x = y_2 + \frac{V_2^2}{2g} + \Delta h_f \tag{15.44}$$

The friction slope S_f is defined as the slope of the EGL, or $\Delta h_f/\Delta x$. Thus, $\Delta h_f = S_f\Delta x$, and defining $\Delta y = y_2 - y_1$, then

$$\frac{V_2^2}{2g} - \frac{V_1^2}{2g} = \frac{d}{dx}\left(\frac{V^2}{2g}\right)\Delta x \tag{15.45}$$

Therefore, Eq. (15.44) becomes

$$\Delta y = S_0\Delta x - S_f\Delta x - \frac{d}{dx}\left(\frac{V^2}{2g}\right)\Delta x$$

Dividing through by Δx and taking the limit as Δx approaches zero gives us

$$\frac{dy}{dx} + \frac{d}{dx}\left(\frac{V^2}{2g}\right) = S_0 - S_f \tag{15.46}$$

The second term is rewritten as $[d(V^2/2g)/dy]\,dy/dx$, so that Eq. (15.46) simplifies to

$$\frac{dy}{dx} = \frac{S_0 - S_f}{1 + d(V^2/2g)/dy} \tag{15.47}$$

To put Eq. (15.47) in a more usable form, the denominator is expressed in terms of the Froude number. This is accomplished by observing that

$$\frac{d}{dy}\left(\frac{V^2}{2g}\right) = \frac{d}{dy}\left(\frac{Q^2}{2gA^2}\right) \tag{15.48}$$

After differentiating the right side of Eq. (15.48), the equation becomes

$$\frac{d}{dy}\left(\frac{V^2}{2g}\right) = \frac{-2Q^2}{2gA^3}\cdot\frac{dA}{dy}$$

But $dA/dy = T$ (top width), and $A/T = D$ (hydraulic depth); therefore,

$$\frac{d}{dy}\left(\frac{V^2}{2g}\right) = \frac{-Q^2}{gA^2D}$$

or

$$\frac{d}{dy}\left(\frac{V^2}{2g}\right) = -\mathrm{Fr}^2$$

Hence, when the expression for $d(V^2/2g)/dy$ is substituted into Eq. (15.47), the result is

$$\frac{dy}{dx} = \frac{S_0 - S_f}{1 - \mathrm{Fr}^2} \tag{15.49}$$

This is the general differential equation for gradually varied flow. It is used to describe the various types of water-surface profiles that occur in open channels. Note that, in the derivation of the equation, S_0 and S_f were taken as positive when the channel and energy grade lines, respectively, were sloping downward in the direction of flow. Also note that y is measured from the bottom of the channel. Therefore, $dy/dx = 0$ if the slope of the water surface is equal to the slope of the channel bottom, and dy/dx is positive if the slope of the water surface is less than the channel slope.

Introduction to Water-Surface Profiles

In the design of projects involving the flow in channels (rivers or irrigation canals, for example), the engineer must often estimate the **water-surface profile** (elevation of the water surface along the channel) for a given discharge. For example, when a dam is being designed for a river project, the water-surface profile in the river upstream must be defined so that the project planners will know how much land to acquire to accommodate the upstream pool. The first step in defining a water-surface profile is to locate a point or points along the channel where the depth can be computed for a given discharge. For example, at a change in slope from mild to steep, critical depth will occur just upstream of the break in grade (see Fig. 15.32). At that point, it is possible to solve for y_c with Eq. (15.25) or (15.27). Also, for flow over the spillway of a dam, there will be a discharge equation for the spillway from which one can calculate the water-surface elevation in the reservoir at the face of the dam. Such points where there is a unique relationship between discharge and water-surface elevation are called **controls**. Once the water-surface elevations at these controls are determined, then the water-surface profile can be extended upstream or downstream from the control points to define the water-surface profile for the entire channel. The completion of the profile is done by numerical integration. However, before this integration is performed, it is usually helpful for the engineer to sketch in the profiles. To assist in the process of sketching the possible profiles, the engineer can refer to different categories of profiles (water-surface profiles have unique characteristics depending on the relationship between normal depth, critical depth, and the actual depth of flow in the channel). This initial sketching of the profiles helps the engineer to scope the problem and to obtain a solution, or solutions, in a minimum amount of time. The next section describes the various types of water-surface profiles.

Types of Water-Surface Profiles

There are 12 different types of water-surface profiles for gradually varied flow in channels, and these are shown schematically in Fig. 15.29. Each profile is identified by a letter and number designator. For example, the first water-surface profile in Fig. 15.29 is identified as an M1 profile. The letter indicates the type of slope of the channel—that is, whether the slope is mild (M), critical (C), steep (S), horizontal (H), or adverse (A). The slope is defined as mild if the uniform flow depth, y_n, is greater than the critical flow depth, y_c. Conversely, if y_n is less than y_c, the channel would be termed steep. If $y_n = y_c$, this is a channel with critical slope. The designation M, S, or C is determined by computing y_n and y_c for the given channel for a given discharge. Equations (15.11) through (15.15) are used to compute y_n, and Eq. (15.27) is used to compute y_c. Fig. 15.30 shows the relationship between y_n and y_c for the H, M, S, C, and A designations. As the name implies, a horizontal slope is one where the channel actually has a zero slope, and an adverse slope is one where the slope of the channel is upward in the direction of flow. Normal depth does not exist for these two cases (for example, water cannot flow at uniform depth in either a horizontal channel or one with adverse slope); therefore, they are given the special designations H and A, respectively.

The number designator for the type of profile relates to the position of the *actual* water surface in relation to the position of the water surface for uniform and critical flow in the channel. If the actual water surface is above that for uniform and critical flow ($y > y_n$; $y > y_c$), then

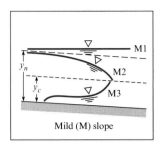

Mild (M) slope

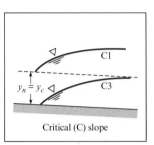

Critical (C) slope

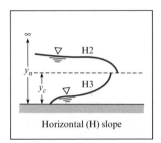

Horizontal (H) slope

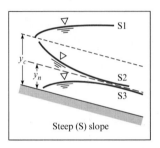

Steep (S) slope

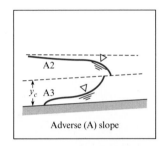

Adverse (A) slope

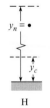

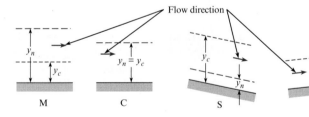

that condition is given a 1 designation; if the actual water surface is between those for uniform and critical flow, then it is given a 2 designation; and if the actual water surface lies below those for uniform and critical flow, then it is given a 3 designation. Fig. 15.31 depicts these conditions for mild and steep slopes.

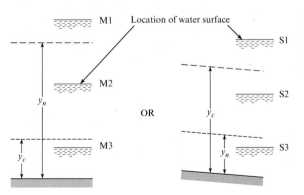

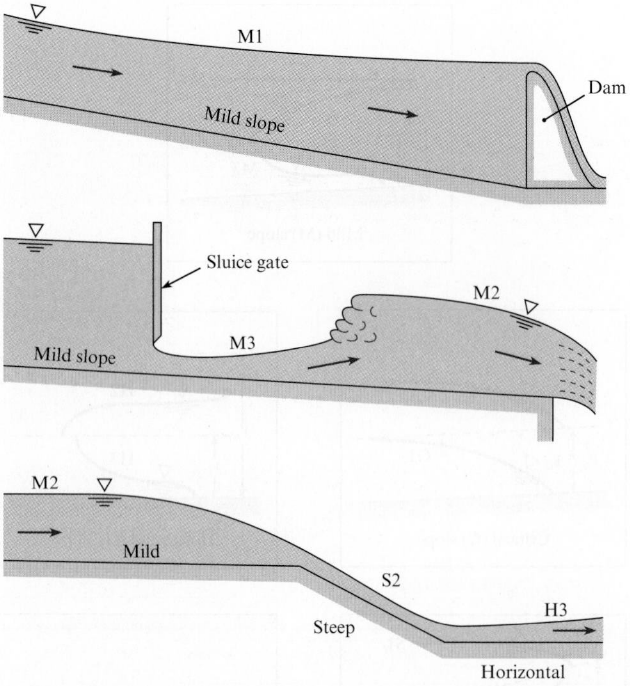

FIGURE 15.32

Water-surface profiles associated with flow behind a dam, flow under a sluice gate, and flow in a channel with a change in grade.

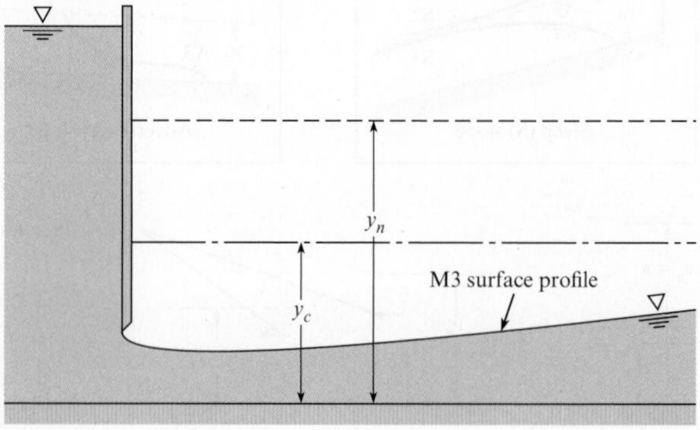

FIGURE 15.33

Water-surface profile, M3 type.

Fig. 15.32 shows how different water-surface profiles can develop in certain field situations. More specifically, if one considers in detail the flow downstream of the sluice gate (see Fig. 15.33), then one can see that the discharge and slope are such that the normal depth is greater than the critical depth; therefore, the slope is mild. The actual depth of flow shown in Fig. 15.33 is less than both y_c and y_n. Hence, a type 3 water-surface profile exists. The complete classification of the profile in Fig. 15.33, therefore, is a mild type 3 profile, or simply an M3 profile. Using these designations, one would categorize the profile upstream of the sluice gate as type M1.

EXAMPLE 15.12

Classification of Water-Surface Profiles

Problem Statement

Classify the water-surface profile for the flow downstream of the sluice gate in Fig. 15.9 if the slope is horizontal and that

for the flow immediately downstream of the break in grade in Fig. 15.15.

Define the Situation

Nonuniform flow is occurring in a channel.

State the Goal

Find the water-surface profile classification for the two different flow situations.

Generate Ideas and Make a Plan

1. Select a number designator based on the location of the actual water surface relative to y_n and y_c (see Fig. 15.31).
2. Select a letter designator to describe the steepness of the slopes, which can also be characterized by the relative size of y_n and y_c (see Fig. 15.30).

Take Action (Execute the Plan)

For Fig. 15.9:

1. The actual depth is less than critical; thus the profile is type 3.

2. The channel is horizontal; hence the profile is designated type H3.

For Fig. 15.15:

1. The actual depth is greater than normal but less than critical, so the profile is type 2.
2. The uniform-flow depth (normal depth y_n) is less than the critical depth; hence the slope is steep. Therefore, the water-surface profile is designated type S2.

Now, refer to Eq. (15.49) to describe the shapes of the profiles. Again, for example, if one considers the M3 profile, it is known that Fr > 1 because the flow is supercritical ($y < y_c$) and that $S_f > S_0$ because the velocity is greater than normal velocity. Hence, a head loss greater than that for normal flow must exist. Inserting these relative values into Eq. (15.49) reveals that both the numerator and the denominator are negative. Thus, dy/dx must be positive (the depth increases in the direction of flow), and as critical depth is approached, the Froude number approaches unity. Hence, the denominator of Eq. (15.49) approaches zero. Therefore, as the depth approaches critical depth, $dy/dx \to \infty$. What actually occurs in cases in which the critical depth is approached in supercritical flow is that a hydraulic jump forms and a discontinuity in profile is thereby produced.

Certain general features of profiles, as shown in Fig. 15.29, are evident. First, as the depth becomes very great, the velocity of flow approaches zero. Hence, Fr $\to$ 0 and $S_f \to$ 0 and dy/dx approaches S_0 because $dy/dx = (S_0 - S_f)(1 - \text{Fr}^2)$. In other words, the depth increases at the same rate at which the channel bottom drops away from the horizontal. Thus, the water surface approaches the horizontal. The profiles that show this tendency are types M1, S1, and C1. A physical example of the M1 type is the water-surface profile upstream of a dam, as shown in Fig. 15.32. The second general feature of several of the profiles is that those that approach normal depth do so asymptotically. This is shown in the S2, S3, M1, and M2 profiles. Also note in Fig. 15.29 that profiles that approach critical depth are shown by dashed lines. This is done because near critical depth either discontinuities develop (hydraulic jump), or the streamlines are very curved (such as near a brink). These profiles cannot be accurately predicted by Eq. (15.49) because this equation is based on one-dimensional flow, which, in these regions, is invalid.

Quantitative Evaluation of the Water-Surface Profile

In practice, most water-surface profiles are generated by numerical integration—that is, by dividing the channel into short reaches and carrying the computation for water-surface elevation from one end of the reach to the other. For one method, called the **direct step method**, the depth and velocity are known at a given section of the channel (one end of the reach), and one arbitrarily chooses the depth at the other end of the reach. Then, the length of the reach is solved for. The applicable equation for quantitative evaluation of the water-surface profile is the energy equation written for a finite reach of channel, Δx:

$$y_1 + \frac{V_1^2}{2g} + S_0 \Delta x = y_2 + \frac{V_2^2}{2g} + S_f \Delta x$$

or

$$\Delta x(S_f - S_0) = \left(y_1 + \frac{V_1^2}{2g}\right) - \left(y_2 + \frac{V_2^2}{2g}\right)$$

or

$$\Delta x = \frac{(y_1 + V_1^2/2g) - (y_2 + V_2^2/2g)}{S_f - S_0} = \frac{(y_1 - y_2) + (V_1^2 - V_2^2)/2g}{S_f - S_0} \qquad (15.50)$$

The procedure for evaluation of a profile starts by ascertaining which type applies to the given reach of channel (using the methods of the preceding subsection). Then, starting from a known depth, compute a finite value of Δx for an arbitrarily chosen change in depth. The process of computing Δx, step by step, up (negative Δx) or down (positive Δx) the channel is repeated until the full reach of channel has been covered. Usually, small changes of y are taken so that the friction slope is approximated by the following equation:

$$S_f = \frac{h_f}{\Delta x} = \frac{fV^2}{8gR_h} \qquad (15.51)$$

Here, V is the mean velocity in the reach, and R_h is the mean hydraulic radius. That is, $V = (V_1 + V_2)/2$, and $R_h = (R_{h1} + R_{h2})/2$. It is obvious that a numerical approach of this type is ideally suited for solution by computer.

EXAMPLE 15.13

Classification and Numerical Analysis of a Water-Surface Profile

Problem Statement

Water discharges from under a sluice gate into a horizontal rectangular channel at a rate of 1 m³/s per meter of width, as shown in the following sketch. What is the classification of the water-surface profile? Quantitatively evaluate the profile downstream of the gate and determine whether it will extend all the way to the abrupt drop 80 m downstream. Make the simplifying assumptions that the resistance factor f is equal to 0.02 and that the hydraulic radius R_h is equal to the depth y.

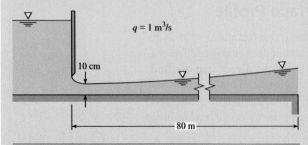

Define the Situation

Water discharges underneath a sluice gate.

Assumptions:

1. Resistance factor f is equal to 0.02.
2. Hydraulic radius R_h is equal to the depth y.

State the Goal

- Classify the downstream profile.
- Determine if increasing slope will prevail all the way to a point of interest 80 m downstream.

Generate Ideas and Make a Plan

1. Determine the letter designation of channel using Fig. 15.30.
2. For flow leaving the sluice gate, determine critical depth y_c, and compare to actual depth of flow. Use this information to refine the classification.
3. Solve for depth versus distance using Eqs. (15.50) and (15.51).

Take Action (Execute the Plan)

1. Channel is horizontal, so the letter designation is H.
2. Determine critical depth y_c using Eq. (15.27):

$$y_c = (q^2/g)^{1/3} = [(1^2 \text{ m}^4/\text{s}^2)/(9.81 \text{ m/s}^2)]^{1/3}$$
$$= \boxed{0.467 \text{ m}}$$

Thus, the depth of flow from the sluice gate is less than the critical depth. Therefore the water-surface profile is classified as

> type H3.

3. To determine depth versus distance along the channel, apply Eqs. (15.50) and (15.51) using the numerical approach given in Table 15.2. Then, plot the results as shown. From the plot, conclude that the

> profile extends to the abrupt drop.

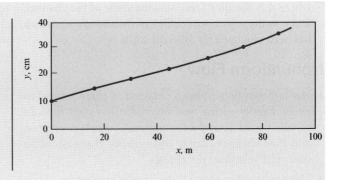

TABLE 15.2 Solution to Example 15.13

Section Number Down-stream of Gate	Depth y, m	Velocity at Section V, m/s	Mean Velocity in Reach, $(V_1 + V_2)/2$	V^2	Mean Hydraulic Radius, $R_m = (y_1 + y_2)/2$	$S_f = \dfrac{fV^2_{\text{mean}}}{8gR_m}$	$\Delta x = \dfrac{(y_1 - y_2) + \dfrac{(V_1^2 - V_2^2)}{2g}}{(S_f - S_0)}$	Distance from Gate x, m
1 (at gate)	0.1	10	...	100	...	...	...	0
	...	...	8.57	73.4	0.12	0.156	15.7	
2	0.14	7.14	...	51.0	...	...	...	15.7
	...	...	6.35	40.3	0.16	0.064	15.3	
3	0.18	5.56	...	30.9	...	...	...	31.0
	...	...	5.05	25.5	0.20	0.032	15.1	
4	0.22	4.54	...	20.6	...	...	...	46.1
	...	...	4.19	17.6	0.24	0.019	13.4	
5	0.26	3.85	...	14.8	...	...	...	59.5
	...	...	3.59	12.9	0.28	0.012	12.4	
6	0.30	3.33	...	11.1	...	...	...	71.9
	...	...	3.13	9.8	0.32	0.008	10.9	
7	0.34	2.94	...	8.6	...	...	...	82.8

15.8 Summarizing Key Knowledge

Describing Open-Channel Flow

- An open channel is one in which a liquid flows with a free surface.
- Steady open-channel flow is classified as either
  - *uniform* (velocity is constant for all points on each streamline) or
  - *nonuniform* (velocity is varying for points along a specific streamline).

Steady and Uniform Flow

- The head loss corresponds to the potential energy change associated with the slope of the channel.
- The discharge is given by the Manning equation:

$$Q = \frac{1}{n} A R_h^{2/3} S_0^{1/2}$$

where A is the flow area, S_0 is the slope of the channel, and n is the resistance coefficient (Manning's n), which has been tabulated for different surfaces.

Nonuniform Flow

- Nonuniform flow in open channels is characterized as either rapidly varied flow or gradually varied flow. In rapidly varied flow, the channel resistance is negligible, and flow changes (depth and velocity changes) occur over relatively short distances.

- The significant π-group is the Froude number:

$$\text{Fr} = \frac{V}{\sqrt{gD_c}}$$

where D_c is the hydraulic depth, A/T. When the Froude number is equal to unity, the flow is critical.

- Subcritical flow occurs when the Froude number is less than unity and supercritical when the Froude number is greater than unity.

Hydraulic Jump

- A hydraulic jump usually occurs when the flow along the channel changes from supercritical to subcritical.

- The governing equation for hydraulic jump in a horizontal, rectangular channel is

$$y_2 = \frac{y_1}{2}\left(\sqrt{1 + 8\,\text{Fr}_1^2} - 1\right)$$

- The corresponding head loss in the hydraulic jump is

$$h_L = \frac{(y_2 - y_1)^3}{4y_1 y_2}$$

- When the flow along the channel changes from subcritical to supercritical flow, the head loss is assumed to be negligible, and the depth and velocity relationship is governed by the change in elevation of the channel bottom and the specific energy, $y + V^2/2g$. Typical cases of this type of flow include the following:

  1. Flow under a sluice gate
  2. An upstep in the channel bottom
  3. Reduction in width of the channel

Gradually Varied Flow

- For gradually varied flow, the governing differential equation is

$$\frac{dy}{dx} = \frac{S_0 - S_f}{1 - \text{Fr}^2}$$

When this equation is integrated along the length of the channel, the depth y is determined as a function of distance x along the channel. This yields the water-surface profile for the reach of the channel.

REFERENCES

1. Limerinos, J.T. "Determination of the Manning Coefficient from Measured Bed Roughness in Natural Channels." Water Supply Paper 1898-B, U.S. Geological Survey, Washington, D.C., 1970.

2. Committee on Hydromechanics of the Hydraulics Division of American Society of Civil Engineers. "Friction Factors in Open Channels." *J. Hydraulics Div., Am. Soc. Civil Eng.* (March 1963).

3. Wolman, M.G. "The Natural Channel of Brandywine Creek, Pennsylvania." Prof. Paper 271, U.S. Geological Survey, Washington D.C., 1954.

4. Yen, B.C. (ed.) *Channel Flow Resistance: Centennial of Manning's Formula.* Littleton, CO: Water Resources Publications, 1992.

5. Chow, Ven Te. *Open Channel Hydraulics.* New York: McGraw-Hill, 1959.

6. Henderson, F.M. *Open Channel Flow.* New York: Macmillan, 1966.

7. American Concrete Pipe Assoc. *Concrete Pipe Design Manual.* Vienna, VA: American Concrete Pipe Assoc., 1980.

8. Parshall, R.L. "The Improved Venturi Flume." *Trans. ASCE,* 89 (1926), 841–851.

9. Roberson, J.A., J.J. Cassidy, and M.H. Chaudhry. *Hydraulic Engineering.* New York: John Wiley, 1988.

10. Hinds, J. "The Hydraulic Design of Flume and Siphon Transitions." *Trans. ASCE,* 92 (1928), pp. 1423–1459.

11. U.S. Bureau of Reclamation. *Design of Small Canal Structures.* U.S. Dept. of Interior, Washington, DC: U.S. Govt. Printing Office, 1978.

12. Rouse, H. (ed.). *Engineering Hydraulics.* New York: John Wiley, 1950.

13. U.S. Bureau of Reclamation. *Hydraulic Design of Stilling Basin and Bucket Energy Dissipators.* Engr. Monograph no. 25, U.S. Supt. of Doc., 1958.

14. Valle, B.L., and G.B. Pasternak. "Submerged and Unsubmerged Natural Hydraulic Jumps in a Bedrock Step-Pool Mountain Channel." *Geomorphology,* 82 (2006), pp. 146-159.

15. Raju, K.G.R. *Flow through Open Channels.* New Delhi: Tata McGraw-Hill, 1981.

PROBLEMS

Describing Open-Channel Flow (§15.1)

15.1 Two channels have the same cross-sectional area, but different geometry, as shown.

  a. Which channel has more wetted surface?

  b. Which channel has the largest wetted perimeter?

  c. Which channel will have more energy loss to friction?

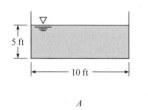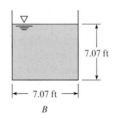

Problem 15.1

Steady Uniform Open-Channel Flow (§15.3)

15.2 A mountain stream flows over a rocky streambed. Apply the Limerinos and Chezy equations to calculate the discharge. The stream has an intermediate rock size d_{84} of 25 cm, an average depth of 1.4 m, a slope of $S = 0.0037$, and a width of 52 m. Choose the closest answer (m³/s): (a) 85, (b) 120, (c) 160, (d) 240, or (e) 410.

15.3 Water is flowing in a rectangular concrete channel. Apply the Manning equation to calculate the discharge. The channel is unfinished concrete 10.1 ft wide and drops 3.5 ft over a run of 1000 ft; the depth of flow is 2.5 ft; $T = 60°F$; and the flow is uniform. Choose the closest answer (cfs): (a) 114, (b) 145, (c) 183, (d) 212, or (e) 565.

15.4 What discharge of water will occur in a trapezoidal channel that has a bottom width of 25 ft and side slopes of 1 vertical to 1 horizontal if the slope of the channel is 2 ft/mile and the depth is 4 ft? The channel is lined with troweled concrete.

15.5 Estimate the discharge in a rock-bedded stream ($d_{84} = 30$ cm) that has an average depth of 1.8 m, a slope of 0.0037, and a width of 52 m. Assume $k_s = d_{84}$.

15.6 A rectangular concrete channel is 14 ft wide and has uniform water flow. If the channel drops 6 ft in a length of 8000 ft, what is the discharge? Assume $T = 60°F$. The depth of flow is 4 ft.

Steady Nonuniform Open-Channel Flow (§15.4)

15.7 Critical flow _____. (Select all of the following that are correct.)

  a. occurs when the discharge is maximum for a given specific energy.

  b. occurs when specific energy is a minimum for a given discharge.

  c. occurs when Fr = 1.

  d. occurs when Fr < 1.

15.8 Water flows at a depth of 18 in. with a velocity of 20 ft/s in a rectangular channel, with width = 3 ft. (a) Is the flow subcritical or supercritical? (b) What is the alternate depth?

15.9 The water discharge in a rectangular channel 20 ft wide is 600 cfs. If the depth of water is 3 ft, is the flow subcritical or supercritical?

15.10 For a rectangular channel 2 m wide and discharge of 12 m³, what is the alternate depth to the 90 cm depth? What is the specific energy for these conditions?

15.11 A broad-crested weir is used to measure discharge in an irrigation ditch. Calculate the discharge. The weir is 5.5 ft long, 2 ft high, and the head on the weir is 2.4 ft. Use Figure 15.13 to find the discharge coefficient. Choose the closest answer: (a) 12, (b) 24, (c) 58, (d) 81, or (e) 105.

15.12 A rectangular channel that is 15 ft wide and has a mild slope ends in a free outfall. If the water depth at the brink is 1.20 ft, what is the discharge in the channel?

15.13 A horizontal rectangular channel 14 ft wide carries a discharge of water of 500 cfs. If the channel ends with a free outfall, what is the depth at the brink?

15.14 What discharge of water will occur over a 3-ft-high, broad-crested weir that is 10 ft long if the head on the weir is 1.8 ft?

15.15 Because of the increased size of ships, the phenomenon called "ship squat" has produced serious problems in harbors where the draft of vessels approaches the depth of the ship channel. When a ship steams up a channel, the resulting flow situation is analogous to open-channel flow in which a constricting flow section exists (the ship reduces the cross-sectional area of the channel). The problem may be analyzed by referencing the water velocity to the ship and applying the energy equation. Thus, at the section of the channel where the ship is located, the relative water velocity in the channel will be greatest, and the water level in the channel will be reduced as dictated by the energy equation. Consequently, the ship itself will be at a lower elevation than if it were stationary; this lowering is referred to as "ship squat." Estimate the squat of the fully loaded supertanker *Bellamya* when it is steaming at 5 kt (1 kt = 0.515 m/s) in a channel that is 35 m deep and 200 m wide. The draft of the *Bellamya* when fully loaded is 29 m. Its width and length are 63 m and 414 m, respectively.

15.16 A small wave is produced in a pond that is 12 in. deep. What is the speed of the wave in the pond?

15.17 A small wave in a pool of water having constant depth travels at a speed of 2 m/s. How deep is the water?

Hydraulic Jumps (§15.6)

15.18 For a hydraulic jump, _____. (Select all of the following that are correct.)

  a. the flow changes from supercritical to subcritical.

  b. the flow changes from subcritical to supercritical.

c. the height of the water abruptly increases from the upstream to the downstream cross section.

d. the downstream and upstream depth are related quantitatively in terms of the upstream Fr.

e. significant energy is lost.

f. the energy equation is a better tool for analysis than the momentum equation.

15.19 The spillway shown has a discharge of 2.5 m^3/s per meter of width occurring over it. What depth y_2 will exist downstream of the hydraulic jump? Assume negligible energy loss over the spillway.

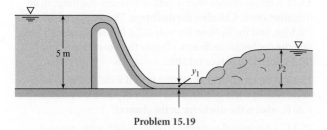

Problem 15.19

15.20 The flow of water downstream from a sluice gate in a horizontal channel has a depth of 35 cm and a flow rate of 10 m^3/s per meter of width. The sluice gate is 2 m wide.

a. Could a hydraulic jump be caused to form downstream of this section?

b. If so, what would be the depth downstream of the jump?

15.21 For the derivation of Eq. (15.28) it is assumed that the bottom shearing force is negligible. For water flowing uniformly at a depth $y_1 = 40$ cm in the concrete channel shown, which is 10 m wide, a sill is installed to force a hydraulic jump to form. Estimate the magnitude of the shearing force F_s associated with the hydraulic jump and then determine F_s/F_H, where F_H is the net hydrostatic force on the hydraulic jump. Assume Manning's n value is $n = 0.012$.

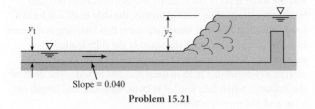

Slope = 0.040

Problem 15.21

Gradually Varied Flow (§15.7)

15.22 Water flows at a rate of 100 ft^3/s in a rectangular channel 10 ft wide. The normal depth in that channel is 2 ft. The actual depth of flow in the channel is 4 ft. The water-surface profile in the channel for these conditions would be classified as (a) M1, (b) M2, (c) S1, or (d) S2.

15.23 The water-surface profile labeled with a question mark is (a) M2, (b) H2, (c), S2, or (d) A2.

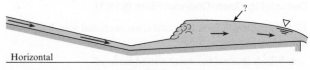

Horizontal

Problem 15.23

15.24 The horizontal rectangular channel downstream of the sluice gate is 10 ft wide, and the water discharge therein is 108 cfs. The water-surface profile was computed by the direct step method. If a 2-ft-high sharp-crested weir were installed at the end of the channel, do you think a hydraulic jump would develop in the channel? If so, approximately where would it be located? Justify your answers by appropriate calculations. Label any water-surface profiles that can be classified.

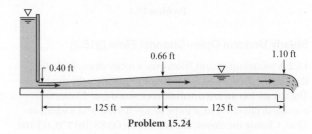

0.40 ft   0.66 ft   1.10 ft

|← 125 ft →|← 125 ft →|

Problem 15.24

15.25 Water flows from under a sluice gate into a horizontal rectangular channel at a rate of 3 m^3/s per meter of width. The channel is concrete, and the depth of water exiting the sluice gate is 20 cm. Apply Eq. (15.42) of §15.6 to construct the water-surface profile up to a depth of 60 cm. In your solution, compute reaches for adjacent pairs of depths given in the following sequence: $d = 20$ cm, 30 cm, 40 cm, 50 cm, and 60 cm. Assume that f is constant with a value of 0.02. Plot your results.

15.26 A horizontal rectangular concrete channel terminates in a free outfall. The channel is 4 m wide and carries a discharge of water of 12 m^3/s. What is the water depth 300 m upstream from the outfall?

15.27 Water flows at a steady rate of 12 cfs per foot of width ($q = 12$ cfs) in the wide rectangular concrete channel shown. Determine the water-surface profile from section 1 to section 2.

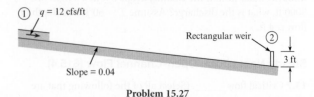

① $q = 12$ cfs/ft

Rectangular weir   ②

3 ft

Slope = 0.04

Problem 15.27

Modeling of Fluid Dynamics Problems

CHAPTER ROAD MAP This chapter describes modeling and introduces two methods that are useful for modeling:

- **Partial Differential Equations (PDEs).** This method involves formulating the governing scientific laws as partial differential equations.
- **Computational Fluid Dynamics (CFD).** This method involves approximating the partial differential equations with algebraic equations and then using a computer algorithm to solve these equations.

FIGURE 16.1

The Eagle X-TS and the workers at the assembly plant where the plane was built. The Eagle X-TS was designed by John Roncz using CFD. Roncz, a world-class designer, is responsible for some portion of 50 aircraft designs. Two of Roncz's designs are on display at the National Air and Space Museum in the United States. (Photo courtesy of John Roncz.)

Roncz describes how he learned fluid mechanics:

"The main advantage I have is that I've never taken a single course in aeronautical engineering. . . . As a result, I've had to figure it all out myself. You understand things better that way." (1)

LEARNING OUTCOMES

MODELING AND PDEs (§16.1, §16.2).
- Describe how engineers build models.
- Explain how engineers apply PDEs in the context of modeling.

MATH TOPICS (§16.2).
- Explain the velocity field.
- Explain Taylor series.
- Explain invariant notation.
- Explain mathematical operators.
- Explain the material derivative.
- Explain the acceleration field.

THE CONTINUITY EQUATION (§16.3).
- List the steps to derive the continuity equation.
- List and describe the various forms of the continuity equation.

THE NAVIER-STOKES EQUATION (§16.4).
- List the steps to derive the Navier-Stokes equation.
- Describe the physics of the Navier-Stokes equation.

CFD (§16.5).
- Describe CFD.
- Describe how engineers select a CFD code.
- Describe how CFD codes work.
- Explain these topics: grid, time step, boundary condition, validation, verification, and turbulence models.

Paths cannot be taught, they can only be taken.

—Traditional Zen saying

In the first chapter of this book, we spoke of your path to success. At the end of this chapter (§16.7), we suggest a path for moving forward.

16.1 Models in Fluid Mechanics

Engineers create models of systems because this process saves money and results in better designs. Modeling involves analyses, experiments, and computer simulations. These topics are introduced in this section.

The Concept of a Model

In engineering, there is something real (e.g., a dam and associated power plant), and there is an idealization (i.e., a model) of this real thing. A model, according to Wang (2), is a *tool to represent a simplified version of reality*. Ford (3) suggests that the model is *a substitute for a real system*. Some examples of models include the following:

- A road map is a model because a map represents a complex array of roads.
- Architects' drawings are models because they represent buildings that will be built.
- A table of contents is a model because it represents the subject matter of a book.

Some examples of models relevant to fluid mechanics are as follows:

- The ideal gas law is a model because it is an idealized (simplified) description of how the variables of density, pressure, and temperature are related.
- A collection of equations can be a model. For example, the energy equation together with the Darcy-Weisbach equation and suitable minor loss coefficients can be used to predict the flow rate for water through a siphon. Using the equations is a substitute for building a system and then correlating experimental data.
- A small-scale car that is used in a wind tunnel to estimate drag acting on a full-scale car is a model.

To advance the discussion of modeling, we next describe an engineering project.

Example of an Engineering Project. The slow sand filter (Fig. 16.2) is a widely used technology for producing clean drinking water. Water enters the filter at the top, and naturally occurring organisms that live in the topmost layer of the filter remove the biological contaminants. This topmost layer, called the *schmutzdecke*, is found in the top few millimeters of the sand layer. The sand and gravel below the schmutzdecke collect dirt and clay particles.

Several years ago, students from the University of Idaho designed a slow sand filter for applications in Kenya. Because slow sand filters do not require chemicals or electricity, this technology is especially suitable to applications in the developing world.

FIGURE 16.2

The slow sand filter.

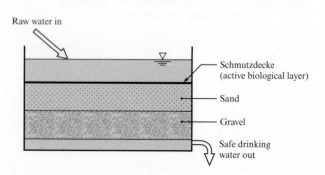

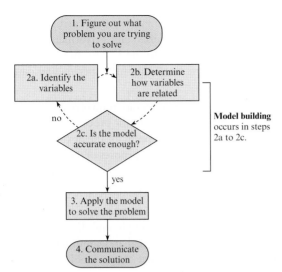

FIGURE 16.3

The model in the context of engineering problem solving.

The team choose to develop various models of the slow sand filter. The model-building process is described in the next subsections.

Summary. A **model** is an idealization or simplified version of reality. Models are valuable when they help engineers and other professionals reach goals in an economical way.

How to Build a Model of a System

The reason for building a model is to solve a problem (Fig. 16.3). The process of model building, according to Montgomery et al. (4), involves identifying relevant variables, determining the relationships between these variables, and then testing the model to ensure that it is accurate (i.e., does the model faithfully capture what happens in reality?). As shown, the process of model building is iterative.

Example. To build a model of a slow sand filter (Fig. 16.2), the modeling process involves the following steps.

- **Step 2a: Identify the variables.** Determine which variables characterize performance. Then, classify the variables into two groups:
  - **Performance variables** characterize how well the product performs. Examples of these variables include the flow rate through the filter, the clarity of the water that leaves the filter, and the time period between maintenance for filter cleaning. Performance variables are dependent variables, meaning that they depend on the values of the design variables.
  - **Design variables** are the factors that engineers can select. Examples of these variables are the depth of water on top of the filter, the thickness of the sand layer, and the distribution of sand and gravel sizes.
- **Step 2b: Determine how the variables are related**. The purpose of this step is to identify cause and effect. For example, if one changes the size of the sand particles, does this make the filter perform better or worse? Why? There are two approaches for determining how the variables are related (4):
  - **Mechanistic models** are based on scientific knowledge of the phenomena. For example, Darcy's law describes flow of fluids through a porous medium such as sand and gravel, and the equation itself tells us the relationship among the variables.
  - **Empirical models** involve relating the variables by using curve fits of experimental data. For example, experiments and correlation could be used to determine the time it takes for the schmutzdecke to develop.

- **Step 2c: Test the model for accuracy**. The result of step 2a is an ability to predict the relationship between design variables (e.g., dimensions, particle sizes) and performance variables (e.g., water quality or flow rate). The purpose of step 2c is to check to see how accurate the predictions are. Much of this time, this step is done by comparing experimental data with predictions.

- **Iterate back to step 2a**. In practice, model building is iterative. Iteration involves repeating a process with the aim of reaching a desired goal. Each repetition of the process is called an iteration, and the results of one iteration are used as the starting point for the next iteration. Iterations are ended when the model has enough accuracy for the purposes of the engineers.

    Example of Iteration (Slow Sand Filter). To build a model of a slow sand filter, one might start out with a model comprised of a few equations and a simple, bench-top experiment. The model would be highly simplified, and the purpose of the first iteration would be to gain experience with modeling and measuring the flow of water through sand. In subsequent iterations, the analytical and experimental models would be developed and continually compared. After analytical models had been developed, the team might create a CFD model to perform parametric studies on the design.

    After the model has been validated through the iterative process, the next steps are to apply the model to solve the problem (step 3 of Fig. 16.3) and to communicate the solution (step 4).

    Summary. Models are built in an iterative process that involves identifying the variables, classifying these variables into performance variables and design variables, and determining how the variables are related. Finally, the model is validated to see if model predictions are accurate enough for the needs of the problem. The most important aspect of model building is to start simple and then use sequential iterations to improve accuracy. Model building was introduced in Chapter 1. When models are based on scientific laws and equations, then the Wales-Woods approaches describes how experts build math models.

Three Methods for Model Building

Model building involves three methods.

Analytical fluid dynamics (AFD) involves knowledge and equations that are commonly found in engineering textbooks and references.

Experimental fluid dynamics (EFD) involves experiments to gather information about variables. EFD is often used to validate calculations, to validate computer solutions, and to determine performance characteristics of systems that are not easily modeled using calculations or computers.

Computational fluid dynamics (CFD) involves computer solutions of the governing partial differential equations. That is, engineers run a computer program to understand how the variables interact.

In real-world applications, model building usually involves an integrated and iterative combination of the preceding approaches. For example, model-building efforts for the slow sand filter might include the following:

- **Darcy's law** (AFD). To predict the rate at which water flows through the sand and gravel, one can apply Darcy's law, which describes flow through a porous medium. This is an example of AFD because it involves a known equation.

- **Measuring permeability** (EFD). To apply Darcy's law, one must estimate the value of the permeability of sand layers. (Permeability is a property of a porous medium that characterizes how easily water flows through the material for a given pressure drop.) To determine permeability, the engineer sets up an experiment and measures the value for various types of sand and gravel.

- **Computational model** (CFD). A commercially available CFD computer model for ground-water flow could be applied to perform parametric studies on the slow sand filter so that engineers could examine many different design variations.
- **Experiments** (EFD). Experiments can measure how long it takes for the schmutzdecke layer to grow.

Summary. In fluid mechanics, there are three approaches to model building: analytical fluid mechanics, experimental fluid mechanics, and computational fluid mechanics. Most models involve two or three of these approaches working synergistically.

Assessing the Value of a Model

Ford (3) asserts that a model is useful when it helps one learn something about the system that the model represents. For example, a road map is useful when it helps one more easily navigate an unfamiliar location, and architects' drawings are useful when they help a builder understand what materials need to be purchased and how the architect intends a house to look.

The value of a model is related to the benefits and the costs (i.e., resources required to produce the model). One way to assess value is to use a ratio of benefits to costs:

$$\begin{pmatrix} \text{value of} \\ \text{a model} \end{pmatrix} = \frac{\text{(benefits provided by the model)}}{\text{(resources required to implement the model)}}$$

To assess value, some questions that engineers might ask include the following:

Benefits:
- Will the model lead to a design that works better?
- Will the model help the team complete the project more quickly?
- Will the model lead to a final design that is lower cost to build? To operate?
- Will the model help the team understand the interactions of the variables?
- Can the model be used for future projects?
- Would the model be beneficial to other engineers who are designing similar systems?

Resources, costs, and risks:
- What is the risk of failure? Can a model that works be developed?
- How accurate will the model be? What accuracy is needed?
- How much engineering time will the model take to build?
- Does software need to be purchased? Experiments built? Other costs?

As shown in Fig. 16.4, the three modeling approaches provide different types of information and have varying levels of cost (time and resources).

- **Algebraic equations**. Applying equations found in textbooks provides estimates (low level of details). Costs are low because estimates usually require a pencil, paper, and a calculator and take about one hour.
- **PDEs**. Finding an existing solution to the governing PDEs provides rich details about the flow. Costs are modest because one has to search the literature, learn the details of the solution, and apply the solution. However, there are only a few solutions in the literature; therefore, this approach is only sometimes useful.
- **CFD**. Developing a CFD solution provides a wealth of details. The costs can be high because one has to obtain a code, learn the code, set up the model, and validate the model.
- **Experiment**. Designing and conducting an experiment provides data from the physical world, which is often used to assess the validity of math-based solutions. Costs for an experiment can range from very low to very high, depending on the scope and nature of the experiment.

FIGURE 16.4

Information provided by a modeling approach versus the cost of the modeling approach.

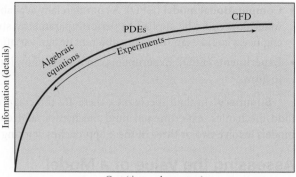

Summary. Three useful methods for model building are analysis, experiment, and computation. Often, these three methods are used in combination via an iterative strategy that involves starting with simple models and then refining these models. There are multiple trade-offs in model building that involve cost, benefits, solution accuracy, and solution detail.

16.2 Foundations for Learning Partial Differential Equations (PDEs)

This section presents:

- Why learning PDEs is useful
- Some mathematical foundations for learning PDEs

Rationale for Learning PDEs

PDEs represent the scientific laws that govern flowing fluids. Solving these equations gives numerical values for the pressure field, the velocity field, or other fields. From these fields, engineers can calculate nearly anything of engineering interest, such as drag force, head loss, and power requirements.

Thus, solving the PDEs is the ultimate solution technique—but there is a catch! In practice, the PDEs have nonlinear terms that prevent direct mathematical solutions, except in a limited number of special cases. These special cases were solved many years ago, and today's engineers do not solve problems by directly solving the PDEs. Nevertheless, there are two benefits to learning PDEs.

Understanding Existing Solutions (Benefit #1). The literature has many existing solutions of PDEs. These solutions classify into two categories:

- **Exact solutions**. An exact solution involves a physical situation in which the equations of motion reduce to equations that can be solved. There are about 100 such solutions in existence. Examples include Poiseuille flow and Couette flow.

- **Idealized solutions**. An idealized solution involves a physical situation in which assumptions are made that allow the governing equations to be simplified and solved mathematically. Two examples of idealized solution are as follows:

  - **Potential flow**. When an external flow around a body is assumed to be inviscid (i.e., frictionless) and irrotational (i.e., the fluid particles are not rotating), the equations reduce to equations that can be solved analytically. This situation is called potential flow.

  - **Laminar boundary layer flow**. When laminar viscous flow near a wall is simplified by making boundary layer assumptions, the equations can be solved. The resulting solution, called the Blasius solution, describes flow in the laminar boundary layer.

Engineers use existing solutions to gain understanding of more complex problems. For example, a bicycle rider was severely injured in a collision caused by a bus that passed too close to him. When a large vehicle passes closely by a cyclist, this causes side forces on the cyclist. To gain insight into these side forces, an engineer used the solution for potential flow around an elliptical body to predict the magnitude and direction of the side force.

A second example involves modeling blood flow in the human abdominal aorta. Sometimes, the aorta loses its structural integrity and bulges out to form an aneurysm. If an aneurysm ruptures, death is common. Thus, the researchers wanted to understand the forces exerted by the flow on the aneurysm walls. Two existing solutions were used to gain insight into this problem: the Poiseuille solution for steady laminar flow in a round tube and the Womersley solution for oscillatory laminar flow in a round tube.

Understanding and Validating CFD (Benefit #2). Because CFD codes solve PDEs, the first step in learning CFD is to learn about the PDEs.

Existing solutions are used to validate CFD codes. For example, when a CFD model of blood flow in an aneurysm was developed, the code was validated in part by modeling an existing analytical solution (i.e., the Womersley solution) and then checking to make sure that the CFD solution matched the analytical solution.

Summary. Three reasons for learning PDEs are (a) to be able to understand and apply existing solutions that are found in the literature, (b) to understand the equations that are being solved by CFD codes, and (c) to validate CFD codes by ensuring that the CFD code can correctly predict the results given by a known classical solution.

The remainder of this section introduces mathematics that are useful in the development of PDEs.

Velocity Field: Cartesian Coordinates

The solution of the equations of motion are fields such as the pressure field, the density field, the temperature field, and the velocity field. Thus, understanding fields is important. This section introduces the velocity field.

In the Cartesian coordinate system, a point in space is identified by specifying coordinates $(x, y, z$; Fig. 16.5). The associated unit vectors are $\mathbf{i}$ in the x direction, $\mathbf{j}$ in the y direction, and $\mathbf{k}$ in the z direction. Notice that the coordinate system is *right-handed,* which means that the cross product of $\mathbf{i}$ and $\mathbf{j}$ is the $\mathbf{k}$ unit vector:

$$\mathbf{i} \times \mathbf{j} = \mathbf{k}$$

The velocity field is given by

$$\mathbf{V} = u(x, y, z, t)\mathbf{i} + v(x, y, z, t)\mathbf{j} + w(x, y, z, t)\mathbf{k} \qquad (16.1)$$

where $u = u(x, y, z, t)$ is the x direction component of the velocity vector, and v and w have similar meanings. The independent variables are position (x, y, z) and time (t).

The next two examples show how to reduce the general form of the velocity field so that it applies to a specific situation. Notice the process steps.

> **EXAMPLE.** Consider steady flow in a plane (Fig. 16.6). Reduce the general equation for the velocity field so that it applies to this situation.
>
> **Ideas/Action.**
>
> **1.** Write the general equation for the velocity field:
>
> $$\mathbf{V} = u(x, y, z, t)\mathbf{i} + v(x, y, z, t)\mathbf{j} + w(x, y, z, t)\mathbf{k}$$
>
> **2.** Analyze the dependent variables. Because the flow is planar, $w = 0$. Thus, the dependent variables reduce to u and v.

FIGURE 16.5

Cartesian coordinates.

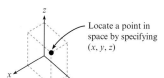

Locate a point in space by specifying (x, y, z)

FIGURE 16.6

Example of velocity components for planar flow.

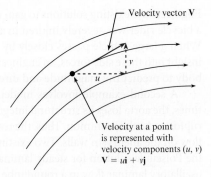

Velocity vector **V**

Velocity at a point
is represented with
velocity components (u, v)
$\mathbf{V} = u\mathbf{i} + v\mathbf{j}$

3. Analyze the independent variables. Because the flow is planar, z is not a parameter. Because the flow is steady, time is not a parameter. Thus, the independent variables are x and y; the velocity field reduces to $\mathbf{V} = u(x,y)\mathbf{i} + v(x, y)\mathbf{j}$.

EXAMPLE. Consider steady flow entering a channel (Fig. 16.7) formed by plates that extend to $\pm\infty$ in the z direction. Such plates are called *infinite plates*. Reduce the general equation for the velocity field so that it applies to this situation.

Ideas/Action

1. Write the general equation for the velocity field:

$$\mathbf{V} = u(x, y, z, t)\mathbf{i} + v(x, y, z, t)\mathbf{j} + w(x, y, z, t)\mathbf{k}$$

2. Analyze the dependent variables. Let $w = 0$ because there is no flow in the z direction.

3. Analyze the independent variables. Because the flow is planar, the velocity does not vary with z. Because the flow is steady, the velocity does not vary this time. Thus, the reduced equation for the velocity field is

$$\mathbf{V} = u(x, y)\mathbf{i} + v(x, y)\mathbf{j}$$

This equation means that both u and v will be nonzero, and both u and v will vary with x and y. The reason is that flow in the entrance to the channel is developing (see Chapter 10). Once the flow is fully developed, then the velocity field will reduce to the form

$$\mathbf{V} = u(y)\mathbf{i}$$

Summary. The general form of the velocity field in Cartesian coordinates is given by Eq. (16.1). To reduce this equation so that it applies to a given situation, analyze the independent and the dependent variables, and eliminate terms that are not relevant or are zero.

Velocity Field: Cylindrical Coordinates

Because cylindrical coordinates are widely used in application, this system is introduced next.

In cylindrical coordinates (Fig. 16.8), a point in space is described by specifying coordinates (r, θ, z). The radius r is measured from the origin, the azimuth angle θ is measured counterclockwise from the x axis, and the height z is measured from the x-y plane.

FIGURE 16.7

Flow between infinite plates.

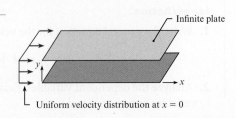

Infinite plate

Uniform velocity distribution at $x = 0$

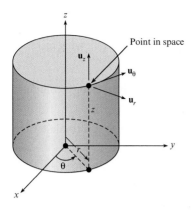

The velocity field in general form is

$$\mathbf{V} = v_r(r, \theta, z, t)\mathbf{u}_r + v_\theta(r, \theta, z, t)\mathbf{u}_\theta + v_z(r, \theta, z, t)\mathbf{u}_z \tag{16.2}$$

EXAMPLE. Fig. 16.9 shows ideal flow over a circular cylinder. Reduce the general form of the velocity field so that it applies to this situation.

Ideas/Action

Represent the velocity vector at a point of interest (see Fig. 16.9).

- **Step 1.** Sketch an x- and y-coordinate axis.
- **Step 2.** Sketch a radius vector of length r.
- **Step 3.** Sketch unit vectors $\mathbf{u}_r$ and $\mathbf{u}_\theta$.
- **Step 4.** Represent the velocity vector with components v_r and v_θ.

Next, do a term-by-analysis of Eq. (16.2). Eliminate z and v_z because the flow is planar. Eliminate t because the flow is steady. Eq. (16.2) reduces to

$$\mathbf{V} = v_r(r, \theta)\mathbf{u}_r + v_\theta(r, \theta)$$

For flow in a plane (e.g., Fig. 16.9), the z direction is not needed; use only the r and θ coordinates. This two-dimensional coordinate system is called **polar coordinates**.

Summary. When cylindrical coordinates are used, the general form of the velocity field is given by Eq. (16.2). For the flow in a plane, the coordinates can be simplified to a 2-D flow, and the coordinates are called polar coordinates.

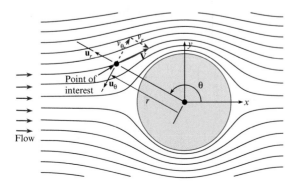

FIGURE 16.9

Using polar coordinates to represent the velocity vector at a point. The pictured flow is ideal flow (i.e., inviscid and irrotational) over a circular cylinder.

Taylor Series

Engineers learn Taylor series because Taylor series are used to perform the following tasks:

- Derive both ordinary and partial differential equations.
- Convert differential equations into algebraic equations that can be solved using a computer algorithm by a CFD program.

A **Taylor series** is a series expansion of a function about a point. The general formula for the function $f(x)$ expanded around the point $x = a$ is

$$f(x) = f(a) + \left(\frac{df}{dx}\right)_a \frac{(x-a)}{1!} + \left(\frac{d^2 f}{dx^2}\right)_a \frac{(x-a)^2}{2!} + \ldots \qquad (16.3)$$

For example, when the function is $f(x) = e^x$, Eq. (16.3) becomes

$$e^x = e^a \left[1 + (x-a) + \frac{(x-a)^2}{2} + \frac{(x-a)^3}{6} + \ldots \right] \qquad (16.4)$$

Kojima et al. (5) suggest that a Taylor series is an *imitation* of an equation, just as equations are imitations of the physical world (see Fig. 16.10).

Taylor series are commonly truncated. This means that higher-order terms are neglected. For example, consider the following Taylor series:

$$\underbrace{f(x) = \frac{1}{1-x}}_{\text{Function}} = \underbrace{1 + x + x^2 + x^3 + x^4 + \ldots \text{ (H.O.T.)}}_{\substack{\text{Taylor series approximation (the acronym} \\ \text{H.O.T. stands for "higher-order terms")}}} \qquad (16.5)$$

When $x = 0.1$, Eq. (16.5) gives

$$\underbrace{f(x) = \frac{1}{1 - 0.1} = 1.11111 \ldots}_{\text{Function}} = \underbrace{1 + 0.1 + 0.01 + 0.001 + 0.0001 \ldots + \ldots \text{ (H.O.T.)}}_{\text{Taylor series approximation}} \qquad (16.6)$$

The effects of neglecting higher-order terms are as follows:

- When two terms are kept, the result is 1.1.
- When three terms are kept, the result is 1.11.
- When four terms are kept, the result is 1.111.

For engineering purposes, it is sometimes useful to modify Eq. (16.3) by changing the independent variables. Change x to $x + \Delta x$ and let $a = x$. The result is

$$f(x + \Delta x) = f(x) + \left(\frac{df}{dx}\right)_x \frac{(\Delta x)}{1!} + \left(\frac{d^2 f}{dx^2}\right)_x \frac{(\Delta x)^2}{2!} + \ldots \text{ (H.O.T.)} \qquad (16.7)$$

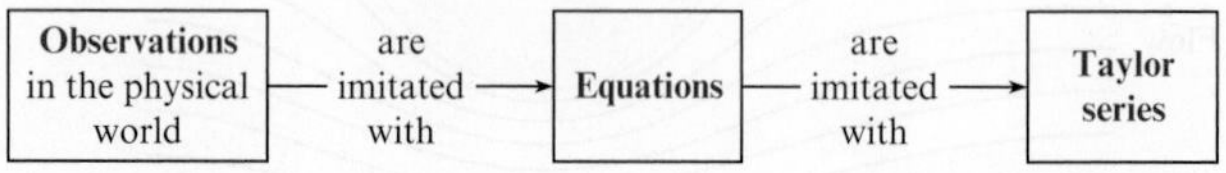

FIGURE 16.10

Data (observations) are idealized with equations. Then, equations are idealized with a Taylor series.

In fluid mechanics, one uses the Taylor series for a function of several variables. The general form of the Taylor series for a function of two variables $f(x, y)$ expanded about point (a, b) is

$$f(x, y) = f(a, b) + \left(\frac{\partial f}{\partial x}\right)_{a,b} \frac{(x - a)}{1!} + \left(\frac{\partial f}{\partial y}\right)_{a,b} \frac{(y - b)}{1!}$$

$$+ \left(\frac{\partial^2 f}{\partial x^2}\right)_{a,b} \frac{(x - a)^2}{2!} + \left(\frac{\partial^2 f}{\partial x \partial y}\right)_{a,b} \frac{2(x - a)(y - b)}{2!} + \left(\frac{\partial^2 f}{\partial y^2}\right)_{a,b} \frac{(y - b)^2}{2!} + \dots$$

(16.8)

To modify Eq. (16.8) so that it is more useful for fluid mechanics, let $x = x + \Delta x$, $y = y$, $a = x$, and $b = y$:

$$f(x + \Delta x, y) = f(x, y) + \left(\frac{\partial f}{\partial x}\right)_{x, y} \frac{\Delta x}{1!} + \left(\frac{\partial^2 f}{\partial x^2}\right)_{x, y} \frac{(\Delta x)^2}{2!} + \left(\frac{\partial^3 f}{\partial x^3}\right)_{x, y} \frac{(\Delta x)^3}{3!} + \dots \quad (16.9)$$

Next, introduce the variables used in fluid mechanics:

$$f(x + \Delta x, y, z, t) = f(x, y, z, t) + \left(\frac{\partial f}{\partial x}\right)_{x, y, z, t} \frac{\Delta x}{1!} + \left(\frac{\partial^2 f}{\partial x^2}\right)_{x, y, z, t} \frac{(\Delta x)^2}{2!} + \dots \text{ H.O.T. } (16.10)$$

Summary. In fluid mechanics, engineers commonly expand functions into a Taylor series, which is a series expansion of about a point. Often, higher-order terms are neglected. A useful form of the Taylor series for fluid mechanics is given in Eq. (16.10).

Mathematical Notation (Invariant Notation and Operators)

In addition to Cartesian, cylindrical, and polar coordinates, engineers use systems such as spherical coordinates, toroidal coordinates, and generalized curvilinear coordinates. Because there is a large amount of detail, engineers sometimes write equations in ways that apply to any coordinate system. **Invariant notation** is a mathematical notation that applies (i.e., generalizes) to any coordinate system.

To introduce invariant notation, consider the gradient of the pressure field (Table 16.1). As shown, the gradient can be written several ways. Also, the mathematical notation can be classified into two categories:

- **Coordinate-specific notation**. Terms in equations are written so that they apply to a specific coordinate system. For example, Table 16.1 shows Cartesian and cylindrical coordinates.

- **Invariant notation**. Terms are written so that they apply to any coordinate system; that is, they generalize.

TABLE 16.1 Alternative Ways to Write the Gradient of the Pressure Field

Category	Description	Mathematical Form
Coordinate-specific notation	Cartesian coordinates	$\frac{\partial p}{\partial x}\mathbf{i} + \frac{\partial p}{\partial y}\mathbf{j} + \frac{\partial p}{\partial z}\mathbf{k}$
	Cylindrical coordinates	$\frac{\partial p}{\partial r}\mathbf{u}_r + \frac{1}{r}\frac{\partial p}{\partial \theta}\mathbf{u}_\theta + \frac{\partial p}{\partial z}\mathbf{u}_z$
Invariant notation	Del notation	∇p
	Gibbs notation	$\text{grad}(p)$
	Indicial notation (Einstein summation convention)	$\frac{\partial p}{\partial x_i}$ or $\partial_i p$

Table 16.1 shows three types of invariant notation:

- **Del notation** is represented by the nabla symbol, ∇. Del notation is the most common approach used in engineering.
- **Gibbs notation** uses words to represent operators—for example, *grad* to represent the gradient. Gibbs notation is common in mathematics.
- **Indicial notation** is a shorthand approach that is common in both engineering and physics.

Mathematical Operators

In mathematics, collections of terms called **operators** are given names because they appear commonly in the equations of mathematical physics. In the equations of fluid mechanics, some common operators include the following:

- **Gradient**. For example, the gradient of the pressure field or the gradient of the velocity field
- **Divergence**. For example, the divergence of the velocity field
- **Curl**. For example, the curl of the velocity field
- **LaPlacian**. For example, the LaPlacian of the velocity field
- **Material derivative**. For example, the time derivative of the temperature field

Each operator has a physical interpretation, and in the next section we show how to develop a physical interpretation by going through the derivation of the partial differential equation. For a thorough introduction to operators, we recommend Schey's book (6).

Summary. A variety of mathematical notations are used. The notations can be classified into invariant and coordinate specific categories. The path that we recommend is to learn each notation (over time) and recognize that the various notations are just different ways to express the same ideas.

The Material Derivative

This subsection introduces an operator called the *material derivative*. This operator has multiple names in the literature, including the (a) substantial derivative, (b) Lagrangian derivative, and (c) derivative following the particle. Whenever you hear any of these labels, recognize that all name the material derivative.

The best way to understand the material derivative is to go through the steps of the derivation, which we do next. We select temperature for the derivation because temperature is easy to visualize.

The purpose of the derivation is to develop an expression for the time rate of change of the temperature of a fluid particle.

Step 1: **Select a fluid particle**. Visualize a fluid particle in a flow that has temperature variations (Fig. 16.11). Notice that as the given fluid particle moves, its temperature will rise because it is being transported from a cooler region to a warmer region.

Step 2: **Apply the definition of the derivative**. The time rate of change of temperature T of the fluid particle is given by the ordinary derivative

$$\frac{dT}{dt} = \lim_{xt \to 0} \frac{T(t + \Delta t) - T(t)}{\Delta t} \tag{16.11}$$

As shown in Fig. 16.12, at time t, the particle is at location (x, y, z).

Thus, the particle's temperature at time t is given by $T = T(x, y, z, t)$. Similarly, the particle's temperature at time $t + \Delta t$ is $T = T(x + \Delta x, y + \Delta y, z + \Delta z, t + \Delta t)$. Substituting into Eq. (16.11) gives

FIGURE 16.11

A fluid particle moving from a region of cooler temperatures to a region of higher temperatures.

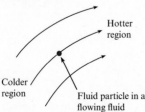

Hotter region

Colder region

Fluid particle in a flowing fluid

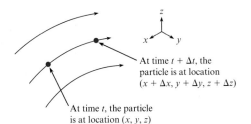

FIGURE 16.12

As a fluid particle is transported by a flowing fluid, its location changes, as shown here.

$$\frac{dT}{dt} = \lim_{\Delta t \to 0} \frac{T(x + \Delta x, y + \Delta y, z + \Delta z, t + \Delta t) - T(x, y, z, t)}{\Delta t} \tag{6.12}$$

Step 3: **Apply Taylor's series.** Expand the numerator in a Taylor's series and neglect higher-order terms:

$$\frac{dT}{dt} = \lim_{\Delta t \to 0} \frac{\left(\frac{\partial T}{\partial x}\right)\frac{\Delta x}{1!} + \left(\frac{\partial T}{\partial y}\right)\frac{\Delta y}{1!} + \left(\frac{\partial T}{\partial z}\right)\frac{\Delta z}{1!} + \left(\frac{\partial T}{\partial t}\right)\frac{\Delta t}{1!}}{\Delta t} \tag{16.13}$$

$$= \lim_{\Delta t \to 0} \left(\frac{\partial T}{\partial x}\right)\frac{\Delta x}{\Delta t} + \left(\frac{\partial T}{\partial y}\right)\frac{\Delta y}{\Delta t} + \left(\frac{\partial T}{\partial z}\right)\frac{\Delta z}{\Delta t} + \left(\frac{\partial T}{\partial t}\right)\frac{\Delta t}{\Delta t}$$

Step 4: **Apply the definition of speed.**

$$u = \lim_{\Delta t \to 0} \frac{\Delta x}{\Delta t}, \quad v = \lim_{\Delta t \to 0} \frac{\Delta y}{\Delta t}, \quad w = \lim_{\Delta t \to 0} \frac{\Delta z}{\Delta t} \tag{16.14}$$

Step 5: **Combine equations.** Insert Eq. (16.14) into Eq. (16.13) to give the final result:

$$\frac{dT}{dt} = \left(\frac{\partial T}{\partial t}\right) + u\left(\frac{\partial T}{\partial x}\right) + v\left(\frac{\partial T}{\partial y}\right) + w\left(\frac{\partial T}{\partial z}\right) \tag{16.15}$$

Step 6: **Interpret the result.** The left side of the equation is the desired result: *the time derivative of a property of a particle.* The right side describes the *mathematical mechanics for doing the derivative when a field is used.* That is, this equation describes how to do the math to obtain a time derivative when a Eulerian description is being used. To summarize:

$$\underbrace{\frac{dT}{dt}}_{\substack{\text{time derivative of} \\ \text{the temperature of} \\ \text{a fluid particle}}} = \underbrace{\left(\frac{\partial T}{\partial t}\right) + u\left(\frac{\partial T}{\partial x}\right) + v\left(\frac{\partial T}{\partial y}\right) + w\left(\frac{\partial T}{\partial z}\right)}_{\substack{\text{mathematics needed to do the derivative when} \\ \text{a field (i.e., a Eulerian approach) is being used}}} \tag{16.16}$$

Step 7: **Generalize the results.** Eq. (16.16) was derived for a specific scalar field (i.e., the temperature field). However, it could have been derived for any scalar field. Thus, let J represent a generic scalar field.

Similarly, Eq. (16.16) could have been derived using any coordinate system. Thus, one can replace the spatial derivatives with an invariant notation. The generalization of Eq. (16.16) is

$$\underbrace{\frac{dJ}{dt}}_{\substack{\text{time derivative} \\ \text{of the property } J \\ \text{of a fluid particle}}} = \underbrace{\left(\frac{\partial J}{\partial t}\right) + \mathbf{V} \cdot \nabla J = \left(\frac{\partial J}{\partial t}\right) + \mathbf{V} \cdot \text{grad}(J)}_{\substack{\text{mathematics needed to do the derivative when} \\ \text{a field (i.e., a Eulerian approach) is being used}}} \tag{16.17}$$

Note that many engineering references write the material derivative using capital letters:

$$\begin{pmatrix} \text{material derivative is} \\ \text{represented using} \end{pmatrix} \Rightarrow \frac{DJ}{Dt} \tag{16.18}$$

In cylindrical coordinates, the material derivative is

$$\frac{dJ}{dt} = \frac{\partial J}{\partial t} + v_r \frac{\partial J}{\partial r} + \frac{v_\theta}{r} \frac{\partial J}{\partial \theta} + v_z \frac{\partial J}{\partial z} \tag{16.19}$$

Summary. The material derivative represents the time rate of change of a property of a fluid particle. As shown in Eq. (16.17), the partial derivative terms (right side) described the mechanics needed to find the derivative. The material derivative can be written in Cartesian coordinates (16.16), cylindrical coordinates (16.19), or in an invariant notation (16.17).

The Acceleration Field

The acceleration field describes the acceleration of each fluid particle:

$$\begin{pmatrix} \text{acceleration at a} \\ \text{point in a field} \end{pmatrix} = \begin{pmatrix} \text{acceleration of} \\ \text{the fluid particle} \\ \text{at this location} \end{pmatrix} = \begin{pmatrix} \text{material derivative} \\ \text{of the} \\ \text{velocity field} \end{pmatrix} \tag{16.20}$$

Therefore, introduce the material derivative to describe the acceleration field:

$$\mathbf{a} = \frac{d\mathbf{V}}{dt} \tag{16.21}$$

To represent Eq. (16.21) in Cartesian coordinates, insert the velocity field from Eq. (16.1):

$$\mathbf{a} = \frac{d}{dt}(u\mathbf{i} + v\mathbf{j} + w\mathbf{k}) = \frac{du}{dt}\mathbf{i} + \frac{dv}{dt}\mathbf{j} + \frac{dw}{dt}\mathbf{k} \tag{16.22}$$

Because du/dt is the material derivative of a scalar field, this term can be evaluated using Eq. (16.16). When this is done for each term on the right side of Eq. (16.22), the acceleration in Cartesian coordinates is given by

$$\begin{pmatrix} \text{acceleration} \\ \text{of a fluid particle} \end{pmatrix} = \mathbf{a} = \frac{d\mathbf{V}}{dt} = \begin{bmatrix} \left\{ \left(\dfrac{\partial u}{\partial t}\right) + u\left(\dfrac{\partial u}{\partial x}\right) + v\left(\dfrac{\partial u}{\partial y}\right) + w\left(\dfrac{\partial u}{\partial z}\right) \right\}\mathbf{i} \\ \left\{ \left(\dfrac{\partial v}{\partial t}\right) + u\left(\dfrac{\partial v}{\partial x}\right) + v\left(\dfrac{\partial v}{\partial y}\right) + w\left(\dfrac{\partial v}{\partial z}\right) \right\}\mathbf{j} \\ \left\{ \left(\dfrac{\partial w}{\partial t}\right) + u\left(\dfrac{\partial w}{\partial x}\right) + v\left(\dfrac{\partial w}{\partial y}\right) + w\left(\dfrac{\partial w}{\partial z}\right) \right\}\mathbf{k} \end{bmatrix} \tag{16.23}$$

When acceleration is derived in cylindrical coordinates, the result is

$$\mathbf{a} = \begin{bmatrix} a_r \\ a_\theta \\ a_z \end{bmatrix} = \begin{bmatrix} \dfrac{dv_r}{dt} - \dfrac{v_\theta^2}{r} \\ \dfrac{dv_\theta}{dt} + \dfrac{v_r v_\theta}{r} \\ \dfrac{dv_z}{dt} \end{bmatrix} = \begin{bmatrix} \dfrac{\partial v_r}{\partial t} + v_r \dfrac{\partial v_r}{\partial r} + \dfrac{v_\theta}{r} \dfrac{\partial v_r}{\partial \theta} + v_z \dfrac{\partial v_r}{\partial z} - \dfrac{v_\theta^2}{r} \\ \dfrac{\partial v_\theta}{\partial t} + v_r \dfrac{\partial v_\theta}{\partial r} + \dfrac{v_\theta}{r} \dfrac{\partial v_\theta}{\partial \theta} + v_z \dfrac{\partial v_\theta}{\partial z} + \dfrac{v_r v_\theta}{r} \\ \dfrac{\partial v_z}{\partial t} + v_r \dfrac{\partial v_z}{\partial r} + \dfrac{v_\theta}{r} \dfrac{\partial v_z}{\partial \theta} + v_z \dfrac{\partial v_z}{\partial z} \end{bmatrix} \tag{16.24}$$

Summary. The acceleration field is given by the material derivative of the velocity field. This can be written in Cartesian coordinates [Eq. (16.23)] or cylindrical coordinates [Eq. (16.24)].

16.3 The Continuity Equation

The continuity equation, according to Frank White (7), is one of five partial differential equations that are needed to model a flowing fluid. The set of five equations is as follows:

- **The continuity equation**. This is the law of conservation of mass applied to a fluid and expressed as a partial differential equation.
- **The momentum equation**. This is Newton's Second Law of Motion applied to fluid. This equation is mostly commonly developed for a Newtonian fluid, and the equation is called the Navier-Stokes equation.
- **The energy equation**. This is the law of conservation of energy applied to a fluid.
- **Equations of state (two equations)**. An equation of state describes how thermodynamic variables are related. For example, an equation of state for density describes how density varies with temperature and pressure.

The continuity equation is described in this section; the Navier-Stokes equation is described in the next section. The other three equations are described in the books by White (7, 8).

In practice, there are multiple ways to write the continuity equation as a partial differential equation. This can be quite confusing when learning. Thus, the main purpose of this section is to introduce the following:

- Various forms of the continuity equation
- Language and ideas for understanding how and why engineers use these different forms

Derivation Using a Control Volume (Conservation Form)

This section introduces one of the ways to derive the continuity equation.

Step 1: **Select a control volume (CV)**. Select a CV (Fig. 16.13) centered around the point (x, y, z). Assume that the CV is stationary and nondeforming.

Let the CV have dimensions $(\Delta x, \Delta y, \Delta z)$, where each dimension is infinitesimal in size. *Infinitesimal* means that dimensions are approaching zero in the sense of the limit in calculus (e.g., limit $\Delta x \to 0$).

Step 2: **Apply conservation of mass**. Apply conservation of mass to the CV. The physics are

$$\text{(rate of accumulation of mass)} + \text{(net outflow of mass)} = \text{(zero)} \qquad \textbf{(16.25)}$$

These physics can represented by this equation:

$$\frac{dm_{cv}}{dt} + \dot{m}_{net} = 0 \qquad \textbf{(16.26)}$$

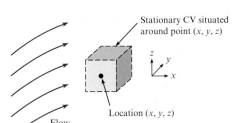
Stationary CV situated around point (x, y, z)

Location (x, y, z)

Flow

FIGURE 16.13

A stationary, nondeforming, infinitesimal CV that is situated at point (x, y, z) in a moving fluid.

FIGURE 16.14

Inflow and outflow of mass through the x faces of the control volume.

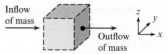

Step 3: **Analyze the accumulation**. The accumulation term is

$$\frac{dm_{cv}}{dt} = \frac{\partial(\text{mass in cv})}{\partial t} = \frac{\partial(\rho \mathcal{V})}{\partial t} = \left(\frac{\partial \rho}{\partial t}\right)\mathcal{V} = \left(\frac{\partial \rho}{\partial t}\right)\Delta x \Delta y \Delta z \qquad \text{(16.27)}$$

Eq. (16.27) uses a partial derivative because the control volume is fixed in place, which means that the variables x, y, and z have fixed values. The volume term was pulled out of the derivative because the volume of the CV is constant with time.

Step 4: **Analyze the outflow**. To analyze $\dot{m}_{net}$, consider flow through the x faces (Fig. 16.14) of the CV. An x face is defined as a face of the cube that is perpendicular to the x axis. As shown, there is outflow through the positive x face and inflow through the negative x face.

The mass flow rates through the x faces are

$$\dot{m}_{\substack{\text{positive} \\ x\text{ face}}} = (\rho A u)_{x + \Delta x/2} = (\rho u)_{x + \Delta x/2}(\Delta y \Delta z)$$

$$\dot{m}_{\substack{\text{negative} \\ x\text{ face}}} = (\rho A u)_{x - \Delta x/2} = (\rho u)_{x - \Delta x/2}(\Delta y \Delta z) \qquad \text{(16.28)}$$

The net flow rate through the x faces is

$$\dot{m}_{net} = \dot{m}_{\substack{\text{positive} \\ x\text{ face}}} - \dot{m}_{\substack{\text{negative} \\ x\text{ face}}}$$

$$= ((\rho u)_{x + \Delta x/2} - ((\rho u)_{x - \Delta x/2}))(\Delta y \Delta z) \qquad \text{(16.29)}$$

Simplify Eq. (16.29) by expanding the derivatives in a Taylor series to give

$$\dot{m}_{\substack{net \\ x\text{ face}}} = \frac{\partial(\rho u)}{\partial x}(\Delta x \Delta y \Delta z) \qquad \text{(16.30)}$$

Repeat the process used to derive Eq. (16.30) for the y face to give

$$\dot{m}_{\substack{net \\ y\text{ face}}} = \frac{\partial(\rho v)}{\partial y}(\Delta x \Delta y \Delta z) \qquad \text{(16.31)}$$

Repeat the process used to derive Eq. (16.30) for the z face to give

$$\dot{m}_{\substack{net \\ z\text{ face}}} = \frac{\partial(\rho w)}{\partial z}(\Delta x \Delta y \Delta z) \qquad \text{(16.32)}$$

To sum the mass flow rates through all faces, add up the terms in Eqs. (16.30) to (16.32):

$$\dot{m}_{net} = \left(\frac{\partial(\rho u)}{\partial x} + \frac{\partial(\rho v)}{\partial y} + \frac{\partial(\rho w)}{\partial z}\right)(\Delta x \Delta y \Delta z) \qquad \text{(16.33)}$$

Step 5: **Combine results**. Insert terms from Eqs. (16.27) and (16.33) into Eq. (16.26):

$$\left(\frac{\partial \rho}{\partial t}\right)(\Delta x \Delta y \Delta z) + \left(\frac{\partial(\rho u)}{\partial x} + \frac{\partial(\rho v)}{\partial y} + \frac{\partial(\rho w)}{\partial z}\right)(\Delta x \Delta y \Delta z) = 0 \qquad \text{(16.34)}$$

Divide through by the volume of the CV to give the final result:

$$\frac{\partial \rho}{\partial t} + \frac{\partial(\rho u)}{\partial x} + \frac{\partial(\rho v)}{\partial y} + \frac{\partial(\rho w)}{\partial z} = 0 \qquad \text{(16.35)}$$

Step 6: **Interpret the physics.** The meaning of Eq. (16.35) is

$$\underbrace{\frac{\partial \rho}{\partial t}}_{\substack{\text{rate of accumulation of mass} \\ \text{in a differential CV divided by} \\ \text{the volume of the CV} \\ \text{(kg/s per m}^3)}} + \underbrace{\frac{\partial(\rho u)}{\partial x} + \frac{\partial(\partial v)}{\partial y} + \frac{\partial(\rho w)}{\partial z}}_{\substack{\text{net rate of mass flow} \\ \text{out of the CV divided by the} \\ \text{volume of the CV} \\ \text{(kg/s per m}^3)}} = 0 \qquad (16.36)$$

Note the dimensions and units of the terms that appear in the continuity equation:

$$\frac{(\text{mass/time})}{(\text{volume})} = \frac{\text{kg/s}}{\text{m}^3} \qquad (16.37)$$

Derivation Using a Fluid Particle (Nonconservation Form)

The literature uses two forms of the continuity equation:

- **Conservation form**. The conservation form, developed in the last subsection, is derived by starting with a differential control volume and applying conservation of mass to this CV. This is a Eulerian approach.

- **Nonconservation form**. The nonconservation form, developed in this subsection, is derived by starting with a differential fluid particle and applying conservation of mass to this particle. This is a Lagrangian approach.

Next, we derive the *nonconservation form* of the continuity equation.

Step 1: **Select a fluid particle.** Select a fluid particle (Fig. 16.15) centered around a point (x, y, z) in space. Because a particle moves with a flowing fluid, this particle is at this location only at a specific instant in time.

Step 2: **Apply conservation of mass.** By definition, the *mass of the particle must stay constant with time*. To say this mathematically:

$$\frac{d(\text{mass})}{dt} = \frac{d[(\text{density})(\text{volume})]}{dt} = \frac{d(\rho \forall)}{dt} = 0 \qquad (16.38)$$

Step 3: **Apply the product rule.** Eq. (16.38) becomes

$$\rho \frac{d\forall}{dt} + \forall \frac{d\rho}{dt} = 0 \qquad (16.39)$$

Step 4: **Analyze the change in volume term.** The change in volume term describes how the volume of the fluid particle changes with time. To analyze this term, apply the definition of the derivative:

$$\frac{d\forall}{dt} = \lim_{\Delta t \to 0} \frac{\forall(t + \Delta t) - \forall(t)}{\Delta t} \qquad (16.40)$$

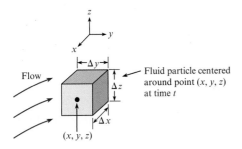

FIGURE 16.15

A fluid particle (infinitesimal in size) located at the point (x, y, z) in a flowing fluid.

In Eq. (16.40), the volume at time t is

$$\Psi(t) = \Delta x \Delta y \Delta z \tag{16.41}$$

and the volume at time $t + \Delta t$ is

$$\Psi(t + \Delta t) = (\Delta x + \Delta x')(\Delta y + \Delta y')(\Delta z + \Delta z') \tag{16.42}$$

where each term of the form $\Delta x'$ represents a change in the length of the side of the particle. Next, multiply out the terms on the right side of Eq. (16.42) and neglect higher-order terms. The equation becomes

$$\Psi(t + \Delta t) \approx \Delta x \Delta y \Delta z + \Delta x' \Delta y \Delta z + \Delta x \Delta y' \Delta z + \Delta x \Delta y \Delta z' \tag{16.43}$$

Next, combine Eqs. (16.41) and (16.43) and apply Taylor's series:

$$\Psi(t + \Delta t) - \Psi(t) \approx \left(\frac{\partial u}{\partial x}\Delta x\right)\Delta y \Delta z \Delta t + \Delta x\left(\frac{\partial v}{\partial y}\Delta y\right)\Delta z \Delta t + \Delta x \Delta y\left(\frac{\partial w}{\partial z}\Delta z\right)\Delta t \tag{16.44}$$

Then, substitute Eq. (16.44) into (16.40) to give

$$\frac{d\Psi}{dt} = \left(\frac{\partial u}{\partial x} + \frac{\partial v}{\partial y} + \frac{\partial w}{\partial z}\right)\Delta x \Delta y \Delta z = \left(\frac{\partial u}{\partial x} + \frac{\partial v}{\partial y} + \frac{\partial w}{\partial z}\right)\Psi \tag{16.45}$$

Step 5: **Combine results.** Substitute Eq. (16.45) into Eq. (16.39), divide each term by the volume of the particle, and rearrange to give

$$\frac{d\rho}{dt} + \rho\left(\frac{\partial u}{\partial x} + \frac{\partial v}{\partial y} + \frac{\partial w}{\partial z}\right) = 0 \tag{16.46}$$

Step 6: **Interpret the physics.** The derivation of Eq. (16.46) reveals two main ideas:

• A change in the density of a fluid particle occurs if, and only if, the volume of the fluid particle is changing with time.

• The volume change of a fluid particle is represented mathematically by the bracketed variables in the second term of Eq. (16.46).

Note that the conservation form [Eq. (16.35)] and the nonconservation form [Eq. (16.46)] are equivalent mathematically because it is possible to start with one of these equations and derive the other.

Summary. Derivation of the conservation and the nonconservation forms of the continuity equation gives equations that are equivalent mathematically. However, these equations have different physical interpretations.

Cylindrical Coordinates

The continuity equation can also be derived in cylindrical coordinates; see Pritchard (9). The result (conservation form) is

$$\underbrace{\frac{\partial \rho}{\partial t}}_{\substack{\text{rate of accumulation of mass}\\ \text{in a differential CV divided by}\\ \text{the volume of the CV}\\ \text{(kg/s per m}^3\text{)}}} + \underbrace{\frac{1}{r}\frac{\partial(r\rho v_r)}{\partial r} + \frac{1}{r}\frac{\partial(\rho v_\theta)}{\partial \theta} + \frac{\partial(\rho v_z)}{\partial z} = 0}_{\substack{\text{net rate of mass flow}\\ \text{out of the CV divided by the}\\ \text{volume of the CV}\\ \text{(kg/s per m}^3\text{)}}} \tag{16.47}$$

One can also derive the continuity equation in spherical coordinates and in other coordinate systems.

Invariant Notation

This subsection shows how to modify the continuity equation to an invariant form. The "del" operator is defined as

$$\nabla \equiv \mathbf{i}\frac{\partial}{\partial x} + \mathbf{j}\frac{\partial}{\partial y} + \mathbf{k}\frac{\partial}{\partial z} \tag{16.48}$$

Thus, start with the continuity equation in Cartesian components and introduce the del operator using the dot product:

$$\frac{\partial \rho}{\partial t} + \frac{\partial(\rho u)}{\partial x} + \frac{\partial(\rho v)}{\partial y} + \frac{\partial(\rho w)}{\partial z} = 0$$

$$\frac{\partial \rho}{\partial t} + \left(\mathbf{i}\frac{\partial}{\partial x} + \mathbf{j}\frac{\partial}{\partial y} + \mathbf{k}\frac{\partial}{\partial z}\right) \cdot ((\rho u)\mathbf{i} + (\rho v)\mathbf{j} + (\rho w)\mathbf{k}) = 0 \tag{16.49}$$

$$\frac{\partial \rho}{\partial t} + \nabla \cdot (\rho \mathbf{V}) = 0$$

The last line in Eq. (16.49) is the continuity equation in an invariant form. The physics are

$$\underbrace{\frac{\partial \rho}{\partial t}}_{\text{accumulation}} + \underbrace{\nabla \cdot (\rho \mathbf{V})}_{\text{net outflow of mass}} = 0 \tag{16.50}$$

The term $\nabla \cdot (\rho \mathbf{V})$ is the divergence. Eq. (16.50) can also be written with the Gibbs notation:

$$\underbrace{\frac{\partial \rho}{\partial t}}_{\text{accumulation}} + \underbrace{\text{div}\,(\rho \mathbf{V})}_{\text{net outflow of mass}} = 0 \tag{16.51}$$

A useful aspect of invariant notion is that it provides a way to describe the physics of a math operator. For example, the physics of the divergence operator can be established from Eq. (16.50):

$$\text{div}\,(\rho \mathbf{V}) = \nabla \cdot (\rho \mathbf{V}) = \frac{\begin{pmatrix} \text{net rate of outflow of mass} \\ \text{from a differential CV centered} \\ \text{about point } (x, y, z) \end{pmatrix}}{(\text{volume of the CV})} \tag{16.52}$$

The physics of the divergence operator can also be found another way. Step 1 is to write Eq. (16.46) in this form:

$$\frac{d\rho}{dt} + \rho(\nabla \cdot \mathbf{V}) = 0$$

$$\frac{d\rho}{dt} + \rho\,\text{div}(\mathbf{V}) = 0 \tag{16.53}$$

Step 2 is to go back through the derivation of Eq. (16.46). This will reveal that

$$\nabla \cdot \mathbf{V} = \text{div}(\mathbf{V}) = \frac{(\text{time rate of change of the volume of a fluid particle})}{(\text{volume of the fluid particle})} \tag{16.54}$$

Summary. The continuity equation can be written in an invariant form. This approach provides a method for developing a physical interpretation of the divergence operator. As shown, the divergence operator has two different physical interpretations.

Continuity for Incompressible (Constant Density) Flow

Because it is common to assume a constant value of density, the continuity equation is often written for the case of constant density. This is usually called incompressible flow.

When density is constant, the continuity equation written in Cartesian coordinates [Eq. (16.36) or Eq. (16.46)] reduces to

$$\frac{\partial u}{\partial x} + \frac{\partial v}{\partial y} + \frac{\partial w}{\partial z} = 0 \tag{16.55}$$

Similarly, the continuity equation for cylindrical coordinates [(Eq. (16.47)] reduces to

$$\frac{1}{r}\frac{\partial(rv_r)}{\partial r} + \frac{1}{r}\frac{\partial v_\theta}{\partial \theta} + \frac{\partial v_z}{\partial z} = 0 \tag{16.56}$$

When density is constant, Eq. (16.51) reduces to

$$\nabla \cdot \mathbf{V} = \text{div}(\mathbf{V}) = 0 \tag{16.57}$$

Summary. When flow is modeled as incompressible, the continuity equation reduces to $\nabla \cdot \mathbf{V} = \text{div}(\mathbf{V}) = 0$, which means that the divergence of the velocity field is zero. This equation can also be written in Cartesian coordinates [Eq. (16.55)] and cylindrical coordinates [Eq. (16.56)].

Summary of the Mathematical Forms of the Continuity Equation

Table 16.2 lists some of the ways to write the continuity equation as a PDE. Recognize that the math simply reflects alternative ways to describe the physics.

As shown in the next example, the continuity equation can be applied in two steps.

• **Step 1: Selection**. From Table 16.2, select an applicable form of the continuity equation.

• **Step 2: Reduction**. Eliminate the variables in the continuity equation that are equal to zero or are negligible.

EXAMPLE. Consider developing laminar flow in a round pipe (Fig. 16.16). At the entrance to the pipe, the velocity profile is uniform. As the flow proceeds down the pipe, the velocity profile becomes fully developed. Assume the flow is steady and constant density. Reduce the general equation for the continuity equation so that it applies to this situation.

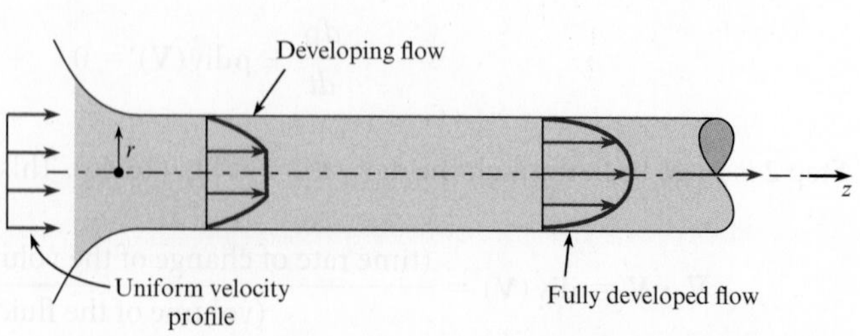

FIGURE 16.16

Developing laminar flow in a round pipe.

TABLE 16.2 Alternative Ways to Write the Continuity Equation as a PDE

	Description	Equation
General equation	Cartesian coordinates (conservation form)	$\dfrac{\partial \rho}{\partial t} + \dfrac{\partial (\rho u)}{\partial x} + \dfrac{\partial (\rho v)}{\partial y} + \dfrac{\partial (\rho w)}{\partial z} = 0$
	Cartesian coordinates (nonconservation form)	$\dfrac{d\rho}{dt} + \rho\left(\dfrac{\partial u}{\partial x} + \dfrac{\partial v}{\partial y} + \dfrac{\partial w}{\partial z}\right) = 0$ $\dfrac{\partial \rho}{\partial t} + \left(u\dfrac{\partial \rho}{\partial x} + v\dfrac{\partial \rho}{\partial y} + w\dfrac{\partial \rho}{\partial z}\right) + \rho\left(\dfrac{\partial u}{\partial x} + \dfrac{\partial v}{\partial y} + \dfrac{\partial w}{\partial z}\right) = 0$
	Cylindrical coordinates (conservation form)	$\dfrac{\partial \rho}{\partial t} + \dfrac{1}{r}\dfrac{\partial (r\rho v_r)}{\partial r} + \dfrac{1}{r}\dfrac{\partial (\rho v_\theta)}{\partial \theta} + \dfrac{\partial (\rho v_z)}{\partial z} = 0$
	Invariant (conservation form)	$\dfrac{\partial \rho}{\partial t} + \nabla \cdot (\rho \mathbf{V}) = 0$
	Invariant (nonconservation form)	$\dfrac{d\rho}{dt} + \rho(\nabla \cdot \mathbf{V}) = 0$
Incompressible flow equation	Invariant form	$\nabla \cdot \mathbf{V} = \operatorname{div}(\mathbf{V}) = 0$
	Cartesian coordinates	$\dfrac{\partial u}{\partial x} + \dfrac{\partial v}{\partial y} + \dfrac{\partial w}{\partial z} = 0$
	Cylindrical coordinates	$\dfrac{1}{r}\dfrac{\partial (rv_r)}{\partial r} + \dfrac{1}{r}\dfrac{\partial v_\theta}{\partial \theta} + \dfrac{\partial v_z}{\partial z} = 0$

Action

Step 1: Selection. Because the flow is constant density and the geometry is a round pipe, select the incompressible flow form of continuity in cylindrical coordinates, Eq. (16.56):

$$\frac{1}{r}\frac{\partial (rv_r)}{\partial r} + \frac{1}{r}\frac{\partial v_\theta}{\partial \theta} + \frac{\partial v_z}{\partial z} = 0$$

Step 2: Reduction. Assume that the flow is symmetric about the z axis. Thus, let $v_\theta = 0$. The continuity equation reduces to

$$\frac{1}{r}\frac{\partial (rv_r)}{\partial r} + \frac{\partial v_z}{\partial z} = 0$$

Review. One could solve this equation to give the velocity field for developing flow in a round pipe. Because this equation has two unknown variables (vr and v_z), one would also need to solve the Navier-Stokes equation.

16.4 The Navier-Stokes Equation

The Navier-Stokes equation, introduced in this section, is widely used in both theory and in application.

The Navier-Stokes equation represents Newton's Second Law of Motion as applied to viscous flow of a Newtonian fluid. In this text, we assume incompressible flow and constant viscosity. In the literature, one can find more general derivations.

FIGURE 16.17

A fluid particle situated in a flowing fluid.

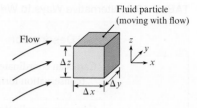

Derivation

Similar to the continuity equations, there are multiple ways to derive the Navier-Stokes equation. This section shows how to derive the equation by starting with a fluid particle and applying Newton's second law. Thus, the result will be the nonconservation form of the equation. Because the derivation is complex, we omit some of the technical details; to access these details, we recommend the text *Viscous Fluid Flow* (8).

Step 1: **Select a fluid particle**. Select a fluid particle in a flowing fluid (Fig. 16.17). Let the particle have the shape of a cube. Assume the dimensions are infinitesimal and that the particle is at the position (x, y, z) at the instant shown.

Step 2: **Apply Newton's second law**.

$$\text{(sum of forces on a fluid particle)} = \text{(mass)(acceleration)}$$

$$\sum \mathbf{F} = m\mathbf{a} = m\frac{d\mathbf{V}}{dt} \tag{16.58}$$

Regarding the forces, the two categories are body forces and surfaces forces. The only possible surface forces are the pressure force and the shear force. Assume that the only body force is the weight $\mathbf{W}$. Eq. (16.58) becomes

$$\text{(weight)} + \text{(pressure force)} + \text{(shear force)} = \text{(density)(volume)(acceleration)}$$

$$\mathbf{W} + \mathbf{F}_p + \mathbf{F}_s = \rho \forall \frac{d\mathbf{V}}{dt} \tag{16.59}$$

The weight is given by

$$\mathbf{W} = \text{(mass)(gravity vector)} = \rho \forall \mathbf{g} \tag{16.60}$$

Insert Eq. (16.60) into Eq. (16.59) to give

$$\rho \forall \mathbf{g} + \mathbf{F}_p + \mathbf{F}_s = \rho \forall \frac{d\mathbf{V}}{dt} \tag{16.61}$$

Step 3: **Analyze the pressure force**. To begin, consider the forces on the x faces of the particle (Fig. 16.18).

The net force due to pressure on the x faces is

$$\mathbf{F}_{\text{pressure}}_{\substack{x \text{ faces}}} = -\left((pA)_{x+\Delta x/2} - (pA)_{x-\Delta x/2}\right)\mathbf{i} = -\left(p_{x+\Delta x/2} - p_{x-\Delta x/2}\right)(\Delta y \Delta z)\mathbf{i} \tag{16.62}$$

FIGURE 16.18

The pressure forces on the x faces of a fluid particle.

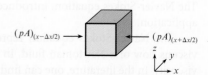

Simplify Eq. (16.62) by applying a Taylor series expansion (twice) and neglecting higher-order terms to give

$$\mathbf{F}_{\substack{\text{pressure} \\ x\text{ faces}}} = \frac{\partial p}{\partial x}(\Delta x \Delta y \Delta z)\mathbf{i} \tag{16.63}$$

Repeat this process for the y and z faces, and combine results to give

$$\mathbf{F}_{\substack{\text{pressure} \\ \text{all faces}}} = -\left(\frac{\partial p}{\partial x}(\Delta x \Delta y \Delta z)\mathbf{i} + \frac{\partial p}{\partial y}(\Delta x \Delta y \Delta z)\mathbf{j} + \frac{\partial p}{\partial z}(\Delta x \Delta y \Delta z)\mathbf{k}\right) \tag{16.64}$$

Simplify Eq. (16.64), and then introduce vector notation to give

$$\mathbf{F}_{\text{pressure}} = -\left(\frac{\partial p}{\partial x}\mathbf{i} + \frac{\partial p}{\partial y}\mathbf{j} + \frac{\partial p}{\partial z}\mathbf{k}\right)(\Delta x \Delta y \Delta z) = -\nabla p(\Delta x \Delta y \Delta z) \tag{16.65}$$

Eq. (16.65) reveals a physical interpretation of the gradient:

$$\begin{pmatrix} \text{gradient of the} \\ \text{pressure field} \\ \text{at a point} \end{pmatrix} = \frac{\begin{pmatrix} \text{net pressure force} \\ \text{on a fluid particle} \end{pmatrix}}{(\text{volume of the particle})} \tag{16.66}$$

Step 4: **Analyze the shear force**. The shear force is the net force on the fluid particle due to shear stresses. Shear stress is caused by viscous effects and is represented mathematically as shown in Fig. 16.19. This figure shows that each face of the fluid particle has three stress components. For example, the positive x face has three stress components, which are τ_{xx}, τ_{xy}, and τ_{xz}. The double subscript notation describes the direction of the stress component and the face on which the component acts. For example:

- τ_{xx} is the shear stress on the x face in the x direction.
- τ_{xy} is the shear stress on the x face in the y direction.
- τ_{xz} is the shear stress on the x face in the z direction.

Shear stress is a type of mathematical entity called a second-order tensor. A *tensor* is analogous to but more general than a vector. Examples: A zeroth-order tensor is a scalar. A first-order tensor is a vector. A second-order tensor has magnitude, direction, and orientation (where orientation describes which face the stress acts on).

To find the net shear force on the particle, each stress component is multiplied by area, and the forces are added. Then, a Taylor series expansion is applied. The result is that

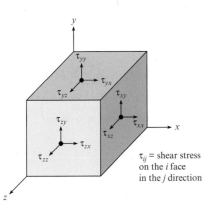

FIGURE 16.19

Shear stresses that act on a fluid particle.

τ_{ij} = shear stress on the i face in the j direction

$$\mathbf{F}_{\text{shear}} = \begin{bmatrix} F_{x,\text{shear}} \\ F_{y,\text{shear}} \\ F_{z,\text{shear}} \end{bmatrix} = \begin{bmatrix} \left(\dfrac{\partial \tau_{xx}}{\partial x} + \dfrac{\partial \tau_{xy}}{\partial x} + \dfrac{\partial \tau_{xz}}{\partial x} \right) \\ \left(\dfrac{\partial \tau_{yx}}{\partial y} + \dfrac{\partial \tau_{yy}}{\partial y} + \dfrac{\partial \tau_{yz}}{\partial y} \right) \\ \left(\dfrac{\partial \tau_{zx}}{\partial z} + \dfrac{\partial \tau_{zy}}{\partial z} + \dfrac{\partial \tau_{zz}}{\partial z} \right) \end{bmatrix} (\Delta x \Delta y \Delta z) \tag{16.67}$$

Eq. (16.67) can be written in invariant notation as

$$\mathbf{F}_{\text{shear}} = (\nabla \cdot \tau)\mathcal{V} = (\text{div}(\tau))\mathcal{V} \tag{16.68}$$

where the terms on the right side represent the divergence of the stress tensor times the volume of the fluid particle.

Eq. (16.68) reveals the physics of the divergence when it operates on the stress tensor. Note that this is the third physical interpretation of the divergence operator in this chapter. This is because the physics of a mathematical operator depends on the context in which the operator is used:

$$\begin{pmatrix} \text{divergence of} \\ \text{the stress tensor} \end{pmatrix} = \dfrac{\begin{pmatrix} \text{net shear force} \\ \text{on a fluid particle} \end{pmatrix}}{(\text{volume of the particle})} \tag{16.69}$$

Step 6: Combine terms. Substitute the shear force, Eq. (16.68), and the pressure force, Eq. (16.65), into Newton's Second Law of Motion, Eq. (16.61). Then, divide by the volume of the fluid particle to give

$$\rho \frac{d\mathbf{V}}{dt} = \rho \mathbf{g} - \nabla p + \nabla \cdot \tau_{ij} \tag{16.70}$$

Eq. (16.70) is the differential form of the linear momentum equation without any assumption about the nature of the fluid. The next step involves modifying this equation so that it applies to a Newtonian fluid.

Step 7: Assume a Newtonian fluid. In 1845, Stokes figured out a way to write the stress tensor in terms of the rate-of-strain tensor of the flowing fluid. The details are omitted here. After Stokes's results are introduced, assume constant density and viscosity. Eq. (16.70) becomes

$$\rho \frac{d\mathbf{V}}{dt} = \rho \mathbf{g} - \nabla p + \mu \nabla^2 \mathbf{V} \tag{16.71}$$

Where $\nabla^2 \mathbf{V}$ is a mathematical operator called the Laplacian of the velocity field. Eq. (16.71) is the final result, the Navier-Stokes equation.

Step 8: Interpret the physics. The physics of the Navier-Stokes equation is

$$\underbrace{\rho \frac{d\mathbf{V}}{dt}}_{\substack{\text{mass of the particle times} \\ \text{acceleration of the particle} \\ \text{divided by the volume of the particle}}} = \underbrace{\rho \mathbf{g}}_{\substack{\text{weight of the} \\ \text{particle divided} \\ \text{by its volume}}} + \underbrace{-\nabla p}_{\substack{\text{net pressure force} \\ \text{on the particle} \\ \text{divided by its volume}}} + \underbrace{\mu \nabla^2 \mathbf{V}}_{\substack{\text{net shear force} \\ \text{on the particle} \\ \text{divided by its volume}}} \tag{16.72}$$

Note the dimensions and units:

$$\text{dimensions} = \frac{\text{force}}{\text{volume}} \sim \frac{\text{N}}{\text{m}^3} = \frac{\text{kg}}{\text{m}^2 \cdot \text{s}^2} \tag{16.73}$$

Cartesian and Cylindrical Coordinates

To write Eq. (16.72) in Cartesian coordinates, find a suitable reference (e.g., the Internet, an advanced fluids text, an engineering handbook) and look up the material derivative ($d\mathbf{V}/dt$), the gradient, and the Laplacian operator in Cartesian coordinates. After substitution, the Navier-Stokes equation (constant properties) for Cartesian coordinates is

$$\rho\left(\frac{\partial u}{\partial t} + u\frac{\partial u}{\partial x} + v\frac{\partial u}{\partial y} + w\frac{\partial u}{\partial z}\right) = \rho g_x - \frac{\partial p}{\partial x} + \mu\left(\frac{\partial^2 u}{\partial x^2} + \frac{\partial^2 u}{\partial y^2} + \frac{\partial^2 u}{\partial z^2}\right)$$

$$\rho\left(\frac{\partial v}{\partial t} + u\frac{\partial v}{\partial x} + v\frac{\partial v}{\partial y} + w\frac{\partial v}{\partial z}\right) = \rho g_y - \frac{\partial p}{\partial y} + \mu\left(\frac{\partial^2 v}{\partial x^2} + \frac{\partial^2 v}{\partial y^2} + \frac{\partial^2 v}{\partial z^2}\right) \quad \textbf{(16.74)}$$

$$\rho\left(\frac{\partial w}{\partial t} + u\frac{\partial w}{\partial x} + v\frac{\partial w}{\partial y} + w\frac{\partial w}{\partial z}\right) = \rho g_z - \frac{\partial p}{\partial z} + \mu\left(\frac{\partial^2 w}{\partial x^2} + \frac{\partial^2 w}{\partial y^2} + \frac{\partial^2 w}{\partial z^2}\right)$$

The Navier-Stokes equation cannot be solved in general because of the nonlinear terms. An example of a nonlinear term is

$$u\frac{\partial u}{\partial x}$$

This term is nonlinear because a dependent variable (u) is multiplied by its first derivative ($\partial u/\partial x$). In general, nonlinear terms in differential equations involve functions of the dependent variables.

The Navier-Stokes equation (constant properties) for cylindrical coordinates is

$$r: \rho\left(\frac{\partial v_r}{\partial t} + v_r\frac{\partial v_r}{\partial r} + \frac{v_\theta}{r}\frac{\partial v_r}{\partial \theta} + v_z\frac{\partial v_r}{\partial z} - \frac{v_\theta^2}{r}\right) = \rho g_r - \frac{\partial p}{\partial r} + \mu\left(\frac{1}{r}\frac{\partial}{\partial r}\left(r\frac{\partial v_r}{\partial r}\right) + \frac{1}{r^2}\frac{\partial^2 v_r}{\partial \theta^2} + \frac{\partial^2 v_r}{\partial z^2} - \frac{v_r}{r^2} - \frac{2}{r^2}\frac{\partial v_\theta}{\partial \theta}\right)$$

$$\theta: \rho\left(\frac{\partial v_\theta}{\partial t} + v_r\frac{\partial v_\theta}{\partial r} + \frac{v_\theta}{r}\frac{\partial v_\theta}{\partial \theta} + v_z\frac{\partial v_\theta}{\partial z} + \frac{v_r v_\theta}{r}\right) = \rho g_\theta - \frac{1}{r}\frac{\partial p}{\partial \theta} + \mu\left(\frac{1}{r}\frac{\partial}{\partial r}\left(r\frac{\partial v_\theta}{\partial r}\right) + \frac{1}{r^2}\frac{\partial^2 v_\theta}{\partial \theta^2} + \frac{\partial^2 v_\theta}{\partial z^2} - \frac{v_\theta}{r^2} + \frac{2}{r^2}\frac{\partial v_\theta}{\partial \theta}\right) \quad \textbf{(16.75)}$$

$$z: \rho\left(\frac{\partial v_z}{\partial t} + v_r\frac{\partial v_z}{\partial r} + \frac{v_\theta}{r}\frac{\partial v_z}{\partial \theta} + v_z\frac{\partial v_z}{\partial z}\right) = \rho g_z - \frac{\partial p}{\partial z} + \mu\left(\frac{1}{r}\frac{\partial}{\partial r}\left(r\frac{\partial v_z}{\partial r}\right) + \frac{1}{r^2}\frac{\partial^2 v_z}{\partial \theta^2} + \frac{\partial^2 v_z}{\partial z^2}\right)$$

Summary. The Navier-Stokes equation represents Newton's Second Law of Motion applied to the viscous flow of a Newtonian fluid. The Navier-Stokes equation has nonlinear terms that prevent an exact mathematical solution for most problems.

16.5 Computational Fluid Dynamics (CFD)

Computational fluid dynamics (CFD) is a method for obtaining approximate solutions to problems in fluid mechanics and heat transfer by using numerical solutions of the governing PDEs. This section describes the following:

- Why CFD is valuable
- How CFD is used in practice
- What CDF programs are and how they work

Why CFD Is Valuable

CFD gives engineers a modeling tool that greatly extends their abilities. For example, there is not a straightforward way to develop and solve equations that will predict the pressure field and streamline patterns for flow around a building. One might use an experimental approach, but this has issues such as matching the Reynolds number and the difficulty of performing parametric studies.

Thus, CFD provides a way to simulate physical phenomena that are impossible for analysis and difficult for experiments. CFD is a useful modeling tool in cases such as the following:

- Complex systems (e.g., ink-jet printers, the human heart, mixing tanks)
- Full-scale simulations (e.g., ships, airplanes, dams)
- Environmental effects (e.g., hurricanes, weather, pollution dispersion)
- Hazards (e.g., explosions, radiation dispersion)
- Physics (e.g., planetary boundary layer, stellar evolution)

CFD is also useful for studying the effects of design perturbations. For example, to design a propeller, one could systemically vary design variables such as the blade profile, blade pitch, and rotation speed and see the effect on performance variables such as efficiency, thrust, and power.

CFD is used in many industries and fields of study: aerospace, automotive, biomedical, chemical processing, HVAC, hydraulics, hydrology, marine, oil and gas, and power generation.

Summary. CFD is valuable to the engineer because

- CFD provides a method for modeling complex problems that cannot effectively be modeled with analytical or experimental fluid mechanics,
- CFD provides a way to consider design perturbations on complex problems such as propeller design and the design of spillways, and
- CFD is widely used in industry.

CFD Codes in Professional Practice

A *code* is engineering lingo for a computer program. In professional practice and most research projects, engineers have the following options:

- **Option 1**. Write their own code. This is rarely done.
- **Option 2**. Apply a code that has been developed by others. This is the most common practice because code development requires years of effort.

This subsection describes three commonly used codes and provides suggestions about selecting a code.

Ground Water Modeling. MODFLOW (10) is a computer program for analyzing groundwater flow. This code has been under development since the early 1980s. MODFLOW is considered the de facto standard for simulation of ground flows. The program is well validated and is considered as legally defensible in U.S. courts.

MODFLOW is available in noncommerical (i.e., free) versions. However, the licensing is limited to government and academic entities. For commercial use, implementations of MODFLOW cost from $1,000 to $7,000 USD (10).

Internal Combustion Engine Modeling. The KIVA codes (11, 12) were originally developed in 1985 by Los Alamos National Laboratory to simulate the processes taking place inside an internal combustion engine. The KIVA programs have become the most widely used CFD programs for multidimensional combustion modeling. KIVA can be applied to understand combustion chemistry processes, such as autoignition of fuels, and to optimize diesel engines for high efficiency and low emissions. Hence, KIVA has been used by engine manufacturers to improve the performance of engines.

Modeling of Flows with Free Surfaces. In 1963, Tony Hirt of the Los Alamos National Laboratory pioneered a computational method called the volume of fluid (VOF) approach, which is useful for tracking and locating a free surface or a fluid–fluid interface. Thus, the VOF method is useful for modeling flows such as flow from a reservoir or the flow of metal into a mold. Dr. Hirt left Los Alamos and founded a company called Flow Science that now markets a code called FLOW-3D.

According to the company's website (13), the following are some examples of applications of FLOW-3D:

- Modeling a coffer dam and spillway of a hydroelectric power plant
- Designing a canoe chute for passage around a low head dam
- Modeling the molding of foamed polyurethane resin, which can expand in volume by more than 30 times during molding

The examples of FLOW-3D, KIVA, and MODFLOW suggest some common ideas:

- **CFD programs can be very useful for applications.** The three codes just described provide technologies for modeling (a) groundwater flow, (b) internal combustion engines, and (c) open-channel flows. There are other codes available that allow one to model other applications. Thus, CFD is a powerful technology for modeling problems that involve fluids.

- **Select a CFD code to match your problem.** CFD codes are developed to solve specific types of problems. FLOW-3D is for open channel flow, whereas KIVA is for internal combustion engines, and MODFLOW is for modeling groundwater flow. Thus, make sure that the CFD code is well suited for the type of problem you want to solve.

- **Use an existing code.** Many codes (e.g., MODFLOW, KIVA, and FLOW-3D) have been under development since the 1980s or earlier. Many years of work have gone into these codes. Thus, it is cost-effective to take advantage of this legacy instead of writing a code from scratch.

Features of CFD Programs

This subsection describes the vocabulary and ideas used by most CFD programs.

Approximation of PDEs. CFD codes apply mathematical methods to develop approximate solutions to the governing PDEs. Approximate solutions (estimates) can be close to reality or far away from reality, depending on the details of how the estimate is made. The accuracy of the estimate is determined in part by how the code was developed. However, most of the accuracy is based on decisions made by the user of the code.

There are many ways to develop approximate solutions of partial differential equations. Three common approaches are the *finite difference method*, the *finite element method,* and the *finite volume method.*

When a partial differential equation is approximated, the result is a set of *many* algebraic equations that are solved at points in space. These points in space are defined using a grid.

Grid Generation. A **grid** (Fig. 16.20) is a set of points in space at which a code solves for values of velocity and other variables of interest. The grid is set up by the user. There are two trade-offs:

- **Accuracy**. If the grid is closely spaced, which is called a fine grid, then the solution is generally more accurate. In the grid shown in Fig. 16.20, notice how the user set a fine grid near the wall of the cylinder.

- **Computational time.** If the grid is coarse (grid lines are widely spaced), then the amount of time for the code to run deceases. Decreasing the computer run time is important because CFD codes can take a long time (i.e., days) to run one simulation.

Grid generation capability is set up by the code developers, and the grid generation itself is done by the user. Wyman (15) describes three approaches available for grid generation:

FIGURE 16.20

A grid used to model subsonic flow past a circular cylinder at a Reynolds number of 10,000. From NASA (14).

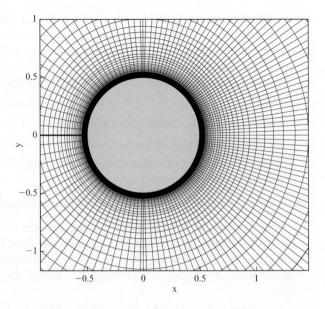

- **Structured grid methods.** With this method, the grid is laid out in a regular repeating pattern called a block (Fig. 16.20). Details (fine grid, coarse grid, etc.) are specified by the user. The advantage of a structured grid is that the user can set up the grid to maximize accuracy while achieving acceptable run time. A drawback of a structured grid is that it can take significant time for the user to input the parameters needed to create the grid. Also, a structured grid requires user expertise for proper layout.

- **Unstructured grid methods.** An unstructured grid is based on a computer algorithm that selects an arbitrary collection of elements to fill the solution domain. Because the elements lack a pattern, the grid is called unstructured. An unstructured grid method is well suited for novices because the grid can be set up easily and quickly and does not require much user expertise. The drawbacks are that the grid may not be good as a structured grid in terms of accuracy and solution time.

- **Hybrid grid methods.** Hybrid grid methods are designed to take advantage of the positive aspects of both structured and unstructured grids. Hybrid grids use a structured grid in local regions while using an unstructured grid in the bulk of the domain.

Time Steps. Because PDEs are being solved by CFD codes, the approximation methods involve solving for variables at specific instances in time. The interval between each solution time is called a **time step**.

Accuracy versus Solution Time. In general, if one selects a fine grid and small time steps, the CFD solution is more accurate. However, fine resolution of space and time drive up the required solution time for the computer. This might seem like a nonissue with today's fast computers, but CFD programs can require days or weeks of solution time. Thus, there is a trade-off between accuracy of a solution and the time that the computer needs for calculations.

Boundary Conditions and Initial Conditions. Solving PDEs, which includes using CFD programs to develop approximate solutions to PDEs, involves the specification of boundary conditions and initial conditions:

- Specifying a **boundary condition** involves *giving numerical values* for the *dependent variables* on the *physical boundaries* that describe that spatial region in which the differential equations are to be solved. For example:

- When flow enters a boundary, the user might specify a known value of velocity at each point. This is known as a *velocity boundary condition*.

- When flow enters a boundary, the user might specify a known value of pressure at each point. This is known as a *pressure boundary condition*.

- Specifying an **initial condition** involves giving numerical values for the *dependent variables* at all spatial points at the starting time of the solutions.

Turbulence (Direct Numerical Simulation). Because most flows of engineering interest involve turbulent flow, CFD codes have methods for analyzing turbulent flow. The most accurate approach, which is called **direct numerical simulation (DNS)**, involves setting the grid and time steps fine enough to resolve the features of the turbulent flow. As a result, Hussan (16) asserts that a DNS solution is very accurate but is also "unrealistic for 99.9% of CFD problems because it is computationally unrealistic." This is because the required computer time is too large for today's computers. Thus, DNS is not used for most problems.

Turbulence Modeling. Turbulence modeling involves the prediction of effects of turbulence by applying simplified equations. These equations are simpler than the full, time-dependent Navier-Stokes equations. CFD Online (www.cfd-online.com) describes 27 turbulence models and is a good source for details.

One of the most widely used turbulence models is called the **k-epsilon model** (or **k-ε model**). This model uses one equation for the turbulent kinetic energy (k) and another equation for the rate of dissipation of the turbulent energy (e). These equations are used together with the **Reynolds-averaged Navier-Stokes (RANS) equations**. The RANS equations are developed by starting with the equations of motion and then taking the time average. According to Hussan (16), the k-ε method "can be very accurate, but it is not suitable for transient flows because the averaging process wipes out most of the important characteristics of a time-dependent solution." The main advantage of the k-ε model is that it is computationally efficient.

Another widely used turbulence model is called **large eddy simulation (LES)**. Large eddy simulation is a compromise between DNS and k-ε. LES uses enough detail to resolve the large-scale structures of the turbulence but uses the k-ε equations to resolve the small scales. The LES method allows one to solve problems that are not well modeled with the k-ε model by using an approach that is more computationally efficient than DNS.

Solver. A solver is the computer algorithm that solves the algebraic equations used by the CFD code. The outputs of the solver are the values of the velocity, pressure, and other relevant fields.

Post Processing. After the solver has generated a solution, the code uses this solution to calculate other parameters of interest. This process is called post processing, and the software that does this work is called the **post processor**. Some common functions of a post processor:

- Calculate derivative variables such as vorticity or shear stress
- Calculate integral variables such as pressure force, shear force, lift, drag, coefficient of lift, and coefficient of drag
- Calculate turbulent quantities such as Reynolds stresses and energy spectra
- Develop plots and other visual representations of data:
  - Plots showing time history—for example, time history of forces or wave heights
  - 2-D contour plots of variables such as pressure, velocity, or vorticity
  - 2-D velocity vector plots
  - 3-D iso-surface plots of parameters such as pressure or vorticity
  - Plots showing streamlines, pathlines, or streaklines
  - Animations of the flow

Verification and Validation

Engineers are very interested in assessing the trustworthiness of solutions. To this end, the CFD community has adopted methods for assessing correctness.

Validation examines the degree to which CFD predictions agree with real-world observations. A common validation strategy is to systematically compare CFD predictions to experimental data or to solutions to well-known problems, called *benchmark solutions*.

Verification examines the degree to which the numerical methods used by the code result in accurate answers. Verification can involve varying the spacing in the grid and ensuring that the predicted results are not dependent on the grid spacing. Similarly, verification can involve varying the time step to ensure that results are time step independent.

16.6 Examples of CFD

This section presents three examples of how professionals apply and think about CFD.

Flow through a Spillway

Problem Definition. This study by Li et al. (18) involved the Canton Dam (see Fig. 16.21), which is located on the North Canadian River in Oklahoma. When the dam was built in 1948, the design was based on maximum flowrate (during a flood) of about 10,000 m³/s. Since this time, improved hydrology data have suggested that the dam should be able to pass a peak flood discharge of 17,700 m³/s. Thus, a new auxiliary spillway was proposed, and the study presents an analysis of the proposed spillway.

Methods. A commercial CFD code, Fluent, was used to solve the time-dependent Reynolds-averaged Navier-Stokes (RANS) equations. The turbulence model was a k-ε model with wall functions. The CFD code was used to develop a tentative design. This design was then built in a 1:54 scale physical model, and the experimental data were used to validate the CFD code.

Results. Li et al. (18, p. 74) stated:

"The physical model results were compared to the CFD model results, and found to be in good agreement. The CFD model was thus validated, which in turn validated the methodology used."

FIGURE 16.21

The Canton Dam showing the proposed new auxiliary spillway.

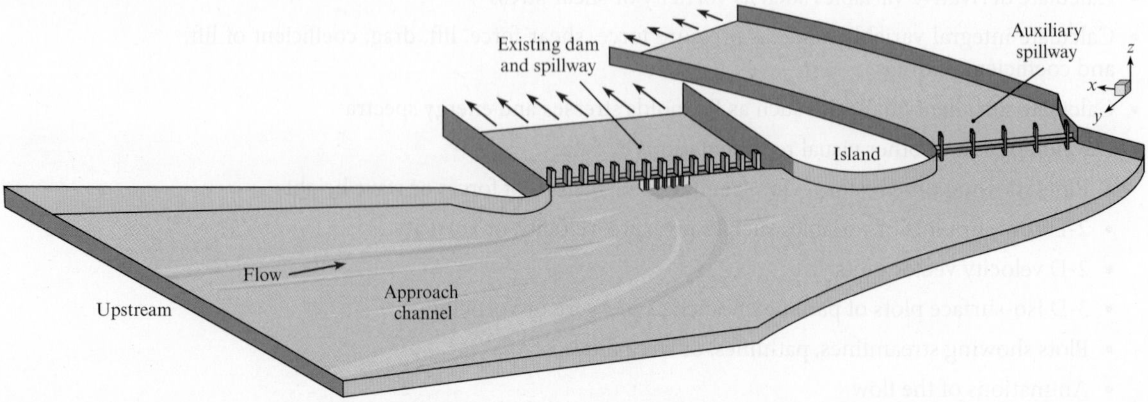

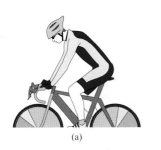  

(a)   (b)   (c)

FIGURE 16.22

Cyclist positions: (a) upright position, (b) dropped position, and (c) time trial position.

In this quote note the following:

- The engineers concluded that the CFD model was trustworthy.
- The engineers suggest that integrating CFD and experimental modeling is a viable approach for the design of hydraulic structures.

Drag on a Cyclist

Problem Definition. Bicycle racers and coaches want to understand how to reduce aerodynamic drag (see Fig. 16.22) because 90% or more of the resistive forces on the cyclist are due to this drag.

However, past CFD studies present issues with how the turbulence models were set up and with the degree of validation of the experiments. Thus, the purposes of the Defraeye et al. (14) study were as follows:

- Evaluate the use of CFD for the analysis of aerodynamic drag of different cycling positions.
- Examine and improve some of the limitations of previous CFD modeling studies for sport applications.

Methods. The experimental method involved wind tunnel experiments to gather pressure data at 30 spatial locations and to provide data on the coefficient of drag. This drag data were measured as the product of coefficient of drag (C_D) and frontal area (A) because accurately measuring frontal area is challenging.

The CFD simulation used both the RANS approach and LES.

Results. The results (Table 16.3) show that the CFD and experimental results differ by about 11% for RANS and about 7% for LES. The authors state that this is considered to be a

TABLE 16.3 Predicted Drag for Cyclists from Defraeye et al. (17)

Cyclist Position (Fig. 16.22)	Turbulence Model	AC_D (m²)	Comparison with Experiment (%)[a]
Upright	RANS	0.219	13
	LES	0.219	13
Dropped	RANS	0.179	7
	LES	0.172	3
Time trial	RANS	0.150	12
	LES	0.142	6

[a]The comparison with the experiment is calculated using this formula:

$$\frac{(AC_D \text{ predicted from CFD}) - (AC_D \text{ measured from experiment})}{(AC_D \text{ measured from experiment})}$$

FIGURE 16.23

The very large optical telescope structure.

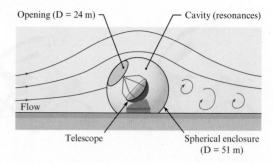

close agreement in CFD studies. The authors report a fair agreement for the predicted surface pressures, especially with LES. Despite the higher accuracy of LES, the authors suggest that the higher computational cost makes RANS more attractive for practical use.

The authors conclude that CFD is a valuable tool for evaluating the drag corresponding to different cyclist positions and for investigating the influence of small adjustments in the cyclist's position. A strong advantage of CFD is that detailed flow field information is obtained that cannot easily be obtained from wind tunnel tests. These details provide insights about the drag force and guidance for position improvements.

Predicting Wind Loads on a Telescope Structure

Problem. Because the next generation of optical telescopes are large, wind loading on the structure becomes more significant. Thus, Mamou et al. (19) conducted a study to investigate the wind loading on the prototype Canadian/U.S. very large optical telescope (VLOT) structure. The study was done during the first phase of design to assess wind loads, vortex shedding, and cavity resonances caused by wind blowing over the opening of the telescope. The structure (Fig. 16.23) is 51 m in diameter, with a 24-m-diameter opening through which the telescope views the sky. The purpose of the study was to assess the capability of a CFD model.

Methods. The code was a fully unsteady Lattice-Boltzmann CFD program. Wind tunnel data were used to validate the code.

Results. The authors noted that cavity resonance due to flow over the opening and vortex shedding from the spherical structure were observed in the wind tunnel experiments and the CFD computations. The CFD code predicted three simultaneously excited cavity modes that were identical to those measured.

16.7 A Path for Moving Forward

For students who want to learn more fluid mechanics, this section gives ideas for moving forward.

Study at the Graduate Level

Some useful graduate courses include partial differential equations, continuum mechanics, numerical methods, fluid mechanics, and computational fluid mechanics. While taking classes, some useful ways to expand one's horizons include the following:

• Read the research literature.

• Read technical books.

• Read on the Internet (for example, see the online CFD community at www.cfd-online.com).

Learn via Application (Jump into the Swimming Pool)

Some ideas for application include the following:

- Find a CFD code and learn to run it.
- Run projects for companies.
- Get involved in research: Take the lead role in writing a research paper.

For students who become involved in research, consider going to conferences and presenting your work. Submit your papers for publication. Sometimes, work will be criticized, but peer reviews are an opportunity for learning.

At research meetings, get to know the members of the community. Most people who attend research meetings have passion for their technical work, and many enjoy helping new people who are becoming engaged in the discipline.

Follow John Roncz's Advice

As John Roncz states (see beginning of chapter), jump in and figure out things yourself. This is really the key to learning anything.

16.8 Summarizing Key Knowledge

Models

- A model is an idealization or simplified version of reality. Models are valuable when they help us reach our goals in an economical way.
- The process of model building is an iterative process including the following steps:
  - Identify the variables.
  - Classify the variables into performance variables (dependent variables) and design variables (independent variables).
  - Determine how to relate the variables. When variables can be related by applying engineering equations, apply the Wales-Woods model. When variables can be related by correlating experimental data, apply regression analysis and other methods from statistics.
  - Validate to determine if model predictions are accurate enough.
- In fluid mechanics, there are three approaches to model building: analytical fluid mechanics, experimental fluid mechanics, and computational fluid mechanics. Most models involve two or three of these approaches working synergistically.
- Model building is best done by starting with simple models and then evolving these models through an iterative process. Multiple trade-offs in model building involve resources, benefits, solution accuracy, and solution detail.

Foundations for Learning Partial Differential Equations (PDEs)

- The PDEs that govern flowing fluids can be solved for only a few special cases because nonlinear terms preclude a general solution. Problems that can be solved are called *exact solutions*. These exact solutions were discovered many years ago.
- Two reasons to learn PDEs are as follows:
  - To understand and apply existing solutions (found in the literature)
  - To understand the equations being solved by CFD codes
- The solution of the PDEs are fields. The general form of a field is exemplified by the velocity field. The velocity field is

Cartesian	$\mathbf{V} = u(x, y, z, t)\mathbf{i} + v(x, y, z, t)\mathbf{j} + w(x, y, z, t)\mathbf{k}$
Cylindrical	$\mathbf{V} = v_r(r, \theta, z, t)\mathbf{u}_r + v_\theta(r, \theta, z, t)\mathbf{u}_\theta + v_z(r, \theta, z, t)\mathbf{u}_z$

- Notice that the velocity field involves the following:
  - *Independent variables.* The independent variables are the three position variables and time.
  - *Dependent variables.* The dependent variables are the three velocity components.
- Taylor series are commonly applied in fluid mechanics for developing derivations and for developing CFD programs. A useful form of the Taylor series is

$$f(x + \Delta x, y, z, t) = f(x, y, z, t) + \left(\frac{\partial f}{\partial x}\right)_{x,y,z,t} \frac{\Delta x}{1!}$$

$$+ \left(\frac{\partial^2 f}{\partial x^2}\right)_{x,y,z,t} \frac{(\Delta x)^2}{2!} + \; \ldots \; \text{H.O.T.}$$

where H.O.T. stands for "higher-order terms." For a small change (i.e., Δx is small), higher-order terms are often neglected.

- PDEs are written in two ways:
  - *Coordinate specific form.* Terms apply to a specific coordinate system. This approach is useful for specific applications.
  - *Invariant form.* Terms apply to any coordinate system; that is, they generalize. This approach is useful for writing (e.g., thesis, research paper) and presentations because the equations are compact, and they illustrate the physics.

- *Invariant notation* is a mathematical notation that applies (i.e., generalizes) to multiple coordinate systems. Three common forms of invariant notation are as follows:
  - *Del notation* uses the nabla symbol ∇.
  - *Gibbs notation* uses words (e.g., grad, div, curl) to represent operators.
  - *Indicial notation* uses subscripted letters to represent vector components and summations.

- An *operator* is a named collection of mathematical terms. Common operators in fluid mechanics equations are as follows:
  - *Gradient* (e.g., the gradient of the pressure field)
  - *Divergence* (e.g., the divergence of the velocity field)
  - *Curl* (e.g., the curl of the velocity field)
  - *LaPlacian* (e.g., the LaPlacian of the velocity field)
  - *Material derivative* (e.g., the time derivative of the temperature field)

- Each operator has one or more physical interpretations. These interpretations can be developed by working through the derivations of the PDEs.

- The *material derivative*
  - has multiple names in the literature (e.g., substantial derivative, Lagrangian derivative, and derivative following the particle),
  - represents the time rate of change of a property of a fluid particle, and
  - is represented in symbols as

$$\frac{dJ}{dt} = \underbrace{\left(\frac{\partial J}{\partial t}\right)}_{\substack{\text{time derivative} \\ \text{of property } J \text{ of} \\ \text{a fluid particle}}} + \underbrace{\mathbf{V} \cdot \nabla J = \left(\frac{\partial J}{\partial t}\right) + \mathbf{V} \cdot \text{grad}(J)}_{\substack{\text{mathematics needed to do the derivative when} \\ \text{a field (i.e., a Eulerian approach) is being used}}}$$

- *Acceleration*, defined at a point in space, means the acceleration of the fluid particle at this point at the given instant in time. Acceleration in Cartesian coordinates is

$$\binom{\text{acceleration}}{\text{of a fluid particle}} = \mathbf{a} = \frac{d\mathbf{V}}{dt} = \begin{bmatrix} \left\{\left(\dfrac{\partial u}{\partial t}\right) + u\left(\dfrac{\partial u}{\partial x}\right) + v\left(\dfrac{\partial u}{\partial y}\right) + w\left(\dfrac{\partial u}{\partial z}\right)\right\}\mathbf{i} \\[2ex] \left\{\left(\dfrac{\partial v}{\partial t}\right) + u\left(\dfrac{\partial v}{\partial x}\right) + v\left(\dfrac{\partial v}{\partial y}\right) + w\left(\dfrac{\partial v}{\partial z}\right)\right\}\mathbf{j} \\[2ex] \left\{\left(\dfrac{\partial w}{\partial t}\right) + u\left(\dfrac{\partial w}{\partial x}\right) + v\left(\dfrac{\partial w}{\partial y}\right) + w\left(\dfrac{\partial w}{\partial z}\right)\right\}\mathbf{k} \end{bmatrix}$$

The Continuity Equation

- Any problem involving a flowing fluid can, in principle, be solved by solving a coupled set of five partial differential equations comprising the continuity equation, the momentum equation, the energy equation, and two equations of state.

- The *conservation form* of the continuity equation is derived by applying the law of conservation of mass to a differential control volume. The resulting equation, in Cartesian coordinates, is

$$\underbrace{\frac{\partial \rho}{\partial t}}_{\substack{\text{rate of accumulation of mass} \\ \text{in a differential CV divided} \\ \text{by the volume of the} \\ \text{CV (kg/s per m}^3\text{)}}} + \underbrace{\frac{\partial(\rho u)}{\partial x} + \frac{\partial(\rho v)}{\partial y} + \frac{\partial(\rho w)}{\partial z} = 0}_{\substack{\text{net rate of mass flow} \\ \text{out of the CV divided by} \\ \text{the volume of the CV} \\ \text{(kg/s per m}^3\text{)}}}$$

- The continuity equation can be expressed using two forms:
  - The *conservation form* is derived by starting with a differential control volume and applying conservation of mass to this CV.

- The *nonconservation form* is derived by starting with a differential fluid particle and applying conservation of mass to this particle.

- The conservation and nonconservation forms are mathematically equivalent because one can start with one form of the equation and derive the other form.

- The nonconservation form of the continuity equation in Cartesian coordinates is

$$\frac{d\rho}{dt} + \rho\left(\frac{\partial u}{\partial x} + \frac{\partial v}{\partial y} + \frac{\partial w}{\partial z}\right) = 0$$

- Derivation of the continuity equation provides two interpretations of the divergence operator:

$$\text{div}(\rho\mathbf{V}) = \nabla \cdot (\rho\mathbf{V}) = \frac{\begin{pmatrix}\text{net rate of outflow of mass} \\ \text{from a differential CV centered} \\ \text{about point } (x, y, z)\end{pmatrix}}{(\text{volume of the CV})}$$

$$\nabla \cdot \mathbf{V} = \text{div}(\mathbf{V})$$
$$= \frac{(\text{time rate of change of the volume of a fluid particle})}{(\text{volume of the fluid particle})}$$

- When density is constant, the flow is called incompressible, and the continuity equation can be written as

Invariant form	$\nabla \cdot \mathbf{V} = \text{div}(\mathbf{V}) = 0$
Cartesian coordinates	$\dfrac{\partial u}{\partial x} + \dfrac{\partial v}{\partial y} + \dfrac{\partial w}{\partial z} = 0$

The Navier-Stokes Equation

- The Navier-Stokes equation is derived by applying Newton's Second Law of Motion to a viscous flow while also assuming that the fluid is Newtonian.

- In invariant form, the Navier-Stokes equation for an incompressible flow with constant density and viscosity is

$$\underbrace{\rho\frac{d\mathbf{V}}{dt}}_{\begin{pmatrix}\text{mass of the particle times}\\\text{acceleration of the particle}\\\text{divided by the volume of the particle}\end{pmatrix}} = \underbrace{\rho\mathbf{g}}_{\begin{pmatrix}\text{weight of the}\\\text{particle divided}\\\text{by its volume}\end{pmatrix}} + \underbrace{-\nabla p}_{\begin{pmatrix}\text{net pressure force}\\\text{on the particle}\\\text{divided by its volume}\end{pmatrix}} + \underbrace{\mu\nabla^2\mathbf{V}}_{\begin{pmatrix}\text{net shear force}\\\text{on the particle}\\\text{divided by its volume}\end{pmatrix}}$$

- Derivation of the Navier-Stokes equation reveals the physics of operators:

  - The gradient of the pressure field describes the net pressure force on a fluid particle divided by the volume of the particle.

  - The divergence of the shear stress tensor describes the net viscous force on a fluid particle divided by the volume of the particle.

FIGURE 16.24

Nonlinear terms in the Navier-Stokes equation contain the product of velocity and its derivative.

- Nonlinear terms (see Fig. 16.24) appear in the acceleration term of the Navier-Stokes equation.

Computational Fluid Dynamics (CFD)

- *Computational fluid dynamics* (CFD) is a method for solving fluid mechanics problems by developing approximate solutions to the governing PDEs. Benefits of learning CFD include the following:

  - CFD can be applied to model complex problems that cannot be modeled effectively with experiment or analysis.

  - CFD provides a way to vary design parameters and learn what happens to the performance of the system under study.

  - CFD is widely used in industry.

- Regarding CFD codes:

  - Engineers typically apply an existing code rather than writing their own code because many excellent codes are available and the process of developing a code takes years of effort.

  - Engineers select codes that fit the type of problem that they are trying to solve (e.g., for modeling groundwater, engineers might select MODFLOW; for modeling an internal combustion engine, engineers might select KIVA).

- CFD codes have an associated language:

  - A *grid* is a set of points in space at which a code solves for values of velocity and other variables of interest.

  - A *time step* is the interval between each solution time.

  - *Boundary conditions* are specified values of the dependent variables (e.g., pressure, velocity) on the physical boundaries of the problem.

- Specifying an *initial condition* involves giving numerical values for the *dependent variables* at all spatial points at the starting time of the solution.

- A *solver* is a label for the computer algorithm that solves the algebraic equations that approximate the PDEs that are being solved by the CFD code.

- A *post processor* is a computer algorithm that uses the solution from the solver to generate plots and calculate parameters such as drag force and shear stress.

- *Validation* assesses the degree to which CFD predictions agree with experimental data.

- *Verification* examines the degree to which the numerical methods used by the code result in accurate answers.

- Three common approaches to modeling turbulent flow are as follows:

  - *Direct numerical simulation* (DNS) involves setting the grid and time steps fine enough to resolve the features of the turbulent flow. DNS is unrealistic for most flows because the required computational time is too large.

  - *Large eddy simulation* (LES) involves direct simulation of the large-scale eddies in the turbulence and approximate simulation of the smaller eddies.

  - The *k-epsilon model* (*k-ε* model) represents turbulence by introducing two extra equations. As compared to DNS and LES, the *k-ε* model is computationally efficient.

REFERENCES

1. Noland, David, Wing Man, *Air & Space: Smithsonian*, 5(5), December 1990/January 1991, p. 34–40.

2. Wang, Herbert, and Mary P. Anderson. *Introduction to Groundwater Modeling : Finite Difference and Finite Element Methods*. San Diego: Academic Press, 1982.

3. Ford, Andrew. *Modeling the Environment*. Washington, DC: Island Press, 2010.

4. Montgomery, Douglas C., George C. Runger, and Norma Faris Hubele. *Engineering Statistics*. Hoboken, NJ: John Wiley, 2011.

5. Kojima, H., S. Togami, and B. Co. *The Manga Guide to Calculus*. San Francisco: No Starch Press, 2009.

6. Schey, H.M. *Div, Grad, Curl, and All That: An Informal Text on Vector Calculus*. New York: W.W. Norton, 2005.

7. White, Frank M. *Fluid Mechanics*. 4e, Boston; London: McGraw-Hill, 2011.

8. White, Frank M. *Viscous Fluid Flow*. New York: McGraw-Hill, 2006.

9. Pritchard, P.J. *Fox and McDonald's Introduction to Fluid Mechanics*, 8e. Hoboken, NJ: John Wiley, 2011.

10. "MODFLOW—Wikipedia, the free encyclopedia." Downloaded on 1/3/12 from http://en.wikipedia.org/wiki/Modflow

11. "KIVA." Downloaded on 1/3/12 from http://www.lanl.gov/orgs/t/t3/codes/kiva.shtml

12. "KIVA (software)—Wikipedia, the free encyclopedia." Downloaded on 1/3/12 from http://en.wikipedia.org/wiki/KIVA_(software)

13. "Computational Fluid Dynamics Software | FLOW-3D from Flow Science, CFD." Downloaded on 1/3/12 from http://www.flow3d.com/

14. "Test Cases." Downloaded on 1/4/12 from http://cfl3d.larc.nasa.gov/Cfl3dv6/cfl3dv6_testcases.html#cylinder

15. Wyman, Nick, "CFD Review | State of the Art in Grid Generation." Downloaded on 1/14/12 from http://www.cfdreview.com/article.pl?sid=01/04/28/2131215

16. http://piv.tamu.edu/CFD/les.htm, downloaded on 2/14/12.

17. Defraeye, Thijs, Bert Blocken, Erwin Koninckx, Peter Hespel, and Jan Carmeliet. "Aerodynamic Study of Different Cyclist Positions: CFD Analysis and Full-Scale Wind-Tunnel Tests." *Journal of Biomechanics* 43, no. 7 (2010).

18. Li, S., S. Cain, M. Wosnik, C. Miller, H. Kocahan, and P.E. Russell Wyckoff. "Numerical Modeling of Probable Maximum Flood Flowing through a System of Spillways." *Journal of Hydraulic Engineering* 137 (2011).

19. Mamou, M., K. Cooper, A. Benmeddour, M. Khalid, J. Fitzsimmons, and R. Sengupta. "Correlation of CFD Predictions and Wind Tunnel Measurements of Mean and Unsteady Wind Loads on a Large Optical Telescope." *Journal of Wind Engineering and Industrial Aerodynamics* 96, no. 6–7 (2008).

PROBLEMS

Models in Fluid Mechanics (§16.1)

16.1 Which of the following could be considered a model? Why? (Select all that apply.)

   a. The ideal gas law

   b. A set of instructions for using a Pitot-static tube to measure velocity

   c. An airplane built from a kit

   d. A computer program to predict the force on a pipe bend

16.2 Apply the modeling building process to the following task. Your team is designing a helium-filled balloon that will travel to at least 80,000 feet elevation in the atmosphere. The balloon will transport a payload comprising a camera and a data acquisition system. Right now, you choose to solve a simpler problem, which is to develop a model that predicts the weight on the earth's surface (at your location) of a helium-filled balloon; the design objective is that the balloon will be neutrally buoyant. This simpler problem can be easily tested with an experiment in your classroom.

a. What are the relevant variables?

b. How are the variables related? What are the relevant equations? How can you apply these equations to develop a single algebraic equation to solve for your goal?

c. What is a simple and low-cost way to test your math model using experimental data?

16.3 Apply the modeling building process to the following task. Your team is designing a two-stage, solid-fuel rocket that is intended to travel to 15,000 feet and then take photos. Right now, you choose to solve a simpler problem, which is to develop a model that predicts the height that a small, low-cost rocket will fly. A small rocket can be purchased from manufacturers such as Estes or Pitsco, and it is relatively easy to measure elevation for such a rocket.

a. What are the relevant variables?

b. How are the variables related? What are the relevant equations?

c. What is a simple and low-cost way to test your math model using experimental data?

Foundations for Learning PDEs (§16.2)

16.4 Why do you think that engineers make the effort to learn partial differential equations? What are the benefits to them?

16.5 Consider the function $f(x) = \dfrac{1}{1-x}$. Show how to find the Taylor series expansion for the function $f(x)$ about the point $x = 0$. Evaluate the numerical value of the Taylor series for $x = 0.1$ using five terms.

16.6 Consider the function $f(x) = \ln(x)$. Show how to find the Taylor series expansion for the function $f(x)$ about the point $x = a$. Then, find the numerical value for $x = 1.5$ and six terms of the Taylor series expansion.

16.7 Consider the flat horizontal plate shown, which is infinite in size in both dimensions. Above the plate is a fluid of viscosity μ. The plate is at rest. Then, at time equals zero seconds, the plate is set in motion to the right with a constant velocity V acting to the right. Consider the velocity field in the fluid above the plate and simplify the general form of the velocity field by answering the following questions.

a. Which velocity components (u, v, w) are zero? Which are nonzero? Why?

b. Which spatial variables (x, y, z) are parameters? Which can be ignored? Why?

c. Is time a parameter? Or, can time be ignored? Why?

d. What is the reduced equation that represents the velocity field?

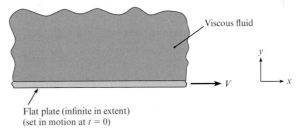

Flat plate (infinite in extent)
(set in motion at $t = 0$)

Problem 16.7

The Continuity Equation (§16.3)

16.8 Compare and contrast the integral form of the continuity equation [(Eq. (5.28)] with the PDE form of the continuity equation [(Eq. (16.36)]. Address the following questions.

a. Are the units and dimensions of each term the same? Or different?

b. How do the physics compare? What is the same? What is different?

c. How do the derivations compare? What is the same? What is different?

d. When would you want to apply the integral form of the continuity equation (Chapter 5)? When would you want to apply the PDE form of the continuity equation (Chapter 16)?

16.9 Start with the conservation form of the continuity equation in Cartesian coordinates and derive the nonconservation form.

16.10 Start with the nonconservation form of the continuity equation in Cartesian coordinates and derive the conservation form.

16.11 Consider water draining out of round hole in the bottom of the round tank as shown. Assume constant density and also assume that the water does not swirl. Then,

a. select the general form of the continuity equation that best applies to this problem, and

b. show how to simplify the general equation from part (a) to develop the reduced form.

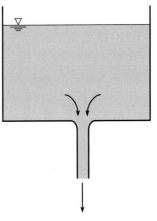

Problem 16.11

The Navier-Stokes Equation (§16.4)

16.12 Answer each question that follows.

a. Is Eq. (16.72) in conservation form or nonconservation form? Why?

b. Is Eq. (16.72) in invariant form or coordinate-specific form? Why?

16.13 What are the physics of the gradient of the pressure field? What are the units? What are the dimensions?

16.14 What are the physics of the divergence of the shear stress tensor? What are the units? What are the dimensions?

16.15 Compare the Navier-Stokes equation to Euler's equation.

a. What are two important similarities?

b. What are two important differences?

16.16 Stress, as introduced in the derivation of the Navier-Stokes equation, is a second-order tensor. Using the Internet, find some articles on tensors and answer the following questions:

a. Why do people use tensors? What are the benefits?

b. What does tensor mean? How is a tensor defined?

c. What are five examples of tensors as they are applied in engineering and physics?

Computational Fluid Dynamics (§16.5)

16.17 If someone asked you why CFD codes are useful for engineers, how would you answer? List your top three reasons in priority order.

16.18 Would you prefer to write your own CFD programs, or would you prefer to use codes that have been written by others? Discuss the advantages and disadvantages of each approach.

16.19 Using the Internet, find one example of a publicly available CFD program (either a commercial or noncommercial code) and describe the code so that others can understand it. Address the following questions in your response.

a. What is the history of the code? When was the code developed? By whom?

b. What is the main purpose of the code? What type of flow is the code well suited for?

c. How much does the code cost?

d. What training and resources are available to help you learn the code?

16.20 Briefly explain each of the following ideas.

a. Grid

b. Time step

c. Solution time for a CFD program versus the accuracy

d. Boundary condition

e. Initial condition

16.21 Briefly explain each of the following ideas.

a. DNS

b. k-epsilon method

c. LES

16.22 Briefly explain each of the following ideas.

a. Post processor

b. Verification

c. Validation

Appendix

For the triangle:
$$A = \frac{bh}{2}$$
$$\bar{I}_{xx} = \frac{bh^3}{36}$$

For the semicircle:
$$A = \frac{\pi r^2}{2}$$
$$\bar{I}_{xx} = 0.110r^4$$
$$\bar{I}_{xx} = \frac{\pi r^4}{8}$$

For the rectangle:
$$A = bh$$
$$\bar{I}_{xx} = \frac{bh^3}{12}$$

For the circle:
$$A = \pi r^2$$
$$\bar{I}_{xx} = \frac{\pi r^4}{4}$$

For the hexagon:
$$A = 2.5981L^2$$
$$\bar{I}_x = 0.5127L^4$$

For the ellipse:
$$A = \pi ab$$
$$\bar{I}_{xx} = \frac{\pi a^3 b}{4}$$

Volume and area formulas:

$$A_{circle} = \pi r^2 = \pi D^2/4$$

$$A_{sphere\ surface} = \pi D^2$$

$$V_{sphere} = \frac{1}{6}\pi D^3 = \frac{4}{3}\pi r^3$$

$$V_{cone} = \frac{1}{12}\pi D^2 h = \frac{1}{3}\pi r^3 h$$

TABLE A.1 Compressible Flow Tables for an Ideal Gas with $k = 1.4$

M or M_1 = local number or Mach number upstream of a normal shock wave; p/p_t = ratio of static pressure to total pressure; ρ/ρ_t = ratio of static density to total density; T/T_t = ratio of static temperature to total temperature; A/A_* = ratio of local cross-sectional area of an isentropic stream tube to cross-sectional area at the point where $M = 1$; M_2 = Mach number downstream of a normal shock wave; p_2/p_1 = static pressure ratio across a normal shock wave; T_2/T_1 = static pressure ratio across a normal shock wave; p_{t_2}/p_{t_1} = total pressure ratio across normal shock wave.

	Subsonic Flow			
M	**p/p_t**	**ρ/ρ_t**	**T/T_t**	**A/A_***
0.00	1.0000	1.0000	1.0000	∞
0.05	0.9983	0.9988	0.9995	11.5914
0.10	0.9930	0.9950	0.9980	5.8218
0.15	0.9844	0.9888	0.9955	3.9103
0.20	0.9725	0.9803	0.9921	2.9630
0.25	0.9575	0.9694	0.9877	2.4027
0.30	0.9395	0.9564	0.9823	2.0351
0.35	0.9188	0.9413	0.9761	1.7780
0.40	0.8956	0.9243	0.9690	1.5901
0.45	0.8703	0.9055	0.9611	1.4487
0.50	0.8430	0.8852	0.9524	1.3398
0.52	0.8317	0.8766	0.9487	1.3034
0.54	0.8201	0.8679	0.9449	1.2703
0.56	0.8082	0.8589	0.9410	1.2403
0.58	0.7962	0.8498	0.9370	1.2130
0.60	0.7840	0.8405	0.9328	1.1882
0.62	0.7716	0.8310	0.9286	1.1657
0.64	0.7591	0.8213	0.9243	1.1452
0.66	0.7465	0.8115	0.9199	1.1265
0.68	0.7338	0.8016	0.9153	1.1097
0.70	0.7209	0.7916	0.9107	1.0944
0.72	0.7080	0.7814	0.9061	1.0806
0.74	0.6951	0.7712	0.9013	1.0681
0.76	0.6821	0.7609	0.8964	1.0570
0.78	0.6691	0.7505	0.8915	1.0471
0.80	0.6560	0.7400	0.8865	1.0382
0.82	0.6430	0.7295	0.8815	1.0305
0.84	0.6300	0.7189	0.8763	1.0237
0.86	0.6170	0.7083	0.8711	1.0179
0.88	0.6041	0.6977	0.8659	1.0129
0.90	0.5913	0.6870	0.8606	1.0089
0.92	0.5785	0.6764	0.8552	1.0056
0.94	0.5658	0.6658	0.8498	1.0031
0.96	0.5532	0.6551	0.8444	1.0014
0.98	0.5407	0.6445	0.8389	1.0003
1.00	0.5283	0.6339	0.8333	1.0000

(Continued)

TABLE A.1 Compressible Flow Tables for an Ideal Gas with $k = 1.4$ (*Continued*)

	Supersonic Flow				Normal Shock Wave			
M_1	p/p_t	ρ/ρ_t	T/T_t	A/A_*	M_2	p_2/p_1	T_2/T_1	p_{t_2}/p_{t_1}
1.00	0.5283	0.6339	0.8333	1.000	1.0000	1.000	1.000	1.0000
1.01	0.5221	0.6287	0.8306	1.000	0.9901	1.023	1.007	0.9999
1.02	0.5160	0.6234	0.8278	1.000	0.9805	1.047	1.013	0.9999
1.03	0.5099	0.6181	0.8250	1.001	0.9712	1.071	1.020	0.9999
1.04	0.5039	0.6129	0.8222	1.001	0.9620	1.095	1.026	0.9999
1.05	0.4979	0.6077	0.8193	1.002	0.9531	1.120	1.033	0.9998
1.06	0.4919	0.6024	0.8165	1.003	0.9444	1.144	1.039	0.9997
1.07	0.4860	0.5972	0.8137	1.004	0.9360	1.169	1.046	0.9996
1.08	0.4800	0.5920	0.8108	1.005	0.9277	1.194	1.052	0.9994
1.09	0.4742	0.5869	0.8080	1.006	0.9196	1.219	1.059	0.9992
1.10	0.4684	0.5817	0.8052	1.008	0.9118	1.245	1.065	0.9989
1.11	0.4626	0.5766	0.8023	1.010	0.9041	1.271	1.071	0.9986
1.12	0.4568	0.5714	0.7994	1.011	0.8966	1.297	1.078	0.9982
1.13	0.4511	0.5663	0.7966	1.013	0.8892	1.323	1.084	0.9978
1.14	0.4455	0.5612	0.7937	1.015	0.8820	1.350	1.090	0.9973
1.15	0.4398	0.5562	0.7908	1.017	0.8750	1.376	1.097	0.9967
1.16	0.4343	0.5511	0.7879	1.020	0.8682	1.403	1.103	0.9961
1.17	0.4287	0.5461	0.7851	1.022	0.8615	1.430	1.109	0.9953
1.18	0.4232	0.5411	0.7822	1.025	0.8549	1.458	1.115	0.9946
1.19	0.4178	0.5361	0.7793	1.026	0.8485	1.485	1.122	0.9937
1.20	0.4124	0.5311	0.7764	1.030	0.8422	1.513	1.128	0.9928
1.21	0.4070	0.5262	0.7735	1.033	0.8360	1.541	1.134	0.9918
1.22	0.4017	0.5213	0.7706	1.037	0.8300	1.570	1.141	0.9907
1.23	0.3964	0.5164	0.7677	1.040	0.8241	1.598	1.147	0.9896
1.24	0.3912	0.5115	0.7648	1.043	0.8183	1.627	1.153	0.9884
1.25	0.3861	0.5067	0.7619	1.047	0.8126	1.656	1.159	0.9871
1.30	0.3609	0.4829	0.7474	1.066	0.7860	1.805	1.191	0.9794
1.35	0.3370	0.4598	0.7329	1.089	0.7618	1.960	1.223	0.9697
1.40	0.3142	0.4374	0.7184	1.115	0.7397	2.120	1.255	0.9582
1.45	0.2927	0.4158	0.7040	1.144	0.7196	2.286	1.287	0.9448
1.50	0.2724	0.3950	0.6897	1.176	0.7011	2.458	1.320	0.9278
1.55	0.2533	0.3750	0.6754	1.212	0.6841	2.636	1.354	0.9132
1.60	0.2353	0.3557	0.6614	1.250	0.6684	2.820	1.388	0.8952
1.65	0.2184	0.3373	0.6475	1.292	0.6540	3.010	1.423	0.8760
1.70	0.2026	0.3197	0.6337	1.338	0.6405	3.205	1.458	0.8557
1.75	0.1878	0.3029	0.6202	1.386	0.6281	3.406	1.495	0.8346
1.80	0.1740	0.2868	0.6068	1.439	0.6165	3.613	1.532	0.8127
1.85	0.1612	0.2715	0.5936	1.495	0.6057	3.826	1.569	0.7902

(*Continued*)

TABLE A.1 Compressible Flow Tables for an Ideal Gas with $k = 1.4$ (*Continued*)

	Supersonic Flow				Normal Shock Wave			
M_1	p/p_t	ρ/ρ_t	T/T_t	A/A_*	M_2	p_2/p_1	T_2/T_1	p_{t_2}/p_{t_1}
1.90	0.1492	0.2570	0.5807	1.555	0.5956	4.045	1.608	0.7674
1.95	0.1381	0.2432	0.5680	1.619	0.5862	4.270	1.647	0.7442
2.00	0.1278	0.2300	0.5556	1.688	0.5774	4.500	1.688	0.7209
2.10	0.1094	0.2058	0.5313	1.837	0.5613	4.978	1.770	0.6742
2.20	0.9352^{-1}†	0.1841	0.5081	2.005	0.5471	5.480	1.857	0.6281
2.30	0.7997^{-1}	0.1646	0.4859	2.193	0.5344	6.005	1.947	0.5833
2.50	0.5853^{-1}	0.1317	0.4444	2.637	0.5130	7.125	2.138	0.4990
2.60	0.5012^{-1}	0.1179	0.4252	2.896	0.5039	7.720	2.238	0.4601
2.70	0.4295^{-1}	0.1056	0.4068	3.183	0.4956	8.338	2.343	0.4236
2.80	0.3685^{-1}	0.9463^{-1}	0.3894	3.500	0.4882	8.980	2.451	0.3895
2.90	0.3165^{-1}	0.8489^{-1}	0.3729	3.850	0.4814	9.645	2.563	0.3577
3.00	0.2722^{-1}	0.7623^{-1}	0.3571	4.235	0.4752	10.330	2.679	0.3283
3.50	0.1311^{-1}	0.4523^{-1}	0.2899	6.790	0.4512	14.130	3.315	0.2129
4.00	0.6586^{-2}	0.2766^{-1}	0.2381	10.72	0.4350	18.500	4.047	0.1388
4.50	0.3455^{-2}	0.1745^{-1}	0.1980	16.56	0.4236	23.460	4.875	0.9170^{-1}
5.00	0.1890^{-2}	0.1134^{-1}	0.1667	25.00	0.4152	29.000	5.800	0.6172^{-1}
5.50	0.1075^{-2}	0.7578^{-2}	0.1418	36.87	0.4090	35.130	6.822	0.4236^{-1}
6.00	0.6334^{-2}	0.5194^{-2}	0.1220	53.18	0.4042	41.830	7.941	0.2965^{-1}
6.50	0.3855^{-2}	0.3643^{-2}	0.1058	75.13	0.4004	49.130	9.156	0.2115^{-1}
7.00	0.2416^{-3}	0.2609^{-2}	0.9259^{-1}	104.1	0.3974	57.000	10.47	0.1535^{-1}
7.50	0.1554^{-3}	0.1904^{-2}	0.8163^{-1}	141.8	0.3949	65.460	11.88	0.1133^{-1}
8.00	0.1024^{-3}	0.1414^{-2}	0.7246^{-1}	190.1	0.3929	74.500	13.39	0.8488^{-2}
8.50	0.6898^{-4}	0.1066^{-2}	0.6472^{-1}	251.1	0.3912	84.130	14.99	0.6449^{-2}
9.00	0.4739^{-4}	0.8150^{-3}	0.5814^{-1}	327.2	0.3898	94.330	16.69	0.4964^{-2}
9.50	0.3314^{-4}	0.6313^{-3}	0.5249^{-1}	421.1	0.3886	105.100	18.49	0.3866^{-2}
10.00	0.2356^{-4}	0.4948^{-3}	0.4762^{-1}	535.9	0.3876	116.500	20.39	0.3045^{-2}

†x^{-n} means $x \cdot 10^{-n}$.

Data source: R. E. Bolz and G. L. Tuve, *The Handbook of Tables for Applied Engineering Sciences*, CRC Press, Inc., Cleveland, 1973. Copyright © 1973 by The Chemical Rubber Co., CRC Press, Inc.

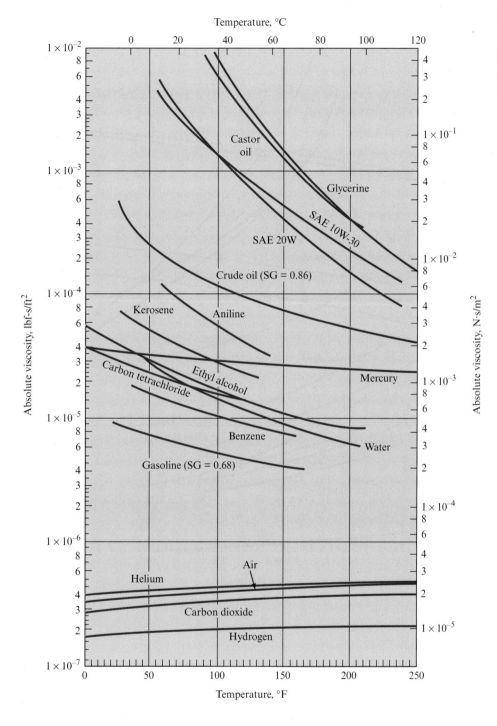

FIGURE A.2

Absolute viscosities of certain gases and liquids (*Data source:* Fluid Mechanics, 5th ed., V. L. Streeter, 1971, McGraw-Hill, New York.)

FIGURE A.3

Kinematic viscosities of certain gases and liquids. The gases are at standard pressure. (*Data source:* Fluid Mechanics, 5th ed., V. L. Streeter, McGraw-Hill, New York.)

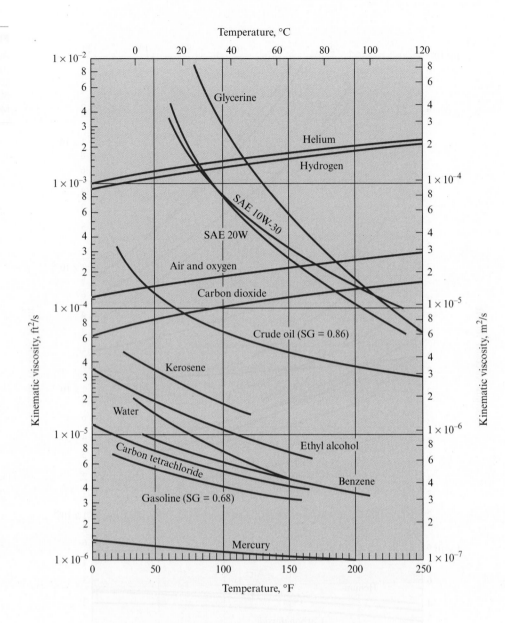

TABLE A.2 Physical Properties of Gases $[T = 15°C\ (59°F),\ p = 1\ \text{atm}]$

Gas	Density kg/m³ (slugs/ft³)	Kinematic Viscosity m²/s (ft²/s)	R Gas Constant J/kg K (ft-lbf/slug-°R)	$\dfrac{c_p}{\dfrac{J}{kg\ K}}\left(\dfrac{Btu}{lbm\text{-}°R}\right)$	$k = \dfrac{c_p}{c_v}$	S Sutherland's Constant K(°R)
Air	1.22 (0.00237)	1.46×10^{-5} (1.58×10^{-4})	287 (1716)	1004 (0.240)	1.40	111 (199)
Carbon dioxide	1.85 (0.0036)	7.84×10^{-6} (8.48×10^{-5})	189 (1130)	841 (0.201)	1.30	222 (400)
Helium	0.169 (0.00033)	1.14×10^{-4} (1.22×10^{-3})	2077 (12,419)	5187 (1.24)	1.66	79.4 (143)
Hydrogen	0.0851 (0.00017)	1.01×10^{-4} (1.09×10^{-3})	4127 (24,677)	14,223 (3.40)	1.41	96.7 (174)
Methane (natural gas)	0.678 (0.0013)	1.59×10^{-5} (1.72×10^{-4})	518 (3098)	2208 (0.528)	1.31	198 (356)
Nitrogen	1.18 (0.0023)	1.45×10^{-5} (1.56×10^{-4})	297 (1776)	1041 (0.249)	1.40	107 (192)
Oxygen	1.35 (0.0026)	1.50×10^{-5} (1.61×10^{-4})	260 (1555)	916 (0.219)	1.40	

Data source: V. L. Streeter (ed.), *Handbook of Fluid Dynamics,* McGraw-Hill Book Company, New York, 1961; also R. E. Bolz and G. L. Tuve, *Handbook of Tables for Applied Engineering Science,* CRC Press, Inc. Cleveland, 1973; and *Handbook of Chemistry and Physics,* Chemical Rubber Company, 1951.

TABLE A.3 Mechanical Properties of Air at Standard Atmospheric Pressure

Temperature	Density	Specific Weight	Dynamic Viscosity	Kinematic Viscosity
	kg/m^3	N/m^3	N·s/m^2	m^2/s
−20°C	1.40	13.70	1.61×10^{-5}	1.16×10^{-5}
−10°C	1.34	13.20	1.67×10^{-5}	1.24×10^{-5}
0°C	1.29	12.70	1.72×10^{-5}	1.33×10^{-5}
10°C	1.25	12.20	1.76×10^{-5}	1.41×10^{-5}
20°C	1.20	11.80	1.81×10^{-5}	1.51×10^{-5}
30°C	1.17	11.40	1.86×10^{-5}	1.60×10^{-5}
40°C	1.13	11.10	1.91×10^{-5}	1.69×10^{-5}
50°C	1.09	10.70	1.95×10^{-5}	1.79×10^{-5}
60°C	1.06	10.40	2.00×10^{-5}	1.89×10^{-5}
70°C	1.03	10.10	2.04×10^{-5}	1.99×10^{-5}
80°C	1.00	9.81	2.09×10^{-5}	2.09×10^{-5}
90°C	0.97	9.54	2.13×10^{-5}	2.19×10^{-5}
100°C	0.95	9.28	2.17×10^{-5}	2.29×10^{-5}
120°C	0.90	8.82	2.26×10^{-5}	2.51×10^{-5}
140°C	0.85	8.38	2.34×10^{-5}	2.74×10^{-5}
160°C	0.81	7.99	2.42×10^{-5}	2.97×10^{-5}
180°C	0.78	7.65	2.50×10^{-5}	3.20×10^{-5}
200°C	0.75	7.32	2.57×10^{-5}	3.44×10^{-5}
	slugs/ft^3	lbf/ft^3	lbf-s/ft^2	ft^2/s
0°F	0.00269	0.0866	3.39×10^{-7}	1.26×10^{-4}
20°F	0.00257	0.0828	3.51×10^{-7}	1.37×10^{-4}
40°F	0.00247	0.0794	3.63×10^{-7}	1.47×10^{-4}
60°F	0.00237	0.0764	3.74×10^{-7}	1.58×10^{-4}
80°F	0.00228	0.0735	3.85×10^{-7}	1.69×10^{-4}
100°F	0.00220	0.0709	3.96×10^{-7}	1.80×10^{-4}
120°F	0.00213	0.0685	4.07×10^{-7}	1.91×10^{-4}
150°F	0.00202	0.0651	4.23×10^{-7}	2.09×10^{-4}
200°F	0.00187	0.0601	4.48×10^{-7}	2.40×10^{-4}
300°F	0.00162	0.0522	4.96×10^{-7}	3.05×10^{-4}
400°F	0.00143	0.0462	5.40×10^{-7}	3.77×10^{-4}

Data source: R. E. Bolz and G. L. Tuve, *Handbook of Tables for Applied Engineering Science*, CRC Press, Inc., Cleveland, 1973. Copyright © 1973 by The Chemical Rubber Co., CRC Press, Inc.

TABLE A.4 Approximate Physical Properties of Common Liquids at Atmospheric Pressure

Liquid and Temperature	Density kg/m^3 (slugs/ft^3)	Specific Gravity	Specific Weight N/m^3 (lbf/ft^3)	Dynamic Viscosity N·s/m^2 (lbf-s/ft^2)	Kinematic Viscosity m^2/s (ft^2/s)	Surface Tension N/m* (lbf/ft)
Ethyl alcohol[1][3] 20°C (68°F)	799 (1.55)	0.79	7,850 (50.0)	1.2×10^{-3} (2.5×10^{-5})	1.5×10^{-6} (1.6×10^{-5})	2.2×10^{-2} (1.5×10^{-3})
Carbon tetrachloride[3] 20°C (68°F)	1,590 (3.09)	1.59	15,600 (99.5)	9.6×10^{-4} (2.0×10^{-5})	6.0×10^{-7} (6.5×10^{-6})	2.6×10^{-2} (1.8×10^{-3})
Glycerine[3] 20°C (68°F)	1,260 (2.45)	1.26	12,300 (78.5)	1.41 (2.95×10^{-2})	1.12×10^{-3} (1.22×10^{-2})	6.3×10^{-2} (4.3×10^{-3})
Kerosene[1][2] 20°C (68°F)	814 (1.58)	0.81	8,010 (51)	1.9×10^{-3} (4.0×10^{-5})	2.37×10^{-6} (2.55×10^{-5})	2.9×10^{-2} (2.0×10^{-3})
Mercury[1][3] 20°C (68°F)	13,550 (26.3)	13.55	133,000 (847)	1.5×10^{-3} (3.1×10^{-5})	1.2×10^{-7} (1.3×10^{-6})	4.8×10^{-1} (3.3×10^{-2})
Sea water 10°C at 3.3% salinity	1,026 (1.99)	1.03	10,070 (64.1)	1.4×10^{-3} (2.9×10^{-5})	1.4×10^{-6} (1.5×10^{-5})	
Oils—38°C (100°F) SAE 10W[4]	870 (1.69)	0.87	8,530 (54.4)	3.6×10^{-2} (7.5×10^{-4})	4.1×10^{-5} (4.4×10^{-4})	
SAE 10W-30[4]	880 (1.71)	0.88	8,630 (55.1)	6.7×10^{-2} (1.4×10^{-3})	7.6×10^{-5} (8.2×10^{-4})	
SAE 30[4]	880 (1.71)	0.88	8,630 (55.1)	1.0×10^{-1} (2.1×10^{-3})	1.1×10^{-4} (1.2×10^{-3})	

*Liquid–air surface tension values.

Data source: (1) V. L. Streeter, *Handbook of Fluid Dynamics*, McGraw-Hill, New York, 1961; (2) V. L. Streeter, *Fluid Mechanics*, 4th ed., McGraw-Hill, New York, 1966; (3) A. A. Newman, *Glycerol*, CRC Press, Cleveland, 1968; (4) R. E. Bolz and G. L. Tuve, *Handbook of Tables for Applied Engineering Sciences*, CRC Press, Cleveland, 1973.

TABLE A.5 Approximate Physical Properties of Water* at Atmospheric Pressure

Temperature	Density	Specific Weight	Dynamic Viscosity	Kinematic Viscosity	Vapor Pressure
	kg/m^3	N/m^3	$N \cdot s/m^2$	m^2/s	N/m^2 abs
0°C	1000	9810	1.79×10^{-3}	1.79×10^{-6}	611
5°C	1000	9810	1.51×10^{-3}	1.51×10^{-6}	872
10°C	1000	9810	1.31×10^{-3}	1.31×10^{-6}	1,230
15°C	999	9800	1.14×10^{-3}	1.14×10^{-6}	1,700
20°C	998	9790	1.00×10^{-3}	1.00×10^{-6}	2,340
25°C	997	9781	8.91×10^{-4}	8.94×10^{-7}	3,170
30°C	996	9771	7.97×10^{-4}	8.00×10^{-7}	4,250
35°C	994	9751	7.20×10^{-4}	7.24×10^{-7}	5,630
40°C	992	9732	6.53×10^{-4}	6.58×10^{-7}	7,380
50°C	988	9693	5.47×10^{-4}	5.53×10^{-7}	12,300
60°C	983	9643	4.66×10^{-4}	4.74×10^{-7}	20,000
70°C	978	9594	4.04×10^{-4}	4.13×10^{-7}	31,200
80°C	972	9535	3.54×10^{-4}	3.64×10^{-7}	47,400
90°C	965	9467	3.15×10^{-4}	3.26×10^{-7}	70,100
100°C	958	9398	2.82×10^{-4}	2.94×10^{-7}	101,300
	$slugs/ft^3$	lbf/ft^3	$lbf\text{-}s/ft^2$	ft^2/s	psia
40°F	1.94	62.43	3.23×10^{-5}	1.66×10^{-5}	0.122
50°F	1.94	62.40	2.73×10^{-5}	1.41×10^{-5}	0.178
60°F	1.94	62.37	2.36×10^{-5}	1.22×10^{-5}	0.256
70°F	1.94	62.30	2.05×10^{-5}	1.06×10^{-5}	0.363
80°F	1.93	62.22	1.80×10^{-5}	0.930×10^{-5}	0.506
100°F	1.93	62.00	1.42×10^{-5}	0.739×10^{-5}	0.949
120°F	1.92	61.72	1.17×10^{-5}	0.609×10^{-5}	1.69
140°F	1.91	61.38	0.981×10^{-5}	0.514×10^{-5}	2.89
160°F	1.90	61.00	0.838×10^{-5}	0.442×10^{-5}	4.74
180°F	1.88	60.58	0.726×10^{-5}	0.385×10^{-5}	7.51
200°F	1.87	60.12	0.637×10^{-5}	0.341×10^{-5}	11.53
212°F	1.86	59.83	0.593×10^{-5}	0.319×10^{-5}	14.70

*Notes: Bulk modulus E_v of water is approximately 2.2 GPa (3.2×10^5 psi).

Data source: R. E. Bolz and G. L. Tuve, *Handbook of Tables for Applied Engineering Science*, CRC Press, Inc., Cleveland, 1973. Copyright © 1973 by The Chemical Rubber Co., CRC Press, Inc.

Answers

Answers to Even-Numbered Problems

Chapter 1

1.6   True

1.8   a. 785 N   b. 35 lbm   c. 1.09 slug   d. 3 N
        e. 0.0933 lbf   f. 3 lbf

1.10 $\rho = 1.33$ kg/m^3; $\rho = 2.57 \times 10^{-3} \frac{\text{slug}}{\text{ft}^3}$

1.12 $\rho = 1.91 \frac{\text{kg}}{\text{m}^3}$

1.14 $\dfrac{m_2}{m_1} = 2.29$

1.16 $m = 26.5$ kg

1.18 $m = 6.98 \times 10^8$ slug $= 1.02 \times 10^{10}$ kg

1.22   a. 17.4 psia   b. 2510 psf   c. 1.20 bar   d. 1.18 atm
        e. 40.2 ft-H$_2$O   f. 35.4 in-Hg

Chapter 2

2.2   1.50%, 2.91 mm

2.4   (a) and (c)

2.6 $\Delta \nu = 6.80 \times 10^{-6}$ m^2/s

2.8   (a) and (c)

2.10   3190 cm^3

2.12   (a)

2.14   (a) $\mu = 1.80 \times 10^{-4} \frac{\text{lbf} \cdot \text{s}}{\text{in}^2} = 2.59 \times 10^{-2} \frac{\text{lbf} \cdot \text{s}}{\text{ft}^2}$
        (b) $\mu = 1.24 \frac{\text{N} \cdot \text{s}}{\text{m}^2}$ (c) more

2.16 $V_8 = 0.179$ m/s, $\tau_8 = 25.5$ Pa; $V_{\text{wall}} = 0$ m/s,
        $\tau_{\text{wall}} = 37.5$ Pa;

2.18 $\mu = 2.22 \times 10^{-2} \frac{\text{N} \cdot \text{s}}{\text{m}^2}$

2.20 $m = 0.625$ g

2.22 $p = 146 \frac{\text{N}}{\text{m}^2}$ gage

2.24   Yes, the water will boil.

Chapter 3

3.2   (a) $\rho_{helium} = 0.173$ kg/m^3
        (b) $\rho_{methane} = 0.418$ kg/m^3

3.4 $p_{abs} = 323$ kPa abs

3.8   (a) $\Delta p = 43.1$ Pa   (b) depth $= 12.0$ m
        (c) $p = 924$ mbar abs   (d) $p = 3.73$ MPa gage
        (e) $p = 589$ kPa gage

3.10 $F_2 = 1680$ N

3.14 $F_{bolts} = 288$ kN (acting downward)

3.16 $V_{added} = 24.6$ in^3

3.20   (a) $p_A - p_B = 1.51$ kPa
        (b) $p_{zA} - p_{zB} = 0.530$ kPa

3.22   (a) $F = 35.4$ kN   (b) Distance from ditch bottom to
        $CP = 0.633$ m

3.24 $F = 6.84$ kN; $y_{cp} - \bar{y} = 25.7$ mm

3.26 $F = 2080$ lbf

3.28 $T = 10{,}900$ ft-lbf

3.30 F per foot $= 6075$ lbf/ft; $F_{bottom\ tie} = 8100$ lbf (tension)

3.32 $R_A = 0.510\gamma W l^2$; Less than

3.34   (a) $F_V = 17.5$ kN   (b) $F_H = 14.7$ kN, $x = 0.467$ m,
        $y_{cp} = 1.56$ m,   (c) $\mathbf{F}_R = 22.9$ kN, $\theta = 40°2'$

3.36 $\mathbf{F}_{\text{Result}} = 5200\mathbf{i} + 8170\mathbf{j}$ lbf

3.40 $SG = 17.0$; Do not bid

3.42   Settle; $\Delta h = 1.04$ ft

3.44 $W_{scrap} = 8270$ N

3.46 $\Delta h = 0.784$ in

3.48 $\Delta h = 0.368$ cm

3.50 $SG = 0.836$

3.52   No; block is unstable with axis vertical

Chapter 4

4.2   b

4.4   a

4.6   d

4.8   d

4.10   c

4.12 $a_\ell = -11.3$ m/s^2; $a_c = 0.900$ m/s^2

4.14 $a_\ell = 3.00$ ft/s^2

4.16 $a_c = \dfrac{32}{27}\left(\dfrac{q_0}{t_0}\right)^2 \dfrac{t^2}{B^3}$

4.18 $a_z = -141$ ft/s^2

4.20 $\dfrac{\partial p}{\partial z} = -65.7$ lbf/ft^3

4.22 $a_s = 116$ ft/s^2

4.24	$h = 22.5$ m
4.26	$V_2 = 9.81$ ft/s
4.28	a and c
4.30	$V = 2.03$ m/s
4.32	$V = 126$ ft/s
4.34	$V = 18.1$ ft/s
4.36	$V = 139$ ft/s
4.38	$V_0 = 2.54$ m/s
4.40	$p_B - p_C = 52.5$ kPa
4.42	$V_{actual} = 60.1$ m/s
4.44	$V_{air} = 13.0$ m/s
4.46	Yes
4.48	$\Delta\theta = \dfrac{4y}{(1 - 4y^2)}$
4.54	$\omega = \sqrt{\dfrac{7.5g}{l}}$

Chapter 5

5.2	$\overline{V} = 1.14$ m/s
5.4	$\dot{m} = 15.5$ kg/s
5.6	$Q = 77.7$ cfs; $Q = 34{,}900$ gpm
5.8	$Q = 1170$ cfs
5.10	$D = 1.80$ m
5.12	$Q = 4.71$ m³/s; $V = 6$ m/s
5.14	$Q = 0.142$ m³/s
5.16	Extensive: mass, volume, energy. Intensive: density, specific energy.
5.18	(b)
5.20	$a_{exit} = 3.40$ m/s² $V_{exit} = 4.40$ m/s
5.22	$V_2 = 15.4$ m/s; Bernoulli Eq. does not apply because density is not contant.
5.24	$V_L = 5.14$ ft/s (upward)
5.26	$t = 50.0$ s
5.28	$V_1 = 3.21$ ft/s; $V_2 = 6.06$ ft/s
5.30	$h_{eq} = 27.3$ ft; Lake dries up when $Q_{in} \leq 49.5 \text{ft}^3/\text{s}$
5.32	$V_{entr} = 6.08$ m/s
5.34	$A = 8.69 \times 10^{-9}$ ft² $= 1.25 \times 10^{-6}$ in²
5.36	$t = \dfrac{2}{d_j^2 \sqrt{2g}} \left[(d^2(h_0^{1/2} - h^{1/2}) + \frac{2}{3}dC_1(h_0^{3/2} - h^{3/2}) + \frac{1}{5}C_1^2(h_0^{5/2} - h^{5/2}) \right]$; $t = 13.8$ s
5.38	$\Delta t = 7.9$ min; $\Delta t_{unpressurized} = 14.9$ min
5.40	$\Delta h = 0.385$ ft
5.42	$V_{out} = 40$ m/s $\left[1 - \exp\left(\dfrac{-t}{10}\right) \right]$
5.44	$V_{cav} = 29.3$ ft/s
5.46	$V_{cav} = 46.0$ ft/s

Chapter 6

6.2	(b), (e), and (f)
6.4	$F = 0.482$ N
6.6	$F_{bot} = 824$ N; $F_{block} = 96.6$ N
6.8	$F_1 = 257$ N; $F_2 = 243$ N
6.10	$\dot{m} = 267$ kg/s; $D = 10.6$ cm
6.12	$F_x = -39.7$ kN (acts to the left); $F_y = -10.4$ kN (acts downward)
6.14	$\mathbf{F} = (-5.82 \text{ lbf})\mathbf{i} + (-21.7 \text{ lbf})\mathbf{j}$
6.16	$F_{\text{per unit width}} = \rho v^2 \sin 45°$ per unit width of wall
6.20	$V_{exit} = 47.6$ ft/s; $F = -168$ lbf
6.22	$F = -3088$ lbf
6.24	F_T per bolt $= 1410$ N
6.26	$F_x = -9.78$ lbf
6.28	$F_x = -1670$ lbf
6.30	$F_y = 5340$ lbf
6.32	$F_x = -15.9$ kN; $F_y = -1.8$ kN
6.34	Thrust $= 124$ kN
6.36	$a = 0.324$ m/s²
6.38	$P = 1.11$ kW

Chapter 7

7.4	b, e, f and g
7.6	b
7.8	$\alpha = \dfrac{1}{4}\left[\dfrac{[(n + 2)(n + 1)]^3}{(3n + 2)(3n + 1)} \right]$; $\alpha = 1.06$
7.10	$\alpha = 1.19$
7.16	$p_A = -187$ psf gage; $V_e = 28.9$ ft/s
7.18	$\frac{p_2}{\gamma} = 79.8$ m
7.20	$p_2 = 18.4$ psig
7.22	$p_1 = 48.3$ Pa gage
7.24	$t = 698$ s $= 11.6$ min
7.26	$P = 45.7$ hp
7.28	$P = 1.21$ MW
7.30	$h_p = 71.5$ m
7.34	$h_L = 1.51$ m
7.36	(a) A to E (b) No (c) Uniform (d) Yes (e) sketch (f) No
7.38	$h_p = 4.08$ m
7.40	$Q = 186$ ft³/s; $p_p = 1170$ lbf/ft²

Chapter 8

8.2	(a) $\frac{M}{LT^2}$ (b) Dimensionless (c) $\frac{ML^2}{T^2}$ (d) $\frac{L}{T}$
8.4	$\dfrac{\dot{m}}{\sqrt{\rho \Delta p D^2}} = f\left(\dfrac{\mu D}{\dot{m}}\right)$

8.6   (a) Fr   (b) We   (c) Re   (d) M
      (e) Re   (f) Fr (g) W   (h) M

8.8 $P_{L,p} = 48$ kW

8.10 $V_w = 5.33$ m/s

8.12 $F_L = 40.1$ kN

8.14 $p = 27.6$ kbar gage

8.16 $\nu_{earth} = 6.71 \times 10^{-5}$ m^2/s

8.18 $V_p = 5.83$ m/s; $\Delta p_p = 2.48$ kPa gage

8.20 $Q_m = 0.101$ m^3/s

8.22 $V_m = 222$ m/s; because M $\geqq$ 0.3, Mach-number effects are important

8.24 $V_m = 1.88$ ft/s; $d_m = 0.312$ ft

8.26   (d)

8.28 $\frac{\Delta p D}{L \rho V^2} = f\left(\frac{\rho V D}{\mu}\right)$; plot results

Chapter 9

9.2 $\mu_{oil} = 8.75 \times 10^{-2}$ N·s/m^2

9.4 $F = 4940$ N

9.6 $P = 0.00780$ W

9.8 $Q = 6.62 \times 10^{-8}$ m^3/s

9.10   (a) False   (b) True   (c) False   (d) False

9.12 $q = 2.45 \times 10^{-3}$ m^2/s

9.14 $\frac{dp}{ds} = -1.61 \times 10^4$ Pa/m; $P = 3.95 \times 10^{-4}$ W

9.16 $V = 0.0150$ m/s

9.18 $V = 0.0150$ m/s

9.20   (a) T   (b) L   (c) L   (d) T   (e) T
      (f) T   (g) L (h) T   (i) L   (j) T

9.22 $F_s = 8.90$ N

9.24 $F_s = 2.82$ N/m; $\frac{du}{dy} = 6.19 \times 10^4$ s^{-1}

9.26 $P = 26.1$ kW

9.28 $P = 103$ hp

Chapter 10

10.2 $L_e = 4.0$ m; Turbulent

10.4   (a) $h_f = 6.21$ ft   (b) $D = 0.227$ m

10.6   (e)

10.8 $\Delta p = 8.43$ kPa

10.12 $f = 0.491$; Re = 131; Write a proof

10.14 $h_f = 95.0$ ft per 100 ft of pipe

10.16 $\Delta p = 0.019$ kPa per 10 m of pipe

10.18   Laminar; plot

10.20 $V = 0.0331$ m/s

10.22 $\Delta p = 628$ Pa

10.24 $f = 0.00946$

10.26 $f = 0.0153$

10.28 $f = 0.024$ (Moody)

10.30   (b)

10.32   (a) $\Delta p = 48.9$ psf   (b) sketch   (c) $P = 349$ hp;

10.34 $D > 0.112$ m

10.36   (a) 0.19   (b) 0.9   (c) 0.2   (d) 0.03   (e) 0.27

10.38   (c)

10.40   (a) Entrance, valve, exit and $f\frac{L}{D}$
      (b) Sketch   (c) $Q = 27.7$ ft^3/s

10.42 $P = 5.61$ W

10.44   (a) $V = 2.14$ m/s   (b) $h = 23.4$ cm

10.46 $925 \leq P \leq 927$ kW depending upon method used for f; sketch

10.48   (c)

10.50 $P = 7.55$ MW

10.52   (e)

10.54 $\frac{V_{trap}}{V_{rect}} = 0.84$

10.56 $Q = 4700$ gpm

10.58   (a)

10.60 $Q_1 = 0.27$ m^3/s; $Q_2 = 0.33$ m^3/s; $h_L = 5.05$ m

Chapter 11

11.2 $C_D = 2.0$

11.4   (a), (b) and (d)

11.8   (e)

11.10 $T = 142$ N

11.14 $M = 9.72$ kN·m

11.16   (a) False   (b) True

11.18 $P = 13.4$ hp

11.20 $V = 117$ mph

11.22   (d)

11.28 $t = 21.1$ min

11.32   (b)

Chapter 12

12.2   M = 27.1

12.4 $c = \sqrt{E_v/\rho}$; $c = 1480$ m/s

12.8   M = 5

12.14 $p_t = 224$ kPa

12.20   (a) $M_e = 2.62$; $T_e = 1,941$ K; $\rho_e = 0.0671$ kg/m^3
      (b) $\dot{m} = 6.78$ kg/s   (c) $T = 18.0$ kN
      (d) $p_t = 691$ kPa $T = 9.86$ kN

12.22 $A/A_* = 2.46$

Chapter 13

13.2 $V = 6.01$ m/s

13.6 $Q = 1.13 \times 10^{-2}$ m³/s

13.8   (a) $\frac{r_m}{D} = 0.224$ (b) $\frac{r_c}{D} = 0.341$ (c) $\dot{m} = 9.96$ kg/s

13.10 $C_c = 0.903$

13.14 $d = 0.601$ m

13.16 $Q = 0.0789$ m³/s

13.18 $Q = 0.232$ m³/s

13.20 $\Delta t = 43.4$ s

Chapter 14

14.2 $F_T = 1370$ N

14.4 $N = 1160$ rpm

14.6 $N = 1170$ rpm

14.10   (a) $Q = 0.32$ m³/s   (b) $P = 13.5$ kW

14.14 $P = 726$ kW

14.16   (a) $\Delta H = 91.3$ m   (b) $Q = 0.878$ m³/s

14.18 $H_{1,600} = 261$ ft

14.20   (a) $Q = 0.0833$ m³/s   (b) Δ pressure head $= 146$ m
       (c) $P = 104$ kW

14.22   (a) $Q = 61.2$ gpm   (b) $Q = 76.4$ gpm
       (c) $Q = 77.4$ gpm

14.24   Mixed flow pump

14.26   Radial flow pump

14.28 $Q = 0.508$ m³/s

14.30   (a) $F_{\text{on bucket}} = 1/2\rho A V_j^2$
       (b) $P = FV_B = (1/2)\rho Q V_j^3/2$
       (c) Operation of more than one bucket

14.32   (a) $\alpha_1 = 6.78°$ (b) $P = 88.9$ MW
       (c) Increase β_2

14.34 $P_{\max} = 3.29$ kW

14.36 $Q = 289$ gpm

Chapter 15

15.2   (b)

15.4 $Q = 608$ cfs

15.6   DW: $Q = 321$ cfs; Manning: $Q = 284$ cfs

15.8   (a) Supercritical   (b) $y_2 = 7.46$ ft

15.10 $E = 3.17$ m; $y_2 = 3.08$ m

15.12 $Q = 187$ cfs

15.14 $Q = 65.7$ cfs

15.16 $V = 5.67$ ft/s

15.18   (a) Yes   (b) No   (c) Yes
       (d) Yes   (e) Yes   (f) No

15.20   (a) Yes   (b) $y_2 = 7.46$ m

15.22   (a)

15.24   Yes; immediately downstream of sluice gate; possibly submerged jump followed by H2

15.26 $d_{300} \approx 1.51$ m

Chapter 16

16.12   True; False; False

Index

Abrupt/sudden expansion, 221–222
Absolute pressure, 16, 59–61
Absolute temperature, 16
Acceleration
    calculating, when velocity field is specified, 109
    centripetal, 108
    convective and local, 108
    defined, 106
    gravitational, 10
    mathematical description of, 107–108
    moving objects and, 192–194
    physical interpretation of, 106–107
Acceleration field, 490–491
Accumulation
    mass, 153
    momentum, 177–178
Action force, 9
Adhesion, 48–49
Adiabatic process, 352, 356
Advance ratio, 408–409
Adverse pressure gradient, 129
Airfoils
    drag and lift on, 336–342
    sound propagation, 354
Airplanes
    drag and lift on, 336–340
    Mach number for, calculating, 356
    total temperature calculation, 358
    wind tunnel applications, 248–249
Alternate depths, 452
Analytical fluid dynamics (AFD), 480
Anderson, J., 266
Anemometers
    cup, 383
    hot-wire or hot-film, 383–384
    laser-Doppler, 384–385
    vane or propeller, 382
Angular momentum equation, 194–197
Apron, 465
Archimedes' principle, 82
Area-averaged velocity, 143
Atmospheric pressure, 58–59
Attached flow, 105, 106
Automobiles
    drag and lift on, 342–345
    model tests for drag force on, 253–254
Average shear stress coefficient, 271, 274
Avogadro's law, 15
Avogadro's number, 17
Axial-flow pumps, 411–415
Axisymmetric bodies, drag and, 323–328

Barometers, 68
Bernoulli equation (in calculations)
    inviscid and irrotational flow, 129
    inviscid flow, 129
Bernoulli equation (theory)
    compared with the energy equation, 221
    continuity equation and, 161
    derivation of, 114
    examples, 117–120
    head form, 114
    irrotational form, 127–128
    physical interpretation—energy is conserved, 115
    physical interpretation—velocity and pressure
        vary, 116
    pressure form, 114
    summary of, 117, 135
Best hydraulic section, 448
Bingham plastics, 43–44
Blasius solution, 268
Body(-ies)
    defined, 6
    floating, 84–87
Body-as-a-particle, 6
Body force, 9, 172–173
Boundary, system, 32
Boundary conditions, 504–505
    for Couette flow, 45
    for Poiseuille flow, 275
Boundary layer, 266–274
    calculations with, 269–274
    description of, 266–267
    laminar, 271
    mixed, 271
    thickness of, 269–270, 274
    tripped, 271
    velocity profiles in, 267–269
Bourdon-tube gage, 68
Boyle's law, 15
Buckingham π theorem, 236
Bulk modulus of elasticity, 34, 38
Buoyant force, 81–88
    as common fluid force, 42
    defined, 81
    floating bodies, 84–87
    immersed bodies, 83–84
Buoyant force equation, 81–82

Capillary action, 49–50
Capillary repulsion, 50
Capture area, 433, 434
Cartesian components, 100

Cartesian coordinates, 483–484, 501
Cavitation, 161–164
  benefits of, 162
  defined, 161
  degradation from, 162
  sites, identifying, 163–164
Center of buoyancy, 83–84
Center of pressure (CP), 73, 76
Centrifugal compressors, 423–425
Centrifugal pumps, 304–305, 415–418
Centripetal acceleration, 108
Centroid, 74
Channel, 264–266
Characteristic curves, 413
Charles's law, 15
Chemical energy, 205–206
Chezy equation, 444, 445, 447
Circulation, 333–335
Closed system (control mass), 147–149, 208
Coefficients. *See* specific coefficients
Coefficient of drag, 271, 321
Coefficient of lift, 335
Coefficient of velocity, 390
Colebrook-White formula, 291
Combined head loss, 282–283, 298–302
Combined head loss equation, 300–301
Component head loss, 282–283, 298–302
Compressible flows/fluids, 351–378
  density, 359
  drag and, 331–333
  isentropic, through a duct with varying area, 366–377
  kinetic, 359–361
  Mach number relationships, 356–361
  normal shock waves, 361–366
  speed of sound, 351–355
  temperature, 357–359
  wave propagation, 351–356
Compressors, centrifugal, 423–425
Computational fluid dynamics (CFD), 477, 480, 501–508
  computer programs (codes), 502–503
  examples of, 506–508
  features of, 503–505
  importance of, 501–502
  validation and verification, 506
Conduits, defined, 279
Conduits, flow in
  centrifugal pumps, 304–305
  classifying, 279–281
  developing versus fully developed, 280–281
  entry or entrance length, 280–281
  flow problems, strategies for solving, 294–298
  laminar flow in round tubes, 286–289
  laminar versus turbulent, 279–280
  nonround, 302–303
  parallel pipes, 3305–306
  pipe head loss, 282–284
  pipe networks, 306–309
  pipe sizes, specifying, 281–282

  stress distributions in pipe flow, 284–286
  turbulent flow, 289–294
Conjugate depth, 463–464
Conservation of energy, 207–208
Conservation of mass. *See* Continuity equation
Consistent Unit Rule, 21
Consistent units, 20–21
Constant density, 64–65, 496
Constant density assumption, 37–38
Constant velocity, 190–192
Continuity equation (in calculations)
  algebraic form, 155
  differential equation form, 497
Continuity equation (theory)
  applications, 154–160
  Bernoulli equation and, 161
  constant density, 496
  cylindrical coordinates, 494
  derivation, 166, 491–494
  differential form, 491–497
  invariant notation, 495–496
  modeling, 491–497
  physical interpretation of, 153–154
  pipe flow form, 159–161
  summary of, 155, 496–497
  units, 155
Continuum assumption, 6–7
Controls in open channel flow, 468
Control mass (closed system), 147–150, 208
Control surface (CS)
  defined, 148
  transport across, 150–151
Control volume (CV) (open system)
  conservation of energy and, 208
  description of, 148
  linear momentum equation for stationary, 180–189
Control volume approach
  closed system (control mass), 147–149
  open system (control volume), 148–149
  properties, intensive and extensive, 148, 149
  Reynolds transport theorem, 151–153
  transport across control surface, 150–151
Convective acceleration, 108
Couette flow
  and viscosity equation, 45–47
  and viscous flow, 263–264
Critical depth, 454
Critical flow, 452–458
Critical-flow flumes, 456–457
Critical mass flow rate, 371
Critical pressure ratio, 372–373
Critical thinking (CT), 2–3
Cross, H., 307
Culverts, uniform flow in, 448–451
Cup anemometers, 383
Curved surface, pressure force on, 78–80
Cyclist, drag on a, 507–508

Cylinders, drag on, 323
Cylindrical coordinates, 484–485, 494, 501

Dam spillways, hydraulic jumps on, 464–466
Darcy-Weisbach equation, 283–284, 443–444
Deforming CV, 148
de Laval nozzles
    flow in, 368–371
    mass flow rate, 371–372, 375–377
    nozzle flow classification by exit conditions, 372–375
    shock waves in, 374–375
    truncated, 375–377
Density, 36–38
    bulk modulus of elasticity, 38
    constant, 64–65, 496
    and constant density assumption, 37–38
    defining, 14
    as fluid property, 34
    specific gravity, 36–37
    total, 359
    variable, 63–64
Depth, conjugate versus sequent, 463–464
Depth ratio, 463
Derivatives, 12–13
Design storm, 449
Developing flow, 280–281
Diagram(s)
    free body, 63
    momentum, 178–180
    Moody, 291–294
    phase, liquid and vapor, 51–52
Differential pressure, 61, 72
Dimensions
    primary and secondary, 22–23
    time as, 22
Dimensional analysis
    approximate, at high Reynolds numbers, 251–254
    Buckingham π theorem, 236
    common π groups, 240–243
    defined, 236
    exponent method, 239–240
    model-prototype performance, 250–251
    need for, 234–236
    open-channel flow, 440–441
    similitude, 243–247
    step-by-step method, 236–239
    variables, selection of significant, 240
Dimensional homogeneity (DH), 23–24
Dimensionality of flow, 104
Dimensionally homogeneous, 23
Direct numerical simulation (DNS), 505
Direct step method, 471
Discharge coefficient, 390, 412–415
Discharge (volume) flow rate, 287–288
    equations for, 142–145
    measuring, 387–402
    in a pipe network, 308–309
Dividing streamline, 96

Doppler effect, 355
Draft tube, 432
Drag curve, standard, 325
Drag force
    airfoil, 336–338
    automobile, 253–254, 342–345
    axisymmetric bodies and, 323–328
    and boundary layer, 271–272
    calculating, 320–324
    coefficient of. *See* coefficient of drag
    as common fluid force, 42
    compressible flows and, 331–332
    on a cyclist, 507–508
    cylinder, 323
    defined, 318
    induced, 338–339
    on a sphere, 327
    on a sphere, calculating, 359–360
    streamlining to reduce, 331
    stress distribution and, 318–320
    surface roughness, 323
    three-dimensional bodies and, 323–328
    two-dimensional bodies and, 321–322
Drag force equation, 320–321
Dynamic similitude, 245–247

Eddy, 267
Efficiency
    equation for, 218
    mechanical, 218–220
    thermal, 218
Efflux, 154
Electrical energy, 206
Electromagnetic flowmeters, 396–397
Elevation, 58–59
Energy
    categories of, 205–206
    conservation of, 207–208
    defined, 205
    specific, 452–453
    technical vocabulary of, 205–207
    units, 205
Energy equation (in calculations), 214
Energy equation (theory)
    applications, 215–216
    compared with the Bernoulli equation, 221
    derivation of, 209, 213–214
    flow and shaft work, 209–211
    kinetic energy correction factor, 211–213
    modeling, 491
    physical interpretation of, 214
    for steady open-channel flow, 442
    summary of, 215
Energy grade line (EGL), 224–227
Engineering notation, 12
Enthalpy
    and thermal energy, 52–53
    total, 357

Entry or entrance length, 280–281
Equation literacy, 9
Equivalent sand roughness, 293
Estimates, 11
Eulerian approach, 99–100
Euler's equation
    calculations involving, 112–113
    derivation of, 109–110
    physical interpretation of, 111–112
    pressure variation, due to changing speed of particle, 111
    pressure variation, normal to curved streamlines, 111–112
    pressure variation, normal to rectilinear streamlines, 111
Expansion, abrupt/sudden, 221–222
Experimental fluid dynamics (EFD), 480
Exponent method, 239–240
Extensive properties, 148, 149

Fan laws, 415
Fanning friction factor, 284
Favorable pressure gradient, 129
Field, 99
Fixed CV, 149
Flat surface
    boundary layer on, 266–267, 274
    pressure force on, 72–80
Floating bodies, 84–86
Flow. *See also* conduits, flow in; open channels, flow in; specific flows
    attached, 105, 106
    classification of, 279–281
    critical, 452–454
    critical mass, 371
    in de Laval nozzles, 368–475
    developing, 280–281
    dimensionality of, 104
    gradually varied, 451, 466–472
    homenergic, 357
    hypersonic, 355
    inviscid, 104
    irrotational, 124, 127–128
    momentum, 176–178
    multidimensional, 104
    nonuniform, 101, 440, 451
    nozzles, 395
    one-dimensional, 104
    rapid, 453
    rapidly varied, 451–452
    subcritical, 452–453
    subsonic, 367
    supercritical, 452
    tranquil, 453
    transonic, 367–368
    uniform, 101, 440
    unsteady, 102
    without free-surface effects, model studies for, 247–249
FLOW-3D, 503
Flow coefficient, 390
Flowing fluids
    acceleration, 106–109

circular cylinders and, 128–129
describing, 100–106, 133
dimensionality, 104
Eulerian and Lagrangian approaches, 99–100
inviscid, 104
laminar, 102–103
pathlines, streaklines, and streamlines, 95–97
regions, 105
rotational motion, 123–125
separation, 105–106
steady and unsteady, 102
turbulent, 102–103
uniform and nonuniform, 101
velocity, 98–100
viscous, 104
Flowing gases, 52–53
Flow measurements. *See* measuring; measuring devices
Flowmeters
    electromagnetic, 396–397
    turbine, 397
    ultrasonic, 397
    vortex, 397
Flow rate
    critical mass, 371
    deriving equations for, 142–144
    differential areas for determining, 146
    example problems, 145–147
    mass, 144, 145, 371–372
    measuring, 387–401
    units, 142, 144
    volume (discharge), 142–145
    working equations for, 145
Flow Science, 503
Flow work, 209–210
Fluids. *See also* flowing fluids
    constant density assumptions for, 37
    defining, 4–5
    ideal, 128
    Newtonian vs. non-Newtonian, 43–44
    shear-thickening, 43–44
    shear-thinning, 43–44
Fluid forces, 41, 42
Fluid in a solid body rotation, 130
Fluid interface rule, 65
Fluid jets, 180–182
Fluid mechanics, 2
Fluid properties, 33–53
    bulk modulus of elasticity, 38
    common fluid forces, 41, 42
    and constant density assumption, 37–38
    looking up, 33–36
    specific gravity, 36–37
    and stress, 38–41
    surface tension, 47–51
    thermal energy in flowing gases, 52–53
    vapor pressure, 51–52
    and viscosity equation, 41–47
Flume, 31

Force(s). *See also* specific forces
    action and reaction, 9
    body and surface, 8–9, 171–172
    defined, 8–9
    between molecules, 4
    relating stress to, 40–41
    seven common fluid forces, 41–42
    summary of, 172
    transitions and, 222–223
Force coefficient, 241, 242
Force units, 10–11
Form drag, 320
Francis turbines, 196–197, 426, 432, 433
Free body diagram (FBD), 63
Free surface, defined, 440
Free-surface effects, model studies for flows without, 247–249
Free-surface model studies, 254–255, 502–503
Friction drag, 320
Friction factor, 284
    laminar flow, 286
    Moody diagram, 291–292
    turbulent flow, 289–290
Friction velocity, 268
Froude number, 242, 243, 440–441
Fully developed flow, 280–281

Gage pressure, 16, 60, 72
Gage pressure rule, 73
Gas(es)
    constant density assumptions for, 37
    defining, 4–5
    real and ideal, 15–16
    thermal energy in flowing, 52–53
Gas constant, specific and universal, 17–18
Gas pressure change rule, 67
Gas turbines, 432
General equation, 11
Geometric similitude, 244–245
Gradually varied flow, 451, 466–473
Gravitational acceleration, 10
Gravitational force, 9
Gravity, specific, 34, 36–37
Grid generation, 503–504
Grid method, 19–20

Hagen-Poiseuille flow, 286
Hardy Cross method, 307
Head
    defined, 115
    piezometric, 65
    pump, 214
    total, 224
    turbine, 214
Head coefficient, 412–415
Head loss
    component/combined, 282–283, 298–302
    defined, 214
    flow problems, strategies for solving, 294–298

    in hydraulic jumps, 464
    laminar flow and, 288–289
    in a nozzle in reverse flow, 252–253
    for orifices, 392
    pipe, 282–284
Heat, specific, 52
Heat transfer, 208
Hirt, T., 502–503
Homenergic flows, 357
Hot-wire or hot-film anemometers, 383–384
HVAC duct, pressure drop in, 303
Hydraulic depth, 453
Hydraulic diameter, 302
Hydraulic grade line (HGL), 224–227
Hydraulic jumps
    dam spillways and, 464–466
    depth relationships, 461–463
    head loss in, 464
    naturally occurring, 466
    occurrences of, 461
    in rectangular channels, 463–464
    submerged, 466
Hydraulic machines, 61–62
Hydraulic radius, 303, 441
Hydrogen bubble visualization, 386
Hydrometers, 82
Hydrophilic surfaces, 49
Hydrophobic surfaces, 49, 50
Hydrostatic condition, 63
Hydrostatic equations, 63–67
    hydrostatic algebraic equation, 64–65
    and hydrostatic condition, 63
    hydrostatic differential equation, 63–64
    working equations, 65–67
Hydrostatic equilibrium, 63. *See also* pressure
Hydrostatic pressure distribution, 73–77
Hypersonic flows, 355

Ideal fluid, 128
Ideal gas, 15–16, 53
Ideal gas law (IGL), 15–18
    units in, 16–17
    universal and specific gas constant, 17–18
    working equations for, 18
Ideally expanded nozzles, 374
Immersed bodies, 83–84
Impulse turbines, 426–429
Induced drag, 339
Induction, 15
Inertial reference frames, 189–190
Inflow, 154
Initial conditions, 504–505
Integrals, 13
Intensive properties, 148, 149
Internal combustion engine modeling, 502
Internal energy, 52
Invariant notation, 487–488, 495–496
Inviscid flow, 104

Irrotational flow, 124, 127–128
Isentropic process
  compressible fluids through a duct with varying area,
    366–377
  defined, 352

Joule, J. P., 207–208

Kaplan turbines, 426
k-epsilon model, 505
Kinematic properties, 33
Kinematic viscosity, 33–35
Kinetic energy correction factor, 211–212
Kinetic pressure, 134, 359–361
KIVA, 502, 503
Kolmogorov length scale, 267
Kutta condition, 337

Lagrangian approach, 99–100
Laminar boundary layer, 271
Laminar flow
  defined, 286
  description of, 379–280
  discharge and mean velocity, 287–288
  head loss and friction factor, 288–289
  kinetic energy correction factor, 212–213
  in round tubes, 286–289
  velocity profile in, 286–287
Laminar velocity profile, 268
Large eddy simulation (LES), 505
Laser-Doppler anemometers (LDAs), 384–385
Length scale, 279
Lift force
  airfoil, 336–342
  automobile, 342–345
  circulation, 333–334
  circulation combined with uniform flow, 334–335
  coefficient, 335. *See* coefficient of lift
  as common fluid force, 42
  defined, 318
  on a rotating sphere, 336
  stress distribution and, 318–320
Lift force equation, 335
Linear momentum equation
  applications, 178–180
  for moving objects, 189–194
  for stationary control volume, 180–189
  summary of, 178
  theory, 175–178
Liquids. *See also* fluids
  constant density assumptions for, 37
  defining, 4–5
Local acceleration, 108
Local Reynolds number, 267
Local shear stress coefficient, 270, 274
Logarithmic velocity distribution, 268
Los Alamos National Laboratory, 502–503

Mach, E., 355
Mach angle, 355
Mach number, 242, 243
  for airplanes, calculating, 356
  area variation and, 366–367
  critical, 332
  defined, 355
  relationships and compressible flows,
    356–361
Mach wave, 355
Macroscopic descriptions, 4
Magnus effect, 335
Manning equation, 445–447
Manning's *n*, 445–447
Manometer, 69–72
Marker methods, 385–387
Mass
  defining, 8
  as dimension, 22
  molar, 17
Mass balance equation, 155
Mass flow rate, 144
  critical, 371
  de Laval nozzles and, 371–372, 375–377
Mass flow rate equation, 145
Mass units, 10–11
Material body, 6
Material derivative, 488–490
Material particle, 6
Material properties, 33
Math models, 25–27
Mean velocity, 143, 287–288
Measuring, 380–401
  flow rate, 387–401
  marker methods, 385–387
  pressure, 67–72, 120–123, 380–387
  velocity, 120–123, 380–387
Measuring devices, 380–385, 389–392
  anemometers, cup, 383
  anemometers, hot-wire or hot-film, 383–384
  anemometers, laser-Doppler, 384–385
  anemometers, vane or propeller, 382
  barometers, 68
  Bourdon-tube gages, 68
  flowmeters, electromagnetic, 396–397
  flowmeters, turbine, 397
  flowmeters, ultrasonic, 397
  flowmeters, vortex, 397
  flow nozzles, 395
  hydrometers, 82
  manometers, 69–72
  orifice meters, 389–394
  piezometers, 68–69
  Pitot-static tube, 121–123, 381
  pressure transducers, 72
  rotameters, 398
  stagnation (Pitot) tube, 380–383

static tube, 381
venturi meters, 394–396
weirs, rectangular, 398–400
weirs, triangular, 401
yaw meters, 382
Mechanical advantage, 61
Mechanical efficiency, 218–220
Mechanical energy, 205
Mechanical work, 206
Mechanics, 2
Metacenter, 84
Metacentric height, 84
Metric system, 12
Microscopic descriptions, 4
Mild slope, 457
Minor loss coefficient, 298–300
Mixed boundary layer, 271
Models (modeling)
    assessing the value of, 481–482
    computational fluid dynamics, 480, 501–508
    continuity equation, 491–497
    defined, 478
    methods for building, 480–481
    Navier-Stokes equation, 491, 497–501
    partial differential equations, 482–491
    process, 478–482
Model testing
    applications, 243
    approximate similitude at high Reynolds numbers,
        251–254
    dynamic similitude, 245–247
    for flows without free-surface effects, 247–249
    free-surface, 254–256
    geometric similitude, 244–245
    model-prototype performance, 250–251
    ship, 244, 256
    spillway, 254–255
MODFLOW, 502, 503
Molar mass, 17
Moles, 16–17
Molecules, forces between, 4
Moment-of-momentum equation, 194–197
Momentum accumulation, 177–178
Momentum diagram, 178–180
Momentum equation (in calculations)
    angular momentum, 194
    linear momentum, 178
Momentum equation (theory)
    angular momentum equation, 194–197
    applications of linear, 178–180
    linear momentum equation, 175–194
    linear, for stationary control volume, 180–189
    Newton's laws of motion, 171–175
    nozzles and, 185–186
    theory of linear, 175–178
    visual solution method, 173–174
Momentum flow, 176–177

Moody diagram, 291–294
Moving objects
    accelerating, 192–194
    constant velocity, 190–192
    linear momentum equations and, 189–194
    reference frames, 189–190
Multidimensional flow, 104

Nardi, A., 25
Navier-Stokes equation, 491
    Cartesian and cylindrical coordinates, 501
    for Couette flow, 264
    derivation, 498–500
    modeling, 497–501
    for Poiseuille flow, 275
    for uniform flow, 262–263
Net positive suction head (NPSH), 420–422
Newton (unit), 10
Newtonian fluids, 43–44
Newton's Laws of Motion, 171–175
Newton's Law of Universal Gravitation (NLUG), 9–10
Nominal Pipe Size (NPS), 282
Noninertial reference frames, 189–190
Non-Newtonian fluids, 43–44
Nonround conduits, 302–303
Nonuniform flow, 101, 440, 451
Normal depth, 440, 445
Normal shock waves
    defined, 361
    in de Laval nozzles, 374–375
    property changes across, 361–363
    in supersonic flows, 364–366
Normal stress, 38–39
Nozzles. *See also* de Laval nozzles
    applications, 184
    flow classification by exit conditions, 372–374
    ideally expanded, 374
    momentum equation and, 185–186
    overexpanded, 374
    underexpanded, 373
Nuclear energy, 206
Numbers, representing, 12

Oblique shock waves, 365
One-dimensional flow, 104
Open channels, defined, 440
Open channels, flow in, 440–442
    best hydraulic section, 448
    Chezy equation, 444, 445, 447
    critical flow, 452–453
    description of, 440–442
    dimensional analysis, 440–442
    energy equation for steady, 442
    gradually varied, 451, 466–472
    hydraulic jump, 461–466
    Manning equation, 445–446
    rapidly varied, 451–452

Open channels (*Continued*)
    Reynolds number, 441, 442
    rock-bedded channels, 444, 445
    steady nonuniform, 451
    steady uniform, 443–451
    transitions, 458–460
    wave celerity, 460–461
Open system. *See* control volume
Orifices, head loss for, 392, 394
Orifice meters, 389–392
Outflow, 154
Overexpanded nozzles, 374

Panel, 72. *See also* flat surface
Panel force working equations, 77
Parallel pipes, 305–306
Partial differential equations (PDEs),
    482–491
    acceleration field, 490–491
    approximation of, 503
    Cartesian coordinates, 483–484
    cylindrical coordinates, 484–485
    material derivative, 488–490
    notation, invariant, 487–488
    operators, 488
    reasons for learning, 482–483
    Taylor series, 486–487
Particle, 6
Particle image velocimetry (PIV),
    386–387
Pathlines, 95–96
Pelton wheel, 426
Performance curves, pumps, 413
Phase diagram, water, 51–52
π-groups
    common, 240–243
    definition, 24
    as dimensionless groups, 24
    exponent method, 239–240
    step-by-step method, 236–239
    use of term, 236
π theorem, Buckingham, 236
Piezometers, 68–69
Piezometric head, 65
Piezometric pressure, 64–65
Pipes
    bends, 186–187
    continuity equation and, 159–161
    expansion, abrupt/sudden, 221–222
    forces on, 222–223
    head loss, 282–284, 294–298
    networks, 306–309
    parallel, 305–306
    sizes, specifying, 281–282
    stress distributions in pipe flow,
        284–286
Pitch angle, 408

Pitot (stagnation tube), 120–121, 380–381
Pitot-static tube, 121–123, 381
Poiseuille flow, 286
    in a channel, 264–266
    in a round pipe, 286–289
    shear stress in a, 45–46
Positive-displacement machines, 406
Post processor, 505
Power
    defined, 206–207
    and rolling resistance, 327–328
Power coefficient, 409
Power equation, 216–218, 273
Power law equation, 290
Power law formula, 269
Prandtl, L., 266, 290, 339
Pressure, 58–80
    absolute, 16, 59–61
    atmospheric, 58–59
    defined, 58
    differential, 61, 72
    fluid properties varied with, 35
    gage, 16, 60, 72
    hydraulic machines, 61–63
    hydrostatic equations, 63–67
    kinetic, 359–361
    measuring, 68–72, 120–123, 380–387
    as normal stress, 38
    piezometric, 64–65
    related to the pressure force, 40
    and shear stress, 38–41
    static, 120
    total, 358
    vacuum, 60–61
Pressure coefficient, 128–129, 250, 257
Pressure distribution, 40
    hydrostatic, 73–74
    for ideal fluid, 128–129
    sketching, 74
    uniform, 73
Pressure field
    circular cylinders and, 128–129
    rotating flow and, 130–132
Pressure force
    as a common fluid force, 42
    on a curved surface, 78–80
    defined, 40–41
    on a flat surface, 72–74
    in viscous flow problems, 262
    related to the pressure distribution, 40
Pressure gradient, favorable and adverse, 129
Pressure ratio, critical, 372–373
Pressure tap, 120
Pressure transducer (PT), 72
Pressure variation. *See* Euler's equation
Pressure vessel force balance, 74
Primary dimensions, 22–23

Problem solving, 24–27
    defining, 24–25
    and math models, 25–27
Process, for a system, 32
Projected area, 320
Propellers, 407–411
Propeller anemometers, 382
Properties. *See also* fluid properties
    defined, 32
    intensive and extensive, 148, 149
    kinematic, 33
    material, 33
    variation of, with temperature and pressure, 35
Pumps
    axial-flow, 411–415
    centrifugal, 304–305, 415–418
    curves, characteristic or performance, 413
    defined and types of, 207
    discharge coefficient, 412–415
    head coefficient, 412–415
    mechanical efficiency, 218–220
    power equation, 216–218
    suction limitations of, 420–422
Pump curve, 304
Pump head, 214

Radial-flow machines, 415–418
Rapidly varied flow, 451–552
Rate of strain, 43
Reaction force, 9
Reaction turbines, 426, 429–430
Real gas, 15–16
Reference frames, 189–190
Relative roughness, 290
Resistance, power and rolling, 327–328
Resistance coefficient, 445
Reynolds, O., 242, 279
Reynolds-averaged Navier-Stokes (RANS) equations, 505
Reynolds number, 242, 243
    approximate similitude at high, 251–254
    and boundary layer, 267
    conduit flow type and, 279–280
    length scale, 279
    open-channel flow and, 442
    similitude for a valve, 249
    similitude for flow over a blimp, 247–248
Reynolds transport theorem, 147, 151–153
Rock-bedded channels, 444, 445
Rolling resistance, 327–328
Rotameters, 398
Rotational motion, 123–125, 130–132
Roughness
    drag and surface, 323–327
    equivalent sand, 293
    relative, 290
    sand roughness height, 290
    type of flow and effects of wall, 291

Roughness coefficient, 446
Runner, turbine, 426

Sand roughness
    equivalent, 293
    height, 290
Saturation pressure, 51
Saturation temperature, 51
Scientific notation, 12
Secondary dimensions, 22–23
Sectional drag coefficient, 321–322
Sequent depth, 464
Sewers, uniform flow in, 448–451
Shaft work, 209–210
Shear force
    and boundary layer, 271
    as common fluid force, 42
    defined, 41
    in viscous flow problems, 263
    related to shear stress, 41
Shear stress
    distributions in pipe flow, 284–286
    in a Poiseuille flow, 45–46
    and pressure, 38–40
    at the wall, 270
    related to shear force, 41
Shear stress coefficient, 242
Shear stress distribution, 41
Shear-thickening fluids, 43–44
Shear-thinning fluids, 43–44
Ship model testing, 244, 256
Shock waves
    in de Laval nozzles, 374–375
    normal. *See* normal shock waves
    oblique, 365
    in supersonic flows, 364–365
Significant figures, 12
Similitude, 243–256
    approximate, at high Reynolds numbers, 251–254
    defined, 243
    dynamic, 245–246
    flows without free-surface effects, 247–249
    free-surface model studies, 254–256
    geometric, 244–245
    model-prototype performance, 250–251
    scope of, 243–244
Skin friction drag, 271–272
Sluice gate, 452
Solids, 5
Solid mechanics, 2
Solver, 505
Sonic booms, 366
Sound
    Doppler effect, 355
    speed of, 351–356
Special-case equation, 11
Specific energy, 452–453

Specific gas constant, 17–18
Specific gravity, 34, 36–37
Specific heat, 52
Specific speed, 418–419, 432
Specific weight, 14–15, 34
Speed of sound, 351–356
Spheres, drag force and
  calculating, 327, 359–361
Spheres, finding π-groups
  using exponent method, 239–240
  using step-by-step method, 237
Spheres, lift on rotating, 336
Spillways
  hydraulic jumps on dam, 464–466
  models, 254–255, 506–507
Stager, R., 25
Stagnation
  point, 96
  tube (Pitot), 120–121, 380–381
  use of term, 359
Stall, 338
Standard atmosphere, 59
Standard Structure of Critical Thinking (SSCT), 3
States, defined, 32–33
Static pressure, 120
Static pressure ratio, 362
Static temperature, 357
Static temperature ratio, 362
Static tube, 381
Steady flow, 102
  energy equation for steady open-channel flow, 442
  nonuniform, 451
  uniform, 448–451
Steady state, 32
Steep slope, 457
Step-by-step method, 236–239
Streaklines, 95–96
Streamlines, 96, 331
Stress. *See also* shear stress and pressure
  defined, 39–40
  distributions and drag and lift, 318–320
  distributions in pipe flow, 284–286
  and fluid properties, 38–41
  normal, 38–39
  relating forces to, 40–41
Strouhal number, 330
Subcritical flow, 452–453
Submerged hydraulic jumps, 466
Subsonic flow, 367
Sudden expansion, 221–222
Supercritical flow, 4452
Supersonic flows
  de Laval nozzles, 368–371
  diffusers, 367
  shock waves in, 364–366
Supersonic wind tunnels
  de Laval nozzles and, 368–371
  flow properties in, 370–371

  mass flow rate in, 372
  test section size in, 370
Surface force, 8, 171–172
Surface roughness, drag and, 323–324
Surface tension
  and adhesion, 48–49
  and capillary action, 49–50
  as fluid property, 34, 47–51
Surface tension force, 42
Surge hydraulic jumps, 466
Surroundings, 32
System(s)
  curve, 304
  defined, 32

Taylor series, 486–487
Temperature
  absolute, 16
  fluid properties varied with, 35
  static, 357
  total, 357, 358
Terminal velocity, 328–329
Theoretical adiabatic power, 423
Theoretical isothermal power, 424
Thermal efficiency, 218
Thermal energy
  defined, 205
  in flowing gases, 52–53
Three-dimensional bodies, drag and,
  323–324
Thrust coefficient, 408–409
Thrust force, 42
Tidal bore, 466
Time, as dimension, 22
Time-averaged velocity, 143
Time steps, 504
Total density, 359
Total enthalpy, 357
Total head tube, 120–121
Total pressure, 358
Total temperature, 357, 358
Tranquil flow, 453
Transition(s)
  abrupt/sudden expansions and, 221–222
  defined, 458
  forces on, 222–223
  open channel, 458–460
  warped-wall, 459
  wedge, 459
Transition region, 267
Transition Reynolds number, 267
Transonic flow, 367–3
Transport
  across control surface, 150–151
  Reynolds transport theorem, 147,
  151–153
Tripped boundary layer, 271
Truncated nozzles, 375–377

Turbines
  defined, 207, 426
  Francis, 196–197, 426, 431, 432
  gas, 432
  impulse, 426–430
  Kaplan, 426
  mechanical efficiency, 218–220
  reaction, 426, 429–430
  specific speed for, 432
  types of, 207
  vane angles, 430–431
  wind, 432–434
Turbine flowmeters, 397
Turbine head, 214
Turbomachinery, 406–438
  categories of, 407
  compressors, centrifugal, 423–424
  defined, 406
  propellers, 407–411
  pumps, axial-flow, 411–415
  pumps, centrifugal, 415–418
  pumps, suction limitations of, 420–422
  radial-flow machines, 415–418
  specific speed, 418–420
  turbines, 426–429
  viscous effects, 422–423
Turbulence modeling, 505
Turbulent flow
  defined, 102, 279–280, 289
  friction factor, 290–291
  Moody diagram, 291–294
  velocity distribution, 290
Turbulent velocity profile, 268–269
Two-dimensional bodies, drag and, 321–322

Ultrasonic flowmeters, 397
Underexpanded nozzles, 373
Uniform flow
  best hydraulic section, 448
  in culverts and sewers, 448–451
  defined, 101, 440
  Navier-Stokes equation for, 262–263
  steady, 443–451
Uniform pressure distribution, 73
U.S. standard atmosphere, 59
Unit prefixes, 12
Units, 10–11, 18–24
  consistent, 20–22
  defined, 19
  and dimensional homogeneity, 23–24
  grid method, 19–20
  in ideal gas law, 16–17
  organized by dimensions, 21–23
  and π-groups, 24
Universal gas constant, 17–18
Unsteady flow, 102
U-tube manometer, 69

Vacuum pressure, 60–61
Validation, 506
Vanes, 182–184, 430–432
Vane anemometers, 382
Vapor pressure, 34, 51–52
Variable density, 63–64
Variable velocity distribution, 188–189
Vector equation, 173–174
Velocity
  area-averaged, 143
  Cartesian components, 100
  coefficient of. *See* coefficient of velocity
  constant, and moving bodies, 190–192
  defined, 98
  Eulerian and Lagrangian approaches, 99–100
  flowing fluids and, 98–100
  friction, 268
  mean, 143, 287–288
  measuring, 120–123, 380–387
  terminal, 328–329
  time-averaged, 103
Velocity distribution
  logarithmic, 268
  measuring, 388
  power-law equation, 290
  turbulent flow, 102
  variable, 188–189
Velocity field, 98–99
  for Couette flow, 274
  for Poiseuille flow, 275
Velocity gradient, 42–43
Velocity profile(s)
  in boundary layer, 267–269
  defined, 43
  laminar, 286–287
  turbulent, 268–269
Vena contracta, 389
Venturi flumes, 457
Venturi meters, 394–396
Venturi nozzles, 119
Verification, 506
Viscosity
  as fluid property, 34
  kinematic, 33
Viscosity equation, 41–47
  and Couette flow, 45–47
  with Newtonian vs. non-Newtonian fluids, 43–44
  reasoning with, 44
  for shear stress in a Poiseuille flow, 45–46
  and velocity gradient, 42–43
Viscous effects, 422–423
Viscous flow, 104, 261–274
  and boundary layer, 266–273
  Couette flow, 263–264
  Navier-Stokes equation for uniform flow, 262–263
  Poiseuille flow in a channel, 264–266
  pressure distribution for, 129

Viscous force, 42
Visual solution method (VSM), 173–174, 178–179
Volume (discharge) flow rate, 142–145
Volume flow rate equation, 145
Vortex flowmeters, 397
Vortex shedding, 105, 330
Vorticity, 125–126

Wales, C., 25
Wales-Woods Model (WWM), 25–27
Wall shear stress, 270
Warped-wall transitions, 459
Water-surface profiles
    defined, 468
    evaluation of, 471–473
    types of, 568–570
Wave celerity, 460–461
Wave celerity equation, 461
Wave propagation, in compressible fluids, 351–356
Weather, and pressure, 59
Weber number, 242, 243

Wedge transitions, 459
Weight, 10
Weight, specific, 14–15, 34
Weirs
    rectangular, 398–400
    triangular, 401
William P. Morgan Cavitation Tunnel, 162
Wind loads on a telescope structure, predicting, 508
Wind tunnels
    applications, 243–244, 248–249
    momentum equation and finding drag force, 189–190
    supersonic. *See* supersonic wind tunnels
Wind turbines, 432–434
Woods, D., 25–26
Work
    defined, 206
    flow and shaft, 209–211
    mechanical, 206
Working equations, 18

Yaw meters, 382